List of the Elements with Their Atomic Symbols and Atomic Weights

Name	Symbol	Atomic Number	Atomic Weight	Name	Symbol	Atomic Number	Atomic Weight
Actinium	Ac	89	(227)*	Manganese	Mn	25	54.938045
Aluminum	Al	13	26.981538	Meitnerium	Mt	109	(276)
Americium	Am	95	(243)	Mendelevium	Md	101	(258)
Antimony	Sb	51	121.760	Mercury	Hg	80	200.59
Argon	Ar	18	39.948	Molybdenum	Mo	42	95.96
Arsenic	As	33	74.92160	Neodymium	Nd	60	144.242
Astatine	At	85	(210)	Neon	Ne	10	20.1797
Barium	Ba	56	137.327	Neptunium	Np	93	(237)
Berkelium	Bk	97	(247)	Nickel	Ni	28	58.6934
Beryllium	Be	4	9.012182	Niobium	Nb	41	92.90638
Bismuth	Bi	83	208.98040	Nitrogen	N	7	14.0067
Bohrium	Bh	107	(272)	Nobelium	No	102	(259)
Boron	B	5	10.811	Osmium	Os	76	190.23
Bromine	Br	35	79.904	Oxygen	O	8	15.9994
Cadmium	Cd	48	112.411	Palladium	Pd	46	106.42
Calcium	Ca	20	40.078	Phosphorus	P	15	30.973762
Californium	Cf	98	(251)	Platinum	Pt	78	195.094
Carbon	C	6	12.0107	Plutonium	Pu	94	(244)
Cerium	Ce	58	140.116	Polonium	Po	84	(209)
Cesium	Cs	55	132.90545	Potassium	K	19	39.0983
Chlorine	Cl	17	35.453	Praseodymium	Pr	59	140.90765
Chromium	Cr	24	51.9961	Promethium	Pm	61	(145)
Cobalt	Co	27	58.933195	Protactinium	Pa	91	231.03588
Copernicium	Cn	112	(285)	Radium	Ra	88	(226)
Copper	Cu	29	63.546	Radon	Rn	86	(222)
Curium	Cm	96	(247)	Rhenium	Re	75	186.207
Darmstadtium	Ds	110	(281)	Rhodium	Rh	45	102.90550
Dubnium	Db	105	(268)	Roentgenium	Rg	111	(280)
Dysprosium	Dy	66	162.500	Rubidium	Rb	37	85.4678
Einsteinium	Es	99	(252)	Ruthenium	Ru	44	101.07
Erbium	Er	68	167.259	Rutherfordium	Rf	104	(265)
Europium	Eu	63	151.964	Samarium	Sm	62	150.36
Fermium	Fm	100	(257)	Scandium	Sc	21	44.955912
Flerovium	Fl	114	(289)	Seaborgium	Sg	106	(271)
Fluorine	F	9	18.998403	Selenium	Se	34	78.96
Francium	Fr	87	(223)	Silicon	Si	14	28.0855
Gadolinium	Gd	64	157.25	Silver	Ag	47	107.8682
Gallium	Ga	31	69.723	Sodium	Na	11	22.989769
Germanium	Ge	32	72.64	Strontium	Sr	38	87.62
Gold	Au	79	196.96657	Sulfur	S	16	32.065
Hafnium	Hf	72	178.49	Tantalum	Ta	73	180.9479
Hassium	Hs	108	(270)	Technetium	Tc	43	(98)
Helium	He	2	4.002602	Tellurium	Te	52	127.60
Holmium	Ho	67	164.93032	Terbium	Tb	65	158.92535
Hydrogen	H	1	1.00794	Thallium	Tl	81	204.3833
Indium	In	49	114.818	Thorium	Th	90	232.0381
Iodine	I	53	126.90447	Thulium	Tm	69	168.93421
Iridium	Ir	77	192.217	Tin	Sn	50	118.710
Iron	Fe	26	55.845	Titanium	Ti	22	47.867
Krypton	Kr	36	83.798	Tungsten	W	74	183.84
Lanthanum	La	57	138.9055	Uranium	U	92	238.02891
Lawrencium	Lr	103	(262)	Vanadium	V	23	50.9415
Lead	Pb	82	207.2	Xenon	Xe	54	131.293
Lithium	Li	3	6.941	Ytterbium	Yb	70	173.054
Livermorium	Lv	116	(293)	Yttrium	Y	39	88.90585
Lutetium	Lu	71	174.9668	Zinc	Zn	30	65.38
Magnesium	Mg	12	24.3050	Zirconium	Zr	40	91.224

*Values in parentheses are the mass numbers of the most common or longest lived isotopes of radioactive elements.

Periodic Table of the Elements

Main groups

1 1A	2 2A	3 3B	4 4B	5 5B	6 6B	7 7B	8	9 8B	10	11 1B	12 2B	13 3A	14 4A	15 5A	16 6A	17 7A	18 8A
1 **H** 1.00794																	2 **He** 4.002602
3 **Li** 6.941	4 **Be** 9.012182											5 **B** 10.811	6 **C** 12.0107	7 **N** 14.0067	8 **O** 15.9994	9 **F** 18.998403	10 **Ne** 20.1797
11 **Na** 22.989769	12 **Mg** 24.3050											13 **Al** 26.981538	14 **Si** 28.0855	15 **P** 30.973762	16 **S** 32.065	17 **Cl** 35.453	18 **Ar** 39.948
19 **K** 39.0983	20 **Ca** 40.078	21 **Sc** 44.955912	22 **Ti** 47.867	23 **V** 50.9415	24 **Cr** 51.9961	25 **Mn** 54.938045	26 **Fe** 55.845	27 **Co** 58.933195	28 **Ni** 58.6934	29 **Cu** 63.546	30 **Zn** 65.38	31 **Ga** 69.723	32 **Ge** 72.64	33 **As** 74.92160	34 **Se** 78.96	35 **Br** 79.904	36 **Kr** 83.798
37 **Rb** 85.4678	38 **Sr** 87.62	39 **Y** 88.90585	40 **Zr** 91.224	41 **Nb** 92.90638	42 **Mo** 95.96	43 **Tc** (98)	44 **Ru** 101.07	45 **Rh** 102.90550	46 **Pd** 106.42	47 **Ag** 107.8682	48 **Cd** 112.411	49 **In** 114.818	50 **Sn** 118.710	51 **Sb** 121.760	52 **Te** 127.60	53 **I** 126.90447	54 **Xe** 131.293
55 **Cs** 132.90545	56 **Ba** 137.327	71 **Lu** 174.9668	72 **Hf** 178.49	73 **Ta** 180.9479	74 **W** 183.84	75 **Re** 186.207	76 **Os** 190.23	77 **Ir** 192.217	78 **Pt** 195.094	79 **Au** 196.96657	80 **Hg** 200.59	81 **Tl** 204.3833	82 **Pb** 207.2	83 **Bi** 208.98040	84 **Po** (209)	85 **At** (210)	86 **Rn** (222)
87 **Fr** (223)	88 **Ra** (226)	103 **Lr** (262)	104 **Rf** (265)	105 **Db** (268)	106 **Sg** (271)	107 **Bh** (272)	108 **Hs** (270)	109 **Mt** (276)	110 **Ds** (281)	111 **Rg** (280)	112 **Cn** (285)	113‡ (284)	114 **Fl** (289)	115‡ (288)	116 **Lv** (293)	117‡ (292)	118‡ (294)

Transition metals

* Lanthanide series

57 **La** 138.9055	58 **Ce** 140.116	59 **Pr** 140.90765	60 **Nd** 144.242	61 **Pm** (145)	62 **Sm** 150.36	63 **Eu** 151.964	64 **Gd** 157.25	65 **Tb** 158.92535	66 **Dy** 162.500	67 **Ho** 164.93032	68 **Er** 167.259	69 **Tm** 168.93421	70 **Yb** 173.054

† Actinide series

89 **Ac** (227)	90 **Th** 232.0381	91 **Pa** 231.03588	92 **U** 238.02891	93 **Np** (237)	94 **Pu** (244)	95 **Am** (243)	96 **Cm** (247)	97 **Bk** (247)	98 **Cf** (251)	99 **Es** (252)	100 **Fm** (257)	101 **Md** (258)	102 **No** (259)

‡ Elements 113, 115, 117, and 118 are currently under review by IUPAC

Chemistry

SEVENTH EDITION

John E. McMurry
Cornell University

Robert C. Fay
Cornell University

Jill K. Robinson
Indiana University

PEARSON

Editor in Chief: Jeanne Zalesky
Acquisitions Editor: Chris Hess
Marketing Manager: Will Moore
Field Marketing Manager: Chris Barker
Program Manager: Jessica Moro
Project Manager: Lisa Pierce
Director of Development: Jennifer Hart
Development Editor: Carol Pritchard-Martinez
Editorial Assistant: Caitlin Falco
Program Management Team Lead: Kristen Flathman
Project Management Team Lead: David Zielonka
Senior Project Manager: Jenna Vittorioso, Lumina Datamatics, Inc.
Compositor: Lumina Datamatics, Inc.
Copyeditor: Lumina Datamatics, Inc.
Photo Research Manager: Maya Gomez
Photo Researcher: Liz Kincaid
Design Manager: Derek Bacchus
Interior Designer: Wanda Espana
Cover Designer: Wanda Espana
Art Coordinator/Illustrator: Mimi Polk, Lachina/Precision Graphics
Text Permissions Manager: William Oplauch
Rights & Permissions Management: Rachel Youdelman
Senior Specialist, Manufacturing: Maura Zaldivar-Garcia
Cover Photo Credit: Dr. Keith Wheeler / Science Source

Library of Congress Cataloging-in-Publication Data

McMurry, John.
 Chemistry/John E. McMurry, Cornell University, Robert C. Fay, Cornell University, Jill K. Robinson, Indiana University.—Seventh edition.
     pages cm
Includes bibliographical references and index.
 ISBN 978-0-321-94317-0 (alk. paper)—ISBN 0-321-94317-1 (alk. paper)
 1. Chemistry—Textbooks.   I. Fay, Robert C., 1936–   II. Robinson, Jill K.
III. Title.
 QD33.2.M36 2014
 540—dc23

                                        2014033963

ISBN-10: 0-321-94317-1; ISBN-13: 978-0-321-94317-0
Printed in the United States of America

3 4 5 6 7 8 9 10—V057—18 17 16 15

www.pearsonhighered.com

ISBN-13: 978-0-321-94317-0
ISBN-10:   0-321-94317-1

Brief Contents

Contents

5 Periodicity and the Electronic Structure of Atoms 154

6 Ionic Compounds: Periodic Trends and Bonding Theory 195

7 Covalent Bonding and Electron-Dot Structures 222

8 Covalent Compounds: Bonding Theories and Molecular Structure 261

14 Chemical Equilibrium 553

15 Aqueous Equilibria: Acids and Bases 603

16 Applications of Aqueous Equilibria 656

17 Thermodynamics: Entropy, Free Energy, and Equilibrium 715

18 Electrochemistry   756

19 Nuclear Chemistry   808

20 Transition Elements and Coordination Chemistry   840

21 Metals and Solid-State Materials   892

22 The Main-Group Elements   927

23 Organic and Biological Chemistry 978

List of Inquiries

About the Authors

John McMurry, educated at Harvard and Columbia, has taught more than 20,000 students in general and organic chemistry over a 40-year period. An emeritus professor of chemistry at Cornell University, Dr. McMurry previously spent 13 years on the faculty at the University of California at Santa Cruz. He has received numerous awards, including the Alfred P. Sloan Fellowship (1969–71), the National Institute of Health Career Development Award (1975–80), the Alexander von Humboldt Senior Scientist Award (1986–87), and the Max Planck Research Award (1991). With the publication of this new edition, he has now authored or coauthored 34 textbooks in various fields of chemistry.

Robert C. Fay, professor emeritus at Cornell University, taught general and inorganic chemistry at Cornell for 45 years beginning in 1962. Known for his clear, well-organized lectures, Dr. Fay was the 1980 recipient of the Clark Distinguished Teaching Award. He has also taught as a visiting professor at Harvard University and the University of Bologna (Italy). A Phi Beta Kappa graduate of Oberlin College, Dr. Fay received his Ph.D. from the University of Illinois. He has been an NSF Science Faculty Fellow at the University of East Anglia and the University of Sussex (England) and a NATO/Heineman Senior Fellow at Oxford University.

Jill K. Robinson received her Ph.D. in Analytical and Atmospheric Chemistry from the University of Colorado at Boulder. She is a Senior Lecturer at Indiana University and teaches general, analytical, and environmental chemistry courses. Her clear and relatable teaching style has been honored with several awards ranging from the Student Choice Award from the University of Wyoming Honors College to the President's Award for Distinguished Teaching at Indiana University. She develops active learning materials for the analytical digital sciences library, promotes nanoscience education in local schools, and serves as advisor for student organizations.

FOR THE STUDENT

Francie came away from her first chemistry lecture in a glow. In one hour she found out t[ ] everything was made up of atoms which were in continual motion. She grasped the idea t[ ] nothing was ever lost or destroyed. Even if something was burned up or rotted away, it [ ] not disappear from the face of the earth; it changed into something else—gases, liquids, a[ ] powders. Everything, decided Francie after that first lecture, was vibrant with life and th[ ] was no death in chemistry. She was puzzled as to why learned people didn't adopt chemis[ ] as a religion.

—**Betty Smith,** *A Tree Grows in Brook[ ]*

OK, not everyone has such a breathless response to their chemistry lectures, and few wo[ ] mistake chemistry as a religion, yet chemistry *is* a subject with great logical beauty. We l[ ] chemistry because it explains the "why" behind many observations of the world around [ ] and we use it every day to help us make informed choices about our health, lifestyle, a[ ] politics. Moreover, chemistry is the fundamental, enabling science that underlies many of t[ ] great advances of the last century that have so lengthened and enriched our lives. Chemist[ ] provides a strong understanding of the physical world and will give you the foundation y[ ] need to go on and make important contributions to science and humanity.

HOW TO USE THIS BOOK

You no doubt have experience using textbooks and know they are not meant to read lik[ ] novel. We have written this book to provide you with a clear, cohesive introduction to che[ ] istry in a way that will help you, as a new student of chemistry, understand and relate to t[ ] subject. While you *could* curl up with this book, you will greatly benefit from continually f[ ] mulating questions and checking your understanding as you *work* through each section. T[ ] way this book is designed and written will help you keep your mind active, thus allowing y[ ] to digest big ideas as you learn some of the many principles of chemistry. Features of this bo[ ] and how you should use them to maximize your learning are described below.

1. **Narrative:** As you read through the text, always challenge yourself to understand t[ ] "why" behind the concept. For example, you will learn that carbon forms four bonds, a[ ] the narrative will give the reason why. By gaining a conceptual understanding, you w[ ] *not need to memorize* a large collection of facts, making learning and retaining importa[ ] principles much easier!

2. **Figures:** Figures are not optional! Most are carefully designed to summarize and conv[ ] important points. *Figure It Out* questions are included to draw your attention to a k[ ] principle. Answer the question by examining the figure and perhaps rereading the relat[ ] narrative. Answers to *Figure It Out* questions are provided near the figure.

3. **Worked Examples:** Numerous worked examples are given throughout the text to sho[ ] the approach for solving a certain type of problem. A stepwise procedure is used with[ ] each worked example.

   Identify—The first step in problem solving is to identify key information and classify [ ] as a known or unknown quantity. This step also involves translating between words a[ ] chemical symbols. Listing knowns on one side and unknowns on the other organizes th[ ] information and makes the process of identifying the correct strategy more visual. Th[ ] *Identify* step will be used in numerical problems.

   Strategy—The strategy describes how to solve the problem without actually solvi[ ] it. Failing to articulate the needed strategy is a common pitfall; too often studen[ ] start manipulating numbers and variables without first identifying key equations or making a plan. Articulating a strategy will develop conceptual understanding and is highly preferable to simply memorizing the steps involved in solving a certain type of problem.

Solution—Once the plan is outlined, the key information can be used and the answer obtained.

Check—A problem is not completed until you have thought about whether the answer makes sense. Use both your practical knowledge of the world and knowledge of chemistry to evaluate your answer. For example, if heat is added to a sample of liquid water and you are asked to calculate the final temperature, you should critically consider your answer: Is the final temperature lower than the original? Shouldn't adding heat raise the temperature? Is the new temperature above 100 °C, the boiling point of water? The *Check* step will be used in problems when the magnitude and sign of a number can be estimated or the physical meaning of the answer verified based on familiar observations.

To test your mastery of the concept explored in worked examples, two problems will follow. **PRACTICE** problems are similar in style and complexity to the worked example and will test your basic understanding. Once you have correctly completed this problem, tackle the **APPLY** problem, in which the concept is used in new situation. Video tutorials explaining some of the APPLY problems illustrate the process of expert thinking and point out how the same principle can be used in multiple ways.

4. **Conceptual Problems:** Conceptual understanding is a primary focus of this book. Conceptual problems are intended to help you with the critical skill of visualizing the structure and interactions of atoms and molecules while probing your understanding of key principles rather than your ability to correctly use numbers in an equation. The time you spend mastering these problems will provide high long-term returns by solidifying main ideas.

5. **Inquiries:** Inquiry sections connect chemistry to the world around you by highlighting useful links in the future careers of many science students. Typical themes are materials, medicine, and the environment. The goal of these sections is to deepen your understanding and aid in retention by tying concepts to memorable applications. These sections can be considered as a capstone for each chapter because *Inquiry* problems review several main concepts and calculations. These sections will also help you prepare for professional exams because they were written in the same style as new versions of these exams. For example, starting in 2015 the MCAT will provide a reading passage about a medical situation and you will be required to apply physical and chemical principles to interpret the system.

6. **End-of-Chapter Study Guide and Problem Sets:** The end-of-chapter study guide can be used either during active study of the chapter or to prepare for an exam. The concept summary provides the central idea for each section, and learning objectives specify key skills needed to solve a variety of problems. Learning objectives are linked to end-of-chapter problems so that you can assess your mastery of that skill.

Working problems is essential for success in chemistry! The number and variety of problems at the end of chapter will give you the practice needed to gain mastery of specific concept. Answers to every other problem are given in the "Answers" section at the back of the book so that you can assess your understanding.

For Instructors

NEW TO THIS EDITION

One of the biggest challenges for general chemistry students is that they are often overwhelmed by the number of topics and massive amount information in the course. Frequently, they do not see connections between new material and previous content, thus creating barriers to learning. *Therefore, the table of contents was revised to create more uniform themes* within chapters and a coherent progression of concepts that build on one another.

- The focus of Chapter 1 has been changed to experimentation and measurements. In this 7th edition, the periodic table and element properties are covered in Chapter 2 (Chemistry Fundamentals: Elements, Molecules, and Ions).

- Coverage of nuclear reactions, radioactivity, and nuclear stability has been consolidated in this edition. Copy on nuclear reactions formerly found in Chapter 2 has been moved to Chapter 19 to keep all nuclear chemistry within one chapter.

- Solution stoichiometry and titrations were moved from Chapter 3 (Mass Relationships in Chemical Reactions) to Chapter 4 (Reactions in Aqueous Solutions).

- At the suggestion of instructors who used the last edition, coverage of redox stoichiometry now appears in the electrochemistry chapter where it is most needed. This change simplifies Chapter 4, which now serves as an introduction to aqueous reactions.

- The new edition features a chapter dedicated to main group chemistry. Main group chemistry sections formerly appearing in Chapter 6: Ionic Compounds: Periodic Trends and Bonding Theory are now incorporated into Chapter 22: The Main Group Elements.

- Covalent bonding and molecular structure are now covered in two chapters (7 and 8) to avoid having to cover an overwhelming amount of material in one chapter. The topic of intermolecular forces was added to Chapter 8 to reinforce its connection to polarity.

- Nuclear chemistry has been moved forward in the table of contents because of its relevance in energy production, medicine, and the environment.

- The chapter on hydrogen and oxygen has been omitted, but key chemical properties and reactions of hydrogen and oxygen are now covered in Chapter 22: The Main Group Elements.

- Chapter 10 dealing with gases now includes content on air pollution and climate change.

- Chapter 23 has been heavily revised to review important general chemistry principles of bonding and structure as they apply to organic and biological molecules. This chapter may be covered as a standalone chapter or sections may be incorporated into earlier chapters if an instructor prefers to cover organic and biological chemistry throughout the year.

NEW! All Worked Examples have been carefully revisited in the context of newly articulated Learning Outcomes.

Worked examples are now tied to Learning Outcomes listed at chapter end and to representative EOC problems so that students can test their own mastery of each skill.

Select worked examples now contain a section called **Identify**, which lays out the known and unknown variables for students. Listing knowns on one side and unknowns on the other organizes the information and makes the process of identifying the correct strategy more visual. The *Identify* step will be used in numerical problems with equations.

Worked examples in the 7th edition now conclude with two problems, one called **Practice** and the other called **Apply**, to help students see how the same principle can be used in different types of problems with different levels of complexity.

To discourage a plug-and-chug approach to problem solving, related Worked Examples from the previous edition have been consolidated, giving students a sense of how different approaches are related.

The number of in-chapter problems has increased by 20% to encourage the students to work problems actively immediately after reading.

NEW! Inquiry Sections have been updated and integrated conceptually into each chapter.

Inquiry sections highlight the importance of chemistry, promote student interest, and deepen the students understanding of the content. The new Inquiry sections include problems that revisit several chapter concepts and can be covered in class, recitation sections, or assigned as homework in MasteringChemistry.

NEW! Chapter Study Guide offers a modern and innovative way for students to review each chapter.

Prepared in a grid format, the main lessons of each chapter are reiterated and linked to learning objectives, associated worked examples, and representative end-of-chapter problems.

NEW! Figure It Out questions promote active learning.

Selected figures are tagged with questions designed to prompt students to look at each illustration more carefully, and interpret graphs and recognize key ideas.

NEW! Looking Forward Notes are now included.

Looking Forward Notes, in addition to Remember Notes, are included to underscore and reiterate connections between topics in different chapters.

NEW! Over 600 new problems have been written for the 7th edition.

New problems ensure there is a way to assess each learning objective in the Study Guide, all of which are suitable for use in MasteringChemistry.

The seventh edition was extensively revised. Here is a list of some of the key changes made in each chapter:

Chapter 1 Chemical Tools: Experimentation and Measurement

- Chapter 1 focuses on experimentation, the scientific method, and measurement and offers a new, robust Inquiry on nanotechnology.
- The scientific method is described in the context of a case study for the development of an insulin drug.

Chapter 2 Atoms, Molecules, and Ions

- Material on the elements and periodic table previously found in Chapter 1 has been relocated here, and nuclear chemistry has been moved to the nuclear chemistry chapter.
- Coverage of the naming of binary molecular compounds was moved to a later point in the chapter to consolidate coverage of the naming of ionic compounds.
- A new Inquiry on green chemistry, focusing on the concept of atom economy, revisits the Law of Conservation of Mass.

Chapter 3 Mass Relationships in Chemical Reactions

- Section 3.2 includes a revised Worked Example on balancing chemical reactions to give students a chance to use the method in simple and complex problems.
- New coverage of mass spectrometry in Section 3.8 explains how molecular weights are measured and mass spectral data

is utilized in problems. The topic of mass spectrometry is connected to crime scene analysis and offers a good example of how the new edition presents chemistry in a modern way.

- A new Inquiry explores CO_2 emissions from various alternative fuels using concepts of stoichiometry.

Chapter 4 Reactions in Aqueous Solution

- Section order and coverage were revised to keep the focus on solution chemistry.
- Problems and worked examples are rearranged so that conceptual worked examples lead off the discussion rather than wrap it up.
- The new Inquiry on sports drinks applies the concepts of electrolytes, solution concentration, and solution stoichiometry.

Chapter 5 Periodicity and Electronic Structure of Atoms

- Section 5.3 on line spectra has been revised to better show how spectral lines of the elements are produced.
- Sections 5.7–5.10 offer a more continuous description of how orbitals can be described using quantum numbers.
- The Inquiry on fluorescent lights was revised to include problems that require students to write electron configurations and interpret line spectra.

Chapter 6 Ionic Compounds: Periodic Trends and Bonding Theory

- As this is the first of three chapters on bonding, it now includes some introduction to topic sequence in Chapters 6–8.

- Every chapter problem is now preceded by a Worked Example and followed by Practice and Apply problems.

- New figures in Section 6.2 help visualize why creating an ion changes the size of an atom.

- Updated Inquiry on ionic liquids includes problems on writing ion electron configurations and relating ion size to properties of the ionic compound.

- Main group chemistry now appears in Chapter 22 (Main Group Chemistry).

Chapter 7 Covalent Bonding and Electron-Dot Structures

- Chapter 7 is now dedicated to covalent bonding using the Lewis electron-dot model. Valence shell electron pair repulsion theory, molecular shape, and molecular orbital theory now appear in Chapter 8.

- Section 7.6 summarizes a general procedure for drawing electron-dot structures and applies the procedure in new Worked Examples.

- The coverage of resonance includes an introduction to the use of curved arrows to denote rearrangement of electrons, a practice that is commonly used in organic chemistry courses.

- The new Inquiry, "How do we make organophosphate insecticides less toxic to humans?," builds on several concepts introduced in this chapter, including polar covalent bonds, electron-dot structures, and resonance.

- The chapter includes many new figures. Much of the new art appears in revised Worked Examples, replacing and/or embellishing Worked Examples appearing in the prior edition.

Chapter 8 Covalent Compounds: Bonding Theories and Molecular Structure

- The focus of Chapter 7 is covalent bonds and electron-dot structures, whereas the focus of Chapter 8 is quantum mechanical theories of covalent bonding, molecular shape, polarity, and intermolecular forces. Polarity and intermolecular forces are a direct extension of molecular shape and have been moved from Chapter 10 to Chapter 8.

- Section 8.1 on the VSEPR model explains use of solid wedges and dashed lines to draw the 3-D structure of molecules.

- Many Worked Examples in this chapter were substantively revised to reflect the chapter's new emphasis. New figures for Worked Examples 8.3 and 8.4 illustrate orbital overlap involved in each type of bond.

- Section 8.5 includes new Figure 8.8: A flowchart to show the strategy for determining molecular polarity. Worked Example 8.6 was revised to follow this flowchart.

- A New Conceptual Worked Example on drawing hydrogen bonds and new end of chapter problems were developed.

- The Inquiry for this chapter was expanded to include intermolecular forces in biomolecular binding. Two new figures were added to illustrate how the mirror image has a different geometric arrangement of atoms and how this can lead to discrimination between these two molecules by a receptor site. New cumulative problems were added that include all topics in the chapter thus far: geometry, hybridization, polarity, intermolecular forces, and mirror images.

Chapter 9 Thermochemistry: Chemical Energy

- Section 9.2, Internal Energy and State Function, includes a new figure to illustrate ΔE in an example of the caloric content of food.

- Section 9.4, Energy and Enthalpy, has a new figure illustrating energy transfer as heat and work in a car's engine to help students grasp the meaning of internal energy.

- Section 9.5, entitled "Thermochemical Equations and the Thermodynamic Standard State," covers all aspects of writing and manipulating thermochemical equations (standard state, stoichiometry, reversibility, and importance of specifying phases).

- Section 9.6 on Enthalpy of Chemical and Physical Change offers improved definitions of endothermic and exothermic phenomena, including new Worked Examples and problems on classifying reactions and identifying direction of heat transfer.

Chapter 10 Gases: Their Properties and Behavior

- Chapter 10 is revised to include three new sections on atmospheric chemistry (air pollution, the greenhouse effect, and climate change) and a new Inquiry on greenhouse gases.

- There are thirty new end-of-chapter problems that require students to describe atmospheric chemistry and utilize many chemistry skills covered thus far in the book.

Chapter 11 Liquids, Solids, and Phase Changes

- Worked Example 11.2 is new and describes how to calculate the energy change associated with heating and phase changes.

- New Section 11.5 now includes two new images to enhance discussion of X-ray diffraction experiments.

- The Inquiry on decaffeination is new and builds on the topics of phase diagrams and energy of phase changes.

Chapter 12 Solutions and Their Properties

- Section 12.2 on Energy Changes and the Solution Process includes a new figure illustrating the hydrogen bonding interactions between solute and solvent (added emphasis on chemical structure and visual explanation of solubility).

- Section 12.3 on Concentration Units for Solutions has refined coverage of concentration units and a new Worked Example on ppm and ppb.

- Section 12.6 on Vapor-Pressure Lowering includes new Worked Examples on the van't Hoff factor and on vapor pressure lowering with a volatile solute.

- The Inquiry on dialysis was expanded and improved through the addition of an illustration of dialysis and follow-up problems dealing with solution concentration and colligative properties.

Chapter 13 Chemical Kinetics

- The first section includes a generic introduction to the concept of a reaction rate, which is now used in problems throughout the chapter instead of reaction rates specific to a reactant or product.
- A new section on Enzyme Catalysis (Section 13.14) has been added, along with new end-of-chapter problems on this topic.
- Coverage of radioactive decay formerly included in this chapter has been moved to the nuclear chemistry chapter.
- The new Inquiry on ozone depletion builds on various kinetics concepts including activation energy determination, calculation of rate, reaction mechanisms, catalysis.

Chapter 14 Chemical Equilibrium

- Section 14.2 on The Equilibrium Constant K_c has an expanded discussion and new Worked Examples dealing with manipulating equations and calculating new values of K_c.
- Section 14.4 on Heterogeneous Equilibria has been revised to clarify when concentrations of pure solids and liquids present in a chemical equation are not included in the equilibrium constant.
- Section 14.5 on Using the Equilibrium Constant has been enhanced by the addition of a new worked example on Judging the Extent of a Reaction.
- Figure 14.6, entitled Steps in Calculating Equilibrium Concentrations, was modified to include the important first step of determining reaction direction.
- The Inquiry on equilibrium and oxygen transport now includes several follow-up problems that give students practice with various equilibrium concepts.

Chapter 15 Aqueous Equilibria: Acids and Bases

- Section 15.3, Factors that Affect Acid Strength, now appears earlier in the chapter to explain *why* chemical structure affects acid strength, and is bolstered by new Worked Example 15.4 entitled 'Evaluating Acid Strength Based Upon Molecular Structure' as well as new end-of-chapter problems.
- Section 15.5 on the pH scale includes new problems exploring environmental issues.
- The Inquiry on acid rain has been updated to include new statistics and a new figure illustrating changes in acid rainfall over time.

Chapter 16 Applications of Aqueous Equilibria

- Coverage of the Henderson-Hasselbalch Equation has been reworked so that students progress from simpler problems to more complex ones.
- Reaction tables are now routinely included in titration problems to help students see what species remain at the end of the neutralization reaction. New Worked Examples are included.
- Section 16.12 on Factors that Affect Solubility has been enhanced with relevant new examples (e.g., tooth decay).
- The new and highly pertinent Inquiry for Chapter 16 on ocean acidification revisits key concepts such as acid-base reactions, buffers, and solubility equilibria in a meaningful environmental context.

Chapter 17 Thermodynamics: Entropy, Free Energy, and Equilibrium

- Section 17.3 on Entropy and Probability is enhanced with a new Worked Example and follow-up problems on the expansion of an ideal gas.
- Different signs of enthalpy and entropy in are broken down on a case-by-case basis in Section 17.7.
- The Inquiry on biological complexity was heavily revised to describe why some biological reactions are spontaneous. The Inquiry now includes concrete examples of the thermodynamics of living systems and four relevant follow-up problems.

Chapter 18 Electrochemistry

- Section on balancing redox reactions using the half-reaction method was taken out of Chapter 4 and placed in Chapter 18 based on reviewer feedback.
- Coverage of fuel cells has been streamlined and incorporated into the Inquiry. New Inquiry problems revisit core thermodynamic and electrochemical concepts.

Chapter 19 Nuclear Chemistry

- All the nuclear chemistry content is now contained in Chapter 19.
- Coverage on balancing a nuclear reaction was revised to more clearly show that mass number and atomic number are equal on both sides of the equation.
- Figure 19.3 was added to illustrate the concept of a radioactive decay series.
- Several improvements were made in Section 19.6 on Fission and Fusion: the difference between nuclear fuel rods used in a reactor and weapons-grade nuclear fuel has been clarified; Figure 19.8 has been updated to include 2013 figures for nuclear energy output.
- New end-of-chapter problems dealing with aspects of nuclear power and nuclear weapons have been added.

Chapter 20 Transition Elements and Coordination Chemistry

- Worked Example 20.5, Identifying Diastereomers, has been revised and moved earlier so that students begin with a conceptual problem.
- Worked Example 20.6, Drawing Diastereomers for Square Planar and Octahedral Complexes, was rewritten to promote conceptual understanding and discourage rote memorization.
- A new Inquiry on the mechanism of action of the antitumor drug cisplatin reinforces several concepts covered in the chapter, including nomenclature, chirality, the formation of coordination compounds, and crystal field theory.

Chapter 21 Metals and Solid-State Materials

- Band theory in metals has been clarified by
  - describing the formation of band from MOs in more detail in the text
  - revising Figure 21.6 to show that bands contain many closely spaced MOs
  - the addition of Figure It Out questions that require extension of band theory to different systems.
- New Figure 21.10 on doping of semiconductors correlates molecular picture with energy level diagrams.

- The connection between LED color and periodic trends is described in Section 21.6. New problems are included.
- The Inquiry on quantum dots was heavily revised to more clearly connect with chapter content on band theory and semiconductors.

Chapter 22 The Main-Group Elements

- Main group chemistry is consolidated into one chapter. The content has been trimmed and key concepts related to periodic trends, bonding, structure, and reactivity are reviewed in the context of main group chemistry.
- The Inquiry Section dealing with barriers to a hydrogen economy describes hydrogen production and storage methods including recent development in photocatalysts.

Chapter 23 Organic and Biological Chemistry

- This chapter was revised so that the focus is on important concepts of structure and bonding that organic chemistry instructors would like students to master in general chemistry.
- Over 50 end-of-chapter problems are completely new.
- Section 23.1 offers an introduction to skeletal structures (line drawings) commonly used as a shorthand method for drawing organic structures.
- Coverage of the alkanes is consolidated in Section 23.1 (the cycloalkanes were formerly covered in Section 23.5 in 6e.).
- Coverage of the naming of organic compounds was shortened in 7e Section 23.3 because the primary focus of the new chapter is on bonding and structure.
- Section 23.4, entitled "Carbohydrates: A Biological Example of Isomers" offers a good example of how the applied chapters at the end of the book explore key concepts (isomerism) in a relevant context (carbohydrates).

- Section 23.4 also offers a good example of how key concepts from other chapters are revisited in the applied chapters at book end. Here chirality is revisited, a subject first presented in the Chapter 8 Inquiry.
- Section 23.5 considers cis-trans isomerism in the context of valence bond theory. Two new Worked Examples are included that describe orbital overlap in organic molecules.
- The theme of cis-trans isomerism is revisited in Section 23.6 with the introduction of the lipids. New Figure 23.6, for example, shows the difference in packing of saturated and unsaturated fats and the role played by intermolecular forces.
- Section 23.5 revisits the concepts of formal charge and resonance first introduced in Ch 7. Problems in this section give students additional practice in the drawing of electron-dot structures and electron "pushing." Common patterns of resonance in organic molecules are introduced as well.
- Section 23.8 is new to the 7th edition, covering conjugated systems in the context of resonance and orbital diagrams. New worked examples tie the section together, offering problems on drawing conjugated π systems, and exploring how to recognize localized vs delocalized electrons.
- Section 23.9, entitled "Proteins: A Biological Example of Conjugation" follows logically from Section 23.8 to look at conjugation in the peptide bond and proteins.
- Section 23.10, new to the 7th edition, considers aromatic compounds in the context of molecular orbital theory. Building on students' understanding of conjugation, molecular orbital theory is invoked to describe the stability of benzene.
- Section 23.11 on the nucleic acids expands on the discussion of aromaticity in describing how aromaticity makes base stacking in the interior of the DNA molecule possible.

ACKNOWLEDGMENTS

Our thanks go to our families and to the many talented people who helped bring this new edition into being. We are grateful to Chris Hess, Acquisitions Editor, for his insight and suggestions that improved the book, to Carol Pritchard-Martinez for her critical review that made the art program and manuscript more understandable for students, to Will Moore, Marketing Manager, who brought new energy to describing features of the seventh edition, to Jenna Vittorioso, Jessica Moro, and Lisa Pierce for their production and editorial efforts. Thank you to Mimi Polk for coordinating art production, and to Liz Kincaid for her photo research efforts. We wish to thank Dr. Ben Burlingham for his contributions in the revision of Chapter 23: Organic and Biological Chemistry. His expertise teaching Organic and Biochemistry led to many improvements that will give students a strong foundation to build upon in future courses.

We are particularly pleased to acknowledge the outstanding contributions of several colleagues who created the many important supplements that turn a textbook into a complete package:

- Charity Lovitt, *University of Washington, Bothell*, and Christine Hermann, *Radford University*, who updated the accompanying Test Bank.
- Joseph Topich, *Virginia Commonwealth University*, who prepared both the full and partial solutions manuals
- Mark Benvenuto, *University of Detroit Mercy*, who contributed valuable content for the Instructor Resource DVD.
- James Zubricky, *The University of Toledo*, who prepared the Student Study Guide to accompany this seventh edition.
- Dennis Taylor, *Clemson University*, who prepared the Instructor Resource Manual
- Sandra Chimon-Peszek, *Calumet College of St. Joseph*, who updated the Laboratory Manual.

Finally, we want to thank all accuracy checkers, text reviewers, our colleagues at so many other institutions who read, criticized, and improved our work.

John McMurry
Robert C. Fay
Jill K. Robinson

REVIEWERS FOR THE SEVENTH EDITION

Jamcs Almy, Golden West College
James Ayers, CO Mesa University
Amina El-Ashmawy, Collin College
Robert Blake, Glendale Community College
Gary Buckley, Cameron University
Ken Capps, Central FL Community College
Joe Casalnuovo, Cal Poly Pomona
Sandra Chimon-Peszek, Calumet College of St. Joseph
Claire Cohen, University of Toledo
David Dobberpuhl, Creighton University
Cheryl Frech, University of Central Oklahoma
Chammi Gamage-Miller, Blinn College–Bryan Campus
Rachel Garcia, San Jacinto College
Carolyn Griffin, Grand Canyon University
Nathanial Grove, UNC Wilmington
Alton Hassell, Baylor University
Sherman Henzel, Monroe Community College
Geoff Hoops, Butler University
Andy Jorgensen, University of Toledo
Jerry Keister, SUNY Buffalo
Angela King, Wake Forest University

Regis Komperda, Wright State University
Peter Kuhlman, Denison University
Don Linn, IUPU Fort Wayne
Rosemary Loza, Ohio State University
Rod Macrae, Marian University
Riham Mahfouz, Thomas Nelson Community College
Jack McKenna, St. Cloud State University
Craig McLauchlan, Illinois State University
Ed Navarre, Southern Illinois University Edwardsville
Christopher Nichols, California State University–Chico
Mya Norman, University of Arkansas
Kris Quinlan, University of Toronto
Betsy Ratcliffe, West Virginia University
Al Rives, Wake Forest University
Richard Roberts, Des Moines Area Community College–Ankeny
Mark Schraf, West Virginia University
Lydia Tien, Monroe Community College
Erik Wasinger, California State University–Chico
Mingming Xu, West Virginia University
James Zubricky, University of Toledo

REVIEWERS OF THE PREVIOUS EDITIONS OF CHEMISTRY

Laura Andersson, Big Bend Community College
David Atwood, University of Kentucky
Mufeed Basti, North Carolina A&T State University
David S. Ballantine, Northern Illinois University
Debbie Beard, Mississippi State University
Ronald Bost, North Central Texas University
Danielle Brabazon, Loyola College
Robert Burk, Carleton University
Myron Cherry, Northeastern State University
Allen Clabo, Francis Marion University
Paul Cohen, University of New Jersey
Katherine Covert, West Virginia University
David De Haan, University of San Diego
Nordulf W. G. Debye, Towson University
Dean Dickerhoof, Colorado School of Mines
Kenneth Dorris, Lamar University
Jon A. Draeger, University of Pittsburgh at Bradford
Brian Earle, Cedar Valley College
Amina El- Ashmawy, Collin County Community College
Joseph W. Ellison, United States Military Academy at West Point
Erik Eriksson, College of the Canyons
Peter M. Fichte, Coker College
Kathy Flynn, College of the Canyons
Joanne Follweiler, Lafayette College

Ted Foster, Folsom Lake College
Cheryl Frech, University of Central Oklahoma
Mark Freilich, University of Memphis
Mark Freitag, Creighton University
Travis Fridgen, Memorial University of Newfoundland
Jack Goldsmith, University of South Carolina Aiken
Thomas Grow, Pensacola Junior College
Katherine Geiser-Bush, Durham Technical Community College
Mildred Hall, Clark State University
Tracy A. Halmi, Pennsylvania State University Erie
Keith Hansen, Lamar University
Lois Hansen-Polcar, Cuyahoga Community College
Wesley Hanson, John Brown University
Michael Hauser, St. Louis Community College–Meramec
M. Dale Hawley, Kansas State University
Patricia Heiden, Michigan Tech University
Thomas Hermann, University of California–San Diego
Thomas Herrington, University of San Diego
Margaret E. Holzer, California State University–Northridge
Todd Hopkins, Baylor University
Narayan S. Hosmane, Northern Illinois University
Jeff Joens, Florida International University
Jerry Keister, University of Buffalo
Chulsung Kim, University of Dubuque

Ranjit Koodali, University of South Dakota
Valerie Land, University of Arkansas Community College
John Landrum, Florida International University
Leroy Laverman, University of California–Santa Barbara
Celestia Lau, Lorain County Community College
Stephen S. Lawrence, Saginaw Valley State University
David Leddy, Michigan Technological University
Shannon Lieb, Butler University
Karen Linscott, Tri-County Technical College
Irving Lipschitz, University of Massachusetts–Lowell
Rudy Luck, Michigan Technological University
Ashley Mahoney, Bethel College
Jack F. McKenna, St. Cloud State University
Iain McNab, University of Toronto
Christina Mewhinney, Eastfield College
David Miller, California State University–Northridge
Rebecca S. Miller, Texas Tech University
Abdul Mohammed, North Carolina A&T State University
Linda Mona, United States Naval Academy
Edward Mottell, Rose-Hulman Institute
Gayle Nicoll, Texas Technological University
Allyn Ontko, University of Wyoming
Robert H. Paine, Rochester Institute of Technology
Cynthia N. Peck, Delta College
Eileen Pérez, University of South Florida

Michael R. Ross, College of St. Benedict/St. John's University
Lev Ryzhkov, Towson University
Svein Saebo, Mississippi State University
John Schreifels, George Mason University
Patricia Schroeder, Johnson County Community College
David Shoop, John Brown University
Penny Snetsinger, Sacred Heart University
Robert L. Snipp, Creighton University
Steven M. Socol, McHenry County College
Thomas E. Sorensen, University of Wisconsin–Milwaukee
L. Sreerama, St. Cloud State University
Keith Stein, University of Missouri–St. Louis
Beth Steiner, University of Akron
Kelly Sullivan, Creighton University
Susan Sutheimer, Green Mountain College
Andrew Sykes, University of South Dakota
Erach Talaty, Wichita State University
Edwin Thall, Florida Community College at Jacksonville
Donald Van Derveer, Georgia Institute of Technology
John B. Vincent, University of Alabama
Steve Watton, Virginia Commonwealth University
Marcy Whitney, University of Alabama
James Wu, Tarrant County Community College
Crystal Lin Yau, Towson University

Showing Students the Connections in Chemistry and Why They Matter

McMurry/Fay/Robinson's Chemistry, Seventh Edition provides a streamlined presentation that blends the quantitative and visual aspects of chemistry, organizes content to highlight connections between topics and emphasizes the application of chemistry to students lives and careers. New content provides a better bridge between organic and biochemistry and general chemistry content, and new and improved pedagogical features make the text a true teaching tool and not just a reference book.

New MasteringChemistry features include conceptual worked examples and integrated Inquiry sections that help make critical connections clear and visible and increase students' understanding of chemistry. The Seventh Edition fully integrates the text with new MasteringChemistry content and functionality to support the learning process before, during, and after class.

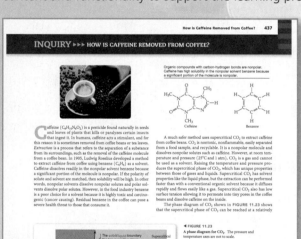

Inquiry Updated inquiry sections now include worked examples and practice problems so that students can apply concepts and skills to scenarios that have relevance to their daily lives. These sections not only highlight the importance of chemistry and promote interest but also deepen students understanding of the content.

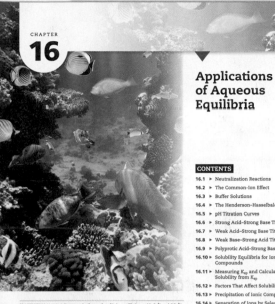

CHAPTER

16

Applications of Aqueous Equilibria

CONTENTS

16.1 ▶ Neutralization Reactions
16.2 ▶ The Common-Ion Effect
16.3 ▶ Buffer Solutions
16.4 ▶ The Henderson-Hasselbalch Equation
16.5 ▶ pH Titration Curves
16.6 ▶ Strong Acid–Strong Base Titrations
16.7 ▶ Weak Acid–Strong Base Titrations
16.8 ▶ Weak Base–Strong Acid Titrations
16.9 ▶ Polyprotic Acid–Strong Base Titrations
16.10 ▶ Solubility Equilibria for Ionic Compounds
16.11 ▶ Measuring K_{sp} and Calculating Solubility from K_{sp}
16.12 ▶ Factors That Affect Solubility
16.13 ▶ Precipitation of Ionic Compounds
16.14 ▶ Separation of Ions by Selective Precipitation
16.15 ▶ Qualitative Analysis

STUDY GUIDE

The limestone ($CaCO_3$) framework of a coral reef is in equilibrium with Ca^{2+} and CO_3^{2-} ions in the ocean.

? What is causing a decrease in the pH of the oceans?

The answer to this question can be found in the **INQUIRY** ▸▸▸ on page 703.

Chapter Openers introduce a topic and question, related to the content in the chapter that will be further discussed in the Inquiry section. The goal is to provide students with a central topic to serve as a waypoint for why the content matters to them beyond what they need to pass the next exam.

CHAPTER

1

Chemical Tools: Experimentation and Measurement

CONTENTS

1.1 ▶ The Scientific Method in a Chemical Context: Improved Pharmaceutical Insulin
1.2 ▶ Experimentation and Measurement
1.3 ▶ Mass and Its Measurement
1.4 ▶ Length and Its Measurement
1.5 ▶ Temperature and Its Measurement
1.6 ▶ Derived Units: Volume and Its Measurement
1.7 ▶ Derived Units: Density and Its Measurement
1.8 ▶ Derived Units: Energy and Its Measurement
1.9 ▶ Accuracy, Precision, and Significant Figures in Measurement
1.10 ▶ Rounding Numbers
1.11 ▶ Calculations: Converting from One Unit to Another

STUDY GUIDE

Instruments for scientific measurements have changed greatly over the centuries. Modern technology has enabled scientists to make images of extremely tiny particles, even individual atoms, using instruments like this atomic force microscope.

? What are the unique properties of nanoscale (1 nm = 10^{-9} m) materials?

The answer to this question can be found in the **INQUIRY** ▸▸▸ on page 23.

Study Guide The end-of-chapter material now includes a Study Guide to help students review each chapter. Prepared in a grid format, the main lessons of each chapter are laid out and linked to learning objectives, associated worked examples, and representative end-of-chapter problems that allow students to assess their comprehension of the material.

26 **CHAPTER 1** Chemical Tools: Experimentation and Measurement

STUDY GUIDE

Section	Concept Summary	Learning Objectives	Test Your Understanding
1.1 ▶ The Scientific Method	The **scientific method** is an iterative process used to perform research. A driving question, often based upon observations, is the first step. Next a **hypothesis** is developed to explain the observation. **Experiments** are designed to test the hypothesis and the results are used to verify or modify the original hypothesis. **Theories** arise when numerous experiments validate a hypothesis and are used to make new predictions. **Models** are simplified representations of complex systems that help make theories more concrete.	1.1 Identify the steps in the scientific method.	Problems 1.28–1.30
		1.2 Differentiate between a qualitative and quantitative measurement.	Problems 1.33–1.35
1.2 ▶ Experimentation and Measurement	Accurate measurement is crucial to scientific experimentation. Scientists use units of measure established by the *Système International* (**SI units**). There are seven fundamental SI units, together with other derived units. (**Table 1.1**)	1.3 Write numbers in scientific notation and use prefixes for multiples of SI units.	Worked Example 1.1; Problems 1.39, 1.49, 1.52, 1.58, and 1.59
1.3 ▶ Mass and Its Measurement	**Mass**, the amount of matter in an object, is measured in the SI unit of **kilograms (kg)**.	1.4 Describe the difference between mass and weight.	Problem 1.36
		1.5 Convert between different prefixes used in mass measurements.	Problem 1.50
1.4 ▶ Length and Its Measurement	**Length** is measured in the SI unit of **meters (m)**.	1.6 Convert between different prefixes used in length measurements.	Problem 1.52 (a) and (b)
1.5 ▶ Temperature and Its Measurement	**Fahrenheit (°F)** is the most common unit for measuring temperature in the United States, whereas **Celsius (°C)** is more common in other parts of the world. **Kelvin (K)** is the standard temperature unit in scientific work.	1.7 Convert between common units of temperature measurements.	Worked Example 1.2; Problems 1.74–1.77
1.6 ▶ Derived Units: Volume and Its Measurement	**Volume**, the amount of space occupied by an object, is measured in SI units by the **cubic meter (m^3)**.	1.8 Convert between SI and metric units of volume.	Problems 1.42 and 1.43, 1.99
		1.9 Convert between different prefixes used in volume measurements.	Problem 1.51
1.7 ▶ Derived Units: Density and Its Measurement	**Density** is a property that relates mass to volume and is measured in the derived SI unit g/cm^3 or g/mL.	1.10 Calculate mass, volume, or density using the formula for density.	Worked Example 1.3; Problems 1.80–1.88, 1.96, 1.100, 1.101
		1.11 Predict whether a substance will float or sink in another substance based on density.	Problem 1.27, 1.97, 1.107
1.8 ▶ Derived Units: Energy and Its Measurement	**Energy** is the capacity to supply heat or do work and is measured in the derived SI unit ($kg \cdot m^2/s^2$), or **joule (J)**. Energy is of two kinds, potential and kinetic. **Kinetic energy (E_K)** is the energy of motion, and **potential energy (E_P)** is stored energy.	1.12 Calculate kinetic energy of a moving object.	Worked Example 1.4; Problem 1.60
		1.13 Convert between common energy units.	Problems 1.94 and 1.95
1.9 ▶ Accuracy,	If measurements are **accurate**, they are close to the	1.14 Specify the number of significant	Worked Example

Helping students relate chemical reasoning to mathematical operations

---

WORKED EXAMPLE 19.3

Using Half-Life to Calculate an Amount Remaining

Phosphorus-32, a radioisotope used in leukemia therapy, has a half-life of 14.26 days. What percent of a sample remains after 35.0 days?

IDENTIFY

Known	Unknown
$t_{1/2} = 14.26$ days	Percent of sample remaining $(N_t/N_0) \times 100$
$t = 35.0$ days	

STRATEGY

The ratio of remaining (N_t) and initial (N_0) amounts of a radioactive sample at time t is given by the equation

$$\ln\left(\frac{N_t}{N_0}\right) = -kt$$

Taking N_0 as 100%, N_t can then be obtained. The value of the rate constant can be found from the equation $k = 0.693/t_{1/2}$.

SOLUTION

The value of the rate constant is calculated using the half-life.

$$k = \frac{0.693}{14.26 \text{ days}} = 4.860 \times 10^{-2} \text{ days}^{-1}$$

Substituting values for t and for k into the equation gives

$$\ln\left(\frac{N_t}{N_0}\right) = (-4.860 \times 10^{-2} \text{ days}^{-1})(35.0 \text{ days}) = -1.70$$

Taking the natural antilogarithm of -1.70 then gives the ratio N_t/N_0:

$$\frac{N_t}{N_0} = \text{antiln}(-1.70) = e^{-1.70} = 0.183$$

Since the initial amount of ^{32}P was 100%, we can set $N_0 = 100\%$ and solve for N_t:

$$\frac{N_t}{100\%} = 0.183 \quad \text{so} \quad N_t = (0.183)(100\%) = 18.3\%$$

After 35.0 days, 18.3% of a ^{32}P sample remains and $100\% - 18.3\% = 81.7\%$ has decayed.

CHECK

We can estimate the answer by considering half-life. Since phosphorus-32 has a half-life of 14.26 days and the time is 35 days, we know the time of the reaction is more than two half-lives. After two half-lives, 75% has reacted and 25% remains. Since the time is over two half-lives, we would estimate that less than 25% remains, which agrees with the answer of 18.3%.

▶ **PRACTICE 19.7** What percentage of $^{14}_{6}$C ($t_{1/2} = 5715$ years) remains in a sample estimated to be 16,230 years old?

▶ **APPLY 19.8** Cesium-137 is a radioactive isotope released as a result of the Fukushima Daiichi nuclear disaster in Japan in 2011. If 89.2% remains after 5.00 years, what is the half-life?

WORKED EXAMPLE 9.2

Calculating Internal Energy Change (ΔE) for a Reaction

The reaction of nitrogen with hydrogen to make ammonia has $\Delta H = -92.2$ kJ. What is the value of ΔE in kilojoules if the reaction is carried out at a constant pressure of 40.0 atm and the volume change is -1.12 L?

$$N_2(g) + 3 H_2(g) \longrightarrow 2 NH_3(g) \quad \Delta H = -92.2 \text{ kJ}$$

IDENTIFY

Known	Unknown
Change in enthalpy ($\Delta H = -92.2$ kJ)	Change in internal energy (ΔE)
Pressure ($P = 40.0$ atm)	
Volume Change ($\Delta V = -1.12$ L)	

STRATEGY

We are given an enthalpy change ΔH, a volume change ΔV, and a pressure P and asked to find an energy change ΔE. Rearrange the equation $\Delta H = \Delta E + P\Delta V$ to the form $\Delta E = \Delta H - P\Delta V$ and substitute the appropriate values for ΔH, P, and ΔV.

SOLUTION

$$\Delta E = \Delta H - P\Delta V$$
where $\Delta H = -92.2$ kJ
$$P\Delta V = (40.0 \text{ atm})(-1.12 \text{ L}) = -44.8 \text{ L} \cdot \text{atm}$$
$$= (-44.8 \text{ L} \cdot \text{atm})\left(101 \frac{\text{J}}{\text{L} \cdot \text{atm}}\right) = -4520 \text{ J} = -4.52 \text{ kJ}$$
$$\Delta E = (-92.2 \text{ kJ}) - (-4.52 \text{ kJ}) = -87.7 \text{ kJ}$$

CHECK

The sign of ΔE is similar in size and magnitude ΔH, which is to be expected because energy transfer as work is usually small compared to heat.

▶ **PRACTICE 9.3** The reaction between hydrogen and oxygen to yield water vapor has $\Delta H = -484$ kJ. How much PV work is done, and what is the value of ΔE in kilojoules for the reaction of 2.00 mol of H_2 with 1.00 mol of O_2 at atmospheric pressure if the volume change is -24.4 L?

$$2 H_2(g) + O_2(g) \longrightarrow 2 H_2O(g) \quad \Delta H° = -484 \text{ kJ}$$

▶ **Conceptual APPLY 9.4** The following reaction has $\Delta E = -186$ kJ/mol.

(a) Is the sign of $P\Delta V$ positive or negative? Explain.
(b) What is the sign and approximate magnitude of ΔH? Explain.

Worked Examples Numerous Worked Examples show the approach for solving different types of problems using a stepwise procedure.

Identify The first step helps students identify key information and classify the known or unknown variables. This step frequently involves translating between words and chemical symbols.

Strategy The strategy describes how to solve the problem without actually solving it. Failing to articulate the needed strategy is a common pitfall; this step involves outlining a plan for solving the problem.

Solution Once the plan is outlined, the key information can be used and the answer obtained.

Check Uses both your practical knowledge of the world and knowledge of chemistry to evaluate your answer.

Conceptual WORKED EXAMPLE 4.7

Visualizing Stoichiometry in Precipitation Reactions

When aqueous solutions of two ionic compounds are mixed, the following results are obtained:

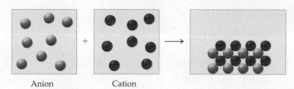

Anion   Cation

(Only the anion of the first compound, represented by blue spheres, and the cation of the second compound, represented by red spheres, are shown.) Which cation and anion combinations are compatible with the observed results?
Anions: NO_3^-, Cl^-, CO_3^{2-}, PO_4^{3-}
Cations: Ca^{2+}, Ag^+, K^+, Cd^{2+}

(a) Identify the limiting and the excess reactant.
(b) How many molecules of excess reactant are left over after the reaction occurs?
(c) How many molecules of product can be made?

STRATEGY

Count the numbers of reactant and product molecules and use coefficients from the balanced equation to relate them to one another.

SOLUTION

(a) Count the number of each type of molecule in the box on the reactant side of the equation. There are 3 ethylene oxide molecules and 5 water molecules. According to the balanced equation the stoichiometry between the reactants is 1:1. Therefore, 5 ethylene oxide molecules would be needed to react with 5 water molecules. Since there are only 3 ethylene oxide molecules, it is the limiting reactant, and water is in excess.

(b) Count the number of water molecules on the product side of the equation. There are 2 water molecules that have not reacted, and water is called the excess reactant.

(c) Count the number of ethylene glycol molecules on the product side of the equation. There are 3 ethylene glycol molecules present.

Therefore, the reaction of 3 ethylene oxide molecules with 5 water molecules results in 3 ethylene glycol molecules with 2 water molecules left over.

3 Ethylene oxide + 5 Water   3 Ethylene glycol + 2 Water

Limiting   Excess   Unreacted
reactant   reactant

▶ **Conceptual PRACTICE 3.13** The following diagram represents the reaction of A (red spheres) with B_2 (blue spheres):

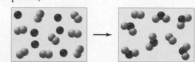

(a) Write a balanced equation for the reaction.
(b) Identify the limiting and excess reactant.
(c) How many molecules of product are made?

Mastering Chemistry

Superior support beyond the classroom with MasteringChemistry®

MasteringChemistry from Pearson is the leading online homework, tutorial, and assessment system, designed to improve results by engaging students before, during, and after class with powerful content. Ensure students arrive ready to learn by assigning educationally effective content before class, and encourage critical thinking and retention with in-class resources such as Learning Catalytics. Students can further master concepts after class through traditional and adaptive homework assignments that provide hints and answer-specific feedback. The Mastering gradebook records scores for all automatically graded assignments in one place, while diagnostic tools give instructors access to rich data to assess student understanding and misconceptions.

Mastering brings learning full circle by continuously adapting to each student and making learning more personal than ever—before, during, and after class.

Before Class

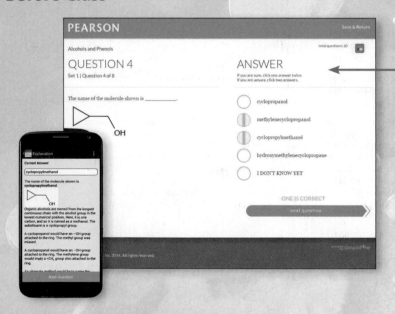

Dynamic Study Modules Now assignable, Dynamic Study Modules (DSMs) enable your students to study on their own to be better prepared to achieve higher test scores. When your students use DSMs outside of class, you can take knowledge transfer out of the classroom, allowing class time to be spent on higher-order learning. Modules can be completed on smartphones, tablets, or computers and assignments will automatically be synced to the MasteringChemistry gradebook.

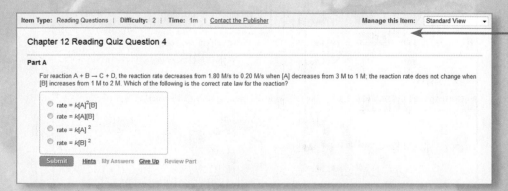

Reading Quizzes give instructors the opportunity to assign reading and test students on their comprehension of chapter content. Reading Quizzes are often useful to provide a common baseline for students prior to coming to class, thereby saving time on lower level content and allowing instructors to use in-class time on more challenging topic.

During Class

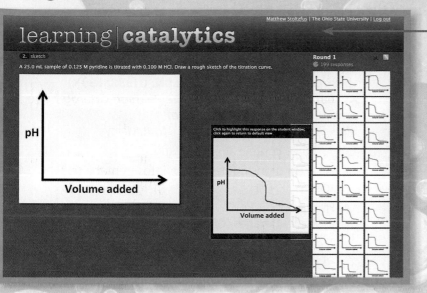

Learning Catalytics Learning Catalytics is a "bring your own device" student engagement, assessment, and classroom intelligence system. With Learning Catalytics you can:

- Assess students in real time, using open-ended tasks to probe student understanding.
- Understand immediately where students are and adjust your lecture accordingly.
- Improve your students' critical-thinking skills.
- Access rich analytics to understand student performance.
- Add your own questions to make Learning Catalytics fit your course exactly.
- Manage student interactions with intelligent grouping and timing.

After Class

Student Tutorials Featuring specific wrong-answer feedback, hints, and a wide variety of educationally effective content, guide your students through the toughest topics in chemistry. The hallmark Hints and Feedback offer scaffolded instruction similar to what students would experience in an office hour, allowing them to learn from their mistakes without being given the answer.

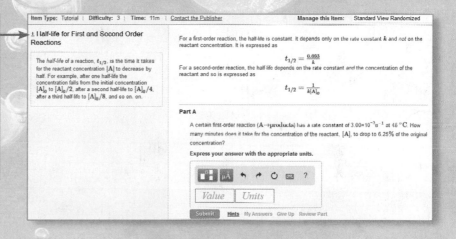

NEW! Adaptive Follow-Up Assignments allow instructors to deliver content to students —automatically personalized for each individual based on the strengths and weaknesses identified by his or her performance on initial Mastering assignments.

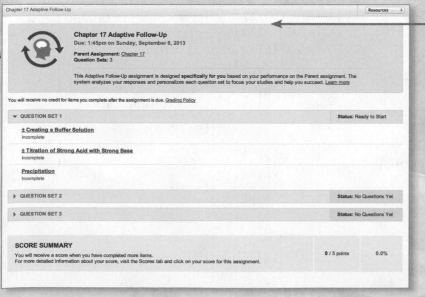

Resources

For Students

Selected Solutions Manual (ISBN: 0133888797)

Joseph Topich, *Virginia Commonwealth University*
This manual contains worked out solutions to all in-chapter problems and even-numbered end-of-chapter problems.

Study Guide (ISBN: 0133888819)

James Zubricky, *University of Toledo*
The Study Guide includes learning goals, an overview, progressive review section with worked examples, and self-tests with answers.

Laboratory Manual (ISBN: 013388662X)

Sandra Chimon-Peszek, *Calumet College of St. Joseph*
The Laboratory Manual contains over 20 experiments that focus on real-world applications. Each experiment corresponds with one or more topics covered in each chapter.

For Instructors

Instructor Resource Center (ISBN: 013388659X)
Available for download on the Pearson catalog page at www.pearsonhighered.com

Mark Benvenuto, *University of Detroit Mercy*

This resource contains the following:

- All illustrations, tables, and photos from the text in JPEG format
- Four pre-built PowerPoint Presentations (lecture, worked examples, images, CRS/ clicker questions)
- TestGen computerized software with the TestGen version of the Testbank
- Word files of the Test Item File

Solutions Manual (ISBN: 0133892298)

Joseph Topich, *Virginia Commonwealth University*
The solutions manual provides worked-out solutions to all in-chapter, conceptual, and end-of-chapter questions and problems. With instructor's permission, this manual may be made available to students.

Instructor Resource Manual (ISBN: 0133886603, Download Only)

Charity Lovitt, *University of Washington, Bothell*
The Instructor manual contains teaching tips, common misconceptions, lecture outlines, and suggested chapter learning goals for students, as well as lecture/laboratory demonstrations and literature references. It also describes the various resources, such as printed test bank questions, animations, and movies that are available to instructors.

Test Bank (ISBN: 0133890694, Download Only)
Available for download on the Pearson catalog page at www.pearsonhighered.com

Charity Lovitt, *University of Washington, Bothell*

Christine Hermann, *Radford University*

The Test Bank contains more than 4,500 multiple choice, true/false, and matching questions.

Chemistry

SEVENTH EDITION

Chemical Tools: Experimentation and Measurement

Instruments for scientific measurements have changed greatly over the centuries. Modern technology has enabled scientists to make images of extremely tiny particles, even individual atoms, using instruments like this atomic force microscope.

 What are the unique properties of nanoscale (1 nm = 10^{-9} m) materials?

The answer to this question can be found in the **INQUIRY** ▸▸▸ on page 23.

CONTENTS

STUDY GUIDE

▲ The sequence of the approximately 5.8 billion nucleic acid units, or *nucleotides*, present in the human genome has been determined using instruments like this automated DNA sequencer.

Life has changed more in the past two centuries than in all the previously recorded span of human history. The Earth's population has increased sevenfold since 1800, and life expectancy has nearly doubled because of our ability to synthesize medicines, control diseases, and increase crop yields. Methods of transportation have changed from horses and buggies to automobiles and airplanes because of our ability to harness the energy in petroleum. Many goods are now made of polymers and ceramics instead of wood and metal because of our ability to manufacture materials with properties unlike any found in nature.

In one way or another, all these changes involve **chemistry**, the study of the composition, properties, and transformations of matter. Chemistry is deeply involved in both the changes that take place in nature and the profound social changes of the past two centuries. In addition, chemistry is central to the current revolution in molecular biology that is revealing the details of how life is genetically regulated. No educated person today can understand the modern world without a basic knowledge of chemistry.

1.1 ▶ THE SCIENTIFIC METHOD IN A CHEMICAL CONTEXT: IMPROVED PHARMACEUTICAL INSULIN

By opening this book, you have already decided that you need to know more about chemistry to pursue your future goals. Perhaps you want to learn how living organisms function, how medicines are made, how human activities can change the environment, how alternative fuels produce clean energy, or how to make materials with novel properties. A good place to start is by learning the experimental approach used by scientists to make new discoveries.

Let's examine the development of Humalog®, a billion-dollar medicine, to illustrate the scientific method and how chemical principles are applied in the pharmaceutical industry. Humalog® was commercialized by Eli Lilly and Company in 1996 and is one of several insulin drugs available to people with diabetes. Do not worry if you do not understand all the details of the chemistry yet as our focus is on the process of modern interdisciplinary research.

Diabetes is caused by inadequate production and/or use of insulin, a hormone involved in the body's metabolism of glucose. High blood glucose levels can lead to severe long-term consequences such as cardiovascular disease, kidney failure, and blindness. Insulin was discovered in 1921 by Dr. Frederick Banting and research associate Charles Best, leading soon after to insulin therapy for diabetic patients. Prior to the commercial availability of insulin in 1923, onset of Type I diabetes meant certain death, and insulin's ability to restore health was so dramatic that it was described as "the raising of the dead." Insulin was initially produced by extracting it from the pancreas glands of pigs and cattle, but beginning in the 1980s recombinant DNA technology was used to make enough human insulin to treat a large number of patients worldwide.

While insulin treatment was once considered a miracle therapy, there are several limitations to the use of human insulin as a drug. **FIGURE 1.1** compares the time profile for insulin

▶ **FIGURE 1.1**

Comparison of insulin profiles. The rise and fall of insulin levels in the blood of a nondiabetic individual and a patient taking an injection of human insulin are shown over time.

Figure It Out

What are the main differences in the time profile for injected insulin when compared to natural insulin release? How do these differences affect treatment of diabetes?

Answer: For injected insulin, the peak concentration is later and the peak shape is more broad than naturally released insulin. These differences lead to high blood sugars initially, but a potential for severe low blood sugar several hours later.

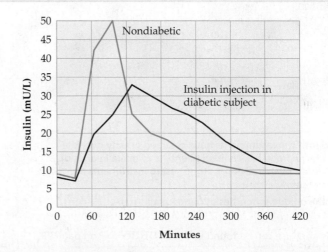

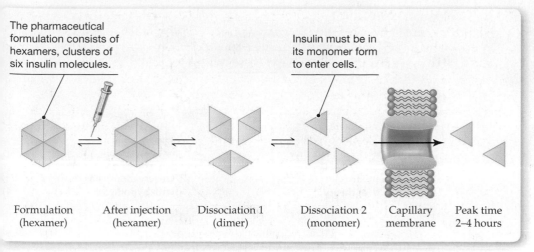

The pharmaceutical formulation consists of hexamers, clusters of six insulin molecules.

Insulin must be in its monomer form to enter cells.

| Formulation (hexamer) | After injection (hexamer) | Dissociation 1 (dimer) | Dissociation 2 (monomer) | Capillary membrane | Peak time 2–4 hours |

◀ **FIGURE 1.2**

Hexamer dissociation after insulin injection. As concentration decreases, insulin hexamers break apart into dimers and then monomers which enter cells through the capillary membrane.

concentration in the blood of a diabetic patient to that of a nondiabetic individual. After a person eats, the peak in insulin concentration in the natural physiological process is sharper and faster than the peak after injection of insulin. This leads to two major problems involving the use of insulin as a drug. Insulin doses depend on the quantity and type of food eaten, but the slow time profile means injections must be given 30 minutes *before* a meal. Failure to adhere to recommended timing can result in large increases in blood glucose (hyperglycemia). Also, because injected insulin will act to lower blood glucose long after food is digested, diabetics must take care to avoid dangerous low blood sugar events (hypoglycemia), which can cause confusion, unconsciousness, and seizures.

Why does the exact same molecule, human insulin, behave differently when produced naturally in the body than when taken in drug form? Differences in the insulin profiles seen in Figure 1.1 are explained by the relatively high concentration of insulin in the pharmaceutical formulation. The drug's shelf life (that is, its stability) is extended when prepared in higher concentrations. Increased stability results from aggregation of insulin *monomers*, single molecules, into *hexamers*, clusters of six insulin molecules. As shown in **FIGURE 1.2**, the hexamers *dissociate*, or break apart, into monomers as insulin becomes diluted in the body. Only the monomeric form can enter the cell by crossing the capillary membrane, causing a time lag in bioavailability. The peak concentration occurs 2 to 4 hours after injection.

Many different principles from chemistry that you will learn about in this book are central to the pharmacological properties of insulin. What, for example, would cause molecules to attract one another and form clusters like the insulin hexamer? In Chapter 8, Bonding Theories and Molecular Structure, you will learn about forces that give molecules like insulin their specific shapes and functions, and cause them to attract one another. Chapter 4, Reactions in Aqueous Solutions, describes how to calculate solution concentrations important in both drug formulations and in the human body. Rates of reactions—such as the time required for hexamer dissociation—and important factors that influence them are explored in Chapter 13, Kinetics. In Chapter 14, on equilibrium, we discuss the control of reversible processes like hexamer formation and the extent to which molecules reside in one state (hexamer) or the other (monomer). We turn now to how knowledge of these and many other scientific concepts was acquired: the scientific method.

The Scientific Method

Dr. Richard DiMarchi, at Eli Lilly and Company, led a team of scientists in the discovery of an improved or "fast acting" insulin, Humalog®. Scientific research begins with a driving question that is frequently based on experimental observations or the desire to learn about the unknown. In this case, measurements of the time profile of injected insulin led to the question, "How can we make a pharmaceutical formulation of human insulin that mimics the body's natural release profile?" A general approach to research is called the **scientific method**. The scientific method is an iterative process involving the formulation of questions and conjectures arising from observations, careful design of experiments, and thoughtful analysis of results. The scientific method involves identifying ways to test the validity of new ideas and

▲ Dr. Richard DiMarchi led a team of scientists in the discovery of the "fast acting" insulin, Humalog®.

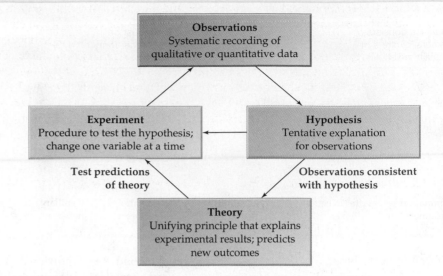

▶ **FIGURE 1.3**

The scientific method. An iterative experimental approach is used in scientific research. Hypotheses and theories are refined based on new experiments and observations.

Figure It Out

What is developed when numerous experimental observations support a hypothesis?

Answer: Theory.

seldom is there only one way to go about it. The main elements of the scientific method, outlined in **FIGURE 1.3**, are the following:

- **Observations** are a systematic recording of natural phenomena and may be **qualitative** descriptive in nature, or **quantitative**, involving measurements.

- A **hypothesis** is a possible explanation for the observation developed based upon facts collected from previous experiments as well as scientific knowledge and intuition. The hypothesis may not be correct, but it must be testable with an experiment.

- An **experiment** is a procedure for testing the hypothesis. Experiments are most useful when they are performed in a *controlled* manner, meaning that only one variable is changed at a time while all others remain constant.

- A **theory** is developed from a hypothesis consistent with experimental data and is a unifying principle that explains experimental results. It also makes predictions about related systems and new experiments are carried out to verify the theory.

Keep in mind as you study chemistry or any other science that theories can never be absolutely proven. There's always the chance that a new experiment might give results that can't be explained by present theory. All a theory can do is provide the best explanation that we can come up with at the present time. Science is an ever-changing field where new observations are made with increasingly sophisticated equipment; it is always possible that existing theories may be modified in the future. Many iterations of the scientific method were required in creating a new analog of insulin that would have a time profile similar to natural insulin. The general hypothesis was that the chemical structure of insulin was responsible for aggregation and modifying it could change properties. In the case of Humalog®, Dr. DiMarchi devised a hypothesis based on observations of the chemical similarity between human insulin and another human hormone called insulin-like growth factor 1 (IGF-1). Both of these hormones are **peptides**, molecules that consist of molecules called **amino acids** linked together in a chain. The structure of IGF-1 was of interest because it exists in solution only in the form of monomers, which results in rapid uptake by cells. A simplified structure of Humalog®, a "fast acting" analog of human insulin is:

LOOKING AHEAD...

The chemical structure of **amino acids** and **peptides** is described in Chapter 23.

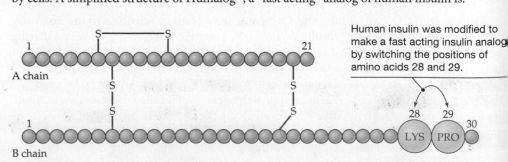

Human insulin was modified to make a fast acting insulin analog by switching the positions of amino acids 28 and 29.

Each amino acid is represented by a single circle, and the overall molecule consists of 51 total amino acids. Both the human insulin and Humalog® molecule consist of two chains, A and B. It was known that in IGF-1, the B chain contained lysine at position 28 and proline at position 29, and that these two amino acids are found in natural human insulin in exactly the reverse order. The hypothesis was that *switching the order of amino acids at positions 28 and 29 on the B chain* would minimize hexamer formation while retaining the biological activity of insulin. Known chemistry was used to synthesize a new analog of insulin, called *Lispro*, in which the two amino acid positions are reversed. Insulin Lispro, marketed as Humalog®, was indeed more "fast acting" than injected human insulin. It aggregated to a lesser extent, resulting in a time profile more closely matching physiological insulin release. Upon successfully concluding clinical trials, including studies for safety and toxicity, doctors worldwide began prescribing Humalog® for treatment of diabetes. Millions of patients in more than 100 countries have benefited from the science that went into its discovery.

Visualizing Chemical Behavior with Molecular Models

How can simply switching the position of two amino acids in an insulin analog result in such drastically different pharmacological properties? A theory to explain this remarkable result was needed. Chemists often make use of molecular models to help develop a theory and to visualize structure–function relationships. **Molecular models** are simplified versions of the way atoms are connected and reveal their three-dimensional arrangement. **FIGURE 1.4** is a ribbon model of Humalog® and human insulin, showing only the position of atoms in the "backbone" of the molecule. A comparison of the structure of these two forms of insulin can help explain the minimized aggregation of the analog. Notice that the configuration at the end of the B chain is significantly different in the two forms of insulin. A bend occurs at the end of the human insulin chain, but not in Humalog®. When the amino acids are switched as in Humalog®, the B chains on adjacent molecules cannot approach closely, preventing some attractive forces from forming. This leads to less self-association and faster dissociation once the insulin analog has been injected. Other than the end of the B chain, the structure of the two molecules is nearly identical, giving Humalog® essentially the same biological activity as human insulin.

In summary, the hypothesis that switching the amino acids at positions 28 and 29 in the B chain of human insulin to create fast-acting insulin was upheld with experimental observations. A theory to explain these observations was developed by experimentally determining

LOOKING AHEAD...

Common types of **molecular models** used to depict molecules will be described in Section 2.10.

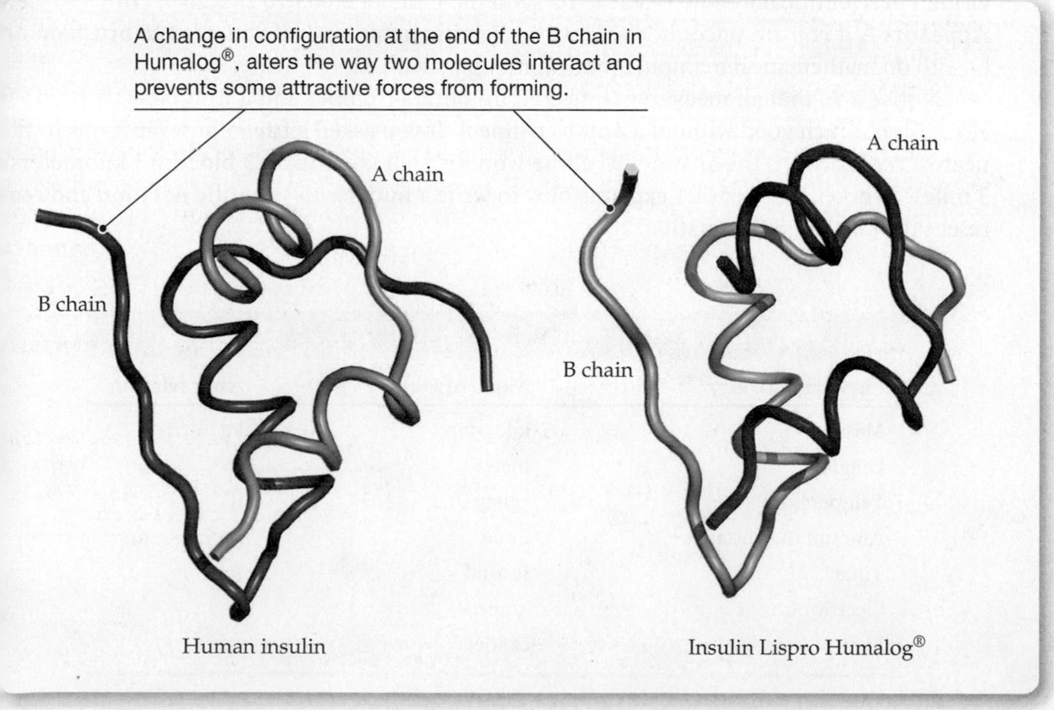

A change in configuration at the end of the B chain in Humalog®, alters the way two molecules interact and prevents some attractive forces from forming.

A chain

A chain

B chain

B chain

Human insulin

Insulin Lispro Humalog®

◀ **FIGURE 1.4**

Ribbon model for human insulin and insulin Lispro. A ribbon model is a useful simplification for depicting how the change in position of lysine and proline alters the backbone configuration at the end of the B chain in Humalog®.

chemical structures and examining molecular models to interpret their meaning. Models showed that the last five amino acids in the B chain were important in aggregation but not biological activity. A prediction could then be made that an insulin analog missing these five amino acids would have a rapid time profile and excellent biological activity. When the analog was prepared, observations supported the prediction, but this analog lacked the stability needed for a drug formulation. Other analogs were prepared, but in the end Humalog® was found to have the most desirable properties. Research laboratories all over the world use the scientific method to discover new phenomena and develop new products.

1.2 ▶ EXPERIMENTATION AND MEASUREMENT

Chemistry is an experimental science. But if our experiments are to be reproducible, we must be able to fully describe the substances we're working with—their amounts, volumes, temperatures, and so forth. Thus, one of the most important requirements in chemistry is that we have a way to measure things.

Under an international agreement concluded in 1960, scientists throughout the world now use the International System of Units for measurement, abbreviated **SI** for the French *Système Internationale d'Unités*. Based on the metric system, which is used in all industrialized countries of the world except the United States, the SI system has seven fundamental units (**TABLE 1.1**). These seven fundamental units, along with others derived from them, suffice for all scientific measurements. We'll look at three of the most common units in this chapter—those for mass, length, and temperature—and will discuss others as the need arises in later chapters.

One problem with any system of measurement is that the sizes of the units often turn out to be inconveniently large or small. For example, a chemist describing the diameter of a sodium atom (0.000 000 000 372 m) would find the meter (m) to be inconveniently large, but an astronomer describing the average distance from the Earth to the Sun (150,000,000,000 m) would find the meter to be inconveniently small. For this reason, SI units are modified through the use of prefixes when they refer to either smaller or larger quantities. Thus, the prefix *milli-* means one-thousandth, and a *milli*meter (mm) is 1/1000 of 1 meter. Similarly, the prefix *kilo-* means one thousand, and a *kilo*meter (km) is 1000 meters. [Note that the SI unit for mass (kilogram) already contains the *kilo-* prefix.] A list of prefixes is shown in **TABLE 1.2**, with the most commonly used ones in red.

Notice how numbers that are either very large or very small are indicated in Table 1.2 using an exponential format called **scientific notation**. For example, the number 55,000 is written in scientific notation as 5.5×10^4, and the number 0.003 20 as 3.20×10^{-3}. Review Appendix A if you are uncomfortable with scientific notation or if you need to brush up on how to do mathematical manipulations on numbers with exponents.

Notice also that all measurements contain both a number and a unit label. A number alone is not much good without a unit to define it. If you asked a friend how far it was to the nearest tennis court, the answer "3" alone wouldn't tell you much, 3 blocks? 3 kilometers? 3 miles? Worked Example 1.1 explains how to write a number in scientific notation and represent the unit in prefix notation.

TABLE 1.1 The Seven Fundamental SI Units of Measure		
Physical Quantity	**Name of Unit**	**Abbreviation**
Mass	kilogram	kg
Length	meter	m
Temperature	kelvin	K
Amount of substance	mole	mol
Time	second	s
Electric current	ampere	A
Luminous intensity	candela	cd

TABLE 1.2   Some Prefixes for Multiples of SI Units. The most commonly used prefixes are shown in red.

Factor	Prefix	Symbol	Example
$1{,}000{,}000{,}000{,}000 = 10^{12}$	tera	T	1 teragram (Tg) = 10^{12} g
$1{,}000{,}000{,}000 = 10^{9}$	**giga**	**G**	1 gigameter (Gm) = 10^{9} m
$1{,}000{,}000 = 10^{6}$	**mega**	**M**	1 megameter (Mm) = 10^{6} m
$1000 = 10^{3}$	**kilo**	**k**	1 kilogram (kg) = 10^{3} g
$100 = 10^{2}$	hecto	h	1 hectogram (hg) = 100 g
$10 = 10^{1}$	deka	da	1 dekagram (dag) = 10 g
$0.1 = 10^{-1}$	**deci**	**d**	1 decimeter (dm) = 0.1 m
$0.01 = 10^{-2}$	**centi**	**c**	1 centimeter (cm) = 0.01 m
$0.001 = 10^{-3}$	**milli**	**m**	1 milligram (mg) = 0.001 g
*$0.000\ 001 = 10^{-6}$	micro	μ	1 micrometer (μm) = 10^{-6} m
*$0.000\ 000\ 001 = 10^{-9}$	nano	n	1 nanosecond (ns) = 10^{-9} s
*$0.000\ 000\ 000\ 001 = 10^{-12}$	pico	p	1 picosecond (ps) = 10^{-12} s
*$0.000\ 000\ 000\ 000\ 001 = 10^{-15}$	femto	f	1 femtomole (fmol) = 10^{-15} mol

*For very small numbers, it is becoming common in scientific work to leave a thin space every three digits to the right of the decimal point, analogous to the comma placed every three digits to the left of the decimal point in large numbers.

━● WORKED EXAMPLE 1.1

Expressing Measurements Using Scientific Notation and SI Units

Express the following quantities in scientific notation and then express the number and unit with the most appropriate prefix.
(a) The diameter of a sodium atom, 0.000 000 000 372 m
(b) The distance from the Earth to the Sun, 150,000,000,000 m

STRATEGY

To write a number in scientific notation, shift the decimal point to the right or left by n places until you obtain a number between 1 and 10. If the decimal is shifted to the right, n is negative and if the decimal is shifted to the left, n is positive. Then multiply the result by 10^{n}. Choose a prefix for the unit that is close to the exponent of the number written in scientific notation.

SOLUTION

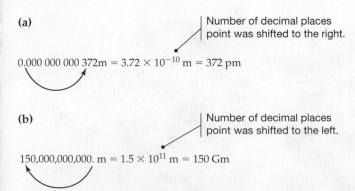

(a) Number of decimal places point was shifted to the right.
$0.000\ 000\ 000\ 372\,\text{m} = 3.72 \times 10^{-10}\ \text{m} = 372\ \text{pm}$

(b) Number of decimal places point was shifted to the left.
$150{,}000{,}000{,}000.\ \text{m} = 1.5 \times 10^{11}\ \text{m} = 150\ \text{Gm}$

▶ **PRACTICE 1.1** Express the following quantities in scientific notation and then express the number and unit with the most appropriate prefix.

(a) The diameter of an insulin molecule, 0.000 000 005 m
(b) The circumference of the Earth at the Equator, 40,075,017 m

▶ **APPLY 1.2** Express the following quantities in scientific notation using fundamental SI units of mass and length given in Table 1.1.

(a) The diameter of a human hair, 70 μm.
(b) The mass of carbon dioxide emitted from a large power plant each year, 20 Tg.

1.3 ▶ MASS AND ITS MEASUREMENT

Mass is defined as the amount of *matter* in an object. **Matter**, in turn, is a catchall term used to describe anything with a physical presence—anything you can touch, taste, or smell. (Stated more scientifically, matter is anything that has mass.) Mass is measured in SI units by the **kilogram** (**kg**; 1 kg = 2.205 U.S. lb). Because the kilogram is too large for many purposes in chemistry, the metric **gram** (**g**; 1 g = 0.001 kg), the **milligram** (**mg**; 1 mg = 0.001 g = 10^{-6} kg), and the **microgram** (μ**g**; 1 μg = 0.001 mg = 10^{-6} g = 10^{-9} kg) are more commonly used. (The symbol μ is the lowercase Greek letter mu.) One gram is a bit less than half the mass of a new U.S. dime.

$$1 \text{ kg} = 1000 \text{ g} = 1,000,000 \text{ mg} = 1,000,000,000 \ \mu\text{g} \quad (2.205 \text{ lb})$$
$$1 \text{ g} = 1000 \text{ mg} = 1,000,000 \ \mu\text{g} \quad (0.035 \ 27 \text{ oz})$$
$$1 \text{ mg} = 1000 \ \mu\text{g}$$

The standard kilogram is set as the mass of a cylindrical bar of platinum–iridium alloy stored in a vault in a suburb of Paris, France. There are 40 copies of this bar distributed throughout the world, with two (Numbers 4 and 20) stored at the U.S. National Institute of Standards and Technology near Washington, D.C.

The terms *mass* and *weight*, although often used interchangeably, have quite different meanings. *Mass* is a physical property that measures the amount of matter in an object, whereas *weight* measures the force with which gravity pulls on an object. Mass is independent of an object's location: your body has the same amount of matter whether you're on Earth or on the moon. Weight, however, *does* depend on an object's location. If you weigh 140 lb on Earth, you would weigh only about 23 lb on the moon, which has a lower gravity than the Earth.

At the same location on Earth, two objects with identical masses experience an identical pull of the Earth's gravity and have identical weights. Thus, the mass of an object can be measured by comparing its weight to the weight of a reference standard of known mass. Much of the confusion between mass and weight is simply due to a language problem. We speak of "weighing" when we really mean that we are measuring mass by comparing two weights. **FIGURE 1.5** shows balances typically used for measuring mass in the laboratory.

1.4 ▶ LENGTH AND ITS MEASUREMENT

The **meter (m)** is the standard unit of length in the SI system. Although originally defined in 1790 as being 1 ten-millionth of the distance from the equator to the North Pole, the meter was redefined in 1889 as the distance between two thin lines on a bar of platinum–iridium alloy stored near Paris, France. To accommodate an increasing need for precision, the meter was redefined again in 1983 as equal to the distance traveled by light through a vacuum in 1/299,792,458 second. Although this new definition isn't as easy to grasp as the distance between two scratches on a bar, it has the great advantage that it can't be lost or damaged.

▲ The mass of a U.S. dime is approximately 2.27 g.

▶ **FIGURE 1.5**

Some balances used for measuring mass in the laboratory.

One meter is 39.37 inches, about 10% longer than an English yard and much too large for most measurements in chemistry. Other more commonly used measures of length are the **centimeter** (**cm**; 1 cm = 0.01 m, a bit less than half an inch), the **millimeter** (**mm**; 1 mm = 0.001 m, about the thickness of a U.S. dime), the **micrometer** (**μm**; 1 μm = 10^{-6} m), the **nanometer** (**nm**; 1 nm = 10^{-9} m), and the **picometer** (**pm**; 1 pm = 10^{-12} m). Thus, a chemist might refer to the diameter of a sodium atom as 372 pm (3.72×10^{-10} m).

$$1 \text{ m} = 100 \text{ cm} = 1000 \text{ mm} = 1{,}000{,}000 \text{ } \mu\text{m} = 1{,}000{,}000{,}000 \text{ nm} \qquad (1.0936 \text{ yd})$$
$$1 \text{ cm} = 10 \text{ mm} = 10{,}000 \text{ } \mu\text{m} = 10{,}000{,}000 \text{ nm} \qquad (0.3937 \text{ in.})$$
$$1 \text{ mm} = 1000 \text{ } \mu\text{m} = 1{,}000{,}000 \text{ nm}$$

1.5 ▶ TEMPERATURE AND ITS MEASUREMENT

Just as the kilogram and the meter are slowly replacing the pound and the yard as common units for mass and length measurement in the United States, the **Celsius degree** (**°C**) is slowly replacing the degree **Fahrenheit** (**°F**) as the common unit for temperature measurement. In scientific work, however, the **kelvin (K)** has replaced both. (Note that we say only "kelvin," not "kelvin degree.")

For all practical purposes, the kelvin and the degree Celsius are the same—both are one-hundredth of the interval between the freezing point of water and the boiling point of water at standard atmospheric pressure. The only real difference between the two units is that the numbers assigned to various points on the scales differ. Whereas the Celsius scale assigns a value of 0 °C to the freezing point of water and 100 °C to the boiling point of water, the Kelvin scale assigns a value of 0 K to the coldest possible temperature, −273.15 °C, sometimes called *absolute zero*. Thus, 0 K = −273.15 °C and 273.15 K = 0 °C. For example, a warm spring day with a Celsius temperature of 25 °C has a Kelvin temperature of 25 + 273.15 = 298 K.

Relationship between the Kelvin and Celsius scales

Temperature in K = Temperature in °C + 273.15

Temperature in °C = Temperature in K − 273.15

In contrast to the Kelvin and Celsius scales, the common Fahrenheit scale specifies an interval of 180° between the freezing point (32 °F) and the boiling point (212 °F) of water. Thus, it takes 180 degrees Fahrenheit to cover the same range as 100 degrees Celsius (or kelvins), and a degree Fahrenheit is therefore only 100/180 = 5/9 as large as a degree Celsius. **FIGURE 1.6** compares the Fahrenheit, Celsius, and Kelvin scales.

▲ The length of the bacteria on the tip of this pin is about 5×10^{-7} m or 500 nm.

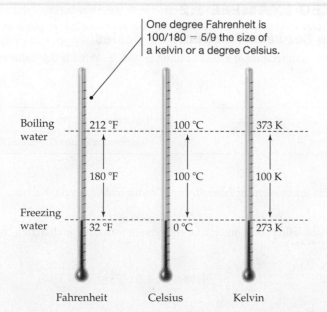

One degree Fahrenheit is 100/180 = 5/9 the size of a kelvin or a degree Celsius.

	Fahrenheit	Celsius	Kelvin
Boiling water	212 °F	100 °C	373 K
	180 °F	100 °C	100 K
Freezing water	32 °F	0 °C	273 K

◀ **FIGURE 1.6**

A comparison of the Fahrenheit, Celsius, and Kelvin temperature scales.

Figure It Out

Which represents the largest increase in temperature: +10 °F, +10 °C, or +10 K?

Answer: Temperature changes of +10° C or +10 K are equal and larger than +10° F.

Two adjustments are needed to convert between Fahrenheit and Celsius scales—one to adjust for the difference in degree size and one to adjust for the difference in zero points. The size adjustment is made using the relationships $1\,°C = (9/5)\,°F$ and $1\,°F = (5/9)\,°C$. The zero-point adjustment is made by remembering that the freezing point of water is higher by 32 on the Fahrenheit scale than on the Celsius scale. Thus, if you want to convert from Celsius to Fahrenheit, you do a size adjustment (multiply $°C$ by $9/5$) and then a zero-point adjustment (add 32). If you want to convert from Fahrenheit to Celsius, you find out how many Fahrenheit degrees there are above freezing (by subtracting 32) and then do a size adjustment (multiply by $5/9$). The following formulas describe the conversions:

Celsius to Fahrenheit	**Fahrenheit to Celsius**
$°F = \left(\dfrac{9\,°F}{5\,°C} \times °C\right) + 32\,°F$	$°C = \dfrac{5\,°C}{9\,°F} \times (°F - 32\,°F)$

Worked Example 1.2 shows how to convert between temperature scales and estimate the answer. Before tackling Worked Example 1.2, we'd like to point out that the Worked Examples in this book suggest a series of steps useful in organizing and analyzing information. While these steps need not always be explicitly shown in your work, they should nevertheless be a standard part of your approach to problem solving.

Problem-Solving Steps In Worked Examples

IDENTIFY

Classify pertinent information as known or unknown. (The quantity needed in the answer will, of course, be unknown.) Specify units and symbols to help identify necessary equations and procedures.

STRATEGY

Find a relationship between the known information and unknown answer, and plan a strategy for getting from one to the other.

SOLUTION

Solve the problem.

CHECK

If possible, make a rough estimate to be sure your calculated answer is reasonable and think about the number and sign to make sure it makes sense.

WORKED EXAMPLE 1.2

Converting Between Temperature Scales

The normal body temperature of a healthy adult is 98.6 °F. What is this value on both Celsius and Kelvin scales?

IDENTIFY

Known	Unknown
Temperature, 98.6 °F	Temperature in units of °C and K

STRATEGY

Use the formulas for converting Fahrenheit to Celsius and Celsius to Kelvin.

SOLUTION

Set up an equation using the temperature conversion formula for changing from Fahrenheit to Celsius:

$$°C = \left(\frac{5\,°C}{9\,°F}\right)(98.6\,°F - 32\,°F) = 37.0\,°C$$

Converting to kelvin gives a temperature of $37.0° + 273.15° = 310.2\,K$.

CHECK

A useful way to double-check a calculation is to estimate the answer. Body temperature in °F is first rounded to the nearest whole number, 99. To account for the difference in zero points of the two scales, 32 is subtracted: $99 - 32 = 67$. Because a degree Fahrenheit is only 5/9 as large as a degree Celsius, the next step is to multiply by 5/9, which can be approximated by dividing by two: $67/2 = 33.5$. The estimate is only slightly lower than the calculated answer (37.0), indicating the mathematical operations have most likely been performed correctly. Estimating Fahrenheit to Celsius conversions is useful as daily temperatures are reported on these two different scales throughout the world.

▸ **PRACTICE 1.3** The melting point of table salt is 1474 °F. What temperature is this on the Celsius and Kelvin scales?

▸ **APPLY 1.4** The metal gallium has a relatively low melting point for a metal, 302.91 K. If the temperature in the cargo compartment carrying a shipment of gallium has a temperature of 88 °F, is the gallium in the solid or liquid state?

▲ The melting point of sodium chloride is 1474 °F.

1.6 ▸ DERIVED UNITS: VOLUME AND ITS MEASUREMENT

Look back at the seven fundamental SI units given in Table 1.1 and you'll find that measures for such familiar quantities as area, volume, density, speed, and pressure are missing. All are examples of *derived* quantities rather than fundamental quantities because they can be expressed using one or more of the seven base units (**TABLE 1.3**).

Volume, the amount of space occupied by an object, is measured in SI units by the **cubic meter (m^3)**, defined as the amount of space occupied by a cube 1 meter on edge (**FIGURE 1.7**).

A cubic meter equals 264.2 U.S. gallons, much too large a quantity for normal use in chemistry. As a result, smaller, more convenient measures are commonly employed. Both the **cubic decimeter (dm^3)** ($1 \, dm^3 = 0.001 \, m^3$), equal in size to the more familiar metric **liter (L)**, and the **cubic centimeter (cm^3)** ($1 \, cm^3 = 0.001 \, dm^3 = 10^{-6} \, m^3$), equal in size to the metric **milliliter (mL)**, are particularly convenient. Slightly larger than 1 U.S. quart, a liter has the volume of a cube 1 dm on edge. Similarly, a milliliter has the volume of a cube 1 cm on edge (Figure 1.7).

$$1 \, m^3 = 1000 \, dm^3 = 1,000,000 \, cm^3 \quad (264.2 \, gal)$$
$$1 \, dm^3 = 1L = 1000 \, mL \quad (1.057 \, qt)$$

FIGURE 1.8 shows some of the equipment frequently used in the laboratory for measuring liquid volume.

TABLE 1.3 Some Derived Quantities		
Quantity	**Definition**	**Derived Unit (Name)**
Area	Length times length	m^2
Volume	Area times length	m^3
Density	Mass per unit volume	kg/m^3
Speed	Distance per unit time	m/s
Acceleration	Change in speed per unit time	m/s^2
Force	Mass times acceleration	$(kg \cdot m)/s^2$ (newton, N)
Pressure	Force per unit area	$kg/(m \cdot s^2)$ (pascal, Pa)
Energy	Force times distance	$(kg \cdot m^2)/s^2$ (joule, J)

▶ **FIGURE 1.7**

Units for measuring volume. A cubic meter is the volume of a cube 1 meter along each edge.

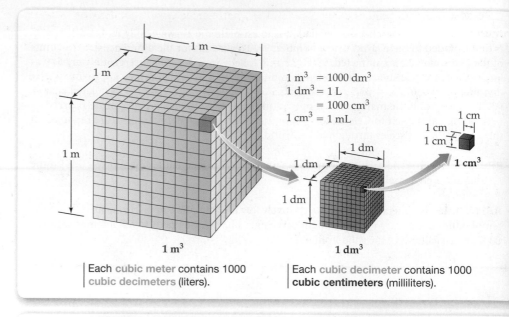

$1 m^3 = 1000 dm^3$
$1 dm^3 = 1 L$
$= 1000 cm^3$
$1 cm^3 = 1 mL$

$1 m^3$

$1 dm^3$

$1 cm^3$

Each cubic meter contains 1000 cubic decimeters (liters).

Each cubic decimeter contains 1000 cubic centimeters (milliliters).

▶ **FIGURE 1.8**

Common items of laboratory equipment used for measuring liquid volume.

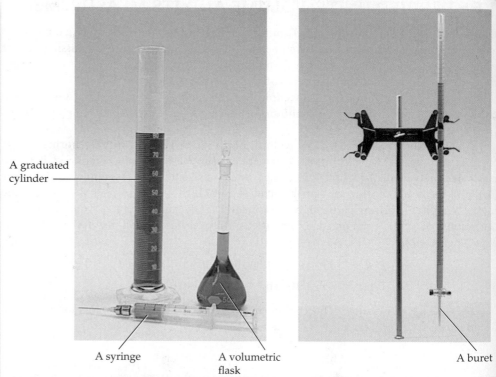

A graduated cylinder

A syringe

A volumetric flask

A buret

▲ Which weighs more, the brass weight or the pillow? Actually, both have identical masses and weights, but the brass has a higher density because its volume is smaller.

1.7 ▶ DERIVED UNITS: DENSITY AND ITS MEASUREMENT

The relationship between the mass of an object and its volume is called *density*. **Density** is calculated as the mass of an object divided by its volume and is expressed in the SI derived unit g/mL for a liquid or g/cm^3 for a solid. The densities of some common materials are given in **TABLE 1.4**.

$$\text{Density} = \frac{\text{Mass (g)}}{\text{Volume (mL or cm}^3)}$$

Because most substances change in volume when heated or cooled, densities are temperature dependent. At 3.98 °C, for example, a 1.0000 mL container holds exactly 1.0000 g of water (density = 1.0000 g/mL). As the temperature is raised, however, the volume occupied by the water expands so that only 0.9584 g fits in the 1.0000 mL container at 100 °C (density = 0.9584 g/mL). *When reporting a density, the temperature must also be specified.*

Although most substances expand when heated and contract when cooled, water behaves differently. Water contracts when cooled from 100 °C to 3.98 °C, but below this temperature it begins to expand again. Thus, the density of liquid water is at its maximum of 1.0000 g/mL at 3.98 °C but decreases to 0.999 87 g/mL at 0 °C (**FIGURE 1.9**). When freezing occurs, the density drops still further to a value of 0.917 g/cm^3 for ice at 0 °C. Ice and any other substance with a density less than that of water will float, but any substance with a density greater than that of water will sink.

Knowing the density of a substance, particularly a liquid, can be very useful because it's often easier to measure a liquid by volume than by mass. Suppose, for example, that you needed 1.55 g of ethyl alcohol. Rather than trying to weigh exactly the right amount, it would be much easier to look up the density of ethyl alcohol (0.7893 g/mL at 20 °C) and measure the correct volume with a syringe as shown in Figure 1.8.

$$\text{Density} = \frac{\text{Mass}}{\text{Volume}} \quad \text{so} \quad \text{Volume} = \frac{\text{Mass}}{\text{Density}}$$

$$\text{Volume} = \frac{1.55 \text{ g ethyl alcohol}}{0.7893 \frac{\text{g}}{\text{mL}}} = 1.96 \text{ mL ethyl alcohol}$$

TABLE 1.4 Densities of Some Common Materials	
Substance	**Density (g/cm³)**
Ice (0 °C)	0.917
Water (3.98 °C)	1.0000
Gold	19.31
Helium (25 °C)	0.000 164
Air (25 °C)	0.001 185
Human fat	0.94
Human muscle	1.06
Cork	0.22–0.26
Balsa wood	0.12
Earth	5.54

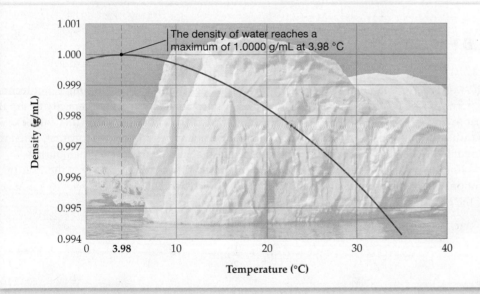

The density of water reaches a maximum of 1.0000 g/mL at 3.98 °C

◀ **FIGURE 1.9**

The density of water at different temperatures.

Figure It Out

Would 10.0 mL of water at 10 °C have the same mass as 10.0 mL of water at 25 °C?

Answer: No, the density of water changes with temperature. 10.0 mL of water at 10 °C would have a higher mass than 10.0 mL of water at 25 °C because the density is higher at 10 °C.

WORKED EXAMPLE 1.3

Relating Mass and Volume Using Density

What is the volume in cm^3 of 201 g of gold?

IDENTIFY

Known	Unknown
Mass of gold (201 g)	Volume of gold (cm³)

continued on next page

How might you determine if this bracelet is pure silver?

STRATEGY

Rearrange the equation for density to solve for volume. Use the known value for the density of gold in Table 1.4 and the mass of gold in the equation.

SOLUTION

$$\text{Volume} = \frac{201 \text{ g gold}}{19.31 \text{ g} / \text{cm}^3} = 10.4 \text{ cm}^3 \text{ gold}$$

CHECK

The mass of gold (201 g) is roughly 10 times larger than the density (19.31 g) so the estimate for volume is 10 cm³, which agrees with the calculated answer.

▶ **PRACTICE 1.5**   Chloroform, a substance once used as an anesthetic, has a density of 1.483 g/mL at 20 °C. How many milliliters would you use if you needed 9.37 g? (1 mL = 1 cm³)

▶ **APPLY 1.6**   You are beachcombing on summer vacation and find a silver bracelet. You take it to the jeweler and he tells you that it is silver plated and will give you $10 for it. You do not want to be swindled so you take the bracelet to your chemistry lab and find its mass on a balance (80.0 g). To measure the volume you place the bracelet in a graduated cylinder (Figure 1.8) containing 10.0 mL of water at 20 °C. The final volume in the graduated cylinder after the bracelet has been added is 17.61 mL. The density of silver at 20 °C is 10.5 g/cm³ and 1 cm³ = 1 mL. What can you conclude about the identity of the metal in the bracelet?

1.8 ▶ DERIVED UNITS: ENERGY AND ITS MEASUREMENT

The word *energy* is familiar to everyone but is surprisingly hard to define in simple, nontechnical terms. A good working definition, however, is to say that **energy** is the capacity to supply heat or do work. The water falling over a dam, for instance, contains energy that can be used to turn a turbine and generate electricity. A tank of propane gas contains energy that, when released in the chemical process of combustion, can heat a house or barbecue a hamburger.

Energy is classified as either *kinetic* or *potential*. **Kinetic energy** (E_K) is the energy of motion. The amount of kinetic energy in a moving object with mass m and velocity v is given by the equation

$$E_K = \frac{1}{2}mv^2$$

The larger the mass of an object and the larger its velocity, the larger the amount of kinetic energy. Thus, water that has fallen over a dam from a great height has a greater velocity and more kinetic energy than the same amount of water that has fallen only a short distance.

Potential energy (E_p), by contrast, is stored energy—perhaps stored in an object because of its height or in a molecule because of chemical reactions it can undergo. The water sitting in a reservoir behind the dam contains potential energy because of its height above the stream at the bottom of the dam. When the water is allowed to fall, its potential energy is converted into kinetic energy. Propane and other substances used as fuels contain potential energy because they can undergo a combustion reaction with oxygen that releases heat. (We'll look at energy in more detail in Chapter 9.)

The units for energy, $(\text{kg} \cdot \text{m}^2)/\text{s}^2$, follow from the expression for kinetic energy, $E_K = 1/2mv^2$. If, for instance, your body has a mass of 50.0 kg (about 110 lb) and you are riding a bicycle at a velocity of 10.0 m/s (about 22 mi/h), your kinetic energy is 2500 $(kg \cdot m^2)/s^2$.

$$E_K = \frac{1}{2}mv^2 = \frac{1}{2}(50.0 \text{ kg})\left(10.0\frac{m}{s}\right)^2 = 2500\frac{kg \cdot m^2}{s^2} = 2500 \text{ J}$$

The SI derived unit for energy $(kg \cdot m^2)/s^2$ is given the name **joule (J)** after the English physicist James Prescott Joule (1818–1889). The joule is a fairly small amount of energy—it takes roughly 100,000 J to heat a coffee cup full of water from room temperature to boiling—so kilojoules (kJ) are more frequently used in chemistry.

In addition to the SI energy unit joule, some chemists and biochemists still use the unit calorie (cal, with a lowercase c). Originally defined as the amount of energy necessary to raise the temperature of 1 g of water by 1 °C (specifically, from 14.5 °C to 15.5 °C), one calorie is now defined as exactly 4.184 J.

$$1 \text{ cal} = 4.184 \text{ J (exactly)}$$

Nutritionists use the somewhat confusing unit Calorie (Cal, with a capital C), which is equal to 1000 calories, or 1 kilocalorie (kcal).

$$1 \text{ Cal} = 1000 \text{ cal} = 1 \text{ kcal} = 4.184 \text{ kJ}$$

The energy value, or caloric content, of food is measured in Calories. Thus, the statement that a banana contains 70 Calories means that 70 Cal (70 kcal, or 290 kJ) of energy is released when the banana is used by the body for fuel.

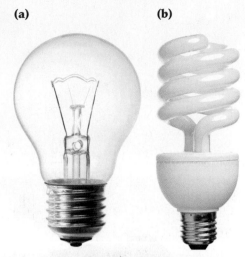

(a)   (b)

▲ **(a)** A 75-watt incandescent bulb uses energy at the rate of 75 J/s. Only about 5% of that energy appears as light, however; the remaining 95% is given off as heat. **(b)** Energy-efficient fluorescent lights are replacing many incandescent bulbs.

WORKED EXAMPLE 1.4

Calculating Kinetic Energy

(a) What is the kinetic energy in joules of a 2360 lb (1070 kg) car moving at 63.3 mi/h (28.3 m/s)? Express the number in scientific notation.
(b) Express the number and unit using an appropriate prefix.

IDENTIFY

Known	Unknown
Mass (1070 kg), velocity (28.3 m/s)	Kinetic energy (E_K) in units of joules (J)

STRATEGY

Use the formula to calculate kinetic energy. If mass is in units of kg and velocity is in units of m/s, then energy will be calculated in units of joules because 1 joule $= \dfrac{1 \text{ kg} \cdot m^2}{s^2}$.

SOLUTION

(a) $E_K = \dfrac{1}{2}mv^2 = \dfrac{1}{2}(1070 \text{ kg})\left(28.3\dfrac{m}{s}\right)^2 = 428,476\dfrac{kg \cdot m^2}{s^2}$

$= 4.28476 \times 10^5 \text{ J}$

(b) Kilo (10^3) and mega (10^6) are both prefixes with exponents similar to the answer in part **(a)**. Therefore, both 428.476 kJ and 0.428 476 MJ are reasonable answers.

CHECK

An answer with a large magnitude should be expected as a car has a very large mass. To estimate the magnitude of kinetic energy, express mass and velocity in the equation using exponents only.

$$E_K \approx (10^3 \text{kg})\left(\frac{10^1 \text{m}}{s}\right)^2 = 10^5\frac{kg \cdot m^2}{s^2}$$

The estimated power of 10 agrees with the detailed calculation.

▶ **PRACTICE 1.7** Some radioactive materials emit a type of radiation called alpha particles at high velocity.

(a) What is the kinetic energy in joules of an alpha particle with a mass of 6.6×10^{-27} kg and a speed of 1.5×10^7 m/s? Express the number in scientific notation.
(b) Express the number and unit using an appropriate prefix.

▶ **APPLY 1.8** A baseball with a mass of 450 g has a kinetic energy of 406 J. Calculate the velocity of the baseball in units of m/s.

1.9 ▶ ACCURACY, PRECISION, AND SIGNIFICANT FIGURES IN MEASUREMENT

Measuring things, whether in cooking, construction, or chemistry, is something that most of us do every day. But how good are those measurements? Any measurement is only as good as the skill of the person doing the work and the reliability of the equipment being used. You've probably noticed, for instance, that you often get slightly different readings when you weigh yourself on a bathroom scale and on a scale at the doctor's office, so there's always some uncertainty about your real weight. The same is true in chemistry—there is always some uncertainty in the value of a measurement.

In talking about the degree of uncertainty in a measurement, we use the words *accuracy* and *precision*. Although most of us use the words interchangeably in daily life, there's actually an important distinction between them. **Accuracy** refers to how close to the true value a given measurement is, whereas **precision** refers to how well a number of independent measurements agree with one another. To see the difference, imagine that you weigh a tennis ball whose true mass is 54.441 778 g. Assume that you take three independent measurements on each of three different types of balance to obtain the data shown in the following table.

▲ This tennis ball has a mass of about 54 g.

Measurement #	Bathroom Scale	Lab Balance	Analytical Balance
1	0.1 kg	54.4 g	54.4418 g
2	0.0 kg	54.5 g	54.4417 g
3	0.1 kg	54.3 g	54.4418 g
(average)	(0.07 kg)	(54.4 g)	(54.4418 g)

If you use a bathroom scale, your measurement (average = 0.07 kg) is neither accurate nor precise. Its accuracy is poor because it measures to only one digit that is far from the true value, and its precision is poor because any two measurements may differ substantially. If you now weigh the ball on an inexpensive laboratory balance, the value you get (average = 54.4 g) has three digits and is fairly accurate, but it is still not very precise because the three readings vary from 54.3 g to 54.5 g, perhaps due to air movements in the room or a sticky mechanism. Finally, if you weigh the ball on an expensive analytical balance like those found in research laboratories, your measurement (average = 54.4418 g) is both precise and accurate. It's accurate because the measurement is very close to the true value, and it's precise because it has six digits that vary little from one reading to another.

To indicate the uncertainty in a measurement, *the value you record should use all the digits you are sure of plus one additional digit that you estimate.* In reading a thermometer that has a mark for each degree, for example, you could be certain about the digits of the nearest mark—say 25 °C—but you would have to estimate between two marks—say between 25 °C and 26 °C—to obtain a value of 25.3 °C.

The total number of digits recorded for a measurement is called the measurement's number of **significant figures**. For example, the mass of the tennis ball as determined on the single-pan balance (54.4 g) has three significant figures, whereas the mass determined on the analytical balance (54.4418 g) has six significant figures. All digits but the last are certain; the final digit is an estimate, which we generally assume to have an error of plus or minus one (± 1).

Finding the number of significant figures in a measurement is usually easy but can be troublesome if zeros are present. Look at the following four quantities:

4.803 cm	Four significant figures: 4, 8, 0, 3
0.006 61 g	Three significant figures: 6, 6, 1
55.220 K	Five significant figures: 5, 5, 2, 2, 0
34,200 m	Anywhere from three (3, 4, 2) to five (3, 4, 2, 0, 0) significant figures

The following rules cover the different situations that arise:

1. **Zeros in the middle of a number are like any other digit; they are always significant.** Thus, 4.803 cm has four significant figures.
2. **Zeros at the beginning of a number are not significant; they act only to locate the decimal point.** Thus, 0.006 61 g has three significant figures. (Note that 0.006 61 g can be rewritten as 6.61×10^{-3} g or as 6.61 mg.)
3. **Zeros at the end of a number and after the decimal point are always significant.** The assumption is that these zeros would not be shown unless they were significant. Thus, 55.220 K has five significant figures. (If the value were known to only four significant figures, we would write 55.22 K.)
4. **Zeros at the end of a number and before the decimal point may or may not be significant.** We can't tell whether they are part of the measurement or whether they just locate the decimal point. Thus, 34,200 m may have three, four, or five significant figures. Often, however, a little common sense is helpful. A temperature reading of 20 °C probably has two significant figures rather than one, since one significant figure would imply a temperature anywhere from 10 °C to 30 °C and would be of little use. Similarly, a volume given as 300 mL probably has three significant figures. On the other hand, a figure of 93,000,000 mi for the distance between the Earth and the Sun probably has only two or three significant figures.

The fourth rule shows why it's helpful to write numbers in scientific notation rather than ordinary notation. Doing so makes it possible to indicate the number of significant figures. Thus, writing the number 34,200 as 3.42×10^4 indicates three significant figures. but writing it as 3.4200×10^4 indicates five significant figures.

One further point about significant figures: certain numbers, such as those obtained when counting objects, are exact and have an effectively infinite number of significant figures. A week has exactly 7 days, for instance, not 6.9 or 7.0 or 7.1, and a foot has exactly 12 inches, not 11.9 or 12.0 or 12.1. In addition, the power of 10 used in scientific notation is an exact number. That is, the number 10^3 is exactly 1000, but the number 1×10^3 has one significant figure.

WORKED EXAMPLE 1.5

Significant Figures

How many significant figures does each of the following measurements have?
(a) 0.036 653 m   (b) 7.2100×10^{-3} g   (c) 72,100 km   (d) $25.03

SOLUTION

(a) 5 (by rule 2)   (b) 5 (by rule 3)
(c) 3, 4, or 5 (by rule 4)   (d) $25.03 is an exact number

▶ **PRACTICE 1.9**   How many significant figures does each of the following quantities have?

(a) 76.600 kJ   (b) $4.502\ 00 \times 10^3$ g   (c) 3000 nm
(d) 0.003 00 mL   (e) 18 students   (f) 3×10^{-5} g
(g) 47.60 mL   (h) 2070 mi

▶ **APPLY 1.10**   Read the volume of the buret and report your answer to the correct number of significant figures. The volume is indicated by the bottom of the meniscus.

▲ What is the volume in this buret (Problem 1.10)?

Conceptual WORKED EXAMPLE 1.6

Determining Precision and Accuracy in a Set of Measurements

Which dartboard represents low accuracy but high precision?

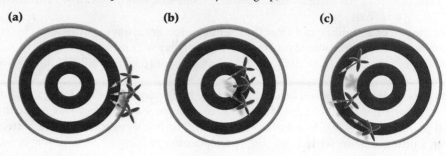

(a)   (b)   (c)

SOLUTION

Precision refers to how close the darts are to one another and accuracy refers to how close they are to the center of the target. Dartboard (**a**) has low accuracy, because the darts are far from the center, but high precision, because the three darts are all in the same location.

▶ Conceptual **PRACTICE 1.11**   Examine the figure in Worked Example 1.6. Which dartboard has

(a) low accuracy and precision
(b) high accuracy and precision

▶ **APPLY 1.12**   A 1.000 mL sample of acetone, a common solvent used as a paint remover, was placed in a small vial whose mass was known to be 4.002 g. The following values were obtained when the acetone-filled vial was weighed: 4.531 g, 4.525 g, and, 4.537 g. How would you characterize the precision and accuracy of these measurements if the true mass of the acetone was 0.7795 g?

▲ Calculators often display more figures than are justified by the precision of the data.

1.10 ▶ ROUNDING NUMBERS

It often happens, particularly when doing arithmetic on a calculator, that a quantity appears to have more significant figures than are really justified. You might calculate the gas mileage of your car, for instance, by finding that it takes 11.70 gallons of gasoline to drive 278 miles:

$$\text{Mileage} = \frac{\text{Miles}}{\text{Gallons}} = \frac{278 \text{ mi}}{11.70 \text{ gal}} = 23.760\,684 \text{ mi/gal (mpg)}$$

Although the answer on the calculator has eight digits, your measurement is really not as precise as it appears. In fact, your answer is precise to only three significant figures and should be **rounded off** to 23.8 mi/gal by removing all nonsignificant figures.

How do you decide how many figures to keep and how many to ignore? For most purposes, a simple procedure using just two rules is sufficient.

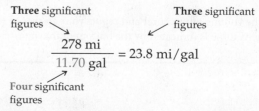

Three significant figures ↘   **Three** significant figures ↙

$$\frac{278 \text{ mi}}{11.70 \text{ gal}} = 23.8 \text{ mi/gal}$$

Four significant figures ↗

1. **In carrying out a multiplication or division, the answer can't have more significant figures than either of the original numbers**. If you think about it, this rule is just common sense. If you don't know the number of miles you drove to better than three significant figures (278 could mean 277, 278, or 279), you certainly can't calculate your mileage to more than the same number of significant figures.

2. **In carrying out an addition or subtraction, the answer can't have more digits to the right of the decimal point than either of the original numbers**. For example, if you have 3.18 L of water and you add 0.013 15 L more, you now have 3.19 L. Again, this rule is just common sense. If you don't know the volume you started with past the second decimal place (it could be 3.17, 3.18, or 3.19), you can't know the total of the combined volumes past the same decimal place.

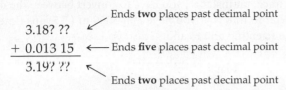

Once you decide how many digits to retain for your answer, the rules for rounding off numbers are as follows:

1. **If the first digit you remove is less than 5, round down by dropping it and all following digits**. Thus, 5.664 525 becomes 5.66 when rounded to three significant figures because the first of the dropped digits (4) is less than 5.

2. **If the first digit you remove is 6 or greater, round up by adding 1 to the digit on the left.** Thus, 5.664 525 becomes 5.7 when rounded to two significant figures because the first of the dropped digits (6) is greater than 5.

3. **If the first digit you remove is 5 and there are more nonzero digits following, round up.** Thus, 5.664 525 becomes 5.665 when rounded to four significant figures because there are nonzero digits (2, 5) after the 5.

4. **If the digit you remove is a 5 with nothing following, round down.** Thus, 5.664 525 becomes 5.664 52 when rounded to six significant figures because there is nothing after the 5.

WORKED EXAMPLE 1.7

Significant Figures in Calculations

It takes 9.25 hours to fly from London, England, to Chicago, Illinois, a distance of 3952 miles. What is the average speed of the airplane in miles per hour?

IDENTIFY

Known	Unknown
Time (9.25 h), distance (3952 mi)	Speed (mi/h)

STRATEGY

Set up a mathematical expression and solve for the answer. Use rules for significant figures in mathematical operations to determine the number of significant figures in the answer.

SOLUTION

First, set up an equation dividing the number of miles flown by the number of hours:

$$\text{Average speed} = \frac{3952 \text{ mi}}{9.25 \text{ h}} = 427.243\,24 \text{ mi/h}$$

Next, decide how many significant figures should be in your answer. Because the problem involves division, and because one of the quantities you started with (9.25 h) has only three significant figures, the answer must also have three significant figures. Finally, round off your answer. The first digit to be dropped (2) is less than 5, so the answer 427.243 24 must be rounded off to 427 mi/h.

In doing this or any other problem, use all figures, significant or not, for the calculation and then round off the final answer. Don't round off at any intermediate step.

▶ **PRACTICE 1.13** Carry out the following calculations, expressing each result with the correct number of significant figures:

(a) 24.567 g + 0.044 78 g = ? g
(b) 4.6742 g ÷ 0.003 71 L = ? g/L
(c) 0.378 mL + 42.3 mL − 1.5833 mL = ? mL

▶ **APPLY 1.14** A sodium chloride solution was prepared in the following manner:

- A 25.0 mL volumetric flask (Figure 1.8) was placed on an analytical balance and found to have a mass of 35.6783 g.
- Sodium chloride was added to flask and the mass of the solid + flask was 36.2365 g.
- The flask was filled to the mark with water and mixed well.

Calculate the concentration of the sodium chloride solution in units of g/mL and give the answer in scientific notation with the correct number of significant figures.

▲ Speed skaters have to convert from laps to meters to find out how far they have gone.

1.11 ▶ CALCULATIONS: CONVERTING FROM ONE UNIT TO ANOTHER

Because so many scientific activities involve numerical calculations—measuring, weighing, preparing solutions, and so forth—it's often necessary to convert a quantity from one unit to another. Converting between units isn't difficult; we all do it every day. If you run 7.5 laps around a 400-meter track, for instance, you have to convert between the distance unit *lap* and the distance unit *meter* to find that you have run 3000 m (7.5 laps times 400 meters/lap). Converting from one scientific unit to another is just as easy.

$$7.5 \text{ laps} \times \frac{400 \text{ meters}}{1 \text{ lap}} = 3000 \text{ meters}$$

The simplest way to carry out calculations that involve different units is to use the **dimensional-analysis method**. In this method, a quantity described in one unit is converted into an equivalent quantity with a different unit by multiplying with a **conversion factor** that expresses the relationship between units.

$$\text{Original quantity} \times \text{Conversion factor} = \text{Equivalent quantity}$$

As an example, we know from Section 1.4 that 1 meter equals 39.37 inches. Writing this relationship as a ratio restates it in the form of a conversion factor, either meters per inch or inches per meter.

Conversion factors between meters and inches

$$\frac{1 \text{ m}}{39.37 \text{ in.}} \quad \text{equals} \quad \frac{39.37 \text{ in.}}{1 \text{ m}} \quad \text{equals} \quad 1$$

Note that this and all other conversion factors are effectively equal to 1 because the quantity above the division line (the numerator) is equal in value to the quantity below the division line (the denominator). Thus, multiplying by a conversion factor is equivalent to multiplying by 1 and so does not change the value of the quantity.

The key to the dimensional-analysis method of problem solving is that units are treated like numbers and can thus be multiplied and divided just as numbers can. The idea when solving a problem is to set up an equation so that unwanted units cancel, leaving only the desired units. Usually it's best to start by writing what you know and then manipulating that known quantity. For example, say you know your height is 69.5 inches and you want to find it in meters. Begin by writing your height in inches and then set up an equation multiplying your height by the conversion factor meters per inch:

$$69.5 \text{ in.} \times \frac{1 \text{ m}}{39.37 \text{ in.}} = 1.77 \text{ m}$$

Starting quantity   Conversion factor   Equivalent quantity

The unit "in." cancels because it appears both above and below the division line, so the only unit that remains is "m."

The dimensional-analysis method gives the right answer only if the conversion factor is arranged so that the unwanted units cancel. If the equation is set up in any other way, the units won't cancel properly and you won't get the right answer. Thus, if you were to multiply your height in inches by an inverted conversion factor of inches per meter rather than meters per inch, you would end up with an incorrect answer expressed in meaningless units.

$$\text{Wrong!} \quad 69.5 \text{ in} \times \frac{39.37 \text{ in.}}{1 \text{ m}} = 2740 \text{ in.}^2/\text{m} \quad ??$$

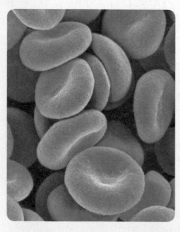

▲ What is the volume of a red blood cell?

The main drawback to using the dimensional-analysis method is that it's easy to get the right answer without really understanding what you're doing. It's therefore best after solving a problem to think through a rough estimate to check your work. If your estimate isn't close to the answer you get from the detailed solution, there's a misunderstanding somewhere and you should think through the problem again.

Even if you don't make an estimate, it's important to be sure that your calculated answer makes sense. If, for example, you were trying to calculate the volume of a human cell and you came up with the answer 5.3 cm³, you should realize that such an answer couldn't possibly be right. Cells are too tiny to be distinguished with the naked eye, but a volume of 5.3 cm³ is about the size of a walnut. Examples 1.8 and 1.9 show how to devise strategies and estimate answers when converting units using dimensional analysis.

▲ The Bugatti Veyron Super Sport has a top speed of 267 mph.

WORKED EXAMPLE 1.8

Unit Conversions Using Significant Figures

The Bugatti Veyron Super Sport is the fastest sports car in the world, with a top speed of 267 miles per hour. What is this speed (reported to the correct number of significant figures) in units of
(a) kilometers per hour?   (b) meters per second?

IDENTIFY

Known	Unknown
Speed (267 mi/h)	Speed (km/h) and speed (m/s)

STRATEGY

(a) Find the conversion factor between km and mi on the inside the back cover of this book, and use the dimensional-analysis method to set up an equation so the "mi" units cancel.

(b) Let's begin with our answer from part (a) with the speed in km/h; the unknown is the speed in m/s. Set up a series of conversion factors so that units of "km" and "h" cancel and you are left with units of "m" in the numerator and "s" in the denominator.

SOLUTION

(a) $\dfrac{267 \text{ mi}}{1 \text{ h}} \times \dfrac{1.609 \text{ km}}{1 \text{ mi}} = 429.603 \dfrac{\text{km}}{\text{h}} = 430. \dfrac{\text{km}}{\text{h}}$

(b) $\dfrac{430. \text{ km}}{1 \text{ h}} \times \dfrac{1000 \text{ m}}{1 \text{ km}} \dfrac{1 \text{ h}}{60 \text{ min}} \times \dfrac{1 \text{ min}}{60 \text{ s}} = 119.444 \dfrac{\text{m}}{\text{s}} = 119 \dfrac{\text{m}}{\text{s}}$

A very fast car!

CHECK

(a) The answer is certainly large, perhaps several hundred kilometers per hour (km/h). A better estimate is to realize that, because 1 mi = 1.609 km, it takes about 1½ times as many kilometers as miles to measure the same distance. Thus, 267 mi is about 400 km, and 267 mi/h is about 400 km/h. The estimate agrees with the detailed solution.

(b) At first glance, the answer makes sense as the speed is very high. A top sprinter can run 100 m in about 10 seconds so the car is roughly ten times faster. This seems reasonable. This is a difficult problem to estimate, however, because it requires several different conversions. It's therefore best to think the problem through one step at a time, writing down the intermediate estimates:

- Because 1 km = 1000 m then the speed is 430,000 m/h or 4.3 × 10⁵ m/h.

- Changing units of time from hours to seconds should decrease the number significantly because the car will travel a shorter distance in 1 second than in 1 hour. Because there are 3600 (3.6 × 10³) seconds in 1 hour, we can estimate by dividing the speed 4.3 × 10⁵ m/h by 3.6 × 10³ s/h

$$\frac{10^5 \text{ m/h}}{10^3 \text{ s/h}} = 10^2 \text{ m/s}$$

Making estimates using powers of 10 is very useful in checking to make sure that your answer is of the correct magnitude. This estimate agrees with the detailed solution.

▶ **PRACTICE 1.15** Gemstones are weighed in *carats*, with 1 carat = 200 mg (exactly). What is the mass in grams of the Hope Diamond, the world's largest blue diamond at 44.4 carats? What is this mass in ounces? (See conversion on the inside back cover.)

▶ **APPLY 1.16** A pure diamond has a density of 3.52 g/cm³. Set up a dimensional-analysis equation to find the volume (cm³) of the Hope Diamond (Problem 1.15).

WORKED EXAMPLE 1.9

Unit Conversions with Squared and Cubed Units

The volcanic explosion that destroyed the Indonesian island of Krakatau on August 27, 1883, released an estimated 4.3 cubic miles (mi^3) of debris into the atmosphere and affected global weather for years. In SI units, how many cubic meters (m^3) of debris were released?

IDENTIFY

Known	Unknown
Volume (4.3 mi^3)	Volume (m^3)

STRATEGY

It's probably simplest to convert first from mi^3 to km^3 and then convert km^3 to m^3. *Notice that the entire conversion factor is cubed.*

SOLUTION

$$4.3 \; \cancel{mi^3} \times \left(\frac{1 \; km}{0.6214 \; \cancel{mi}} \right)^3 = 17.92 \; km^3$$

$$17.92 \; \cancel{km^3} \times \left(\frac{1000 \; m}{1 \; \cancel{km}} \right)^3 = 1.792 \times 10^{10} \; m^3$$

$$= 1.8 \times 10^{10} \; m^3 \quad \text{Rounded off}$$

CHECK

One meter is much less than 1 mile, so it takes a large number of cubic meters to equal 1 mi^3, and the answer is going to be very large. Because 1 km is about 0.6 mi, 1 km^3 is about $(0.6)^3 = 0.2$ times as large as 1 mi^3. Thus, each mi^3 contains about 5 km^3, and 4.3 mi^3 contains about 20 km^3. Each km^3, in turn, contains $(1000 \; m)^3 = 10^9 \; m^3$. Thus, the volume of debris from the Krakatau explosion was about $20 \times 10^9 \; m^3$, or $2 \times 10^{10} \; m^3$. The estimate agrees with the detailed solution.

▶ PRACTICE 1.17

(a) A 12-inch pizza has an area of 113.112 square inches (in^2). What is the area in units of cm^2? (1 inch = 2.54 cm, exactly)

(b) A can of soda has a volume of 355 mL. What is the volume expressed in SI units of m^3?

▶ APPLY 1.18 How large, in cubic centimeters, is the volume of a red blood cell (in cm^3) if the cell has a cylindrical shape with a diameter of 6×10^{-6} m and a height of 2×10^{-6} m? What is the volume in pL?

▲ Volcano on Krakatau in Indonesia

INQUIRY ▶▶▶ WHAT ARE THE UNIQUE PROPERTIES OF NANOSCALE MATERIALS?

Imagine a world of new lightweight replacements for metals, synthetic scaffolds on which bones can be regrown, drugs that target and kill cancer cells with a minimum of side effects, and faster, smaller computers. Those are but a few of the developments that might emerge from *nanotechnology*, one of the hottest research areas in science today.

Nanotechnology is the study and production of materials and structures that have at least one dimension between 1 nm and 100 nm, where one nanometer is one billionth of a meter. Nanotechnology is an explosively growing, multidisciplinary enterprise, spanning the fields of chemistry, physics, biology, medicine, materials science, environmental science, and engineering.

To appreciate the extremely small size of nanomaterials, it is useful to compare the sizes of objects on different scales. **Macroscale** items are large enough to be observed with the human eye and are measured with instruments such as rulers and calipers. The **microscale** is a smaller size regime and is so named because dimensions of materials are in the micrometer range ($1 \mu m = 1 \times 10^{-6}$ m). Microscale objects, such as cells, cannot be seen with the human eye and therefore must be imaged with an optical microscope. One thousand times smaller than the microscale is the **nanoscale**, representing particles with nanometer-sized (1 nm $= 1 \times 10^{-9}$ m) dimensions. Atoms and molecules are nanoscale entities that use specialized instruments such as electron microscopes and atomic force microscopes for imaging. **FIGURE 1.10** depicts the scale regimes and representative objects.

The atomic force microscope (AFM, seen in the chapter-opening photo), invented in 1985, creates images of nanoscale objects by using a sharp tip to "feel" the surface. The tip is attached to a flexible cantilever that acts as a spring as it moves over hills and valleys in a sample (see **FIGURE 1.11**). The vertical deflection of the

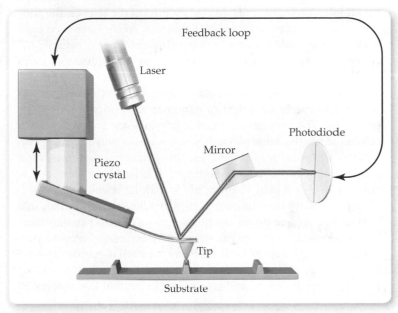

▲ FIGURE 1.11

Schematic of an atomic force microscope.

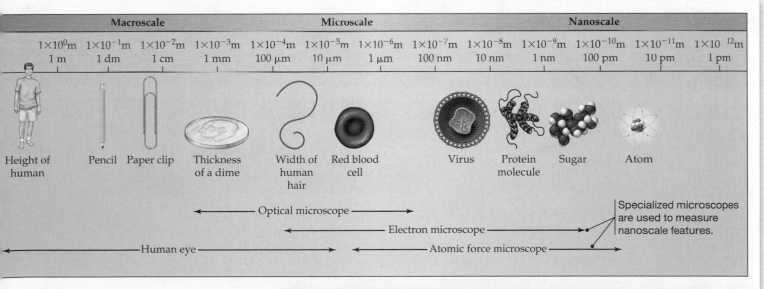

▲ FIGURE 1.10

Size scale of macroscopic and microscopic objects.

Figure It Out

What objects can be seen with a microscope but not the human eye? What particles can be imaged with an atomic force microscope, but not an optical microscope?

Answer: Cells can be seen with a microscope, but not the eye. Molecules and viruses can be seen with an atomic force microscope, but not an optical microscope. Some very high resolution images of individual atoms have been recorded with an atomic force microscope.

continued on next page

continued from previous page

tip is measured using laser light that reflects off the back of the cantilever and is sensed with a *photodiode* (light detector). A *piezoelectric crystal* (material that expands when a voltage is applied) is used to move the tip back and forth across the surface, thus "scanning" the sample. A computer records the position of the laser on the photodiode and creates an image of the surface. AFMs have been used to image nanoscale features in many different types of samples including computer chips, catalysts, viruses, and individual molecules like DNA.

Nanotechnology is an exciting research frontier because reducing the size of an object to nanometer proportions alters its properties. You may have learned in previous science classes that a substance has the same properties regardless of how much is present. For instance, gold is a yellow, shiny material that has a high melting point and conducts electricity. It is relatively *inert*, or unreactive, which is why it is useful for making jewelry or money. Gold has exactly the same properties in a large bar as it does in a tiny flake. But these properties do not extend to gold nanoparticles. Gold nanoparticles are chemically reactive and range in color from red to purple. In general, *nanoparticles have unique properties that vary with size and composition.* They tend to have lower melting points, different colors, and greater reactivity than the material in bulk.

▲ Gold and silver nanoparticles create the red and yellow colors in this stained-glass art from Medieval times.

 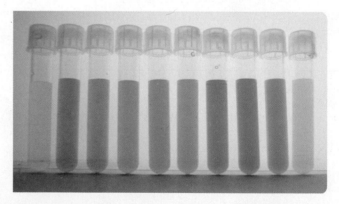

▲ A bar of gold (left) and solutions of various-sized gold nanoparticles (right). Gold does not retain its characteristic color at the nanoscale. The red solution in the vial on the left contains gold nanoparticles with diameters of 3–30 nm, and particle size increases going to the right. The violet color on the right is characteristic of gold particles hundreds of nanometers in diameter.

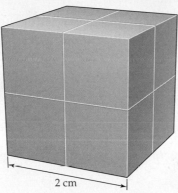

$SA = 6 \times (1 \text{ cm} \times 1 \text{ cm}) = 6 \text{ cm}^2$

$V = 1 \text{ cm} \times 1 \text{ cm} \times 1 \text{ cm} = 1 \text{ cm}^3$

$\dfrac{SA}{V} = 6$

$SA = 6 \times (2 \text{ cm} \times 2 \text{ cm}) = 24 \text{ cm}^2$

$V = 2 \text{ cm} \times 2 \text{ cm} \times 2 \text{ cm} = 8 \text{ cm}^3$

$\dfrac{SA}{V} = 3$

The *surface area to volume ratio* of a particle is a measure that can explain why some properties change with particle size. In small particles, a substantial portion of atoms are on the surface which makes them more reactive. **FIGURE 1.12** illustrates how the size of a cubic object influences this quantity. The surface area (SA) is calculated by multiplying the number of sides by the area of each side, $A = (l \times w)$, and the volume (V) is calculated using the formula $V = l \times w \times h$. Both the surface area and volume on the cube with the side length of 2 cm are larger than the cube with a 1 cm side. However, the surface area to volume ratio is smaller for the larger cube.

PROBLEM 1.19   Using Figure 1.10, estimate in powers of 10

   (a) How many times larger the diameter of a human hair is than a 10 nm gold nanoparticle.

   (b) How many times larger a red blood cell is than a sugar molecule.

PROBLEM 1.20   Calculate the percentage of atoms on the surface of a cubic nanoparticle if the diameter of the atoms is 250 pm and the edge length of the particle is

   (a) 5.0 nm

   (b) 10.0 nm

PROBLEM 1.21   Catalytic converters use nanoscale particles of precious metals such as platinum to change pollutants in automobile exhaust into less harmful gases. Calculate the following quantities for two different spherical particles of platinum with diameters of 5.0 nm and 5.0 μm.

   (a) surface area in units of μm^2 ($SA = 4\pi r^2$)

   (b) volume in units of μm^3 $\left(V = \dfrac{4}{3}\pi r^3\right)$

   (c) surface area to volume ratio in units of μm^{-1}

   (d) How many times larger is the surface area to volume ratio of the 5 nm particle than the 5 μm particle?

PROBLEM 1.22   Platinum is an expensive and rare metal used in catalytic converters and other industrial applications. Much research has been devoted to maximizing reactive properties of other metals by shrinking them to nanoparticles in the hope that someday they will be an efficient and economic alternative to platinum.

   (a) Explain why changing the size of a metal particle influences its reactivity.

   (b) What is the economic benefit of using small particles?

   (c) What properties other than reactivity might you expect to change as the size of particle approaches the nanoscale?

STUDY GUIDE

Section	Concept Summary	Learning Objectives	Test Your Understanding
1.1 ▶ The Scientific Method	The **scientific method** is an iterative process used to perform research. A driving question, often based upon observations, is the first step. Next a **hypothesis** is developed to explain the observation. **Experiments** are designed to test the hypothesis and the results are used to verify or modify the original hypothesis. **Theories** arise when numerous experiments validate a hypothesis and are used to make new predictions. **Models** are simplified representations of complex systems that help make theories more concrete.	**1.1** Identify the steps in the scientific method.	Problems 1.28–1.30
		1.2 Differentiate between a qualitative and quantitative measurement.	Problems 1.33–1.35
1.2 ▶ Experimentation and Measurement	Accurate measurement is crucial to scientific experimentation. Scientists use units of measure established by the *Système Internationale* (**SI units**). There are seven fundamental SI units, together with other derived units. (**Table 1.1**)	**1.3** Write numbers in scientific notation and use prefixes for multiples of SI units.	Worked Example 1.1; Problems 1.39, 1.49, 1.52, 1.58, and 1.59
1.3 ▶ Mass and Its Measurement	**Mass**, the amount of matter in an object, is measured in the SI unit of **kilograms (kg)**.	**1.4** Describe the difference between mass and weight.	Problem 1.36
		1.5 Convert between different prefixes used in mass measurements.	Problem 1.50
1.4 ▶ Length and Its Measurement	**Length** is measured in the SI unit of **meters (m)**.	**1.6** Convert between different prefixes used in length measurements.	Problem 1.52 (a) and (b)
1.5 ▶ Temperature and Its Measurement	**Fahrenheit (°F)** is the most common unit for measuring temperature in the United States, whereas **Celsius (°C)** is more common in other parts of the world. **Kelvin (K)** is the standard temperature unit in scientific work.	**1.7** Convert between common units of temperature measurements.	Worked Example 1.2; Problems 1.74–1.77
1.6 ▶ Derived Units: Volume and Its Measurement	Volume, the amount of space occupied by an object, is measured in SI units by the **cubic meter (m^3)**.	**1.8** Convert between SI and metric units of volume.	Problems 1.42 and 1.43, 1.99
		1.9 Convert between different prefixes used in volume measurements.	Problem 1.51
1.7 ▶ Derived Units: Density and Its Measurement	**Density** is a property that relates mass to volume and is measured in the derived SI unit g/cm^3 or g/mL.	**1.10** Calculate mass, volume, or density using the formula for density.	Worked Example 1.3; Problems 1.80–1.88, 1.96, 1.100, 1.101
		1.11 Predict whether a substance will float or sink in another substance based on density.	Problem 1.27, 1.97, 1.107
1.8 ▶ Derived Units: Energy and Its Measurement	**Energy** is the capacity to supply heat or do work and is measured in the derived SI unit ($kg \cdot m^2/s^2$), or **joule (J)**. Energy is of two kinds, potential and kinetic. **Kinetic energy (E_K)** is the energy of motion, and **potential energy (E_P)** is stored energy.	**1.12** Calculate kinetic energy of a moving object.	Worked Example 1.4; Problem 1.60
		1.13 Convert between common energy units.	Problems 1.94 and 1.95
1.9 ▶ Accuracy, Precision, and Significant Figures in Measurement	If measurements are **accurate**, they are close to the true value, and if measurements are **precise** they are reproducible or close to one another.	**1.14** Specify the number of significant figures in a measurement.	Worked Example 1.5; Problems 1.54 and 1.55
		1.15 Evaluate the level of accuracy and precision in a data set.	Worked Example 1.6; Problem 1.12
		1.16 Report a measurement to the appropriate number of significant figures.	Problems 1.25 and 1.26

Section	Concept Summary	Learning Objectives	Test Your Understanding
1.10 ▸ Rounding Numbers	It's important when measuring physical quantities or carrying out calculations to indicate the precision of the measurement by **rounding off** the result to the correct number of **significant figures**.	**1.17** Report the answer of mathematical calculations to the correct number of significant figures.	Worked Example 1.7; Problems 1.62–1.65
		1.18 Round a measurement to a specified number of significant figures.	Problems 1.56 and 1.57
1.11 ▸ Unit Conversions	Because many experiments involve numerical calculations, it's often necessary to manipulate and convert different units of measure. The simplest way to carry out such conversions is to use the **dimensional-analysis method**, in which an equation is set up so that unwanted units cancel and only the desired units remain	**1.19** Change a measurement into different units using appropriate conversion factors.	Worked Examples 1.8 and 1.9, Problems 1.66–1.73, 1.108, 1.103

KEY TERMS

accuracy *16*
amino acids *4*
Celsius degree (°C) *9*
centimeter (cm) *9*
chemistry *2*
conversion factor *20*
cubic centimeter (cm^3) *11*
cubic decimeter (dm^3) *11*
cubic meter (m^3) *11*
density *12*
dimensional-analysis method *20*
energy *14*

experiment *4*
Fahrenheit (°F) *9*
gram (g) *8*
hypothesis *4*
joule (J) *15*
kelvin (K) *9*
kilogram (kg) *8*
kinetic energy (E_K) *14*
liter (L) *11*
macroscale *23*
microscale *23*
mass *8*

matter *8*
meter (m) *8*
microgram (μg) *8*
micrometer (μm) *9*
milligram (mg) *8*
milliliter (mL) *11*
millimeter (mm) *9*
molecular model *5*
nanometer (nm) *9*
nanoscale *23*
nanotechnology *23*
observation *4*

peptide *4*
picometer (pm) *9*
potential energy (E_P) *14*
precision *16*
rounding off *27*
scientific method *3*
scientific notation *6*
SI unit *26*
significant figure *16*
theory *4*
qualitative *4*
quantitative *4*

KEY EQUATIONS

• **Relationship between the Kelvin and Celsius Scales (Section 1.5)**

Temperature in K = Temperature in °C + 273.15
Temperature in °C = Temperature in K − 273.15

• **Converting between Celsius and Fahrenheit temperatures (Section 1.5)**

$$°F = \left(\frac{9\,°F}{5\,°C} \times °C \right) + 32\,°F \quad °C = \frac{5\,°C}{9\,°F} \times (°F - 32\,°F)$$

• **Calculating density (Section 1.7)**

$$\text{Density} = \frac{\text{Mass (g)}}{\text{Volume (mL or cm}^3)}$$

• **Calculating kinetic energy (Section 1.8)**

$$E_K = \frac{1}{2}mv^2$$

CONCEPTUAL PROBLEMS

*Problems at the end of each chapter begin with a section called "Concep-
tual Problems." The problems in this section are visual or abstract rather
than numerical and are intended to probe your understanding rather than
your facility with numbers and formulas. Answers to even-numbered problems (in color) can be found at the end of the book following the appendices.
Problems 1.1–1.22 appear within the chapter.*

1.23 Which block in each of the following drawings of a balance is
more dense, red or green? Explain.

(a)   (b)

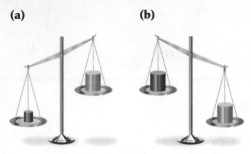

1.24 What is the temperature reading on the following Celsius thermometer? How many significant figures do you have in your
answer?

1.25 How many milliliters of water does the graduated cylinder in
(a) contain, and how tall in centimeters is the paper clip in (b)?
How many significant figures do you have in each answer?

(a)   (b)

1.26 Assume that you have two graduated cylinders, one with a capacity of 5 mL (a) and the other with a capacity of 50 mL (b). Draw
a line in each, showing how much liquid you would add if you

needed to measure 2.64 mL of water. Which cylinder will give the
more accurate measurement? Explain.

(a)   (b)

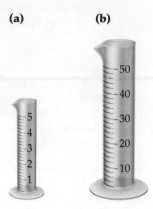

1.27 The following cylinder contains three liquids that do not mix
with one another: water (density = 1.0 g/mL), vegetable oil
(density = 0.93 g/mL), and mercury (density = 13.5 g/mL).
Which liquid is which?

SECTION PROBLEMS

*The Section Problems at the end of each chapter cover specific topics from the
various sections of the chapter. These problems are presented in pairs, with
each even-numbered problem followed by an odd-numbered one requiring
similar skills. These paired problems are followed by unpaired Chapter Problems that draw on various parts of the chapter. Even-numbered problems
(in color) are answered at the end of the book following the appendixes.*

Scientific Method (Section 1.1)

1.28 The following statements pertain to the development of the theory of combustion by the French chemist Lavoisier in the eigh-
teenth century. Match the statement with the appropriate step
(observation, hypothesis, experiment designed to test hypothesis)
in the scientific method.

(a) A metal is burned in a closed container and the change in
mass of the solid and volume of the gas is measured.

(b) Oxygen gas combines with a substance during its combustion.

(c) Combustion of a metal in a closed container ceases after a length of time.

1.29 The following statements pertain to the development of the theory of the structure of DNA. Match the statement with the appropriate step (observation, hypothesis, experiment designed to test hypothesis) in the scientific method.

(a) Two strands of DNA wind around one another in a helical structure.

(b) In a sample of DNA, there are equal amounts of the bases A and T and equal amounts of the bases C and G.

(c) Direct X-rays at a sample of crystallized DNA and interpret the diffraction pattern for structural information.

1.30 What is the difference between a hypothesis and theory?

(a) A hypothesis provides an explanation for a phenomenon, but a theory does not.

(b) A theory provides an explanation for a phenomenon, but hypothesis does not.

(c) Both a theory and a hypothesis provide an explanation for a phenomenon, but a theory has been upheld by experimental observations.

1.31 Dr. DiMarchi proposed that switching the positions of two amino acids in human insulin would create an analog with a faster time profile. Is this a hypothesis or theory?

1.32 How do molecular models help chemists develop theories?

1.33 Label the following statements about insulin therapy for diabetics as quantitative or qualitative observations:

(a) Insulin injections allow a diabetic patient to survive.

(b) The peak in the time release profile for natural insulin is 90 minutes while the peak for injected insulin is 150 minutes.

1.34 Label the following statements about the world's largest gold bar as quantitative or qualitative observations. (This gold bar was worth approximately \$10.25 million in 2013.)

(a) The melting point of gold is 1064.2 °C.

(b) The volume of the gold bar is 15,730 cm^3.

(c) Gold metal is a conductor of electricity.

(d) The mass of the gold bar is 250 kg.

(e) The gold bar is yellow and shiny.

1.35 Label the following statements as quantitative or qualitative observations.

(a) An object weighs less on the moon than on Earth.

(b) An object that weighs 50 pounds on Earth only weighs 8.3 pounds on the moon.

(c) The freezing point of water is cold.

(d) The freezing point of water is 0 °C.

Units and Significant Figures (Sections 1.2–1.10)

1.36 What is the difference between mass and weight?

1.37 What is the difference between a derived SI unit and a fundamental SI unit? Give an example of each.

1.38 What SI units are used for measuring the following quantities? For derived units, express your answers in terms of the six fundamental units.

(a) Mass   (b) Length

(c) Temperature   (d) Volume

(e) Energy   (f) Density

1.39 What SI prefixes correspond to the following multipliers?

(a) 10^3 (b) 10^{-6}

(c) 10^9 (d) 10^{-12}

(e) 10^{-2}

1.40 Which is larger, a Fahrenheit degree or a Celsius degree? By how much?

1.41 What is the difference between a kelvin and a Celsius degree?

1.42 What is the difference between a cubic decimeter (SI) and a liter (metric)?

1.43 What is the difference between a cubic centimeter (SI) and a milliliter (metric)?

1.44 Which of the following statements use exact numbers?

(a) 1 ft = 12 in.

(b) 1 cal = 4.184 J

(c) The height of Mt. Everest is 29,035 ft.

(d) The world record for the 1 mile run, set by Morocco's Hicham el Guerrouj in July, 1999, is 3 minutes, 43.13 seconds.

1.45 What is the difference in mass between a nickel that weighs 4.8 g and a nickel that weighs 4.8673 g?

1.46 Bottles of wine sometimes carry the notation "Volume = 75 cL." What does the unit cL mean?

1.47 What do the following abbreviations stand for?

(a) dL   (b) dm

(c) μm   (d) nL

(e) MJ

1.48 Which quantity in each of the following pairs is larger?

(a) 5.63×10^6 cm or 6.02×10^1 km

(b) 46 μs or 3.2×10^{-2} ms

(c) 200,098 g or 17×10^1 kg

1.49 Which quantity in each of the following pairs is smaller?

(a) 154 pm or 7.7×10^{-9} cm

(b) 1.86×10^{11} μm or 2.02×10^2 km

(c) 2.9 GA or 3.1×10^{15} μA

1.50 How many picograms are in 1 mg? In 35 ng?

1.51 How many microliters are in 1 L? In 20 mL?

1.52 Carry out the following conversions:

(a) 5 pm = _____ cm = _____ nm

(b) 8.5 cm^3 = _____ m^3 = _____ mm^3

(c) 65.2 mg = _____ g = _____ pg

1.53 Which is larger, and by approximately how much?

(a) A liter or a quart

(b) A mile or a kilometer

(c) A gram or an ounce

(d) A centimeter or an inch

1.54 How many significant figures are in each of the following measurements?

(a) 35.0445 g   (b) 59.0001 cm

(c) 0.030 03 kg   (d) 0.004 50 m

(e) 67,000 m^2 (f) 3.8200×10^3 L

1.55 How many significant figures are in each of the following measurements?

(a) \$130.95   (b) 2000.003 g

(c) 5 ft 3 in.   (d) 510 J

(e) 5.10×10^2 J   (f) 10 students

1.56 The Vehicle Assembly Building at the John F. Kennedy Space Center in Cape Canaveral, Florida, is the largest building in the world, with a volume of 3,666,500 m^3. Round off this quantity to four significant figures; and then to two significant figures. Express the answers in scientific notation.

1.57 The diameter of the Earth at the equator is 7926.381 mi. Round off this quantity to four significant figures; then to two significant figures. Express the answers in scientific notation.

1.58 Express the following measurements in scientific notation:

 (a) 453.32 mg

 (b) 0.000 042 1 mL

 (c) 667,000 g

1.59 Convert the following measurements from scientific notation to standard notation:

 (a) 3.221×10^{-3} mm   (b) 8.940×10^5 m

 (c) $1.350\,82 \times 10^{-12}$ m^3 (d) 6.4100×10^2 km

1.60 Round off the following quantities to the number of significant figures indicated in parentheses:

 (a) 35,670.06 m (4, 6)   (b) 68.507 g (2, 3)

 (c) 4.995×10^3 cm (3)   (d) $2.309\,85 \times 10^{-4}$ kg (5)

1.61 Round off the following quantities to the number of significant figures indicated in parentheses:

 (a) 7.0001 kg (4)   (b) 1.605 km (3)

 (c) 13.2151 g/cm^3 (3)   (d) 2,300,000.1 (7)

1.62 Express the results of the following calculations with the correct number of significant figures:

 (a) 4.884×2.05 (b) $94.61 \div 3.7$

 (c) $3.7 \div 94.61$ (d) $5502.3 + 24 + 0.01$

 (e) $86.3 + 1.42 - 0.09$ (f) 5.7×2.31

1.63 Express the results of the following calculations with the correct number of significant figures:

 (a) $\dfrac{3.41 - 0.23}{5.233} \times 0.205$ (b) $\dfrac{5.556 \times 2.3}{4.223 - 0.08}$

1.64 The world record for the women's outdoor 20,000-meter run, set in 2000 by Tegla Loroupe, is 1:05:26.6 (seconds are given to the nearest tenth). What was her average speed, expressed in miles per hour with the correct number of significant figures? (Assume that the race distance is accurate to 5 significant figures.)

1.65 In the United States, the emissions limit for carbon monoxide in motorcycle engine exhaust is 12.0 g of carbon monoxide per kilometer driven. What is this limit expressed in mg per mile with the correct number of significant figures?

Unit Conversions (Section 1.11)

1.66 Carry out the following conversions:

 (a) How many grams of meat are in a quarter-pound hamburger (0.25 lb)?

 (b) How tall in meters is the Willis Tower, formerly called the Sears Tower, in Chicago (1454 ft)?

 (c) How large in square meters is the land area of Australia (2,941,526 mi^2)?

1.67 Convert the following quantities into SI units with the correct number of significant figures:

 (a) 5.4 in.   (b) 66.31 lb

 (c) 0.5521 gal   (d) 65 mi/h

 (e) 978.3 yd^3 (f) 2.380 mi^2

1.68 The volume of water used for crop irrigation is measured in acre feet, where 1 acre-foot is the amount of water needed to cover 1 acre of land to a depth of 1 ft.

 (a) If there are 640 acres per square mile, how many cubic feet of water are in 1 acre-foot?

 (b) How many acre-feet are in Lake Erie (total volume = 116 mi^3)?

1.69 The height of a horse is usually measured in *hands* instead of in feet, where 1 hand equals 1/3 ft (exactly).

 (a) How tall in centimeters is a horse of 18.6 hands?

 (b) What is the volume in cubic meters of a box measuring $6 \times 2.5 \times 15$ hands?

1.70 Weights in England are commonly measured in *stones*, where 1 stone = 14 lb. What is the weight in pounds of a person who weighs 8.65 stones?

1.71 Concentrations of substances dissolved in solution are often expressed as mass per unit volume. For example, normal human blood has a cholesterol concentration of about 200 mg/100 mL. Express this concentration in the following units:

 (a) mg/L   (b) $\mu g/mL$

 (c) g/L   (d) $ng/\mu L$

 (e) How much total blood cholesterol in grams does a person have if the normal blood volume in the body is 5 L?

1.72 Administration of digitalis, a drug used to control atrial fibrillation in heart patients, must be carefully controlled because even a modest overdose can be fatal. To take differences between patients into account, drug dosages are prescribed in terms of mg/kg body weight. Thus, a child and an adult differ greatly in weight, but both receive the same dosage per kilogram of body weight. At a dosage of 20 $\mu g/kg$ body weight, how many milligrams of digitalis should a 160 lb patient receive?

1.73 Among many alternative units that might be considered as a measure of time is the *shake* rather than the second. Based on the expression "faster than a shake of a lamb's tail," we'll define 1 shake as equal to 2.5×10^{-4} s. If a car is traveling at 55 mi/h, what is its speed in cm/shake?

Temperature (Section 1.5)

1.74 The normal body temperature of a goat is 39.9 °C, and that of an Australian spiny anteater is 22.2 °C. Express these temperatures in degrees Fahrenheit.

1.75 Of the 90 or so naturally occurring elements, only four are liquid near room temperature: mercury (melting point = −38.87 °C), bromine (melting point = −7.2 °C), cesium (melting point = 28.40 °C), and gallium (melting point = 29.78 °C). Convert these melting points to degrees Fahrenheit.

1.76 Suppose that your oven is calibrated in degrees Fahrenheit but a recipe calls for you to bake at 175 °C. What oven setting should you use?

1.77 Tungsten, the element used to make filaments in lightbulbs, has a melting point of 6192 °F. Convert this temperature to degrees Celsius and to kelvin.

1.78 Suppose you were dissatisfied with both Celsius and Fahrenheit units and wanted to design your own temperature scale based on ethyl alcohol (ethanol). On the Celsius scale, ethanol has a melting point of −117.3 °C and a boiling point of 78.5 °C, but on your new scale calibrated in units of degrees ethanol, °E, you define ethanol to melt at 0 °E and boil at 200 °E.

 (a) How does your ethanol degree compare in size with a Celsius degree?

(b) How does an ethanol degree compare in size with a Fahrenheit degree?

(c) What are the melting and boiling points of water on the ethanol scale?

(d) What is normal human body temperature (98.6 °F) on the ethanol scale?

(e) If the outside thermometer reads 130 °E, how would you dress to go out?

1.79 Answer parts **(a)**–**(d)** of Problem 1.78 assuming that your new temperature scale is based on ammonia, NH_3. On the Celsius scale, ammonia has a melting point of -77.7 °C and a boiling point of -33.4 °C, but on your new scale calibrated in units of degrees ammonia, °A, you define ammonia to melt at 0 °A and boil at 100 °A.

Density (Section 1.7)

1.80 What is the density of glass in g/cm^3 if a sample weighing 27.43 g has a volume of 12.40 cm^3?

1.81 What is the density of lead in g/cm^3 if a sample weighing 206.77 g has a volume of 15.50 cm^3?

1.82 A vessel contains 4.67 L of bromine, whose density is 3.10 g/cm^3. What is the mass of the bromine in the vessel (in kilograms)?

1.83 Aspirin has a density of 1.40 g/cm^3. What is the volume in cubic centimeters of an aspirin tablet weighing 250 mg? Of a tablet weighing 500 lb?

1.84 Gaseous hydrogen has a density of 0.0899 g/L at 0 °C, and gaseous chlorine has a density of 3.214 g/L at the same temperature. How many liters of each would you need if you wanted 1.0078 g of hydrogen and 35.45 g of chlorine?

1.85 The density of silver is 10.5 g/cm^3. What is the mass (in kilograms) of a cube of silver that measures 0.62 m on each side?

1.86 What is the density of lead in g/cm^3 if a rectangular bar measuring 0.50 cm in height, 1.55 cm in width, and 25.00 cm in length has a mass of 220.9 g?

1.87 What is the density of lithium metal in g/cm^3 if a cylindrical wire with a diameter of 2.40 mm and a length of 15.0 cm has a mass of 0.3624 g?

1.88 You would like to determine if a set of antique silverware is pure silver. The mass of a small fork was measured on a balance and found to be 80.56 g. The volume was found by dropping the fork into a graduated cylinder initially containing 10.0 mL of water. The volume after the fork was added was 15.90 mL. Calculate the density of the fork. If the density of pure silver at the same temperature is 10.5 g/cm^3, is the fork pure silver?

1.89 An experiment is performed to determine if pennies are made of pure copper. The mass of 10 pennies was measured on a balance and found to be 24.656 g. The volume was found by dropping the 10 pennies into a graduated cylinder initially containing 10.0 mL of water. The volume after the pennies were added was 12.90 mL. Calculate the density of the pennies. If the density of pure copper at the same temperature is 8.96 g/cm^3, are the pennies made of pure copper?

Energy (Section 1.8)

1.90 Which has more kinetic energy, a 1400 kg car moving at 115 km/h or a 12,000 kg truck moving at 38 km/h?

1.91 Assume that the kinetic energy of a 1400 kg car moving at 115 km/h (Problem 1.90) is converted entirely into heat. How many calories of heat are released, and what amount of water in liters could be heated from 20.0 °C to 50.0 °C by the car's energy? (One calorie raises the temperature of 1 mL of water by 1 °C.)

1.92 The combustion of 45.0 g of methane (natural gas) releases 2498 kJ of heat energy. How much energy in kilocalories (kcal) would combustion of 0.450 ounces of methane release?

1.93 Sodium (Na) metal undergoes a chemical reaction with chlorine (Cl) gas to yield sodium chloride, or common table salt. If 1.00 g of sodium reacts with 1.54 g of chlorine, 2.54 g of sodium chloride is formed and 17.9 kJ of heat is released. How much sodium and how much chlorine in grams would have to react to release 171 kcal of heat?

1.94 A Big Mac hamburger from McDonald's contains 540 Calories.

(a) How many kilojoules does a Big Mac contain?

(b) For how many hours could the amount of energy in a Big Mac light a 100 watt lightbulb? (1 watt = 1 J/s)

1.95 A 20 fluid oz. soda contains 238 Calories.

(a) How many kilojoules does the soda contain?

(b) For how many hours could the amount of energy in the soda light a 75 watt lightbulb? (1 watt = 1 J/s)

CHAPTER PROBLEMS

1.96 When an irregularly shaped chunk of silicon weighing 8.763 g was placed in a graduated cylinder containing 25.00 mL of water, the water level in the cylinder rose to 28.76 mL. What is the density of silicon in g/cm^3?

1.97 Lignum vitae is a hard, durable, and extremely dense wood used to make ship bearings. A sphere of this wood with a diameter of 7.60 cm has a mass of 313 g.

(a) What is the density of the lignum vitae sphere?

(b) Will the sphere float or sink in water?

(c) Will the sphere float or sink in chloroform? (The density of chloroform is 1.48 g/mL.)

1.98 Sodium chloride has a melting point of 1074 K and a boiling point of 1686 K. Convert these temperatures to degrees Celsius and to degrees Fahrenheit.

1.99 A large tanker truck for carrying gasoline has a capacity of 3.4×10^4 L.

(a) What is the tanker's capacity in gallons?

(b) If the retail price of gasoline is $3.00 per gallon, what is the value of the truck's full load of gasoline?

1.100 The density of chloroform, a widely used organic solvent, is 1.4832 g/mL at 20 °C. How many milliliters would you use if you wanted 112.5 g of chloroform?

1.101 More sulfuric acid (density = 1.8302 g/cm^3) is produced than any other chemical—approximately 3.6×10^{11} lb/yr worldwide. What is the volume of this amount in liters?

1.102 Answer the following questions:

(a) An old rule of thumb in cooking says: "A pint's a pound the world around." What is the density in g/mL of a substance for which 1 pt = 1 lb exactly?

(b) There are exactly 640 acres in 1 square mile. How many square meters are in 1 acre?

(c) A certain type of wood has a density of 0.40 g/cm^3. What is the mass of 1.0 cord of this wood in kg, where 1 cord is 128 cubic feet of wood?

(d) A particular sample of crude oil has a density of 0.85 g/mL. What is the mass of 1.00 barrel of this crude oil in kg, where a barrel of oil is exactly 42 gallons?

(e) A gallon of ice cream contains exactly 32 servings, and each serving has 165 Calories, of which 30.0% are derived from fat. How many Calories derived from fat would you consume if you ate one half-gallon of ice cream?

1.103 A 1.0 ounce piece of chocolate contains 15 mg of caffeine, and a 6.0 ounce cup of regular coffee contains 105 mg of caffeine. How much chocolate would you have to consume to get as much caffeine as you would from 2.0 cups of coffee?

1.104 A bag of Hershey's Kisses contains the following information:

Serving size: 9 pieces = 41 g

Calories per serving: 230

Total fat per serving: 13 g

(a) The bag contains 2.0 lbs of Hershey's Kisses. How many Kisses are in the bag?

(b) The density of a Hershey's Kiss is 1.4 g/mL. What is the volume of a single Hershey's Kiss?

(c) How many Calories are in one Hershey's Kiss?

(d) Each gram of fat yields 9 Calories when metabolized. What percent of the calories in Hershey's Kisses are derived from fat?

1.105 Vinaigrette salad dressing consists mainly of oil and vinegar. The density of olive oil is 0.918 g/cm^3, the density of vinegar is 1.006 g/cm^3, and the two do not mix. If a certain mixture of olive oil and vinegar has a total mass of 397.8 g and a total volume of 422.8 cm^3, what is the volume of oil and what is the volume of vinegar in the mixture?

1.106 At a certain point, the Celsius and Fahrenheit scales "cross," giving the same numerical value on both. At what temperature does this crossover occur?

1.107 Imagine that you place a cork measuring 1.30 cm × 5.50 cm × 3.00 cm in a pan of water and that on top of the cork you place a small cube of lead measuring 1.15 cm on each edge. The density of cork is 0.235 g/cm^3, and the density of lead is 11.35 g/cm^3. Will the combination of cork plus lead float or sink?

1.108 The density of polystyrene, a plastic commonly used to make CD cases and transparent cups is 0.037 lbs/in^3. Calculate the density in units of g/cm^3.

1.109 The density of polypropylene, a plastic commonly used to make bottle caps, yogurt containers, and carpeting, is 0.55 oz/in^3. Calculate the density in units of g/cm^3.

1.110 A 125 mL sample of water at 293.2 K was heated for 8 min, 25 s so as to give a constant temperature increase of 3.0 °F/min. What is the final temperature of the water in degrees Celsius?

1.111 A calibrated flask was filled to the 25.00 mL mark with ethyl alcohol. By weighing the flask before and after adding the alcohol, it was determined that the flask contained 19.7325 g of alcohol. In a second experiment, 25.0920 g of metal beads were added to the flask, and the flask was again filled to the 25.00 mL mark with ethyl alcohol. The total mass of the metal plus alcohol in the flask was determined to be 38.4704 g. What is the density of the metal in g/mL?

1.112 Brass is a copper–zinc alloy. What is the mass in grams of a brass cylinder having a length of 1.62 in. and a diameter of 0.514 in. if the composition of the brass is 67.0% copper and 33.0% zinc by mass? The density of copper is 8.92 g/cm^3, and the density of zinc is 7.14 g/cm^3. Assume that the density of the brass varies linearly with composition.

1.113 Ocean currents are measured in *Sverdrups* (sv) where 1 sv = 10^9 m^3/s. The Gulf Stream off the tip of Florida, for instance, has a flow of 35 sv.

(a) What is the flow of the Gulf Stream in milliliters per minute?

(b) What mass of water in the Gulf Stream flows past a given point in 24 hours? The density of seawater is 1.025 g/mL.

(c) How much time is required for 1 petaliter (PL; 1 PL = 10^{15} L) of seawater to flow past a given point?

1.114 The element gallium (Ga) has the second largest liquid range of any element, melting at 29.78 °C and boiling at 2204 °C at atmospheric pressure.

(a) What is the density of gallium in g/cm^3 at 25 °C if a 1 in cube has a mass of 0.2133 lb?

(b) Assume that you construct a thermometer using gallium as the fluid instead of mercury, and that you define the melting point of gallium as 0 °G and the boiling point of gallium as 1000 °G. What is the melting point of sodium chloride (801 °C) on the gallium scale?

1.115 Distances over land are measured in *statute miles* (5280 ft) but distances over water are measured in *nautical miles*, where 1 nautical mile was originally defined as 1 minute of arc along an Earth meridian, or 1/21,600 of the Earth's circumference through the poles. A ship's speed through the water is measured in *knots*, where 1 knot = 1 nautical mile per hour. Historically, the unit *knot* derived from the practice of measuring a ship's speed by throwing a log tied to a knotted line over the side. The line had a knot tied in it at intervals of 47 ft. 3 in., and the number of knots run out in 28 seconds was counted to determine speed.

(a) How many feet are in a nautical mile? How many meters?

(b) The northern bluefin tuna can weigh up to 1500 pounds and can swim at speeds up to 48 miles per hour. How fast is this in knots?

(c) A *league* is defined as 3 nautical miles. The Mariana Trench, with a depth of 35,798 feet, is the deepest point in the ocean. How many leagues deep is this?

(d) By international agreement, the nautical mile is now defined as exactly 1852 meters. By what percentage does this current definition differ from the original definition, and by what percentage does it differ from a statute mile?

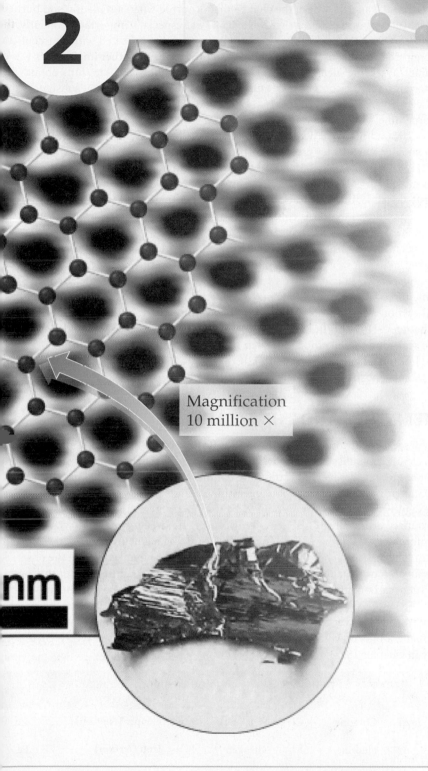

Magnification
10 million ×

nm

Atoms, Molecules, and Ions

If you could take a large piece of a pure element such as carbon and cut it into ever smaller and smaller pieces, you would find that it is made of a vast number of tiny fundamental units that we call *atoms*. A scanning tunneling microscope generated this image of individual carbon atoms in the small flake of highly ordered pyrolytic graphite shown.

 How is the principle of atom economy used to minimize waste in a chemical synthesis?

The answer to this question can be found in the **INQUIRY** ▶▶▶ on page 66.

People have always been fascinated by changes, particularly by those that are dramatic or useful. In the ancient world, the change that occurred when a stick of wood burned, gave off heat, and turned into a small pile of ash was especially important. Similarly, the change that occurred when a reddish lump of rock (iron ore) was heated with charcoal and produced a gray metal (iron) useful for making weapons, tools, and other implements was of enormous value. Observing such changes eventually caused philosophers to think about what different materials might be composed of and led to the idea of fundamental substances that we today call elements.

At the same time, philosophers were pondering the question of elements, they were also thinking about related matters: What makes up an element? Is matter continuously divisible into ever smaller and smaller pieces, or is there an ultimate limit? Can you cut a piece of gold in two, take one of the pieces and cut *it* in two, and so on infinitely, or is there a point at which you must stop? Most thinkers, including Plato and Aristotle, believed that matter is continuously divisible, but the Greek philosopher Democritus (460–370 B.C.) disagreed. Democritus proposed that matter is composed of tiny, discrete particles, which we now call *atoms*, from the Greek word *atomos*, meaning "indivisible." Little else was learned about elements and atoms until the birth of modern experimental science some 2000 years later.

Although atomic theory existed for many years, atoms were never directly "seen" until the invention of specialized instruments such as the scanning tunneling microscope (STM) and the atomic force microscope (AFM) in the 1980s. An STM was used to create the image of individual carbon atoms in the chapter opening photo and an AFM is described in the Inquiry in Chapter 1. Now, for the first time, matter can be directly manipulated and studied at the atomic scale. Individual atoms can even be moved in an attempt to build materials in novel ways. The control and study of matter at the atomic level has important implications in computing, biochemistry, and industrial products such as batteries, catalysts, and plastics.

2.1 ▶ CHEMISTRY AND THE ELEMENTS

Everything you see around you is formed from one or more of 118 presently known *elements*. An **element** is a fundamental substance that can't be chemically changed or broken down into anything simpler. Mercury, silver, and sulfur are common examples, as listed in **TABLE 2.1**.

Actually, the previous statement about everything being made of one or more of 118 elements is an exaggeration because only about 90 of the 118 occur naturally. The remaining 28 have been produced artificially by nuclear chemists using high-energy particle accelerators.

Furthermore, only 83 of the 90 or so naturally occurring elements are found in any appreciable abundance. Hydrogen is thought to account for approximately 75% of the observed mass in the universe; oxygen and silicon together account for 75% of the mass of the Earth's crust; and oxygen, carbon, hydrogen, and nitrogen make up more than 95% of the mass of the human body (**FIGURE 2.1**). By contrast, there is probably less than 20 grams of the element francium (Fr) dispersed over the entire Earth at any one time. Francium is an unstable radioactive element, atoms of which are continually being formed and destroyed. We'll discuss **radioactivity** in Chapter 18.

▲ Samples of mercury, silver, and sulfur (clockwise from top left).

LOOKING AHEAD...

As we'll see in Chapter 18, **radioactivity** is the spontaneous decay of certain unstable atoms with emission of some form of radiation.

TABLE 2.1 **Names and Symbols of Some Common Elements.** Latin names from which the symbols of some elements are derived are shown in parentheses.

Aluminum	**Al**	Chlorine	**Cl**	Manganese	**Mn**	Copper (*cuprum*)	**Cu**
Argon	**Ar**	Fluorine	**F**	Nitrogen	**N**	Iron (*ferrum*)	**Fe**
Barium	**Ba**	Helium	**He**	Oxygen	**O**	Lead (*plumbum*)	**Pb**
Boron	**B**	Hydrogen	**H**	Phosphorus	**P**	Mercury (*hydrargyrum*)	**Hg**
Bromine	**Br**	Iodine	**I**	Silicon	**Si**	Potassium (*kalium*)	**K**
Calcium	**Ca**	Lithium	**Li**	Sulfur	**S**	Silver (*argentum*)	**Ag**
Carbon	**C**	Magnesium	**Mg**	Zinc	**Zn**	Sodium (*natrium*)	**Na**

(a) Relative abundance on Earth

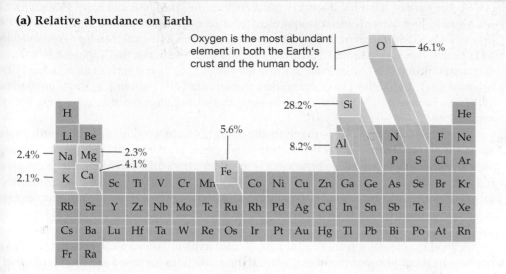

Oxygen is the most abundant element in both the Earth's crust and the human body.

(b) Relative abundance in the human body

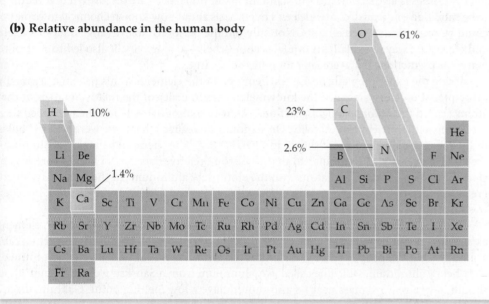

◄ **FIGURE 2.1**

Estimated elemental composition by mass percent of (a) the Earth's crust and (b) the human body. Only the major constituents are shown in each case; small amounts of many other elements are also present.

Figure It Out

What is the most abundant element in both the Earth's crust and the human body? What is the second most abundant element in **(a)** the Earth's crust and **(b)** the human body?

Answer: In both the Earth's crust and the human body, the most abundant element is oxygen. Silicon is the second, most abundant element in the Earth's crust, and carbon is the second most abundant element in the human body.

For simplicity, chemists refer to specific elements using one- or two-letter symbols. As shown by the examples in Table 2.1, the first letter of an element's symbol is always capitalized and the second letter, if any, is lowercase. Many of the symbols are just the first one or two letters of the element's English name: H = hydrogen, C = carbon, Al = aluminum, and so forth. Other symbols derive from Latin or other languages: Na = sodium (Latin, *natrium*), Pb = lead (Latin, *plumbum*), and W = tungsten (German, *wolfram*). The names, symbols, and other information about all 118 known elements are given inside the front cover of this book, organized in a format you've undoubtedly seen before called the **periodic table**.

2.2 ▶ ELEMENTS AND THE PERIODIC TABLE

Ten elements have been known since the beginning of recorded history: antimony (Sb), carbon (C), copper (Cu), gold (Au), iron (Fe), lead (Pb), mercury (Hg), silver (Ag), sulfur (S), and tin (Sn). The first "new" element to be found in several thousand years was arsenic (As), discovered in about 1250. In fact, only 24 elements were known when the United States was founded in 1776.

As the pace of scientific discovery quickened in the late 1700s and early 1800s, chemists began to look for similarities among elements that might allow general conclusions to be drawn. Particularly important among the early successes was Johann Döbereiner's observation in 1829 that there were several *triads*, or groups of three elements, that appeared to behave similarly. Calcium (Ca), strontium (Sr), and barium (Ba) form one such triad; chlorine (Cl), bromine (Br), and iodine (I) form another; and lithium (Li), sodium (Na), and potassium (K) form a third. By 1843, 16 such triads were known and chemists were searching for an explanation.

Numerous attempts were made in the mid-1800s to account for the similarities among groups of elements, but the breakthrough came in 1869 when the Russian chemist Dmitri Mendeleev created the forerunner of the modern periodic table. Mendeleev's creation is an ideal example of how a scientific theory develops. At first, there is only disconnected information—a large number of elements and many observations about their properties and behavior. As more and more facts become known, people try to organize the data in ways that make sense until ultimately a consistent hypothesis emerges.

A good hypothesis must do two things: It must explain known facts, and it must make predictions about phenomena yet unknown. If the predictions are tested and found true, then the hypothesis is a good one and will stand until additional facts are discovered that require it to be modified or discarded. Mendeleev's hypothesis about how known chemical information could be organized passed all tests. Not only did the periodic table arrange data in a useful and consistent way to explain known facts about chemical reactivity, it also led to several remarkable predictions that were later found to be accurate.

Using the experimentally observed chemistry of the elements as his primary organizing principle, Mendeleev arranged the known elements in order of the relative masses of their atoms (called their *atomic weights*, Section 2.9), with hydrogen = 1, and then grouped them according to their chemical reactivity. On so doing, he realized that there were several "holes" in the table, some of which are shown in **FIGURE 2.2**. The chemical behavior of aluminum (relative mass ≈ 27.3) is similar to that of boron (relative mass ≈ 11), but there was no element known at the time that fits into the slot below aluminum. In the same way, silicon (relative mass ≈ 28) is similar in many respects to carbon (relative mass ≈ 12), but there was no element known that fits below silicon.

Looking at the holes in the table, Mendeleev predicted that two then-unknown elements existed and might be found at some future time. Furthermore, he predicted with remarkable accuracy what the properties of these unknown elements would be. The element immediately below aluminum, which he called *eka*-aluminum from a Sanskrit word meaning "first," should have a relative mass near 68 and should have a low melting point. Gallium, discovered in 1875, has exactly these properties. The element below silicon, which Mendeleev called *eka*-silicon, should have a relative mass near 72 and should be dark gray in color. Germanium, discovered in 1886, fits the description perfectly (**TABLE 2.2**).

▲ Left to right, samples of chlorine, bromine, and iodine, one of Döbereiner's triads of elements with similar chemical properties.

▲ Gallium is a shiny, low-melting metal.

▲ Germanium is a hard, gray semimetal.

▶ **FIGURE 2.2**

A portion of Mendeleev's periodic table. The table shows the relative masses of atoms as known at the time and some of the holes representing unknown elements.

H = 1							
Li = 7	Be = 9.4		B = 11	C = 12	N = 14	O = 16	F = 19
Na = 23	Mg = 24		Al = 27.3	Si = 28	P = 31	S = 32	Cl = 35.5
K = 39	Ca = 40	?, Ti, V, Cr, Mn, Fe, Co, Ni, Cu, Zn	? = 68	? = 72	As = 75	Se = 78	Br = 80

There is an unknown element, which turns out to be gallium (Ga), beneath aluminum (Al)...

...and another unknown element, which turns out to be germanium (Ge), beneath silicon (Si).

Figure It Out

In the 1800s, Li, Na, and K were grouped as a triad due to their similar reactivity. How is this triad represented in Mendeleev's periodic table?

Answer: Elements with similar reactivity were placed in vertical groups.

TABLE 2.2 A Comparison of Predicted and Observed Properties for Gallium (*eka*-Aluminum) and Germanium (*eka*-Silicon)

Element	Property	Mendeleev's Prediction	Observed Property
Gallium	Relative mass	68	69.7
	Density	5.9 g/cm^3	5.91 g/cm^3
	Melting point	Low	29.8 °C
Germanium	Relative mass	72	72.6
	Density	5.5 g/cm^3	5.35 g/cm^3
	Color	Dark gray	Light gray

In the modern periodic table, shown in **FIGURE 2.3**, elements are placed on a grid with seven horizontal rows, called **periods**, and 18 vertical columns, called **groups**. When organized in this way, *the elements in a given group have similar chemical properties*. Lithium, sodium, potassium, and the other metallic elements in group 1A behave similarly. Beryllium, magnesium, calcium, and the other elements in group 2A behave similarly. Fluorine, chlorine, bromine, and the other elements in group 7A behave similarly, and so on throughout the table. (Mendeleev, by the way, was completely unaware of the existence of the group 8A elements—He, Ne, Ar, Kr, Xe, and Rn—because none were known when he constructed his table. All are colorless, odorless gases with little or no chemical reactivity, and none were discovered until 1894, when argon was first isolated.)

The overall form of the periodic table is well accepted, but chemists in different countries have historically used different conventions for labeling the groups. To resolve these difficulties, an international standard calls for numbering the groups from 1 to 18 going left to right. This standard has not yet found complete acceptance, however, and we'll continue to use the U.S. system of numbers and capital letters—group 3B instead of group 3 and group 7A instead of group 17, for example. Labels for the newer system are also shown in Figure 2.3.

One further note: There are actually 32 groups in the periodic table rather than 18, but to make the table fit manageably on a page, the 14 elements beginning with lanthanum (the *lanthanides*) and the 14 beginning with actinium (the *actinides*) are pulled out and shown below the others. These groups are not numbered.

We'll see repeatedly throughout this book that the periodic table of the elements is the most important organizing principle in chemistry. The time you take now to familiarize yourself with the layout and organization of the periodic table will pay off later on. Notice in Figure 2.3, for instance, that there is a regular progression in the size of the seven periods (rows). The first period has only 2 elements, hydrogen (H) and helium (He); the second and third periods have 8 elements each; the fourth and fifth periods have 18 elements each; and the sixth and seventh periods, which include the lanthanides and actinides, have 32 elements each. We'll see in Chapter 5 that this regular progression in the periodic table reflects a similar regularity in the structure of atoms.

Notice also that not all groups in the periodic table have the same number of elements. The two larger groups on the left and the six larger groups on the right are called the **main groups**. Most of the elements on which life is based—carbon, hydrogen, nitrogen, oxygen, and phosphorus, for instance—are main-group elements. The 10 smaller groups in the middle of the table are called the **transition metal groups**. Most of the metals you're probably familiar with—iron, copper, zinc, and gold, for instance—are transition metals. And the 14 groups shown separately at the bottom of the table are called the **inner transition metal groups**.

LOOKING AHEAD...

We'll see in Chapter 5 that the detailed structure of a given atom is related to its position in the periodic table.

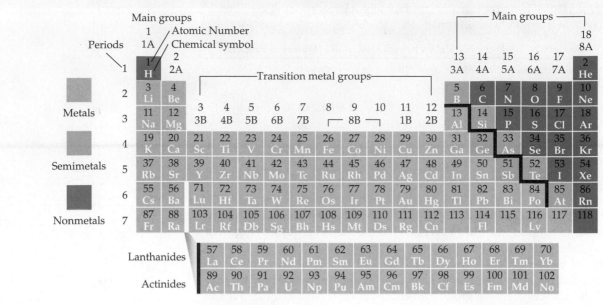

▲ **FIGURE 2.3**

The periodic table. Each element is identified by a one- or two-letter symbol and is characterized by an atomic number (Section 2.8). Elements are organized into 18 vertical columns, or *groups*, and seven horizontal rows, or *periods*. The two groups on the left and the six on the right are the *main groups* (1, 2, and 13–18); the 10 in the middle are the *transition metal groups* (3–12). The 14 elements beginning with lanthanum are the *lanthanides*, and the 14 elements beginning with actinium are the *actinides*. Together, the lanthanides and actinides are known as the *inner transition metal groups*. Two systems for numbering the groups are shown above the top row and explained in the text.

Those elements (except hydrogen) on the left side of the zigzag line running from boron (B) to astatine (At) are metals; those elements (plus hydrogen) to the right of the line are nonmetals; and seven of the nine elements abutting the line are metalloids, or semimetals.

Figure It Out

What is the symbol for the element located in Period 5, Group 16? Classify the element as a metal, semimetal, or nonmetal.

Answer: Te, semimetal

▲ Addition of a solution of silver nitrate to a solution of sodium chloride yields a white precipitate of solid silver chloride.

2.3 ▶ SOME COMMON GROUPS OF ELEMENTS AND THEIR PROPERTIES

Any characteristic that can be used to describe or identify matter is called a **property**. Examples include volume, amount, odor, color, and temperature. Still other properties include such characteristics as melting point, solubility, and chemical behavior. For example, we might list some properties of sodium chloride (table salt) by saying that it melts at 1474 °F (or 801 °C), dissolves in water, and undergoes a chemical reaction when it comes into contact with a silver nitrate solution.

Properties can be classified as either *intensive* or *extensive*, depending on whether the value of the property changes with the amount of the sample. **Intensive properties**, like temperature and melting point, have values that do not depend on the amount of sample: A small ice cube might have the same temperature as a massive iceberg. **Extensive properties**, like length and volume, have values that *do* depend on the sample size: An ice cube is much smaller than an iceberg.

Properties can also be classified as either *physical* or *chemical*, depending on whether the property involves a change in the chemical makeup of a substance. **Physical properties** are characteristics that do not involve a change in a sample's chemical makeup, whereas **chemical properties** are characteristics that *do* involve a change in chemical makeup. The melting point of ice, for instance, is a physical property because melting causes the water to change only in form, from solid to liquid, but not in chemical makeup. Water has the exact same chemical composition (H_2O) in both states of matter. The rusting of an iron (Fe) bicycle left in the rain is a chemical property, however, because iron combines with oxygen and moisture from the air to give the new substance, rust (which has the formula Fe_2O_3). **TABLE 2.3** lists other examples of both physical and chemical properties.

As noted previously, the elements in a group of the periodic table often show remarkable similarities in their chemical properties. Look at the following groups, for instance, to see some examples:

- **Group 1A—Alkali metals** Lithium (Li), sodium (Na), potassium (K), rubidium (Rb), and cesium (Cs) are soft, silvery metals. All react rapidly, often violently, with water to form products that are highly alkaline, or *basic*—hence the name *alkali metals*. Because of their high reactivity, the alkali metals are never found in nature in the pure state but only in combination with other elements. Francium (Fr) is also an alkali metal but, as noted previously, it is so rare that little is known about it.

  Note that group 1A also contains hydrogen (H) even though, as a colorless gas, it is completely different in appearance and behavior from the alkali metals. We'll see the reason for this classification in Section 5.13.

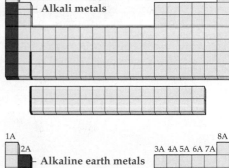

- **Group 2A—Alkaline earth metals** Beryllium (Be), magnesium (Mg), calcium (Ca), strontium (Sr), barium (Ba), and radium (Ra) are also lustrous, silvery metals but are less reactive than their neighbors in group 1A. Like the alkali metals, the alkaline earths are never found in nature in the pure state.

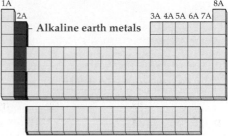

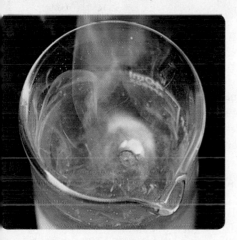

▲ Sodium, one of the alkali metals, reacts violently with water to yield hydrogen gas and an alkaline (basic) solution.

▲ Magnesium, one of the alkaline earth metals, burns in air.

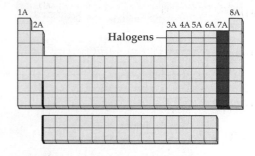

- **Group 7A—Halogens** Fluorine (F), chlorine (Cl), bromine (Br), and iodine (I), are colorful, corrosive nonmetals. They are found in nature only in combination with other elements, such as with sodium in table salt (sodium chloride, NaCl). In fact, the group name *halogen* is taken from the Greek word *hals*, meaning "salt." Astatine (At) is also a halogen, but it exists in such tiny amounts that little is known about it.

TABLE 2.3   Some Examples of Physical and Chemical Properties

Physical Properties		Chemical Properties
Temperature	Amount	Rusting (of iron)
Color	Odor	Combustion (of gasoline)
Melting point	Solubility	Tarnishing (of silver)
Electrical conductivity	Hardness	Cooking (of an egg)

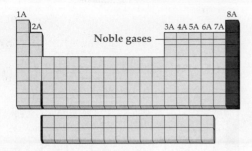

- **Group 8A—Noble gases** Helium (He), neon (Ne), argon (Ar), krypton (Kr), xenon (Xe), and radon (Rn) are colorless gases with very low chemical reactivity. Helium and neon don't combine with any other element; argon, krypton, and xenon combine with very few.

As indicated in Figure 2.3, the elements of the periodic table are often divided into three major categories: metals, nonmetals, and semimetals.

▲ Bromine, a halogen, is a corrosive dark red liquid at room temperature.

▲ Neon, one of the noble gases, is used in neon lights and signs.

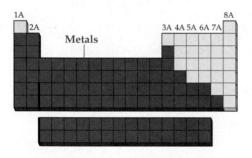

- **Metals** Metals, the largest category of elements, are found on the left side of the periodic table, bounded on the right by a zigzag line running from boron (B) at the top to astatine (At) at the bottom. The metals are easy to characterize by their appearance. All except mercury are solid at room temperature, and most have the silvery shine we normally associate with metals. In addition, metals are generally malleable rather than brittle, can be twisted and drawn into wires without breaking, and are good conductors of heat and electricity.

- **Nonmetals** Except for hydrogen, nonmetals are found on the right side of the periodic table and, like metals, are easy to characterize by their appearance. Eleven of the seventeen nonmetals are gases, one is a liquid (bromine), and only five are solids at room temperature (carbon, phosphorus, sulfur, selenium, and iodine). None are silvery in appearance, and several are brightly colored. The solid nonmetals are brittle rather than malleable and are poor conductors of heat and electricity.

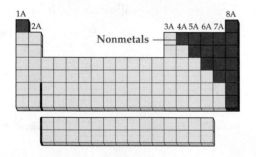

▲ Lead, aluminum, copper, gold, iron, and silver (clockwise from left) are typical metals. All conduct electricity and can be drawn into wires.

▲ Bromine, carbon, phosphorus, and sulfur (clockwise from top left) are typical nonmetals. None conduct electricity or can be made into wires.

- **Semimetals** Seven of the nine elements adjacent to the zigzag boundary between metals and nonmetals—boron, silicon, germanium, arsenic, antimony, tellurium, and astatine—are called semimetals because their properties are intermediate between those of their metallic and nonmetallic neighbors. Although most are silvery in appearance and all are solid at room temperature, semimetals are brittle rather than malleable and tend to be poor conductors of heat and electricity. Silicon, for example, is a widely used *semiconductor*, a substance whose electrical conductivity is intermediate between that of a metal and an insulator.

Conceptual WORKED EXAMPLE 2.1

Classifying Elements Based Upon Properties and Location in the Periodic Table

An element is a colorless gas at room temperature and pressure. It is highly reactive and therefore only found in nature combined with other elements. Classify the element as a metal, nonmetal, or semimetal and specify its general location in the periodic table.

STRATEGY

Match the description of the element to the properties of the main categories of elements and the four specific groups: alkali metals, alkaline earth metals, halogens, and noble gases.

SOLUTION

The element is a nonmetal because it is a gas. All metals (with the exception of mercury) and semimetals are solids at room temperature and pressure. Its high reactivity indicates it is a halogen, but it cannot be bromine or iodine since these elements are not gases at room temperature.

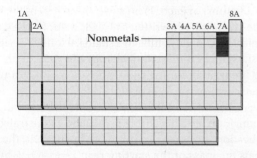

Conceptual PRACTICE 2.1 An element is a shiny, silver-colored solid at room temperature and pressure. It conducts electricity and can be found in nature in its pure form. Classify the element as a metal, nonmetal, or semimetal and specify its general location in the periodic table.

Conceptual APPLY 2.2 An element is indicated in the periodic table.

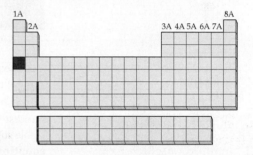

(a) Give the symbol and name of the element, and classify it as a metal, nonmetal, or semimetal.
(b) Predict the appearance of the element at room temperature and pressure.
(c) What would happen if you were to hit a pure sample of this element with a hammer?
(d) Could this element be used to make a wire to conduct electricity in your home?

2.4 ▶ OBSERVATIONS SUPPORTING ATOMIC THEORY: THE CONSERVATION OF MASS AND THE LAW OF DEFINITE PROPORTIONS

The notion that matter was composed of atoms was an accepted scientific theory long before the technology to "see" individual atoms was invented. Hundreds of years ago, scientists designed clever experiments and made careful measurements to obtain results that supported atomic theory. The Englishman Robert Boyle (1627–1691) is generally credited with being the first to study chemistry as a separate intellectual discipline and the first to carry out rigorous chemical experiments. Through a careful series of experiments into the nature and behavior of gases, Boyle provided clear evidence for the atomic makeup of matter. In addition, Boyle was the first to clearly define an element as a substance that cannot be chemically broken down further and to suggest that a substantial number of different elements might exist. Atoms of these different elements, in turn, can join together in different ways to yield a vast number of different substances we call **chemical compounds**

Progress in chemistry was slow in the decades following Boyle, and it was not until the work of Joseph Priestley (1733–1804) that the next great leap was made. Priestley prepared and isolated the gas oxygen in 1774 by heating the compound mercury oxide (HgO) according to the chemical equation we would now write as $2 \, HgO \rightarrow 2 \, Hg + O_2$.

▶ Heating the red powder HgO causes it to decompose into the silvery liquid mercury and the colorless gas oxygen.

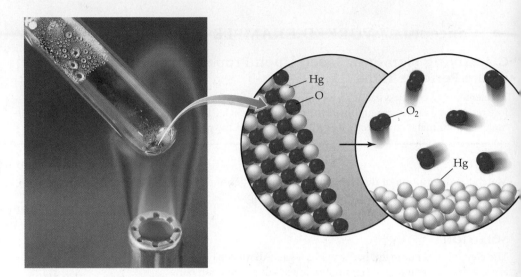

Hg
O
O_2
Hg

In this standard format for writing chemical transformations, each compound is described by its **chemical formula**, which lists the symbols of its constituent elements and uses subscripts to indicate the number of atoms of each. If no subscript is given, the number 1 is understood. Thus, sodium chloride (table salt) is written as NaCl, water as H_2O, and sucrose (table sugar) as $C_{12}H_{22}O_{11}$. A chemical reaction is written in a standard format called a **chemical equation**, in which the reactant substances undergoing change are written on the left, the product substances being formed are written on the right, and an arrow is drawn between them to indicate the direction of the chemical transformation.

Soon after Priestley's discovery, Antoine Lavoisier (1743–1794) showed that oxygen is the key substance involved in combustion. Furthermore, Lavoisier demonstrated with careful measurements that when combustion is carried out in a closed container, the mass of the combustion products exactly equals the mass of the starting reactants. When hydrogen gas burns and combines with oxygen to yield water (H_2O), for instance, the mass of the water formed is equal to the mass of the hydrogen and oxygen consumed. Called the **law of mass conservation**, this principle is a cornerstone of chemical science.

> **Law of mass conservation** Mass is neither created nor destroyed in chemical reactions.

It's easy to demonstrate the law of mass conservation by carrying out an experiment like that shown in **FIGURE 2.4**. If 3.25 g of mercury nitrate [$Hg(NO_3)_2$] and 3.32 g of potassium iodide (KI) are each dissolved in water and the solutions are mixed, an immediate chemical reaction occurs leading to the formation of the insoluble orange solid mercury iodide (HgI_2). Filtering the reaction mixture gives 4.55 g of mercury iodide, and evaporation of the water from the remaining solution leaves 2.02 g of potassium nitrate (KNO_3). Thus, the combined mass of the reactants (3.25 g + 3.32 g = 6.57 g) is exactly equal to the combined mass of the products (4.55 g + 2.02 g = 6.57 g).

Known amounts of solid KI and solid $Hg(NO_3)_2$ are weighed and then dissolved in water.

The solutions are mixed to give **solid HgI_2**, which is removed by filtration.

The solution that remains is evaporated to give solid KNO_3. On weighing, the combined masses of the products equal the combined masses of the reactants.

▲ **FIGURE 2.4**

An illustration of the law of mass conservation. In any chemical reaction, the combined mass of the final products equals the combined mass of the starting reactants.

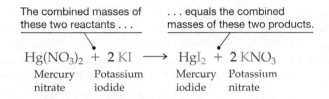

The combined masses of these two reactants . . .

. . . equals the combined masses of these two products.

$$Hg(NO_3)_2 + 2 KI \longrightarrow HgI_2 + 2 KNO_3$$

Mercury Potassium   Mercury Potassium
nitrate   iodide   iodide   nitrate

Further investigations in the decades following Lavoisier led the French chemist Joseph Proust (1754–1826) to formulate a second fundamental chemical principle that we now call the **law of definite proportions**:

> **Law of definite proportions** Different samples of a pure chemical compound always contain the same proportion of elements by mass.

Every sample of water (H_2O) contains 1 part hydrogen and 8 parts oxygen by mass; every sample of carbon dioxide (CO_2) contains 3 parts carbon and 8 parts oxygen by mass; and so on. *Elements combine in specific proportions, not in random proportions.*

2.5 ▶ THE LAW OF MULTIPLE PROPORTIONS AND DALTON'S ATOMIC THEORY

At the same time that Proust was formulating the law of definite proportions, the English schoolteacher John Dalton (1766–1844) was exploring along similar lines. His work led him to propose what has come to be called the **law of multiple proportions**:

> **Law of multiple proportions** Elements can combine in different ways to form different chemical compounds, whose mass ratios are simple whole-number multiples of each other.

▲ The brown gas NO_2 is an important component of urban pollution.

The key to Dalton's proposition was his realization that the *same* elements sometimes combine in different ratios to give *different* chemical compounds. For example, oxygen and nitrogen can combine either in a 7:8 mass ratio to make the compound we know today as nitric oxide (NO) or in a 7:16 mass ratio to make the compound we know as nitrogen dioxide (NO$_2$). The second compound contains exactly twice as much oxygen as the first.

NO: 7 g nitrogen per 8 g oxygen   N:O mass ratio = 7:8

NO$_2$:   7 g nitrogen per 16 g oxygen   N:O mass ratio = 7:16

Comparison of N:O ratios
in NO and NO$_2$ $$\frac{\text{N:O mass ratio in NO}}{\text{N:O mass ratio in NO}_2} = \frac{(7\,\text{g N})/(8\,\text{g O})}{(7\,\text{g N})/(16\,\text{g O})} = 2$$

This result makes sense only if we assume that matter is composed of discrete atoms that have characteristic masses and combine with one another in specific and well-defined ways (**FIGURE 2.5**).

Taking all three laws together—the law of mass conservation, the law of definite proportions, and the law of multiple proportions—ultimately led Dalton to propose a new theory of matter. He reasoned as follows:

- *Elements are made up of tiny particles called **atoms**.* Although Dalton didn't know what atoms were like, he nevertheless felt they were necessary to explain why there were so many different elements.

- *Each element is characterized by the mass of its atoms. Atoms of the same element have the same mass, but atoms of different elements have different masses.* Dalton realized that there must be some feature that distinguishes the atoms of one element from those of another. Because Proust's law of definite proportions showed that elements always combine in specific mass ratios, Dalton reasoned that the distinguishing feature among atoms of different elements must be mass.

- *The chemical combination of elements to make different chemical compounds occurs when whole numbers of atoms join in fixed proportions.* Only if whole numbers of atoms combine will different samples of a pure chemical compound always contain the same proportion of elements by mass (the law of definite proportions and the law of multiple proportions). Fractional parts of atoms are never involved in chemical reactions.

- *Chemical reactions only rearrange how atoms are combined in chemical compounds; the atoms themselves don't change.* Dalton realized that atoms must be chemically indestructible for the law of mass conservation to be valid. If the same numbers and kinds of atoms are present in both reactants and products, then the masses of reactants and products must also be the same.

Not everything that Dalton proposed was correct. He thought, for instance, that water had the formula HO rather than H$_2$O. Nevertheless, his atomic theory of matter was ultimately accepted and came to form a cornerstone of modern chemical science.

▲ These samples of sulfur and carbon have different masses but contain the same number of atoms.

▶ **FIGURE 2.5**

An illustration of Dalton's law of multiple proportions.

Figure It Out

Does a compound with the formula N$_2$O$_5$ follow the Law of Multiple Proportions?

Answer: Yes, any chemical formula with whole number subscripts follows the Law of Multiple Proportions.

Atoms of nitrogen and oxygen can combine in specific proportions to make either NO or NO$_2$.

NO$_2$ contains exactly twice as many atoms of oxygen per atom of nitrogen as NO does.

WORKED EXAMPLE 2.2

Using the Law of Multiple Proportions

Methane and propane are both constituents of natural gas. A sample of methane contains 5.70 g of carbon atoms and 1.90 g of hydrogen atoms combined in a certain way, whereas a sample of propane contains 4.47 g of carbon atoms and 0.993 g of hydrogen atoms combined in a different way. Show that the two compounds obey the law of multiple proportions.

STRATEGY

Find the C:H mass ratio in each compound, and then compare the ratios to see whether they are simple multiples of each other.

SOLUTION

$$\text{Methane:} \quad \text{C:H mass ratio} = \frac{5.70 \text{ g C}}{1.90 \text{ g H}} = 3.00$$

$$\text{Propane:} \quad \text{C:H mass ratio} = \frac{4.47 \text{ g C}}{0.993 \text{ g H}} = 4.50$$

$$\frac{\text{C:H mass ratio in methane}}{\text{C:H mass ratio in propane}} = \frac{3.00}{4.50} = \frac{2}{3}$$

▶ **PRACTICE 2.3** Compounds A and B are colorless gases obtained by combining sulfur with oxygen. Compound A results from combining 6.00 g of sulfur with 5.99 g of oxygen, and compound B results from combining 8.60 g of sulfur with 12.88 g of oxygen. Show that the mass ratios in the two compounds are simple multiples of each other.

▶ **APPLY 2.4** If the chemical formula of compound A in Problem 2.3 is SO_2, what is the chemical formula of compound B?

▲ Sulfur burns with a bluish flame to yield colorless SO_2 gas.

2.6 ▶ ATOMIC STRUCTURE: ELECTRONS

Dalton's atomic theory is fine as far as it goes, but it leaves unanswered the obvious question: What makes up an atom? Dalton himself had no way of answering this question, and it was not until nearly a century later that experiments by the English physicist J. J. Thomson (1856–1940) provided some clues. Thomson's experiments involved the use of *cathode-ray tubes* (CRTs), early predecessors of the tubes found in older televisions and computer displays.

As shown in **FIGURE 2.6A**, a cathode-ray tube (CRT) is a sealed glass vessel from which the air has been removed and in which two thin pieces of metal, called *electrodes*, have been sealed. When a sufficiently high voltage is applied across the electrodes, an electric current flows through the tube from the negatively charged electrode (the *cathode*) to the positively charged electrode (the *anode*). If the tube is not fully evacuated but still contains a small amount of air or other gas, the flowing current is visible as a glow called a *cathode ray*.

Experiments by a number of physicists in the 1890s had shown that cathode rays can be deflected by bringing either a magnet or an electrically charged plate near the tube (**FIGURE 2.6B**). Because the beam is produced at a negative electrode and is deflected toward a positive plate, Thomson proposed that cathode rays must consist of tiny, negatively charged particles, which we now call **electrons**. Furthermore, because electrons are emitted from electrodes made of many different metals, all these different metals must contain electrons.

Thomson reasoned that the amount of deflection of the electron beam in a cathode-ray tube due to a nearby magnetic or electric field should depend on three factors:

1. *The strength of the deflecting magnetic or electric field.* The stronger the magnet or the higher the voltage on the charged plate, the greater the deflection.
2. *The size of the negative charge on the electron.* The larger the charge on the particle, the greater its interaction with the magnetic or electric field and the greater the deflection.
3. *The mass of the electron.* The lighter the particle, the greater its deflection (just as a Ping-Pong ball is more easily deflected than a bowling ball).

(a) The electron beam ordinarily travels in a straight line.

(b) The beam is deflected by either a magnetic field or an electric field.

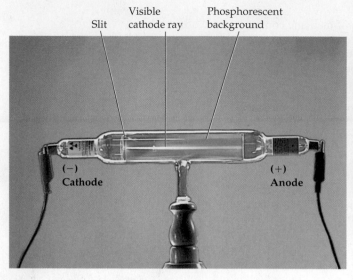

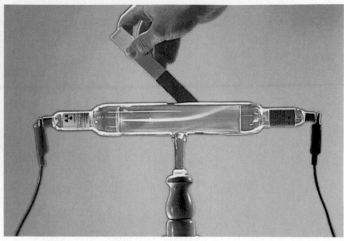

▲ **FIGURE 2.6**

A cathode-ray tube. In a cathode-ray tube, a stream of electrons emitted from the negatively charged cathode passes through a slit, moves toward the positively charged anode, and is detected by a phosphorescent strip.

By carefully measuring the amount of deflection caused by electric and magnetic fields of known strength, Thomson was able to calculate the ratio of the electron's electric charge to its mass: its *charge-to-mass ratio, e/m*. The modern value is

$$\frac{e}{m} = 1.758\ 820 \times 10^8\ \text{C/g}$$

where e is the magnitude of the charge on the electron in coulombs (C) and m is the mass of the electron in grams. (We'll say more about **coulombs** and **electrical charge** in Chapter 18.) Note that because e is defined as a positive quantity, the actual (negative) charge on the electron is $-e$.

Thomson was able to measure only the ratio of charge to mass, not charge or mass itself, and it was left to the American R. A. Millikan (1868–1953) to devise a method for measuring the mass of an electron (**FIGURE 2.7**). In Millikan's experiment, a fine mist of oil was sprayed into a chamber, and the tiny droplets were allowed to fall between two horizontal plates. Observing the droplets through a telescopic eyepiece made it possible to determine how rapidly they fell through the air, which in turn allowed their masses to be calculated. The droplets were then given a negative charge by irradiating them with X rays. X rays knocked electrons from gas molecules in the surrounding air, and the electrons stuck to the oil droplet. By applying a voltage to the plates, with the upper plate positive, it was possible to counteract the downward fall of the charged droplets and keep them suspended.

With the voltage on the plates and the mass of the droplets known, Millikan was able to show that the charge on a given droplet was always a small whole-number multiple of e, whose modern value is $1.602\ 176 \times 10^{-19}$ C. Substituting the value of e into Thomson's charge-to-mass ratio then gives the mass m of the electron as $9.109\ 382 \times 10^{-28}$ g:

$$\text{Because} \quad \frac{e}{m} = 1.758\ 820 \times 10^8\ \text{C/g}$$

$$\text{then} \quad m = \frac{e}{1.758\ 820 \times 10^8\ \text{C/g}} = \frac{1.602\ 176 \times 10^{-19}\ \cancel{\text{C}}}{1.758\ 820 \times 10^8\ \cancel{\text{C}}/\text{g}}$$

$$= 9.109\ 382 \times 10^{-28}\ \text{g}$$

LOOKING AHEAD...

We'll see in Chapter 18 that **coulombs** and **electrical charge** are fundamental to electrochemistry—the area of chemistry involving batteries, fuel cells, and electroplating of metals.

An alpha particle
(relative mass = 7000;
charge = $+2e$)

An electron
(relative mass = 1;
charge = $-1e$)

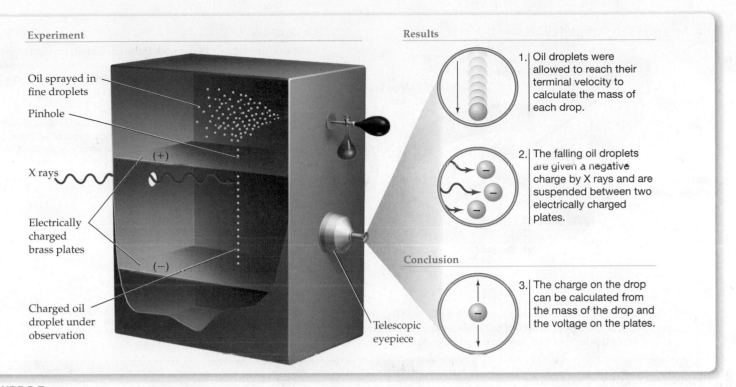

Experiment

Oil sprayed in
fine droplets

Pinhole

X rays

(+)

Electrically
charged
brass plates

(−)

Charged oil
droplet under
observation

Telescopic
eyepiece

Results

1. Oil droplets were allowed to reach their terminal velocity to calculate the mass of each drop.

2. The falling oil droplets are given a negative charge by X rays and are suspended between two electrically charged plates.

Conclusion

3. The charge on the drop can be calculated from the mass of the drop and the voltage on the plates.

▲ **FIGURE 2.7**

Millikan's oil drop experiment.

2.7 ▶ ATOMIC STRUCTURE: PROTONS AND NEUTRONS

Think about the consequences of Thomson's cathode-ray experiments. Because matter is electrically neutral overall, the fact that the atoms in an electrode can give off negatively charged particles (electrons) must mean that those same atoms also contain positively charged particles for electrical balance. The search for those positively charged particles and for an overall picture of atomic structure led to a landmark experiment published in 1911 by the New Zealand physicist Ernest Rutherford (1871–1937).

Rutherford's work involved the use of *alpha* (α) *particles*, a type of emission previously found to be given off by a number of naturally occurring radioactive elements, including radium, polonium, and radon. Rutherford knew that alpha particles are about 7000 times more massive than electrons and that they have a positive charge that is twice the magnitude of the charge on an electron but opposite in sign.

When Rutherford directed a beam of alpha particles at a thin gold foil, he found that almost all the particles passed through the foil undeflected. A very small number, however (about 1 of every 20,000), were deflected at an angle, and a few actually bounced back toward the particle source (**FIGURE 2.8**). Rutherford described his reaction to the results as follows:

It was quite the most incredible event that has ever happened to me in my life. It was almost as incredible as if you fired a 15-inch shell at a piece of tissue paper and it came back and hit you.

Rutherford explained his results by proposing that a metal atom must be almost entirely empty space and have its mass concentrated in a tiny central core that he called the **nucleus**. If the nucleus contains the atom's positive charges and most of its mass, and if the electrons are a relatively large distance away, then it is clear why the observed scattering results are obtained: most alpha particles encounter empty space as they fly through the foil. Only when a positive alpha particle chances to come near a small but massive positive nucleus is it repelled strongly enough to make it bounce backward.

Modern measurements show that an atom has a diameter of roughly 10^{-10} m and that a nucleus has a diameter of about 10^{-15} m. It's difficult to imagine from these numbers alone, though, just how small a nucleus really is. For comparison purposes, if an atom were the size of a large domed stadium, the nucleus would be approximately the size of a small pea in the center of the playing field.

▲ The relative size of the nucleus in an atom is roughly the same as that of a pea in the middle of this huge stadium.

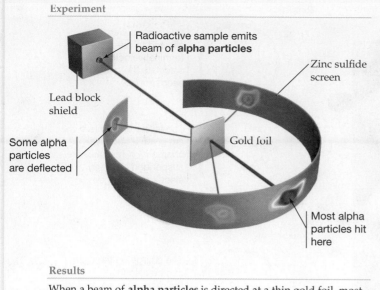

Experiment

Radioactive sample emits beam of **alpha particles**

Lead block shield

Some alpha particles are deflected

Zinc sulfide screen

Gold foil

Most alpha particles hit here

Results

When a beam of **alpha particles** is directed at a thin gold foil, most particles pass through undeflected, but some are deflected at large angles and a few bounce back toward the particle source.

▲ **FIGURE 2.8**

Rutherford's scattering experiment.

Conclusion

Because the majority of particles are not deflected, the gold atoms must be almost entirely empty space. The atom's mass is concentrated in a tiny dense core, which deflects the occasional alpha particle.

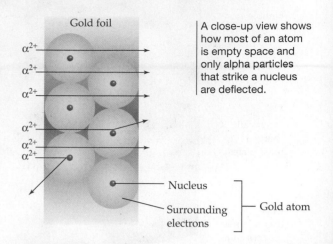

Gold foil

A close-up view shows how most of an atom is empty space and only alpha particles that strike a nucleus are deflected.

Nucleus

Surrounding electrons

Gold atom

Figure It Out

What was the surprising result in the gold foil experiment? What conclusions about atomic structure were drawn from this observation?

Answer: Some alpha particles bounced backward. This observation could only be explained by the alpha particle interacting with a dense, positively charged region in a gold atom (the nucleus).

TABLE 2.4   A Comparison of Subatomic Particles

Particle	Mass		Charge	
	Grams	**u***	**Coulombs**	**e**
Electron	$9.109\,382 \times 10^{-28}$	$5.485\,799 \times 10^{-4}$	$-1.602\,176 \times 10^{-19}$	-1
Proton	$1.672\,622 \times 10^{-24}$	$1.007\,276$	$+1.602\,176 \times 10^{-19}$	$+1$
Neutron	$1.674\,927 \times 10^{-24}$	$1.008\,665$	0	0

*The unified atomic mass unit (u) is defined in Section 2.9.

Further experiments by Rutherford and others between 1910 and 1930 showed that a nucleus is composed of two kinds of particles, called *protons* and *neutrons*. **Protons** have a mass of $1.672\,622 \times 10^{-24}$ g (about 1836 times that of an electron) and are positively charged. Because the charge on a proton is opposite in sign but equal in size to that on an electron, the numbers of protons and electrons in a neutral atom are equal. **Neutrons** ($1.674\,927 \times 10^{-24}$ g) are almost identical in mass to protons but carry no charge, and the number of neutrons in a nucleus is not directly related to the numbers of protons and electrons. **TABLE 2.4** compares the three fundamental subatomic particles, and **FIGURE 2.9** gives an overall view of the atom.

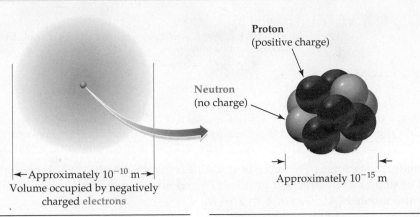

Proton
(positive charge)

Neutron
(no charge)

←Approximately 10^{-10} m→
Volume occupied by negatively
charged **electrons**

Approximately 10^{-15} m

A number of **electrons** equal to the number of protons move about the nucleus and account for most of the atom's volume.	The **protons** and **neutrons** in the nucleus take up very little volume but contain essentially all the atom's mass.

WORKED EXAMPLE 2.3

Calculations Using Atomic Size

Ordinary "lead" pencils actually are made of a form of carbon called *graphite*. If a pencil line is 0.35 mm wide and the diameter of a carbon atom is 1.5×10^{-10} m, how many atoms wide is the line?

IDENTIFY

Known	Unknown
Diameter of a C atom $\left(\dfrac{1.5 \times 10^{-10}\,\text{m}}{1\,\text{atom}}\right)$	Number of C atoms in width of line
Pencil line = 0.35 mm	

STRATEGY

Begin with the known information that is not a conversion factor (width of the pencil line). Next set up an equation using appropriate conversion factors so that the unwanted units cancel. Conversion factors between prefixes are also needed in this problem.

SOLUTION

$$\text{Atoms} = 0.35\,\cancel{\text{mm}} \times \frac{1\,\cancel{\text{m}}}{1000\,\cancel{\text{mm}}} \times \frac{1\,\text{atom}}{1.5 \times 10^{-10}\,\cancel{\text{m}}} = 2.3 \times 10^6\,\text{atoms}$$

CHECK

A single carbon atom is about 10^{-10} m across, so it takes 10^{10} carbon atoms placed side by side to stretch 1 m, 10^7 carbon atoms to stretch 1 mm, and about 0.3×10^7 (or 3×10^6; 3 *million*) carbon atoms to stretch 0.35 mm. The estimate agrees with the solution.

▶ **PRACTICE 2.5** The gold foil that Rutherford used in his scattering experiment had a thickness of approximately 0.005 mm. If a single gold atom has a diameter of 2.9×10^{-10} m, how many atoms thick was Rutherford's foil?

▶ **APPLY 2.6** A small speck of carbon, the size of a pinhead, contains about 10^{19} atoms, the diameter of a carbon atom is 1.5×10^{-10} m, and the circumference of the Earth at the equator is 40,075 km. How many times around the Earth would the atoms from this speck of carbon extend if they were laid side by side?

2.8 ▶ ATOMIC NUMBERS

Thus far, we've described atoms only in general terms and have not yet answered the most important question: What is it that makes one atom different from another? How, for example, does an atom of gold differ from an atom of carbon? The answer turns out to be quite simple. *Elements differ from one another according to the number of protons in the nucleus*, a value called the element's **atomic number** (Z). That is, all atoms of a given element contain the same number of **protons** in their nuclei. All hydrogen atoms, atomic number 1, have 1 proton; all helium atoms, atomic number 2, have 2 protons; all carbon atoms, atomic number 6, have 6 protons; and so on. In addition, every neutral atom contains a number of electrons equal to its number of protons.

REMEMBER...

The periodic table is arranged by increasing atomic number of the elements (Figure 2.3). The atomic number is defined as the number of **protons** in the nucleus and gives the identity of the element.

Atomic number (Z)

= Number of protons in an atom's nucleus
= Number of electrons around the nucleus in a neutral atom

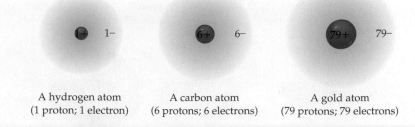

| A hydrogen atom | A carbon atom | A gold atom |
| (1 proton; 1 electron) | (6 protons; 6 electrons) | (79 protons; 79 electrons) |

In addition to protons, the nuclei of all atoms (other than hydrogen) also contain neutrons. The sum of the numbers of protons (Z) and neutrons (N) in an atom is called the atom's **mass number** (A). That is, $A = Z + N$.

Mass number (A) = Number of protons (Z) + number of neutrons (N)

Most hydrogen atoms have 1 proton and no neutrons, so their mass number is $A = 1 + 0 = 1$. Most helium atoms have 2 protons and 2 neutrons, so their mass number is $A = 2 + 2 = 4$. Most carbon atoms have 6 protons and 6 neutrons, so their mass number is $A = 6 + 6 = 12$. Except for hydrogen, stable atoms always contain at least as many neutrons as protons, although there is no simple way to predict how many neutrons a given atom will have.

Notice that we said *most* hydrogen atoms have mass number 1, *most* helium atoms have mass number 4, and *most* carbon atoms have mass number 12. In fact, different atoms of the same element can have different mass numbers depending on how many neutrons they have. Atoms with identical atomic numbers but different mass numbers are called **isotopes**. Hydrogen, for example, has three isotopes.

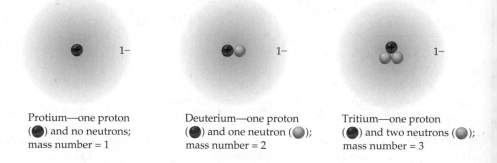

Protium—one proton	Deuterium—one proton	Tritium—one proton
(⬤) and no neutrons;	(⬤) and one neutron (◯);	(⬤) and two neutrons (◯);
mass number = 1	mass number = 2	mass number = 3

All hydrogen atoms have 1 proton in their nucleus (otherwise they wouldn't be hydrogen), but 99.985% of them have no neutrons. These hydrogen atoms, called *protium*, have mass number 1. In addition, 0.015% of hydrogen atoms, called *deuterium*, have 1 neutron and mass number 2. Still other hydrogen atoms, called *tritium*, have 2 neutrons and mass number 3. An unstable, radioactive isotope, tritium occurs only in trace amounts on Earth but is made artificially in nuclear reactors. As other examples, there are 15 known isotopes of nitrogen, only 2 of which occur naturally on Earth, and 25 known isotopes of uranium, only 3 of which occur naturally. In total, more than 3600 isotopes of the 118 known elements have been identified.

A specific isotope is represented by its element symbol accompanied by its mass number as a left superscript and its atomic number as a left subscript. Thus, protium is represented as $^{1}_{1}\text{H}$, deuterium as $^{2}_{1}\text{H}$, and tritium as $^{3}_{1}\text{H}$. Similarly, the two naturally occurring isotopes of nitrogen are represented as $^{14}_{7}\text{N}$ (spoken as "nitrogen-14") and $^{15}_{7}\text{N}$ (nitrogen-15). The number of neutrons in an isotope is not given explicitly but can be calculated by subtracting the atomic number (subscript) from the mass number (superscript). For example, subtracting the atomic number 7 from the mass number 14 indicates that a $^{14}_{7}\text{N}$ atom has 7 neutrons.

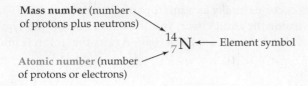

Mass number (number
of protons plus neutrons)

$^{14}_{7}$N ← Element symbol

Atomic number (number
of protons or electrons)

The number of neutrons in an atom has relatively little effect on the atom's chemical properties. The chemical behavior of an element is determined almost entirely by the number of electrons it has, which in turn is determined by the number of protons in its nucleus. All three isotopes of hydrogen therefore behave similarly (although not identically) in their chemical reactions.

WORKED EXAMPLE 2.4

Interpreting Isotope Symbols

The isotope of uranium used to generate nuclear power is $^{235}_{92}$U. How many protons, neutrons, and electrons does an atom of $^{235}_{92}$U have?

STRATEGY

The atomic number (subscript 92) in the symbol $^{235}_{92}$U indicates the number of protons and electrons in the atom. The number of neutrons is the difference between the mass number (superscript 235) and the atomic number (92).

SOLUTION

An atom of $^{235}_{92}$U has 92 protons, 92 electrons, and $235 - 92 = 143$ neutrons.

▶ **PRACTICE 2.7**　The isotope $^{75}_{34}$Se is used medically for the diagnosis of pancreatic disorders. How many protons, neutrons, and electrons does an atom of $^{75}_{34}$Se have?

▶ **APPLY 2.8**　Element X is toxic to humans in high concentration but is essential to life in low concentrations. Identify element X, whose atoms contain 24 protons, and write the symbol for the isotope of X that has 28 neutrons. (Refer to Figure 2.3, the periodic table.)

▲ Uranium-235 is used as fuel in this nuclear-powered icebreaker.

2.9 ▶ ATOMIC WEIGHTS AND THE MOLE

Pick up a pencil, and look at the small amount of tip visible. How many atoms (pencil lead is made of carbon) do you think are in the tip? One thing is certain: Atoms are so tiny that the number needed to make a visible sample is enormous. In fact, even the smallest speck of dust visible to the naked eye contains at least 10^{17} atoms. Thus, the mass in grams of a single atom is much too small a number for convenience so chemists use a unit called an *atomic mass unit* (*amu*), or, more correctly, the **unified atomic mass unit (u)** (also known as a *dalton* [Da] in biological work). One unified atomic mass unit is defined as exactly 1/12 the mass of an atom of $^{12}_{6}$C and is equal to $1.660\,539 \times 10^{-24}$ g.

$$\blacktriangleright \ \mathbf{1\,u} = \frac{\text{Mass of one } ^{12}_{6}\text{C atom}}{12} = 1.660\,539 \times 10^{-24}\,\text{g}$$

Because the mass of an atom's electrons is negligible compared to the mass of its protons and neutrons, defining 1 u as 1/12 the mass of a $^{12}_{6}$C atom means that protons and neutrons both have a mass of almost exactly 1 u (Table 2.4). Thus, the mass of a specific atom in atomic mass units—called the atom's **atomic mass**—is numerically close to the atom's mass number. A $^{1}_{1}$H atom, for instance, has a mass of 1.007 825; a $^{235}_{92}$U atom has an atomic mass of 235.043 930; and so forth. In practice, atomic masses are taken to be dimensionless, and the unit u is understood rather than specified.

Most elements occur naturally as a mixture of different isotopes. Thus, if you look at the periodic table inside the front cover, you'll see listed below the symbol for each element a value called the element's *atomic weight*. Again, the unit u is understood but not specified.

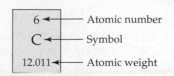

6 ← —— Atomic number
C ← —— Symbol
12.011 ← —— Atomic weight

An element's **atomic weight** is the weighted average of the isotopic masses of the element's naturally occurring isotopes. Carbon, for example, occurs on Earth as a mixture of two major isotopes, $^{12}_{6}C$ (98.89% natural abundance) and $^{13}_{6}C$ (1.11% natural abundance). Although the isotopic mass of any individual carbon atom is either exactly 12 (a carbon-12 atom) or 13.0034 (a carbon-13 atom), the *average* atomic mass—that is, the atomic weight—of a large collection of carbon atoms is 12.011. A third carbon isotope, $^{14}_{6}C$, also exists, but its natural abundance is so small that it can be ignored when calculating atomic weight.

$$\text{Atomic weight of C} = (\text{Mass of } {}^{12}_{6}C)(\text{Abundance of } {}^{12}_{6}C) + (\text{Mass of } {}^{13}_{6}C)(\text{Abundance of } {}^{13}_{6}C)$$
$$= (12)(0.9889) + (13.0034)(0.0111)$$
$$= 11.867 + 0.144 = 12.011$$

WORKED EXAMPLE 2.5

Calculating an Atomic Weight

Chlorine has two naturally occurring isotopes: $^{35}_{17}Cl$, with a natural abundance of 75.76% and an isotopic mass of 34.969, and $^{37}_{17}Cl$, with a natural abundance of 24.24% and an isotopic mass of 36.966. What is the atomic weight of chlorine?

IDENTIFY

Known	Unknown
Isotope masses and abundances	Atomic weight of chlorine
(75.76% of isotope with mass 34.969 and 24.24% of isotope with mass 36.966)	

STRATEGY

The atomic weight of an element is the weighted average of the isotopic masses, which equals the sum of the masses of each isotope times the natural abundance of that isotope:

$$\text{Atomic weight} = (\text{Mass of } {}^{37}_{17}Cl)(\text{Abundance of } {}^{37}_{17}Cl)$$
$$+ (\text{Mass of } {}^{37}_{17}Cl)(\text{Abundance of } {}^{37}_{17}Cl)$$

SOLUTION
Atomic weight $= (34.969)(0.7576) + (36.966)(0.2424) = 35.45$

CHECK

The atomic weight is somewhere between 35 and 37, the masses of the two individual isotopes. It is closer to 35 since this is the mass of the more abundant isotope.

▶ **PRACTICE 2.9** Copper metal has two naturally occurring isotopes: copper-63 (69.15%; isotopic mass = 62.93) and copper-65 (30.85%; isotopic mass = 64.93). Calculate the atomic weight of copper, and check your answer in a periodic table.

▶ **APPLY 2.10** Gallium has two naturally occurring isotopes: gallium-69 (isotopic mass = 68.9256) and gallium-71 (isotopic mass = 70.9247). The atomic weight of gallium is 69.7231.

(a) Without doing any calculations, state which isotope has the greater abundance.

(b) Calculate the percent abundance of each isotope.

Atomic weights are extremely useful because they are conversion factors between numbers of atoms and masses; that is, they allow us to *count* a large number of atoms by *weighing* a sample of the substance. For instance, knowing that carbon has an atomic weight of 12.011

ets us calculate that a small pencil tip made of carbon and weighing 15 mg (1.5×10^{-2} g) contains 7.5×10^{20} atoms:

$$(1.5 \times 10^{-2}\,\cancel{g})\left(\frac{1\,\cancel{u}}{1.6605 \times 10^{-24}\,\cancel{g}}\right)\left(\frac{1\,\text{C atom}}{12.011\,\cancel{u}}\right) = 7.5 \times 10^{20}\,\text{C atoms}$$

When referring to the enormous numbers of atoms that make up the visible amounts we typically deal with, chemists use the fundamental SI unit for amount called a *mole*, abbreviated *mol*. One **mole** of any element is the amount whose mass in grams, called its **molar mass**, is numerically equal to its atomic weight. One mole of carbon atoms has a mass of 12.011 g, and one mole of silver atoms has a mass of 107.868 g. Molar mass thus acts as a conversion factor that lets you convert between mass in grams and number of atoms. *Whenever you have the same number of moles of different elements, you also have the same number of atoms.*

How many atoms are there in a mole? Experiments show that one mole of any element contains $6.022\,141 \times 10^{23}$ atoms, a value called **Avogadro's number**, abbreviated N_A, after the Italian scientist who first recognized the importance of the mass/number relationship. Avogadro's number of atoms of any element—that is, one mole—has a mass in grams equal to the element's atomic weight.

It's hard to grasp the magnitude of a quantity as large as Avogadro's number, but some comparisons might give you a sense of scale: The age of the universe in seconds (13.7 billion years, or 4.32×10^{17} s) is less than a millionth the size of Avogadro's number. The number of liters of water in the world's oceans (1.3×10^{21} L) is less than one-hundredth the size of Avogadro's number. The mass of the Earth in kilograms (5.98×10^{24} kg) is only 10 times Avogadro's number. We'll return to the mole throughout the book and see some of its many uses in Chapter 3.

▲ These samples of helium, sulfur, copper, and mercury each contain 1 mol. Do they have the same mass?

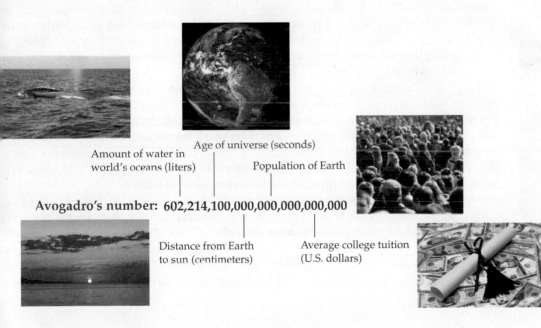

Amount of water in world's oceans (liters)

Age of universe (seconds)

Population of Earth

Avogadro's number: 602,214,100,000,000,000,000,000

Distance from Earth to sun (centimeters)

Average college tuition (U.S. dollars)

WORKED EXAMPLE 2.6

Converting Between Mass and Numbers of Moles and Atoms

How many moles and how many atoms of silicon are in a sample weighing 10.53 g? The atomic weight of silicon is 28.0855.

IDENTIFY

Known	Unknown
Mass of sample 10.53 g	Moles and number of atoms of silicon
Atomic weight of silicon (28.0855)	

STRATEGY

The molar mass (28.0855 g/mol) is numerically equivalent to the atomic weight. Use molar mass to convert between mass and number of moles, and then use Avogadro's number to convert between moles and number of atoms.

continued on next page

SOLUTION

$$(10.53 \text{ g Si})\left(\frac{1 \text{ mol Si}}{28.0855 \text{ g Si}}\right) = 0.3749 \text{ mol Si}$$

$$(0.3749 \text{ mol Si})\left(\frac{6.022 \times 10^{23} \text{ atoms Si}}{1 \text{ mol Si}}\right) = 2.258 \times 10^{23} \text{ atoms Si}$$

CHECK

A mass of 10.53 g of silicon is a bit more than one-third the molar mass of silicon (28.0855 g/mol), so the sample contains a bit more than 0.33 mol. This number of moles, in turn, contains a bit

more than one-third of Avogadro's number of atoms, or about 2×10^{23} atoms.

▶ **PRACTICE 2.11**   How many moles and how many atoms of platinum are in a ring with a mass of 9.50 g? Use the periodic table on the inside front cover of the book to look up the atomic weight of platinum.

▶ **APPLY 2.12**   If 2.26×10^{22} atoms of element Y have a mass of 1.50 g, what is the identity of Y?

2.10 ▶ MIXTURES AND CHEMICAL COMPOUNDS; MOLECULES AND COVALENT BONDS

Although only 90 elements occur naturally, there are far more than 90 different substances on Earth. Water, sugar, protein in food, and cotton or rayon in clothing are familiar substances that are not pure elements. Matter, anything that has mass and occupies volume, can be classified as either mixtures or pure substances (**FIGURE 2.10**). Pure substances, in turn, can be either elements or chemical compounds.

A **mixture** is simply a blend of two or more substances added together in some arbitrary proportion without chemically changing the individual substances themselves. Thus, the constituent units in the mixture are not all the same, and the proportion of the units is variable. Hydrogen gas and oxygen gas, for instance, can be mixed in any ratio without changing them (as long as there is no flame nearby to initiate reaction), just as a spoonful of sugar and a spoonful of salt can be mixed.

A chemical compound, in contrast to a mixture, is a pure substance that is formed when atoms of different elements combine in a specific way to create a new material with properties completely unlike those of its constituent elements. A chemical compound has a constant composition throughout, and its constituent units are all identical. For example, when atoms of sodium (a soft, silvery metal) combine with atoms of chlorine (a toxic, yellow-green gas), the familiar white solid called *sodium chloride* (table salt) is formed. Similarly, when two atoms of hydrogen combine with one atom of oxygen, water is formed.

To see how a chemical compound is formed, imagine what must happen when two atoms approach each other at the beginning of a chemical reaction. Because the electrons of an atom occupy a much greater volume than the nucleus, it's the electrons that actually make the contact when atoms collide. Thus, it's the electrons that form the connections, or **chemical bonds**, that join atoms together in compounds. Chemical bonds between atoms are usually classified as either *covalent* or *ionic*. As a general rule, covalent bonds occur primarily

▲ The crystalline quartz sand on this beach is a pure compound (SiO_2), but the seawater is a liquid mixture of many compounds dissolved in water.

▶ **FIGURE 2.10**

A scheme for the classification of matter.

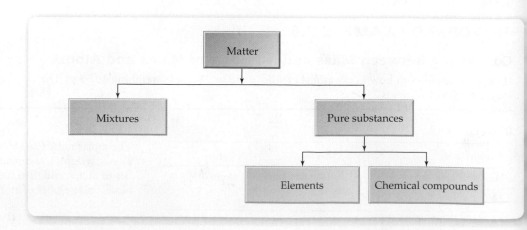

The two teams are joined together because both are tugging on the same rope.

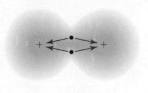

Similarly, two atoms are joined together when both **nuclei (+)** tug on the same **electrons (dots)**.

▲ **FIGURE 2.11**

A covalent bond between atoms is analogous to a tug-of-war.

between nonmetal atoms, while ionic bonds occur primarily between metal and nonmetal atoms. Let's look briefly at both kinds, beginning with covalent bonds.

A **covalent bond**, the most common kind of chemical bond, results when two atoms *share* several (usually two) electrons. A simple way to think about a covalent bond is to imagine it as a tug-of-war. If two people pull on the same rope, they are effectively joined together. Neither person can escape from the other as long as both hold on. Similarly with atoms: When two atoms both hold on to some shared electrons, the atoms are bonded together (**FIGURE 2.11**).

The unit of matter that results when two or more atoms are joined by covalent bonds is called a **molecule**. A hydrogen chloride (HCl) molecule results when a hydrogen atom and a chlorine atom share two electrons. A water (H_2O) molecule results when each of two hydrogen atoms shares two electrons with a single oxygen atom. An ammonia (NH_3) molecule results when each of three hydrogen atoms shares two electrons with a nitrogen atom, and so on. To visualize these and other molecules, it helps to imagine the individual atoms as spheres joined together to form molecules with specific three-dimensional shapes, as shown in **FIGURE 2.12**. *Ball-and-stick* models specifically indicate the covalent bonds between atoms, while *space-filling* models accurately portray overall molecular shape but don't explicitly show covalent bonds.

◀ **FIGURE 2.12**

Molecular models. Drawings such as these help in visualizing molecules.

Ball-and-stick models show atoms (spheres) joined together by covalent bonds (sticks).

Space-filling models portray the overall molecular shape but don't explicitly show covalent bonds.

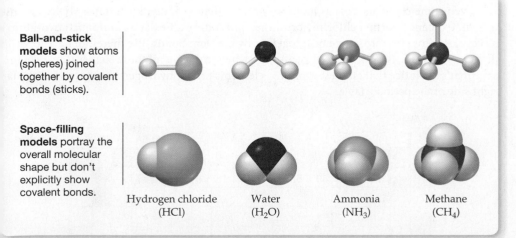

Hydrogen chloride (HCl)   Water (H_2O)   Ammonia (NH_3)   Methane (CH_4)

Chemists normally represent a molecule by giving its **structural formula**, which shows the specific connections between atoms and therefore gives much more information than the chemical formula alone. Ethyl alcohol, for example, has the chemical formula C_2H_6O and the following structural formula:

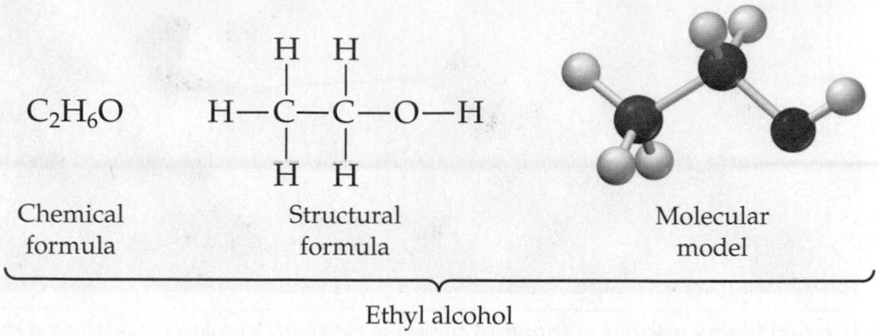

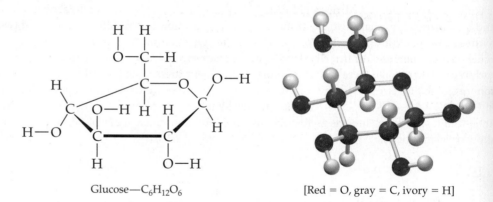

C_2H_6O

Chemical
formula

Structural
formula

Molecular
model

Ethyl alcohol

A structural formula uses lines between atoms to indicate the covalent bonds. Thus, the two carbon atoms in ethyl alcohol are covalently bonded to each other, the oxygen atom is bonded to one of the carbon atoms, and the six hydrogen atoms are distributed three to one carbon, two to the other carbon, and one to the oxygen.

Structural formulas are particularly important in *organic chemistry*—the chemistry of carbon compounds—where the behavior of large, complex molecules is almost entirely governed by their structure. Take even a relatively simple substance like glucose, for instance. The molecular formula of glucose, $C_6H_{12}O_6$, tells nothing about how the atoms are connected. In fact, you could probably imagine a great many different ways in which the 24 atoms might be connected. The structural formula for glucose, however, shows that 5 carbons and 1 oxygen form a ring of atoms, with the remaining 5 oxygens each bonded to 1 hydrogen and distributed on different carbons.

Glucose—$C_6H_{12}O_6$

[Red = O, gray = C, ivory = H]

Even some elements exist as molecules rather than as individual atoms. Hydrogen, nitrogen, oxygen, fluorine, chlorine, bromine, and iodine all exist as *diatomic* (two-atom) molecules whose two atoms are held together by covalent bonds. We therefore have to write them as such—H_2, N_2, O_2, F_2, Cl_2, Br_2, and I_2—when using any of these elements in a chemical equation. Notice that all these diatomic elements except hydrogen cluster toward the far right side of the periodic table.

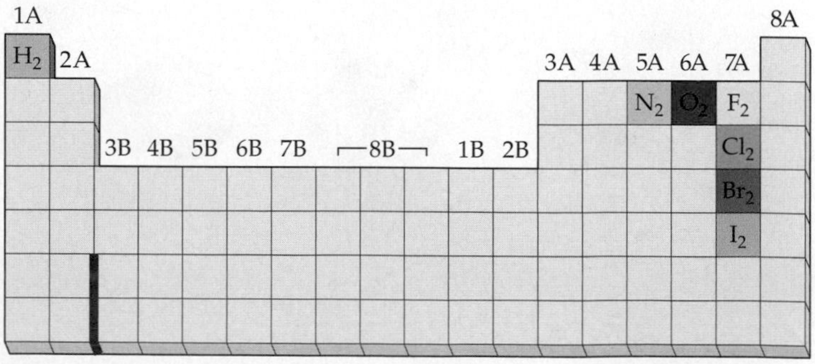

Conceptual WORKED EXAMPLE 2.7

Visual Representations of Mixtures and Compounds

Which of the following drawings represents a mixture, which a pure compound, and which an element?

(a) **(b)** **(c)**

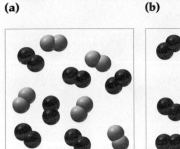  

STRATEGY

Most people (professional chemists included) find chemistry easier to grasp when they can visualize the behavior of atoms, thereby turning symbols into pictures. The Conceptual Problems in this text are intended to help you do that, frequently representing atoms and molecules as collections of spheres. Don't take the pictures literally; focus instead on interpreting what they represent. An element contains only one kind of atom while a compound contains two or more different elements bonded together. A pure substance contains only one type of element or compound while mixture contains two or more substances.

SOLUTION

Drawing **(a)** represents a mixture of two diatomic elements, one composed of two red atoms and one composed of two blue atoms. Drawing **(b)** represents molecules of a pure diatomic element because all atoms are identical. Drawing **(c)** represents molecules of a pure compound composed of one red and one blue atom.

▶ Conceptual **PRACTICE 2.13**   Which of the following drawings represents a pure sample of hydrogen peroxide (H_2O_2) molecules? The red spheres represent oxygen atoms, and the ivory spheres represent hydrogen.

(a) **(b)** **(c)** **(d)**

▶ Conceptual **APPLY 2.14**   Red and blue spheres represent atoms of different elements.

(a) Which drawing(s) illustrate a pure substance?
(b) Which drawing(s) illustrate a mixture?
(c) Which two drawings illustrate the law of multiple proportions?

(a) **(b)**

(c) **(d)**

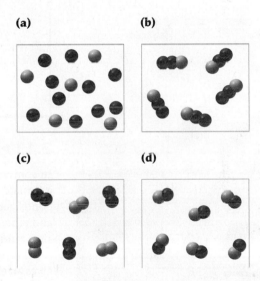

Conceptual WORKED EXAMPLE 2.8

Converting Between Structural and Molecular Formulas

Propane, C_3H_8, has a structure in which the three carbon atoms are bonded in a row, each end carbon is bonded to three hydrogens, and the middle carbon is bonded to two hydrogens. Draw the structural formula, using lines between atoms to represent covalent bonds.

SOLUTION

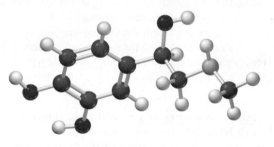

Propane

▶ **Conceptual** **PRACTICE 2.15**   Draw the structural formula of methylamine, CH_5N, a substance responsible for the odor of rotting fish. The carbon atom is bonded to the nitrogen atom and to three hydrogens. The nitrogen atom is bonded to the carbon and two hydrogens.

▶ **Conceptual** **APPLY 2.16**   Adrenaline, the so-called flight-or-fight hormone, can be represented by the following ball-and-stick model. What is the chemical formula of adrenaline? (Gray = C, ivory = H, red = O, blue = N)

2.11 ▶ IONS AND IONIC BONDS

In contrast to a covalent bond, an **ionic bond** results not from a sharing of electrons but from a transfer of one or more electrons from one atom to another. As noted previously, ionic bonds generally form between a metal and a nonmetal. Metals, such as sodium, magnesium, and zinc, tend to give up electrons, whereas nonmetals, such as oxygen, nitrogen, and chlorine, tend to accept electrons.

For example, when sodium metal comes in contact with chlorine gas, a sodium atom gives an electron to a chlorine atom, resulting in the formation of two charged particles, called **ions**. Because a sodium atom loses one electron, it loses one negative charge and becomes an Na^+ ion with a charge of $+1$. Such positive ions are called **cations** (pronounced **cat**-ions). Conversely, because a chlorine atom gains an electron, it gains a negative charge and becomes a Cl^- ion with a charge of -1. Such negative ions are called **anions** (**an**-ions).

▲ Chlorine is a toxic green gas, sodium is a reactive metal, and sodium chloride is a harmless white solid.

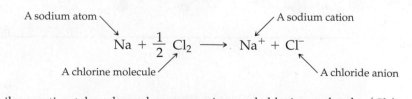

A similar reaction takes place when magnesium and chlorine molecules (Cl_2) come in contact to form $MgCl_2$. A magnesium atom transfers an electron to each of two chlorine atoms, yielding the doubly charged Mg^{2+} cation and two Cl^- anions.

$$Mg + Cl_2 \longrightarrow Mg^{2+} + Cl^- + Cl^-(MgCl_2)$$

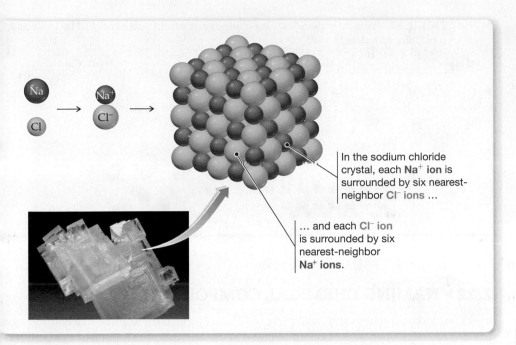

The arrangement of Na⁺ and Cl⁻ ions in a crystal of sodium chloride. There is no discrete "molecule" of NaCl. Instead, the entire crystal is an ionic solid.

In the sodium chloride crystal, each **Na⁺** ion is surrounded by six nearest-neighbor **Cl⁻ ions** ...

... and each **Cl⁻ ion** is surrounded by six nearest-neighbor **Na⁺ ions**.

Because opposite charges attract, positively charged cations such as Na^+ and Mg^{2+} experience a strong electrical attraction to negatively charged anions like Cl^-, an attraction that we call an ionic bond. Unlike what happens when covalent bonds are formed, though, we can't really talk about discrete Na^+Cl^- *molecules* under normal conditions. We can speak only of an **ionic solid**, in which equal numbers of Na^+ and Cl^- ions are packed together in a regular way (**FIGURE 2.13**). In a crystal of table salt, for instance, each Na^+ ion is surrounded by six nearby Cl^- ions, and each Cl^- ion is surrounded by six nearby Na^+ ions, but we can't specify what pairs of ions "belong" to each other as we can with atoms in covalent molecules.

Charged, covalently bonded groups of atoms, called **polyatomic ions**, are also common—ammonium ion (NH_4^+), hydroxide ion (OH^-), nitrate ion (NO_3^-), and the doubly charged sulfate ion (SO_4^{2-}) are examples (**FIGURE 2.14**). You can think of these polyatomic ions as charged molecules because they consist of specific numbers and kinds of atoms joined together by covalent bonds, with the overall unit having a positive or negative charge. When writing the formulas of substances that contain more than one of these ions, parentheses are placed around the entire polyatomic unit. The formula $Ba(NO_3)_2$, for instance, indicates a substance made of Ba^{2+} cations and NO_3^- polyatomic anions in a 1:2 ratio. We'll learn how to name compounds with these ions in Section 2.12.

⎯● **WORKED EXAMPLE 2.9**

Identifying Ionic and Molecular Compounds

Which of the following compounds would you expect to be ionic and which molecular (covalent)?

(a) BaF_2 (b) SF_4 (c) PH_3 (d) CH_3OH

STRATEGY

Remember that covalent bonds generally form between nonmetal atoms, while ionic bonds form between metal and nonmetal atoms.

SOLUTION

Compound **(a)** is composed of a metal (barium) and a nonmetal (fluorine) and is likely to be ionic. Compounds **(b)**–**(d)** are composed entirely of nonmetals and therefore are probably molecular.

continued on next page

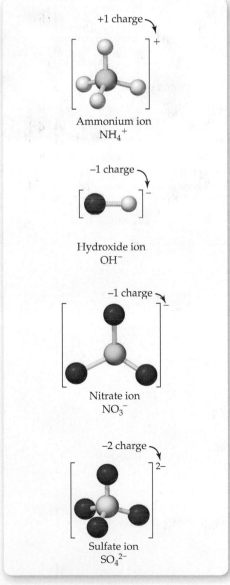

+1 charge

Ammonium ion
NH_4^+

−1 charge

Hydroxide ion
OH^-

−1 charge

Nitrate ion
NO_3^-

−2 charge

Sulfate ion
SO_4^{2-}

▲ **FIGURE 2.14**

Molecular models of some polyatomic ions.

▶ **PRACTICE 2.17** Which of the following compounds would you expect to be ionic and which molecular (covalent)?

(a) LiBr **(b)** $SiCl_4$ **(c)** BF_3 **(d)** CaO

▶ **Conceptual APPLY 2.18** Which of the following drawings most likely represents an ionic compound and which a molecular (covalent) compound? Explain.

(a) **(b)**

2.12 ▶ NAMING CHEMICAL COMPOUNDS

In the early days of chemistry, when few pure substances were known, newly discovered compounds were often given fanciful names—morphine, quicklime, potash, and barbituric acid (said to be named by its discoverer in honor of his friend Barbara) to cite a few. Today, with more than 40 million pure compounds known, there would be chaos without a systematic method for naming compounds. Every chemical compound must be given a name that not only defines it uniquely but also allows chemists (and computers) to know its chemical structure.

Different kinds of compounds are named by different rules. Ordinary table salt, for instance, is named *sodium chloride* because of its formula NaCl, but common table sugar ($C_{12}H_{22}O_{11}$) is named *β-D-fructofuranosyl-α-D-glucopyranoside* because of special rules for carbohydrates. (Organic compounds often have quite complex structures and correspondingly complex names, though we'll not discuss them in this text.) We'll begin by seeing how to name simple ionic compounds and then introduce additional rules in later chapters as the need arises.

Naming Binary Ionic Compounds

Binary ionic compounds—those made of only two elements—are named by identifying first the positive ion and then the negative ion. The positive ion takes the same name as the element, while the negative ion takes the first part of its name from the element and then adds the ending *-ide*. For example, KBr is named potassium bromide: *potassium* for the K^+ ion and *brom*ide for the negative Br^- ion derived from the element *brom*ine.

LiF	$CaBr_2$	$AlCl_3$
Lithium fluoride	Calcium bromide	Aluminum chloride

FIGURE 2.15 shows some common main-group ions, and **FIGURE 2.16** shows some common transition metal ions. There are several interesting points about Figure 2.15. Notice, for instance, that metals tend to form cations and nonmetals tend to form anions. Also note that elements within a given group of the periodic table form ions with the same charge and that the charge is related to the group number. Main-group metals usually form cations whose charge is equal to the group number. For example, Group 1A elements form singly positive ions (M^+, where M is a metal), group 2A elements form doubly positive ions (M^{2+}), and group 3A elements form triply positive ions (M^{3+}). Main-group nonmetals usually form anions whose charge is equal to the group number in the U.S. system minus eight. Thus, group 6A elements form doubly negative ions ($6 - 8 = -2$), group 7A elements form singly negative ions ($7 - 8 = -1$), and group 8A elements form no ions at all ($8 - 8 = 0$). We'll see the reason for this behavior in Chapter 6.

▲ Morphine, a pain-killing agent found in the opium poppy, was named after Morpheus, the Greek god of dreams.

◀ **FIGURE 2.15**

Main-group cations (blue) and anions (red). A *cation* bears the same name as the element it is derived from; an *anion* name has an *-ide* ending.

1 +1 1A	+2					+3	+4	-3	-2	-1	18 8A
H^+ H^- Hydride	2 2A					13 3A	14 4A	15 5A	16 6A	17 7A	
Li^+	Be^{2+}							N^{3-} Nitride	O^{2-} Oxide	F^- Fluoride	
Na^+	Mg^{2+}					Al^{3+}			S^{2-} Sulfide	Cl^- Chloride	
K^+	Ca^{2+}					Ga^{3+}			Se^{2-} Selenide	Br^- Bromide	
Rb^+	Sr^{2+}					In^{3+}	Sn^{2+} Sn^{4+}		Te^{2-} Telluride	I^- Iodide	
Cs^+	Ba^{2+}					Tl^+ Tl^{3+}	Pb^{2+} Pb^{4+}				

◀ **FIGURE 2.16**

Common transition metal ions. Only ions that exist in aqueous solution are shown.

3 3B	4 4B	5 5B	6 6B	7 7B	8	9 — 8B —	10	11 1B	12 2B
Sc^{3+}	Ti^{3+}	V^{2+} V^{3+}	Cr^{2+} Cr^{3+}	Mn^{2+}	Fe^{2+} Fe^{3+}	Co^{2+}	Ni^{2+}	Cu^+ Cu^{2+}	Zn^{2+}
Y^{3+}					Ru^{3+}	Rh^{3+}	Pd^{2+}	Ag^+	Cd^{2+}
									Hg^{2+} $(Hg_2)^{2+}$

Notice also, in both Figures 2.15 and 2.16 that some metals form more than one kind of cation. Iron, for instance, forms both the doubly charged Fe^{2+} ion and the triply charged Fe^{3+} ion. In naming these ions, we distinguish between them by using a Roman numeral in parentheses to indicate the number of charges. Thus, $FeCl_2$ is named iron(II) chloride and $FeCl_3$ is iron(III) chloride. Alternatively, an older method distinguishes between the ions by using the Latin name of the element (*ferrum* in the case of iron) together with the ending *-ous* for the ion with lower charge and *-ic* for the ion with higher charge. Thus, $FeCl_2$ is sometimes called ferrous chloride and $FeCl_3$ is called ferric chloride. Although still in use, this older naming system is being phased out and we'll rarely use it in this book.

Fe^{2+}	Fe^{3+}	Sn^{2+}	Sn^{4+}
Iron(II) ion	Iron(III) ion	Tin(II) ion	Tin(IV) ion
Ferrous ion	Ferric ion	Stannous ion	Stannic ion
(From the Latin *ferrum* = iron)		(From the Latin *stannum* = tin)	

In any neutral compound, the total number of positive charges must equal the total number of negative charges. Thus, you can always figure out the number of positive charges on a metal cation by counting the number of negative charges on the associated anion(s).

▲ Crystals of iron(II) chloride tetrahydrate are greenish, and crystals of iron(III) chloride hexahydrate are brownish yellow.

In $FeCl_2$, for example, the iron ion must be Fe(II) because there are two Cl^- ions associated with it. Similarly, in $TiCl_3$ the titanium ion is Ti(III) because there are three Cl^- anions associated with it. As a general rule, a Roman numeral is needed for transition-metal compounds to avoid ambiguity. In addition, the main-group metals tin (Sn), thallium (Tl), and lead (Pb) can form more than one kind of ion and need Roman numerals for naming their compounds. Metals in group 1A and group 2A form only one cation, however, so Roman numerals are not needed.

WORKED EXAMPLE 2.10

Converting Between Names and Formulas for Binary Ionic Compounds

Give systematic names for the following compounds:

(a) $BaCl_2$ (b) $CrCl_3$ (c) PbS   (d) Fe_2O_3

STRATEGY

Name the cation with the name of the element and the anion using the first part of the element name + "ide." If the cation is a transition metal, then the charge is specified with Roman numerals. Figure out the number of positive charges on each transition metal cation by counting the number of negative charges on the associated anion(s). Refer to Figures 2.15 and 2.16 as necessary.

SOLUTION

(a) Barium chloride   No Roman numeral is necessary because barium, a group 2A element, forms only Ba^{2+}.

(b) Chromium(III) chloride   The Roman numeral III is necessary to specify the +3 charge on chromium (a transition metal).

(c) Lead(II) sulfide   The sulfide anion (S^{2-}) has a double negative charge, so the lead cation must be doubly positive.

(d) Iron(III) oxide   The three oxide anions (O^{2-}) have a total negative charge of −6, so the two iron cations must have a total charge of +6. Thus, each is Fe(III).

▶ **PRACTICE 2.19**   Write formulas for the following compounds:

(a) Magnesium fluoride
(b) Tin(IV) oxide
(c) Iron(III) sulfide

▶ **Conceptual APPLY 2.20**   Three binary ionic compounds are represented on the following periodic table: red with red, green with green, and blue with blue. Name each, and write its likely formula.

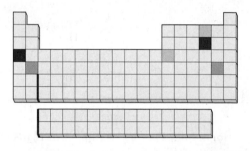

Naming Compounds with Polyatomic Ions

Ionic compounds that contain polyatomic ions are named in the same way as binary ionic compounds: First the cation is identified and then the anion. For example, $Ba(NO_3)_2$ is called *barium nitrate* because Ba^{2+} is the cation and the NO_3^- polyatomic anion has the name *nitrate*. Unfortunately, there is no simple systematic way of naming the polyatomic ions themselves, so it's necessary to memorize the names, formulas, and charges of the most common ones, listed in **TABLE 2.5**. The ammonium ion (NH_4^+) is the only cation on the list; all the others are anions.

Several points about the ions in Table 2.5 need special mention. First, note that the names of most polyatomic anions end in *-ite* or *-ate*. Only hydroxide (OH^-), cyanide (CN^-), and peroxide (O_2^{2-}) have the *-ide* ending. Second, note that several of the ions form a series of **oxoanions**, binary polyatomic anions in which an atom of a given element is combined with different numbers of oxygen atoms—hypochlorite (ClO^-), chlorite (ClO_2^-), chlorate (ClO_3^-), and perchlorate (ClO_4^-), for example. When there are only two oxoanions in a

TABLE 2.5	Some Common Polyatomic Ions		

Formula	Name	Formula	Name
Cation		Singly charged anions (continued)	
NH_4^+	Ammonium	NO_2^-	Nitrite
		NO_3^-	Nitrate
Singly charged anions		Doubly charged anions	
$CH_3CO_2^-$	Acetate	CO_3^{2-}	Carbonate
CN^-	Cyanide	CrO_4^{2-}	Chromate
ClO^-	Hypochlorite	$Cr_2O_7^{2-}$	Dichromate
ClO_2^-	Chlorite	O_2^{2-}	Peroxide
ClO_3^-	Chlorate	HPO_4^{2-}	Hydrogen phosphate
ClO_4^-	Perchlorate	SO_3^{2-}	Sulfite
$H_2PO_4^-$	Dihydrogen phosphate	SO_4^{2-}	Sulfate
HCO_3^-	Hydrogen carbonate (or bicarbonate)	$S_2O_3^{2-}$	Thiosulfate
HSO_4^-	Hydrogen sulfate (or bisulfate)	**Triply charged anion**	
OH^-	Hydroxide	PO_4^{3-}	Phosphate
MnO_4^-	Permanganate		

series, as with sulfite (SO_3^{2-}) and sulfate (SO_4^{2-}), the ion with fewer oxygens takes the *-ite* ending and the ion with more oxygens takes the *-ate* ending.

SO_3^{2-} Sul*fite* ion (fewer oxygens) SO_4^{2-} Sul*fate* ion (more oxygens)

NO_2^- Nit*rite* ion (fewer oxygens) NO_3^- Nit*rate* ion (more oxygens)

When there are more than two oxoanions in a series, the prefix *hypo-* (meaning "less than") is used for the ion with the fewest oxygens, and the prefix *per-* (meaning "more than") is used for the ion with the most oxygens.

ClO^- *Hypo*chlorite ion (less oxygen than chlorite)

ClO_2^- Chlorite ion

ClO_3^- Chlorate ion

ClO_4^- *Per*chlorate iron (more oxygen than chlorate)

Third, note that several pairs of ions are related by the presence or absence of a hydrogen ion. The hydrogen carbonate anion (HCO_3^-) differs from the carbonate anion (CO_3^{2-}) by the presence of H^+, and the hydrogen sulfate anion (HSO_4^-) differs from the sulfate anion (SO_4^{2-}) by the presence of H^+. The ion that has the additional hydrogen is sometimes referred to using the prefix *bi-*, although this usage is now discouraged; for example, $NaHCO_3$ is sometimes called sodium bicarbonate.

HCO_3^- Hydrogen carbonate (*bi*carbonate) ion CO_2^{-2} Carbonate iron

HSO_4^- Hydrogen sulfate (*bi*sulfate) ion SO_4^{2-} Sulfate ion

WORKED EXAMPLE 2.11

Converting Between Names and Formulas for Compounds with Polyatomic Ions

Give systematic names for the following compounds:

(a) $LiNO_3$ (b) $KHSO_4$ (c) $CuCO_3$ (d) $Fe(ClO_4)_3$

continued on next page

STRATEGY

Name the cation first and the anion second. Unfortunately, there is no alternative: The names and charges of the common polyatomic ions must be memorized. Refer to Table 2.5 if you need help.

SOLUTION

(a) Lithium nitrate — Lithium (group 1A) forms only the Li^+ ion and does not need a Roman numeral.

(b) Potassium hydrogen sulfate — Potassium (group 1A) forms only the K^+ ion.

(c) Copper(II) carbonate — The carbonate ion has a -2 charge, so copper must be $+2$. A Roman numeral is needed because copper, a transition metal, can form more than one ion.

(d) Iron(III) perchlorate — There are three perchlorate ions, each with a -1 charge, so the iron must have a $+3$ charge.

▶ **PRACTICE 2.21**   Write formulas for the following compounds:

(a) Potassium hypochlorite
(b) Silver(I) chromate
(c) Iron(III) carbonate

▶ **Conceptual APPLY 2.22**   The following drawings are those of solid ionic compounds, with red spheres representing the cations and blue spheres representing the anions in each.

(1)   **(2)**

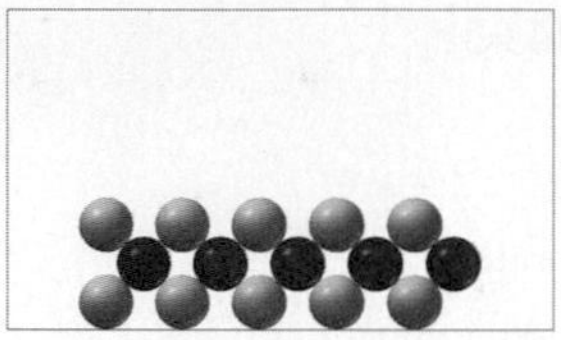

Which of the following formulas are consistent with each drawing?

(a) LiBr   (b) $NaNO_2$ (c) $CaCl_2$ (d) K_2CO_3 (e) $Fe_2(SO_4)_3$

Naming Binary Molecular Compounds

Binary molecular compounds—those made of only two covalently bonded elements—are named in much the same way as binary ionic compounds. One of the elements in the compound is more electron-poor, or *cationlike*, and the other element is more electron-rich, or *anionlike*. As with ionic compounds, the cationlike element takes the name of the element itself, and the anionlike element takes an *-ide* ending. The compound HF, for example, is called *hydrogen fluoride*.

HF   Hydrogen is more cationlike because it is farther left in the periodic table, and fluoride is more anionlike because it is farther right. The compound is therefore named *hydrogen fluoride*.

We'll see a quantitative way to decide which element is more cationlike and which is more anionlike in Section 7.3 but you might note for now that it's usually possible to decide by looking at the relative positions of the elements in the periodic table. The farther left and toward the bottom of the periodic table an element occurs, the more likely it is to be cationlike; the farther right and toward the top an element occurs (except for the noble gases), the more likely it is to be anionlike.

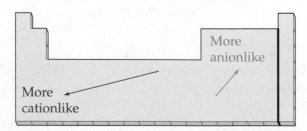

The following examples show how this generalization applies:

CO   Carbon monoxide (C is in group 4A; O is in group 6A)
CO_2   Carbon dioxide
PCl_3   Phosphorus trichloride (P is in group 5A; Cl is in group 7A)
SF_4   Sulfur tetrafluoride (S is in group 6A; F is in group 7A)
N_2O_4   Dinitrogen tetroxide (N is in group 5A; O is in group 6A)

Because nonmetals often combine with one another in different proportions to form different compounds, numerical prefixes are usually included in the names of binary molecular compounds to specify the numbers of each kind of atom present. The compound CO, for example, is called carbon *mono*xide, and CO_2 is called carbon *di*oxide. TABLE 2.6 lists the most common numerical prefixes. Note that when the prefix ends in *a* or *o* (but not *i*) and the anion name begins with a vowel (*oxide*, for instance); the *a* or *o* on the prefix is dropped to avoid having two vowels together in the name. Thus, we write carbon *mono*xide rather than carbon *mono*oxide for CO and dinitrogen *tetr*oxide rather than dinitrogen *tetra*oxide for N_2O_4. A *mono-* prefix is not used for the atom named first: CO_2 is called carbon dioxide rather than monocarbon dioxide.

TABLE 2.6 Numerical Prefixes for Naming Compounds	
Prefix	**Meaning**
mono-	1
di-	2
tri-	3
tetra-	4
penta-	5
hexa-	6
hepta-	7
octa-	8
nona-	9
deca-	10

WORKED EXAMPLE 2.12

Converting Between Names and Formulas for Binary Molecular Compounds

Give systematic names for the following compounds:

(a) PCl_3 (b) N_2O_3 (c) P_4O_7 (d) BrF_3

STRATEGY

Look at a periodic table to see which element in each compound is more cationlike (located farther to the left or lower) and which is more anionlike (located farther to the right or higher). Then name the compound using the appropriate numerical prefix to specify the number of atoms.

SOLUTION

(a) Phosphorus trichloride   (b) Dinitrogen trioxide
(c) Tetraphosphorus heptoxide   (d) Bromine trifluoride

PRACTICE 2.23 Write formulas for compounds with the following names:

(a) Disulfur dichloride   (b) Iodine monochloride
(c) Nitrogen triiodide

Conceptual APPLY 2.24   Give systematic names for the following compounds:

(a)

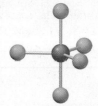

(b)

Purple = P, green = Cl   Blue = N, red = O

INQUIRY ▶▶▶ HOW IS THE PRINCIPLE OF ATOM ECONOMY USED TO MINIMIZE WASTE IN A CHEMICAL SYNTHESIS?

Chemical synthesis, combining atoms of different elements to make new compounds, is central to the global economy and a source of many products that enhance our lives. Dyes, fertilizers, plastics, synthetic fabrics, medicines, and electronic components are familiar examples of substances produced by chemical reactions. In the past, rapid and economic production methods have taken precedence over environmental considerations. Many chemical processes use large amounts of energy; non-renewable, petroleum-based feedstocks; and hazardous materials that pollute the environment. However, as dangers of commonly used chemicals have been discovered, scientists have begun to change their approach to chemical synthesis.

Green chemistry is the design of chemical products and processes that reduce or eliminate the use or generation of hazardous substances. It is different than remediation in that it aims to eliminate pollution by *preventing* it from happening in the first place. Green chemistry principles focus on using more efficient reactions with benign starting materials, using renewable resources, conserving energy, and creating waste materials that can be reused,

recycled, or biodegraded. Adoption of green chemistry technologies provides economic benefits, improved safety, and the promise of a sustainable future.

Chemists use green chemistry principles to design processes at the atomic level to prevent the formation of pollutants and waste. **Atom economy** is a concept conceived by Stanford chemistry professor Barry Trost, which states that it is *best to have all or most starting atoms end up in the desired product rather than in waste by-products*. It can be thought of as the efficiency of the reaction in terms of number of atoms and can be calculated as follows:

$$\text{Percent Atom Economy} = \frac{\Sigma \text{Atomic weight}_{\text{(atoms in all reactants)}}}{\Sigma \text{Atomic weight}_{\text{(atoms in desired product)}}} \times$$

where Σ (Epsilon) means "sum."

The numerator is the sum of the atomic weights of all atoms in the reactants and the denominator is the sum of the atomic weights of all atoms in the desired product. Worked Example 2.12 shows how to calculate the atom economy of a reaction.

● WORKED EXAMPLE 2.12

Calculating Atom Economy of a Reaction

Calculate the percent atom economy in the reaction between salicylic acid and acetic anhydride in the synthesis of aspirin.

Salicylic acid + Acetic anhydride → Aspirin + Acetic acid

IDENTIFY

Known	Unknown
Structural formulas for reactants and products	Percent atom economy

STRATEGY

Step 1. Using the structural formulas, count the number of each type of atom in reactant and product molecules to determine molecular formulas. Compute the sum of atomic weights by multiplying the number of each type of atom by its atomic weight and adding them all together.

Step 2. Calculate percent atom economy from the sums found in Step 1 and the formula provided.

SOLUTION

Step 1. The chemical formula of the two reactant molecules are $C_7H_6O_3$ (salicylic acid) and $C_4H_6O_3$ (acetic anhydride). Therefore in the reactants, there are a total of **11** carbon atoms, **12** hydrogen atoms, and **6** oxygen atoms. The chemical formula of the desired product is $C_9H_8O_4$ (aspirin) and there are **9** carbon atoms, **8** hydrogen atoms, and **4** oxygen atoms.

The sum of atomic weights of atoms in the reactants and desired product is calculated as follows:

Reactants

$11\,C = (11)(12.0) = 132.0$	
$12\,H = (12)(1.0) = 12.0$	
$6\,O = (6)(16.0) = 96.0$	

Sum of atomic weights of atoms in reactants $= 240.0$

Desired Product:

9 C = (9)(12.0) = 108.0

8 H = (8)(1.0) = 8.0

4 O = (4)(16.0) = 64.0

Sum of atomic weights of atoms in desired product = 180.0

Step 2. The formula for percent atom economy can be applied:

$$\text{Percent Atom Economy} = \frac{(180.0)}{(240.0)} \times 100 = 75.0\%$$

This means that the reaction is 75.0% efficient in its utilization of matter. The molecule acetic acid (CH_3COOH) is a "by-product" as it is not desired in the synthesis. Thus, 2 carbon atoms, 4 hydrogen atoms, and 2 oxygen atoms are considered to be waste in the production of one aspirin molecule.

PROBLEM 2.25 The Law of Conservation of Mass states that "mass is neither created nor destroyed in chemical reactions." How does the green chemistry principle of atom economy illustrate this law?

PROBLEM 2.26 Propene is a raw material for a wide variety of products including the polymer polypropylene used in plastic wrap and Styrofoam cups. Calculate the atom economy for the synthesis of propene from propanol. (Note: Sulfuric acid, H_2SO_4, is a catalyst that can be recovered so it is not considered in atom economy calculations.)

$$\underset{\text{Propanol}}{H-\overset{\overset{\displaystyle H}{|}}{\underset{\underset{\displaystyle H}{|}}{C}}-\overset{\overset{\displaystyle H}{|}}{\underset{\underset{\displaystyle H}{|}}{C}}-\overset{\overset{\displaystyle H}{|}}{\underset{\underset{\displaystyle H}{|}}{C}}-O-H} \xrightarrow[\text{Heat}]{H_2SO_4}$$

$$\underset{\text{Propene}}{\overset{H}{\underset{H}{>}}C=\overset{\overset{\displaystyle H}{|}}{C}-\overset{\overset{\displaystyle H}{|}}{\underset{\underset{\displaystyle H}{|}}{C}}-H} \quad + \quad \underset{\text{Water}}{H-\overset{O}{\diagdown}H}$$

PROBLEM 2.27 Examine the two reactions important in chemical synthesis of organic compounds.

Reaction 1: An Addition Reaction (combination of two or more molecules to form a larger molecule)

$$\overset{H}{\underset{H}{>}}C=C\overset{H}{\underset{H}{<}} \quad + \quad Cl_2 \quad \longrightarrow \quad \underset{\text{Desired product}}{H-\overset{\overset{\displaystyle Cl}{|}}{\underset{\underset{\displaystyle H}{|}}{C}}-\overset{\overset{\displaystyle Cl}{|}}{\underset{\underset{\displaystyle H}{|}}{C}}-H}$$

Reaction 2: A Substitution Reaction (an atom or group of atoms is replaced by a different atom)

$$\underset{}{H-\overset{\overset{\displaystyle Cl}{|}}{\underset{\underset{\displaystyle H}{|}}{C}}-H} \quad + \quad Br^- \quad \longrightarrow \quad \underset{\text{Desired product}}{H-\overset{\overset{\displaystyle Br}{|}}{\underset{\underset{\displaystyle H}{|}}{C}}-H} \quad + \quad Cl^-$$

(a) Without performing any calculations, state which reaction has the higher percent atom economy.

(b) Calculate the percent atom economy for both reactions.

PROBLEM 2.28 Ibuprofen (the active ingredient in the over-the-counter drugs Advil and Motrin) is a molecule that alleviates pain and reduces fever and swelling. Use the ball-and-stick model of ibuprofen to determine the molecular formula. (Gray = C, ivory = H, red = O)

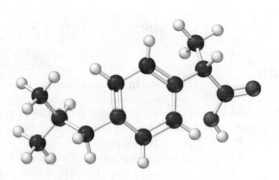

PROBLEM 2.29 Ibuprofen was initially synthesized by a process developed by Boots Co. in the 1960s. Six reaction steps were utilized, and the percent atom economy was 40%. For every one mole of ibuprofen produced, 1 mole of Na, 23 moles of H, 1 mole of N, 7 moles of C, 8 moles of O, and 1 mole of Cl were unused and considered as waste.

(a) Calculate the mass (g) of each element wasted for every one mole of ibuprofen produced.

(b) Calculate the total mass (g) wasted for every one mole of ibuprofen produced.

(c) Yearly production of ibuprofen is approximately 30 million lbs, which is equivalent to 6.6×10^7 moles. Calculate the total mass (kg) of matter wasted in the annual production ibuprofen.

PROBLEM 2.30 In the 1990s, BHC Co. developed a three-step synthesis for ibuprofen with a percent atom economy of 77.5%. This synthesis is "greener" than the original Boots Co. synthesis (Problem 2.29) because only 4 mol of H, 2 mol of C, and 2 mol of O are wasted for every mole of ibuprofen produced.

(a) Calculate the mass (g) of each element wasted for every one mole of ibuprofen produced.

(b) Calculate the total mass (g) wasted for every one mole of ibuprofen produced.

(c) Yearly production of ibuprofen is approximately 30 million lbs., which is equivalent to 6.6×10^7 moles. Calculate the total mass wasted in the annual production ibuprofen by the BHC Co. synthesis.

(d) What are the savings of waste (kg) of the BHC Co. ibuprofen synthesis over the Boots Co. synthesis in the yearly production of ibuprofen (Problem 2.29)?

STUDY GUIDE

Section	Concept Summary	Learning Objectives	Test Your Understanding
2.1 ▶ Chemistry and the Elements	All matter is formed from one or more of the 118 presently known **elements**–fundamental substances that can't be chemically broken down. Elements are symbolized by one- or two-letter abbreviations.	**2.1** Use symbols to represent element names.	Problems 2.46–2.51
2.2 ▶ Elements and the Periodic Table	Elements are organized into a **periodic table** with **groups** (columns) and **periods** (rows). Elements in the same groups show similar chemical behavior. Elements are classified as **metals**, **nonmetals**, or **semimetals**.	**2.2** Identify the location of metals, nonmetals, and semimetals on the periodic table.	Problems 2.32, 2.35, 2.58, 2.59
		2.3 Indicate the atomic number, group number, and period number for an element whose position in the periodic is given.	Problems 2.34, 2.53, 2.54
		2.4 Identify groups as main group, transition metal group, or inner transition metal group.	Problems 2.55–2.59
2.3 ▶ Some Common Groups of Elements and Their Properties	The characteristics, or **properties**, that are used to describe matter can be classified in several ways. **Physical properties** are those that can be determined without changing the chemical composition of the sample, whereas **chemical properties** are those that do involve a chemical change. Intensive properties are those whose values do not depend on the size of the sample, whereas **extensive properties** are those that do depend on sample amount.	**2.5** Specify the location and give examples of elements in the alkali metal, alkaline earth metal, halogen, and noble gas groups.	Problems 2.31, 2.61–2.63
		2.6 Use the properties of an element to classify it as metal, nonmetal, or semimetal and give its location in the periodic table.	Worked Example 2.1; Problems 2.33, 2.64–2.67
2.4 ▶ The Conservation of Mass and the Law of Definite Proportions	Elements join together in different ways to make **chemical compounds** and a pure compound always has the same proportion of elements by mass. During a chemical reaction, the **law of mass conservation** applies and the mass of reactants is the same as the mass of products.	**2.7** Determine the mass of the products in a reaction using the law of mass conservation.	Problem 2.72–2.73
2.5 ▶ The Law of Multiple Proportions and Dalton's Atomic Theory	**Elements** are made of tiny particles called **atoms**, which can combine in simple numerical ratios according to the **law of multiple proportions**.	**2.8** Demonstrate the law of multiple proportions using mass composition of two compounds of the same elements.	Worked Example 2.2; Problems 2.74–2.81
2.6 ▶ Atomic Structure: Electrons	Atoms are composed of three fundamental particles: **protons** are positively charged, **electrons** are negatively charged, and **neutrons** are neutral.	**2.9** Describe Thomson's cathode-ray experiment and what it contributed to the current model of atomic structure. (Figure 2.3)	Problems 2.82–2.84
		2.10 Describe Millikan's oil drop experiment and what it contributed to the current model of atomic structure. (Figure 2.4)	Problems 2.85, 2.86
2.7 ▶ Atomic Structure: Proton and Neutrons	According to the nuclear model of an atom proposed by Ernest Rutherford, protons and neutrons are clustered into a dense core called the **nucleus**, while electrons move around the nucleus at a relatively great distance.	**2.11** Describe Rutherford's gold foil experiment and what it contributed to the current model of atomic structure. (Figure 2.5)	Problems 2.87, 2.88
		2.12 Describe the structure and size of the atom. (Figure 2.6)	Problem 2.89
		2.13 Calculate the number of atoms in a sample given the size of the atom.	Problem 2.90, 2.91

Section	Concept Summary	Learning Objectives	Test Your Understanding
2.8 ▶ Atomic Numbers	Elements differ from one another according to how many protons their atoms contain, a value called the **atomic number (Z)** of the element. The sum of an atom's protons and neutrons is its **mass number (A)**. Although all atoms of a specific element have the same atomic number, different atoms of an element can have different mass numbers, depending on how many neutrons they have. Atoms with identical atomic numbers but different mass numbers are called **isotopes**.	**2.14** Determine the mass number, atomic number, and number of protons, neutrons, and electrons from an isotope symbol.	Worked Example 2.4, Problem 2.100–2.105
		2.15 Write isotope symbols for elements.	Problems 2.94–2.99
2.9 ▶ Atomic Weights and the Mole	Atomic weights are measured using the **unified atomic mass unit (u)**, defined as 1/12 the mass of a ^{12}C atom. Because both protons and neutrons have a mass of approximately 1, the mass of an atom in atomic mass units (the isotopic mass) is numerically close to the atom's mass number. The element's **atomic weight** is a weighted average of the mass of its naturally occurring isotopes. When referring to the enormous numbers of atoms that make up visible amounts of matter, the fundamental SI unit called a *mole* is used. One **mole** is the amount whose mass in grams, called its **molar mass**, is numerically equal to the atomic weight. Numerically, one mole of any element contains 6.022×10^{23} atoms, a value called **Avogadro's number** (N_A).	**2.16** Calculate atomic weight given the fractional abundance and mass of each isotope.	Worked Example 2.5; Problems 2.106–2.112
		2.17 Convert between grams and numbers of moles or atoms using molar mass and Avogadro's number.	Worked Example 2.6; Problems 2.113–2.115
		2.18 Identify an element given the mass and number of atoms or moles.	Problem 2.116–2.119
2.10 ▶ Mixtures and Chemical Compounds; Molecules and Covalent Bonds	Most substances are **chemical compounds**, formed when atoms of two or more elements combine in a **chemical reaction**. The atoms in a compound are held together by one of two kinds of **chemical bonds**. **Covalent bonds** form when two atoms share electrons to give a new unit of matter called a **molecule**.	**2.19** Classify molecular representations of matter as a mixture, pure substance, element, or compound.	Worked Example 2.7; Problems 2.13, 2.14
		2.20 Convert between structural formulas, ball-and-stick models, and chemical formulas.	Worked Example 2.8, Problems 2.39–2.41, 2.126–2.131
2.11 ▶ Ions and Ionic Bonds	**Ionic bonds** form when one atom completely transfers one or more electrons to another atom, resulting in the formation of **ions**. Positively charged ions (**cations**) are strongly attracted to negatively charged ions (**anions**) by electrical forces.	**2.21** Classify bonds as ionic or covalent.	Problems 2.120, 2.121
		2.22 Determine the number of electrons and protons from chemical symbol and charge.	Problems 2.42, 2.124, 2.125
		2.23 Match the molecular representation of an ionic compound with its chemical formula.	Problem 2.43
2.12 ▶ Naming Chemical Compounds	Chemical compounds are named systematically by following a series of rules. Binary ionic compounds are named by identifying first the positive ion and then the negative ion. Binary molecular compounds are similarly named by identifying the cationlike and anionlike elements. Naming compounds with **polyatomic ions** involves memorizing the names and formulas of the most common ones.	**2.24** Convert between name and formula for binary ionic compounds.	Worked Example 2.10; Problems 2.132–2.135
		2.25 Convert between formula and name for ionic compounds with polyatomic ions.	Worked Example 2.11; Problems 2.136–2.138
		2.26 Convert between name and formula for binary molecular compounds.	Worked Example 2.12; Problems 2.140–2.149

KEY TERMS

anion 58	chemical property 38	isotope 50	oxoanion 62
atom 44	covalent bond 55	law of definite proportions 43	period 37
atom economy 66	electron 45	law of mass conservation 42	periodic table 35
atomic mass 51	element 34	law of multiple proportions 43	physical property 38
atomic number (Z) 49	extensive property 38	main group 37	polyatomic ion 59
atomic weight 52	green chemistry 66	mass number 50	property 38
Avogadro's number 53	group 37	mass number 50	proton 48
cation 58	inner transition metal group 37	molar mass 53	structural formula 56
chemical bond 54	intensive property 38	mole 53	transition metal group 37
chemical compound 42	ion 58	molecule 55	unified atomic mass unit (u) 51
chemical equation 42	ionic bond 58	neutron 48	
chemical formula 42	ionic solid 59	nucleus 47	

CONCEPTUAL PROBLEMS

Problems 2.1–2.30 appear within the chapter.

2.31 Where on the following outline of a periodic table are the indicated elements or groups of elements?

(a) Alkali metals

(b) Halogens

(c) Alkaline earth metals

(d) Transition metals

(e) Hydrogen

(f) Helium

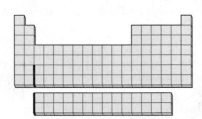

2.32 Where on the following outline of a periodic table does the dividing line between metals and nonmetals fall?

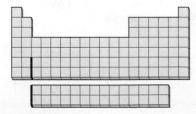

2.33 Is the red element on the following periodic table likely to be a gas, a liquid, or a solid? What is the atomic number of the blue element? What is the group number of the green, blue, and red element? Name at least one other element that is chemically similar to the green element.

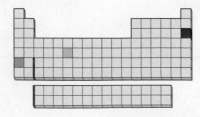

2.34 The element indicated on the following periodic table is used in smoke detectors. Identify it, give its atomic number, and tell what kind of group it's in.

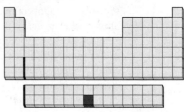

2.35 Identify the three elements indicated on the periodic table and give the group that they are in. Classify these elements as metals, nonmetals, or semimetals. Would you expect these elements to have similar or different chemical reactivity?

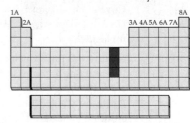

2.36 If yellow spheres represent sulfur atoms and red spheres represent oxygen atoms, which of the following drawings shows a collection of sulfur dioxide (SO_2) units?

(a)

(b)

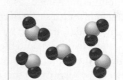

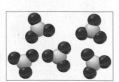

(c)

(d)

2.37 Assume that the mixture of substances in drawing (a) undergoes a reaction. Which of the drawings (b)–(d) represents a product mixture consistent with the law of mass conservation?

(a)

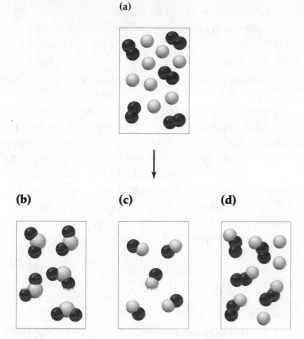

(b)   (c)   (d)

2.38 In the following drawings, red spheres represent protons and blue spheres represent neutrons. Which of the drawings represent different isotopes of the same element, and which represents a different element altogether?

(a)   (b)   (c)

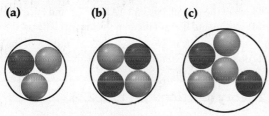

2.39 Methionine, one of the 20 amino acid building blocks from which proteins are made, has the following structure. What is the chemical formula of methionine? In writing the formula, list the element symbols in alphabetical order and give the number of each element as a subscript.

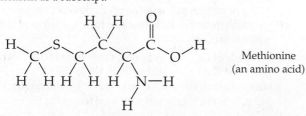

Methionine (an amino acid)

2.40 Thymine, one of the four bases in deoxyribonucleic acid (DNA), has the following structure. What is the chemical formula of thymine? In writing the formula, list the element symbols in alphabetical order and give the number of each element as a subscript.

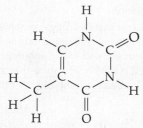

2.41 Give molecular formulas corresponding to each of the following ball-and-stick molecular representations (red = O, gray = C, blue = N, ivory = H). In writing the formula, list the elements in alphabetical order.

(a) Alanine (an amino acid)

(b) Ethylene glycol (automobile antifreeze)

(c) Acetic acid (vinegar)

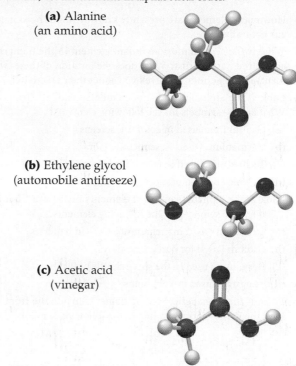

2.42 Which of the following three drawings represents a neutral Na atom, which represents a Ca atom with two positive electrical charges (Ca^{2+}), and which represents an F atom with one minus charge (F^-)?

(a)   (b)   (c)

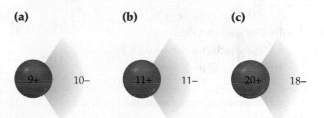

2.43 In the following drawings, red spheres represent cations and blue spheres represent anions. Match each of the drawings (a)–(d) with the following ionic compounds:

(i) $Ca_3(PO_4)_2$ (ii) Li_2CO_3 (iii) $FeCl_2$ (iv) $MgSO_4$

(a)   (b)

(c)   (d)

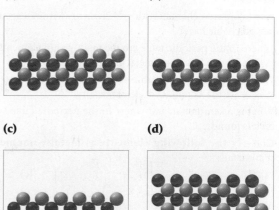

SECTION PROBLEMS

Elements and the Periodic Table (Sections 2.1–2.3)

2.44 How many elements are presently known? About how many occur naturally?

2.45 What is the second most abundant element in the human body and in the Earth's crust? What does the location of these two elements in the periodic table tell you about their reactivity?

2.46 Look at the alphabetical list of elements inside the front cover. What are the symbols for the following elements?

(a) Gadolinium (used in color TV screens)

(b) Germanium (used in semiconductors)

(c) Technetium (used in biomedical imaging)

(d) Arsenic (used in pesticides)

2.47 Look at the alphabetical list of elements inside the front cover. What are the symbols for the following elements?

(a) Cadmium (used in rechargeable Ni–Cd batteries)

(b) Iridium (used for hardening alloys)

(c) Beryllium (used in the space shuttle)

(d) Tungsten (used in light bulbs)

2.48 Look at the alphabetical list of elements inside the front cover. Give the names corresponding to the following symbols:

(a) Te   (b) Re   (c) Be

(d) Ar   (e) Pu

2.49 Look at the alphabetical list of elements inside the front cover. Give the names corresponding to the following symbols:

(a) B   (b) Rh   (c) Cf

(d) Os   (e) Ga

2.50 What is wrong with each of the following statements?

(a) The symbol for tin is Ti.

(b) The symbol for manganese is Mg.

(c) The symbol for potassium is Po.

(d) The symbol for helium is HE.

2.51 What is wrong with each of the following statements?

(a) The symbol for carbon is ca.

(b) The symbol for sodium is So.

(c) The symbol for nitrogen is Ni.

(d) The symbol for chlorine is Cr.

2.52 In the periodic table, (a) what are the rows called and (b) what are the columns called?

2.53 How many groups are there in the periodic table? How are they labeled?

2.54 What common characteristics do elements within a group of the periodic table have?

2.55 Where in the periodic table are the main-group elements found? Where are the transition metal groups found?

2.56 Where in the periodic table are the metallic elements found? Where are the nonmetallic elements found?

2.57 What is a semimetal, and where in the periodic table are semimetals found?

2.58 Classify the following elements as metals, nonmetals, or semimetals:

(a) Ti   (b) Te   (c) Se

(d) Sc   (e) Si

2.59 Classify the following elements as metals, nonmetals, or semimetals:

(a) Ar   (b) Sb   (c) Mo

(d) Cl   (e) N   (f) Mg

2.60 List several general properties of the following groups:

(a) Alkali metals   (b) Noble gases

(c) Halogens

2.61 (a) Without looking at a periodic table, list as many alkali metals as you can. (There are five common ones.)

(b) Without looking at a periodic table, list as many alkaline earth metals as you can. (There are five common ones.)

2.62 Without looking at a periodic table, list as many halogens as you can. (There are four common ones.)

2.63 Without looking at a periodic table, list as many noble gases as you can. (There are six common ones.)

2.64 At room temperature, a certain element is found to be a soft silver-colored solid that reacts violently with water and is a good conductor of electricity. Is the element likely to be a metal, a nonmetal, or a semimetal?

2.65 At room temperature, a certain element is found to be shiny silver-colored solid that is a poor conductor of electricity. When a sample of the element is hit with a hammer, it shatters. Is the element likely to be a metal, a nonmetal, or a semimetal?

2.66 At room temperature, a certain element is yellow crystalline solid. It does not conduct electricity, and when hit with a hammer it shatters. Is the element likely to be a metal, a nonmetal, or a semimetal?

2.67 At room temperature, a certain element is a colorless, unreactive gas. Is the element likely to be a metal, a nonmetal, or a semimetal?

2.68 In which of the periodic groups 1A, 2A, 5A, and 7A is the first letter of all elements' symbol the same as the first letter of their name?

2.69 For which elements in groups 1A, 2A, 5A, and 7A of the periodic table does the first letter of their symbol differ from the first letter of their name?

Atomic Theory (Sections 2.4 and 2.5)

2.70 How does Dalton's atomic theory account for the law of mass conservation and the law of definite proportions?

2.71 What is the law of multiple proportions, and how does Dalton's atomic theory account for it?

2.72 A sample of mercury with a mass of 114.0 g was combined with 12.8 g of oxygen gas, and the resulting reaction gave 123.1 g of mercury(II) oxide. How much oxygen was left over after the reaction was complete?

2.73 A sample of $CaCO_3$ was heated, causing it to form CaO and CO_2 gas. Solid CaO remained behind, while the CO_2 escaped to the atmosphere. If the $CaCO_3$ weighed 612 g and the CaO weighed 343 g, how many grams of CO_2 were formed in the reaction?

2.74 In methane, one part hydrogen combines with three parts carbon by mass. If a sample of a compound containing only carbon and hydrogen contains 32.0 g of carbon and 8.0 g of hydrogen, could the sample be methane? If the sample is not methane, show that the law of multiple proportions is followed for methane and this other substance.

2.75 In borane, one part hydrogen combines with 3.6 parts boron by mass. A compound containing only hydrogen and boron contains 6.0 g of hydrogen and 43.2 g of boron. Could this compound be borane? If it is not borane, show that the law of multiple proportions is followed for borane and this other substance.

2.76 Benzene, ethane, and ethylene are just three of a large number of *hydrocarbons*—compounds that contain only carbon and hydrogen. Show how the following data are consistent with the law of multiple proportions.

Compound	Mass of carbon in 5.00 g sample	Mass of hydrogen in 5.00 g sample
Benzene	4.61 g	0.39 g
Ethane	4.00 g	1.00 g
Ethylene	4.29 g	0.71 g

2.77 The atomic weight of carbon (12.011) is approximately 12 times that of hydrogen (1.008).

   (a) Show how you can use this knowledge to calculate possible formulas for benzene, ethane, and ethylene (Problem 2.76).

   (b) Show how your answer to part **(a)** is consistent with the actual formulas for benzene (C_6H_6), ethane (C_2H_6), and ethylene (C_2H_4).

2.78 Two compounds containing carbon and oxygen have the following percent composition by mass.

   Compound 1: 42.9% carbon and 57.1% oxygen
   Compound 2: 27.3% carbon and 72.7% oxygen

   Show that the law of multiple proportions is followed. If the formula of the first compound is CO, what is the formula of the second compound?

2.79 In addition to carbon monoxide (CO) and carbon dioxide (CO_2), there is a third compound of carbon and oxygen called *carbon suboxide*. If a 2.500 g sample of carbon suboxide contains 1.32 g of C and 1.18 g of O, show that the law of multiple proportions is followed. What is a possible formula for carbon suboxide?

2.80 A compound of zinc and sulfur contains 67.1% zinc by mass. What is the ratio of zinc and sulfur atoms in the compound?

2.81 There are two compounds of titanium and chlorine. One compound contains 31.04% titanium by mass, and the other contains 74.76% chlorine by mass. What are the ratios of titanium and chlorine atoms in the two compounds?

Elements and Atoms (Sections 2.6–2.8)

2.82 The results from Thomson's cathode-ray tube experiment led to the discovery of which subatomic particle?

2.83 What affects the magnitude of the deflection of the cathode ray in Thomson's experiment?

2.84 What property was measured in Thomson's cathode-ray experiment?

   (a) The mass of the electron.

   (b) The mass of the proton.

   (c) The charge of the electron.

   (d) The mass to charge ratio of the electron.

   (e) Both the mass and charge of the electron.

2.85 What property was measured in Millikan's oil drop experiment?

   (a) The mass of the electron.

   (b) The mass of the proton.

   (c) The charge of the electron.

   (d) The mass to charge ratio of the electron.

   (e) Both the mass and charge of the electron.

2.86 Which of the following charges is NOT possible for the overall charge on an oil droplet in Millikan's experiment? For this problem we'll round the currently accepted charge of an electron to 1.602×10^{-19} C.

   (a) -1.010×10^{-18} C

   (b) -8.010×10^{-19} C

   (c) -2.403×10^{-18} C

2.87 What discovery about atomic structure was made from the results of Rutherford's gold foil experiment?

2.88 Prior to Rutherford's gold foil experiment, the "plum pudding" model of the atom represented atomic structure. In this model, the atom is composed of electrons interspersed within a positive cloud of charge. If this were the correct model of the atom, how would the results of Rutherford's experiment have been different?

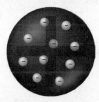

   (a) The alpha particles would pass right through the gold foil with little to no deflection.

   (b) Most of the alpha particles would be deflected back toward the source.

   (c) Most of the alpha particles would be absorbed by the atom and would not pass through or be deflected from the gold foil.

2.89 A period at the end of sentence written with a graphite pencil has a diameter of 1 mm. If the period represented the nucleus, approximately how large is the diameter of the entire atom in units of m?

2.90 A ¼ inch thick lead sheet is used for protection from medical X-rays. If a single lead atom has a diameter 350 pm, how many atoms thick is the lead sheet?

2.91 A period at the end of sentence written with a graphite pencil has a diameter of 1 mm. How many carbon atoms would it take to line up across the period, if a single carbon atom has a diameter of 150 pm?

2.92 What is the difference between an atom's atomic number and its mass number?

2.93 What is the difference between an element's atomic number and its atomic mass?

2.94 The subscript giving the atomic number of an atom is often left off when writing an isotope symbol. For example, $^{13}_{6}C$ is often written simply as ^{13}C. Why is this allowed?

2.95 Iodine has a *lower* atomic weight than tellurium (126.90 for iodine, 127.60 for tellurium) even though it has a *higher* atomic number (53 for iodine, 52 for tellurium). Explain.

2.96 Give the names and symbols for the following elements:

   (a) An element with atomic number 6

   (b) An element with 18 protons in its nucleus

   (c) An element with 23 electrons

2.97 The radioactive isotope cesium-137 was produced in large amounts in fallout from the 1985 nuclear power plant disaster at Chernobyl, Ukraine. Write the symbol for this isotope in standard format.

2.98 Write symbols for the following isotopes:

   (a) Radon-220

   (b) Polonium-210

   (c) Gold-197

2.99 Write symbols for the following isotopes:

(a) $Z = 58$ and $A = 140$

(b) $Z = 27$ and $A = 60$

2.100 How many protons, neutrons, and electrons are in each of the following atoms?

(a) $^{15}_{7}N$

(b) $^{60}_{27}Co$

(c) $^{131}_{53}I$

(d) $^{148}_{58}Ce$

2.101 How many protons and neutrons are in the nucleus of the following atoms?

(a) ^{27}Al

(b) ^{32}S

(c) ^{64}Zn

(d) ^{207}Pb

2.102 Identify the following elements:

(a) $^{24}_{12}X$

(b) $^{58}_{28}X$

(c) $^{104}_{46}X$

(d) $^{183}_{74}X$

2.103 Identify the following elements:

(a) $^{202}_{80}X$

(b) $^{195}_{78}X$

(c) $^{184}_{76}X$

(d) $^{209}_{83}X$

2.104 Which of the following isotope symbols can't be correct?

$^{18}_{9}F$ $^{12}_{5}C$ $^{33}_{35}Br$ $^{18}_{8}O$ $^{11}_{5}Bo$

2.105 Which of the following isotope symbols can't be correct?

$^{14}_{7}Ni$ $^{131}_{54}Xe$ $^{54}_{26}Fe$ $^{73}_{23}Ge$ $^{1}_{2}He$

Atomic Weight and Moles (Section 2.9)

2.106 Copper has two naturally occurring isotopes, including ^{65}Cu. Look at the periodic table and tell whether the second isotope is ^{63}Cu or ^{66}Cu.

2.107 Sulfur has four naturally occurring isotopes, including ^{33}S, ^{34}S, and ^{36}S. Look at the periodic table and tell whether the fourth isotope is ^{32}S or ^{35}S.

2.108 Naturally occurring boron consists of two isotopes: ^{10}B (19.9%) with an isotopic mass of 10.0129 and ^{11}B (80.1%) with an isotopic mass of 11.009 31. What is the atomic weight of boron? Check your answer by looking at a periodic table.

2.109 Naturally occurring silver consists of two isotopes: ^{107}Ag (51.84%) with an isotopic mass of 106.9051 and ^{109}Ag (48.16%) with an isotopic mass of 108.9048. What is the atomic weight of silver? Check your answer in a periodic table.

2.110 Magnesium has three naturally occurring isotopes: ^{24}Mg (23.985) with 78.99% abundance, ^{25}Mg (24.986) with 10.00% abundance, and a third with 11.01% abundance. Look up the atomic weight of magnesium, and then calculate the mass of the third isotope.

2.111 A sample of naturally occurring silicon consists of ^{28}Si (27.9769), ^{29}Si (28.9765), and ^{30}Si (29.9738). If the atomic weight of silicon is 28.0855 and the natural abundance of ^{29}Si is 4.68%, what are the natural abundances of ^{28}Si and ^{30}Si?

2.112 Copper metal has two naturally occurring isotopes: copper-63 (69.15%; isotopic mass = 62.93) and copper-65 (30.85%; isotopic mass 64.93). Calculate the atomic weight of copper and check your answer in the periodic table.

2.113 Based on your answer to Problem 2.112, how many atoms of copper are in an old penny made of pure copper and weighing 2.15 g?

2.114 What is the mass in grams of each of the following samples?

(a) 1.505 mol of Ti

(b) 0.337 mol of Na

(c) 2.583 mol of U

2.115 How many moles are in each of the following samples?

(a) 11.51 g of Ti

(b) 29.127 g of Na

(c) 1.477 kg of U

2.116 If the atomic weight of an element is x, what is the mass in grams of 6.02×10^{23} atoms of the element? How does your answer compare numerically with the atomic weight of element x?

2.117 If the atomic weight of an element is x, what is the mass in grams of 3.17×10^{20} atoms of the element?

2.118 If 6.02×10^{23} atoms of element Y have a mass of 83.80 g, what is the identity of Y?

2.119 If 4.61×10^{21} atoms of element Z have a mass of 0.815 g, what is the identity of Z?

Chemical Compounds (Sections 2.10 and 2.11)

2.120 What is the difference between a covalent bond and an ionic bond?

2.121 Which of the following bonds are likely to be covalent and which ionic? Explain.

(a) B $\cdots$ Br

(b) Na $\cdots$ Br

(c) Br $\cdots$ Cl

(d) O $\cdots$ Br

2.122 The symbol CO stands for carbon monoxide, but the symbol Co stands for the element cobalt. Explain.

2.123 Correct the error in each of the following statements:

(a) The formula of ammonia is NH3.

(b) Molecules of potassium chloride have the formula KCl.

(c) Cl^- is a cation.

(d) CH_4 is a polyatomic ion.

2.124 How many protons and electrons are in each of the following ions?

(a) Be^{2+}

(b) Rb^+

(c) Se^{2-}

(d) Au^{3+}

2.125 What is the identity of the element X in the following ions?

(a) X^{2+}, a cation that has 36 electrons

(b) X^-, an anion that has 36 electrons

2.126 The structural formula of isopropyl alcohol, better known as "rubbing alcohol," is shown. What is the chemical formula of isopropyl alcohol?

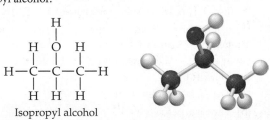

Isopropyl alcohol

2.127 Lactic acid, a compound found both in sour milk and in tired muscles, has the structure shown. What is its chemical formula?

Lactic acid

2.128 Butane, the fuel used in disposable lighters, has the formula C_4H_{10}. The carbon atoms are connected in the sequence C—C—C—C, and each carbon has four covalent bonds. Draw the structural formula of butane.

2.129 Cyclohexane, C_6H_{12}, is an important starting material used in the industrial synthesis of nylon. Each carbon has four covalent bonds, two to hydrogen and two to other carbons. Draw the structural formula of cyclohexane.

2.130 Isooctane, the substance in gasoline from which the term *octane rating* derives, has the formula C_8H_{18}. Each carbon has four covalent bonds, and the atoms are connected in the sequence shown. Draw the complete structural formula of isooctane.

2.131 Fructose, $C_6H_{12}O_6$, is the sweetest naturally occurring sugar and is found in many fruits and berries. Each carbon has four covalent bonds, each oxygen has two covalent bonds, each hydrogen has one covalent bond, and the atoms are connected in the sequence shown. Draw the complete structural formula of fructose.

Naming Compounds (Section 2.12)

2.132 Give systematic names for the following binary compounds:
(a) CsF
(b) K_2O
(c) CuO

2.133 Give systematic names for the following binary compounds:
(a) BaS
(b) $BeBr_2$
(c) $FeCl_3$

2.134 Write formulas for the following binary compounds:
(a) Potassium chloride
(b) Tin(II) bromide
(c) Calcium oxide
(d) Barium chloride
(e) Aluminum hydride

2.135 Write formulas for the following binary compounds:
(a) Vanadium(III) chloride
(b) Manganese(IV) oxide
(c) Copper(II) sulfide
(d) Aluminum oxide

2.136 Write formulas for the following compounds:
(a) Calcium acetate
(b) Iron(II) cyanide
(c) Sodium dichromate
(d) Chromium(III) sulfate
(e) Mercury(II) perchlorate

2.137 Write formulas for the following compounds:
(a) Lithium phosphate
(b) Magnesium hydrogen sulfate
(c) Manganese(II) nitrate
(d) Chromium(III) sulfate

2.138 Give systematic names for the following compounds:
(a) $Ca(ClO)_2$
(b) $Ag_2S_2O_3$
(c) NaH_2PO_4
(d) $Sn(NO_3)_2$
(e) $Pb(CH_3CO_2)_4$
(f) $(NH_4)_2SO_4$

2.139 Name the following ions:
(a) Ba^{2+}
(b) Cs^+
(c) V^{3+}
(d) HCO_3^-
(e) NH_4^+
(f) Ni^{2+}
(g) NO_2^-
(h) ClO_2^-
(i) Mn^{2+}
(j) ClO_4^-

2.140 What are the formulas of the compounds formed from the following ions?
(a) Ca^{2+} and Br^-
(b) Ca^{2+} and SO_4^{2-}
(c) Al^{3+} and SO_4^{2-}

2.141 What are the formulas of the compounds formed from the following ions?
(a) Na^+ and NO_3^-
(b) K^+ and SO_4^{2-}
(c) Sr^{2+} and Cl^-

2.142 Write formulas for compounds of calcium with each of the following:
(a) Chlorine
(b) Oxygen
(c) Sulfur

2.143 Write formulas for compounds of rubidium with each of the following:
(a) Bromine
(b) Nitrogen
(c) Selenium

2.144 Give the formulas and charges of the following ions:
(a) Sulfite ion
(b) Phosphate ion
(c) Zirconium(IV) ion
(d) Chromate ion
(e) Acetate ion
(f) Thiosulfate ion

2.145 What are the charges on the positive ions in the following compounds?
(a) $Zn(CN)_2$
(b) $Fe(NO_2)_3$
(c) $Ti(SO_4)_2$
(d) $Sn_3(PO_4)_2$
(e) Hg_2S
(f) MnO_2
(g) KIO_4
(h) $Cu(CH_3CO_2)_2$

2.146 Name the following binary molecular compounds:
(a) CCl_4
(b) ClO_2
(c) N_2O
(d) N_2O_3

2.147 Give systematic names for the following compounds:
(a) NCl_3
(b) P_4O_6
(c) S_2F_2

2.148 Name the following binary compounds of nitrogen and oxygen:
(a) NO
(b) N_2O
(c) NO_2
(d) N_2O_4
(e) N_2O_5

2.149 Name the following binary compounds of sulfur and oxygen:
(a) SO
(b) S_2O_2
(c) S_5O
(d) S_7O_2
(e) SO_3

2.150 Fill in the missing information to give formulas for the following compounds:
(a) $Na_?SO_4$
(b) $Ba_?(PO_4)_?$
(c) $Ga_?(SO_4)_?$

2.151 Write formulas for each of the following compounds:

(a) Sodium peroxide

(b) Aluminum bromide

(c) Chromium(III) sulfate

CHAPTER PROBLEMS

2.152 Germanium has five naturally occurring isotopes: ^{70}Ge, 20.5%, 69.924; ^{72}Ge, 27.4%, 71.922; ^{73}Ge, 7.8%, 72.923; ^{74}Ge, 36.5%, 73.921; and ^{76}Ge, 7.8%, 75.921. What is the atomic weight of germanium?

2.153 Fluorine occurs naturally as a single isotope. How many protons, neutrons, and electrons are present in deuterium fluoride (^2HF)? (Deuterium is ^2H.)

2.154 Ammonia (NH_3) and hydrazine (N_2H_4) are both compounds of nitrogen and hydrogen. Based on the law of multiple proportions, how many grams of hydrogen would you expect 2.34 g of nitrogen to combine with to yield ammonia? To yield hydrazine?

2.155 If 3.670 g of nitrogen combines with 0.5275 g of hydrogen to yield compound X, how many grams of nitrogen would combine with 1.575 g of hydrogen to make the same compound? Is X ammonia (NH_3) or hydrazine (N_2H_4)?

2.156 Identify the following atoms:

(a) A halogen with 53 electrons

(b) A noble gas with $A = 84$

2.157 Hydrogen has three isotopes (^1H, ^2H, and ^3H), and chlorine has two isotopes (^{35}Cl and ^{37}Cl). How many isotopic kinds of HCl are there? Write the formula for each, and tell how many protons, neutrons, and electrons each contains.

2.158 Prior to 1961, the atomic mass unit was defined as 1/16 the mass of the atomic weight of oxygen; that is, the atomic weight of oxygen was defined as exactly 16. What was the mass of a ^{12}C atom prior to 1961 if the atomic weight of oxygen on today's scale is 15.9994?

2.159 What was the mass in atomic mass units of a ^{40}Ca atom prior to 1961 if its mass on today's scale is 39.9626? (See Problem 2.158.)

2.160 The *molecular weight* of a compound is the sum of the atomic weights of all atoms in the molecule. What is the molecular mass of acetaminophen ($C_8H_9NO_2$), the active ingredient in Tylenol?

2.161 The *mass percent* of an element in a compound is the mass of the element (total mass of the element's atoms in the compound) divided by the mass of the compound (total mass of all atoms in the compound) times 100%. What is the mass percent of each element in acetaminophen? (See Problem 2.160.)

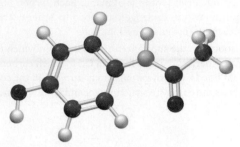

Acetaminophen ($C_8H_9NO_2$)

2.162 Tetrahydrofuran, an organic substance used as a solvent in many pharmaceutical processes, has the formula C_4H_8O. In tetrahydrofuran, the four C atoms are bonded in a row, each C atom is bonded to two H atoms, each H atom is bonded to one C atom, and the O atom is bonded to two C atoms. Write a structural formula for tetrahydrofuran.

2.163 In an alternate universe, the smallest negatively charged particle, analogous to our electron, is called a blorvek. To determine the charge on a single blorvek, an experiment like Millikan's with charged oil droplets was carried out and the following results were recorded:

Droplet Number	Charge (C)
1	7.74×10^{-16}
2	4.42×10^{-16}
3	2.21×10^{-16}
4	4.98×10^{-16}
5	6.64×10^{-16}

(a) Based on these observations, what is the largest possible value for the charge on a blorvek?

(b) Further experiments found a droplet with a charge of 5.81×10^{-16} C. Does this new result change your answer to part (a)? If so, what is the new largest value for the blorvek's charge?

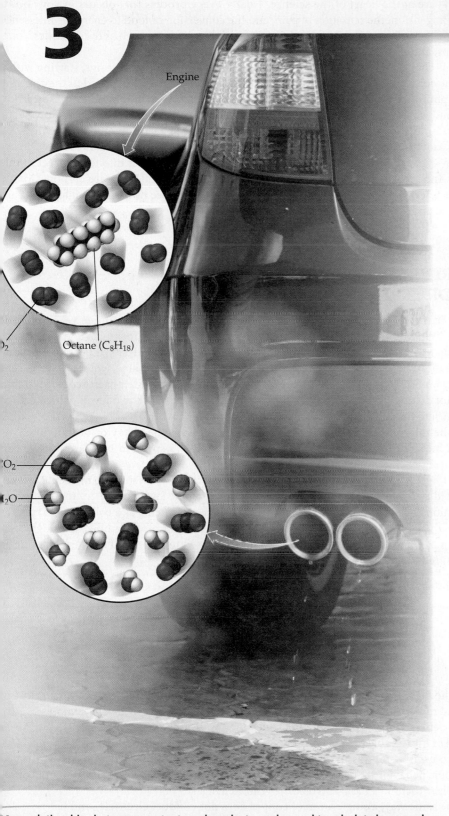

Engine

O_2

Octane (C_8H_{18})

O_2

H_2O

Mass Relationships in Chemical Reactions

Mass relationships between reactants and products can be used to calculate how much carbon dioxide is released when gasoline or other carbon-containing fuels are burned.

 Can Alternative Fuels Decrease CO₂ Emissions?

The answer to this question can be found in the **INQUIRY** ▶▶▶ on page 101.

CONTENTS

STUDY GUIDE

It's important to realize that chemical *reactions*—the change of one substance into another—are at the heart of the science. Nearly every process that occurs in your body including vision, the sensation of pain, and the conversion of food to energy, is in essence a series of chemical reactions. Likewise, products we use every day such as dyes, plastics, computer chips, and metals are manufactured by chemical reactions.

In this chapter, we'll begin learning how to describe chemical reactions, starting with a look at the conventions for writing chemical equations like the one depicted in the chapter-opening photo that shows how octane in a car engine reacts with oxygen to make carbon dioxide and water. Next, we'll examine mass relationships between reactants and products an understanding of which allows us to determine how much carbon dioxide is emitted by burning gasoline in our car's engine, an important inquiry because carbon dioxide is a *greenhouse gas*—a gas that traps heat in the atmosphere. We delve deeper into this subject in this chapter's Inquiry section, which describes how to calculate and compare carbon dioxide emissions for various types of alternative fuels. Finally, we'll see how chemical formulas are determined and how molecular weights are measured.

3.1 ▶ REPRESENTING CHEMISTRY ON DIFFERENT LEVELS

Before starting the main subject of this chapter, let's first answer a simple yet important question: What do numbers and symbols represent in chemical formulas and equations? Answering this question isn't as easy as it sounds because a chemical symbol can have different meanings under different circumstances. Chemists use the same symbols to represent chemistry on both a small-scale, microscopic level and a large-scale, macroscopic level and tend to not distinguish between what is happening on each of the two levels, which can be very confusing to newcomers to the field.

On the microscopic level, chemical symbols represent the behavior of individual atoms and molecules. Atoms and molecules are much too small to be seen, but we can nevertheless describe their microscopic behavior. For example, we can read the equation $2\,H_2 + O_2 \rightarrow 2\,H_2O$ to mean "Two molecules of hydrogen react with one molecule of oxygen to yield two molecules of water." It's on the microscopic level that we try to understand how reactions occur. Although simplistic, we visualize a molecule as a collection of spheres stuck together. In trying to understand how H_2 reacts with O_2, for example, you might picture H_2 and O_2 molecules as made of two spheres pressed together and a water molecule as made of three spheres.

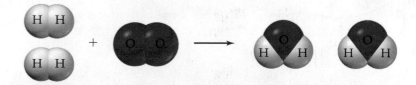

On the macroscopic level, formulas and equations represent the large-scale behaviors of atoms and molecules that give rise to visible properties. In other words, the symbols H_2, O_2, and H_2O represent not just single molecules but vast numbers of molecules that together have a set of measurable physical properties. A collection of a large number of H_2O molecules appears to us as a colorless liquid that freezes at 0 °C and boils at 100 °C. Clearly, it's this macroscopic behavior we deal with in the laboratory when we weigh specific amounts of reactants, place them in a flask, and observe visible changes.

In the same way, a single atom of copper does not conduct electricity and has no color on a microscopic level. On a macroscopic level, however, a large collection of copper atoms appears to us as a shiny, reddish-brown solid that can be drawn into electrical wires or made into coins.

A chemical formula or equation can be read either on the macroscopic level or on the microscopic level. The symbol H_2O can be interpreted either as one tiny, invisible molecule or as a vast collection of molecules large enough to swim in. You will learn to interpret a chemical formula or equation differently depending upon the context in which it is presented.

3.2 ▶ BALANCING CHEMICAL EQUATIONS

Preceding chapters provided several examples of reactions: hydrogen reacting with oxygen to yield water, sodium reacting with chlorine to yield sodium chloride, and mercury(II) nitrate reacting with potassium iodide to yield mercury(II) iodide:

4 H and 2 O atoms on this side   4 H and 2 O atoms on this side

$$2\,H_2 + O_2 \longrightarrow 2\,H_2O$$

2 Na and 2 Cl atoms on this side   2 Na and 2 Cl atoms on this side

$$2\,Na + Cl_2 \longrightarrow 2\,NaCl$$

1 Hg, 2 N, 6 O, 2 K, and 2 I atoms on this side   1 Hg, 2 N, 6 O, 2 K, and 2 I atoms on this side

$$Hg(NO_3)_2 + 2\,KI \longrightarrow HgI_2 + 2\,KNO_3$$

Look carefully at how these equations are written. Because hydrogen, oxygen, and chlorine exist as covalent diatomic molecules, we must write them as H_2, O_2, and Cl_2, rather than as isolated atoms (Section 2.10). Now, look at the atoms on each side of the reaction arrow. Although we haven't explicitly stated it yet, chemical equations are always written so that they are **balanced**; that is, the numbers and kinds of atoms on both sides of the reaction arrow are the same. This requirement is a consequence of the **law of mass conservation** (Section 2.4). Because atoms are neither created nor destroyed in chemical reactions, their numbers and kinds must remain the same in both products and reactants.

Balancing a chemical equation involves finding out how many *formula units* of each different substance take part in the reaction. A **formula unit**, as its name implies, is one unit—whether atom, ion, or molecule—corresponding to a given formula. One formula unit of NaCl is one Na^+ ion and one Cl^- ion, one formula unit of $MgBr_2$ is one Mg^{2+} ion and two Br^- ions, and one formula unit of H_2O is one H_2O molecule.

Complicated equations generally need to be balanced using a systematic method, as will be shown in later chapters, whereas simpler equations can often be balanced using a mixture of common sense and trial and error:

1. *Write an unbalanced equation using the correct chemical formula unit for each reactant and product.* For the reaction of ammonia (NH_3) with oxygen to form nitrogen monoxide and water, we begin by writing:

$$NH_3 + O_2 \longrightarrow NO + H_2O \qquad \text{Unbalanced}$$

2. *Find suitable **coefficients**—the numbers placed before formulas to indicate how many formula units of each substance are required to balance the equation.* Only these coefficients can be changed when balancing an equation; the formulas themselves can't be changed. It is best to begin by balancing the elements that appear in only two species in the equation. In the reaction of ammonia with oxygen, the element H appears in NH_3 and H_2O and the element N appears in NH_3 and NO. To balance H, a coefficient of 2 is placed in front of NH_3 and a coefficient of 3 is placed in front of H_2O. There are now 6 H atoms in both reactants and products.

$$2\,NH_3 + O_2 \longrightarrow NO + 3\,H_2O \qquad \text{Balanced for H}$$

To balance N, a coefficient of 2 is needed in front of NO because there are 2 N atoms in the reactants.

$$2\,NH_3 + O_2 \longrightarrow 2\,NO + 3\,H_2O \qquad \text{Balanced for H and N}$$

REMEMBER...

According to the **law of mass conservation**, mass is neither created nor destroyed in chemical reactions. (Section 2.4)

Elements that are not combined with other elements should be balanced last as changing the coefficient will not impact other species in the equation. In this case, adding a coefficient of 5/2 in front of O_2 in the reactants will balance O_2 because there are 5 O_2 atoms in the products.

$$2\,NH_3 + \frac{5}{2}O_2 \longrightarrow 2\,NO + 3\,H_2O \qquad \text{Balanced for H and N and O}$$

3. *Report coefficients to their smallest whole-number values.* The equation for the reaction of ammonia with oxygen is now balanced, but it is common to use whole-number coefficients in balanced equations. Therefore, multiply all the coefficients by 2 to get the final balanced equation.

$$4\,NH_3 + 5\,O_2 \longrightarrow 4\,NO + 6\,H_2O \qquad \text{Balanced}$$

During the trial-and-error process for balancing, you may arrive at an equation that has coefficients that need to be reduced. If you had arrived at the balanced equation

$$8\,NH_3 + 10\,O_2 \longrightarrow 8\,NO + 12\,H_2O$$

it would be necessary to divide by a common divisor to give the smallest whole-number values. Dividing all the coefficients by 2 would result in the correct balanced equation.

4. *Check your answer by making sure that the numbers and kinds of atoms are the same on both sides of the equation.*

$$\underset{\substack{\text{4 N, 12 H, and 10 O}\\\text{atoms on this side}}}{\diagdown} \qquad \underset{\substack{\text{4 N, 12 H, and 10 O}\\\text{atoms on this side}}}{\diagup}$$
$$4\,NH_3 + 5\,O_2 \longrightarrow 4\,NO + 6\,H_2O$$

Let's work through some additional examples.

Conceptual WORKED EXAMPLE 3.1

Visualizing Atoms and Molecules in a Chemical Reaction

Write a balanced equation for the reaction of element A (red spheres) with element B (blue spheres) as represented below:

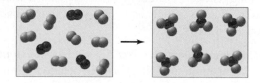

STRATEGY

Balancing the reactions shown in this molecular representation is just a matter of counting the numbers of reactant and product formula units. In this example, the reactant box contains three red A_2 molecules and nine blue B_2 molecules, whereas the product box contains six AB_3 molecules with no reactant left over.

SOLUTION

$$3\,A_2 + 9\,B_2 \longrightarrow 6\,AB_3 \quad \text{dividing by 3 reduces the equation to } A_2 + 3\,B_2 \longrightarrow 2\,AB_3$$

CHECK

In any balanced equation, the numbers and kinds of atoms must be the same on both sides.

$$\underset{\substack{\text{2 A and 6 B atoms}\\\text{on this side}}}{\diagdown} \qquad \underset{\substack{\text{2 A and 6 B atoms}\\\text{on this side}}}{\diagup}$$
$$A_2 + 3\,B_2 \longrightarrow 2\,AB_3$$

▶ **Conceptual** PRACTICE 3.1   Write a balanced equation for the reaction of element A (red spheres) with element B (green spheres) as represented below:

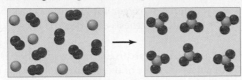

▶ **Conceptual** APPLY 3.2   Draw a molecular representation of reactants and products for the following reaction, which plays a significant role in atmospheric pollution. Nitrogen monoxide reacts with oxygen to produce nitrogen dioxide. Represent each atom as a sphere labeled with the elemental symbol N or O. Make sure that your drawing represents a *balanced* chemical reaction.

●─ WORKED EXAMPLE 3.2

Balancing a Chemical Equation

Propane, C_3H_8, is a colorless, odorless gas often used as a heating and cooking fuel in campers and rural homes. Write a balanced equation for the combustion reaction of propane with oxygen to yield carbon dioxide and water.

STRATEGY AND SOLUTION

Follow the four steps described in the text:

Step 1. Write the unbalanced equation using correct chemical formulas for all substances:

$$C_3H_8 + O_2 \longrightarrow CO_2 + H_2O \qquad \text{Unbalanced}$$

Step 2. Find coefficients to balance the equation. Begin by balancing the elements that appear in two species; in this reaction, these are C and H. Look at the unbalanced equation, and note that there are 3 carbon atoms on the left side of the equation but only 1 on the right side. If we add a coefficient of 3 to CO_2 on the right, the carbons balance:

$$C_3H_8 + O_2 \longrightarrow 3\,CO_2 + H_2O \qquad \text{Balanced for C}$$

Next, look at the number of hydrogen atoms. There are 8 hydrogens on the left but only 2 (in H_2O) on the right. By adding a coefficient of 4 to the H_2O on the right, the hydrogens balance:

$$C_3H_8 + O_2 \longrightarrow 3\,CO_2 + 4\,H_2O \qquad \text{Balanced for C and H}$$

Find the coefficient for O_2 last as oxygen is not combined with other elements. Look at the number of oxygen atoms. There are 2 on the left but 10 on the right. By adding a coefficient of 5 to the O_2 on the left, the oxygens balance:

$$C_3H_8 + 5\,O_2 \longrightarrow 3\,CO_2 + 4\,H_2O \qquad \text{Balanced for C, H, and O}$$

Step 3. Make sure that the coefficients are reduced to their smallest whole-number values. In fact, our answer is already correct, but we might have arrived at a different answer through trial and error:

$$2\,C_3H_8 + 10\,O_2 \longrightarrow 6\,CO_2 + 8\,H_2O$$

Although the preceding equation is balanced, the coefficients are not the smallest whole numbers. It would be necessary to divide all coefficients by 2 to reach the final equation. A coefficient of 1 is never written but is implied if no other coefficient is given.

$$C_3H_8 + 5\,O_2 \longrightarrow 3\,CO_2 + 4\,H_2O$$

Step 4. Check the answer by counting the numbers and kinds of atoms on both sides of the equation to make sure that they're the same:

3 C, 8 H, and 10 O atoms on this side | 3 C, 8 H, and 10 O atoms on this side

$$C_3H_8 + 5\,O_2 \longrightarrow 3\,CO_2 + 4\,H_2O$$

continued on next page

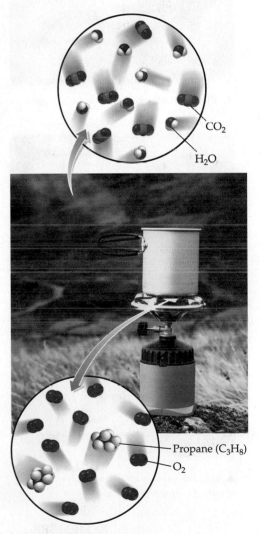

CO_2

H_2O

Propane (C_3H_8)

O_2

▲ Propane is used as a fuel in camp stoves and rural homes.

▲ Violent reaction of potassium chlorate with table sugar.

▶ **PRACTICE 3.3** Balance the following equations:
(a) $C_6H_{12}O_6 \longrightarrow C_2H_6O + CO_2$ (fermentation of sugar to yield ethyl alcohol)
(b) $NaClO_3 \longrightarrow NaCl + O_2$ (source of oxygen for breathing masks in airliners)
(c) $NH_3 + Cl_2 \longrightarrow N_2H_4 + NH_4Cl$ (synthesis of hydrazine for rocket fuel)

▶ **APPLY 3.4** The major ingredient in ordinary safety matches is potassium chlorate, $KClO_3$, a substance that can act as a source of oxygen in combustion reactions. Its reaction with ordinary table sugar (sucrose, $C_{12}H_{22}O_{11}$), for example, occurs violently to yield potassium chloride, carbon dioxide, and water. Write a balanced equation for the reaction.

3.3 ▶ CHEMICAL ARITHMETIC: STOICHIOMETRY

Imagine a laboratory experiment—perhaps the reaction of ethylene, C_2H_4, with hydrogen chloride, HCl, to prepare ethyl chloride, C_2H_5Cl, a colorless, low-boiling liquid that doctors and athletic trainers use as a spray-on anesthetic for minor injuries. You might note that in writing this and other equations, the designations (g) for gas, (l) for liquid, (s) for solid, and (aq) for aqueous solutions are often appended to the symbols of reactants and products to show their physical state. We'll do this frequently from now on.

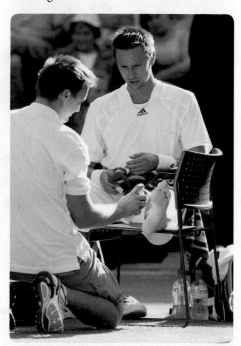

$C_2H_4(g)$ + $HCl(g)$ ⟶ $C_2H_5Cl(l)$
Ethylene   Hydrogen chloride   Ethyl chloride (an anesthetic)

How much ethylene and how much hydrogen chloride should you use for your experiment? According to the coefficients of the balafnced equation, a 1:1 numerical ratio of the two reactants is needed. But because you can't count the reactant molecules, you have to weigh them. That is, you must convert a *number* ratio of reactant molecules, as given by coefficients in the balanced equation, into a *mass* ratio to be sure that you are using the right amounts.

Mass ratios are determined by using the *molecular weights* of the substances involved in a reaction. Just as the **unified atomic mass unit (u)** of an element is the average mass of the element's *atoms* (Section 2.9), the **molecular weight** of a substance is the average mass of the substance's *molecules*. Numerically, molecular weight (or, more generally, **formula weight** to include both ionic and molecular substances) equals the sum of the atomic weights of all atoms in the molecule.

▲ Ethyl chloride is often used as a spray-on anesthetic for athletic injuries.

REMEMBER...

The atomic weight of an element is the weighted average mass of the element's naturally occurring isotopes. Although atomic weight is usually written as dimensionless, the unit of atomic weight is the **unified atomic mass unit (u)**. (Section 2.9)

> **Molecular weight** Sum of atomic weights of all atoms in a molecule.
> **Formula weight** Sum of atomic weights of all atoms in a formula unit of any compound, molecular or ionic.

As examples, the molecular weight of ethylene is 28.0, the molecular weight of hydrogen chloride is 36.5, and the molecular weight of ethyl chloride is 64.5. (These numbers are rounded off to one decimal place for convenience; the actual values are known more precisely.)

For ethylene, C_2H_4:

Atomic weight of 2 C	$= (2)(12.0) = 24.0$
Atomic weight of 4 H	$= (4)(1.0) = 4.0$
Molecular weight of C_2H_4	$= 28.0$

For hydrogen chloride, HCl:

Atomic weight of H	$= 1.0$
Atomic weight of Cl	$= 35.5$
Molecular weight of HCl	$= 36.5$

For ethyl chloride, C_2H_5Cl:

Atomic weight of 2 C	$= (2)(12.0) = 24.0$
Atomic weight of 5 H	$= (5)(1.0) = 5.0$
Atomic weight of Cl	$= 35.5$
Molecular weight of C_2H_5Cl	$= 64.5$

How do we use molecular weights? We saw in Section 2.9 that one **mole** of any element is the amount whose mass in grams, or *molar mass*, is numerically equal to the element's atomic weight. In the same way, one mole of any chemical compound is the amount whose mass in grams is numerically equal to the compound's molecular weight (or formula weight) and contains **Avogadro's number** of formula units (6.022×10^{23}). Thus, 1 mol of ethylene has a mass of 28.0 g, 1 mol of HCl has a mass of 36.5 g, and 1 mol of C_2H_5Cl has a mass of 64.5 g.

Mol. wt. of HCl = 36.5	Molar mass of HCl = 36.5 g/mol	1 mol of HCl = 6.022×10^{23} HCl molecules
Mol. wt. of C_2H_4 = 28.0	Molar mass of C_2H_4 = 28.0 g/mol	1 mol of C_2H_4 = 6.022×10^{23} C_2H_4 molecules
Mol. wt. of C_2H_5Cl = 64.5	Molar mass of C_2H_5Cl = 64.5 g/mol	1 mol of C_2H_5Cl = 6.022×10^{23} C_2H_5Cl molecules

WORKED EXAMPLE 3.3

Calculating a Molecular Weight

What is the molecular weight of glucose ($C_6H_{12}O_6$), and what is its molar mass in grams per mole?

IDENTIFY

Known	Unknown
Chemical formula of glucose ($C_6H_{12}O_6$)	Molecular weight and molar mass of glucose

STRATEGY

The molecular weight of a substance is the sum of the atomic weights of the constituent atoms. List the elements present in the molecule, and look up the atomic weight of each (we'll round off to one decimal place for convenience):

$$C (12.0) \quad H (1.0) \quad O (16.0)$$

Then, multiply the atomic weight of each element by the number of times that element appears in the chemical formula, and total the results.

SOLUTION

$$
\begin{aligned}
C_6\, (6 \times 12.0) &= 72.0 \\
H_{12}\, (12 \times 1.0) &= 12.0 \\
O_6\, (6 \times 16.0) &= 96.0 \\
\hline
\text{Mol. wt. of } C_6H_{12}O_6 &= 180.0
\end{aligned}
$$

Because one *molecule* of glucose has a mass of 180.0 u, 1 *mol* of glucose has a mass of 180.0 g. Thus, the molar mass of glucose is 180.0 g/mol.

▶ **PRACTICE 3.5** Calculate the formula weight or molecular weight of the following substances:

(a) Fe_2O_3 (rust)
(b) H_2SO_4 (sulfuric acid)
(c) $C_6H_8O_7$ (citric acid)
(d) $C_{16}H_{18}N_2O_4S$ (penicillin G)

▶ **Conceptual APPLY 3.6** Use the structural formula of sucrose to determine its molecular weight and molar mass in grams per mole. (C = gray, O = red, H = ivory.)

REMEMBER...

The **mole** is the fundamental SI unit for measuring the amount of matter. One mole of any substance—atom, ion, or molecule—is the amount whose mass in grams is numerically equal to the substance's atomic or formula weight. One mole contains **Avogadro's number** (6.022×10^{23}) of formula units. (Section 2.9)

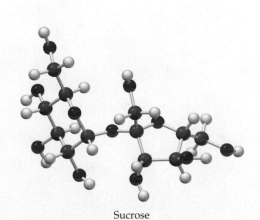

Sucrose

WORKED EXAMPLE 3.4

Interconverting Mass and Moles

How many moles of glucose, which is used to treat low blood sugar, are in a tablet containing 2.00 g? (The molar mass of glucose, $C_6H_{12}O_6$, was calculated in Worked Example 3.3.)

IDENTIFY

Known	Unknown
Mass of glucose (2.00 g)	Moles of glucose

STRATEGY

The known quantity (grams) can be converted to the unknown quantity (moles) using the molar mass of glucose as a conversion factor. Set up an equation so that the unwanted unit cancels.

SOLUTION

$$2.00 \text{ g glucose} \times \frac{1 \text{ mol glucose}}{180.0 \text{ g glucose}} = 0.0111 \text{ mol glucose}$$

$$= 1.11 \times 10^{-2} \text{ mol glucose}$$

CHECK

Because the molar mass of glucose is 180.0 g/mol, 1 mol of glucose has a mass of 180.0 g. Thus, 2.00 g of glucose is a bit more than one-hundredth of a mole, or 0.01 mol. The estimate agrees with the detailed solution.

▶ **PRACTICE 3.7** How many moles are in 5.26 g of $NaHCO_3$, the main ingredient in Alka-Seltzer tablets?

▶ **APPLY 3.8** When a diabetic experiences low blood glucose, possibly due to an excess of insulin or increased levels of exercise, the treatment is consumption of glucose tablets.

(a) How many grams of glucose are in the recommended amount for treatment of an adult, 0.0833 mol glucose?
(b) A typical tablet contains 3.75 g of glucose. How many tablets should be eaten?
(c) How many molecules of glucose are in 0.0833 mol?

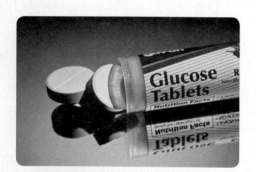

In any balanced chemical equation, the coefficients tell the number of formula units, and thus the number of moles, of each substance in the reaction. You can then use molar masses as conversion factors to calculate reactant masses. If you saw the following balanced equation for the industrial synthesis of ammonia, for instance, you would know that 3 mol of $H_2(g)$ (3 mol × 2.0 g/mol = 6.0 g) is needed for reaction with 1 mol of $N_2(g)$ (28.0 g) to yield 2 mol of $NH_3(g)$ (2 mol × 17.0 g/mol = 34.0 g).

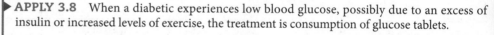

This number of moles of hydrogen...	...reacts with this number of moles of nitrogen...	...to yield this number of moles of ammonia.

$$3 \text{ H}_2(g) + 1 \text{ N}_2(g) \longrightarrow 2 \text{ NH}_3(g)$$

In referring to the chemical arithmetic needed for mole–mass conversions, we use the word **stoichiometry** (stoy-key-*ahm*-uh-tree; from the Greek *stoicheion*, "element," and *metron*, "measure"). Let's look again at the reaction of ethylene with HCl and assume that we have 15.0 g of ethylene and need to know how many grams of HCl to use in the reaction.

$$C_2H_4(g) + HCl(g) \longrightarrow C_2H_5Cl(l)$$

According to the coefficients in the balanced equation, 1 molecule of HCl reacts with molecule of ethylene, so 1 mol of HCl is needed for reaction with each mole of ethylene. To find out how many grams of HCl are needed to react with 15.0 g of ethylene, we first have to find out how many moles of ethylene are in 15.0 g. We do this gram-to-mole conversion by calculating the molar mass of ethylene and using that value as a conversion factor:

Mol. wt. of $C_2H_4 = (2 \times 12.0) + (4 \times 1.0) = 28.0$

Molar mass of $C_2H_4 = 28.0$ g/mol

Moles of $C_2H_4 = 15.0 \text{ g ethylene} \times \dfrac{1 \text{ mol ethylene}}{28.0 \text{ g ethylene}} = 0.536 \text{ mol ethylene}$

Now that we know how many moles of ethylene we have (0.536 mol), we also know from the balanced equation how many moles of HCl we need (0.536 mol), and we have to do a mole-to-gram conversion to find the mass of HCl required. Once again, the conversion is done by calculating the molar mass of HCl and using that value as a conversion factor:

Mol. wt. of HCl $= 1.0 + 35.5 = 36.5$

Molar mass of HCl $= 36.5$ g/mol

Grams of HCl $= 0.536 \text{ mol } C_2H_4 \times \dfrac{1 \text{ mol HCl}}{1 \text{ mol } C_2H_4} \times \dfrac{36.5 \text{ g HCl}}{1 \text{ mol HCl}} = 19.6 \text{ g HCl}$

Thus, 19.6 g of HCl is needed to react with 15.0 g of ethylene.

Look carefully at the sequence of steps in the calculation just completed. *Moles* (numbers of molecules) are given by the coefficients in the balanced equation, but *grams* are used to weigh reactants in the laboratory. Moles tell us *how many molecules* of each reactant are needed, whereas grams tell us *how much mass* of each reactant is needed.

Moles $\longrightarrow$ Numbers of molecules or formula units

Grams $\longrightarrow$ Mass

The flow diagram in **FIGURE 3.1** illustrates the necessary conversions. Note again that you can't go directly from the number of grams of one reactant to the number of grams of another reactant. You *must* first convert to moles.

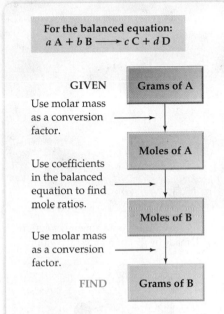

For the balanced equation:
$a\,\text{A} + b\,\text{B} \longrightarrow c\,\text{C} + d\,\text{D}$

GIVEN Use molar mass as a conversion factor.	**Grams of A**
Use coefficients in the balanced equation to find mole ratios.	**Moles of A**
Use molar mass as a conversion factor.	**Moles of B**
FIND	**Grams of B**

▲ **FIGURE 3.1**

Stoichiometry: Conversions between moles and grams for a chemical reaction. The numbers of moles tell how many molecules of each reactant are needed, as given by the coefficients of the balanced equation; the numbers of grams tell what mass of each reactant is needed.

Figure It Out

Why does a gram-to-mole conversion need to be done when relating the mass of one reactant to another?

Answer: Relative amounts of reactants and products for a reaction can be found from coefficients in the balanced equation. These coefficients represent moles or number of atoms/molecules, not mass.

WORKED EXAMPLE 3.5

Relating the Masses of Reactants and Products

Aqueous solutions of sodium hypochlorite (NaOCl), best known as household bleach, are prepared by reaction of sodium hydroxide with chlorine. How many grams of NaOH are needed to react with 25.0 g of Cl_2?

$$2\,\text{NaOH}(aq) + Cl_2(g) \longrightarrow \text{NaOCl}(aq) + \text{NaCl}(aq) + H_2O(l)$$

IDENTIFY

Known	Unknown
Mass of Cl_2 (25.0 g)	Mass of NaOH (g)
Balanced reaction	

STRATEGY

The goal is to relate the known amount of one reactant (Cl_2) with the other reactant (NaOH). Finding the relationships between quantities of reactants requires working in moles and using the balanced equation to relate amounts. Molar masses are used to interconvert between moles and grams. Use the general strategy outlined in Figure 3.1.

SOLUTION

Step 1. Convert grams of Cl_2 to moles of Cl_2. This gram-to-mole conversion is done in the usual way, using the molar mass of Cl_2 (70.9 g/mol) as the conversion factor:

$$25.0 \text{ g } Cl_2 \times \dfrac{1 \text{ mol } Cl_2}{70.9 \text{ g } Cl_2} = 0.353 \text{ mol } Cl_2$$

continued on next page

▲ Household bleach is an aqueous solution of NaOCl, made by reaction of NaOH with Cl_2.

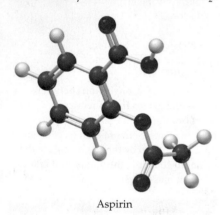

Aspirin

Step 2. Convert moles of Cl_2 to moles of NaOH. The coefficients in the balanced equation show that each mole of Cl_2 reacts with 2 mol of NaOH.

$$0.353 \text{ mol } Cl_2 \times \frac{2 \text{ mol NaOH}}{1 \text{ mol } Cl_2} = 0.706 \text{ mol NaOH}$$

Step 3. Convert moles of NaOH to grams of NaOH. Carry out a mole-to-gram conversion using the molar mass of NaOH (40.0 g/mol) as a conversion factor to find that 28.2 g of NaOH is required for the reaction:

$$0.706 \text{ mol NaOH} \times \frac{40.0 \text{ g NaOH}}{1 \text{ mol NaOH}} = 28.2 \text{ g NaOH}$$

The problem can also be worked by combining the steps and setting up one large equation:

$$\text{Grams of NaOH} = 25.0 \text{ g } Cl_2 \times \frac{1 \text{ mol } Cl_2}{70.9 \text{ g } Cl_2} \times \frac{2 \text{ mol NaOH}}{1 \text{ mol } Cl_2} \times \frac{40.0 \text{ g NaOH}}{1 \text{ mol NaOH}}$$

$$= 28.2 \text{ g NaOH}$$

CHECK

The molar mass of NaOH is about half that of Cl_2, and 2 mol of NaOH is needed per 1 mol of Cl_2. Thus, the needed mass of NaOH will be similar to that of Cl_2, or about 25 g.

▶ **PRACTICE 3.9** Aspirin is prepared by reaction of salicylic acid ($C_7H_6O_3$) with acetic anhydride ($C_4H_6O_3$) according to the following equation:

$$C_7H_6O_3(s) + C_4H_6O_3(l) \longrightarrow C_9H_8O_4(s) + CH_3CO_2H(l)$$

Salicylic   Acetic   Aspirin   Acetic acid
acid   anhydride

How many grams of acetic anhydride are needed to react with 4.50 g of salicylic acid?

▶ **APPLY 3.10** Refer to the balanced reaction for the synthesis of aspirin in Problem 3.9.

(a) How many grams of salicylic acid are needed to make 10.0 g of aspirin?
(b) How many grams of acetic acid are formed as a by-product when 10.0 g of aspirin are synthesized?

3.4 ▶ YIELDS OF CHEMICAL REACTIONS

In the stoichiometry examples worked out in the preceding section, we made the unstated assumption that all reactions "go to completion." That is, we assumed that all reactant molecules are converted to products. In fact, few reactions behave so nicely. More often, a large majority of molecules react as expected, but other processes, or *side reactions*, also occur. Thus, the amount of product actually formed, called the **yield** of the reaction, is usually less than the amount predicted by calculations.

The amount of product actually formed in a reaction divided by the amount theoretically possible and multiplied by 100% is the reaction's **percent yield**. For example, if a given reaction *could* provide 6.9 g of a product according to its stoichiometry but actually provides only 4.7 g, then its percent yield is 4.7/6.9 × 100% = 68%.

REMEMBER...

When performing a series of calculations in the laboratory, all the significant figures should be retained until the final answer is rounded to the correct number of significant figures (Section 1.10). However, in the multi-step problems in the remainder of this book, we will round the number at each step to save space and to keep the focus on the new concept in the calculation.

$$\text{Percent yield} = \frac{\text{Actual yield of product}}{\text{Theoretical yield of product}} \times 100\%$$

Worked Example 3.6 shows how to calculate and use percent yield.

WORKED EXAMPLE 3.6

Calculating Percent Yield

Methyl *tert*-butyl ether (MTBE, $C_5H_{12}O$), a gasoline additive now being phased out in many places because of health concerns, can be made by reaction of isobutylene (C_4H_8) with methanol (CH_4O). What is the percent yield of the reaction if 32.8 g of MTBE is obtained from reaction of 26.3 g of isobutylene with sufficient methanol?

$$C_4H_8(g) + CH_4O(l) \longrightarrow C_5H_{12}O(l)$$
Isobutylene   Methyl *tert*-butyl ether (MTBE)

Methyl *tert*-butyl ether

IDENTIFY

Known	Unknown
Balanced reaction	Percent yield of reaction
Mass of isobutylene (26.3 g)	
Methanol (sufficient amount to react with isobutylene)	
Mass of MTBE (32.8 g) (actual yield of product)	

STRATEGY

To calculate the unknown quantity (percent yield), the actual yield and the theoretical yield must be known. Since the actual yield (32.8 g MTBE) is given in the problem, we need to calculate the theoretical yield, which is the amount of MTBE that could be produced from the complete reaction of 26.3 g of isobutylene.

Step 1. Calculate the molar masses of reactants and products to use as conversion factors.

Step 2. Find the theoretical amount of product, MTBE, using the coefficients from the balanced equation and molar masses as conversion factors.

Step 3. Use the equation for percent yield.

SOLUTION

Step 1. Calculation of Molar Masses

Isobutylene, C_4H_8:
Mol. wt. = $(4 \times 12.0) + (8 \times 1.0) = 56.0$
Molar mass of isobutylene = 56.0 g/mol

MTBE, $C_5H_{12}O$:
Mol. wt. = $(5 \times 12.0) + (12 \times 1.0) + 16.0 = 88.0$
Molar mass of MTBE = 88.0 g/mol

Step 2. Theoretical Amount of Product
To calculate the amount of MTBE that could theoretically be produced from 26.3 g of isobutylene, we first have to find the number of moles of reactant, using molar mass as the conversion factor:

$$26.3 \text{ g isobutylene} \times \frac{1 \text{ mol isobutylene}}{56.0 \text{ g isobutylene}} = 0.470 \text{ mol isobutylene}$$

According to the balanced equation, 1 mol of product is produced per mole of reactant, so we know that 0.470 mol of isobutylene can theoretically yield 0.470 mol of MTBE. Finding the mass of this MTBE requires a mole-to-mass conversion:

$$0.470 \text{ mol isobutylene} \times \frac{1 \text{ mol MTBE}}{1 \text{ mol isobutylene}} \times \frac{88.0 \text{ g MTBE}}{1 \text{ mol MTBE}}$$
$$= 41.4 \text{ g MTBE}$$

Step 3. Percent Yield Equation
Dividing the actual amount by the theoretical amount and multiplying by 100% gives the percent yield:

$$\frac{32.8 \text{ g MTBE}}{41.4 \text{ g MTBE}} \times 100\% = 79.2\%$$

CHECK

It is difficult to estimate a magnitude of the number for percent yield as the calculation involves multiple steps. However, the value for percent yield must be between 0% and 100%; therefore, the answer of 79.2% is in the correct range.

▶ **PRACTICE 3.11** Ethyl alcohol is prepared industrially by the reaction of ethylene, C_2H_4, with water. What is the percent yield of the reaction if 4.6 g of ethylene gives 4.7 g of ethyl alcohol?

$$C_2H_4(g) + H_2O(l) \longrightarrow C_2H_6O(l)$$
Ethylene   Ethyl alcohol

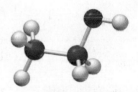

Ethyl alcohol

▶ **APPLY 3.12**

(a) Diethyl ether ($C_4H_{10}O$), the "ether" used medically as an anesthetic, is prepared commercially by treatment of ethyl alcohol (C_2H_6O) with an acid. How many grams of diethyl ether would you obtain from 40.0 g of ethyl alcohol if the percent yield of the reaction is 87.0%?

$$2 C_2H_6O(l) \xrightarrow{\text{Acid}} C_4H_{10}O(l) + H_2O(l)$$
Ethyl alcohol   Diethyl ether

(b) How many grams of ethyl alcohol would be needed to produce 100.0 g of diethyl ether if the percent yield of reaction is 87.0%?

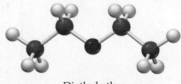

Diethyl ether

3.5 ▶ REACTIONS WITH LIMITING AMOUNTS OF REACTANTS

Because chemists usually write balanced equations, it's easy to get the impression that reactions are always carried out using exactly the right proportions of reactants. In fact, this is often not the case. Many reactions are carried out using an excess amount of one reactant—more than is actually needed according to stoichiometry. Look, for instance, at the industrial synthesis of ethylene glycol, $C_2H_6O_2$, a substance used both as automobile antifreeze and as a starting material for the preparation of polyester polymers. Approximately 18 million metric tons of ethylene glycol are prepared each year worldwide by reaction of ethylene oxide, C_2H_4O, with water at high temperature (1 metric ton = 1000 kg = 2205 lb).

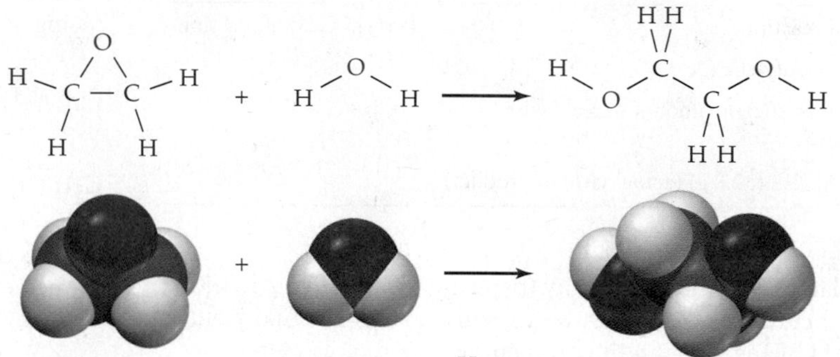

| C_2H_4O (g) | + | H_2O (l) | Heat→ | $C_2H_6O_2$ (l) |
| Ethylene oxide | | Water | | Ethylene glycol |

Because water is so cheap and so abundant, it doesn't make sense to worry about using exactly 1 mol of water for each mole of ethylene oxide. Rather, it's much easier to use an excess of water to be certain that enough is present to entirely consume the more valuable ethylene oxide reactant. Of course, when an excess of water is present, only the amount required by stoichiometry undergoes reaction. The excess water does not react and remains unchanged.

Whenever the ratios of reactant molecules used in an experiment are different from those given by the coefficients of the balanced equation, a surplus of one reactant is left over after the reaction is finished. Thus, the extent to which a chemical reaction takes place depends on the reactant that is present in limiting amount—the **limiting reactant**. The other reactant is said to be the *excess reactant*.

The situation with excess reactants and limiting reactants is analogous to what sometimes happens with people and chairs. If there are five people in a room but only three chairs, then only three people can sit while the other two stand because the number of people sitting is limited by the number of available chairs. The chairs are analogous to the limiting reactant, whereas the people are the excess reactant. Worked Example 3.7 shows how to visualize a limiting reactant problem.

Conceptual WORKED EXAMPLE 3.7

Identifying a Limiting Reactant from a Molecular Representation

Examine the balanced reaction for the production of ethylene glycol from ethylene oxide and water and the graphical molecular representation shown below.

$$C_2H_4O \ + \ H_2O \ \xrightarrow{Heat} \ C_2H_6O_2$$
Ethylene oxide   Water   Ethylene glycol

Identify the limiting and the excess reactant.
How many molecules of excess reactant are left over after the reaction occurs?
How many molecules of product can be made?

RATEGY

nt the numbers of reactant and product molecules and use coefficients from the balanced
tion to relate them to one another.

OLUTION

Count the number of each type of molecule in the box on the reactant side of the equation.
There are 3 ethylene oxide molecules and 5 water molecules. According to the balanced
equation the stoichiometry between the reactants is 1:1. Therefore, 5 ethylene oxide molecules would be needed to react with 5 water molecules. Since there are only 3 ethylene
oxide molecules, it is the limiting reactant, and water is in excess.
Count the number of water molecules on the product side of the equation. There are
2 water molecules that have not reacted, and water is called the excess reactant.
Count the number of ethylene glycol molecules on the product side of the equation. There
are 3 ethylene glycol molecules present.

Therefore, the reaction of 3 ethylene oxide molecules with 5 water molecules results in
ylene glycol molecules with 2 water molecules left over.

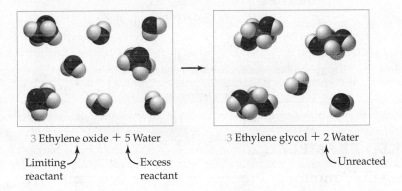

3 Ethylene oxide + 5 Water   3 Ethylene glycol + 2 Water

Limiting   Excess   Unreacted
reactant   reactant

Conceptual PRACTICE 3.13 The following diagram represents the reaction of A (red
eres) with B_2 (blue spheres):

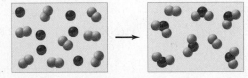

Write a balanced equation for the reaction.
Identify the limiting and excess reactant.
How many molecules of product are made?

Conceptual APPLY 3.14 Draw a diagram similar to the one shown in Problem 3.13 for
following reaction, when 8 molecules of AB react with 6 molecules of B_2. Represent each
n as a sphere labeled with the symbol A or B. Specify the limiting and excess reactant.

$$2 AB + B_2 \longrightarrow 2 AB_2$$

FIGURE 3.2 is a flow chart that summarizes the steps needed for identifying a limiting reactant and calculating the theoretical mass of product formed when specified amounts of reactants are mixed. The process involves calculating the amount of product that can be made if each reactant is completely used up. The limiting reactant can be determined by comparing the amount of product formed from each reactant. The limiting reactant will form the *lowest* amount of product, which is the *theoretical yield*. Just as in previous stoichiometry examples, amounts of reactants and products are related from the balanced equation and number of moles. Worked Example 3.8 shows how to determine the limiting reactant and how to calculate the amount of product and excess reactant.

▶ **FIGURE 3.2**

Steps for identifying a limiting reactant and calculating theoretical yield.

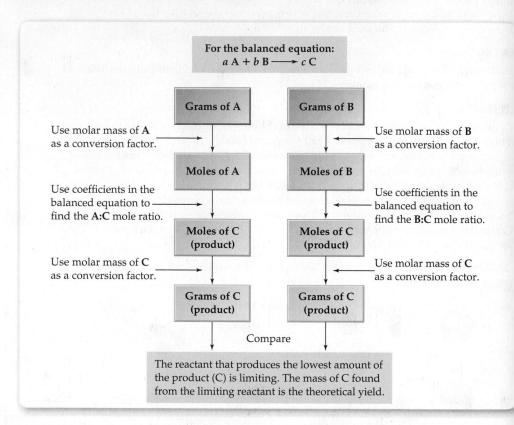

For the balanced equation:
$$a\,A + b\,B \longrightarrow c\,C$$

Grams of A

Use molar mass of **A** as a conversion factor.

Moles of A

Use coefficients in the balanced equation to find the **A:C** mole ratio.

Moles of C (product)

Use molar mass of **C** as a conversion factor.

Grams of C (product)

Grams of B

Use molar mass of **B** as a conversion factor.

Moles of B

Use coefficients in the balanced equation to find the **B:C** mole ratio.

Moles of C (product)

Use molar mass of **C** as a conversion factor.

Grams of C (product)

Compare

The reactant that produces the lowest amount of the product (C) is limiting. The mass of C found from the limiting reactant is the theoretical yield.

● WORKED EXAMPLE 3.8

Calculating the Amount of Product or Excess Reactant when One Reactant Is Limiting

Cisplatin, an anticancer agent used for the treatment of solid tumors, is prepared by the reaction of ammonia with potassium tetrachloroplatinate. Assume that 10.0 g of K_2PtCl_4 and 10.0 g of NH_3 are allowed to react.

$$\underset{\text{Potassium tetrachloroplatinate}}{K_2PtCl_4(aq)} + 2\,NH_3(aq) \longrightarrow \underset{\text{Cisplatin}}{Pt(NH_3)_2Cl_2(s)} + 2\,KCl(aq)$$

(a) Which reactant is limiting, and which is in excess? How many grams of cisplatin are formed?
(b) How many grams of the excess reactant are consumed, and how many grams remain?

IDENTIFY

Known	Unknown
Balanced reaction	Limiting reactant
Mass of NH_3 (10.0 g)	Mass of $Pt(NH_3)_2Cl_2$ (g)
Mass of K_2PtCl_4 (10.0 g)	Mass of excess reactant consumed (g)
	Mass of excess reactant left over (g)

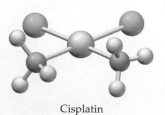

Cisplatin

STRATEGY

(a) Because the amount of each reactant is known, the procedure outlined in Figure 3.2 can be used to determine which reactant is limiting. Use stoichiometry to calculate the amount of product that can be made from the amount of each reactant given. The reactant that results in the least amount of product is the limiting reactant and determines the maximum amount of product that can be made.

(b) Stoichiometry can also be used to relate the amount of limiting reactant to the amount of excess reactant.

SOLUTION

(a) Step 1. Calculation of Molar Masses
Finding the molar amounts of reactants always begins by calculating formula weight and using molar masses as conversion factors:

Form. weight of $K_2PtCl_4 = (2 \times 39.1) + 195.1$
$$+ (4 \times 35.5) = 415.3$$

Molar mass of $K_2PtCl_4 = 415.3$ g/mol

Mol. wt. of $NH_3 = 14.0$ amu $+ (3 \times 1.0) = 17.0$
Molar mass of $NH_3 = 17.0$ g/mol

Mol. wt. of $Pt(NH_3)_2Cl_2 = 195.1 + (2 \times 17.0)$
$$+ (2 \times 35.5) = 300.1$$

Molar mass of $Pt(NH_3)_2Cl_2 = 300.1$ g/mol

Step 2. Amount of Product Made from Each Reactant
Taking the given amount of *each* reactant, use stoichiometry to find the amount of product that can be made if the entire amount reacted.

$$10.0 \text{ g } K_2PtCl_4 \times \frac{1 \text{ mol } K_2PtCl_4}{415.3 \text{ g } K_2PtCl_4} \times \frac{1 \text{ mol } Pt(NH_3)_2Cl_2}{1 \text{ mol } K_2PtCl_4}$$

$$\times \frac{300.1 \text{ g } Pt(NH_3)_2Cl_2}{1 \text{ mol } Pt(NH_3)_2Cl_2} = 7.23 \text{ g } Pt(NH_3)_2Cl_2$$

$$10.0 \text{ g } NH_3 \times \frac{1 \text{ mol } NH_3}{17.0 \text{ g } NH_3} \times \frac{1 \text{ mol } Pt(NH_3)_2Cl_2}{2 \text{ mol } NH_3}$$

$$\times \frac{300.1 \text{ g } Pt(NH_3)_2Cl_2}{1 \text{ mol } Pt(NH_3)_2Cl_2} = 88.3 \text{ g } Pt(NH_3)_2Cl_2$$

These calculations tell us that K_2PtCl_4 is the limiting reactant because it produces the fewest grams of product. The excess reactant is NH_3 because its consumption would produce a larger

amount of product. Only 7.23 g of $Pt(NH_3)_2Cl_2$ can be produced, given the initial amount of K_2PtCl_4.

(b) With the identities of the excess reactant and limiting reactant known, we can use stoichiometry to find out how much NH_3 reacts and how much is left over. We know that all the limiting reactant (K_2PtCl_4) is used up; stoichiometry allows us to calculate the amount of NH_3 that reacts as follows:

$$10.0 \text{ g } K_2PtCl_4 \times \frac{1 \text{ mol } K_2PtCl_4}{415.3 \text{ g } K_2PtCl_4} \times \frac{2 \text{ mol } NH_3}{1 \text{ mol } K_2PtCl_4}$$

$$\times \frac{17.0 \text{ g } NH_3}{1 \text{ mol } NH_3} = 0.819 \text{ g } NH_3$$

Grams of unreacted $NH_3 = (10.0 \text{ g} - 0.819 \text{ g}) = 9.2 \text{ g } NH_3$

CHECK

It is reasonable that K_2PtCl_4 (415.3 g/mol) is the limiting reactant because it has a much higher molar mass than NH_3 (17.0 g/mol) and the reaction started with an equal number of grams of each. The amount of product, $Pt(NH_3)_2Cl_2$, can be estimated from reaction stoichiometry with the limiting reactant and relative molar masses. Because 1 mol of K_2PtCl_4 will produce 1 mol of $Pt(NH_3)_2Cl_2$ and the molar mass of $Pt(NH_3)_2Cl_2$ (300.1 g/mol) is about $\frac{3}{4}$ of the molar mass of K_2PtCl_4 (415.3 g/mol), then the amount produced (7.23 g) should be about $\frac{3}{4}$ the mass of the initial amount (10.0 g).

▶ **PRACTICE 3.15** Lithium oxide is used aboard the space shuttle to remove water from the air supply.

(a) If 80.0 kg of water is to be removed and 65 kg of Li_2O is available, which reactant is limiting?
(b) How many kilograms of the excess reactant remain?
(c) How many kilograms of the product (LiOH) will be produced?

$$Li_2O(s) + H_2O(g) \longrightarrow 2 \text{ LiOH}(s)$$

▶ **APPLY 3.16** After lithium hydroxide is produced aboard the space shuttle by reaction of Li_2O with H_2O (Problem 3.15), it is used to remove exhaled carbon dioxide from the air supply. Initially 400.0 g of LiOH were present and 500.0 g of $LiHCO_3$ have been produced. Can the reaction remove any additional CO_2 from the air? If so, how much?

$$LiOH(s) + CO_2(g) \longrightarrow LiHCO_3(s)$$

3.6 ▶ PERCENT COMPOSITION AND EMPIRICAL FORMULAS

All the substances we've dealt with thus far have had known formulas. When a new compound is made in the laboratory or found in nature, however, its formula must be experimentally determined.

Determining the formula of a new compound begins with analyzing the substance to discover what elements it contains and how much of each element is present—that is, to find its *composition*. The **percent composition** of a compound is expressed by identifying the elements present and giving the mass percent of each. For example, we might express the percent composition of a certain colorless liquid found in gasoline by saying that it contains 84.1% carbon and 15.9% hydrogen by mass. In other words, a 100.0 g sample of the compound contains 84.1 g of carbon atoms and 15.9 g of hydrogen atoms.

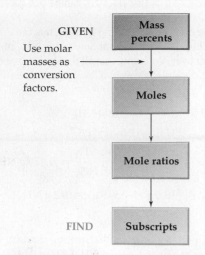

GIVEN

Use molar masses as conversion factors.

Mass percents

↓

Moles

↓

Mole ratios

↓

FIND

Subscripts

▲ **FIGURE 3.3**

Calculating the formula of a compound from its percent composition.

Figure It Out

A mole ratio is the number of moles of one element divided by the number of moles of another element. How do mole ratios relate to subscripts in a chemical formula?

Answer: The subscripts in a chemical formula represent moles. For example in 1 mole of C_2H_4, there are two moles of C atoms and four moles of hydrogen atoms. The mole ratio is the same as the ratio of subscripts.

Knowing a compound's percent composition makes it possible to calculate the compound's chemical formula. As shown in **FIGURE 3.3**, the strategy is to find the relative number of moles of each element in the compound and then use those numbers to establish the mole ratios of the elements. The mole ratios, in turn, correspond to the subscripts in the chemical formula.

Let's use for our example the colorless liquid whose composition is 84.1% carbon and 15.9% hydrogen by mass. Arbitrarily taking 100 g of the substance to make the calculation easier, we find by using molar masses as conversion factors that the 100 g contains:

$$84.1 \text{ g C} \times \frac{1 \text{ mol C}}{12.01 \text{ g C}} = 7.00 \text{ mol C}$$

$$15.9 \text{ g H} \times \frac{1 \text{ mol H}}{1.008 \text{ g H}} = 15.8 \text{ mol H}$$

With the relative numbers of moles of C and H known, we find the mole ratio by dividing both by the smaller number (7.00):

$$C_{\left(\frac{7.00}{7.00}\right)} H_{\left(\frac{15.8}{7.00}\right)} = C_1 H_{2.26}$$

The C:H mole ratio of 1:2.26 means that we can write $C_1H_{2.26}$ as a temporary formula for the liquid. Multiplying the subscripts by small integers in a trial-and-error procedure until whole numbers are found then gives the **empirical formula**, which gives the smallest whole-number ratios of atoms in the compound. In the present instance, we need to multiply the subscripts by 4 to obtain the empirical formula C_4H_9. (The subscripts may not always be *exact* integers because of small errors in the data, but the discrepancies should be small.)

$$C_{(1 \times 4)} H_{(2.26 \times 4)} = C_4 H_{9.04} = C_4 H_9$$

An empirical formula determined from percent composition tells only the *ratios* of atoms in a compound. The **molecular formula**, which tells the *actual numbers* of atoms in a molecule, can be either the same as the empirical formula or a multiple of it. To determine the molecular formula, it's necessary to know the molecular weight of the substance. In the present instance, the molecular weight of our compound (octane) is 114.2, which is a simple multiple of the empirical molecular weight for C_4H_9 (57.1).

To find the multiple, divide the molecular weight by the empirical formula weight:

$$\text{Multiple} = \frac{\text{Molecular weight}}{\text{Empirical formula weight}} = \frac{114.2}{57.1} = 2.00$$

Then multiply the subscripts in the empirical formula by this multiple to obtain the molecular formula. In our example, the molecular formula of octane is $C_{(4 \times 2)} H_{(9 \times 2)}$, or C_8H_{18}.

Just as we can find the empirical formula of a substance from its percent composition, we can also find the percent composition of a substance from its empirical (or molecular) formula. The strategies for the two kinds of calculations are exactly opposite. Aspirin, for example, has the molecular formula $C_9H_8O_4$ and thus has a C:H:O mole ratio of 9:8:4. We can convert this mole ratio into a mass ratio, and thus into percent composition, by carrying out mole-to-gram conversions.

Let's assume we start with 1 mol of compound to simplify the calculation:

$$1 \text{ mol aspirin} \times \frac{9 \text{ mol C}}{1 \text{ mol aspirin}} \times \frac{12.0 \text{ g C}}{1 \text{ mol C}} = 108 \text{ g C}$$

$$1 \text{ mol aspirin} \times \frac{8 \text{ mol H}}{1 \text{ mol aspirin}} \times \frac{1.01 \text{ g H}}{1 \text{ mol H}} = 8.08 \text{ g H}$$

$$1 \text{ mol aspirin} \times \frac{4 \text{ mol O}}{1 \text{ mol aspirin}} \times \frac{16.0 \text{ g O}}{1 \text{ mol O}} = 64.0 \text{ g O}$$

Dividing the mass of each element by the total mass and multiplying by 100% then gives the percent composition:

$$\text{Total mass of 1 mol aspirin} = 108 \text{ g} + 8.08 \text{ g} + 64.0 \text{ g} = 180 \text{ g}$$

$$\% \ C = \frac{108 \ g \ C}{180 \ g} \times 100\% = 60.0\%$$

$$\% \ H = \frac{8.08 \ g \ H}{180 \ g} \times 100\% = 4.49\%$$

$$\% \ O = \frac{64.0 \ g \ O}{180 \ g} \times 100\% = 35.6\%$$

The answer can be checked by confirming that the sum of the mass percentages is within a rounding error of 100%: 60.0% + 4.49% + 35.6% = 100.1%.

Worked Example 3.10 further illustrates conversions between percent composition and empirical formulas.

WORKED EXAMPLE 3.9

Calculating Empirical Formulas, Molecular Formulas, and Percent Composition

(a) Vitamin C (ascorbic acid) contains 40.92% C, 4.58% H, and 54.50% O by mass. What is the empirical formula of ascorbic acid?

Conceptual **(b)** Use the structure of ascorbic acid to determine the molecular formula for ascorbic acid. (Gray = C, red = O, ivory = H.) By what integer should the empirical formula be multiplied to convert it to the molecular formula?

(a) **Empirical formula of ascorbic acid:**

IDENTIFY

Known	Unknown
Percent composition of ascorbic acid	Empirical formula of ascorbic acid
(40.92% C, 4.58% H, and 54.50% O by mass)	

STRATEGY

Figure 3.3 outlines the procedure for converting from percent composition to empirical formula. Assume 100.0 g of ascorbic acid, convert grams of each element to moles, and divide by the smallest number of moles to obtain mole ratios for subscripts.

SOLUTION

(a) **Step 1.** Per Figure 3.3, our first step is to convert the mass of each element in the sample to moles.

$$40.92 \ g \ C \times \frac{1 \ mol \ C}{12.0 \ g \ C} = 3.41 \ mol \ C$$

$$4.58 \ g \ H \times \frac{1 \ mol \ H}{1.01 \ g \ H} = 4.53 \ mol \ H$$

$$54.50 \ g \ O \times \frac{1 \ mol \ O}{16.0 \ g \ O} = 3.41 \ mol \ O$$

Step 2. Find mole ratios. Dividing each of the three numbers by the smallest (3.41 mol) gives a C:H:O mole ratio of 1:1.33:1 and a temporary formula of $C_1H_{1.33}O_1$.

Step 3. Determine what whole-number subscripts belong in the empirical formula. Multiplying the subscripts by small integers, such as 1, 2, 3, or 4, in a trial-and-error procedure until whole numbers are found gives the empirical formula: $C_{(3 \times 1)}H_{(3 \times 1.33)}O_{(3 \times 1)} = C_3H_4O_3$.

(b) **To find the integer multiple that converts the empirical formula into the molecular formula:**

IDENTIFY

Known	Unknown
Empirical formula from part **(a)**	Molecular formula
Ball-and-stick model for ascorbic acid	

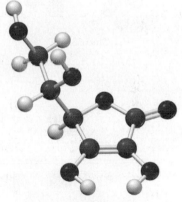

Ascorbic acid

continued on next page

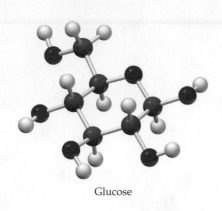

Glucose

STRATEGY

Find the molecular formula by counting atoms in the structural formula provided. Determine the multiplication factor needed to turn subscripts in the empirical formula into the molecular formula.

SOLUTION

Counting the atoms in the structural formula for ascorbic acid gives a molecular formula of $C_6H_8O_6$. If the subscripts in the empirical formula weight $(C_3H_4O_3)$ are multiplied by 2, then the molecular formula is obtained.

CHECK

The empirical formula can be multiplied by a whole number to obtain the molecular formula. This is strong evidence that the empirical formula has been determined correctly.

▶ **PRACTICE 3.17**   What is the empirical formula of the ingredient in Bufferin tablets that has the percent composition 14.25% C, 56.93% O, and 28.83% Mg by mass?

▶ **PRACTICE 3.18**   Dimethylhydrazine, a colorless liquid used as a rocket fuel, is 40.0% C, 13.3% H, and 46.7% N. What is the empirical formula? What is the molecular formula if the molecular weight is 60.0 g/mol?

▶ **Conceptual APPLY 3.19**   Use the structural formula for glucose to determine the molecular formula. What is the empirical formula, and what is the percent composition of each atom in glucose?

3.7 ▶ DETERMINING EMPIRICAL FORMULAS: ELEMENTAL ANALYSIS

One of the most common methods used to determine percent composition and empirical formulas, particularly for organic compounds containing carbon and hydrogen, is *combustion analysis*. In this method, a compound of unknown composition is burned with oxygen to produce the volatile combustion products CO_2 and H_2O, which are separated and have their amounts determined by an automated instrument. Methane (CH_4), for instance, burns according to the balanced equation:

$$CH_4(g) + 2\,O_2(g) \longrightarrow CO_2(g) + 2\,H_2O(g)$$

With the amounts of the carbon-containing product (CO_2) and hydrogen-containing product (H_2O) established, the strategy is to calculate the number of moles of carbon and hydrogen in the products, from which we can find the C:H mole ratio of the starting compound. This information, in turn, provides the chemical formula, as outlined by the flow diagram in **FIGURE 3.4**.

As an example of how combustion analysis works, imagine that we have a sample of a pure substance—say, naphthalene, which is often used for household moth balls. We weigh a known amount of the sample, burn it in pure oxygen, and then analyze the products. Let's say that 0.330 g of naphthalene reacts with O_2 and that 1.133 g of CO_2 and 0.185 g of H_2O are formed. The first thing to find out is the number of moles of carbon and hydrogen in the CO_2 and H_2O products so that we can calculate the number of moles of each element originally present in the naphthalene sample.

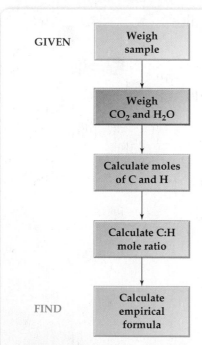

GIVEN

Weigh sample

Weigh CO_2 and H_2O

Calculate moles of C and H

Calculate C:H mole ratio

FIND

Calculate empirical formula

▲ **FIGURE 3.4**

Determining an empirical formula from combustion analysis of a compound containing C and H.

Figure It Out

Generally explain how the mass of CO_2 and H_2O produced during combustion analysis is useful in determining the empirical formula of the hydrocarbon.

Answer: All of the CO_2 comes from carbon in the compound. Similarly, all of the H_2O comes from hydrogen in the original compound. Therefore, the mass of these compounds can be used to find the number of moles of C and H in the original compound.

$$\text{Moles of C in 1.133 g } CO_2 = 1.133 \text{ g } CO_2 \times \frac{1 \text{ mol } CO_2}{44.01 \text{ g } CO_2} \times \frac{1 \text{ mol C}}{1 \text{ mol } CO_2}$$

$$= 0.02574 \text{ mol C}$$

$$\text{Moles of H in 0.185 g } H_2O = 0.185 \text{ g } H_2O \times \frac{1 \text{ mol } H_2O}{18.02 \text{ g } H_2O} \times \frac{2 \text{ mol H}}{1 \text{ mol } H_2O}$$

$$= 0.0205 \text{ mol H}$$

Although it's not necessary in this instance because naphthalene contains only carbon and hydrogen, we can make sure that all the mass is accounted for and that no other elements

re present. To do so, we carry out mole-to-gram conversions to find the number of grams of
C and H in the starting sample:

$$\text{Mass of C} = 0.02574 \ \text{mol C} \times \frac{12.01 \ \text{g C}}{1 \ \text{mol C}} = 0.3091 \ \text{g C}$$

$$\text{Mass of H} = 0.0205 \ \text{mol H} \times \frac{1.01 \ \text{g H}}{1 \ \text{mol H}} = 0.0207 \ \text{g H}$$

$$\text{Total mass of C and H} = 0.3091 \ \text{g} + 0.0207 \ \text{g} = 0.3298 \ \text{g}$$

Because the total mass of the C and H in the products (0.3298 g) is the same as the mass
of the starting sample (0.330 g), we know that no other elements are present in naphthalene.

With the relative number of moles of C and H in naphthalene known, divide the larger
number of moles by the smaller number to get the formula $C_{1.26}H_1$:

$$C_{\left(\frac{0.02574}{0.0205}\right)}H_{\left(\frac{0.0205}{0.0205}\right)} = C_{1.26}H_1$$

Then multiply the subscripts by small integers in a trial-and-error procedure until whole
numbers are found to obtain the whole-number formula C_5H_4:

Multiply subscripts by 2: $C_{(1.26 \times 2)}H_{(1 \times 2)} = C_{2.52}H_2$

Multiply subscripts by 3: $C_{(1.26 \times 3)}H_{(1 \times 3)} = C_{3.78}H_3$

Multiply subscripts by 4: $C_{(1.26 \times 4)}H_{(1 \times 4)} = C_{5.04}H_4 = C_5H_4$ (Both subscripts are integers)

Elemental analysis provides only an empirical formula. To determine the molecular
formula, it's also necessary to know the substance's molecular weight. The next section will
show one common method for experimentally measuring molecular weight, mass spectrometry. In the present problem, the molecular weight of naphthalene is 128.2, or twice
the empirical formula weight of C_5H_4 (64.1). Thus, the molecular formula of naphthalene is
$C_{(2 \times 5)}H_{(2 \times 4)} = C_{10}H_8$.

Worked Example 3.11 shows an example of combustion analysis when the sample contains oxygen in addition to carbon and hydrogen.

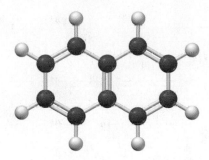

Naphthalene

WORKED EXAMPLE 3.10

Calculating an Empirical Formula and a Molecular Formula from a Combustion Analysis

Caproic acid, the substance responsible for the aroma of goats, dirty socks, and old shoes,
contains carbon, hydrogen, and oxygen. On combustion analysis, a 0.450 g sample of
caproic acid gives 0.418 g of H_2O and 1.023 g of CO_2. What is the empirical formula
of caproic acid? If the molecular weight of caproic acid is 116.2, what is the molecular
formula?

IDENTIFY

Known	Unknown
Mass of caproic acid (0.450 g)	Empirical formula
Mass of H_2O (0.418 g) and CO_2 (1.023 g)	Molecular formula
Mol. wt. of caproic acid (116.2)	

STRATEGY

Use the procedure outlined in Figure 3.4 to turn combustion analysis data into an empirical formula. This molecule also contains oxygen, and because oxygen yields no combustion
products, its presence in a molecule can't be directly detected by combustion analysis. Rather,
the presence of oxygen must be inferred by subtracting the calculated masses of C and H from
the total mass of the sample. (See Steps 2 and 3 illustrating how the amount of oxygen can be
determined.)

continued on next page

Caproic acid

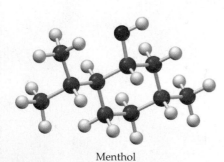

Menthol

SOLUTION

Step 1. Find the molar amounts of C and H in the sample:

$$\text{Moles of C} = 1.023 \text{ g } CO_2 \times \frac{1 \text{ mol } CO_2}{44.01 \text{ g } CO_2} \times \frac{1 \text{ mol C}}{1 \text{ mol } CO_2} = 0.023\,24 \text{ mol C}$$

$$\text{Moles of H} = 0.418 \text{ g } H_2O \times \frac{1 \text{ mol } H_2O}{18.02 \text{ g } H_2O} \times \frac{2 \text{ mol H}}{1 \text{ mol } H_2O} = 0.0469 \text{ mol H}$$

Step 2. Find the number of grams of each element in the sample:
Start with C and H as these can be determined from the number of moles found in Step 1.

$$\text{Mass of C} = 0.023\,24 \text{ mol C} \times \frac{12.01 \text{ g C}}{1 \text{ mol C}} = 0.2791 \text{ g C}$$

$$\text{Mass of H} = 0.0464 \text{ mol H} \times \frac{1.01 \text{ g H}}{1 \text{ mol H}} = 0.0469 \text{ g H}$$

Subtracting the masses of C and H from the mass of the starting sample indicates that 0.124 g is unaccounted for:

$$0.450 \text{ g} - (0.2791 \text{ g} + 0.0469 \text{ g}) = 0.124 \text{ g}$$

Step 3. Find the moles of oxygen:
Because we are told that oxygen is also present in the sample, the "missing" mass must be due to oxygen, which can't be detected by combustion. We therefore need to find the number of moles of oxygen in the sample:

$$\text{Moles of O} = 0.124 \text{ g } O \times \frac{1 \text{ mol O}}{16.00 \text{ g } O} = 0.007\,75 \text{ mol O}$$

Step 4. Find the mole ratios of the elements:
Knowing the relative numbers of moles of all three elements, C, H, and O, we divide the three numbers of moles by the smallest number (0.007 75 mol of oxygen) to arrive at a C:H:O ratio of 3:6:1.

$$C_{\left(\frac{0.02324}{0.00775}\right)}H_{\left(\frac{0.0464}{0.00775}\right)}O_{\left(\frac{0.00775}{0.00775}\right)} = C_3H_6O$$

Step 5. Find the molecular formula:
The empirical formula of caproic acid is, therefore, C_3H_6O, and the empirical formula weight is 58.1. Because the molecular weight of caproic acid is 116.2, or twice the empirical formula weight, the molecular formula of caproic acid must be $C_{(2\times3)}H_{(2\times6)}O_{(2\times1)} = C_6H_{12}O_2$.

CHECK

If a simple empirical formula is obtained and the molecular weight is a whole-number multiple of the empirical formula weight, then the formulas have most likely been correctly determined.

▶ **PRACTICE 3.20** Menthol, a flavoring agent obtained from peppermint oil, contains carbon, hydrogen, and oxygen. On combustion analysis, 1.00 g of menthol yields 1.161 g of H_2O and 2.818 g of CO_2. What is the empirical formula of menthol? Check your answer with the structural formula provided.

▶ **PRACTICE 3.21** Combustion analysis is performed on 0.50 g of a hydrocarbon, and 1.55 g of CO_2 and 0.697 g of H_2O are produced. What is the empirical formula of the hydrocarbon? If the molar mass is 142.0 g/mol, what is the molecular formula?

▶ **APPLY 3.22**
(a) Polychlorinated biphenyls (PCBs) were compounds used as coolants in transformers and capacitors, but their production was banned by the U.S. Congress in 1979 because they are highly toxic and persist in the environment. When 1.0 g of a PCB containing carbon, hydrogen, and chlorine was subjected to combustion analysis, 1.617 g of CO_2 and 0.138 g of H_2O were produced. What is the empirical formula?
(b) If the molecular weight is 326.26, what is the molecular formula?
(c) Can combustion analysis be used to determine the empirical formula of a compound containing carbon, hydrogen, oxygen, and chlorine?

3.8 ▶ DETERMINING MOLECULAR WEIGHTS: MASS SPECTROMETRY

As we saw in the previous section, determining a compound's molecular formula requires knowledge of its molecular weight. But how is molecular weight determined?

The most common method of determining both atomic and molecular weights is with an instrument called a *mass spectrometer*. More than 20 different kinds of mass spectrometers are commercially available, depending on the intended application, but the electron-impact, magnetic-sector instrument, shown in **FIGURE 3.5a**, is particularly common. In this instrument, the sample is vaporized and injected as a dilute gas into an evacuated chamber, where it is bombarded with a beam of high-energy electrons. The electron beam knocks other electrons from the sample molecules, which become positively charged ions. A reaction for ionization of a molecule (M) is shown below:

$$M(g) + e^-_{\text{high energy}} \longrightarrow M^+(g) + 2\,e^-$$

Ionization is necessary as electric and magnetic fields will only exert a force on a charged species, not a neutral molecule. Some of these ionized molecules survive, and others fragment

(a) A mass spectrometer

The resulting ions are passed between the poles of a magnet and are deflected according to their mass-to-charge ratio.

Magnet

Slits

Accelerating grid

The deflected ions pass through a slit into a detector assembly.

Heated filament

Slit

Detector

Molecules are **ionized** by collision with a high-energy electron beam.

Sample inlet

To vacuum pump

Electron beam

(b) A mass spectrum for naphthalene

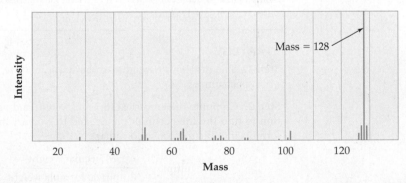

Intensity

Mass = 128

20   40   60   80   100   120

Mass

▲ **FIGURE 3.5**

Mass spectrometry. **(a)** Schematic illustration of an electron-impact, magnetic-sector mass spectrometer. **(b)** A mass spectrum of naphthalene, molecular weight = 128, showing peaks of different masses on the horizontal axis.

Figure It Out

Compare two ions with a +1 charge. How will the path of a light ion differ from the path of a heavy ion in the curved tube?

Answer: The light ion will make sharper turn than the heavier ion because a force from the magnetic field will be able to influence the path of the lighter ion to a greater degree.

into smaller ions. Collisions with the electron beam have sufficient energy to not only ionize the molecule but also break bonds. The various ions of different masses are then accelerated by an electric field and passed between the poles of a strong magnet, which deflects them through a curved, evacuated pipe.

The radius of deflection of a charged ion, M^+, as it passes between the magnet poles depends on its mass, with lighter ions deflected more strongly than heavier ones. This is similar to vehicles rounding a turn at high speed. A small, light race car will be able to make a tighter turn than a heavy semitruck. By varying the strength of the magnetic field, it's possible to focus ions of different masses through a slit at the end of the curved pipe and onto a detector assembly. The mass spectrum that results is plotted as a graph of intensity versus ion mass.

Although a typical mass spectrum contains ions of many different masses, the heaviest ion is generally due to the ionized molecule itself, the so-called molecular ion. By measuring the mass of this molecular ion, the molecular weight of the molecule can be determined. Naphthalene, for example, gives rise to an intense peak at mass 128 in its spectrum, consistent with a molecular formula of $C_{10}H_8$ (**FIGURE 3.5b**). There is a small peak at mass 129 in the spectrum that arises from the presence of the carbon-13 **isotope** in a naturally occurring sample of napthalene, $^{13}C_{10}\,^1H_8$. The intensity is lower because the abundance of carbon-13 is low, approximately 1%. Most naturally occurring carbon is the isotope carbon-12.

Modern mass spectrometers are so precise that molecular weights can often be measured to seven significant figures. A $^{12}C_{10}\,^1H_8$ molecule of naphthalene has a molecular weight of 128.0626 as measured by mass spectrometry. High mass accuracy is often needed to make an identification of a compound. For example, two compounds with different molecular formulas can have very similar masses, $C_5H_8O = 84.0570$ and $C_6H_{12} = 84.0934$. Highly accurate mass measurements are frequently used to confirm the identity of molecules synthesized in laboratories, such as natural products used as pharmaceuticals. A confirmation of the correct molecular formula and structure of a drug is of utmost importance.

> **REMEMBER...**
>
> Atoms with identical atomic numbers but with different mass numbers are called **isotopes**. Carbon has several isotopes, of which only ^{12}C and ^{13}C are stable. (Section 2.8)

WORKED EXAMPLE 3.11

Determination of Molecular Formula from Combustion Analysis and a Mass Spectrum

A compound has an empirical formula of CH as determined from combustion analysis. The mass spectrum for the compound is shown in **FIGURE 3.6**. What is the molecular formula?

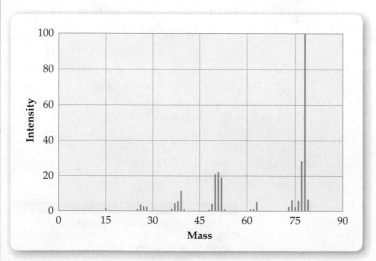

▲ **FIGURE 3.6**
The mass spectrum for a hydrocarbon.

IDENTIFY

Known	Unknown
Empirical formula (CH)	Molecular formula
Mass spectrum (used to find molecular weight)	

STRATEGY

Step 1. The molecular weight can be found by interpreting the mass spectrum.

Step 2. Compute the empirical formula weight and use the equation to find the integer multiple needed to convert the empirical formula into the molecular formula.

$$\text{Multiple} = \frac{\text{Molecular weight}}{\text{Empirical formula weight}}$$

SOLUTION

Step 1. The molecular weight is 78 as determined from the most intense peak with the highest mass. The peak at 79 has the greatest mass but a very low intensity. This peak is most likely due to the abundance of carbon-13 isotope in the natural sample.

Step 2. The empirical formula weight of CH is 13. The whole-number multiple for converting the empirical formula into the molecular formula can be found by substituting values into the equation.

$$\text{Multiple} = \frac{\text{Molecular weight}}{\text{Empirical formula weight}} = \frac{78}{13} = 6$$

The molecular formula can be found by multiplying subscripts of the empirical formula by 6:

$$C_{(1\times6)}H_{(1\times6)} = C_6H_6$$

▶ **PRACTICE 3.23**   A compound has an empirical formula of C_6H_5 as determined from combustion analysis. The mass spectrum for the compound is shown in **FIGURE 3.7**. What is the molecular formula?

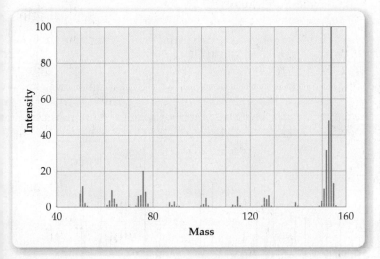

▲ **FIGURE 3.7**

The mass spectrum of a compound with empirical formula C_6H_5.

▶ **APPLY 3.24**   Combustion analysis was performed on 1.00 g of a compound containing C, H, and N, and 2.79 g of CO_2 and 0.57 g of H_2O were produced. Given the mass spectrum for the compound in **FIGURE 3.8**, what is the empirical formula? What is the molecular formula?

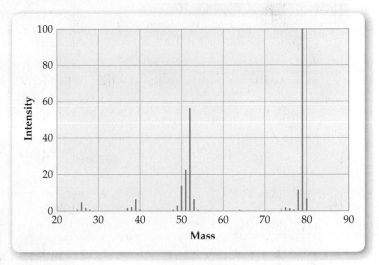

▲ **FIGURE 3.8**

The mass spectrum for a compound containing C, H, and N.

After viewing the mass spectra for several compounds, you may have begun to notice that different compounds have a unique pattern of masses with specific intensities. The mass spectrum is sometimes referred to as a "fingerprint" and can be used to identify an unknown substance, just as an actual fingerprint can identify a person. Popular crime solving shows on TV often show laboratory technicians performing an analysis to identify an unknown substance using mass spectrometry. Electron-impact ionization produces highly reproducible mass spectra, and a database containing thousands of compounds is available for comparison. The mass spectrum for the unknown can be compared to the mass spectrum of standards found in the database. Identification occurs when the mass of peaks and their relative intensity in the sample match a known substance. **FIGURE 3.9** shows the mass spectra for two illegal drugs.

▶ **FIGURE 3.9**

Electron-impact ionization mass spectra.
Two illicit drugs, cocaine and heroin, have
different mass spectra, which can be used
in identifying unknown substances.

Figure It Out

How are fragment ions useful in identifying
an unknown compound using its mass
spectrum?

Answer: Fragment ion masses give multiple points of
matching for a given compound, making it highly improbable
that another compound would have exactly the same ions.

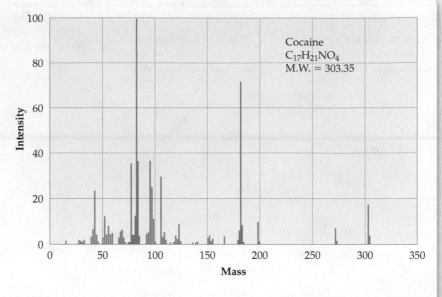

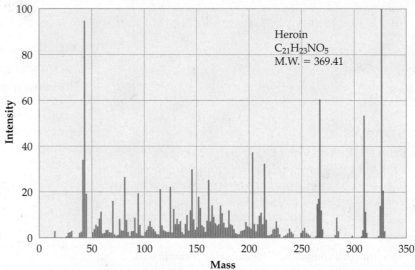

NQUIRY ▶▶▶ CAN ALTERNATIVE FUELS DECREASE CO$_2$ EMISSIONS?

Imagine that you are buying a new car and would like to learn more about the environmental benefits of alternative fuels. The U.S. Department of Energy provides information on several types of alternative fuels including biodiesel, electric, ethanol (E85), compressed natural gas (CNG), liquefied natural gas (LNG), hydrogen, and liquefied petroleum gas (propane). One important criterion is that the fuel must be readily available; the four alternative fuels that are easiest to obtain are electric, propane, ethanol, and CNG. Energy independence and sustainability are other important factors and all four alternative fuels are domestically produced, but only ethanol and a small fraction of natural gas is renewable. Other factors include cost of the fuel, cost of the vehicle, safety, driving distance, and emissions. For this example, we will evaluate only *the contribution to greenhouse gas emissions, specifically carbon dioxide*. **Greenhouse gases** are molecules that absorb heat radiation emitted from the earth and cause warming effect on climate. Carbon dioxide is a greenhouse gas of particular concern because its concentration in the atmosphere has increased approximately 40% since the start of the Industrial Revolution as a result of burning fossil fuels.

ELECTRIC VEHICLES

Electric vehicles are sold by many automakers; examples include the Chevy Volt and the Nissan Leaf. Electric cars are powered by batteries and have zero tailpipe emissions. However, the batteries are charged by plugging into a source of electricity, and emissions may be produced by the source of electrical power, typically a power plant. Most of the electricity in the United States is produced by steam turbine generators at power plants from primary resources such as coal, natural gas, and nuclear energy. Electricity generated from fossil fuels adds CO$_2$ to the atmosphere. According to the U.S. Energy Information Administration, in 2011, 42% of the nation's electricity was generated by coal, 25% by natural gas, 19% by nuclear energy, and less than 1% by petroleum. About 13% of electricity is produced from renewable sources of energy, including hydropower, biomass, wind, geothermal, and solar power.

PROPANE

Propane, also known as liquefied petroleum gas (LPG), has been used as a vehicle fuel for decades. Gasoline-to-propane conversion kits can be purchased for most cars and trucks, and liquid propane fueling stations are widespread. Propane (C$_3$H$_8$) is stored under pressure inside a tank. As pressure is released, the liquid propane vaporizes and turns into a colorless, odorless gas that is used for combustion. Propane is produced as a by-product of natural gas processing and crude oil refining.

▲ Propane-powered police vehicle

ETHANOL (E85)

Ethanol (C$_2$H$_5$OH) is a renewable fuel made from various plant materials. An important source is the fermentation of corn, which is why ethanol fueling stations are concentrated in the midwestern United States. More recently, research efforts are focusing on breaking down cellulose, a component in the cell wall of green plants and algae, as a source of ethanol. Flexible fuel vehicles (FFVs) have engines that run on either gasoline or E85, a gasoline–ethanol blend containing 51–83% ethanol. More than 70 vehicles are classified as FFVs and most major automakers, including Ford, Chevrolet, Toyota, and Honda, produce them. In fact, more than 95% of U.S. gasoline contains a small percentage of ethanol to oxygenate the fuel and reduce air pollution. The potential of ethanol to become an important vehicle fuel was recognized by Henry Ford and other early automakers, and the first FFV was Ford's Model T produced from 1908 to 1927.

▲ An E85-powered 2009 Chevrolet HHR

continued on next page

continued from previous page

COMPRESSED NATURAL GAS

Many traditional vehicles can be converted to run on compressed natural gas (CNG); the cost of a conversion kit and installation runs from $3000 to $4000. Natural gas, predominantly methane (CH_4), is a domestically produced fuel that has both fossil and renewable sources. Sources of natural gas include wells, extraction during crude oil production, and mining from subsurface porous rock reservoirs in a process called hydraulic fracturing or fracking. Renewable sources are emerging and arise from decaying organic materials such as waste from plants, landfills, wastewater, and livestock. Natural gas is easily transported and stored through existing utility infrastructure as it is commonly supplied to residential and industrial buildings for heating and other energy uses. Because of the gaseous nature of this fuel, it must be stored onboard a vehicle in either a compressed gaseous (CNG) or

a liquefied (LNG) state. To provide adequate driving range, CNG is stored onboard a vehicle in cylinders at a pressure of 3000–360 pounds per square inch.

EVALUATING CARBON DIOXIDE EMISSIONS FROM COMBUSTION

Stoichiometry can be used to calculate and compare the amount of CO_2 produced from combustion of gasoline and alternative fuels. For our comparison, let's calculate the amount of CO_2 emitted by driving a car for 250 miles. A typical sedan has a gas mileage of 25 miles per gallon (mpg) in the city, and therefore the total amount of gasoline is 10.0 gallons. Worked Example 3.9 shows the calculation of CO_2 released by the combustion of gasoline, and Problem 3.25 provides information for calculating CO_2 emissions from alternative fuels (electric, ethanol, propane, and CNG).

●— WORKED EXAMPLE 3.12

Calculating the Amount of Carbon Dioxide Emitted from Combustion of Fuels

Calculate the amount of carbon dioxide (in kilograms) emitted from burning 10.0 gallons of gasoline. Gasoline is composed of hydrocarbons, molecules containing carbon and hydrogen, where the number of carbon atoms ranges from 4 to 12. For ease of calculation, let's assume gasoline is primarily octane (C_8H_{18}). The density of gasoline is 0.74 kg/L and 1 L = 0.2642 gal.

IDENTIFY

Known	Unknown
Volume of gasoline (10.0 gallons)	Mass of CO_2 produced (kg)
Chemical formula for octane (C_8H_{18})	
Combustion of octane is the chemical reaction	
Density of gasoline (0.74 kg/L)	
1 L = 0.2642 gallons	

STRATEGY

The amount of gasoline (a reactant) is known. The unknown (mass of CO_2 product) can be found by writing and balancing a reaction for combustion and using stoichiometry to relate the amount of reactant to product.

Step 1. Write a reaction for the combustion of octane and balance it.

Step 2. Convert the amount of octane from volume in gallons to mass in kilograms.

Step 3. Use stoichiometry to calculate the amount of CO_2 produced.

SOLUTION

Step 1. In combustion, octane reacts with oxygen to produce carbon dioxide and water. The reaction can be balanced by

systematically changing coefficients so that the number of type of atoms is equal on both sides.

$$2\,C_8H_{18}(l) + 25\,O_2(g) \longrightarrow 16\,CO_2(g) + 18\,H_2O(g)$$

Step 2. The mass of octane in 10 gallons of gasoline can be calculated using the conversion factors provided.

$$10.0 \text{ gal} \times \frac{\text{L}}{0.2642 \text{ gal}} \times \frac{0.74 \text{ kg}}{\text{L}} = 28 \text{ kg } C_8H_{18}$$

Step 3. Stoichiometry can be used to find the amount of CO_2 produced from the initial mass of octane.

$$28 \text{ kg } C_8H_{18} \times \frac{1000 \text{ g}}{1 \text{ kg}} \times \frac{1 \text{ mol } C_8H_{18}}{114 \text{ g } C_8H_{18}} \times \frac{16 \text{ mol } CO_2}{2 \text{ mol } C_8H_{18}}$$

$$\times \frac{44.0 \text{ g } CO_2}{1 \text{ mol } CO_2} \times \frac{1 \text{ kg}}{1000 \text{ g}} = 87 \text{ kg } CO_2$$

CHECK

The mass of carbon dioxide is higher than the initial mass of octane because each carbon is now combined with two oxygen in CO_2 instead of approximately two hydrogen in the octane molecule, C_8H_{18}. The mass increases because oxygen has a larger atomic mass than hydrogen.

(Data in the following problem was obtained from the Department of Energy's Office for Alternative Fuels.)

PROBLEM 3.25 Calculate the amount of carbon dioxide (in kilograms) emitted when four alternative fuels are burned to provide the same amount of energy as 10.0 gallons of gasoline. Compare the carbon dioxide emissions from alternative fuels to gasoline (Worked Example 3.12).

(a) **Ethanol (C_2H_5OH)** Although E85 is a blend of ethanol and gasoline, let's use pure ethanol in our calculation for simplification. A gallon of ethanol contains 68% of the energy of a gallon of gas, so 14.7 gallons of ethanol provides the same amount of energy as 10.0 gallons of gasoline. The density of ethanol is 0.79 kg/L and 1 L = 0.2642 gal.

(b) **Liquefied Petroleum Gas/Propane (C_3H_8)** A gallon of propane contains 73% of the energy of a gallon of gas, so 13.7 gallons of propane provides the same amount of energy as 10.0 gallons of gasoline. The density of liquefied propane is 0.49 kg/L and 1 L = 0.2642 gal.

(c) **Compressed Natural Gas (CH_4)** It takes 25.7 kg of natural gas, methane (CH_4), to provide the same amount of energy as 10.0 gallons of gasoline.

(d) **Electricity from a Coal-Burning Power Plant** An electric power plant using bituminous coal produces 0.94 kg CO_2/kWh, and an electric vehicle uses 35 kWh per 100 miles. (A kilowatt-hour, symbolized kWh, is a common unit of electrical energy equivalent to 3.6 megajoules.)

(e) **Electricity from a Natural Gas–Burning Power Plant** An electric power plant using natural gas produces 0.55 kg CO_2/kWh, and an electric vehicle uses 35 kWh per 100 miles.

(f) Which fuel produces the least amount of CO_2 when burned to provide energy for a car: gasoline, ethanol, propane, CNG, electricity from coal, or electricity from natural gas?

LIFE CYCLE ANALYSIS

In determining the impact of alternative fuels on CO_2 emissions, factors other than direct combustion must be considered. *Life cycle analysis* is a technique used to assess the environmental impacts of all stages of a product's life, including raw material extraction, processing, manufacturing, and distribution. A study by Argonne National Laboratory found that when the entire fuel life cycles are considered, using corn-based ethanol instead of gasoline reduces emissions by 19–52%, depending on the source of energy used during ethanol production. Using cellulosic ethanol provides an even greater benefit and could reduce emissions by up to 86%. This large decrease in CO_2 is not due to differences in direct combustion of the two fuels but due to the CO_2 captured when the plant used to make the ethanol is grown. In photosynthesis, plants use carbon dioxide and water to produce glucose and oxygen, ultimately converting CO_2 into organic plant matter. The unbalanced reaction for photosynthesis is shown.

$$CO_2\,(g)\;+\;H_2O\,(l)\;\longrightarrow\;C_6H_{12}O_6\,(s)\;+\;O_2\,(g)$$

PROBLEM 3.26
(a) Balance the reaction for photosynthesis.
(b) If one acre of corn absorbs 1400 lb of carbon dioxide, how many kilograms of glucose are produced during photosynthesis? (1 kg = 2.2 lb)

COAL SCRUBBING

Electric vehicles may not offer much advantage in regions that depend heavily on coal for electricity generation. However, in geographic areas with relatively low-polluting energy plants (such as nuclear or hydroelectric), significantly less carbon dioxide emissions are derived from the use of electric vehicles than from conventional vehicles running on gasoline or diesel. However, the many advantages of coal as a source of energy in electric power plants will likely lead to its continued use in the future. There is a large domestic supply of coal, along with an established infrastructure for mining and transportation, and relatively low cost.

One "cleaner" option for the continued use of coal is to remove or "scrub" CO_2 from the effluent released from power plants. The most promising technology involves the use of amines, compounds containing the ($-NH_2$) group, to bind CO_2 and remove it. To date, this technology has not been implemented to a large extent due to the high cost of installation and operation. Small pilot operations and further research are, however, under way to improve methods and bring down cost. The reaction between carbon dioxide and monoethanolamine is shown:

$$CO_2(g)\;+\;2\,HOCH_2CH_2NH_2(aq)\;\longrightarrow$$
$$HOCH_2CH_2NH_3{}^+(aq)\;+\;HOCH_2CH_2NHCO_2{}^-(aq)$$

PROBLEM 3.27
(a) A large power plant in the United States emits approximately 20 million tons of CO_2 annually. How many tons of monoethanolamine are required to absorb all the CO_2? (1 ton = 907.2 kg)
(b) If 25 million tons of monoethanolamine were used, how many tons of CO_2 would escape to the atmosphere?

PROBLEM 3.28 Would you consider using an alternative fuel? Base your decision on availability of fuel in your region, sustainability, and CO_2 emissions. Explain your reasoning. (More information can be found on the U.S. Department of Energy's Alternative Fuels Data Center website.)

STUDY GUIDE

Section	Concept Summary	Learning Objectives	Test Your Understanding
3.1 and 3.2 ▸ Balancing Chemical Equations	Because mass is neither created nor destroyed in chemical reactions, all chemical equations must be **balanced**—that is, the numbers and kinds of atoms on both sides of the reaction arrow must be the same. A balanced equation tells the number ratio of reactant and product **formula units** in a reaction. The **coefficients** represent the number of moles or number of molecules/atoms of a reactant or product.	**3.1** Visualize bonds broken and formed in a chemical reaction and relate numbers of molecules or atoms to the balanced reaction.	Worked Example 3.1; Problems 3.29, 3.30
		3.2 Balance a chemical reaction, given the formulas of reactants and products.	Worked Example 3.2; Problems 3.38, 3.40
3.3 ▸ Chemical Arithmetic: Stoichiometry	Just as atomic weight is the mass of an atom, **molecular weight** is the mass of a molecule. The analogous term **formula weight** is used for ionic and other nonmolecular substances. Molecular weight is the sum of the atomic masses of all atoms in the molecule. One mole of a substance is the amount whose mass in grams is numerically equal to the substance's molecular or formula mass. Carrying out chemical calculations using mass–mole relationships is called **stoichiometry** and is done using molar masses and ratios of coefficients in the balanced equation as conversion factors.	**3.3** Calculate formula weight, molecular weight, and molar mass, given a chemical formula or structure.	Worked Example 3.3; Problems 3.32, 3.42, 3.44, 3.45
		3.4 Interconvert between mass, moles, and molecules or atoms of a substance.	Worked Example 3.4; Problems 3.46, 3.48, 3.52, 3.54
		3.5 Relate the amount (moles or mass) of reactants and products in a balanced equation using stoichiometry.	Worked Example 3.5; Problems 3.60, 3.62, 3.64
3.4 ▸ Yields of Chemical Reactions	The amount of product actually formed in a reaction—the reaction's **yield**—is often less than the amount theoretically possible. Dividing the actual amount by the theoretical amount and multiplying by 100% gives the reaction's **percent yield**.	**3.6** Calculate percent yield, given amounts of reactants and products.	Worked Example 3.6; Problems 3.11, 3.12
3.5 ▸ Reactions with Limiting Amounts of Reactants	Often, reactions are carried out with an excess of one reactant beyond that called for by the balanced equation. In such cases, the extent to which the reaction takes place depends on the reactant present in limiting amount, the **limiting reactant**.	**3.7** Visualize the relative amounts of atoms or molecules in the reactants and products of a balanced reaction.	Worked Example 3.7; Problem 3.34
		3.8 Determine which reactant is limiting and calculate the theoretical yield of the product and the amount of excess reactant.	Worked Example 3.8; Problems 3.70, 3.72
		3.9 Calculate percent yield when one reactant is limiting.	Problems 3.74–3.81
3.6 and 3.7 ▸ Percent Composition, Empirical Formulas, and Combustion Analysis	The chemical makeup of a substance is described by its **percent composition**—the percentage of the substance's mass due to each of its constituent elements. Elemental analysis is used to calculate a substance's **empirical formula**, which gives the smallest whole-number ratio of atoms of the elements in the compound. To determine the **molecular formula**, which may be a simple multiple of the empirical formula, it's also necessary to know the substance's molecular weight.	**3.10** Calculate the percent composition, given a chemical formula or structure.	Problems 3.33, 3.82, 3.83
		3.11 Determine the empirical and molecular formula, given the mass percent composition and molecular weight of a compound.	Worked Example 3.9; Problems 3.84–3.87, 3.92
		3.12 Determine the empirical and molecular formula, given combustion analysis data and molecular weight.	Worked Example 3.10; Problems 3.88, 3.89, 3.94, 3.95
3.8 ▸ Determining Molecular Weights: Mass Spectrometry	Molecular weights are experimentally measured by mass spectrometry. A mass spectrometer separates ions of different masses by altering the direction they travel using a magnetic field.	**3.13** Determine the molecular weight of a substance, given a mass spectrum. Determine empirical and molecular formula using both mass and combustion analysis data.	Worked Example 3.11; Problem 3.100, 3.101
		3.14 Identify a compound using a molecular weight measured with a high accuracy mass spectrometer.	Problems 3.98, 3.99

KEY TERMS

balanced equation 79
coefficient 79
empirical formula 92
formula unit 79

formula weight 82
greenhouse gases 101
limiting reactant 88

molecular formula 92
molecular weight 82
percent composition 91

percent yield 86
stoichiometry 84
yield 86

KEY EQUATIONS

Molecular and Formula Weight (Section 3.3)

Molecular Weight Sum of atomic weights of all atoms in a molecule.

Formula Weight Sum of atomic weights of all atoms in a formula unit of any compound, molecular or ionic.

Percent yield (Section 3.4)

$$\text{Percent yield} = \frac{\text{Actual yield of product}}{\text{Theoretical yield of product}} \times 100\%$$

CONCEPTUAL PROBLEMS

Problems 3.1–3.28 appear within the chapter.

3.29 The reaction of A (red spheres) with B (blue spheres) is shown in the following diagram:

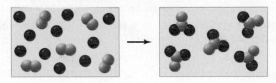

Which equation best describes the stoichiometry of the reaction?
(a) $A_2 + 2B \longrightarrow A_2B_2$ (b) $10A + 5B_2 \longrightarrow 5A_2B_2$
(c) $2A + B_2 \longrightarrow A_2B_2$ (d) $5A + 5B_2 \longrightarrow 5A_2B_2$

3.30 If blue spheres represent nitrogen atoms and red spheres represent oxygen atoms, which box represents reactants and which represents products for the reaction $2NO(g) + O_2(g) \longrightarrow 2NO_2(g)$?

(a)   **(b)**

(c)   **(d)**

3.31 Cytosine, a constituent of deoxyribonucleic acid (DNA), can be represented by the following molecular model. If 0.001 mol of cytosine is submitted to combustion analysis, how many moles of CO_2 and how many moles of H_2O would be formed? (Gray = C, red = O, blue = N, ivory = H.)

Cytosine

3.32 Fluoxetine, marketed as an antidepressant under the name Prozac, can be represented by the following ball-and-stick molecular model. Write the molecular formula for fluoxetine, and calculate its molecular weight (red = O, gray = C, blue = N, yellow-green = F, ivory = H).

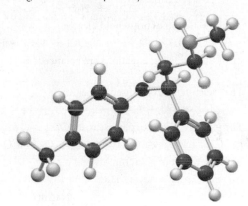

Fluoxetine

3.33 What is the percent composition of cysteine, one of the 20 amino acids commonly found in proteins? (Gray = C, red = O, blue = N, yellow = S, ivory = H.)

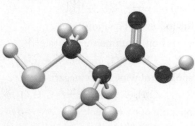

Cysteine

3.34 The following diagram represents the reaction of A_2 (red spheres) with B_2 (blue spheres):

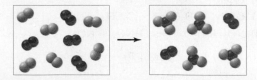

(a) Write a balanced equation for the reaction, and identify the limiting reactant.

(b) How many moles of product can be made from 1.0 mol of A_2 and 1.0 mol of B_2?

3.35 A hydrocarbon of unknown formula C_xH_y was submitted to combustion analysis with the following results. What is the empirical formula of the hydrocarbon?

$= H_2O$ $= CO_2$

SECTION PROBLEMS

Balancing Equations (Section 3.2)

3.36 Which of the following equations are balanced?

(a) The development reaction in silver-halide photography:

$$2\,AgBr + 2\,NaOH + C_6H_6O_2 \longrightarrow$$
$$2\,Ag + H_2O + 2\,NaBr + C_6H_4O_2$$

(b) The preparation of household bleach:

$$2\,NaOH + Cl_2 \longrightarrow NaOCl + NaCl + H_2O$$

3.37 Which of the following equations are balanced? Balance any that need it.

(a) The thermite reaction, used in welding:

$$Al + Fe_2O_3 \longrightarrow Al_2O_3 + Fe$$

(b) The photosynthesis of glucose from CO_2:

$$6\,CO_2 + 6\,H_2O \longrightarrow C_6H_{12}O_6 + 6\,O_2$$

(c) The separation of gold from its ore:

$$Au + 2\,NaCN + O_2 + H_2O \longrightarrow$$
$$NaAu(CN)_2 + 3\,NaOH$$

3.38 Balance the following equations:

(a) $Mg + HNO_3 \longrightarrow H_2 + Mg(NO_3)_2$

(b) $CaC_2 + H_2O \longrightarrow Ca(OH)_2 + C_2H_2$

(c) $S + O_2 \longrightarrow SO_3$

(d) $UO_2 + HF \longrightarrow UF_4 + H_2O$

3.39 Balance the following equations:

(a) The explosion of ammonium nitrate:

$$NH_4NO_3 \longrightarrow N_2 + O_2 + H_2O$$

(b) The spoilage of wine into vinegar:

$$C_2H_6O + O_2 \longrightarrow C_2H_4O_2 + H_2O$$

(c) The burning of rocket fuel:

$$C_2H_8N_2 + N_2O_4 \longrightarrow N_2 + CO_2 + H_2O$$

3.40 Balance the following equations:

(a) $SiCl_4 + H_2O \longrightarrow SiO_2 + HCl$

(b) $P_4O_{10} + H_2O \longrightarrow H_3PO_4$

(c) $CaCN_2 + H_2O \longrightarrow CaCO_3 + NH_3$

(d) $NO_2 + H_2O \longrightarrow HNO_3 + NO$

3.41 Balance the following equations:

(a) $VCl_3 + Na + CO \longrightarrow V(CO)_6 + NaCl$

(b) $RuI_3 + CO + Ag \longrightarrow Ru(CO)_5 + AgI$

(c) $CoS + CO + Cu \longrightarrow Co_2(CO)_8 + Cu_2S$

Molecular Weights and Stoichiometry (Section 3.3)

3.42 What are the molecular (formula) weights of the following substances?

(a) Hg_2Cl_2 (calomel, used at one time as a bowel purgative)

(b) $C_4H_8O_2$ (butyric acid, responsible for the odor of rancid butter

(c) CF_2Cl_2 (a chlorofluorocarbon that destroys the stratospheri ozone layer)

3.43 What are the formulas of the following substances?

(a) $PCl_?$; Mol. wt. = 137.3

(b) Nicotine, $C_{10}H_{14}N_?$; Mol. wt. = 162.2

3.44 What are the molecular weights of the following pharmaceuticals

(a) $C_{33}H_{35}FN_2O_5$ (atorvastatin, lowers blood cholesterol)

(b) $C_{22}H_{27}F_3O_4S$ (fluticasone, anti-inflammatory)

(c) $C_{16}H_{16}ClNO_2S$ (clopidogrel, inhibits blood clots)

3.45 What are the molecular weights of the following herbicides?

(a) $C_6H_6Cl_2O_3$ (2,4-dichlorophenoxyacetic acid, effective o broadleaf plants)

(b) $C_{15}H_{22}ClNO_2$ (metolachlor, pre-emergent herbicide)

(c) $C_8H_6Cl_2O_3$ (dicamba, effective on broadleaf plants)

3.46 How many grams are in a mole of each of the following substances?

(a) Ti   (c) Hg

(b) Br_2 (d) H_2O

3.47 How many moles are in a gram of each of the following substances?

(a) Cr   (c) Au

(b) Cl_2 (d) NH_3

3.48 How many moles of ions are in 27.5 g of $MgCl_2$?

3.49 How many moles of anions are in 35.6 g of AlF_3?

3.50 What is the molecular weight of chloroform if 0.0275 mol weigh 3.28 g?

3.51 What is the molecular weight of cholesterol if 0.5731 mol weigh 221.6 g?

3.52 Iron(II) sulfate, $FeSO_4$, is prescribed for the treatment of anemia How many moles of $FeSO_4$ are present in a standard 300 mg tab let? How many iron(II) ions?

3.53 The "lead" in lead pencils is actually almost pure carbon, and th mass of a period mark made by a lead pencil is about 0.0001 g How many carbon atoms are in the period?

3.54 An average cup of coffee contains about 125 mg of caffeine, $C_8H_{10}N_4O_2$. How many moles of caffeine are in a cup? How many molecules of caffeine?

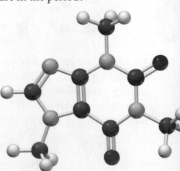

Caffeine

3.55 What is the mass in grams of each of the following samples?

(a) 0.0015 mol of sodium

(b) 0.0015 mol of lead

(c) 0.0015 mol of diazepam (Valium), $C_{16}H_{13}ClN_2O$

3.56 A sample that weighs 25.12 g contains 6.022×10^{23} particles. If 25.00% of the total number of particles are argon atoms and 75.00% are another element, what is the chemical identity of the other constituent?

3.57 A sample that weighs 107.75 g is a mixture of 30% helium atoms and 70% krypton atoms. How many particles are present in the sample?

3.58 Titanium metal is obtained from the mineral rutile, TiO_2. How many kilograms of rutile are needed to produce 100.0 kg of Ti?

3.59 Iron metal can be produced from the mineral hematite, Fe_2O_3, by reaction with carbon. How many kilograms of iron are present in 105 kg of hematite?

3.60 In the preparation of iron from hematite, Fe_2O_3 reacts with carbon:

$$Fe_2O_3 + C \longrightarrow Fe + CO_2 \qquad \text{Unbalanced}$$

(a) Balance the equation.

(b) How many moles of carbon are needed to react with 525 g of hematite?

(c) How many grams of carbon are needed to react with 525 g of hematite?

3.61 An alternative method for preparing pure iron from Fe_2O_3 is by reaction with carbon monoxide:

$$Fe_2O_3 + CO \longrightarrow Fe + CO_2 \qquad \text{Unbalanced}$$

(a) Balance the equation.

(b) How many grams of CO are needed to react with 3.02 g of Fe_2O_3?

(c) How many grams of CO are needed to react with 1.68 mol of Fe_2O_3?

3.62 Magnesium metal burns in oxygen to form magnesium oxide, MgO.

(a) Write a balanced equation for the reaction.

(b) How many grams of oxygen are needed to react with 25.0 g of Mg? How many grams of MgO will result?

(c) How many grams of Mg are needed to react with 25.0 g of O_2? How many grams of MgO will result?

3.63 Ethylene gas, C_2H_4, reacts with water at high temperature to yield ethyl alcohol, C_2H_6O.

(a) How many grams of ethylene are needed to react with 0.133 mol of H_2O? How many grams of ethyl alcohol will result?

(b) How many grams of water are needed to react with 0.371 mol of ethylene? How many grams of ethyl alcohol will result?

3.64 Pure oxygen was first made by heating mercury(II) oxide:

$$HgO \xrightarrow{\text{Heat}} Hg + O_2 \qquad \text{Unbalanced}$$

(a) Balance the equation.

(b) How many grams of mercury and how many grams of oxygen are formed from 45.5 g of HgO?

(c) How many grams of HgO would you need to obtain 33.3 g of O_2?

3.65 Titanium dioxide (TiO_2), the substance used as the pigment in white paint, is prepared industrially by reaction of $TiCl_4$ with O_2 at high temperature:

$$TiCl_4 + O_2 \xrightarrow{\text{Heat}} TiO_2 + 2 Cl_2$$

How many kilograms of TiO_2 can be prepared from 5.60 kg of $TiCl_4$?

3.66 Silver metal reacts with chlorine (Cl_2) to yield silver chloride. If 2.00 g of Ag reacts with 0.657 g of Cl_2, what is the empirical formula of silver chloride?

3.67 Aluminum reacts with oxygen to yield aluminum oxide. If 5.0 g of Al reacts with 4.45 g of O_2, what is the empirical formula of aluminum oxide?

3.68 The industrial production of hydriodic acid takes place by treatment of iodine with hydrazine (N_2H_4):

$$2 I_2 + N_2H_4 \longrightarrow 4 HI + N_2$$

(a) How many grams of I_2 are needed to react with 36.7 g of (N_2H_4)?

(b) How many grams of HI are produced from the reaction of 115.7 g of N_2H_4 with excess iodine?

3.69 An alternative method for producing hydriodic acid is the reaction of iodine with hydrogen sulfide:

$$H_2S + I_2 \longrightarrow 2 HI + S$$

(a) How many grams of I_2 are needed to react with 49.2 g of H_2S?

(b) How many grams of HI are produced from the reaction of 95.4 g of H_2S with excess I_2?

Limiting Reactants and Reaction Yield (Sections 3.4 and 3.5)

3.70 Assume that you have 1.39 mol of H_2 and 3.44 mol of N_2. How many grams of ammonia (NH_3) can you make, and how many grams of which reactant will be left over?

$$3 H_2 + N_2 \longrightarrow 2 NH_3$$

3.71 Hydrogen and chlorine react to yield hydrogen chloride: $H_2 + Cl_2 \rightarrow 2 HCl$. How many grams of HCl are formed from reaction of 3.56 g of H_2 with 8.94 g of Cl_2? Which reactant is limiting?

3.72 How many grams of the dry-cleaning solvent 1,2-dichloroethane (also called ethylene chloride), $C_2H_4Cl_2$, can be prepared by reaction of 15.4 g of ethylene, C_2H_4, with 3.74 g of Cl_2?

$$C_2H_4 + Cl_2 \longrightarrow C_2H_4Cl_2$$

1,2-Dichloroethane
(ethylene chloride)

3.73 How many grams of each product result from the following reactions, and how many grams of which reactant are left over?

(a) $(1.3 \text{ g NaCl}) + (3.5 \text{ g AgNO}_3) \longrightarrow$
$$(x \text{ g AgCl}) + (y \text{ g NaNO}_3)$$

(b) $(2.65 \text{ g BaCl}_2) + (6.78 \text{ g H}_2\text{SO}_4) \longrightarrow$
$$(x \text{ g BaSO}_4) + (y \text{ g HCl})$$

3.74 Nickel(II) sulfate, used for nickel plating, is prepared by treatment of nickel(II) carbonate with sulfuric acid:

$$NiCO_3 + H_2SO_4 \longrightarrow NiSO_4 + CO_2 + H_2O$$

(a) How many grams of H_2SO_4 are needed to react with 14.5 g of $NiCO_3$?

(b) How many grams of $NiSO_4$ are obtained if the yield is 78.9%?

3.75 Hydrazine, N_2H_4, once used as a rocket propellant, reacts with oxygen:

$$N_2H_4 + O_2 \longrightarrow N_2 + 2 H_2O$$

(a) How many grams of O_2 are needed to react with 50.0 g of N_2H_4?

(b) How many grams of N_2 are obtained if the yield is 85.5%?

3.76 Limestone ($CaCO_3$) reacts with hydrochloric acid according to the equation $CaCO_3 + 2 HCl \rightarrow CaCl_2 + H_2O + CO_2$. If 1.00 mol of CO_2 has a volume of 22.4 L under the reaction conditions, how many liters of gas can be formed by reaction of 2.35 g of $CaCO_3$ with 2.35 g of HCl? Which reactant is limiting?

3.77 Sodium azide (NaN_3) yields N_2 gas when heated to 300 °C, a reaction used in automobile air bags. If 1.00 mol of N_2 has a volume of 47.0 L under the reaction conditions, how many liters of gas can be formed by heating 38.5 g of NaN_3? The reaction is

$$2 NaN_3(s) \longrightarrow 3 N_2(g) + 2 Na(s)$$

3.78 Acetic acid (CH_3CO_2H) reacts with isopentyl alcohol ($C_5H_{12}O$) to yield isopentyl acetate ($C_7H_{14}O_2$), a fragrant substance with the odor of bananas. If the yield from the reaction of acetic acid with isopentyl alcohol is 45%, how many grams of isopentyl acetate are formed from 3.58 g of acetic acid and 4.75 g of isopentyl alcohol? The reaction is

$$CH_3CO_2H + C_5H_{12}O \longrightarrow C_7H_{14}O_2 + H_2O$$

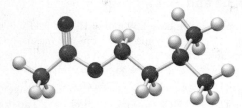

Isopentyl acetate

3.79 Cisplatin $[Pt(NH_3)_2Cl_2]$, a compound used in cancer treatment, is prepared by reaction of ammonia with potassium tetrachloroplatinate:

$$K_2PtCl_4 + 2 NH_3 \longrightarrow 2 KCl + Pt(NH_3)_2Cl_2$$

How many grams of cisplatin are formed from 55.8 g of K_2PtCl_4 and 35.6 g of NH_3 if the reaction takes place in 95% yield based on the limiting reactant?

3.80 If 1.87 g of acetic acid (CH_3COOH) reacts with 2.31 g of isopentyl alcohol ($C_5H_{12}O$) to give 2.96 g of isopentyl acetate ($C_7H_{14}O_2$), what is the percent yield of the reaction?

3.81 If 3.42 g of K_2PtCl_4 and 1.61 g of NH_3 give 2.08 g of cisplatin (Problem 3.79), what is the percent yield of the reaction?

Formulas and Elemental Analysis (Sections 3.6 and 3.7)

3.82 Urea, a substance commonly used as a fertilizer, has the formula CH_4N_2O. What is its percent composition by mass?

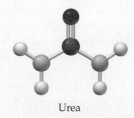

Urea

3.83 Calculate the mass percent composition of each of the following substances:

(a) Malachite, a copper-containing mineral: $Cu_2(OH)_2CO_3$

(b) Acetaminophen, a headache remedy: $C_8H_9NO_2$

(c) Prussian blue, an ink pigment: $Fe_4[Fe(CN)_6]_3$

3.84 An unknown liquid is composed of 5.57% H, 28.01% Cl, and 66.42% C. The molecular weight found by mass spectrometry is 126.58. What is the molecular formula of the compound?

3.85 An unknown liquid is composed of 34.31% C, 5.28% H, and 60.41% I. The molecular weight found by mass spectrometry is 210.06. What is the molecular formula of the compound?

3.86 What is the empirical formula of stannous fluoride, the first fluoride compound added to toothpaste to protect teeth against decay? Its mass percent composition is 24.25% F and 75.75% Sn.

3.87 What are the empirical formulas of each of the following substances?

(a) Ibuprofen, a headache remedy: 75.69% C, 15.51% O, 8.80% H

(b) Magnetite, a naturally occurring magnetic mineral: 72.36% Fe, 27.64% O

(c) Zircon, a mineral from which cubic zirconia is made: 34.91% O, 15.32% Si, 49.77% Zr

3.88 Combustion analysis of 45.62 mg of toluene, a commonly used solvent, gives 35.67 mg of H_2O and 152.5 mg of CO_2. What is the empirical formula of toluene?

3.89 Coniine, a toxic substance isolated from poison hemlock, contains only carbon, hydrogen, and nitrogen. Combustion analysis of a 5.024 mg sample yields 13.90 mg of CO_2 and 6.048 mg of H_2O. What is the empirical formula of coniine?

3.90 Cytochrome c is an iron-containing enzyme found in the cells of all aerobic organisms. If cytochrome c is 0.43% Fe by mass, what is its minimum molecular weight?

3.91 Nitrogen fixation in the root nodules of peas and other leguminous plants is carried out by the molybdenum-containing enzyme *nitrogenase*. What is the molecular weight of nitrogenase if the enzyme contains two molybdenum atoms and is 0.0872% Mo by mass?

3.92 Disilane, Si_2H_x, is analyzed and found to contain 90.28% silicon by mass. What is the value of x?

3.93 A certain metal sulfide, MS_2, is used extensively as a high-temperature lubricant. If MS_2 is 40.06% sulfur by mass, what is the identity of the metal M?

3.94 Combustion analysis of a 31.472 mg sample of the widely used flame retardant Decabrom gave 1.444 mg of CO_2. Is the molecular formula of Decabrom $C_{12}Br_{10}$ or $C_{12}Br_{10}O$?

3.95 The stimulant amphetamine contains only carbon, hydrogen, and nitrogen. Combustion analysis of a 42.92 mg sample of amphetamine gives 37.187 mg of H_2O and 125.75 mg of CO_2. If the molar mass of amphetamine is less than 160 g/mol, what is its molecular formula?

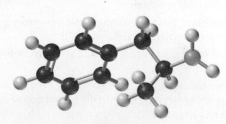

Amphetamine

Mass Spectrometry (Section 3.8)

3.96 Describe the path of a neutral molecule in the mass spectrometer. Why is ionization a necessary first step?

3.97 Why is high precision and accuracy in molecular weight measurement needed in identifying the formula of molecules made in the lab?

3.98 The molecular weight of an organic compound was found to be 70.042 11 by mass spectrometry. Is the sample C_5H_{10}, C_4H_6O, or $C_3H_6N_2$? Exact masses of elements are 1.007 825 (^1H); 12.000 00 (^{12}C); 14.003 074 (^{14}N); and 15.994 915 (^{16}O).

3.99 The mass of an organic compound was found to be 58.077 46 by mass spectrometry. Is the sample C_4H_{10}, C_3H_6O, or $C_2H_6N_2$? Exact masses of elements are 1.007 825 (^1H); 12.000 00 (^{12}C); 14.003 074 (^{14}N); and 15.994 915 (^{16}O).

3.100 **(a)** Combustion analysis of 50.0 mg of benzene, a commonly used solvent composed of carbon and hydrogen, gives 34.6 mg of H_2O and 169.2 mg of CO_2. What is the empirical formula of benzene?

(b) Given the mass spectrum of benzene, identify the molecular weight and give the molecular formula.

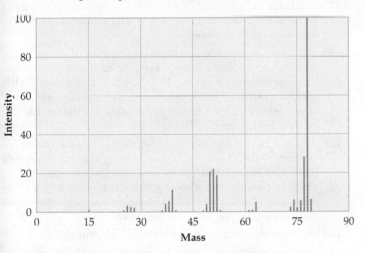

3.101 **(a)** Combustion analysis of 150.0 mg of 1,2,3,benzenetriol, a compound composed of carbon, hydrogen, and oxygen, gives 64.3 mg of H_2O and 314.2 mg of CO_2. What is the empirical formula of 1,2,3,benzenetriol?

(b) Given the mass spectrum of 1,2,3,benzenetriol, identify the molecular weight and give the molecular formula.

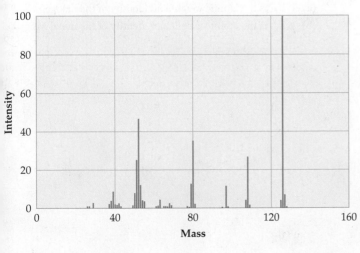

CHAPTER PROBLEMS

3.102 *Ringer's solution*, used in the treatment of burns and wounds, is prepared by dissolving 4.30 g of NaCl, 0.150 g of KCl, and 0.165 g of $CaCl_2$ in water and diluting to a volume of 500.0 mL. What is the molarity of each of the component ions in the solution?

3.103 Balance the following equations:

(a) $C_6H_5NO_2 + O_2 \longrightarrow CO_2 + H_2O + NO_2$

(b) $Au + H_2SeO_4 \longrightarrow Au_2(SeO_4)_3 + H_2SeO_3 + H_2O$

(c) $NH_4ClO_4 + Al \longrightarrow Al_2O_3 + N_2 + Cl_2 + H_2O$

3.104 Silver sulfide, the tarnish on silverware, comes from reaction of silver metal with hydrogen sulfide (H_2S):

$$Ag + H_2S + O_2 \longrightarrow Ag_2S + H_2O \quad \text{Unbalanced}$$

(a) Balance the equation.

(b) If the reaction were used intentionally to prepare Ag_2S, how many grams would be formed from 496 g of Ag, 80.0 g of H_2S, and 40.0 g of O_2 if the reaction takes place in 95% yield based on the limiting reactant?

3.105 Give the percent composition of each of the following substances:

(a) Glucose, $C_6H_{12}O_6$

(b) Sulfuric acid, H_2SO_4

(c) Potassium permanganate, $KMnO_4$

(d) Saccharin, $C_7H_5NO_3S$

3.106 What are the empirical formulas of substances with the following mass percent compositions?

(a) Aspirin: 4.48% H, 60.00% C, 35.52% O

(b) Ilmenite (a titanium-containing ore): 31.63% O, 31.56% Ti, 36.81% Fe

(c) Sodium thiosulfate (photographic "fixer"): 30.36% O, 29.08% Na, 40.56% S

3.107 The reaction of tungsten hexachloride (WCl_6) with bismuth gives hexatungsten dodecachloride (W_6Cl_{12}):

$$WCl_6 + Bi \longrightarrow W_6Cl_{12} + BiCl_3 \quad \text{Unbalanced}$$

(a) Balance the equation.

(b) How many grams of bismuth react with 150.0 g of WCl_6?

(c) When 228 g of WCl_6 react with 175 g of Bi, how much W_6Cl_{12} is formed based on the limiting reactant?

3.108 Sodium borohydride, $NaBH_4$, a substance used in the synthesis of many pharmaceutical agents, can be prepared by reaction of NaH with B_2H_6 according to the equation $2\,NaH + B_2H_6 \longrightarrow 2\,NaBH_4$.

(a) How many grams of $NaBH_4$ can be prepared by reaction between 8.55 g of NaH and 6.75 g of B_2H_6?

(b) Which reactant is limiting, and how many grams of the excess reactant will be left over?

3.109 Ferrocene, a substance proposed for use as a gasoline additive, has the percent composition 5.42% H, 64.56% C, and 30.02% Fe. What is the empirical formula of ferrocene?

3.110 The molar mass of HCl is 36.5 g/mol, and the average mass per HCl molecule is 36.5 u. Use the fact that $1\,u = 1.6605 \times 10^{-24}\,g$ to calculate Avogadro's number.

3.111 Ethylene glycol, commonly used as automobile antifreeze, contains only carbon, hydrogen, and oxygen. Combustion analysis of a 23.46 mg sample yields 20.42 mg of H_2O and 33.27 mg of CO_2.

What is the empirical formula of ethylene glycol? What is its molecular formula if it has a molecular weight of 62.0?

3.112 The molecular weight of ethylene glycol is 62.0689 when calculated using the atomic weights found in a standard periodic table, yet the molecular weight determined experimentally by high-resolution mass spectrometry is 62.0368. Explain the discrepancy.

3.113 Balance the following equations:

(a) $CO(NH_2)_2(aq) + HOCl(aq) \longrightarrow$
$$NCl_3(aq) + CO_2(aq) + H_2O(l)$$

(b) $Ca_3(PO_4)_2(s) + SiO_2(s) + C(s) \longrightarrow$
$$P_4(g) + CaSiO_3(s) + CO(g)$$

3.114 Assume that gasoline has the formula C_8H_{18} and has a density of 0.703 g/mL. How many pounds of CO_2 are produced from the complete combustion of 1.00 gal of gasoline?

3.115 Compound X contains only carbon, hydrogen, nitrogen, and chlorine. When 1.00 g of X is dissolved in water and allowed to react with excess silver nitrate, $AgNO_3$, all the chlorine in X reacts and 1.95 g of solid AgCl is formed. When 1.00 g of X undergoes complete combustion, 0.900 g of CO_2 and 0.735 g of H_2O are formed. What is the empirical formula of X?

3.116 A pulverized rock sample believed to be pure calcium carbonate, $CaCO_3$, is subjected to chemical analysis and found to contain 51.3% Ca, 7.7% C, and 41.0% O by mass. Why can't this rock sample be pure $CaCO_3$?

3.117 A certain alcoholic beverage contains only ethanol (C_2H_6O) and water. When a sample of this beverage undergoes combustion, the ethanol burns but the water simply evaporates and is collected along with the water produced by combustion. The combustion reaction is

$$C_2H_6O(l) + 3 O_2(g) \longrightarrow 2 CO_2(g) + 3 H_2O(g)$$

When a 10.00 g sample of this beverage is burned, 11.27 g of water is collected. What is the mass in grams of ethanol, and what is the mass of water in the original sample?

3.118 A mixture of FeO and Fe_2O_3 with a mass of 10.0 g is converted to 7.43 g of pure Fe metal. What are the amounts in grams of FeO and Fe_2O_3 in the original sample?

3.119 A compound of formula XCl_3 reacts with aqueous $AgNO_3$ to yield solid AgCl according to the following equation:

$$XCl_3(aq) + 3 AgNO_3(aq) \longrightarrow X(NO_3)_3(aq) + 3 AgCl(s)$$

When a solution containing 0.634 g of XCl_3 was allowed to react with an excess of aqueous $AgNO_3$, 1.68 g of solid AgCl was formed. What is the identity of the atom X?

3.120 When eaten, dietary carbohydrates are digested to yield glucose ($C_6H_{12}O_6$), which is then metabolized to yield carbon dioxide and water:

$$C_6H_{12}O_6 + O_2 \longrightarrow CO_2 + H_2O \qquad \text{Unbalanced}$$

Balance the equation, and calculate both the mass in grams and the volume in liters of the CO_2 produced from 66.3 g of glucose, assuming that 1 mol of CO_2 has a volume of 25.4 L at normal body temperature.

3.121 A copper wire having a mass of 2.196 g was allowed to react with an excess of sulfur. The excess sulfur was then burned, yielding SO_2 gas. The mass of the copper sulfide produced was 2.748 g.

(a) What is the percent composition of copper sulfide?

(b) What is its empirical formula?

(c) Calculate the number of copper ions per cubic centimeter if the density of the copper sulfide is 5.6 g/cm^3.

3.122 Element X, a member of group 5A, forms two chlorides, XCl_3 and XCl_5. Reaction of an excess of Cl_2 with 8.729 g of XCl_3 yields 13.233 g of XCl_5. What is the atomic weight and the identity of the element X?

3.123 A mixture of XCl_3 and XCl_5 weighing 10.00 g contains 81.04% Cl by mass. How many grams of XCl_3 and how many grams of XCl_5 are present in the mixture?

3.124 Ammonium nitrate, a potential ingredient of terrorist bombs, can be made nonexplosive by addition of diammonium hydrogen phosphate, $(NH_4)_2HPO_4$. Analysis of such a $NH_4NO_3 - (NH_4)_2HPO_4$ mixture showed the mass percent of nitrogen to be 30.43%. What is the mass ratio of the two components in the mixture?

3.125 Window glass is typically made by mixing soda ash (Na_2CO_3), limestone ($CaCO_3$), and silica sand (SiO_2) and then heating to 1500 °C to drive off CO_2 from the (Na_2CO_3) and $CaCO_3$. The resultant glass consists of about 12% Na_2O by mass, 13% CaO by mass, and 75% SiO_2 by mass. How much of each reactant would you start with to prepare 0.35 kg of glass?

3.126 An unidentified metal M reacts with an unidentified halogen X to form a compound MX_2. When heated, the compound decomposes by the reaction:

$$2 MX_2(s) \longrightarrow 2 MX(s) + X_2(g)$$

When 1.12 g of MX_2 is heated, 0.720 g of MX is obtained, along with 56.0 mL of X_2 gas. Under the conditions used, 1.00 mol of the gas has a volume of 22.41 L.

(a) What is the atomic weight and identity of the halogen X?

(b) What is the atomic weight and identity of the metal M?

CHAPTER

4

Reactions in Aqueous Solution

Many chemical reactions in your body occur in aqueous solution and fall into the general categories of precipitation, acid-base, and redox reactions. Vigorous exercise depletes fluids and electrolytes; upsetting the body's delicate chemical balance.

 How do sports drinks replenish the chemicals lost in sweat?

The answer to this question can be found in the **INQUIRY** ▶▶▶ on page 142.

CONTENTS

O ur world is based on water. Approximately 71% of the Earth's surface is covered by water, and another 3% is covered by ice; 66% of the mass of an adult human body is water, and water is needed to sustain all living organisms. It's therefore not surprising that a large amount of important chemistry, including all those reactions that happen in our bodies, takes place in water—that is, in *aqueous solution*.

We saw in the previous chapter how chemical reactions are described and how the specific mass relationships among reactant and product substances in reactions can be calculated. In this chapter, we'll continue the study of chemical reactions by describing some ways that chemical reactions can be classified and by exploring some general ways in which reactions take place.

4.1 ▶ SOLUTION CONCENTRATION: MOLARITY

For a chemical reaction to occur, the reacting molecules or ions must come into contact. This means that the reactants must be mobile, which in turn means that most chemical reactions are carried out in the liquid state or in solution rather than in the solid state. It's therefore necessary to have a standard means to describe exact quantities of substances in solution.

As we've seen, stoichiometry calculations for chemical reactions always require working in moles. Thus, the most generally useful means of expressing a solution's concentration is **molarity (M)**, the number of moles of a substance, or **solute**, dissolved in enough solvent to make 1 liter of solution. For example, a solution made by dissolving 1.00 mol (58.5 g) of NaCl in enough water to give 1.00 L of solution has a concentration of 1.00 mol/L, or 1.00 M. The molarity of any solution is found by dividing the number of moles of solute by the number of liters of solution:

$$\text{Molarity (M)} = \frac{\text{Moles of solute}}{\text{Liters of solution}}$$

Note that it's the final volume of the *solution* that's important, not the starting volume of the *solvent* used. The final volume of the solution might be a bit larger than the volume of the solvent because of the additional volume of the solute. In practice, a solution of known molarity is prepared by weighing an appropriate amount of solute and placing it in a container called a *volumetric flask*, as shown in **FIGURE 4.1**. Enough solvent is added to dissolve

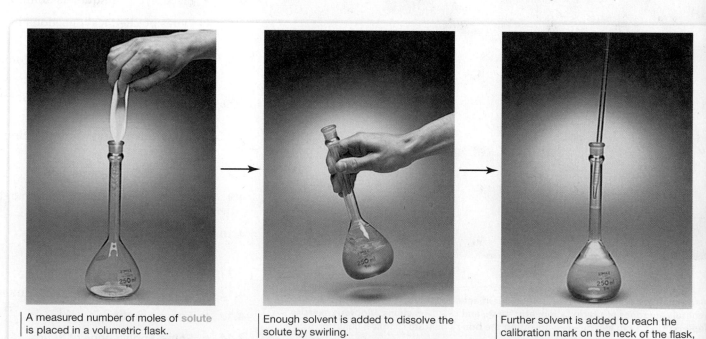

A measured number of moles of solute is placed in a volumetric flask. | Enough solvent is added to dissolve the solute by swirling. | Further solvent is added to reach the calibration mark on the neck of the flask, and the solution is mixed until uniform.

▲ **FIGURE 4.1**

Preparing a solution of known molarity.

the solute, and further solvent is added until an accurately calibrated final volume is reached. The solution is then gently mixed to reach a uniform concentration.

Molarity can be used as a conversion factor to relate a solution's volume to the number of moles of solute. If we know the molarity and volume of a solution, we can calculate the number of moles of solute. If we know the number of moles of solute and the molarity of the solution, we can find the solution's volume. Worked Examples 4.1 and 4.2 show how the calculations are done.

$$\text{Molarity} = \frac{\text{Moles of solute}}{\text{Volume of solution(L)}}$$

$$\text{Moles of solute} = \text{Molarity} \times \text{Volume of solution} \qquad \text{Volume of solution} = \frac{\text{Moles of solute}}{\text{Molarity}}$$

WORKED EXAMPLE 4.1

Calculating the Molarity of a Solution

What is the molarity of a solution made by dissolving 2.355 g of sulfuric acid (H_2SO_4) in water and diluting to a final volume of 50.0 mL?

IDENTIFY

Known	Unknown
Mass of solute (2.355 g of sulfuric acid)	Molarity ($M = mol/L$)
Volume of solution (50.0 mL)	

STRATEGY

Molarity is the number of moles of solute per liter of solution. Thus, it's necessary to find the number of moles of sulfuric acid in 2.355 g and then divide by the volume of the solution in liters.

SOLUTION

$$\text{Mol. wt. } H_2SO_4 = (2 \times 1.0) + (32.1) + (4 \times 16.0) = 98.1$$

$$\text{Molar mass of } H_2SO_4 = 98.1 \text{ g/mol}$$

$$2.355 \text{ g } H_2SO_4 \times \frac{1 \text{ mol } H_2SO_4}{98.1 \text{ g } H_2SO_4} = 0.0240 \text{ mol } H_2SO_4$$

$$\frac{0.0240 \text{ mol } H_2SO_4}{0.0500 \text{ L}} = 0.480 \text{ M}$$

The solution has a sulfuric acid concentration of 0.480 M.

▶ **PRACTICE 4.1**   A sweetened iced tea beverage of 355 mL contains 43.0 g of sucrose ($C_{12}H_{22}O_{11}$). Calculate the molarity of sucrose in the drink.

▶ **APPLY 4.2**   The procedure in an experiment calls for a 1.00 M solution of KCl. A student prepares the solution by adding 37.3 g to KCl to 500 mL of water. Does the resulting solution have the correct molarity? Explain.

WORKED EXAMPLE 4.2

Calculating the Number of Moles of Solute in a Solution

Hydrochloric acid (HCl) is sold commercially as a 12.0 M aqueous solution. How many moles of HCl are in 300.0 mL of 12.0 M solution?

IDENTIFY

Known	Unknown
Volume of solution (300.0 mL)	Moles of solute (mol HCl)
Molarity of solution (12.0 M)	

continued on next page

STRATEGY

The number of moles of solute is calculated by multiplying the molarity of the solution by its volume.

SOLUTION

$$\text{Moles of HCl} = (\text{Molarity of solution}) \times (\text{Volume of solution})$$

$$= \frac{12.0 \text{ mol HCl}}{1 \text{ L}} \times 0.3000 \text{ L} = 3.60 \text{ mol HCl}$$

There are 3.60 mol of HCl in 300.0 mL of 12.0 M solution.

CHECK

One liter of 12.0 M HCl solution contains 12 mol of HCl, so 300 mL (0.3 L) of solution contains $0.3 \times 12 = 3.6$ mol.

▶ **PRACTICE 4.3**   How many moles of solute are present in the following solutions?

(a) 125 mL of 0.20 M $NaHCO_3$
(b) 650.0 mL of 2.50 M H_2SO_4

▶ **PRACTICE 4.4**   How many grams of solute would you use to prepare the following solutions?

(a) 500.0 mL of 1.25 M NaOH
(b) 1.50 L of 0.250 M glucose $(C_6H_{12}O_6)$

▶ **APPLY 4.5**   The concentration of cholesterol $(C_{27}H_{46}O)$ in normal blood is approximately 0.005 M.

(a) How many grams of cholesterol are in 750 mL of blood?
(b) A medical test for cholesterol requires 25 mg. How many milliliters of blood are needed to perform the test?

Cholesterol

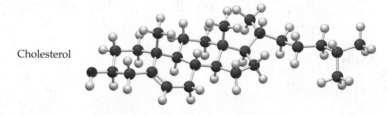

4.2 ▶ DILUTING CONCENTRATED SOLUTIONS

For convenience, chemicals are sometimes bought and stored as concentrated solutions, which are then diluted before use. Aqueous hydrochloric acid, for example, is sold commercially as a 12.0 M solution, yet it is most commonly used in the laboratory after dilution with water to a final concentration of either 6.0 M or 1.0 M.

$$\text{Concentrated solution} + \text{Solvent} \longrightarrow \text{Dilute solution}$$

The main thing to remember when diluting a concentrated solution is that the number of moles of solute is constant; only the volume of the solution is changed by adding more solvent. Because the number of moles of solute can be calculated by multiplying molarity times volume, we can set up the following equation:

Moles of solute (constant) $= \text{Molarity} \times \text{Volume}$
$$= M_i \times V_i = M_f \times V_f$$

where M_i is the initial molarity, V_i is the initial volume, M_f is the final molarity, and V_f is the final volume after dilution. Rearranging this equation into a more useful form shows that

▲ Just as the frozen orange juice concentrate must be diluted before use by adding water, many chemical solutions must also be diluted.

the molar concentration after dilution (M_f) can be found by multiplying the initial concentration (M_i) by the ratio of initial and final volumes (V_i/V_f):

$$M_f = M_i \times \frac{V_i}{V_f}$$

Suppose, for example, that we dilute 50.0 mL of a solution of 2.00 M H_2SO_4 to a volume of 200.0 mL. The solution volume *increases* by a factor of 4 (from 50 mL to 200 mL), so the concentration of the solution must *decrease* by a factor of 4 (from 2.00 M to 0.500 M):

$$M_f = 2.00\ M \times \frac{50.0\ \text{mL}}{200.0\ \text{mL}} = 0.500\ M$$

In practice, dilutions are usually carried out as shown in **FIGURE 4.2**. The volume to be diluted is withdrawn using a calibrated tube called a *pipet*, placed in an empty volumetric flask of the chosen volume, and diluted to the calibration mark on the flask. The one common exception to this order of steps is when diluting a strong acid such as H_2SO_4, where a large amount of heat is released. In such instances, it is much safer to add the acid slowly to the water rather than adding water to the acid.

← Calibration mark

The volume to be diluted is placed in an empty volumetric flask.

Solvent is added to a level just below the calibration mark, and the flask is shaken.

More solvent is added to reach the calibration mark, and the flask is again shaken.

▲ **FIGURE 4.2**
The procedure for diluting a concentrated solution.

WORKED EXAMPLE 4.3

Diluting a Solution

What volume of 1.000 M NaOH solution is required to prepare 500.0 mL of 0.2500 M NaOH?

IDENTIFY

Known	Unknown
Final volume (V_f = 500.0 mL)	Initial volume (V_i)
Final concentration (M_f = 0.2500 M)	
Initial concentration (M_i = 1.000 M)	

continued on next page

STRATEGY

The initial volume (V_i) can be found by rearranging the equation $M_i \times V_i = M_f \times V_f$ to solve for the unknown, $V_i = (M_f/M_i) \times V_f$.

SOLUTION

$$V_i = \frac{M_f}{M_i} \times V_f = \frac{0.2500\ M}{1.000\ M} \times 500.0\ mL = 125.0\ mL$$

CHECK

Because the concentration decreases by a factor of 4 after dilution (from 1.000 M to 0.2500 M), the volume must increase by a factor of 4. Thus, to prepare 500.0 mL of solution, we should start with $500.0/4 = 125.0$ mL.

▶ **PRACTICE 4.6** What is the final concentration if 75.0 mL of a 3.50 M glucose solution is diluted to a volume of 400.0 mL?

▶ **APPLY 4.7** Sulfuric acid is normally purchased at a concentration of 18.0 M. How would you prepare 250.0 mL of 0.500 M aqueous H_2SO_4? (Remember to add acid to water rather than water to acid.)

4.3 ▶ ELECTROLYTES IN AQUEOUS SOLUTION

Before beginning a study of reactions in water, it's helpful to look at some general properties of aqueous solutions. We all know from experience that both sugar (sucrose) and table salt (NaCl) dissolve in water. The solutions that result, though, are quite different. When sucrose, a molecular substance, dissolves in water, the resulting solution contains neutral sucrose *molecules* surrounded by water. When NaCl, an ionic substance, dissolves in water, the solution contains separate Na^+ and Cl^- *ions* surrounded by water. Because of the presence of the charged ions, the NaCl solution conducts an electric current, but the sucrose solution does not.

$$C_{12}H_{22}O_{11}(s) \xrightarrow{H_2O} C_{12}H_{22}O_{11}(aq)$$
$$\text{Sucrose}$$
$$NaCl(s) \xrightarrow{H_2O} Na^+(aq) + Cl^-(aq)$$

The electrical conductivity of an aqueous NaCl solution is easy to demonstrate using a battery, a light bulb, and several pieces of wire, connected as shown in **FIGURE 4.3**. When the wires are dipped into an aqueous NaCl solution, the positively charged Na^+ ions move through the solution toward the wire connected to the negatively charged terminal of the battery and the negatively charged Cl^- ions move toward the wire connected to the positively charged terminal of the battery. The resulting movement of electrical charges allows a current to flow, so the bulb lights. When the wires are dipped into an aqueous sucrose solution, however, there are no ions to carry the current, so the bulb remains dark.

Substances such as NaCl or KBr, which dissolve in water to produce conducting solutions of ions, are called **electrolytes**. Substances such as sucrose or ethyl alcohol, which do not produce ions in aqueous solution, are **nonelectrolytes**. Most electrolytes are ionic compounds, but some are molecular. Hydrogen chloride, for instance, is a gaseous molecular compound when pure but **dissociates**, or splits apart, to give H^+ and Cl^- ions when it dissolves in water.

$$HCl(g) \xrightarrow{H_2O} H^+(aq) + Cl^-(aq)$$

Compounds that dissociate to a large extent (70–100%) into ions when dissolved in water are said to be **strong electrolytes**, while compounds that dissociate to only a small extent are **weak electrolytes**. Potassium chloride and most other ionic compounds, for instance, are largely dissociated in dilute solution and are thus strong electrolytes. Acetic acid (CH_3CO_2H), by contrast, dissociates only to the extent of about 1.3% in a 0.10 M solution and is a weak electrolyte. As a result, a 0.10 M solution of acetic acid is only weakly conducting and the bulb in Figure 4.3 would only light dimly.

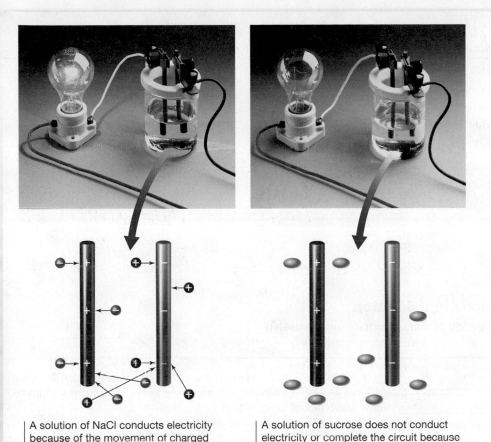

◄ **FIGURE 4.3**
Testing the conductivity of aqueous solutions.

Figure It Out

Describe the appearance of the light bulb if it were placed in 1.0 M solutions of (a) HNO₃, (b) HF, and (c) C₂H₅OH. Refer to Table 4.1 for classification of these compounds.

Answer: (a) The bulb would be bright since HNO₃ is a strong electrolyte and dissociates nearly completely into ions. (b) The bulb would light dimly because HF is a weak electrolyte and dissociates only to a small extent. (c) The bulb will not light because ethanol is a nonelectrolyte.

A solution of NaCl conducts electricity because of the movement of charged particles (ions), thereby completing the circuit and allowing the bulb to light.

A solution of sucrose does not conduct electricity or complete the circuit because it contains no mobile charged particles. The bulb therefore remains dark.

For 0.10 M solutions:

$$KCl(aq) \rightleftharpoons K^+(aq) + Cl^-(aq) \quad \text{Strong electrolyte}$$
$$(2\%) \qquad\qquad (98\%)$$

$$CH_3CO_2H(aq) \rightleftharpoons H^+(aq) + CH_3CO_2^-(aq) \quad \text{Weak electrolyte}$$
$$(99\%) \qquad\qquad (1\%)$$

Note that a forward-and-backward double arrow ($\rightleftharpoons$) is used in the dissociation equation to indicate that the reaction takes place simultaneously in both directions. That is, dissociation is a dynamic process in which an *equilibrium* is established between the forward and reverse reactions. Dissociation of acetic acid takes place in the forward direction, while recombination of H^+ and $CH_3CO_2^-$ ions takes place in the reverse direction. The size of the equilibrium arrow indicates whether the equilibrium reaction forms mostly products or mostly reactants. Ultimately, the concentrations of the reactants and products reach constant values and no longer change with time. We'll learn much more about **chemical equilibria** in Chapter 14.

A brief list of some common substances classified according to their electrolyte strength is given in **TABLE 4.1**. Note that pure water is a nonelectrolyte because it does not dissociate appreciably into H^+ and OH^- ions. We'll explore the dissociation of water in more detail in Section 15.4.

LOOKING AHEAD...

A **chemical equilibrium**, as we'll see in Chapters 13 and 14, is the state in which a reaction takes place in both forward and backward directions so that the concentrations of products and reactants remain constant over time.

Strong- completely dissoedhes
size how many Oxygens

TABLE 4.1 **Electrolyte Classification of Some Common Substances**

Strong Electrolytes	Weak Electrolytes	Nonelectrolytes
HCl, HBr, HI	CH_3CO_2H	H_2O
$HClO_4$	HF	CH_3OH (methyl alcohol)
HNO_3	HCN	C_2H_5OH (ethyl alcohol)
H_2SO_4		$C_{12}H_{22}O_{11}$ (sucrose)
KBr		Most compounds of carbon (organic compounds)
NaCl		
NaOH, KOH		
Other soluble ionic compounds		

WORKED EXAMPLE 4.4

Calculating the Concentration of Ions in a Solution

What is the total molar concentration of ions in a 0.350 M solution of the strong electrolyte Na_2SO_4, assuming complete dissociation?

IDENTIFY

Known	Unknown
Molar concentration of compound (0.350 M Na_2SO_4)	Total molar concentration of ions

STRATEGY

The ions come from the dissociation of Na_2SO_4, and therefore a reaction for dissolving Na_2SO_4 in water is written. The balanced reaction shows that 3 mol of ions are formed from 1 mol of the compound: 2 mol of Na^+ and 1 mol of SO_4^{2-}.

$$Na_2SO_4(s) \xrightarrow{H_2O} 2\,Na^+(aq) + SO_4^{2-}(aq)$$

SOLUTION

Assuming complete dissociation, the total molar concentration of ions is three times the molarity of Na_2SO_4, or 1.05 M:

$$\frac{0.350 \text{ mol Na}_2\text{SO}_4}{1 \text{ L}} \times \frac{3 \text{ mol ions}}{1 \text{ mol Na}_2\text{SO}_4} = 1.05 \text{ M}$$

▶ **PRACTICE 4.8** What is the molar concentration of Br^- ions in a 0.225 M aqueous solution of $FeBr_3$, assuming complete dissociation?

▶ **Conceptual APPLY 4.9** Three different substances, A_2X, A_2Y, and A_2Z, are dissolved in water, with the following results. (Water molecules are omitted for clarity.)

(a) Which of the substances is the strongest electrolyte, and which is the weakest? Explain.

(b) What is the molar concentration of A ions and Y ions in a 0.350 M solution of A_2Y?

(c) What is the percent ionization of A_2X indicated in the graphical representation?

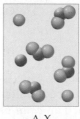

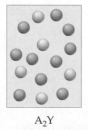

A_2X A_2Y A_2Z

4.4 ▶ TYPES OF CHEMICAL REACTIONS IN AQUEOUS SOLUTION

Many common chemical reactions that take place in aqueous solution fall into one of three general categories: *precipitation reactions*, *acid–base neutralization reactions*, and *oxidation–reduction reactions*. Let's look briefly at an example of each before looking at them in more detail in subsequent sections.

- **Precipitation reactions** are processes in which soluble ionic reactants (strong electrolytes) yield an insoluble solid product called a *precipitate*, which falls out of the solution thereby removing some of the dissolved ions. Most precipitations take place when the anions and cations of two ionic compounds change partners. For example, an aqueous solution of lead(II) nitrate reacts with an aqueous solution of potassium iodide to yield an aqueous solution of potassium nitrate plus an insoluble yellow precipitate of lead(II) iodide:

$$Pb(NO_3)_2(aq) + 2\,KI(aq) \longrightarrow 2\,KNO_3(aq) + PbI_2(s)$$

- **Acid–base neutralization reactions** are processes in which an acid reacts with a base to yield water plus an ionic compound called a *salt*. Acids are compounds that produce H^+ ions when dissolved in water and bases are compounds that produce OH^- ions when dissolved in water. Thus, a neutralization reaction removes H^+ and OH^- ions from solution, just as a precipitation reaction removes metal and nonmetal ions. The reaction between hydrochloric acid and aqueous sodium hydroxide to yield water plus aqueous sodium chloride is a typical example:

$$HCl(aq) + NaOH(aq) \longrightarrow H_2O(l) + NaCl(aq)$$

- **Oxidation–reduction reactions**, or **redox reactions**, are processes in which one or more electrons are transferred between reaction partners (atoms, molecules, or ions). As a result of this electron transfer, the charges on atoms in the various reactants change. When metallic magnesium reacts with aqueous hydrochloric acid, for instance, a magnesium atom gives an electron to each of two H^+ ions, forming an Mg^{2+} ion and a H_2 molecule. The charge on the magnesium changes from 0 to +2, and the charge on each hydrogen changes from +1 to 0:

$$Mg(s) + 2\,HCl(aq) \longrightarrow MgCl_2(aq) + H_2(g)$$

▲ Reaction of aqueous lead(II) nitrate with aqueous potassium iodide gives a yellow precipitate of lead(II) iodide. $Pb(NO_3)_2(aq) + 2\,KI(aq) \longrightarrow PbI_2(s) + 2\,KNO_3(aq)$

4.5 ▶ AQUEOUS REACTIONS AND NET IONIC EQUATIONS

The equations we've been writing up to this point have all been **molecular equations**. That is, all the substances involved in the reactions have been written using their complete formulas as if they were *molecules*. In the previous section, for instance, we wrote the precipitation reaction of lead(II) nitrate with potassium iodide to yield solid PbI_2 using only the parenthetical **state abbreviation** (*aq*) to indicate that the substances are dissolved in aqueous solution. Nowhere in the equation was it indicated that ions are involved. There are, however, more accurate ways of writing equations to reflect the chemical changes that occur during reactions, as the following series of examples illustrates.

REMEMBER …

The physical state of a substance in a chemical reaction is often indicated with a parenthetical **state abbreviation** (*s*) for solid, (*l*) for liquid, (*g*) for gas, and (*aq*) for aqueous solution. (Section 3.3)

A Molecular Equation

$$Pb(NO_3)_2(aq) + 2\,KI(aq) \longrightarrow 2\,KNO_3(aq) + PbI_2(s)$$

This equation implies that molecules are interacting. It is the case, however, that lead nitrate, potassium iodide, and potassium nitrate are strong electrolytes that dissolve in water to yield solutions of ions. Thus, it's more accurate to write the precipitation reaction as an **ionic equation**, in which all the ions are explicitly shown.

An Ionic Equation

$$Pb^{2+}(aq) + 2\,NO_3^-(aq) + 2\,K^+(aq) + 2\,I^-(aq) \longrightarrow 2\,K^+(aq) + 2\,NO_3^-(aq) + PbI_2(s)$$

This ionic equation shows that the NO_3^- and K^+ ions undergo no change during the reaction. Instead, they appear on both sides of the reaction arrow and act merely as **spectator ions**, whose only role is to balance the charge. A **net ionic equation** gives only the species that react (the Pb^{2+} and I^- ions in this instance) because spectator ions are cancelled from both sides of the equation.

An Ionic Equation

$$Pb^{2+}(aq) + 2\,\cancel{NO_3^-}(aq) + 2\,\cancel{K^+}(aq) + 2\,I^-(aq) \longrightarrow 2\,\cancel{K^+}(aq) + 2\,\cancel{NO_3^-}(aq) + PbI_2(s)$$

A Net Ionic Equation

$$Pb^{2+}(aq) + 2\,I^-(aq) \longrightarrow PbI_2(s)$$

Leaving the spectator ions out of a net ionic equation doesn't mean that their presence is irrelevant. If a reaction occurs by mixing a solution of Pb^{2+} ions with a solution of I^- ions, then those solutions must also contain additional ions to balance the charge in each. That is, the Pb^{2+} solution must also contain an anion, and the I^- solution must also contain a cation. Leaving these other ions out of the net ionic equation only implies that these ions do not undergo a chemical reaction. Any nonreactive spectator ion could serve to balance charge.

WORKED EXAMPLE 4.5

Writing a Net Ionic Equation

Aqueous hydrochloric acid reacts with zinc metal to yield hydrogen gas and aqueous zinc chloride. Write a net ionic equation for the process, and identify the spectator ions.

$$2\,HCl(aq) + Zn(s) \longrightarrow H_2(g) + ZnCl_2(aq)$$

STRATEGY

First, write the ionic equation, listing all the species present in solution. Table 4.1 tells us that both HCl (a molecular compound) and $ZnCl_2$ (a soluble ionic compound) are strong electrolytes that exist as ions in solution. Then find the ions that are present on both sides of the reaction arrow—the spectator ions—and cancel them to leave the net ionic equation.

SOLUTION

Ionic Equation

$$2\,H^+(aq) + 2\,\cancel{Cl^-}(aq) + Zn(s) \longrightarrow H_2(g) + Zn^{2+}(aq) + 2\,\cancel{Cl^-}(aq)$$

Net Ionic Equation

$$2\,H^+(aq) + Zn(s) \longrightarrow H_2(g) + Zn^{2+}(aq)$$

Chloride ion is the only spectator ion in this reaction.

▲ Zinc metal reacts with aqueous hydrochloric acid to give hydrogen gas and aqueous Zn^{2+} ions.

▶ **PRACTICE 4.10** Write net ionic equations for the following reactions:

(a) $2\,AgNO_3(aq) + Na_2CrO_4(aq) \longrightarrow Ag_2CrO_4(s) + 2\,NaNO_3(aq)$
(b) $H_2SO_4(aq) + MgCO_3(s) \longrightarrow H_2O(l) + CO_2(g) + MgSO_4(aq)$
(c) $Hg(NO_3)_2(aq) + 2\,NH_4I(aq) \longrightarrow HgI_2(s) + 2\,NH_4NO_3(aq)$

▶ **Conceptual** **APPLY 4.11** Two clear, colorless solutions are mixed and a white precipitate forms and settles to the bottom of the container. The diagram on the right represents the ions in solution and the solid after the reaction has occurred. Identify the spectator ions and give the balanced net ionic equation and molecular equation for the reaction.

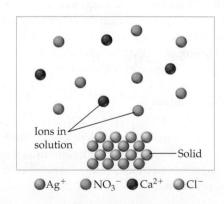

Ions in solution

Solid

●Ag^+ ●NO_3^- ●Ca^{2+} ●Cl^-

4.6 ▶ PRECIPITATION REACTIONS AND SOLUBILITY GUIDELINES

How can you predict whether a precipitation reaction will occur on mixing aqueous solutions of two substances? To do that, you must know the **solubility** of each potential product—how much of each compound will dissolve in a given amount of solvent at a given temperature. If a substance has a high solubility in water, no precipitate will form. If a substance has a low solubility in water, it's likely to precipitate from an aqueous solution.

Solubility is a complex matter, and it's not always possible to make correct predictions. Furthermore, solubility depends on the concentrations of the reactant ions, and the very words *soluble* and *insoluble* are imprecise. A substance is considered soluble if it dissolves to give a concentration of 0.01 M or greater. A compound is soluble if it meets either (or both) of the following criteria:

1. **A compound is soluble if it contains one of the following *cations*:**

   - Li^+, Na^+, K^+, Rb^+, Cs^+ (group 1A cations)
   - NH_4^+ (ammonium ion)

   That is, essentially all ionic compounds containing an alkali metal cation or ammonium cation are soluble in water and will not precipitate, regardless of the anions present.

2. **A compound is soluble if it contains one of the following *anions*:**

   - Cl^-, Br^-, I^- (halide)
       except: Ag^+, Hg_2^{2+}, and Pb^{2+} halides
   - NO_3^- (nitrate), ClO_4^- (perchlorate), $CH_3CO_2^-$ (acetate), and SO_4^{2-} (sulfate)
       except: Sr^{2+}, Ba^{2+}, Hg_2^{2+}, and Pb^{2+} sulfates

   That is, most ionic compounds containing a halide, nitrate, perchlorate, acetate, or sulfate anion are soluble in water and will not precipitate regardless of the cations present. The exceptions that *will* precipitate are silver(I), mercury(I) and lead(II) halides, and strontium, barium, mercury(I), and lead(II) sulfates.

On the other hand, a compound that does *not* contain one of the cations or anions listed above is *not* soluble. Thus, carbonates (CO_3^{2-}), sulfides (S^{2-}), phosphates (PO_4^{3-}), and hydroxides (OH^-) are generally not soluble unless they contain an alkali metal or ammonium cation. The main exceptions are the sulfides and hydroxides of Ca^{2+}, Sr^{2+}, and Ba^{2+}. Solubility guidelines are summarized in TABLE 4.2.

You might notice that most of the ions that impart solubility to compounds are singly charged—either singly positive (Li^+, Na^+, K^+, Rb^+, Cs^+, NH_4^+) or singly negative (Cl^-, Br^-, I^-, NO_3^-, ClO_4^-, $CH_3CO_2^-$). Very few doubly charged ions or triply charged ions form soluble compounds. This solubility behavior arises because of the relatively strong ionic bonds in compounds containing ions with multiple charges. The greater the strength of the

TABLE 4.2 Solubility Guidelines for Ionic Compounds in Water

Soluble Compounds	Common Exceptions
Li^+, Na^+, K^+, Rb^+, Cs^+ (group 1A cations)	None
NH_4^+ (ammonium ion)	None
Cl^-, Br^-, I^- (halide) *when paired with*	Halides of Ag^+, Hg_2^{2+}, Pb^{2+} *= insoluble*
NO_3^- (nitrate)	None
ClO_4^- (perchlorate)	None
$CH_3CO_2^-$ (acetate)	None
SO_4^{2-} (sulfate) *when paired with*	Sulfates of Sr^{2+}, Ba^{2+}, Hg_2^{2+}, Pb^{2+} *= insoluble*

Insoluble Compounds	Common Exceptions
CO_3^{2-} (carbonate)	Carbonates of group 1A cations, NH_4^+ *= soluble*
S^{2-} (sulfide)	Sulfides of group 1A cations, NH_4^+, Ca^{2+}, Sr^{2+}, and Ba^{2+} *= soluble*
PO_4^{3-} (phosphate)	Phosphates of group 1A cations, NH_4^+ *= soluble*
OH^- (hydroxide)	Hydroxides of group 1A cations, NH_4^+, Ca^{2+}, Sr^{2+}, and Ba^{2+} *= soluble*

▲ Reaction of aqueous AgNO$_3$ with aqueous Na$_2$CO$_3$ gives a white precipitate of Ag$_2$CO$_3$.

ionic bonds holding ions together in a crystal, the more difficult it is to break those bonds apart during the solution process. We'll return to this topic in Section 6.8.

Using the solubility guidelines makes it possible not only to predict whether a precipitate will form when solutions of two ionic compounds are mixed but also to prepare a specific compound by purposefully carrying out a precipitation. If, for example, you wanted to prepare a sample of solid silver carbonate, Ag$_2$CO$_3$, you could mix a solution of AgNO$_3$ with a solution of Na$_2$CO$_3$. Both starting compounds are soluble in water, as is NaNO$_3$. Silver carbonate is the only insoluble combination of ions and will therefore precipitate from solution.

$$2\,AgNO_3(aq) + Na_2CO_3(aq) \longrightarrow Ag_2CO_3(s) + 2\,NaNO_3(aq)$$

WORKED EXAMPLE 4.6

Predicting the Product of a Precipitation Reaction

Will a precipitation reaction occur when aqueous solutions of CdCl$_2$ and (NH$_4$)$_2$S are mixed? If so, write the net ionic equation.

STRATEGY

Determine the possible products of the reaction by combining the cation from one reactant with the anion from the other reactant.

$$CdCl_2(aq) + (NH_4)_2S(aq) \longrightarrow CdS(?) + 2\,NH_4Cl(?)$$

Next predict the solubility of each product using the guidelines in Table 4.2.

SOLUTION

Of the two possible products, the solubility guidelines predict that CdS, a sulfide, is insoluble and that NH$_4$Cl, an ammonium compound and a halide, is soluble. Thus, a precipitation reaction will likely occur:

$$Cd^{2+}(aq) + S^{2-}(aq) \longrightarrow CdS(s)$$

▶ **PRACTICE 4.12** Predict whether a precipitation reaction will occur in each of the following situations. Write a net ionic equation for each reaction that occurs.

(a) NiCl$_2$(aq) + (NH$_4$)$_2$S(aq) $\longrightarrow$?
(b) Na$_2$CrO$_4$(aq) + Pb(NO$_3$)$_2$(aq) $\longrightarrow$?
(c) AgClO$_4$(aq) + CaBr$_2$(aq) $\longrightarrow$?
(d) ZnCl$_2$(aq) + K$_2$CO$_3$(aq) $\longrightarrow$?

▶ **APPLY 4.13** How might you use a precipitation reaction to prepare a sample of Ca$_3$(PO$_4$)$_2$? Write the net ionic equation.

Conceptual WORKED EXAMPLE 4.7

Visualizing Stoichiometry in Precipitation Reactions

When aqueous solutions of two ionic compounds are mixed, the following results are obtained:

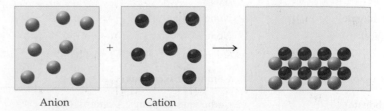

Anion   Cation

(Only the anion of the first compound, represented by blue spheres, and the cation of the second compound, represented by red spheres, are shown.) Which cation and anion combinations are compatible with the observed results?
Anions: NO$_3^-$, Cl$^-$, CO$_3^{2-}$, PO$_4^{3-}$
Cations: Ca^{2+}, Ag$^+$, K$^+$, Cd^{2+}

STRATEGY

The drawing represents a precipitation reaction because it shows that ions in solution fall to the bottom of the container in an ordered arrangement. Counting the spheres shows that the cation and anion react in equal numbers (8 of each), so they must have the same number of charges—either both singly charged or both doubly charged. (There is no triply charged cation in the list.) Look at all the possible combinations, and decide which would precipitate.

SOLUTION

Possible combinations of singly charged ions: $AgNO_3$, KNO_3, $AgCl$, KCl

Possible combinations of doubly charged ions: $CaCO_3$, $CdCO_3$

Of the possible combinations, $AgCl$, $CaCO_3$, and $CdCO_3$ are insoluble. Therefore, the precipitate could arise from three possible combinations of anions and cations (1) Ag^+ and Cl^-, (2) Ca^{2+} and CO_3^{2-}, or (3) Cd^{2+} and CO_3^{2-}.

▶ **Conceptual PRACTICE 4.14** An aqueous solution containing an anion, represented by blue spheres, is added to another solution containing a cation, represented by red spheres, and the following result is obtained.

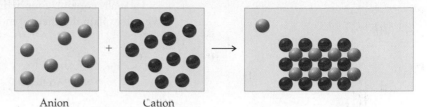

Anion   Cation

Which cation and anion combinations are compatible with the observed results?

Anions: S^{2-}, PO_4^{3-}, ClO_4^-
Cations: Mg^{2+}, Fe^{3+}, NH_4^+, Zn^{2+}

▶ **Conceptual APPLY 4.15** A solution containing the compounds $Pb(NO_3)_2$, KBr, and $Ba(CH_3CO_2)_2$ is prepared and a white precipitate forms. Identify the precipitate and draw a diagram similar to the one in Worked Example 4.7 to represent the precipitation reaction.

4.7 ▶ ACIDS, BASES, AND NEUTRALIZATION REACTIONS

In 1777, the French chemist Antoine Lavoisier proposed that all acids contain a common element: oxygen. In fact, the word *oxygen* is derived from a Greek phrase meaning "acid former." Lavoisier's idea had to be modified, however, when the English chemist Sir Humphrey Davy (1778–1829) showed in 1810 that muriatic acid (now called hydrochloric acid) contains only hydrogen and chlorine but no oxygen. Davy's studies thus suggested that the common element in acids is *hydrogen*, not oxygen.

The relationship between acidic behavior and the presence of hydrogen in a compound was clarified in 1887 by the Swedish chemist Svante Arrhenius (1859–1927). Arrhenius proposed that an **acid** is a substance that dissociates in water to give hydrogen ions (H^+) and a **base** is a substance that dissociates in water to give hydroxide ions (OH^-):

$$\textbf{An acid} \quad HA(aq) \longrightarrow H^+(aq) + A^-(aq)$$
$$\textbf{A base} \quad MOH(aq) \longrightarrow M^+(aq) + OH^-(aq)$$

In these equations, HA is a general formula for an acid—for example, HCl or HNO_3—and MOH is a general formula for a metal hydroxide—for example, $NaOH$ or KOH.

Although convenient to use in equations, the symbol $H^+(aq)$ does not really represent the structure of the ion present in aqueous solution. As a bare hydrogen nucleus—a proton—with no electron nearby, H^+ is much too reactive to exist by itself. Rather, the H^+ bonds to the oxygen atom of a water molecule and forms the more stable **hydronium ion, H_3O^+**. We'll sometimes write $H^+(aq)$ for convenience, particularly when balancing equations, but will more often write $H_3O^+(aq)$ to represent an aqueous acid solution. Hydrogen chloride, for instance, gives $Cl^-(aq)$ and $H_3O^+(aq)$ when it dissolves in water.

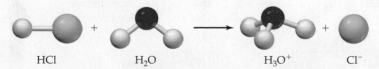

HCl H_2O H_3O^+ Cl^-

Different acids dissociate to different extents in aqueous solution. Acids that dissociate to a large extent are strong electrolytes and **strong acids**, whereas acids that dissociate to only a small extent are weak electrolytes and **weak acids**. We've already seen in Table 4.1, for instance, that HCl, $HClO_4$, HNO_3, and H_2SO_4 are strong electrolytes and therefore strong acids, while CH_3CO_2H and HF are weak electrolytes and therefore weak acids. You might note that acetic acid actually contains four hydrogens, but only the one bonded to the oxygen atom dissociates. We will explain the effect of molecular structure on acid dissociation in Chapter 15.

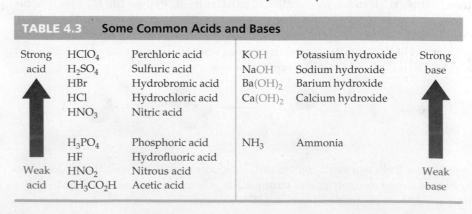

Acetic acid

Nonacidic

Acidic

Different acids can have different numbers of acidic hydrogens and yield different numbers of H_3O^+ ions in solution. Hydrochloric acid (HCl) is said to be a **monoprotic acid** because it provides only one H^+ ion, but sulfuric acid (H_2SO_4) is a **diprotic acid** because it can provide two H^+ ions. Phosphoric acid (H_3PO_4) is a **triprotic acid** and can provide three H^+ ions. With sulfuric acid, the first dissociation of an H^+ is complete—all H_2SO_4 molecules lose one H^+—but the second dissociation is incomplete, as indicated by the double arrow in the following equation:

$$\text{Sulfuric acid:} \quad H_2SO_4(aq) + H_2O(l) \longrightarrow HSO_4^-(aq) + H_3O^+(aq)$$
$$HSO_4^-(aq) + H_2O(l) \rightleftharpoons SO_4^{2-}(aq) + H_3O^+(aq)$$

With phosphoric acid, none of the three dissociations is complete:

$$\text{Phosphoric acid:} \quad H_3PO_4(aq) + H_2O(l) \rightleftharpoons H_2PO_4^-(aq) + H_3O^+(aq)$$
$$H_2PO_4^-(aq) + H_2O(l) \rightleftharpoons HPO_4^{2-}(aq) + H_3O^+(aq)$$
$$HPO_4^{2-}(aq) + H_2O(l) \rightleftharpoons PO_4^{3-}(aq) + H_3O^+(aq)$$

Bases, like acids, can also be either strong or weak, depending on the extent to which they produce OH^- ions in aqueous solution. Most metal hydroxides, such as NaOH and $Ba(OH)_2$, are strong electrolytes and **strong bases**, but ammonia (NH_3) is a weak electrolyte and a **weak base**. Ammonia is a weak base because it reacts to a small extent with water to yield NH_4^+ and OH^- ions. In fact, aqueous solutions of ammonia are often called *ammonium hydroxide*, although this is really a misnomer because the concentrations of NH_4^+ and OH^- ions are low.

$$NH_3(g) + H_2O(l) \rightleftharpoons NH_4^+(aq) + OH^-(aq)$$

As with the dissociation of acetic acid, discussed in Section 4.3, the reaction of ammonia with water takes place only to a small extent (about 1%). Most of the ammonia remains unreacted, and we therefore write the reaction with a double arrow to show that a dynamic equilibrium exists between the forward and reverse reactions.

TABLE 4.3 summarizes the names, formulas, and classification of some common acids and bases.

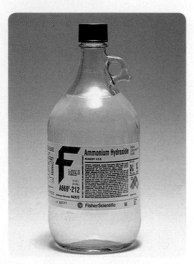

▲ Shouldn't this bottle be labeled "Aqueous Ammonia" rather than "Ammonium Hydroxide"?

REMEMBER...

Oxoanions are polyatomic anions in which an atom of a given element is combined with different numbers of oxygen atoms. (Section 2.12)

Naming Acids

Most acids are **oxoacids**; that is, they contain oxygen in addition to hydrogen and other elements. When dissolved in water, an oxoacid yields one or more H^+ ions and an **oxoanion** like one of those listed in **TABLE 4.4** and discussed previously in Section 2.12.

TABLE 4.3	Some Common Acids and Bases				
Strong acid	$HClO_4$	Perchloric acid	KOH	Potassium hydroxide	Strong base
	H_2SO_4	Sulfuric acid	NaOH	Sodium hydroxide	
	HBr	Hydrobromic acid	$Ba(OH)_2$	Barium hydroxide	
	HCl	Hydrochloric acid	$Ca(OH)_2$	Calcium hydroxide	
	HNO_3	Nitric acid			
	H_3PO_4	Phosphoric acid	NH_3	Ammonia	
	HF	Hydrofluoric acid			
Weak acid	HNO_2	Nitrous acid			Weak base
	CH_3CO_2H	Acetic acid			

TABLE 4.4   Common Oxoacids and Their Anions

Oxoacid		Oxoanion	
HNO_2	Nitrous acid	NO_2^-	Nitrite ion
HNO_3	Nitric acid	NO_3^-	Nitrate ion
H_3PO_4	Phosphoric acid	PO_4^{3-}	Phosphate ion
H_2SO_3	Sulfurous acid	SO_3^{2-}	Sulfite ion
H_2SO_4	Sulfuric acid	SO_4^{2-}	Sulfate ion
$HClO$	Hypochlorous acid	ClO^-	Hypochlorite ion
$HClO_2$	Chlorous acid	ClO_2^-	Chlorite ion
$HClO_3$	Chloric acid	ClO_3^-	Chlorate ion
$HClO_4$	Perchloric acid	ClO_4^-	Perchlorate ion

The names of oxoacids are related to the names of the corresponding oxoanions, with the -ite or -ate ending of the anion name replaced by -ous acid or -ic acid, respectively. In other words, the acid with fewer oxygens has an -ous ending, and the acid with more oxygens has an -ic ending. The compound HNO_2, for example, is called *nitrous acid* because it has fewer oxygens and yields the nitrite ion (NO_2^-) when dissolved in water, while HNO_3 is called *nitric acid* because it has more oxygens and yields the nitrate ion (NO_3^-) when dissolved in water.

Nitrous Acid Gives Nitrite Ion

$$HNO_2(aq) + H_2O(l) \xrightarrow{\text{Dissolve in water}} H_3O^+(aq) + NO_2^-(aq)$$

Nitric Acid Gives Nitrate Ion

$$HNO_3(aq) + H_2O(l) \xrightarrow{\text{Dissolve in water}} H_3O^+(aq) + NO_3^-(aq)$$

In a similar way, hypochlorous acid yields the hypochlorite ion, chlorous acid yields the chlorite ion, chloric acid yields the chlorate ion, and perchloric acid yields the perchlorate ion (Table 4.4).

In addition to the oxoacids, there are a small number of other common acids, such as HCl, that do not contain oxygen. For such compounds, the prefix *hydro-* and the suffix -ic acid are used for the aqueous solution.

Hydrogen Chloride Gives *Hydrochloric Acid*

$$HCl(g) + H_2O(l) \xrightarrow{\text{Dissolve in water}} H_3O^+(aq) + Cl^-(aq)$$

Hydrogen Cyanide Gives *Hydrocyanic Acid*

$$HCN(g) + H_2O(l) \xrightarrow{\text{Dissolve in water}} H_3O^+(aq) + CN^-(aq)$$

WORKED EXAMPLE 4.8

Naming Acids

Name the following acids:
(a) $HBrO(aq)$
(b) $H_2S(aq)$

STRATEGY

To name an acid, look at its formula and decide whether the compound is an oxoacid. If so, the name must reflect the number of oxygen atoms, according to Table 4.4. If the compound is not an oxoacid, it is named using the prefix *hydro-* and the suffix -ic acid.

continued on next page

SOLUTION

(a) This compound is an oxoacid that yields hypobromite ion (BrO^-) when dissolved in water. Its name is *hypobromous acid*.

(b) This compound is not an oxoacid but yields sulfide ion when dissolved in water. As a pure gas, H_2S is named hydrogen sulfide. In water solution, it is called *hydrosulfuric acid*.

▶ **PRACTICE 4.16**   Name the following acids:

(a) HI   (b) $HBrO_2$ (c) H_2CrO_4

▶ **APPLY 4.17**   Give likely chemical formulas corresponding to the following names:

(a) Phosphorous acid   (b) Hydroselenic acid

Neutralization Reactions

When an acid and a base are mixed in the right stoichiometric proportions, both acidic and basic properties disappear because of a neutralization reaction that produces water and an ionic **salt**. The anion of the salt (A^-) comes from the acid, and the cation of the salt (M^+) comes from the base:

A neutralization reaction

$$HA(aq) + MOH(aq) \longrightarrow H_2O(l) + MA(aq)$$

$$\text{Acid} \qquad \text{Base} \qquad \qquad \text{Water} \qquad \text{A salt}$$

Because salts are generally strong electrolytes in aqueous solution, we can write the neutralization reaction of a strong acid with a strong base as an ionic equation:

$$H^+(aq) + A^-(aq) + M^+(aq) + OH^-(aq) \longrightarrow H_2O(l) + M^+(aq) + A^-(aq)$$

Canceling the ions that appear on both sides of the ionic equation, A^- and M^+, gives the net ionic equation, which describes the reaction of any strong acid with any strong base in water.

Net Ionic Equation

$$H^+(aq) + OH^-(aq) \longrightarrow H_2O(l)$$
$$\text{or} \quad H_3O^+(aq) + OH^-(aq) \longrightarrow 2\,H_2O(l)$$

For the reaction of a weak acid with a strong base, a similar neutralization occurs, but we must write the molecular formula of the acid rather than simply $H^+(aq)$, because the dissociation of the acid in water is incomplete. Instead, the acid exists primarily as the neutral molecule. In the reaction of the weak acid HF with the strong base KOH, for example, we write the net ionic equation as

$$HF(aq) + OH^-(aq) \longrightarrow H_2O(l) + F^-(aq)$$

● WORKED EXAMPLE 4.9

Writing Ionic and Net Ionic Equations for an Acid–Base Reaction

Write both an ionic equation and a net ionic equation for the neutralization reaction of aqueous HBr and aqueous $Ba(OH)_2$.

STRATEGY

Hydrogen bromide is a strong acid whose aqueous solution contains H^+ ions and Br^- ions. Barium hydroxide is a strong base whose aqueous solution contains Ba^{2+} and OH^- ions. Thus, we have a mixture of four different ions on the reactant side. Write the neutralization reaction as an ionic equation, and then cancel spectator ions to give the net ionic equation.

SOLUTION

Ionic Equation:

$$2\,H^+(aq) + 2\,Br^-(aq) + Ba^{2+}(aq) + 2\,OH^-(aq) \longrightarrow 2\,H_2O(l) + 2\,Br^-(aq) + Ba^{2+}(aq)$$

Net Ionic Equation

$$2H^+(aq) + 2OH^-(aq) \longrightarrow 2H_2O(l)$$
$$\text{or} \quad H^+(aq) + OH^-(aq) \longrightarrow H_2O(l)$$

The reaction of HBr with $Ba(OH)_2$ involves the combination of a proton (H^+) from the acid with OH^- from the base to yield water and an aqueous salt ($BaBr_2$).

▶ **PRACTICE 4.18** Write a balanced ionic equation and net ionic equation for each of the following acid–base reactions:

(a) $2\,CsOH(aq) + H_2SO_4(aq) \rightarrow ?$
(b) $Ca(OH)_2(aq) + 2\,CH_3CO_2H(aq) \rightarrow ?$

▶ **APPLY 4.19** Milk of magnesia (active ingredient: magnesium hydroxide) is used as an antacid to treat indigestion and heartburn. Write a balanced ionic equation and net ionic equation for the reaction of magnesium hydroxide with stomach acid, hydrochloric acid.

4.8 ▶ SOLUTION STOICHIOMETRY

We remarked in Section 4.1 that molarity is a conversion factor between numbers of moles of solute and the volume of a solution. Thus, if we know the volume and molarity of a solution, we can calculate the number of moles of solute. If we know the number of moles of solute and molarity, we can find the volume.

As indicated by the flow diagram in **FIGURE 4.4**, using molarity is critical for carrying out stoichiometry calculations on substances in solution. Molarity makes it possible to calculate the volume of one solution needed to react with a given volume of another solution. This sort of calculation is particularly important in the chemistry of acids and bases, as shown in Worked Example 4.10.

● WORKED EXAMPLE 4.10

Reaction Stoichiometry in Solution

Stomach acid, a dilute solution of HCl in water, can be neutralized by reaction with sodium hydrogen carbonate, $NaHCO_3$, according to the equation

$$HCl(aq) + NaHCO_3(aq) \longrightarrow NaCl(aq) + H_2O(l) + CO_2(g)$$

How many milliliters of 0.125 M $NaHCO_3$ solution are needed to neutralize 18.0 mL of 0.100 M HCl?

IDENTIFY

Known	Unknown
Volume of HCl (18.0 mL)	Volume of $NaHCO_3$ (mL)
Concentration of HCl (0.100 M)	
Concentration of $NaHCO_3$ (0.125 M)	
Balanced reaction	

STRATEGY

Solving stoichiometry problems always requires finding the number of moles of one reactant, and using the coefficients of the balanced equation to find the number of moles of the other reactant. Molarity is used as a conversion factor between volume and moles. Figure 4.4 outlines the steps for relating the volumes of reactants or products.

continued on next page

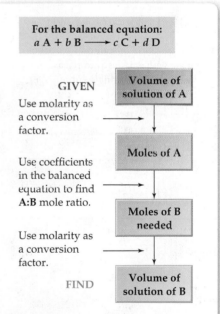

▲ **FIGURE 4.4**

Using molarity as a conversion factor between moles and volume in stoichiometry calculations.

Figure It Out

Show the conversion factors needed to convert from liters of solution A to liters of solution B. Assume the molarity of solution A = X, the stoichiometric coefficients for A and B are *a* and *b*, respectively, and the molarity of solution B = Y.

Answer: $L_A \times \dfrac{X \text{ mol } A}{1 L} \times \dfrac{b \text{ mol } B}{a \text{ mol } A} \times \dfrac{1L}{Y \text{ mol } B} = L_B$

▲ Neutralization of sodium hydrogen carbonate with acid leads to release of CO_2 gas, visible in this fizzing solution.

SOLUTION

We first have to find how many moles of HCl are in 18.0 mL of a 0.100 M solution by multiplying volume times molarity:

$$\text{Moles of HCl} = 18.0 \text{ mL} \times \frac{1 \text{ L}}{1000 \text{ mL}} \times \frac{0.100 \text{ mol}}{1 \text{ L}} = 1.80 \times 10^{-3} \text{ mol HCl}$$

Next, find the moles of $NaHCO_3$ using the coefficients from the balanced equation.

$$1.80 \times 10^{-3} \text{ mol HCl} \times \frac{1 \text{ mol NaHCO}_3}{1 \text{ mol HCl}} = 1.80 \times 10^{-3} \text{ mol NaHCO}_3$$

Find the volume of $NaHCO_3$ by using the molarity to convert between moles and volume.

$$1.80 \times 10^{-3} \text{ mol NaHCO}_3 \times \frac{1 \text{ L solution}}{0.125 \text{ mol NaHCO}_3} \times \frac{1000 \text{ mL}}{1 \text{ L solution}} = 14.4 \text{ mL solution}$$

Thus, 14.4 mL of the 0.125 M $NaHCO_3$ solution is needed to neutralize 18.0 mL of the 0.100 M HCl solution.

CHECK

The balanced equation shows that HCl and $NaHCO_3$ react in a 1 : 1 molar ratio, and we are told that the concentrations of the two solutions are about the same. Thus, the volume of the $NaHCO_3$ solution must be about the same as that of the HCl solution.

▶ **PRACTICE 4.20**   What volume of 0.250 M H_2SO_4 is needed to react with 50.0 mL of 0.100 M NaOH? The equation is

$$H_2SO_4(aq) + 2 NaOH(aq) \longrightarrow Na_2SO_4(aq) + 2 H_2O(l)$$

▶ **APPLY 4.21**   What is the molarity of a HNO_3 solution if 68.5 mL is needed to react with 25.0 mL of 0.150 M KOH solution? The equation is

$$HNO_3(aq) + KOH(aq) \longrightarrow KNO_3(aq) + H_2O(l)$$

4.9 ▶ MEASURING THE CONCENTRATION OF A SOLUTION: TITRATION

Measuring the concentration of solutions is important for a variety of reasons. For example, the concentration of ions such as H^+, OH^-, Fe^{3+}, Mg^{2+}, Ca^{2+}, and Cl^- are important in the quality of water used for drinking, irrigation, and industrial processes. Furthermore, solutions prepared by the methods described in Section 4.1 may not result in the exact molarity calculated by dividing moles of solute by volume of solution. This discrepancy arises because chemicals often cannot be purchased in their pure form or they react with other chemicals in the solvent or in the air. A technique frequently used for determining a solution's exact molarity is called a *titration*.

Titration is a procedure for determining the concentration of a solution by allowing a measured volume of that solution to react with a second solution of another substance (the *standard solution*) whose concentration is known. By finding the volume of the standard solution that reacts with the measured volume of the first solution, the concentration of the first solution can be calculated. (It's necessary, though, that the reaction go to completion and have a yield of 100%.)

To see how titration works, let's imagine that we have an HCl solution (an acid) whose concentration we want to find by allowing it to react with NaOH (a base) in an acid–base neutralization reaction. The balanced equation is

$$NaOH(aq) + HCl(aq) \longrightarrow NaCl(aq) + H_2O(l)$$

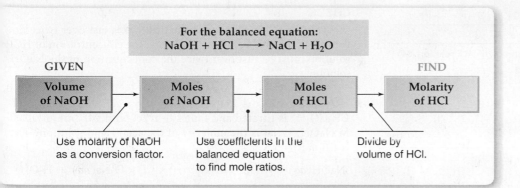

For the balanced equation:
$$NaOH + HCl \longrightarrow NaCl + H_2O$$

GIVEN

| Volume of NaOH | → | Moles of NaOH | → | Moles of HCl | → | Molarity of HCl |

FIND

Use molarity of NaOH as a conversion factor.

Use coefficients in the balanced equation to find mole ratios.

Divide by volume of HCl.

◀ **FIGURE 4.5**

A flow diagram for an acid–base titration. The calculations needed to determine the concentration of an HCl solution by titration with an NaOH standard solution are summarized.

A measured volume of acid solution is placed in a flask, and phenolphthalein indicator is added.

Base solution of known concentration is added from a buret until the indicator changes **color**. Reading the volume of base from the buret allows calculation of the acid concentration.

◀ **FIGURE 4.6**

Titration of an acid solution of unknown concentration with a base solution of known concentration.

Figure It Out

What does the color change indicate in the chemical reaction between HCl and NaOH?

Answer: The HCl is completely reacted and there is excess base in solution. The first faint, permanent color change indicates the end of the titration.

We'll begin the titration by measuring out a known volume of the HCl solution and adding a small amount of an *indicator*, a compound that undergoes a color change during the course of the reaction. The compound phenolphthalein, for instance, is colorless in acid solution but turns red in base solution. Next, we fill a calibrated glass tube called a *buret* with an NaOH standard solution of known concentration and slowly add the NaOH to the HCl. When the phenolphthalein just begins to turn pink, all the HCl has completely reacted and the solution now has a tiny amount of excess NaOH. By then reading from the buret to find the volume of the NaOH standard solution that has been added to react with the known volume of HCl solution, we can calculate the concentration of the HCl. The strategy is summarized in **FIGURE 4.5**, and the procedure is shown in **FIGURE 4.6**.

WORKED EXAMPLE 4.11

Determining the Concentration of a Solution Using a Titration Procedure

A 20.0 mL sample of hydrochloric acid (HCl) is titrated and found to react with 42.6 mL of 0.100 M NaOH. What is the molarity of the hydrochloric acid solution?

continued on next page

IDENTIFY

Known	Unknown
Volume of HCl solution (20.0 mL)	Concentration of HCl solution (M)
Volume of NaOH solution (42.6 mL)	
Concentration of NaOH solution (0.100 M)	
Balanced reaction	

STRATEGY

Solving stoichiometry problems always requires finding the number of moles of one reactant, and using the coefficients of the balanced equation to find the number of moles of the other reactant. Molarity is used as a conversion factor between volume and moles. Figure 4.5 outlines the steps for finding concentration in a titration experiment.

SOLUTION

Using the molarity of the NaOH standard solution as a conversion factor, we can calculate the number of moles of NaOH undergoing reaction:

$$\text{Moles of NaOH} = 0.0426 \text{ L NaOH} \times \frac{0.100 \text{ mol NaOH}}{1 \text{ L NaOH}}$$

$$= 0.004\,26 \text{ mol NaOH}$$

According to the balanced equation, the number of moles of HCl is the same as that of NaOH:

$$\text{Moles of HCl} = 0.004\,26 \text{ mol NaOH} \times \frac{1 \text{ mol HCl}}{1 \text{ mol NaOH}} = 0.004\,26 \text{ mol HCl}$$

Dividing the number of moles of HCl by the volume then gives the molarity of the HCl:

$$\text{HCl molarity} = \frac{0.004\,26 \text{ mol HCl}}{0.0200 \text{ L HCl}} = 0.213 \text{ M HCl}$$

CHECK

The volume of NaOH used in the titration is just over twice the volume of the HCl solution. Therefore, the concentration of HCl solution must be just over twice the concentration of the NaOH solution.

▶ **PRACTICE 4.22** A 25.0 mL sample of vinegar (dilute acetic acid, CH_3CO_2H) is titrated and found to react with 94.7 mL of 0.200 M NaOH. What is the molarity of the acetic acid solution? The reaction is

$$NaOH(aq) + CH_3CO_2H(aq) \longrightarrow CH_3CO_2Na(aq) + H_2O(l)$$

▶ **Conceptual APPLY 4.23** Assume that the buret contains H^+ ions, the flask contains OH^- ions, and each has a volume of 100 mL. How many milliliters would you need to add from the buret to the flask to neutralize all the OH^- ions in a titration procedure? The equation is $H^+(aq) + OH^-(aq) \longrightarrow H_2O(l)$.

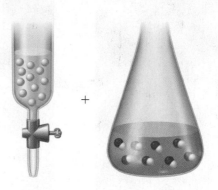

4.10 ▶ OXIDATION–REDUCTION (REDOX) REACTIONS

Purple aqueous permanganate ion, MnO_4^-, reacts with aqueous Fe^{2+} ion to yield Fe^{3+} and pale pink Mn^{2+}:

$$MnO_4^-(aq) + 5 Fe^{2+}(aq) + 8 H^+(aq) \longrightarrow Mn^{2+}(aq) + 5 Fe^{3+}(aq) + 4 H_2O(l)$$

Magnesium metal burns in air with an intense white light to form solid magnesium oxide:

$$2 Mg(s) + O_2(g) \longrightarrow 2 MgO(s)$$

Red phosphorus reacts with liquid bromine to form liquid phosphorus tribromide:

$$2 P(s) + 3 Br_2(l) \longrightarrow 2 PBr_3(l)$$

Although these and many thousands of other reactions appear unrelated, and many don't even take place in aqueous solution, all are oxidation–reduction (redox) reactions.

Historically, the word *oxidation* referred to the combination of an element with oxygen to yield an oxide, and the word *reduction* referred to the removal of oxygen from an oxide to yield the element. Such oxidation–reduction processes have been crucial to the development of human civilization and still have enormous commercial value. The oxidation (rusting) of iron metal by reaction with moist air has been known for millennia and is still a serious problem that causes enormous structural damage to buildings, boats, and bridges. The reduction

Aqueous potassium permanganate, deep purple in color, is frequently used as an *oxidizing agent*, as described in the text. Magnesium metal burns in air to give MgO. Elemental phosphorus reacts spectacularly with bromine to give PBr_3.

of iron ore (Fe_2O_3) with charcoal (C) to make iron metal has been carried out since prehistoric times and is still used today in the initial stages of steelmaking.

$$4\, Fe(s) + 3\, O_2(g) \longrightarrow 2\, Fe_2O_3(s) \qquad \text{Rusting of iron: an \textbf{oxidation} of Fe}$$

$$2\, Fe_2O_3(s) + 3\, C(s) \longrightarrow 4\, Fe(s) + 3\, CO_2(g) \quad \text{Manufacture of iron: a \textbf{reduction} of } Fe_2O_3$$

Today, the words *oxidation* and *reduction* have taken on a much broader meaning. An **oxidation** is now defined as the loss of one or more electrons by a substance, whether element, compound, or ion, and a **reduction** is the gain of one or more electrons by a substance. Thus, an oxidation–reduction, or redox, reaction is any process in which electrons are transferred from one substance to another.

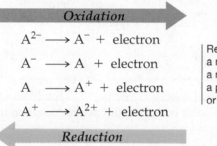

Oxidation

$$A^{2-} \longrightarrow A^- + \text{electron}$$
$$A^- \longrightarrow A + \text{electron}$$
$$A \longrightarrow A^+ + \text{electron}$$
$$A^+ \longrightarrow A^{2+} + \text{electron}$$

Reduction

Reactant A might be:
a neutral atom,
a monatomic ion,
a polyatomic ion,
or a molecule.

How can you tell when a redox reaction takes place? The answer is that you assign to each atom in a compound a value called an **oxidation number** (or *oxidation state*), which indicates whether the atom is neutral, electron-rich, or electron-poor. By comparing the oxidation number of an atom before and after reaction, you can tell whether the atom has gained or lost electrons. Note that oxidation numbers don't necessarily imply ionic charges; they are just a convenient device to help keep track of electrons during redox reactions.

The rules for assigning oxidation numbers are as follows:

. **An atom in its elemental state has an oxidation number of 0. For example:**

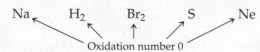

$$Na \qquad H_2 \qquad Br_2 \qquad S \qquad Ne$$

Oxidation number 0

. **An atom in a monatomic ion has an oxidation number identical to its charge.** Review Section 2.12 to see the charges on some common ions. For example:

$$Na^+ \qquad Ca^{2+} \qquad Al^{3+} \qquad Cl^- \qquad O^{2-}$$
$$+1 \qquad\;\; +2 \qquad\;\; +3 \qquad\;\; -1 \qquad\;\; -2$$

3. **An atom in a polyatomic ion or in a molecular compound usually has the same oxida-tion number it would have if it were a monatomic ion.** In the hydroxide ion (OH^-), for instance, the hydrogen atom has an oxidation number of +1, as if it were H^+, and the oxy-gen atom has an oxidation number of -2, as if it were a monatomic O^{2-} ion.

In general, the farther left an element is in the periodic table, the more probable that it will be "cationlike." Metals, therefore, usually have positive oxidation numbers. The far-ther right an element is in the periodic table, the more probable that it will be "anionlike." Nonmetals, such as O, N, and the halogens, usually have negative oxidation numbers. We'll see the reasons for these trends in Sections 6.3–6.5.

Nonmetals;
"anionlike"
Negative oxidation numbers

Metals;
"cationlike"
Positive oxidation numbers

(a) **Hydrogen can be either +1 or −1.** When bonded to a metal, such as Na or Ca, hydro-gen has an oxidation number of -1. When bonded to a nonmetal, such as C, N, O, or Cl, hydrogen has an oxidation number of +1.

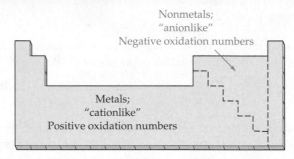

(b) **Oxygen usually has an oxidation number of** -2. The major exception is in com-pounds called *peroxides*, which contain either the O_2^{2-} ion or an O–O covalent bond in a molecule. Both oxygen atoms in a peroxide have an oxidation number of -1.

$$\underset{+1 \quad -2 \quad +1}{H-O-H} \qquad \underset{+1 \quad -1 \quad -1 \quad +1}{H-O-O-H} \qquad \underset{-1 \qquad -1}{[O-O]^{2-}}$$

(c) **Halogens usually have an oxidation number of** -1. The major exception is in com-pounds of chlorine, bromine, or iodine in which the halogen atom is bonded to oxy-gen. In such cases, the oxygen has an oxidation number of -2, and the halogen has a positive oxidation number. In Cl_2O, for instance, the O atom has an oxidation number of -2 and each Cl atom has an oxidation number of +1.

$$\underset{+1 \quad -2 \quad +1}{Cl-O-Cl} \qquad \underset{+1 \quad -2 \quad +1}{H-O-Br}$$

4. **The sum of the oxidation numbers is 0 for a neutral compound and is equal to the net charge for a polyatomic ion.** This rule is particularly useful for finding the oxidation number of an atom in difficult cases. The general idea is to assign oxidation numbers to the "easy" atoms first and then find the oxidation number of the "difficult" atom by subtrac-tion. For example, suppose we want to know the oxidation number of the sulfur atom in sulfuric acid, H_2SO_4. Since each H atom is +1 and each O atom is -2, the S atom must have an oxidation number of +6 for the compound to have no net charge:

$$\underset{+1 \quad ? \quad -2}{H_2SO_4} \qquad 2(+1) + (?) + 4(-2) = 0 \text{ net charge}$$
$$? = 0 - 2(+1) - 4(-2) = +6$$

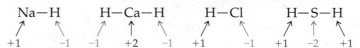

To find the oxidation number of the chlorine atom in the perchlorate anion (ClO_4^-), we know that each oxygen is -2, so the Cl atom must have an oxidation number of $+7$ for there to be a net charge of -1 on the ion:

$$ClO_4^-$$
$$? \quad -2$$
$$? + 4(-2) = -1 \text{ net charge}$$
$$? = -1 - 4(-2) = +7$$

To find the oxidation number of the nitrogen atom in the ammonium cation (NH_4^+), we know that each H atom is $+1$, so the N atom must have an oxidation number of -3 for the ion to have a net charge of $+1$:

$$NH_4^+$$
$$? \quad +1$$
$$? + 4(+1) = +1 \text{ net charge}$$
$$? = +1 - 4(+1) = -3$$

● WORKED EXAMPLE 4.12

Assigning Oxidation Numbers

Assign oxidation numbers to each atom in the following substances:
(a) CdS   **(b)** AlH_3 **(c)** $S_2O_3^{2-}$ **(d)** $Na_2Cr_2O_7$

STRATEGY

(a) The sulfur atom in S^{2-} has an oxidation number of -2, so Cd must be $+2$.

(b) H bonded to a metal has the oxidation number -1, so Al must be $+3$.

(c) O usually has the oxidation number -2, so S must be $+2$ for the anion to have a net charge of -2: for $(2\,S^{+2})\,(3\,O^{-2})$, $2(+2) + 3(-2) = -2$ net charge.

(d) Na is always $+1$, and oxygen is -2, so Cr must be $+6$ for the compound to be neutral: for $(2\,Na^+)\,(2\,Cr^{+6})\,(7\,O^{-2})$, $2(+1) + 2(+6) + 7(-2) = 0$ net charge.

SOLUTION

(a) CdS
$\uparrow \; \uparrow$
$+2 \; -2$

(b) AlH_3
$\uparrow \; \uparrow$
$+3 \; -1$

(c) $S_2O_3^{2-}$
$\uparrow \; \uparrow$
$+2 \; -2$

(d) $Na_2Cr_2O_7$
$\uparrow \quad \uparrow \quad \uparrow$
$+1 \quad +6 \quad -2$

▶ **PRACTICE 4.24**   Assign an oxidation number to each atom in the following compounds:

(a) $SnCl_4$ **(b)** CrO_3 **(c)** $VOCl_3$
(d) V_2O_3 **(e)** HNO_3 **(f)** $FeSO_4$

▶ **APPLY 4.25**   Chlorine can have several different oxidation numbers ranging in value from -1 to $+7$.

(a) Write the formula and give the name of the chlorine oxide compound in which chlorine has an oxidation number of $+2, +3, +6$, and $+7$.

(b) Based on oxidation numbers, which chlorine oxide from part (a) cannot react with molecular oxygen?

4.11 ▶ IDENTIFYING REDOX REACTIONS

Once oxidation numbers are assigned, it's clear why all the reactions mentioned in the previous section are redox processes. Take the rusting of iron, for example. Two of the reactants, Fe and O_2, are neutral elements and have oxidation numbers of 0. In the product, however, the oxygen atoms have an oxidation number of -2 and the iron atoms have an oxidation number of $+3$. Thus, Fe has undergone a change from 0 to $+3$ (a loss of electrons, or oxidation), and O has undergone a change from 0 to -2 (a gain of electrons, or reduction). Note that the total number of electrons given up by the atoms being oxidized ($4\,Fe \times 3\,\text{electrons}/Fe = 12$ electrons) is the same as the number gained by the atoms being reduced ($6\,O \times 2\,\text{electrons}/O = 12$ electrons).

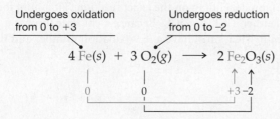

Undergoes oxidation from 0 to $+3$ Undergoes reduction from 0 to -2

$$4\,Fe(s) + 3\,O_2(g) \longrightarrow 2\,Fe_2O_3(s)$$
$$0 \qquad\qquad 0 \qquad\qquad +3\ -2$$

▲ The iron used in this prehistoric dagger handle was made by the reduction of iron ore with charcoal.

A similar analysis can be carried out for the production of iron metal from its ore. The iron atom is reduced because it goes from an oxidation number of +3 in the reactant (Fe_2O_3) to 0 in the product (Fe). At the same time, the carbon atom is oxidized because it goes from an oxidation number of 0 in the reactant (C) to +4 in the product (CO_2). The oxygen atoms undergo no change because they have an oxidation number of −2 in both reactant and product. The total number of electrons given up by the atoms being oxidized ($3\,C \times 4$ electrons/C = 12 electrons) is the same as the number gained by the atoms being reduced ($4\,Fe \times 3$ electrons/Fe = 12 electrons).

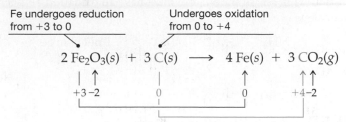

$$2\,Fe_2O_3(s) + 3\,C(s) \longrightarrow 4\,Fe(s) + 3\,CO_2(g)$$

As these examples show, oxidation and reduction reactions, called **half-reactions**, always occur together. A redox reaction consists of two half-reactions; one oxidation half-reaction and one reduction half-reaction. Whenever one atom loses one or more electrons, another atom must gain those electrons. The substance that *causes* a reduction by giving up electrons—the iron atom in the reaction of Fe with O_2 and the carbon atom in the reaction of C with Fe_2O_3—is called a **reducing agent**. The substance that causes an oxidation by accepting electrons—the oxygen atom in the reaction of Fe with O_2 and the iron atom in the reaction of C with Fe_2O_3—is called an **oxidizing agent**. The reducing agent is itself oxidized when it gives up electrons and the oxidizing agent is itself reduced when it accepts electrons.

Reducing agent
- Causes reduction
- Loses one or more electrons
- Undergoes oxidation
- Oxidation number of atom increases

Oxidizing agent
- Causes oxidation
- Gains one or more electrons
- Undergoes reduction
- Oxidation number of atom decreases

We'll see in later chapters that redox reactions are common for almost every element in the periodic table except for the noble gas elements of group 8A. In general, metals give up electrons and act as reducing agents, while reactive nonmetals such as O_2 and the halogens accept electrons and act as oxidizing agents.

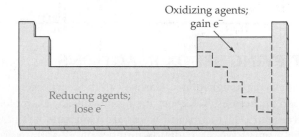

Different metals can give up different numbers of electrons in redox reactions. Lithium, sodium, and the other group 1A elements give up only one electron and become monopositive ions with oxidation numbers of +1. Beryllium, magnesium, and the other group 2A elements, however, typically give up two electrons and become dipositive ions. The transition metals in the middle of the periodic table can give up a variable number of electrons to yield more than one kind of ion depending on the exact reaction. Titanium, for example, can react with chlorine to yield either $TiCl_3$ or $TiCl_4$. Because a chloride ion has a −1 oxidation number, the titanium atom in $TiCl_3$ must have a +3 oxidation number and the titanium atom in $TiCl_4$ must be +4.

●→ WORKED EXAMPLE 4.13

Identifying Oxidizing and Reducing Agents

Assign oxidation numbers to all atoms, tell in each case which substance is undergoing oxidation and which reduction, and identify the oxidizing and reducing agents.

(a) $Ca(s) + 2 H^+(aq) \longrightarrow Ca^{2+}(aq) + H_2(g)$

(b) $2 Fe^{2+}(aq) + Cl_2(aq) \longrightarrow 2 Fe^{3+}(aq) + 2 Cl^-(aq)$

STRATEGY AND SOLUTION

(a) The elements Ca and H_2 have oxidation numbers of 0; Ca^{2+} is +2 and H^+ is +1. Ca is oxidized, because its oxidation number increases from 0 to +2, and H^+ is reduced, because its oxidation number decreases from +1 to 0. The reducing agent is the substance that gives away electrons, thereby going to a higher oxidation number, and the oxidizing agent is the substance that accepts electrons, thereby going to a lower oxidation number. In the present case, calcium is the reducing agent and H^+ is the oxidizing agent.

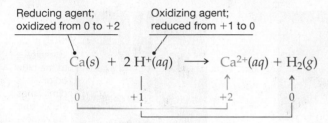

Reducing agent; oxidized from 0 to +2

Oxidizing agent; reduced from +1 to 0

$$Ca(s) + 2 H^+(aq) \longrightarrow Ca^{2+}(aq) + H_2(g)$$

$\quad 0 \qquad\quad +1 \qquad\qquad +2 \qquad\quad 0$

(b) Atoms of the neutral element Cl_2 have an oxidation number of 0; the monatomic ions have oxidation numbers equal to their charge:

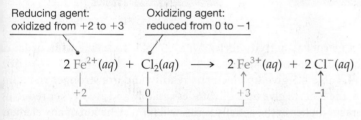

Reducing agent: oxidized from +2 to +3

Oxidizing agent: reduced from 0 to −1

$$2 Fe^{2+}(aq) + Cl_2(aq) \longrightarrow 2 Fe^{3+}(aq) + 2 Cl^-(aq)$$

$\quad +2 \qquad\quad 0 \qquad\qquad +3 \qquad\quad -1$

Fe^{2+} is oxidized because its oxidation number increases from +2 to +3, and Cl_2 is reduced because its oxidation number decreases from 0 to −1. Fe^{2+} is the reducing agent, and Cl_2 is the oxidizing agent.

▶ **PRACTICE 4.26** In each of the following reactions, tell which substance is undergoing an oxidation and which a reduction, and identify the oxidizing and reducing agents.

(a) $SnO_2(s) + 2 C(s) \longrightarrow Sn(s) + 2 CO(g)$

(b) $Sn^{2+}(aq) + 2 Fe^{3+}(aq) \longrightarrow Sn^{4+}(aq) + 2 Fe^{2+}(aq)$

(c) $4 NH_3(g) + 5 O_2(g) \longrightarrow 4 NO(g) + 6 H_2O(l)$

▶ **APPLY 4.27** Police often use a Breathalyzer test to determine the ethanol (C_2H_5OH) content in a person's blood. The test involves a redox reaction that produces a color change. Potassium dichromate is reddish orange and chromium(III) sulfate is green. The balanced reaction is:

$$2 K_2Cr_2O_7(aq) + 3 C_2H_5OH(g) + 8 H_2SO_4(aq) \longrightarrow$$
$$2 Cr_2(SO_4)_3(aq) + 2 K_2SO_4(aq)$$
$$+ 3 CH_3COOH(aq) + 11 H_2O(l)$$

(a) Identify the element that gets oxidized and the element that gets reduced.

(b) Give the oxidizing agent and the reducing agent.

▲ A Breathalyzer test measures alcohol concentration in exhaled breath using a redox reaction.

4.12 ▶ THE ACTIVITY SERIES OF THE ELEMENTS

The reaction of an aqueous cation, usually a metal ion, with a free element to give a different cation and a different element is among the simplest of all redox processes. Aqueous copper(II) ion reacts with iron metal, for example, to give iron(II) ion and copper metal (**FIGURE 4.7**):

$$Fe(s) + Cu^{2+}(aq) \longrightarrow Fe^{2+}(aq) + Cu(s)$$

Similarly, aqueous acid reacts with magnesium metal to yield magnesium ion and hydrogen gas:

$$Mg(s) + 2 H^+(aq) \longrightarrow Mg^{2+}(aq) + H_2(g)$$

Whether a reaction occurs between a given ion and a given element depends on the relative ease with which the various substances gain or lose electrons—that is, on how easily each substance is reduced or oxidized. By noting the results from a succession of different

▶ **FIGURE 4.7**

The redox reaction of iron with aqueous copper(II) ion.

Figure It Out

What is happening at the atomic level on the surface of the nail?

Answer: Cu²⁺ ions are gaining 2 electrons from Fe(s) atoms, forming a coating of Cu(s) on the surface of the nail. Fe(s) in the nail dissolves into solution as Fe²⁺ ions.

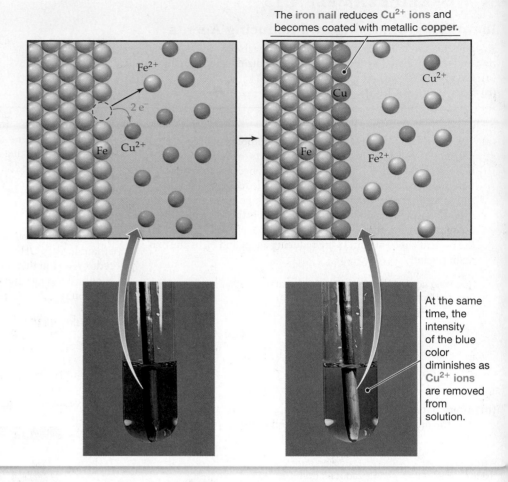

The iron nail reduces Cu^{2+} ions and becomes coated with metallic copper.

At the same time, the intensity of the blue color diminishes as Cu^{2+} ions are removed from solution.

reactions, it's possible to construct an **activity series**, which ranks the elements in order of their reducing ability in aqueous solution (**TABLE 4.5**).

Elements at the top of Table 4.5 give up electrons readily and are stronger reducing agents, whereas elements at the bottom give up electrons less readily and are weaker reducing agents. As a result, any element higher in the activity series will reduce the ion of any element lower in the activity series. Because copper is above silver, for example, copper metal gives electrons to Ag^+ ions (**FIGURE 4.8**).

$$Cu(s) + 2\,Ag^+(aq) \longrightarrow Cu^{2+}(aq) + 2\,Ag(s)$$

Conversely, because gold is below silver in the activity series, gold metal does not give electrons to Ag^+ ions.

$$Au(s) + 3\,Ag^+(aq) \not\longrightarrow Au^{3+}(aq) + 3\,Ag(s) \qquad \textit{Does not occur}$$

The position of hydrogen in the activity series is particularly important because it indicates which metals react with aqueous acid (H^+) to release H_2 gas. The metals at the top of the series—the alkali metals of group 1A and alkaline earth metals of group 2A—are such powerful reducing agents that they react even with pure water, in which the concentration of H^+ is very low:

Oxidized   Reduced   Unchanged

$$2\,Na(s) + 2\,H_2O(l) \longrightarrow 2\,Na^+(aq) + 2\,OH^-(aq) + H_2(g)$$

$$0 \qquad\quad +1\,-2 \qquad\qquad +1 \qquad\quad -2\,+1 \qquad\quad 0$$

TABLE 4.5   **A Partial Activity Series of the Elements**

Oxidation Reaction

Strongly reducing ↑

These elements react rapidly with aqueous H^+ ions (acid) or with liquid H_2O to release H_2 gas.	$Li \rightarrow Li^+ + e^-$ $K \rightarrow K^+ + e^-$ $Ba \rightarrow Ba^{2+} + 2 e^-$ $Ca \rightarrow Ca^{2+} + 2 e^-$ $Na \rightarrow Na^+ + e^-$
These elements react with aqueous H^+ ions or with steam to release H_2 gas.	$Mg \rightarrow Mg^{2+} + 2 e^-$ $Al \rightarrow Al^{3+} + 3 e^-$ $Mn \rightarrow Mn^{2+} + 2 e^-$ $Zn \rightarrow Zn^{2+} + 2 e^-$ $Cr \rightarrow Cr^{3+} + 3 e^-$ $Fe \rightarrow Fe^{2+} + 2 e^-$
These elements react with aqueous H^+ ions to release H_2 gas.	$Co \rightarrow Co^{2+} + 2 e^-$ $Ni \rightarrow Ni^{2+} + 2 e^-$ $Sn \rightarrow Sn^{2+} + 2 e^-$
	$H_2 \rightarrow 2 H^+ + 2 e^-$
These elements do not react with aqueous H^+ ions to release H_2.	$Cu \rightarrow Cu^{2+} + 2 e^-$ $Ag \rightarrow Ag^+ + e^-$ $Hg \rightarrow Hg^{2+} + 2 e^-$ $Pt \rightarrow Pt^{2+} + 2 e^-$ $Au \rightarrow Au^{3+} + 3 e^-$

Weakly reducing

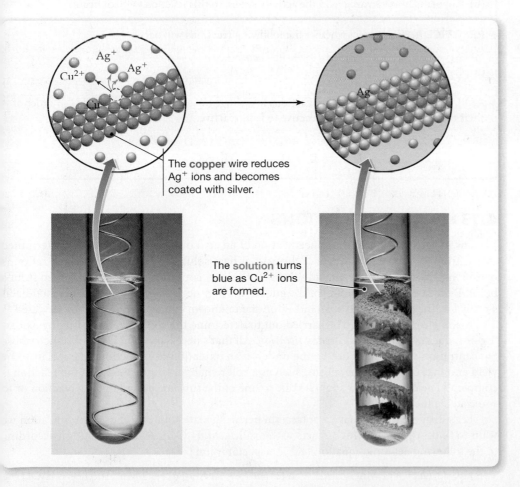

◄ **FIGURE 4.8**

The redox reaction of copper with aqueous Ag^+ ion.

Ag^+

Cu^{2+} Ag^+

Cu

Ag

The **copper** wire reduces Ag^+ ions and becomes coated with silver.

The **solution** turns blue as Cu^{2+} ions are formed.

In contrast, the metals in the middle of the series react with aqueous acid but not with water, and the metals at the bottom of the series react with neither aqueous acid nor water:

$$Fe(s) + 2\,H^+(aq) \longrightarrow Fe^{2+}(aq) + H_2(g)$$
$$Ag(s) + H^+(aq) \longrightarrow \text{No reaction}$$

Notice that the most easily oxidized metals—those at the top of the activity series—are on the left of the periodic table. Conversely, the least easily oxidized metals—those at the bottom of the activity series—are in the transition metal groups closer to the right side of the table.

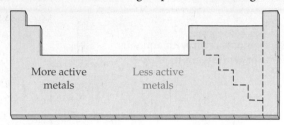

More active
metals

Less active
metals

WORKED EXAMPLE 4.14

Predicting the Products of a Redox Reaction

Predict whether the following redox reactions will occur:
(a) $Hg^{2+}(aq) + Zn(s) \longrightarrow Hg(l) + Zn^{2+}(aq)$
(b) $2\,H^+(aq) + Cu(s) \longrightarrow H_2(g) + Cu^{2+}(aq)$

STRATEGY

Look at Table 4.5 to find the relative reactivities of the elements.

SOLUTION

(a) Zinc is above mercury in the activity series, so this reaction will occur.
(b) Copper is below hydrogen in the activity series, so this reaction will not occur.

▶ **PRACTICE 4.28** Predict whether the following reactions will occur:

(a) $2\,H^+(aq) + Pt(s) \longrightarrow H_2(g) + Pt^{2+}(aq)$
(b) $Ca^{2+}(aq) + Mg(s) \longrightarrow Ca(s) + Mg^{2+}(aq)$

▶ **APPLY 4.29** Use the following reactions to arrange the elements **A**, **B**, **C**, and **D** in order of their redox reactivity from most reactive to least reactive.

$$A + D^+ \longrightarrow A^+ + D \qquad C^+ + D \longrightarrow C + D^+$$
$$B^+ + D \longrightarrow B + D^+ \qquad B + C^+ \longrightarrow B^+ + C$$

4.13 ▶ REDOX TITRATIONS

REMEMBER...

The reaction used for a **titration** must go to completion and have a yield of 100%.
(Section 4.9)

We saw in Section 4.9 that the concentration of an acid or base solution can be determined by **titration**. A measured volume of the acid or base solution of unknown concentration is placed in a flask, and a base or acid solution of known concentration is slowly added from a buret. By measuring the volume of the added solution necessary for a complete reaction, as signaled by the color change of an indicator, the unknown concentration can be calculated.

A similar procedure can be carried out to determine the concentration of many oxidizing or reducing agents using a *redox titration*. All that's necessary is that the substance whose concentration you want to determine undergo an oxidation or reduction reaction in 100% yield and that there be some means, such as a color change, to indicate when the reaction is complete. The color change might be due to one of the substances undergoing reaction or to some added indicator.

Let's imagine that we have a potassium permanganate solution whose concentration we want to find. Aqueous $KMnO_4$ reacts with oxalic acid, $H_2C_2O_4$, in acidic solution according to the following net ionic equation (K^+ is a spectator ion):

$$5\,H_2C_2O_4(aq) + 2\,MnO_4^-(aq) + 6\,H^+(aq) \rightarrow 10\,CO_2(g) + 2\,Mn^{2+}(aq) + 8\,H_2O(l)$$

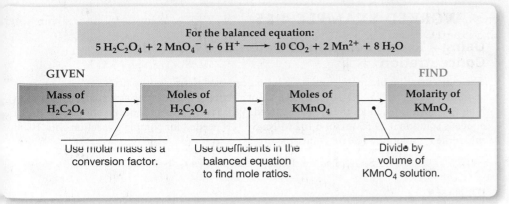

For the balanced equation:
$$5\,H_2C_2O_4 + 2\,MnO_4^- + 6\,H^+ \longrightarrow 10\,CO_2 + 2\,Mn^{2+} + 8\,H_2O$$

GIVEN

| Mass of $H_2C_2O_4$ | Moles of $H_2C_2O_4$ | Moles of $KMnO_4$ | Molarity of $KMnO_4$ |

FIND

Use molar mass as a conversion factor.

Use coefficients in the balanced equation to find mole ratios.

Divide by volume of $KMnO_4$ solution.

▲ **FIGURE 4.9**

A summary of calculations for determining the concentration of a $KMnO_4$ solution by redox titration of $H_2C_2O_4$.

The reaction goes to completion with 100% yield and is accompanied by a sharp color change when the intense purple color of the MnO_4^- ion disappears.

The strategy used is outlined in **FIGURE 4.9**. As with acid–base titrations, the general idea is to measure a known amount of one substance—in this case, $H_2C_2O_4$—and use mole ratios from the balanced equation to find the number of moles of the second substance—in this case, $KMnO_4$—necessary for complete reaction. With the molar amount of $KMnO_4$ thus known, titration gives the volume of solution containing that amount. Dividing the number of moles by the volume gives the concentration.

As an example of how the procedure works, let's carefully weigh some amount of $H_2C_2O_4$—say, 0.2585 g—and dissolve it in approximately 100 mL of 0.5 M H_2SO_4. The exact volume isn't important because we're concerned only with the amount of dissolved $H_2C_2O_4$, not with its concentration. Next, we place an aqueous $KMnO_4$ solution of unknown concentration in a buret and slowly add it to the $H_2C_2O_4$ solution. The purple color of the added MnO_4^- initially disappears as reaction occurs, but we continue the addition until a faint color persists, indicating that all the $H_2C_2O_4$ has reacted and that MnO_4^- ion is no longer being reduced. At this equivalence point, or *end point*, of the titration, we might find that 22.35 mL of the $KMnO_4$ solution has been added (**FIGURE 4.10**).

To calculate the molarity of the $KMnO_4$ solution, we need to find the number of moles of $KMnO_4$ present in 22.35 mL of solution used for titration. We do this by following the procedure outlined in Figure 4.9 and first calculating the number of moles of oxalic acid that react with the permanganate ion. A gram-to-mole conversion is done using the molar mass of $H_2C_2O_4$ as the conversion factor:

$$\text{Moles of } H_2C_2O_4 = 0.2585\ \text{g } H_2C_2O_4 \times \frac{1\ \text{mol } H_2C_2O_4}{90.03\ \text{g } H_2C_2O_4}$$
$$= 2.871 \times 10^{-3}\ \text{mol } H_2C_2O_4$$

According to the balanced equation, 5 mol of oxalic acid react with 2 mol of permanganate ion. Thus, we can calculate the number of moles of $KMnO_4$ that react with 2.871×10^{-3} mol of $H_2C_2O_4$:

$$\text{Moles of } KMnO_4 = 2.871 \times 10^{-3}\ \text{mol } H_2C_2O_4 \times \frac{2\ \text{mol } KMnO_4}{5\ \text{mol } H_2C_2O_4}$$
$$= 1.148 \times 10^{-3}\ \text{mol } KMnO_4$$

Knowing both the number of moles of $KMnO_4$ that react (1.148×10^{-3} mol) and the volume of the $KMnO_4$ solution (22.35 mL), we can calculate the molarity:

$$\text{Molarity} = \frac{1.148 \times 10^{-3}\ \text{mol } KMnO_4}{22.35\ \text{mL}} \times \frac{1000\ \text{mL}}{1\ \text{L}} = 0.051\,36\ \text{M}$$

The molarity of the $KMnO_4$ solution is 0.051 36 M.

A precise amount of oxalic acid is weighed and dissolved in aqueous H_2SO_4.

Aqueous $KMnO_4$ of unknown concentration is added from a buret until …

… the purple color persists, indicating that all of the oxalic acid has reacted.

▲ **FIGURE 4.10**

The redox titration of oxalic acid, $H_2C_2O_4$, with $KMnO_4$.

Figure It Out

Why does the solution in the flask remain clear until the end point of the titration is reached?

Answer: The purple titrant MnO_4^- is reduced to colorless Mn^{2+} by $H_2C_2O_4$. As soon as the $H_2C_2O_4$ is used up, additional MnO_4^- added to the flask generates the purple end point.

WORKED EXAMPLE 4.15

Using a Redox Reaction to Determine a Solution's Concentration

The concentration of an aqueous I_3^- solution can be determined by titration with aqueous sodium thiosulfate, $Na_2S_2O_3$, in the presence of a starch indicator. The starch turns from deep blue to colorless when all the I_3^- has reacted. What is the molar concentration of I_3^- in an aqueous solution if 24.55 mL of 0.102 M $Na_2S_2O_3$ is needed for complete reaction with 10.00 mL of the I_3^- solution? The net ionic equation is

$$2\,S_2O_3^{2-}(aq) + I_3^-(aq) \longrightarrow S_4O_6^{2-}(aq) + 3\,I^-(aq)$$

IDENTIFY

Known	Unknown
Volume of I_3^- solution (10.00 mL)	Molarity of I_3^- solution
Volume of $Na_2S_2O_3$ solution (24.55 mL)	
Molarity of $Na_2S_2O_3$ solution (0.102 M)	

STRATEGY

The procedure is similar to that outlined in Figure 4.9 except that volume of the $Na_2S_2O_3$ solution can be used to find moles of $S_2O_3^{2-}$ instead of a gram-to-mole conversion.

SOLUTION

We first need to find the number of moles of thiosulfate ion used for the titration:

$$24.55\ \text{mL} \times \frac{1\ \text{L}}{1000\ \text{mL}} \times \frac{0.102\ \text{mol S}_2\text{O}_3^{2-}}{1\ \text{L}} = 2.50 \times 10^{-3}\ \text{mol S}_2\text{O}_3^{2-}$$

According to the balanced equation, 2 mol of $S_2O_3^{2-}$ ion react with 1 mol of I_3^- ion. Thus, we can find the number of moles of I_3^- ion:

$$2.50 \times 10^{-3}\ \text{mol S}_2\text{O}_3^{2-} \times \frac{1\ \text{mol I}_3^-}{2\ \text{mol S}_2\text{O}_3^{2-}} = 1.25 \times 10^{-3}\ \text{mol I}_3^-$$

Knowing both the number of moles of I_3^- (1.25×10^{-3} mol) and the volume of the I_3^- solution (10.00 mL), let us calculate molarity:

$$\frac{1.25 \times 10^{-3}\ \text{mol I}_3^-}{10.00\ \text{mL}} \times \frac{10^3\ \text{mL}}{1\ \text{L}} = 0.125\ \text{M}$$

The molarity of the I_3^- solution is 0.125 M.

CHECK

According to the balanced equation, the amount of $S_2O_3^{2-}$ needed for the reaction (2 mol) is twice the amount of I_3^- (1 mol). The titration results indicate that the volume of the $S_2O_3^{2-}$ solution (24.55 mL) is a little over twice the volume of the I_3^- solution (10.00 mL). Thus, the concentrations of the two solutions must be about the same—approximately 0.1 M.

▶ **PRACTICE 4.30**   What is the molar concentration of Fe^{2+} ion in an aqueous solution if 31.50 mL of 0.105 M $KBrO_3$ is required for complete reaction with 10.00 mL of the Fe^{2+} solution? The net ionic equation is:

$$6\,Fe^{2+}(aq) + BrO_3^-(aq) + 6\,H^+(aq) \longrightarrow 6\,Fe^{3+}(aq) + Br^-(aq) + 3\,H_2O(l)$$

▶ **APPLY 4.31**   Iron(II) sulfate is a soluble ionic compound added as a source of iron in vitamin tablets. Determine the mass of iron (mg) in one tablet that has been dissolved in 10.0 mL of water and titrated with 14.92 mL of 0.0100 M $K_2Cr_2O_7$ solution. The net ionic equation is:

$$Cr_2O_7^{2-}(aq) + 6\,Fe^{2+}(aq) + 14\,H^+(aq) \longrightarrow 2\,Cr^{3+}(aq) + 6\,Fe^{3+}(aq) + 7\,H_2O(l)$$

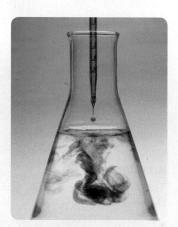

▲ The reddish I_3^- *solution* turns a *deep blue* color when it is added to a solution containing a small amount of starch.

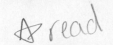 *read*

4.14 ▸ SOME APPLICATIONS OF REDOX REACTIONS

Redox reactions take place with every element in the periodic table except helium and neon and occur in a vast number of processes throughout nature, biology, and industry. Here are just a few examples:

- **Combustion.** *Combustion* is the burning of a fuel by oxidation with oxygen in air. Gasoline, fuel oil, natural gas, wood, paper, and other organic substances of carbon and hydrogen are the most common fuels. Even some metals, such as magnesium and calcium, will burn in air.

$$CH_4(g) + 2\,O_2(g) \longrightarrow CO_2(g) + 2\,H_2O(l)$$
$$\text{Methane}$$
$$\text{(Natural gas)}$$

- **Bleaching.** *Bleaching* uses redox reactions to decolorize or lighten colored materials. Dark hair is bleached to turn it blond, clothes are bleached to remove stains, wood pulp is bleached to make white paper, and so on. The exact oxidizing agent used depends on the situation—hydrogen peroxide (H_2O_2) is used for hair, sodium hypochlorite (NaOCl) is used for clothes, and ozone or chlorine dioxide is used for wood pulp—but the principle is always the same. In all cases, colored impurities are destroyed by reaction with a strong oxidizing agent.

▲ Dark hair can be bleached by a redox reaction with hydrogen peroxide.

- **Batteries.** Although they come in many types and sizes, redox reactions power all types of batteries. In a typical redox reaction carried out in the laboratory—say, the reaction of zinc metal with Ag^+ to yield Zn^{2+} and silver metal—the reactants are simply mixed in a flask and electrons are transferred by direct contact between them. In a battery, however, the two reactants are kept in separate compartments and the electrons are transferred through a wire running between them.

   The inexpensive alkaline battery commonly used in flashlights and other small household items uses a thin steel can containing zinc powder and a paste of potassium hydroxide as one reactant, separated by paper from a paste of powdered carbon and manganese dioxide as the other reactant. A graphite rod with a metal cap sticks into the MnO_2 to provide electrical contact. When the can and the graphite rod are connected by a wire, zinc sends electrons flowing through the wire toward the MnO_2 in a redox reaction. The resultant electrical current can be used to light a bulb or power a small electronic device. The reaction is

$$Zn(s) + 2\,MnO_2(s) \longrightarrow ZnO(s) + Mn_2O_3(s)$$

We'll look at the chemistry of batteries in more detail in Section 18.10.

- **Metallurgy.** *Metallurgy*, the extraction and purification of metals from their ores, makes use of numerous redox processes. We'll see in Section 21.2, for example, that metallic zinc is prepared by reduction of ZnO with *coke*, a form of carbon:

$$ZnO(s) + C(s) \longrightarrow Zn(s) + CO(g)$$

- **Corrosion.** *Corrosion* is the deterioration of a metal by oxidation, such as the rusting of iron in moist air. The economic consequences of rusting are enormous: It has been estimated that up to one-fourth of the iron produced in the United States is used to replace bridges, buildings, and other structures that have been destroyed by corrosion. (The raised dot in the formula $Fe_2O_3 \cdot H_2O$ for rust indicates that one water molecule is associated with each Fe_2O_3 in an unspecified way.)

$$4\,Fe(s) + 3\,O_2(g) \xrightarrow{\text{H}_2\text{O}} 2\,Fe_2O_3 \cdot H_2O(s)$$
$$\text{Rust}$$

- **Respiration.** The term *respiration* refers to the processes of breathing and using oxygen for the many biological redox reactions that provide the energy needed by living organisms. The energy is released from food molecules slowly and in complex, multistep pathways, but the overall result of respiration is similar to that of a combustion reaction. For example, the simple sugar glucose ($C_6H_{12}O_6$) reacts with O_2 to give CO_2 and H_2O according to the following equation:

$$C_6H_{12}O_6 + 6\,O_2 \rightarrow 6\,CO_2 + 6\,H_2O + \text{energy}$$
$$\text{Glucose}$$
$$\text{(a carbohydrate)}$$

INQUIRY ▶▶▶ HOW DO SPORTS DRINKS REPLENISH THE CHEMICALS LOST IN SWEAT?

Sports drinks such as Gatorade and Powerade are commonly consumed by both professional and amateur athletes during exercise. How do these drinks help athletes perform better and recover more quickly? The development of sports drinks started in 1965 when University of Florida football coaches noted that players became extremely fatigued, lost significant amounts of weight (often more than 10 pounds), and seldom needed to urinate after exercising in the heat. The team consulted with Robert Cade, a kidney specialist at the University of Florida's College of Medicine, who speculated that electrolytes lost in sweat were upsetting the body's delicate chemical balance. Sodium and potassium ions were of primary concern due to their importance in nerve and muscle function, regulation of body heat, distribution of water, and transport of solutes such as glucose for energy.

To test his hypothesis, Cade and a team of researchers received permission from the coaching staff to study the fluids of freshman players before and after exercising vigorously in the heat. The results were staggering; after exercise the players had an electrolyte imbalance, low blood sugar, and decreased total blood volume. These factors led to diminished physical performance and in some cases extreme heat exhaustion. With data in hand, Cade's team began pursuing a remedy to address all these issues. A drink to replace the fluids and electrolytes lost through sweat and the carbohydrates burned for energy was created. The first batch contained water, salt, and sugar, but tasted so terrible that no one could drink it until Cade's wife suggested adding lemon juice.

By 1966, the drink known as *Gatorade* became a staple for the team and hospitalizations of players due to heat exhaustion became almost nonexistent. Only one player who had not drunk any Gatorade had a problem and the Gators advanced to the prestigious Orange Bowl for the first time in the school's history. The university released an official statement about Gatorade in late December 1966 that the *Florida Times-Union* summed up with this headline: "One Lil' Swig of That Kickapoo Juice and Biff, Bam, Sock—It's Gators, 8-2."

▲ Gatorade and other sports drinks contain electrolytes and conduct electricity.

PROBLEM 4.32 A vitamin-fortified brand of a sports beverage contains sodium chloride (NaCl), sodium citrate ($NaC_6H_7O_7$), and potassium dihydrogen phosphate (KH_2PO_4), as well as the substances whose structures are seen below.

Citric acid ($C_6H_8O_7$)

Fructose ($C_6H_{12}O_6$)

Vitamin B3 ($C_6NH_5O_2$)

(a) Use Table 4.1 (Electrolyte Classification of Some Common Substances) to classify the components of the sports drink as a strong electrolyte, weak electrolyte, or nonelectrolyte.

(b) Identify which substances replenish electrolytes with important biological functions.

PROBLEM 4.33 The nutritional label on Powerade specifies that there are 150 mg of sodium and 35 mg of potassium in 360 mL of the beverage. Calculate the concentration of sodium and potassium ions in units of molarity.

PROBLEM 4.34 The concentration of sodium ions in Powerade is 0.416 mg/mL. Imagine that you want to prepare your own drink with the same concentration of sodium ions. How many grams of sodium chloride are needed to prepare 0.500 L of solution?

Questions 4.35 and 4.36 refer to the analysis of chloride ion in a sports beverage by precipitation with silver.

PROBLEM 4.35 Many sports drinks contain phosphate ion (PO_4^{3-}) which will also precipitate with silver, thus interfering with the chloride measurement. The phosphate ion can be removed by precipitation prior to the analysis of chloride.

  (a) Use the solubility guidelines (Table 4.2) to choose a cation from the list below that would form a precipitate with phosphate, but not with chloride.

  $K^+, Ba^{2+}, Pb^{2+}, NH_4^+$

(b) Write the net ionic reaction for the precipitation reaction from part (a).

PROBLEM 4.36 Following the removal of phosphate by precipitation, an excess of silver ion was added to 100.0 mL of a sports beverage. A white precipitate of silver chloride was isolated by filtration, dried, and found to have a mass of 172 mg. Calculate the concentration of chloride ion in the drink in units of molarity.

PROBLEM 4.37 The flavor of the first batch of Gatorade was improved by adding lemon juice, which contains citric acid $(H_3C_6H_5O_7)$. Citric acid is still added as flavoring to sports drinks today. The concentration of citric acid in a beverage was determined by titration with sodium hydroxide according to the reaction:

$$H_3C_6H_5O_7(aq) + 3\,NaOH(aq) \longrightarrow Na_3C_6H_5O_7(aq) + 3\,H_2O(l)$$

If 25.0 mL of the beverage required 35.6 mL of 0.0400 M NaOH for a complete reaction, calculate the molarity of citric acid.

STUDY GUIDE

Section	Concept Summary	Learning Objectives	Test Your Understanding
4.1 ▶ Concentrations in Solution: Molarity	The concentration of a substance in solution is usually expressed as **molarity (M)**, defined as the number of moles of a substance (**solute**) dissolved per liter of solution. A solution's molarity acts as a conversion factor between solution volume and number of moles of solute, making it possible to carry out stoichiometry calculations on solutions.	**4.1** Calculate the molarity of a solution given the mass of solute and total volume.	Worked Example 4.1; Problem 4.52
		4.2 Calculate the a amount of solute in a given volume of solution with a known molarity.	Worked Example 4.2; Problems 4.48, 4.50, 4.127
		4.3 Describe the proper technique for preparing solutions of known molarity.	Problems 4.2, 4.58
4.2 ▶ Diluting Concentrated Solutions	When carrying out a dilution, only the volume is changed by adding solvent; the amount of solute is unchanged.	**4.4** Calculate the concentration of a solution that has been diluted.	Worked Example 4.3; Problems 4.38, 4.54, 4.56
		4.5 Describe the proper technique for diluting solutions.	Problem 4.59
4.3 ▶ Electrolytes in Aqueous Solution	Many reactions take place in aqueous solution. Substances whose aqueous solutions contain ions conduct electricity and are called **electrolytes**. Ionic compounds, such as NaCl, and molecular compounds that **dissociate** substantially into ions when dissolved in water are **strong electrolytes**. Substances that dissociate to only a small extent are **weak electrolytes**, and substances that do not produce ions in aqueous solution are **nonelectrolytes**.	**4.6** Classify a substance as a strong, weak, or nonelectrolyte. (Table 4.1)	Problems 4.39, 4.60, 4.64
		4.7 Calculate the concentration of ions in a strong electrolyte solution.	Worked Example 4.4; Problem 4.66, 4.67, 4.126, 4.128
4.4 ▶ Types of Chemical Reactions in Aqueous Solution	Aqueous reactions can be classified into three major groups. **Precipitation reactions** occur when solutions of two ionic substances are mixed and a precipitate falls from solution. **Acid–base neutralization reactions** occur when an acid is mixed with a base, yielding water and an ionic **salt**. **Oxidation–reduction reactions**, or **redox reactions**, are processes in which one or more electrons are transferred between reaction partners.	**4.8** Classify a reaction as a precipitation, acid–base neutralization, or oxidation–reduction (redox) reaction.	Problems 4.68, 4.69
4.5 ▶ Aqueous Reactions and Net Ionic Equations	Aqueous ionic compounds exist as cations and anions in solution. An **ionic equation** shows all the ions in a reaction and a **net ionic equation** shows only the ions that take part in a reaction. **Spectator ions** are present to balance charge but do not take part in the chemical reaction.	**4.9** Write a net ionic equation and identify spectator ions given the molecular equation.	Worked Example 4.5; Problems 4.70, 4.71
4.6 ▶ Precipitation Reactions and Solubility Guidelines	Solubility guidelines (Table 4.2) are used to predict which combinations of anions and cations in ionic compounds will be soluble and insoluble. To predict whether a precipitate will form in a reaction, write the formula of possible products and determine **solubility**.	**4.10** Use the solubility guidelines (Table 4.2) to predict the solubility of an ionic compound in water.	Problems 4.72, 4.73
		4.11 Predict whether a precipitation reaction will occur and write the ionic and net ionic equations.	Worked Example 4.6 and 4.7; Problems 4.40, 4.41, 4.74, 4.76, 4.141
4.7 ▶ Acids, Bases, and Neutralization Reactions	An acid is a substance that dissociates in water to give hydrogen (H^+) ions and a base is a substance that dissociates to give hydroxide ions (OH^-). The neutralization of a strong acid with a strong base can be written as a net ionic equation, in which nonparticipating, spectator ions are not specified: $$H^+(aq) + OH^-(aq) \longrightarrow H_2O(l)$$	**4.12** Convert between name and formula for an acid.	Worked Example 4.8; Problem 4.16, 4.17
		4.13 Classify acids as strong or weak based on the molecular picture of dissociation.	Problem 4.42
		4.14 Write the ionic equation and net ionic equation for an acid–base neutralization reaction.	Worked Example 4.9; Problems, 4.88, 4.90

Section	Concept Summary	Learning Objectives	Test Your Understanding
4.8 ▶ Solution Stoichiometry	Stoichiometry calculations are performed by relating amounts of reactants and products in a balanced equation in units of moles since stoichiometric coefficients refer to moles. Molarity is a conversion factor between numbers of moles of solute and the volume of a solution.	**4.15** Convert between moles and volume using molarity in stoichiometry calculations.	Worked Example 4.10; Problems 4.92, 4.94, 4.129
4.9 ▶ Titrations	**Titration** is a technique used to find the exact concentration of a solution. A fixed volume of solution with unknown concentration is added to a flask. A solution with a known concentration (titrant) is added from a buret until the reaction is complete. The measured volume of titrant and reaction stoichiometry is used to calculate the concentration of the solution in the flask.	**4.16** Determine the concentration of a solution using titration data.	Worked Example 4.11; Problems 4.29, 4.96, 4.98
		4.17 Visualize the substances present in solution during a titration procedure.	Problems 4.23, 4.44, 4.45
4.10 ▶ Oxidation–Reduction (Redox) Reactions	An **oxidation** is the loss of one or more electrons; a **reduction** is the gain of one or more electrons. Redox reactions can be identified by assigning to each atom in a substance an **oxidation number**, which provides a measure of whether the atom is neutral, electron rich, or electron poor. Comparing the oxidation numbers of an atom before and after reaction shows whether the atom has gained or lost electrons.	**4.18** Assign oxidation numbers to atoms in a compound.	Worked Example 4.12; Problems 4.104, 4.106, 4.108, 4.133
4.11 ▶ Identifying Redox Reactions	Oxidations and reductions must occur together. Whenever one substance loses one or more electrons (is oxidized), another substance gains the electrons (is reduced). The substance that causes a reduction by giving up electrons is called a **reducing agent**. The substance that causes an oxidation by accepting electrons is called an **oxidizing agent**. The reducing agent is itself oxidized when it gives up electrons, and the oxidizing agent is itself reduced when it accepts electrons.	**4.19** Identify redox reactions, oxidizing agents, and reducing agents.	Worked Example 4.13; Problems 4.110, 4.111
4.12 ▶ The Activity Series of the Elements	Among the simplest of redox processes is the reaction of an aqueous cation, usually a metal ion, with a free element to give a different ion and a different element. Noting the results from a succession of different reactions makes it possible to organize an **activity series**, which ranks the elements in order of their reducing ability in aqueous solution	**4.20** Use the location of elements in the periodic table and activity series to predict if a redox reaction will occur.	Worked Example 4.14; Problems 4.46, 4.47, 4.112, 4.113
		4.21 Develop an activity series and predict if a redox reaction will occur based on experimental data provided.	Problems 4.114, 4.115, 4.134
4.13 ▶ Redox Titrations	The concentration of an oxidizing agent or a reducing agent in solution can be determined by a redox titration.	**4.22** Use a redox titration to determine the concentration of an oxidizing or reducing agent in solution.	Worked Example 4.15; Problems 4.116, 4.118, 4.120, 4.124

KEY TERMS

KEY EQUATIONS

- **Molarity (Section 4.1)**

$$\text{Molarity (M)} = \frac{\text{Moles of solute}}{\text{Liters of solution}}$$

- **Dilution (Section 4.2)**

$$M_i \times V_i = M_f \times V_f$$

CONCEPTUAL PROBLEMS

Problems 4.1–4.37 appear within the chapter.

4.38 Box **(a)** represents 1.0 mL of a solution of particles at a given concentration. Which of the boxes **(b)**–**(d)** represents 1.0 mL of the solution that results after **(a)** has been diluted by doubling the volume of its solvent?

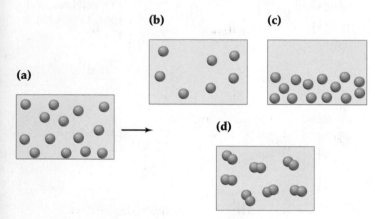

4.39 Three different substances, AX_3, AY_3, and AZ_3, are dissolved in water with the following results. (Water molecules are omitted for clarity.)

(a) Which of the substances is the strongest electrolyte, and which is the weakest?

(b) What is the molar concentration of A ions and Y ions in a 0.500 M solution of AX_3?

(c) What is the percent ionization of AZ_3 indicated in the graphical representation?

4.40 Assume that an aqueous solution of a cation, represented as a red sphere, is allowed to mix with a solution of an anion, represented as a yellow sphere. Three possible outcomes are represented by boxes (1)–(3):

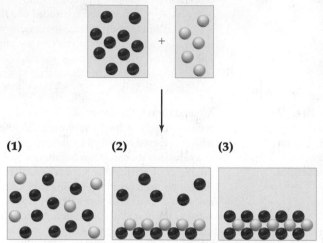

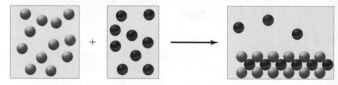

Which outcome corresponds to each of the following reactions?

(a) $2 Na^+(aq) + CO_3^{2-}(aq) \rightarrow$

(b) $Ba^{2+}(aq) + CrO_4^{2-}(aq) \rightarrow$

(c) $2 Ag^+(aq) + SO_3^{2-}(aq) \rightarrow$

4.41 Assume that an aqueous solution of a cation, represented as a blue sphere, is allowed to mix with a solution of an anion, represented as a red sphere, and that the following result is obtained:

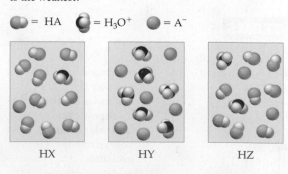

Which combinations of cation and anion, chosen from the following lists, are compatible with the observed results? Explain.

Cations: $Na^+, Ca^{2+}, Ag^+, Ni^{2+}$

Anions: $Cl^-, CO_3^{2-}, CrO_4^{2-}, NO_3^-$

4.42 The following pictures represent aqueous solutions of three acids HA (A = X, Y, or Z), with surrounding water molecules omitted for clarity. Which of the three is the strongest acid, and which is the weakest?

= HA   = H_3O^+ = A^-

HX   HY   HZ

4.43 Assume that an aqueous solution of OH^-, represented as a blue sphere, is allowed to mix with a solution of an acid H_nA, represented as a red sphere. Three possible outcomes are depicted by boxes (1)–(3), where the green spheres represent A^{n-}, the anion of the acid:

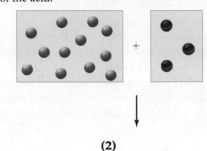 +

(1)   **(2)**   **(3)**

Which outcome corresponds to each of the following reactions?

(a) $HF + OH^- \rightarrow H_2O + F^-$

(b) $H_2SO_3 + 2\,OH^- \rightarrow 2\,H_2O + SO_3^{2-}$

(c) $H_3PO_4 + 3\,OH^- \rightarrow 3\,H_2O + PO_4^{3-}$

4.44 The concentration of an aqueous solution of NaOCl (sodium hypochlorite; the active ingredient in household bleach) can be determined by a redox titration with iodide ion in acidic solution:

$$OCl^-(aq) + 2\,I^-(aq) + 2\,H^+(aq) \rightarrow Cl^-(aq) + I_2(aq) + H_2O(l)$$

Assume that the blue spheres in the buret represent I^- ions, the red spheres in the flask represent OCl^- ions, the concentration of the I^- ions in the buret is 0.120 M, and the volumes in the buret and the flask are identical. What is the concentration of NaOCl in the flask? What percentage of the I^- solution in the buret must be added to the flask to react with all the OCl^- ions?

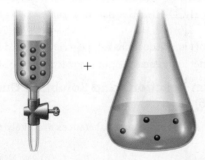

4.45 Assume that the electrical conductivity of a solution depends on the total concentration of dissolved ions and that you measure the conductivity of three different solutions while carrying out titration procedures:

(a) Begin with 1.00 L of 0.100 M KCl, and titrate by adding 0.100 M $AgNO_3$.

(b) Begin with 1.00 L of 0.100 M HF, and titrate by adding 0.100 M KOH.

(c) Begin with 1.00 L of 0.100 M $BaCl_2$, and titrate by adding 0.100 M Na_2SO_4.

Which of the following graphs corresponds to which titration?

(1)   **(2)**

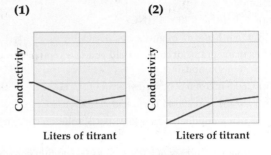

(3)

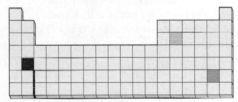

4.46 Based on the positions in the periodic table, which of the following reactions would you expect to occur?

(a) $Red^+ + Green \rightarrow Red + Green^+$

(b) $Blue + Green^+ \rightarrow Blue^+ + Green$

(c) $Red + Blue^+ \rightarrow Red^+ + Blue$

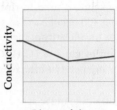

4.47 The following two redox reactions occur between aqueous cations and solid metals. Will a solution of green cations react with solid blue metal? Explain.

(a)

(b)

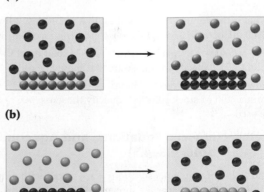

SECTION PROBLEMS

Molarity and Dilution (Sections 4.1 and 4.2)

4.48 How many moles of solute are present in each of the following solutions?
(a) 35.0 mL of 1.200 M HNO_3
(b) 175 mL of 0.67 M glucose ($C_6H_{12}O_6$)

4.49 How many grams of solute would you use to prepare each of the following solutions?
(a) 250.0 mL of 0.600 M ethyl alcohol (C_2H_6O)
(b) 167 mL of 0.200 M boric acid (H_3BO_3)

4.50 How many milliliters of a 0.45 M $BaCl_2$ solution contain 15.0 g of $BaCl_2$?

4.51 How many milliliters of a 0.350 M KOH solution contain 0.0171 mol of KOH?

4.52 The sterile saline solution used to rinse contact lenses can be made by dissolving 400 mg of NaCl in sterile water and diluting to 100 mL. What is the molarity of the solution?

4.53 The concentration of glucose ($C_6H_{12}O_6$) in normal blood is approximately 90 mg per 100 mL. What is the molarity of the glucose?

4.54 Copper reacts with dilute nitric acid according to the following equation:

$$3\,Cu(s) + 8\,HNO_3(aq) \rightarrow 3\,Cu(NO_3)_2(aq) + 2\,NO(g) + 4\,H_2O(l)$$

If a copper penny weighing 3.045 g is dissolved in a small amount of nitric acid and the resultant solution is diluted to 50.0 mL with water, what is the molarity of the $Cu(NO_3)_2$?

4.55 Pennies minted after 1982 are mostly zinc (97.5%) with a copper cover. If a post-1982 penny is dissolved in a small amount of nitric acid, the copper coating reacts as in Problem 4.54 and the exposed zinc reacts according to the following equation:

$$Zn(s) + 2\,HNO_3(aq) \rightarrow Zn(NO_3)_2(aq) + H_2(g)$$

For a penny that weighs 2.482 g, what is the molarity of the $Zn(NO_3)_2$ if the resultant solution is diluted to 250.0 mL with water?

4.56 A bottle of 12.0 M hydrochloric acid has only 35.7 mL left in it. What will the HCl concentration be if the solution is diluted to 250.0 mL?

4.57 What is the volume of the solution that would result by diluting 70.00 mL of 0.0913 M NaOH to a concentration of 0.0150 M?

4.58 How would you prepare 500 mL of 0.33 M solution of $CaCl_2$ from solid $CaCl_2$? Specify the glassware that should be used.

4.59 How would you prepare 250 mL of 0.150 M solution of $CaCl_2$ from a 3.00 M stock solution? Specify the glassware that should be used.

Electrolytes, Net Ionic Equations, and Aqueous Reactions (Sections 4.3–4.5)

4.60 The following aqueous solutions were tested with a light bulb conductivity apparatus, as shown in Figure 4.1. What result—dark, dim, or bright—do you expect from each?
(a) 0.10 M potassium chloride
(b) 0.10 M methanol
(c) 0.10 M acetic acid

4.61 The following aqueous solutions were tested with a light bulb conductivity apparatus, as shown in Figure 4.3. What result—dark, dim, or bright—do you expect from each?
(a) 0.10 M hydrofluoric acid
(b) 0.10 M sodium chloride
(c) 0.10 M glucose ($C_6H_{12}O_6$)

4.62 Individual solutions of $Ba(OH)_2$ and H_2SO_4 both conduct electricity, but the conductivity disappears when equal molar amounts of the solutions are mixed. Explain.

4.63 A solution of HCl in water conducts electricity, but a solution of HCl in chloroform, $CHCl_3$, does not. What does this observation tell you about how HCl exists in water and how it exists in chloroform?

4.64 Classify each of the following substances as either a strong electrolyte, weak electrolyte, or nonelectrolyte:
(a) HBr
(b) HF
(c) $NaClO_4$
(d) $(NH_4)_2CO_3$
(e) NH_3
(f) Ethyl alcohol

4.65 Is it possible for a molecular substance to be a strong electrolyte? Explain.

4.66 What is the total molar concentration of ions in each of the following solutions, assuming complete dissociation?
(a) A 0.750 M solution of K_2CO_3
(b) A 0.355 M solution of $AlCl_3$

4.67 What is the total molar concentration of ions in each of the following solutions?
(a) A 1.250 M solution of CH_3OH
(b) A 0.225 M solution of $HClO_4$

4.68 Classify each of the following reactions as a precipitation, acid–base neutralization, or oxidation–reduction:
(a) $Hg(NO_3)_2(aq) + 2\,NaI(aq) \longrightarrow 2\,NaNO_3(aq) + HgI_2(s)$
(b) $2\,HgO(s) \xrightarrow{heat} 2\,Hg(l) + O_2(g)$
(c) $H_3PO_4(aq) + 3\,KOH(aq) \longrightarrow K_3PO_4(aq) + 3\,H_2O(l)$

4.69 Classify each of the following reactions as a precipitation, acid–base neutralization, or oxidation–reduction:
(a) $S_8(s) + 8\,O_2(g) \rightarrow 8\,SO_2(g)$
(b) $NiCl_2(aq) + Na_2S(aq) \rightarrow NiS(s) + 2\,NaCl(aq)$
(c) $2\,CH_3CO_2H(aq) + Ba(OH)_2(aq) \rightarrow (CH_3CO_2)_2Ba(aq) + 2\,H_2O(l)$

4.70 Write net ionic equations for the reactions listed in Problem 4.68.

4.71 Write net ionic equations for the reactions listed in Problem 4.69.

Precipitation Reactions and Solubility Guidelines (Section 4.6)

4.72 Which of the following substances are likely to be soluble in water?
(a) $PbSO_4$
(b) $Ba(NO_3)_2$
(c) $SnCO_3$
(d) $(NH_4)_3PO_4$

4.73 Which of the following substances are likely to be soluble in water?
(a) ZnS
(b) $Au_2(CO_3)_3$
(c) $PbCl_2$
(d) Na_2S

4.74 Predict whether a precipitation reaction will occur when aqueous solutions of the following substances are mixed. For those that form a precipitate, write the net ionic reaction.
(a) $NaOH + HClO_4$
(b) $FeCl_2 + KOH$
(c) $(NH_4)_2SO_4 + NiCl_2$
(d) $CH_3CO_2Na + HCl$

4.75 Predict whether a precipitation reaction will occur when aqueous solutions of the following substances are mixed. For those that form a precipitate, write the net ionic reaction.

(a) $MnCl_2 + Na_2S$ (b) $HNO_3 + CuSO_4$

(c) $Hg(NO_3)_2 + Na_3PO_4$ (d) $Ba(NO_3)_2 + KOH$

4.76 Which of the following solutions will not form a precipitate when added to 0.10 M $BaCl_2$?

(a) 0.10 M $LiNO_3$ (b) 0.10 M K_2SO_4

(c) 0.10 M $AgNO_3$

4.77 Which of the following solutions will not form a precipitate when added to 0.10 M NaOH?

(a) 0.10 M $MgBr_2$ (b) 0.10 M NH_4Br

(c) 0.10 M $FeCl_2$

4.78 How would you prepare the following substances by a precipitation reaction?

(a) $PbSO_4$ (b) $Mg_3(PO_4)_2$

(c) $ZnCrO_4$

4.79 How would you prepare the following substances by a precipitation reaction?

(a) $Al(OH)_3$ (b) FeS

(c) $CoCO_3$

4.80 What is the mass and the identity of the precipitate that forms when 30.0 mL of 0.150 M HCl reacts with 25.0 mL of 0.200 M $AgNO_3$?

4.81 What is the mass and the identity of the precipitate that forms when 55.0 mL of 0.100 M $BaCl_2$ reacts with 40.0 mL of 0.150 M Na_2CO_3?

4.82 Assume that you have an aqueous mixture of $NaNO_3$ and $AgNO_3$. How could you use a precipitation reaction to separate the two metal ions?

4.83 Assume that you have an aqueous mixture of $BaCl_2$ and $CuCl_2$. How could you use a precipitation reaction to separate the two metal ions?

4.84 Assume that you have an aqueous solution of an unknown salt. Treatment of the solution with dilute NaOH, Na_2SO_4, and KCl produces no precipitate. Which of the following cations might the solution contain?

(a) Ag^+ (b) Cs^+

(c) Ba^{2+} (d) NH_4^+

4.85 Assume that you have an aqueous solution of an unknown salt. Treatment of the solution with dilute $BaCl_2$, $AgNO_3$, and $Cu(NO_3)_2$ produces no precipitate. Which of the following anions might the solution contain?

(a) Cl^- (b) NO_3^-

(c) OH^- (d) SO_4^{2-}

Acids, Bases, and Neutralization Reactions (Section 4.7)

4.86 Assume that you are given a solution of an unknown acid or base. How can you tell whether the unknown substance is acidic or basic?

4.87 Why do we use a double arrow $\rightleftharpoons$ to show the dissociation of a weak acid or weak base in aqueous solution?

4.88 Write balanced ionic equations for the following reactions:

(a) Aqueous perchloric acid is neutralized by aqueous calcium hydroxide.

(b) Aqueous sodium hydroxide is neutralized by aqueous acetic acid.

4.89 Write balanced ionic equations for the following reactions:

(a) Aqueous hydrobromic acid is neutralized by aqueous calcium hydroxide.

(b) Aqueous barium hydroxide is neutralized by aqueous nitric acid.

4.90 Write balanced net ionic equations for the following reactions:

(a) $LiOH(aq) + HI(aq) \rightarrow$?

(b) $HBr(aq) + Ca(OH)_2(aq) \rightarrow$?

4.91 Write balanced net ionic equations for the following reactions. Note that $HClO_3$ is a strong acid.

(a) $Fe(OH)_3(s) + H_2SO_4(aq) \rightarrow$?

(b) $HClO_3(aq) + NaOH(aq) \rightarrow$?

Solution Stoichiometry and Titration (Sections 4.8 and 4.9)

4.92 A flask containing 450 mL of 0.500 M HBr was accidentally knocked to the floor. How many grams of K_2CO_3 would you need to put on the spill to neutralize the acid according to the following equation?

$$2 HBr(aq) + K_2CO_3(aq) \longrightarrow 2 KBr(aq) + CO_2(g) + H_2O(l)$$

4.93 The odor of skunks is caused by chemical compounds called *thiols*. These compounds, of which butanethiol ($C_4H_{10}S$) is a representative example, can be deodorized by reaction with household bleach (NaOCl) according to the following equation:

$$2 C_4H_{10}S(l) + NaOCl(aq) \longrightarrow C_8H_{18}S_2(l) + NaCl(aq) + H_2O(l)$$

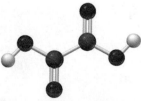

.089 g

Butanethiol

How many grams of butanethiol can be deodorized by reaction with 5.00 mL of 0.0985 M NaOCl?

4.94 Potassium permanganate ($KMnO_4$) reacts with oxalic acid ($H_2C_2O_4$) in aqueous sulfuric acid according to the following equation:

$$2 KMnO_4(aq) + 5 H_2C_2O_4(aq) + 3 H_2SO_4(aq) \longrightarrow$$
$$2 MnSO_4(aq) + 10 CO_2(g) + 8 H_2O(l) + K_2SO_4(aq)$$

How many milliliters of a 0.250 M $KMnO_4$ solution are needed to react completely with 3.225 g of oxalic acid?

4.95 Oxalic acid, $H_2C_2O_4$, is a toxic substance found in spinach leaves. What is the molarity of a solution made by dissolving 12.0 g of oxalic acid in enough water to give 400.0 mL of solution? How many milliliters of 0.100 M KOH would you need to titrate 25.0 mL of the oxalic acid solution according to the following equation?

$$H_2C_2O_4(aq) + 2 KOH(aq) \longrightarrow K_2C_2O_4(aq) + 2 H_2O(l)$$

Oxalic acid

4.96 How many milliliters of 1.00 M KOH must be added to neutralize the following solutions?

(a) A mixture of 0.240 M LiOH (25.0 mL) and 0.200 M HBr (75.0 mL)

(b) A mixture of 0.300 M HCl (45.0 mL) and 0.250 M NaOH (10.0 mL)

4.97 How many milliliters of 2.00 M HCl must be added to neutralize the following solutions?

(a) A mixture of 0.160 M HNO_3 (100.0 mL) and 0.100 M KOH (400.0 mL)

(b) A mixture of 0.120 M NaOH (350.0 mL) and 0.190 M HBr (150.0 mL)

4.98 If the following solutions are mixed, is the resulting solution acidic, basic, or neutral?

(a) 50.0 mL of 0.100 M HBr and 30.0 mL of 0.200 M KOH

(b) 100.0 mL of 0.0750 M HCl and 75.0 mL of 0.100 M $Ba(OH)_2$

4.99 If the following solutions are mixed, is the resulting solution acidic, basic, or neutral?

(a) 65.0 mL of 0.0500 M $HClO_4$ and 40.0 mL of 0.0750 M NaOH

(b) 125.0 mL of 0.100 M HNO_3 and 90.0 mL of 0.0750 M $Ca(OH)_2$

Redox Reactions, Oxidation Numbers, and Activity Series (Sections 4.10–4.12)

4.100 Where in the periodic table are the best reducing agents found? The best oxidizing agents?

4.101 Where in the periodic table are the most easily reduced elements found? The most easily oxidized?

4.102 In each of the following instances, tell whether the substance gains electrons or loses electrons in a redox reaction:

(a) An oxidizing agent

(b) A reducing agent

(c) A substance undergoing oxidation

(d) A substance undergoing reduction

4.103 Tell for each of the following substances whether the oxidation number increases or decreases in a redox reaction:

(a) An oxidizing agent

(b) A reducing agent

(c) A substance undergoing oxidation

(d) A substance undergoing reduction

4.104 Assign oxidation numbers to each element in the following compounds:

(a) NO_2 (b) SO_3

(c) $COCl_2$ (d) CH_2Cl_2

(e) $KClO_3$ (f) HNO_3

4.105 Assign oxidation numbers to each element in the following compounds:

(a) $VOCl_3$ (b) $CuSO_4$

(c) CH_2O (d) Mn_2O_7

(e) OsO_4 (f) H_2PtCl_6

4.106 Assign oxidation numbers to each element in the following ions:

(a) ClO_3^- (b) SO_3^{2-}

(c) $C_2O_4^{2-}$ (d) NO_2^-

(e) BrO^- (f) AsO_4^{3-}

4.107 Assign oxidation numbers to each element in the following ions:

(a) $Cr(OH)_4^-$ (b) $S_2O_3^{2-}$

(c) NO_3^- (d) MnO_4^{2-}

(e) HPO_4^{2-} (f) $V_2O_7^{4-}$

4.108 Nitrogen can have several different oxidation numbers ranging in value from −3 to +5.

(a) Write the formula and give the name of the nitrogen oxide compound in which nitrogen has an oxidation number of +1, +2, +4, and +5.

(b) Based on oxidation numbers, which nitrogen oxide from par (a) cannot react with molecular oxygen?

4.109 Phosphorus can have several different oxidation numbers ranging in value from −3 to +5.

(a) When phosphorus burns in air or oxygen, it yields eithe tetraphosphorus hexoxide or tetraphorus decoxide Write the formula and give the oxidation number in each compound.

(b) Based on oxidation numbers, which phosphorus oxide com- pound from part (a) was formed by combustion with a lim- ited supply of oxygen?

4.110 Which element is oxidized and which is reduced in each of the following reactions?

(a) $Ca(s) + Sn^{2+}(aq) \longrightarrow Ca^{2+}(aq) + Sn(s)$

(b) $ICl(s) + H_2O(l) \longrightarrow HCl(aq) + HOI(aq)$

4.111 Which element is oxidized and which is reduced in each of the following reactions?

(a) $Si(s) + 2 Cl_2(g) \longrightarrow SiCl_4(l)$

(b) $Cl_2(g) + 2 NaBr(aq) \longrightarrow Br_2(aq) + 2 NaCl(aq)$

4.112 Use the activity series of metals (Table 4.5) to predict the outcome of each of the following reactions. If no reaction occurs, write N.R

(a) $Na^+(aq) + Zn(s) \longrightarrow ?$

(b) $HCl(aq) + Pt(s) \longrightarrow ?$

(c) $Ag^+(aq) + Au(s) \longrightarrow ?$

(d) $Au^{3+}(aq) + Ag(s) \longrightarrow ?$

4.113 Neither strontium (Sr) nor antimony (Sb) is shown in the activity series of Table 4.5. Based on their positions in the periodic table, which would you expect to be the better reducing agent? Will the following reaction occur? Explain.

$$2 Sb^{3+}(aq) + 3 Sr(s) \longrightarrow 2 Sb(s) + 3 Sr^{2+}(aq)$$

4.114 (a) Use the following reactions to arrange the elements **A, B, C,** and **D** in order of their decreasing ability as reducing agents:

$A + B^+ \longrightarrow A^+ + B$ $C^+ + D \longrightarrow$ no reaction

$B + D^+ \longrightarrow B^+ + D$ $B + C^+ \longrightarrow B^+ + C$

(b) Which of the following reactions would you expect to occur according to the activity series you established in part (a)?

(1) $A^+ + C \longrightarrow A + C^+$

(2) $A^+ + D \longrightarrow A + D^+$

4.115 (a) Use the following reactions to arrange the elements **A, B, C,** and **D** in order of their decreasing ability as reducing agents:

$2 A + B^{2+} \longrightarrow 2 A^+ + B$ $B + D^{2+} \longrightarrow B^{2+} + D$

$A^+ + C \longrightarrow$ no reaction $2 C + B^{2+} \longrightarrow 2 C^+ + B$

(b) Which of the following reactions would you expect to occur according to the activity series you established in part (a)?

(1) $2 A^+ + D \longrightarrow 2 A + D^{2+}$

(2) $D^{2+} + 2 C \longrightarrow D + 2 C^+$

Redox Titrations (Section 4.13)

4.116 Iodine, I_2, reacts with aqueous thiosulfate ion in neutral solution according to the balanced equation

$$I_2(aq) + 2 S_2O_3^{2-}(aq) \longrightarrow S_4O_6^{2-}(aq) + 2 I^-(aq)$$

How many grams of I_2 are present in a solution if 35.20 mL of 0.150 M $Na_2S_2O_3$ solution is needed to titrate the I_2 solution?

4.117 How many milliliters of 0.250 M $Na_2S_2O_3$ solution is needed for complete reaction with 2.486 g of I_2 according to the equation in Problem 4.116?

4.118 Dichromate ion, $Cr_2O_7^{2-}$, reacts with aqueous iron(II) ion in acidic solution according to the balanced equation

$$Cr_2O_7^{2-}(aq) + 6\,Fe^{2+}(aq) + 14\,H^+(aq) \longrightarrow$$
$$2\,Cr^{3+}(aq) + 6\,Fe^{3+}(aq) + 7\,H_2O(l)$$

What is the concentration of Fe^{2+} if 46.99 mL of 0.2004 M $K_2Cr_2O_7$ is needed to titrate 50.00 mL of the Fe^{2+} solution?

4.119 A volume of 18.72 mL of 0.1500 M $K_2Cr_2O_7$ solution was required to titrate a sample of $FeSO_4$ according to the equation in Problem 4.118. What is the mass of the sample?

4.120 What is the molar concentration of As(III) in a solution if 22.35 mL of 0.100 M $KBrO_3$ is needed for complete reaction with 50.00 mL of the As(III) solution? The balanced equation is:

$$3\,H_3AsO_3(aq) + BrO_3^-(aq) \longrightarrow Br^-(aq) + 3\,H_3AsO_4(aq)$$

4.121 Standardized solutions of $KBrO_3$ are frequently used in redox titrations. The necessary solution can be made by dissolving $KBrO_3$ in water and then titrating it with an As(III) solution. What is the molar concentration of a $KBrO_3$ solution if 28.55 mL of the solution is needed to titrate 1.550 g of As_2O_3? See Problem 4.120 for the balanced equation. (As_2O_3 dissolves in aqueous acid solution to yield H_3AsO_3: $As_2O_3 + 3\,H_2O \longrightarrow 2\,H_3AsO_3$.)

4.122 The metal content of iron in ores can be determined by a redox procedure in which the sample is first oxidized with Br_2 to convert all the iron to Fe^{3+} and then titrated with Sn^{2+} to reduce the Fe^{3+} to Fe^{2+}. The balanced equation is:

$$2\,Fe^{3+}(aq) + Sn^{2+}(aq) \longrightarrow 2\,Fe^{2+}(aq) + Sn^{4+}(aq)$$

What is the mass percent Fe in a 0.1875 g sample of ore if 13.28 mL of a 0.1015 M Sn^{2+} solution is needed to titrate the Fe^{3+}?

4.123 The concentration of the Sn^{2+} solution used in Problem 4.122 can be found by letting it react with a known amount of Fe^{2+}. What is the molar concentration of an Sn^{2+} solution if 23.84 mL is required for complete reaction with 1.4855 g of Fe_2O_3?

4.124 Alcohol levels in blood can be determined by a redox reaction with potassium dichromate according to the balanced equation

$$C_2H_5OH(aq) + 2\,Cr_2O_7^{2-}(aq) + 16\,H^+(aq) \longrightarrow$$
$$2\,CO_2(g) + 4\,Cr^{3+}(aq) + 11\,H_2O(l)$$

What is the blood alcohol level in mass percent if 8.76 mL of 0.049 88 M $K_2Cr_2O_7$ is required for complete reaction with a 10.002 g sample of blood?

4.125 Calcium levels in blood can be determined by adding oxalate ion to precipitate calcium oxalate, CaC_2O_4, followed by dissolving the precipitate in aqueous acid and titrating the resulting oxalic acid ($H_2C_2O_4$) with $KMnO_4$:

$$5\,H_2C_2O_4(aq) + 2\,MnO_4^-(aq) + 6\,H^+(aq) \longrightarrow$$
$$10\,CO_2(g) + 2\,Mn^{2+}(aq) + 8\,H_2O(l)$$

How many milligrams of Ca^{2+} are present in 10.0 mL of blood if 21.08 mL of 0.000 988 M $KMnO_4$ solution is needed for the titration?

CHAPTER PROBLEMS

4.126 *Ringer's solution*, used in the treatment of burns and wounds, is prepared by dissolving 4.30 g of NaCl, 0.150 g of KCl, and 0.165 g of $CaCl_2$ in water and diluting to a volume of 500.0 mL. What is the molarity of each of the component ions in the solution?

4.127 The estimated concentration of gold in the oceans is 1.0×10^{-11} g/mL.
   (a) Express the concentration in mol/L.
   (b) Assuming that the volume of the oceans is 1.3×10^{21} L, estimate the amount of dissolved gold in grams in the oceans.

4.128 What is the molarity of each ion in a solution prepared by dissolving 0.550 g of Na_2SO_4, 1.188 g of Na_3PO_4, and 0.223 g of Li_2SO_4 in water and diluting to a volume of 100.00 mL?

4.129 Assume that you have 1.00 g of a mixture of benzoic acid (Mol. wt. = 122) and gallic acid (Mol. wt. = 170), both of which contain one acidic hydrogen that reacts with NaOH. On titrating the mixture with 0.500 M NaOH, 14.7 mL of base is needed to completely react with both acids. What mass in grams of each acid is present in the original mixture?

4.130 The concentration of a solution of potassium permanganate, $KMnO_4$, can be determined by titration against a known amount of oxalic acid, $H_2C_2O_4$, according to the following equation:

$$5\,H_2C_2O_4(aq) + 2\,KMnO_4(aq) + 3\,H_2SO_4(aq) \longrightarrow$$
$$10\,CO_2(g) + 2\,MnSO_4(aq) + K_2SO_4(aq) + 8\,H_2O(l)$$

What is the concentration of a $KMnO_4$ solution if 22.35 mL reacts with 0.5170 g of oxalic acid?

4.131 A compound with the formula $XOCl_2$ reacts with water, yielding HCl and another acid H_2XO_3, which has two acidic hydrogens that react with NaOH. When 0.350 g of $XOCl_2$ was added to 50.0 mL of water and the resultant solution was titrated, 96.1 mL of 0.1225 M NaOH was required to react with all the acid.
   (a) Write a balanced equation for the reaction of $XOCl_2$ with H_2O.
   (b) What are the atomic mass and identity of element X?

4.132 An alternative procedure to that given in Problem 4.122 for determining the amount of iron in a sample is to convert the iron to Fe^{2+} and then titrate it with a solution of $Ce(NH_4)_2(NO_3)_6$:

$$Fe^{2+}(aq) + Ce^{4+}(aq) \longrightarrow Fe^{3+}(aq) + Ce^{3+}(aq)$$

What is the mass percent of iron in a sample if 1.2284 g of the sample requires 54.91 mL of 0.1018 M $Ce(NH_4)_2(NO_3)_6$ for complete reaction?

4.133 Assign oxidation numbers to each atom in the following substances:
   (a) Ethane, C_2H_6, a constituent of natural gas
   (b) Borax, $Na_2B_4O_7$, a mineral used in laundry detergents
   (c) $Mg_2Si_2O_6$, a silicate mineral

4.134 (a) Use the following reactions to arrange the elements **A**, **B**, **C**, and **D** in order of their decreasing ability as reducing agents:

$$C + B^+ \longrightarrow C^+ + B \qquad\qquad A^+ + D \longrightarrow \text{No reaction}$$
$$C^+ + A \longrightarrow \text{No reaction} \qquad\qquad D + B^+ \longrightarrow D^+ + B$$

   (b) Which of the following reactions would you expect to occur according to the activity series you established in part (a)?
   (1) $A^+ + C \longrightarrow A + C^+$
   (2) $A^+ + B \longrightarrow A + B^+$

4.135 Some metals occur naturally in their elemental state while others occur as compounds in ores. Gold, for instance, is found as the free metal; mercury is obtained by heating mercury(II) sulfide ore in oxygen; and zinc is obtained by heating zinc(II) oxide ore with coke (carbon). Judging from their positions in the activity series, which of the metals silver, platinum, and chromium would probably be obtained by
   (a) finding it in its elemental state?
   (b) heating its sulfide with oxygen?
   (c) heating its oxide with coke?

4.136 A sample weighing 14.98 g and containing a small amount of copper was treated to give a solution containing aqueous Cu^{2+} ions. Sodium iodide was then added to yield solid copper(I) iodide plus I_3^- ion, and the I_3^- was titrated with thiosulfate, $S_2O_3^{2-}$. The titration required 10.49 mL of 0.100 M $Na_2S_2O_3$ for complete reaction. What is the mass percent copper in the sample? The balanced equations are

$$2 Cu^{2+}(aq) + 5 I^-(aq) \longrightarrow 2 CuI(s) + I_3^-(aq)$$

$$I_3^-(aq) + 2 S_2O_3^{2-}(aq) \longrightarrow 3 I^-(aq) + S_4O_6^{2-}(aq)$$

4.137 The solubility of an ionic compound can be described quantitatively by a value called the *solubility product constant*, K_{sp}. For the general solubility process

$A_aB_b \rightleftharpoons a A^{n+} + b B^{m-}$, $K_{sp} = [A^{n+}]^a [B^{m-}]^b$. The brackets refer to concentrations in moles per liter.

    (a) Write the expression for the solubility product constant of Ag_2CrO_4.

    (b) If $K_{sp} = 1.1 \times 10^{-12}$ for Ag_2CrO_4, what are the molar concentrations of Ag^+ and CrO_4^{2-} in solution?

4.138 Write the expression for the solubility product constant of MgF_2 (see Problem 4.137). If $[Mg^{2+}] = 2.6 \times 10^{-4} \, mol/L$ in a solution, what is the value of K_{sp}?

4.139 Succinic acid, an intermediate in the metabolism of food molecules, has molecular weight = 118.1. When 1.926 g of succinic acid was dissolved in water and titrated, 65.20 mL of 0.5000 M NaOH solution was required to neutralize the acid. How many acidic hydrogens are there in a molecule of succinic acid?

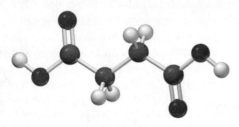

Succinic acid

4.140 How could you use a precipitation reaction to separate each of the following pairs of cations? Write the formula for each reactant you would add, and write a balanced net ionic equation for each reaction.

    (a) K^+ and Hg_2^{2+} **(b)** Pb^{2+} and Ni^{2+}

    (c) Ca^{2+} and NH_4^+ **(d)** Fe^{2+} and Ba^{2+}

4.141 How could you use a precipitation reaction to separate each of the following pairs of anions? Write the formula for each reactant you would add, and write a balanced net ionic equation for each reaction.

    (a) Cl^- and NO_3^- **(b)** S^{2-} and SO_4^{2-}

    (c) SO_4^{2-} and CO_3^{2-} **(d)** OH^- and ClO_4^-

4.142 A 100.0 mL solution containing aqueous HCl and HBr was titrated with 0.1235 M NaOH. The volume of base required to neutralize the acid was 47.14 mL. Aqueous $AgNO_3$ was then added to precipitate the Cl^- and Br^- ions as AgCl and AgBr. The mass of the silver halides obtained was 0.9974 g. What are the molarities of the HCl and HBr in the original solution?

4.143 A mixture of CuO and Cu_2O with a mass of 10.50 g is reduced to give 8.66 g of pure Cu metal. What are the amounts in grams of CuO and Cu_2O in the original mixture?

4.144 When 75.0 mL of a 0.100 M lead(II) nitrate solution is mixed with 100.0 mL of a 0.190 M potassium iodide solution, a yellow–orange precipitate of lead(II) iodide is formed.

    (a) What mass in grams of lead(II) iodide is formed, assuming the reaction goes to completion?

    (b) What is the molarity of each of the ions Pb^{2+}, K^+, NO_3^-, and I^- in the resulting solution?

4.145 A sample of metal (**M**) reacted with both steam and aqueous HCl to release H_2, but did not react with water at room temperature.

When 1.000 g of the metal was burned in oxygen, it formed 1.890 g of a metal oxide, M_2O_3. What is the identity of the metal?

4.146 An unknown metal (**M**) was found not to react with either water or steam, but its reactivity with aqueous acid was not investigated. When a 1.000 g sample of the metal was burned in oxygen and the resulting metal oxide converted to a metal sulfide, 1.504 g of sulfide was obtained. What is the identity of the metal?

4.147 A mixture of acetic acid (CH_3CO_2H; monoprotic) and oxalic acid ($H_2C_2O_4$; diprotic) requires 27.15 mL of 0.100 M NaOH to neutralize it. When an identical amount of the mixture is titrated, 15.05 mL of 0.0247 M $KMnO_4$ is needed for complete reaction. What is the mass percent of each acid in the mixture? (Acetic acid does not react with MnO_4^-. The equation for the reaction of oxalic acid with MnO_4^- was given in Problem 4.125.)

4.148 Iron content in ores can be determined by a redox procedure in which the sample is first reduced with Sn^{2+}, as in Problem 4.122 and then titrated with $KMnO_4$ to oxidize the Fe^{2+} to Fe^{3+}. The balanced equation is

$$MnO_4^-(aq) + 5 Fe^{2+}(aq) + 8 H^+(aq) \longrightarrow$$
$$Mn^{2+}(aq) + 5 Fe^{3+}(aq) + 4 H_2O(l)$$

What is the mass percent Fe in a 2.368 g sample if 48.39 mL of a 0.1116 M $KMnO_4$ solution is needed to titrate the Fe^{3+}?

4.149 A mixture of $FeCl_2$ and NaCl is dissolved in water, and addition of aqueous silver nitrate then yields 7.0149 g of a precipitate. When an identical amount of the mixture is titrated with MnO_4^-, 14.28 mL of 0.198 M $KMnO_4$ is needed for complete reaction. What are the mass percents of the two compounds in the mixture? (Na^+ and Cl^- do not react with MnO_4^-. The equation for the reaction of Fe^{2+} with MnO_4^- was given in Problem 4.148.)

MULTICONCEPT PROBLEMS

4.150 Salicylic acid, used in the manufacture of aspirin, contains only the elements C, H, and O and has only one acidic hydrogen that reacts with NaOH. When 1.00 g of salicylic acid undergoes complete combustion, 2.23 g CO_2 and 0.39 g H_2O are obtained. When 1.00 g of salicylic acid is titrated with 0.100 M NaOH, 72.4 mL of base is needed for complete reaction. What are the empirical and molecular formulas of salicylic acid?

4.151 Compound X contains only the elements C, H, O, and S. A 5.00 g sample undergoes complete combustion to give 4.83 g of CO_2, 1.48 g of H_2O, and a certain amount of SO_2 that is further oxidized to SO_3 and dissolved in water to form sulfuric acid, H_2SO_4. On titration of the H_2SO_4, 109.8 mL of 1.00 M NaOH is needed for complete reaction. (Both H atoms in sulfuric acid are acidic and react with NaOH.)

    (a) What is the empirical formula of X?

    (b) When 5.00 g of X is titrated with NaOH, it is found that X has two acidic hydrogens that react with NaOH and that 54.9 mL of 1.00 M NaOH is required to completely neutralize the sample. What is the molecular formula of X?

4.152 A 1.268 g sample of a metal carbonate (MCO_3) was treated with 100.00 mL of 0.1083 M sulfuric acid (H_2SO_4), yielding CO_2 gas and an aqueous solution of the metal sulfate (MSO_4). The solution was boiled to remove all the dissolved CO_2 and was then titrated with 0.1241 M NaOH. A 71.02 mL volume of NaOH was required to neutralize the excess H_2SO_4.

    (a) What is the identity of the metal M?

    (b) How many liters of CO_2 gas were produced if the density of CO_2 is 1.799 g/L?

4.153 Element M is prepared industrially by a two-step procedure according to the following (unbalanced) equations:

(1) $M_2O_3(s) + C(s) + Cl_2(g) \longrightarrow MCl_3(l) + CO(g)$

(2) $MCl_3(l) + H_2(g) \longrightarrow M(s) + HCl(g)$

Assume that 0.855 g of M_2O_3 is submitted to the reaction sequence. When the HCl produced in Step (2) is dissolved in water and titrated with 0.511 M NaOH, 144.2 mL of the NaOH solution is required to neutralize the HCl.

(a) Balance both equations.

(b) What is the atomic mass of element M, and what is its identity?

(c) What mass of M in grams is produced in the reaction?

4.154 Assume that you dissolve 10.0 g of a mixture of NaOH and $Ba(OH)_2$ in 250.0 mL of water and titrate with 1.50 M hydrochloric acid. The titration is complete after 108.9 mL of the acid has been added. What is the mass in grams of each substance in the mixture?

4.155 The following three solutions are mixed: 100.0 mL of 0.100 M Na_2SO_4, 50.0 mL of 0.300 M $ZnCl_2$, and 100.0 mL of 0.200 M $Ba(CN)_2$.

(a) What ionic compounds will precipitate out of solution?

(b) What is the molarity of each ion remaining in the solution assuming complete precipitation of all insoluble compounds?

4.156 A 250.0 g sample of a white solid is known to be a mixture of KNO_3, $BaCl_2$, and NaCl. When 100.0 g of this mixture is dissolved in water and allowed to react with excess H_2SO_4, 67.3 g of a white precipitate is collected. When the remaining 150.0 g of the mixture is dissolved in water and allowed to react with excess $AgNO_3$, 197.6 g of a second precipitate is collected.

(a) What are the formulas of the two precipitates?

(b) What is the mass of each substance in the original 250 g mixture?

4.157 Four solutions are prepared and mixed in the following order:

(1) Start with 100.0 mL of 0.100 M $BaCl_2$

(2) Add 50.0 mL of 0.100 M $AgNO_3$

(3) Add 50.0 mL of 0.100 M H_2SO_4

(4) Add 250.0 mL of 0.100 M NH_3

Write an equation for any reaction that occurs after each step, and calculate the concentrations of Ba^{2+}, Cl^-, NO_3^-, NH_3, and NH_4^+ in the final solution, assuming that all reactions go to completion.

4.158 To 100.0 mL of a solution that contains 0.120 M $Cr(NO_3)_2$ and 0.500 M HNO_3 is added 20.0 mL of 0.250 M $K_2Cr_2O_7$. The dichromate and chromium(II) ions react to give chromium(III) ions.

(a) Write a balanced net ionic equation for the reaction.

(b) Calculate the concentrations of all ions in the solution after reaction. Check your concentrations to make sure that the solution is electrically neutral.

4.159 Sodium nitrite, $NaNO_2$, is frequently added to processed meats as a preservative. The amount of nitrite ion in a sample can be determined by acidifying to form nitrous acid (HNO_2), letting the nitrous acid react with an excess of iodide ion, and then titrating the I_3^- ion that results with thiosulfate solution in the presence of a starch indicator. The unbalanced equations are

(1) $HNO_2 + I^- \longrightarrow NO + I_3^-$ (in acidic solution)

(2) $I_3^- + S_2O_3^{2-} \longrightarrow I^- + S_4O_6^{2-}$

(a) Balance the two redox equations.

(b) When a nitrite-containing sample with a mass of 2.935 g was analyzed, 18.77 mL of 0.1500 M $Na_2S_2O_3$ solution was needed for the reaction. What is the mass percent of NO_2^- ion in the sample?

4.160 Brass is an approximately 4:1 alloy of copper and zinc, along with small amounts of tin, lead, and iron. The mass percents of copper and zinc can be determined by a procedure that begins with

dissolving the brass in hot nitric acid. The resulting solution of Cu^{2+} and Zn^{2+} ions is then treated with aqueous ammonia to lower its acidity, followed by addition of sodium thiocyanate (NaSCN) and sulfurous acid (H_2SO_3) to precipitate copper(I) thiocyanate (CuSCN). The solid CuSCN is collected, dissolved in aqueous acid, and treated with potassium iodate (KIO_3) to give iodine, which is then titrated with aqueous sodium thiosulfate ($Na_2S_2O_3$). The filtrate remaining after CuSCN has been removed is neutralized by addition of aqueous ammonia, and a solution of diammonium hydrogen phosphate (($NH_4)_2HPO_4$) is added to yield a precipitate of zinc ammonium phosphate ($ZnNH_4PO_4$). Heating the precipitate to 900 °C converts it to zinc pyrophosphate ($Zn_2P_2O_7$), which is weighed. The equations are

(1) $Cu(s) + NO_3^-(aq) \longrightarrow Cu^{2+}(aq) + NO(g)$ (in acid)

(2) $Cu^{2+}(aq) + SCN^-(aq) + HSO_3^-(aq) \longrightarrow$
$\qquad\qquad CuSCN(s) + HSO_4^-(aq)$ (in acid)

(3) $Cu^+(aq) + IO_3^-(aq) \longrightarrow Cu^{2+}(aq) + I_2(aq)$ (in acid)

(4) $I_2(aq) + S_2O_3^{2-}(aq) \longrightarrow I^-(aq) + S_4O_6^{2-}(aq)$ (in acid)

(5) $ZnNH_4PO_4(s) \longrightarrow Zn_2P_2O_7(s) + H_2O(g) + NH_3(g)$

(a) Balance all equations.

(b) When a brass sample with a mass of 0.544 g was subjected to the preceding analysis, 10.82 mL of 0.1220 M sodium thiosulfate was required for the reaction with iodine. What is the mass percent copper in the brass?

(c) The brass sample in part (b) yielded 0.246 g of $Zn_2P_2O_7$. What is the mass percent zinc in the brass?

4.161 A certain metal sulfide, MS_n (where n is a small integer), is widely used as a high-temperature lubricant. The substance is prepared by reaction of the metal pentachloride (MCl_5) with sodium sulfide (Na_2S). Heating the metal sulfide to 700 °C in air gives the metal trioxide (MO_3) and sulfur dioxide (SO_2), which reacts with Fe^{3+} ion under aqueous acidic conditions to give sulfate ion (SO_4^{2-}). Addition of aqueous $BaCl_2$ then forms a precipitate of $BaSO_4$. The unbalanced equations are:

(1) $MCl_5(s) + Na_2S(s) \longrightarrow MS_n(s) + S(l) + NaCl(s)$

(2) $MS_n(s) + O_2(g) \longrightarrow MO_3(s) + SO_2(g)$

(3) $SO_2(g) + Fe^{3+}(aq) \longrightarrow Fe^{2+}(aq) + SO_4^{2-}(aq)$ (in acid)

(4) $SO_4^{2-}(aq) + Ba^{2+}(aq) \longrightarrow BaSO_4(s)$

Assume that you begin with 4.61 g of MCl_5 and that reaction (1) proceeds in 91.3% yield. After oxidation of the MS_n product, oxidation of SO_2, and precipitation of sulfate ion, 7.19 g of $BaSO_4(s)$ is obtained.

(a) How many moles of sulfur are present in the MS_n sample?

(b) Assuming several possible values for n ($n = 1, 2, 3 \ldots$), what is the atomic weight of **M** in each case?

(c) What is the likely identity of the metal **M**, and what is the formula of the metal sulfide MS_n?

(d) Balance all equations.

4.162 On heating a 0.200 g sample of a certain semimetal **M** in air, the corresponding oxide M_2O_3 was obtained. When the oxide was dissolved in aqueous acid and titrated with $KMnO_4$, 10.7 mL of 0.100 M MnO_4^- was required for complete reaction. The unbalanced equation is

$H_3MO_3(aq) + MnO_4^-(aq) \longrightarrow H_3MO_4(aq)$
$\qquad\qquad\qquad + Mn^{2+}(aq)$ (in acid)

(a) Balance the equation.

(b) How many moles of oxide were formed, and how many moles of semimetal were in the initial 0.200 g sample?

(c) What is the identity of the semimetal M?

Periodicity and the Electronic Structure of Atoms

The color emitted by elements heated in a flame arises from electrons moving between different energy levels in an atom. Atoms emit characteristic colors because each has its own unique energy levels, referred to as electronic structure.

 How does knowledge of atomic emission spectra help us build more efficient light bulbs?

The answer to this question can be found in the **INQUIRY** ▶▶▶ on page 184.

The periodic table, introduced in Section 2.2, is the most important organizing principle in chemistry. If you know the properties of any one element in a group, or column, of the periodic table, you can make a good guess at the properties of every other element in the same group and even of the elements in neighboring groups. Although the periodic table was originally constructed from empirical observations, its scientific underpinnings have long been established and are well understood.

To see why it's called the *periodic* table, look at the graph of atomic radius versus atomic number in **FIGURE 5.1**, which shows a periodic rise-and-fall pattern. Beginning on the left with atomic number 1 (hydrogen), the size of the atoms increases to a maximum at atomic number 3 (lithium), then decreases to a minimum, then increases again to a maximum at atomic number 11 (sodium), then decreases, and so on. It turns out that all the maxima occur for atoms of group 1A elements—Li, Na, K, Rb, Cs, and Fr—and that the minima occur for atoms of the group 7A elements—F, Cl, Br, and I.

There's nothing unique about the periodicity of atomic radii shown in Figure 5.1. Any of several dozen other physical or chemical properties could be plotted in a similar way with similar results. We'll look at several examples of such periodicity in this chapter and the next.

5.1 ▶ THE NATURE OF RADIANT ENERGY AND THE ELECTROMAGNETIC SPECTRUM

What fundamental property of atoms is responsible for the periodic variations we observe in atomic radii and in so many other characteristics of the elements? This question occupied the thoughts of chemists for more than 50 years after Mendeleev, and it was not until well into the 1920s that the answer was established. To understand how the answer slowly emerged, it's necessary to look first at the nature of visible light and other forms of *radiant energy*. *Spectroscopy*, the study of the interaction of radiant energy with matter, has provided immense insight into atomic structure.

Although they appear quite different to our senses, visible light, infrared radiation, microwaves, radio waves, X rays, and so on are all different forms of *electromagnetic radiation* or **radiant energy**. Collectively, they make up the **electromagnetic spectrum**, shown in **FIGURE 5.2**.

Electromagnetic energy traveling through a vacuum behaves in some ways like ocean waves traveling through water. Like ocean waves, electromagnetic energy is characterized by a *frequency*, a *wavelength*, and an *amplitude*. If you could stand in one place and look at a sideways, cutaway view of an ocean wave moving through the water, you would see a regular rise-and-fall pattern like that in **FIGURE 5.3**.

▲ Ocean waves, like electromagnetic waves, are characterized by a wavelength, a frequency, and an amplitude.

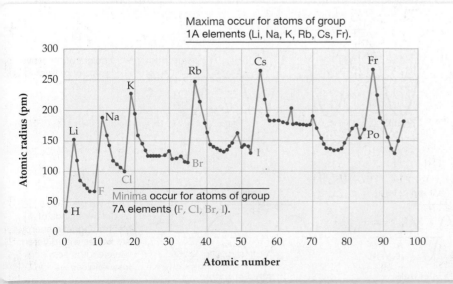

Maxima occur for atoms of group 1A elements (Li, Na, K, Rb, Cs, Fr).

◀ **FIGURE 5.1**

A graph of atomic radius in picometers (pm) versus atomic number. A clear rise-and-fall pattern of periodicity is evident. (Accurate data are not available for the group 8A elements.)

▶ **FIGURE 5.2**

The electromagnetic spectrum. The spectrum consists of a continuous range of wavelengths and frequencies, from radio waves at the low-frequency end to gamma rays at the high-frequency end.

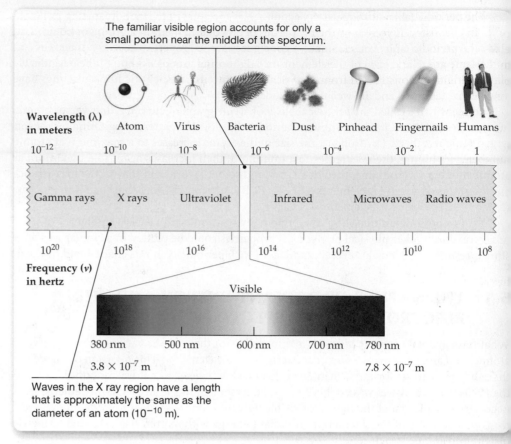

The **frequency** (ν, Greek nu) of a wave is simply the number of wave peaks that pass by a given point per unit time, usually expressed in units of reciprocal seconds, or **hertz** (Hz, 1 Hz = 1 s^{-1}). The **wavelength** (λ, Greek lambda) of the wave is the distance from one wave peak to the next, and the **amplitude** of the wave is the height of the wave, measured from the center line between peak and trough. Physically, what we perceive as the intensity of electromagnetic energy is proportional to the square of the wave amplitude. A faint beam and a blinding glare of light may have the same wavelength and frequency, but they differ greatly in amplitude.

▶ **FIGURE 5.3**

The nature of electromagnetic waves. The waves are characterized by a wavelength, a frequency, and an amplitude.

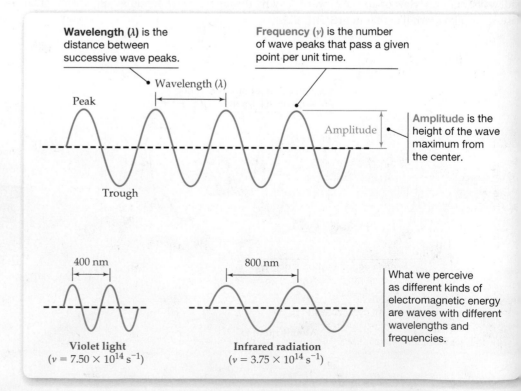

Multiplying the wavelength of a wave in meters (m) by its frequency in reciprocal seconds (s^{-1}) gives the speed of the wave in meters per second (m/s). The rate of travel of all electromagnetic energy in a vacuum is a constant value, commonly called the speed of light and abbreviated c. Its numerical value is defined as exactly $2.997\,924\,58 \times 10^8$ m/s, usually rounded off to 3.00×10^8 m/s:

$$\text{Wavelength} \times \text{Frequency} = \text{Speed}$$
$$\lambda \, (\text{m}) \times \nu \, (\text{s}^{-1}) = c \, (\text{m/s})$$

which can be rewritten as

$$\lambda = \frac{c}{\nu} \quad \text{or} \quad \nu = \frac{c}{\lambda}$$

This equation says that frequency and wavelength are inversely related: Electromagnetic energy with a longer wavelength has a lower frequency, and energy with a shorter wavelength has a higher frequency. Worked Example 5.1 demonstrates how to convert between the wavelength and frequency of electromagnetic radiation.

WORKED EXAMPLE 5.1

Calculating a Frequency from a Wavelength

The light blue glow given off by mercury streetlamps has a wavelength of 436 nm. What is its frequency in hertz?

IDENTIFY

Known	Unknown
Wavelength (λ)	Frequency (ν)

STRATEGY

Use the equation that relates wavelength and frequency. Don't forget to convert from nanometers to meters as units on the speed of light (c) are m/s.

SOLUTION

$$\nu = \frac{c}{\lambda} = \frac{\left(3.00 \times 10^8 \, \dfrac{\text{m}}{\text{s}}\right)}{(436 \, \text{nm})\left(\dfrac{1 \, \text{m}}{10^9 \, \text{nm}}\right)}$$
$$= 6.88 \times 10^{14} \, \text{s}^{-1} = 6.88 \times 10^{14} \, \text{Hz}$$

The frequency of the light is $6.88 \times 10^{14} \, \text{s}^{-1}$, or 6.88×10^{14} Hz.

▶ **PRACTICE 5.1** What is the wavelength in meters of an FM radio wave with frequency $\nu = 102.5$ MHz? Of a medical X ray with $\nu = 9.55 \times 10^{17}$ Hz?

▶ **Conceptual APPLY 5.2** Two electromagnetic waves are represented below.

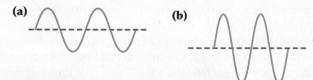

(a) Which wave has the higher frequency?
(b) Which wave represents a more intense (brighter) beam of light?
(c) Which wave represents blue light, and which represents red light?

▲ Does the blue glow from this mercury lamp correspond to a longer or shorter wavelength than the yellow glow from a sodium lamp?

5.2 ▶ PARTICLELIKE PROPERTIES OF RADIANT ENERGY: THE PHOTOELECTRIC EFFECT AND PLANCK'S POSTULATE

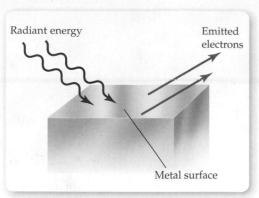

Radiant energy → Emitted electrons

Metal surface

▲ **FIGURE 5.4**

The photoelectric effect. Irradiation of a clean metal surface with radiant energy above a threshold frequency causes electrons to be emitted.

▲ Night-vision goggles are based on the photoelectric effect of metal alloys that emit electrons when irradiated with infrared light.

One important step toward developing a model of atomic structure came in 1905, when Albert Einstein (1879–1955) proposed an explanation of the **photoelectric effect**. Scientists had known since the late 1800s that irradiating a clean metal surface with light causes electrons to be ejected from the metal (**FIGURE 5.4**). Furthermore, the frequency of the light used for the irradiation must be above some threshold value, which is different for every metal. Blue light ($\nu \approx 6.5 \times 10^{14}$ Hz) causes metallic sodium to emit electrons, for example, but red light ($\nu \approx 4.5 \times 10^{14}$ Hz) has no effect on sodium.

Einstein explained the photoelectric effect by assuming that a beam of light behaves as if it were a stream of small particles, called **photons**, whose energy (E) is related to their frequency, ν (or wavelength, λ), by an equation called **Planck's postulate** after the German physicist Max Planck (1858–1947).

> **Planck's postulate** $\quad E = h\nu = \dfrac{hc}{\lambda}$

The proportionality constant h represents a fundamental physical constant that we now call Planck's constant and that has the value $h = 6.626 \times 10^{-34}$ J·s. For example, one photon of red light with a frequency $\nu = 4.62 \times 10^{14}$ s^{-1} (wavelength $\lambda = 649$ nm) has an energy of 3.06×10^{-19} J. [Recall from Section 1.8 that the SI unit for energy is the joule (J), where $1\,\text{J} = 1\,(\text{kg} \cdot \text{m}^2)/\text{s}^2$.]

$$E = h\nu = (6.626 \times 10^{-34}\,\text{J·s})(4.62 \times 10^{14}\,\text{s}^{-1}) = 3.06 \times 10^{-19}\,\text{J}$$

You might also recall from Section 2.9 that 1 mole (mol) of anything is the amount that contains Avogadro's number (6.022×10^{23}) of entities. Thus, it's often convenient to express electromagnetic energy on a per-mole basis rather than a per-photon basis. Multiplying the per-photon energy of 3.06×10^{-19} J by Avogadro's number gives an energy of 184 kJ/mol.

$$\left(3.06 \times 10^{-19}\,\frac{\text{J}}{\text{photon}}\right)\left(6.022 \times 10^{23}\,\frac{\text{photon}}{\text{mol}}\right) = 1.84 \times 10^5\,\frac{\text{J}}{\text{mol}}$$

$$\left(1.84 \times 10^5\,\frac{\text{J}}{\text{mol}}\right)\left(\frac{1\,\text{kJ}}{1000\,\text{J}}\right) = 184\,\text{kJ/mol}$$

Higher frequencies and shorter wavelengths correspond to higher energy radiation, while lower frequencies and longer wavelengths correspond to lower energy. Blue light ($\lambda \approx 450$ nm), for instance, has a shorter wavelength and is more energetic than red light ($\lambda \approx 650$ nm). Similarly, an X ray ($\lambda \approx 1$ nm) has a shorter wavelength and is more energetic than an FM radio wave ($\lambda \approx 10^{10}$ nm, or 10 m).

If the frequency (or energy) of the photon striking a metal is below a minimum value, no electron is ejected. Above the threshold level, however, sufficient energy is transferred from the photon to an electron to overcome the attractive forces holding the electron to the metal (**FIGURE 5.5**). The amount of energy necessary to eject an electron is called the work function (Φ) of the metal and is lowest for the group 1A and group 2A elements. That is, elements on the left side of the periodic table hold their electrons less tightly than other metals and lose them more readily (**TABLE 5.1**).

Note again that the energy of an individual photon depends only on its frequency (or wavelength), not on the intensity of the light beam. The intensity of a light beam is a measure of the *number* of photons in the beam, whereas frequency is a measure of the *energies* of those photons. A low-intensity beam of high-energy photons might easily knock a few electrons loose from a metal, but a high-intensity beam of low-energy photons might not be able to knock loose a single electron. As a rough analogy, think of throwing balls of different masses at a glass window. A thousand ping-pong balls (lower energy) would only bounce off the

TABLE 5.1 Work Functions of Some Common Metals	
Element	**Work Function (Φ) (kJ/mol)**
Cs	188
K	221
Na	228
Ca	277
Mg	353
Cu	437
Fe	451

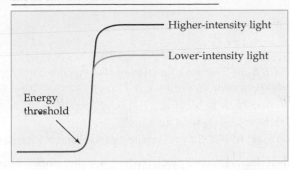

A plot of the number of electrons ejected from a metal surface versus light frequency shows a threshold value.

Higher-intensity light

Lower-intensity light

Energy threshold

Frequency →

Number of electrons ejected

Increasing the intensity of the light while keeping the frequency constant increases the number of ejected electrons but does not change the threshold value.

◀ **FIGURE 5.5**

Dependence of the photoelectric effect on frequency.

Figure It Out

Would electrons be ejected from a metal surface if (**a**) very bright light with a frequency below the threshold is used, or (**b**) very dim light with a frequency above the threshold is used? How does this observation support the theory that light has particlelike properties?

Answer: (**a**) No electrons ejected. (**b**) Low number of electrons ejected. Only light above a certain frequency has enough energy to eject an electron. This is analogous to a "particle" such as a baseball being able to break a window, whereas a ping-pong ball cannot do so.

window, but a single baseball (higher energy) would break the glass. In the same way, low-energy photons bounce off the metal surface, but a single photon at or above the threshold energy can "break" the metal and dislodge an electron.

The main conclusion from Einstein's work was that the behavior of light and other forms of electromagnetic energy is more complex than had been formerly believed. In addition to behaving as waves, *light energy can also behave as small particles.* The idea might seem strange at first but becomes less so if you think of light as analogous to matter. Both are said to be **quantized**, meaning that both matter and electromagnetic energy occur only in discrete amounts. Just as there can be either 1 or 2 hydrogen atoms but not 1.5 or 1.8, there can be 1 or 2 photons of light but not 1.5 or 1.8. A **quantum** is the smallest possible unit of a quantity, just like an atom is the smallest possible quantity of an element. The quantum of energy corresponding to one photon of light is almost inconceivably small, just as the amount of matter in one atom is inconceivably small, but the idea is the same. Planck's postulate and Einstein's photons with particlelike properties were a dramatic departure from the laws of classical physics at the time and ultimately led to a revolution in the way that scientists thought about the structure of the atom.

▲ A glass window can be broken by a single baseball, but a thousand ping-pong balls would only bounce off.

─● WORKED EXAMPLE 5.2

Calculating the Energy of a Photon

What is the energy in kilojoules per mole of radar waves with $\nu = 3.35 \times 10^8$ Hz?

IDENTIFY

Known	Unknown
Frequency (ν)	Energy (E)

STRATEGY

The energy of a photon with frequency ν can be calculated with the equation $E_{photon} = h\nu$. To find the energy per mole of photons, the energy of one photon must be multiplied by Avogadro's number (Section 2.9).

SOLUTION

$$E = h\nu = (6.626 \times 10^{-34}\,J \cdot s)(3.35 \times 10^8\,s^{-1}) = 2.22 \times 10^{-25}\,J$$

$$\left(2.22 \times 10^{-25}\,\frac{J}{photon}\right)\left(6.022 \times 10^{23}\,\frac{photon}{mol}\right) = 0.134\,J/mol$$

$$= 1.34 \times 10^{-4}\,kJ/mol$$

▶ **PRACTICE 5.3** The biological effects of a given dose of electromagnetic energy generally become more serious as the energy of the radiation increases: Infrared radiation has a pleasant warming effect; ultraviolet radiation causes tanning and burning; and X rays can cause considerable tissue damage. What energies in kilojoules per mole are associated with the following wavelengths: infrared radiation with $\lambda = 1.55 \times 10^{-6}$ m, ultraviolet light with $\lambda = 250$ nm, and X rays with $\lambda = 5.49$ nm?

▶ **APPLY 5.4** It requires 74 kJ to heat a cup of water from room temperature to boiling in the microwave oven. If the wavelength of microwave radiation is 2.3×10^{-3} m, how many moles of photons are required to heat the water?

WORKED EXAMPLE 5.3

Calculating the Frequency for a Photoelectric Effect Given a Work Function

The work function of lithium metal is $\Phi = 283$ kJ/mol. What is the minimum frequency of light needed to eject electrons from lithium?

IDENTIFY

Known	Unknown
Work Function (Φ) = Energy (E)	Frequency (ν)

STRATEGY

The work function of a metal is the minimum energy needed to eject electrons: $E = 283$ kJ/mol for lithium. Convert the work function from kJ/mol to J/photon to find the energy of a single photon, and rearrange Planck's equation $E = h\nu$ to solve for frequency, $\nu = E/h$.

SOLUTION

$$E = \left(\frac{283 \text{ kJ}}{\text{mol}}\right)\left(\frac{1000 \text{ J}}{1 \text{ kJ}}\right)\left(\frac{1 \text{ mol}}{6.022 \times 10^{23} \text{ photons}}\right) = 4.70 \times 10^{-19} \frac{\text{J}}{\text{photon}}$$

$$\nu = \frac{E}{h} = \frac{4.70 \times 10^{-19} \text{ J}}{(6.626 \times 10^{-34} \text{ J} \cdot \text{s})} = 7.09 \times 10^{14} \text{ s}^{-1} \text{ or } 7.09 \times 10^{14} \text{ Hz}$$

▶ **PRACTICE 5.5** The work function of zinc metal is 350 kJ/mol. Will photons of violet light with $\lambda = 390$ nm cause electrons to be ejected from a sample of zinc?

▶ **Conceptual APPLY 5.6** Compare the two elements Rb and Ag.
(a) Which element do you predict to have a higher work function?
(b) Which element exhibits the photoelectric effect at a longer wavelength of light?

5.3 ▶ THE INTERACTION OF RADIANT ENERGY WITH ATOMS: LINE SPECTRA

Now that we understand that electromagnetic radiation has both wavelike and particlelike properties, we can examine how its interaction with matter provides clues about atomic structure. Let's compare the emission of light from an atom with the emission of light from a light bulb. The light that we see from the Sun or from a typical light bulb is "white" light, meaning that it consists of an essentially continuous distribution of wavelengths spanning the entire visible region of the electromagnetic spectrum. That white light actually consists of a spectrum of many colors of light is made evident when a narrow beam of white light is passed through a glass prism to produce a "rainbow" of colors (**FIGURE 5.6a**). This happens

▶ **FIGURE 5.6**

Separation of white light into its constituent colors. (a) When a narrow beam of ordinary white light is passed through a glass prism, different wavelengths travel through the glass at different rates and appear as different colors. A similar effect occurs when light passes through water droplets in the air, forming a rainbow, or (b) through ice crystals in clouds, causing an unusual weather phenomenon called a *parhelion*, or sundog.

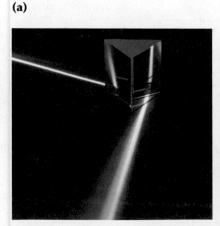

because the different wavelengths contained in white light travel through the glass at different rates. The prism separates the white light into its component colors, ranging from red at the long-wavelength end of the spectrum (780 nm) to violet at the short-wavelength end (380 nm). This separation into colors also occurs when light travels through water droplets in the air, forming a rainbow, or through oriented ice crystals in clouds, causing a *parhelion*, or sundog (**FIGURE 5.6b**).

What do visible light and other kinds of electromagnetic energy have to do with atomic structure? It turns out that atoms give off light when heated or otherwise energetically excited, thereby providing a clue to their atomic makeup. Unlike the white light from the Sun, though, an energetically excited atom emits light not in a continuous distribution of wavelengths but only at a certain specific wavelengths. When passed first through a narrow slit and then through a prism, the light emitted by an excited atom is found to consist of only a few wavelengths rather than a full rainbow of colors, giving a series of discrete lines on an otherwise dark background—a **line spectrum** that is unique for each element. If hydrogen atoms are electrically excited in a discharge tube, they give off a pinkish light made of several different colors (**FIGURE 5.7a**). Other elements also produce line spectra upon excitation such as neon in signs or sodium salts in a flame (**FIGURE 5.7b**). In fact, the brilliant colors of fireworks are produced by mixtures of metal atoms that have been heated by explosive powder.

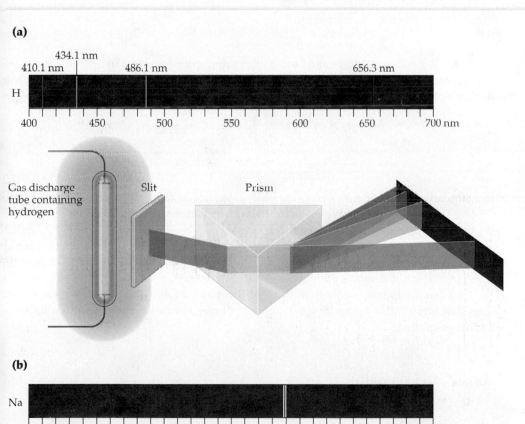

(a)

H

434.1 nm
410.1 nm   486.1 nm   656.3 nm

400   450   500   550   600   650   700 nm

Gas discharge tube containing hydrogen   Slit   Prism

(b)

Na

400   450   500   550   600   650   700 nm

Ne

400   450   500   550   600   650   700 nm

Wavelength, λ (nm)

◀ **FIGURE 5.7**

Atomic line spectra. (a) When light from a hydrogen discharge tube is passed through a narrow slit and then a prism, the colors of light in its emission spectrum are separated. **(b)** Each element, such as sodium and neon, has its own characteristic line spectrum that can be used to identify or quantify the amount of an element in a sample.

Figure It Out

How is the emission spectrum for an atom different than the emission spectrum of the Sun?

Answer: The emission spectrum from the Sun consists of all visible wavelengths of electromagnetic radiation. The emission spectra from atoms consist of discrete visible wavelengths, a line spectrum.

▶ Electronically excited hydrogen atoms give off pink light and neon atoms emit orange light in discharge tubes. Thermally excited sodium atoms emit yellow light in a flame.

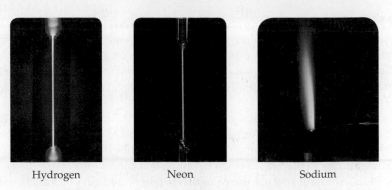

Hydrogen   Neon   Sodium

Soon after the discovery that energetic atoms emit light of specific wavelengths, chemists began cataloging the line spectra of various elements. They rapidly found that each element has its own unique spectral "signature," and they began using the results to identify the elements present in minerals and other substances. Not until the work of the Swiss schoolteacher Johann Balmer in 1885, though, was a pattern discovered in atomic line spectra. It was known at the time that hydrogen produced a spectrum with four lines, as shown in Figure 5.7a. The wavelengths of the four lines are 656.3 nm (red), 486.1 nm (blue–green), 434.1 nm (blue), and 410.1 nm (indigo).

Thinking about the hydrogen spectrum and trying by trial and error to organize the data in various ways, Balmer discovered that the wavelengths of the four lines in the hydrogen spectrum can be expressed by the equation

$$\frac{1}{\lambda} = R_\infty \left[\frac{1}{2^2} - \frac{1}{n^2} \right] \quad \text{or} \quad \nu = R_\infty \cdot c \left[\frac{1}{2^2} - \frac{1}{n^2} \right]$$

where R_∞ is a constant (now called the *Rydberg constant*) equal to $1.097 \times 10^{-2} \, \text{nm}^{-1}$ and n is an integer greater than 2. The red spectral line at 656.3 nm, for example, results from Balmer's equation when $n = 3$:

$$\frac{1}{\lambda} = \left[1.097 \times 10^{-2} \, \text{nm}^{-1} \right] \left[\frac{1}{2^2} - \frac{1}{3^2} \right] = 1.524 \times 10^{-3} \, \text{nm}^{-1}$$

$$\lambda = \frac{1}{1.524 \times 10^{-3} \, \text{nm}^{-1}} = 656.3 \, \text{nm}$$

Similarly, a value of $n = 4$ gives the blue–green line at 486.1 nm, a value of $n = 5$ gives the blue line at 434.1 nm, and so on. Solve Balmer's equation yourself to make sure.

Subsequent to the discovery of the Balmer series of lines in the visible region of the electromagnetic spectrum, it was found that many other spectral lines are also present in nonvisible regions of the spectrum. Hydrogen, for instance, shows a series of spectral lines in the ultraviolet region and several other series in the infrared region.

By adapting Balmer's equation, the Swedish physicist Johannes Rydberg was able to show that every line in the entire spectrum of hydrogen can be fit by a generalized **Balmer–Rydberg equation**:

Balmer–Rydberg equation $\dfrac{1}{\lambda} = R_\infty \left[\dfrac{1}{m^2} - \dfrac{1}{n^2} \right] \quad \text{or} \quad \nu = R_\infty \cdot c \left[\dfrac{1}{m^2} - \dfrac{1}{n^2} \right]$

where m and n represent integers with $n > m$. If $m = 1$, then the ultraviolet series of lines results. If $m = 2$, then Balmer's series of visible lines results. If $m = 3$, an infrared series is described, and so forth for still larger values of m. Some of these other spectral lines are calculated in Worked Example 5.4.

5.4 ▶ THE BOHR MODEL OF THE ATOM: QUANTIZED ENERGY

As often happens in science, experimental results are obtained before a theory to explain them is developed. The discovery of the line spectrum for hydrogen came before the Balmer–Rydberg mathematical model, and a theory of atomic structure to explain line spectra did not arise for several decades. During the time period 1900–1911, scientists conducted key experiments and developed new theories that helped to solve the puzzle of line spectra. Among them were Planck's postulate of quantized energy (1900), Einstein's concept of photons and the photoelectric effect (1905), and **Rutherford's nuclear model** of the atom (1911). Building upon these seminal discoveries, Niels Bohr (1885–1962), a Danish physicist working in Rutherford's lab, proposed an atomic model that predicted the existence of line spectra. His model of the hydrogen atom described a small, positively charged nucleus with an electron circling around it, much as a planet orbits the Sun. Bohr postulated that the energy levels of the orbits are *quantized* so that only certain specific orbits corresponding to certain specific energies for the electron are available. You might think of the quantized nature of orbits in terms of an analogy: climbing stairs versus a ramp. The height of a ramp changes continuously, but stairs change height only in discrete amounts; the height reached by climbing each stair is thus quantized.

The success of the Bohr model was that it was able to explain the line spectrum observed for hydrogen. Each orbit has its own radius, referred to as *n*, which is directly related to energy. As the radius increases, the energy also increases. Thus, $n = 2$ has greater energy than $n = 1$ and $n = 3$ has greater energy than $n = 2$, and so on. In Bohr's model, there is no change in energy when an electron moves within its orbit, but when an electron falls into a lower orbit, it emits a photon whose energy equals the difference in the energies of the two orbits (**FIGURE 5.8**).

$$\Delta E = E_{\text{final}} - E_{\text{initial}} = h\nu$$

Quantization of energy in the atom arises from the fact that an electron cannot reside between orbits in the Bohr model, just like you cannot stand between steps on the stairs.

▲ A ramp changes height continuously, but stairs are quantized, changing height only in discrete amounts. In the same way, electromagnetic energy is not continuous but is emitted only in discrete amounts.

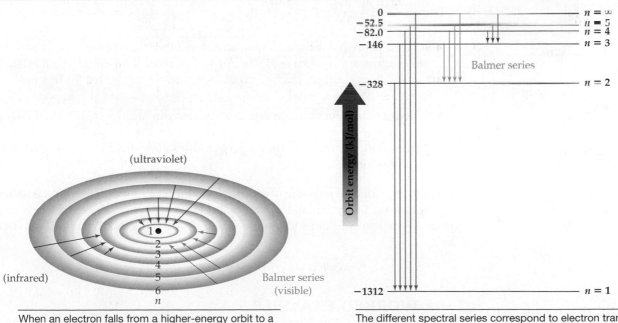

When an electron falls from a higher-energy orbit to a lower-energy orbit, it emits electromagnetic radiation whose frequency corresponds to the energy difference between the orbits.

The different spectral series correspond to electron transitions from higher-energy to lower-energy orbits.

▲ **FIGURE 5.8**

The Bohr model for hydrogen predicts atomic line spectra.

Figure It Out

Which electron transition corresponds to the emission of light with the most energy: $n = 5$ to $n = 3$, $n = 5$ to $n = 2$, $n = 5$, to $n = 1$?

Answer: $n = 5$ to $n = 1$ as shown by the difference in energy between the orbits.

When energy is absorbed by an atom, an electron moves from a lower to higher energy orbit ($n_{initial} < n_{final}$). Conversely, when an electron falls from a higher to lower energy orbit, energy is released ($n_{inital} > n_{final}$). In the hydrogen atom, electrons moving from $n = 6, 5, 4, 3$ to $n = 2$ result in emission of wavelengths in the visible region of the electromagnetic spectrum (Balmer series). As seen in Figure 5.8, electrons moving from $n = 6, 5, 4, 3$ to $n = 1$ correspond to greater energy than the Balmer series and result in emission of photons in the ultraviolet region of the spectrum. Similarly, electronic transitions from $n = 6, 5, 4$ to $n = 3$ are lower in energy than the Balmer series and correlate to spectral lines in the infrared.

The variables m and n in the Balmer–Rydberg equation for hydrogen (Section 5.3) represent the energy levels of the orbits in the Bohr model. The variable n corresponds to the n value of the higher-energy orbit, and the variable m corresponds to the n value of the lower energy orbit closer to the nucleus.

$$\frac{1}{\lambda} = R_\infty \left[\frac{1}{m^2} - \frac{1}{n^2} \right]$$

n level of lower n level of higher
energy orbit   energy orbit

Notice in Figure 5.8 that as n becomes larger and approaches infinity, the energy difference between $n = \infty$ and $n = 1$ converges to a value of 1312 kJ/mol. That is, 1312 kJ is released when electrons come from a great distance (the "infinite" level) and add to H^+ to give a mole of hydrogen atoms, each with an electron in its lowest energy state:

$$H^+ + e^- \longrightarrow H + \text{Energy} \qquad (1312 \text{ kJ/mol})$$

Because the energy released upon adding an electron to H^+ is equal to the energy absorbed on removing an electron from a hydrogen atom, we can also say that 1312 kJ/mol is required to remove the electron from a hydrogen atom. We'll see in the next chapter that the amount of energy necessary to remove an electron from a given atom provides an important clue about that element's chemical reactivity.

Although the Bohr model was very successful in accounting for the line spectrum of hydrogen, it suffered from several limitations.

- It failed to predict the spectrum of any atom other than hydrogen and only works for *one-electron species* such as H, He^+, or Li^{2+}. If more than one electron exists, interactions such as electron–electron repulsions must be accounted for in a more complex model.

- It does not give an accurate depiction of electron location. *Electrons do not move in fixed, defined orbits.* In fact, we can never know the precise location of an electron in the atom and can only define probabilities of an electron existing within a given volume of space.

However, the fundamental idea of quantized energy levels for the electron was an important theory for which Bohr was awarded the Nobel Prize in Physics in 1922. The next several sections will expand on Bohr's ideas and further describe the modern model for electrons in an atom.

WORKED EXAMPLE 5.4

Relating the Bohr Model and the Balmer–Rydberg Equation

(a) Use the Balmer-Rydberg equation to calculate the wavelength of the photon emitted when an electron falls from the $n = 4$ level to the $n = 1$ level in the hydrogen atom.

(b) Calculate the energy of the photon in units of kJ/mol. Check that the energy of the photon corresponds to the difference in energy between the $n = 4$ and $n = 1$ level in Figure 5.8.

(a) STRATEGY

Use the Balmer–Rydberg equation and identify the values of m and n. n represents the highest energy orbit, $n = 4$, and m represents the lowest energy orbit, $n = 1$. Therefore, in the Rydberg equation $n = 4$ and $m = 1$.

SOLUTION

Solving the equation for wavelength gives

$$\frac{1}{\lambda} = R_\infty \left[\frac{1}{m^2} - \frac{1}{n^2} \right] = (1.097 \times 10^{-2}\, nm^{-1}) \left[\frac{1}{1^2} - \frac{1}{4^2} \right]$$

$$= 1.028 \times 10^{-2}\, nm^{-1}$$

$$\text{or} \quad \lambda = \frac{1}{1.028 \times 10^{-2}\, nm^{-1}} = 97.3\, nm$$

(b) STRATEGY

Use Planck's postulate to convert between wavelength and energy. To use this equation, first convert the units on wavelength to meters because the speed of light is expressed in units of (m/s). Planck's equation gives the energy in units of J/photon, which must be converted to kJ/mol.

SOLUTION

$$E = \frac{hc}{\lambda} = \frac{(6.626 \times 10^{-34}\, J\cdot s)\left(3.00 \times 10^8\, \dfrac{m}{s} \right)}{(97.2\, nm)\left(\dfrac{1\, m}{10^9\, nm} \right)} = 2.05 \times 10^{-18}\, \frac{J}{photon}$$

$$\left(2.05 \times 10^{-18}\, \frac{J}{photon} \right)\left(6.022 \times 10^{23}\, \frac{photon}{mol} \right) = 1.23 \times 10^6\, J/mol$$

$$= 1.23 \times 10^3\, kJ/mol$$

CHECK

The wavelength of the photon released in the $n = 4$ to $n = 1$ transition is in the ultraviolet region of the spectrum which is consistent with theory. Figure 5.8 gives the energy in kJ/mol for each n level and thus $\Delta E = E_{final} - E_{initial} = [-1312\, kJ/mol - (-82\, kJ/mol)] = -1.23 \times 10^3\, kJ/mol$. The magnitude is consistent with the photon energy and the negative sign indicates energy was emitted from the atom because the electron fell from a higher energy to a lower energy orbit. A positive sign of ΔE would mean that energy was absorbed and represents an electron moving from a lower energy to a higher energy orbit.

▶ **PRACTICE 5.7** The Balmer equation can be extended beyond the visible portion of the electromagnetic spectrum to include lines in the ultraviolet. What is the wavelength in nanometers and energy in kJ/mol of ultraviolet light in the Balmer series corresponding to a value of $n = 7$?

▶ **APPLY 5.8**

(a) What is the longest-wavelength line in nanometers in the infrared series for hydrogen where $m = 3$?

(b) What is the shortest-wavelength line in nanometers in the infrared series for hydrogen where $m = 3$? (Hint: $n = \infty$)

5.5 ▶ WAVELIKE PROPERTIES OF MATTER: DE BROGLIE'S HYPOTHESIS

The analogy between matter and radiant energy developed in the early 1900s was further extended in 1924 by the French physicist Louis de Broglie (1892–1987). de Broglie suggested that if *light* can behave in some respects like *matter*, then perhaps *matter* can behave in some respects like *light*. That is, perhaps matter is wavelike as well as particlelike.

In developing his theory about the wavelike behavior of matter, de Broglie focused on the inverse relationship between energy and wavelength for photons:

$$\text{Since} \quad E = \frac{hc}{\lambda} \quad \text{then} \quad \lambda = \frac{hc}{E}$$

Using the famous equation $E = mc^2$ proposed in 1905 by Einstein as part of his special theory of relativity, and substituting for E, then gives

$$\lambda = \frac{hc}{E} = \frac{hc}{mc^2} = \frac{h}{mc}$$

de Broglie suggested that a similar equation might be applied to moving particles like electrons by replacing the speed of light, c, by the speed of the particle, v. The resultant **de Broglie equation** allows calculation of a "wavelength" of an electron or of any other particle or object of mass m moving at velocity v:

> **de Broglie equation** $\quad \lambda = \dfrac{h}{mv}$

Worked Example 5.5 demonstrates how to use the de Broglie equation to calculate the wavelength of a particle in motion.

WORKED EXAMPLE 5.5

Calculating the de Broglie Wavelength of Moving Particles

Calculate the wavelength of an electron in a hydrogen atom with a mass of 9.11×10^{-31} kg and a velocity v of 2.2×10^6 m/s (about 1% of the speed of light).

STRATEGY

Since mass and velocity are known, the de Broglie equation can be used to calculate wavelength. Note that mass and velocity must be in SI units of kg and m/s.

SOLUTION

Planck's constant, which is usually expressed in units of joule seconds (J·s), is expressed for the present purposes in units of $(kg \cdot m^2)/s [1 J = 1 (kg \cdot m^2)/s^2]$.

$$\lambda = \frac{h}{mv} = \frac{6.626 \times 10^{-34} \dfrac{kg \cdot m^2}{s}}{(9.11 \times 10^{-31} \, kg)\left(2.2 \times 10^6 \, \dfrac{m}{s}\right)} = 3.3 \times 10^{-10} \, m$$

▶ **PRACTICE 5.9** What is the de Broglie wavelength in meters of a small car with a mass of 1150 kg traveling at a velocity of 55.0 mi/h (24.6 m/s)? Is this wavelength longer or shorter than the diameter of an atom (approximately 200 pm)?

▶ **APPLY 5.10** Electron microscopes make use of the de Broglie wavelength of high-speed electrons to image tiny objects. The magnification of an electron microscope ($\approx 10^6$ times) is much greater than an optical microscope using visible light ($\approx 10^3$ times) because the wavelengths of high-speed electrons are much smaller than the wavelengths of visible light. To produce a high-quality image, the wavelength of the electron must be about 10 times less than the diameter of the particle. What electron velocity is required if a 1 nm particle is to be imaged?

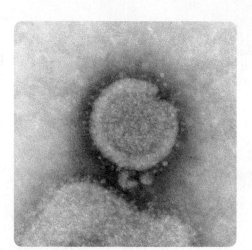

▲ An image of an avian flu virus created by a transmission electron microscope. Electron microscopes are able to image much smaller objects than optical microscopes by using a beam of high-speed electrons with short de Broglie wavelengths.

What does it mean to say that light and matter act both as waves and as particles? On the everyday human scale, the answer is "not much." We have never directly observed the wave nature of familiar objects such as a baseball with our eyes because the de Broglie wavelengths of macroscopic objects are much too small. The problem in trying to understand the dual wave/particle description of light and matter is that our common sense isn't up to the task. Our intuition has been developed from personal experiences, using our eyes and other senses to tell us how light and matter are "supposed" to behave. It is difficult to imagine a particle simultaneously behaving as a wave as we have never observed such a phenomenon.

In contrast, on the atomic scale, where distances and masses are so tiny, light and matter behave in a manner different from what we're used to. Experiments have been performed that support the theory that subatomic particles such as electrons exhibit wave properties. In 1927, C. Davisson and L. Germer, working at Bell Labs, directed a beam of electrons at a nickel crystal generating a diffraction pattern (**FIGURE 5.9**). Electrons traveling as waves were diffracted, or bent, around nuclei in the crystal and the bright and dark areas are produced by constructive and destructive interference of the waves. The diffraction pattern provides direct experimental evidence for de Broglie's theory because diffraction and interference are characteristics of waves not particles. The wave nature of electrons is now used to study the structure of materials and produce images with atomic resolution in electron microscopes.

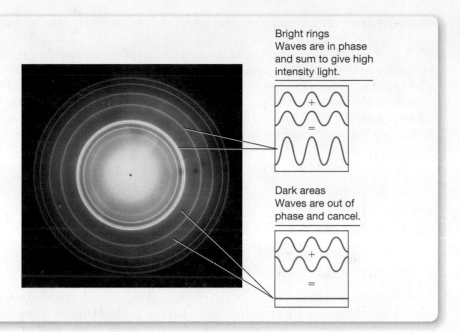

Bright rings
Waves are in phase
and sum to give high
intensity light.

Dark areas
Waves are out of
phase and cancel.

◀ **FIGURE 5.9**

Diffraction pattern generated by directing a beam of electrons at a nickel crystal.

5.6 ▶ THE QUANTUM MECHANICAL MODEL OF THE ATOM: HEISENBERG'S UNCERTAINTY PRINCIPLE

With the particlelike nature of energy and the wavelike nature of matter established, let's return to the problem of atomic structure. Several models of atomic structure were proposed in the late nineteenth and early twentieth centuries, such as the Bohr model described in Section 5.4. The Bohr model was important historically because of its conclusion that electrons have only specific energy levels available to them, but as we noted the model fails for atoms with more than one electron.

The breakthrough in understanding atomic structure came in 1926, when the Austrian physicist Erwin Schrödinger (1887–1961) proposed what has come to be called the **quantum mechanical model** of the atom. The fundamental idea behind the model is that it's best to abandon the notion of an electron as a small particle moving around the nucleus in a defined path and to concentrate instead on the electron's wavelike properties. In fact, it was shown in 1927 by Werner Heisenberg (1901–1976) that it is *impossible* to know precisely where an electron is and what path it follows—a statement called the **Heisenberg uncertainty principle**.

The Heisenberg uncertainty principle can be understood by imagining what would happen if we tried to determine the position of an electron at a given moment. For us to "see" the electron, light photons of an appropriate frequency would have to interact with and bounce off the electron. But such an interaction would transfer energy from the photon to the electron, thereby increasing the energy of the electron and making it move faster. Thus, the very act of determining the electron's position would make that position change.

In mathematical terms, Heisenberg's principle states that the uncertainty in the electron's position, Δx, times the uncertainty in its momentum, Δmv, is equal to or greater than the quantity $h/4\pi$:

Heisenberg uncertainty principle $(\Delta x)(\Delta mv) \geq \dfrac{h}{4\pi}$

According to this equation, we can never know both the position and the velocity of an electron (or of any other object) beyond a certain level of precision. If we know the velocity with a high degree of certainty (Δmv is small), then the *position* of the electron must be uncertain (Δx must be large). Conversely, if we know the position of the electron exactly (Δx is small), then we can't know its velocity (Δmv must be large). As a result, an electron

▲ Even the motion of very fast objects such as bullets can be captured in daily life. On the atomic scale, however, velocity and position can't both be known precisely.

will always appear as something of a blur whenever we attempt to make any physical mea surements of its position and velocity.

A brief calculation can help make the conclusions of the uncertainty principle cleare As mentioned in the previous section, the mass m of an electron is 9.11×10^{-31} kg and th velocity v of an electron in a hydrogen atom is 2.2×10^6 m/s. If we assume that the velocit is known to within 10%, or 0.2×10^6 m/s, then the uncertainty in the electron's position in hydrogen atom is greater than 3×10^{-10} m, or 300 pm. But since the diameter of a hydroge atom is only 240 pm, *the uncertainty in the electron's position is similar in size to the atom itself!*

$$\text{If } (\Delta x)(\Delta mv) \geq \frac{h}{4\pi} \quad \text{then } (\Delta x) \geq \frac{h}{(4\pi)(\Delta mv)}$$

$$\Delta x \geq \frac{6.626 \times 10^{-34} \frac{\text{kg} \cdot \text{m}^2}{\text{s}}}{(4)(3.1416)(9.11 \times 10^{-31} \text{ kg})\left(0.2 \times 10^6 \frac{\text{m}}{\text{s}}\right)}$$

$$\Delta x \geq 3 \times 10^{-10} \text{ m} \quad \text{or} \quad 300 \text{ pm}$$

When the mass m of an object is relatively large, as in daily life, then both Δx and Δv in th Heisenberg relationship are very small, so we have no problem in measuring both positio and velocity for visible objects. The problem arises only on the atomic scale.

5.7 ▶ THE QUANTUM MECHANICAL MODEL OF THE ATOM: ORBITALS AND QUANTUM NUMBERS

Schrödinger's quantum mechanical model of atomic structure is framed in the form of mathematical expression called a *wave equation* because it is similar in form to the equatio used to describe the motion of ordinary waves in fluids. The solutions to the wave equation ar called **wave functions**, or **orbitals**, and are represented by the symbol ψ (Greek psi). The bes way to think about an electron's wave function is to regard it as an expression whose square ψ^2, defines the probability of finding the electron within a given volume of space around th nucleus. It is important to distinguish the difference between "orbits" in the Bohr model o the atom and "orbitals" in the quantum mechanical model. An orbit defines a specific loca tion for the electron while the orbital is a mathematical equation. As Heisenberg showed, w can never be completely certain about an electron's position. A wave function, however, tell where the electron will most probably be found.

$$\underset{\text{equation}}{\text{Wave}} \xrightarrow{\text{Solve}} \underset{\text{or orbital }(\psi)}{\text{Wave function}} \longrightarrow \underset{\substack{\text{electron in a region} \\ \text{of space }(\psi^2)}}{\text{Probability of finding}}$$

A wave function is characterized by three parameters called **quantum numbers**, repre sented as n, l, and m_l, which describe the energy level of the orbital and the three-dimensiona shape of the region in space occupied by a given electron.

- **The principal quantum number (n)** is a positive integer ($n = 1, 2, 3, 4, \ldots$) on whicl the size and energy level of the orbital primarily depend. For hydrogen and other one electron atoms, such as He^+, the energy of an orbital depends only on n. For atoms witl more than one electron, the energy level of an orbital depends both on n and on the quantum number.

  As the value of n increases, the number of allowed orbitals increases and the size o those orbitals becomes larger, thus allowing an electron to be farther from the nucleus Because it takes energy to separate a negative charge from a positive charge, this increase distance between the electron and the nucleus means that the energy of the electron i the orbital increases as the quantum number n increases.

  We often speak of orbitals as being grouped according to the principal quantun number n into successive layers, or **shells**, around the nucleus. Those orbitals witl $n = 3$, for example, are said to be in the third shell.

- **The angular-momentum quantum number (l)** defines the three-dimensional shape of the orbital. For an orbital whose principal quantum number is n, the angular-momentum quantum number l can have any integral value from 0 to $n - 1$. Thus, within each shell, there are n different shapes for orbitals.

$$\text{If } n = 1, \text{then } l = 0$$
$$\text{If } n = 2, \text{then } l = 0 \text{ or } 1$$
$$\text{If } n = 3, \text{then } l = 0, 1, \text{ or } 2$$
$$\ldots \text{ and so on}$$

Just as it's convenient to think of orbitals as being grouped into shells according to the principal quantum number n, we often speak of orbitals within a shell as being further grouped into **subshells** according to the angular-momentum quantum number l. Different subshells are usually designated by letters rather than by numbers, following the order s, p, d, f, g. (Historically, the letters $s, p, d,$ and f arose from the use of the words *sharp*, *principal*, *diffuse*, and *fundamental* to describe various lines in atomic spectra.) After f, successive subshells are designated alphabetically: $g, h,$ and so on.

Quantum number l:	0	1	2	3	4	$\ldots$
Subshell notation:	s	p	d	f	g	$\ldots$

As an example, an orbital with $n = 3$ and $l = 2$ is a $3d$ orbital: 3 to represent the third shell and d to represent the $l = 2$ subshell.

- **The magnetic quantum number (m_l)** defines the spatial orientation of the orbital with respect to a standard set of coordinate axes. For an orbital whose angular-momentum quantum number is l, the magnetic quantum number m_l can have any integral value from $-l$ to $+l$. Thus, within each subshell—orbitals with the same shape, or value of l— there are $2l + 1$ different spatial orientations for those orbitals. We'll explore this point further in the next section.

$$\text{If } l = 0, \text{then } m_l = 0$$
$$\text{If } l = 1, \text{then } m_l = -1, 0, \text{ or } +1$$
$$\text{If } l = 2, \text{then } m_l = -2, -1, 0, +1, \text{ or } +2$$
$$\ldots \text{ and so forth}$$

A summary of the allowed combinations of quantum numbers for the first four shells is given in **TABLE 5.2**.

The energy levels of various orbitals are shown in **FIGURE 5.10**. As noted earlier in this section, the energy levels of different orbitals in a hydrogen atom depend only on the

TABLE 5.2 Allowed Combinations of Quantum Numbers n, l, and m_l for the First Four Shells

n	l	m_l	Orbital Notation	Number of Orbitals in Subshell	Number of Orbitals in Shell
1	0	0	$1s$	1	1
2	0	0	$2s$	1	4
	1	$-1, 0, +1$	$2p$	3	
3	0	0	$3s$	1	9
	1	$-1, 0, +1$	$3p$	3	
	2	$-2, -1, 0, +1, +2$	$3d$	5	
4	0	0	$4s$	1	16
	1	$-1, 0, +1$	$4p$	3	
	2	$-2, -1, 0, +1, +2$	$4d$	5	
	3	$-3, -2, -1, 0, +1, +2, +3$	$4f$	7	

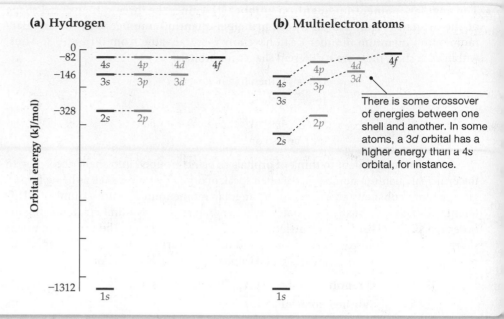

▶ **FIGURE 5.10**

Energy levels of atomic orbitals: (a) hydrogen and (b) a typical multielectron atom. The differences between energies of various subshells in (b) are exaggerated for clarity.

(a) Hydrogen

(b) Multielectron atoms

There is some crossover of energies between one shell and another. In some atoms, a 3d orbital has a higher energy than a 4s orbital, for instance.

Figure It Out

Rank the orbitals (4s, 4p, 4d, and 4f) from lowest to highest energy in a hydrogen atom and a multielectron atom.

Answer: In hydrogen the energies of 4s, 4p, 4d, and 4f are all equivalent. In a multielectron atom the order of energy from lowest to highest is 4s > 4p > 4d > 4f.

principal quantum number n, but the energy levels of orbitals in multielectron atoms depend on both n and l. In other words, the orbitals in a given shell all have the same energy for hydrogen but have slightly different energies for other atoms, depending on their subshell. In fact, there is even some crossover of energies between one shell and another. A 3d orbital in some multielectron atoms has a higher energy than a 4s orbital, for instance.

● WORKED EXAMPLE 5.6

Assigning Quantum Numbers to an Orbital

Give the possible combinations of quantum numbers for a 4p orbital.

STRATEGY

The principal quantum number n gives the shell number, which is 4 in this case. The angular-momentum quantum number l gives the subshell designation. For a p orbital $l = 1$. The magnetic quantum number m_l is related to the spatial orientation of the orbital and can have any of the three values from $-l$ to $+l$. For $l = 1$, values for $m_l = -1, 0,$ or $+1$.

SOLUTION

The allowable combinations are

$$n = 4, l = 1, m_l = -1 \qquad n = 4, l = 1, m_l = 0 \qquad n = 4, l = 1, m_l = +1$$

▶ **PRACTICE 5.11** Give orbital notations for electrons in orbitals with the following quantum numbers:

(a) $n = 2, l = 1, m_l = 1$ **(b)** $n = 4, l = 3, m_l = -2$ **(c)** $n = 3, l = 2, m_l = -1$

▶ **APPLY 5.10** Extend Table 5.2 to show allowed combinations of quantum numbers when $n = 5$. How many orbitals are in the fifth shell?

5.8 ▶ THE SHAPES OF ORBITALS

We said in the previous section that the square of a wave function, or orbital, describes the probability of finding the electron within a specific region of space. The shape of that spatial region is defined by the angular-momentum quantum number l, with $l = 0$ called an s orbital, $l = 1$ a p orbital, $l = 2$ a d orbital, and so forth. Of the various possibilities, s, p, d, and *f* orbitals are the most important because these are the only ones actually occupied in known elements. Let's look at each of the four individually.

s Orbitals

All *s* orbitals are spherical, meaning that the probability of finding an *s* electron depends only on distance from the nucleus, not on direction. Furthermore, because there is only one possible orientation of a sphere in space, an *s* orbital has $m_l = 0$ and there is only one *s* orbital per shell.

As shown in **FIGURE 5.11**, the value of ψ^2 for an *s* orbital is greatest near the nucleus and then drops off rapidly as the distance from the nucleus increases, although it never goes all the way to zero, even at a large distance. As a result, there is no definite boundary to the atom and no definite size. For purposes like that of Figure 5.11, however, we usually imagine a boundary surface enclosing the volume where an electron spends most (say, 95%) of its time.

Although all *s* orbitals are spherical, there are significant differences among the *s* orbitals in different shells. For one thing, the size of the *s* orbital increases in successively higher shells, implying that an electron in an outer-shell *s* orbital is farther from the nucleus on average than an electron in an inner-shell *s* orbital. For another thing, the electron distribution in an outer-shell *s* orbital has more than one region of high probability. As shown in Figure 5.11, a 2*s* orbital is essentially a sphere within a sphere and has two regions of high probability, separated by a surface of zero probability called a **node**. Similarly, a 3*s* orbital has three regions of high probability and two spherical nodes.

The concept of an orbital node—a surface of zero electron probability separating regions of nonzero probability—is difficult to grasp because it raises the question "How does an electron get from one region of the orbital to another if it's not allowed to be at the node?" The question is misleading, though, because it assumes particlelike behavior for the electron rather than wavelike behavior.

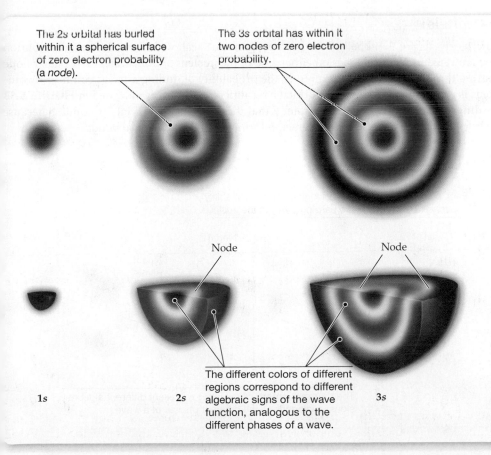

The 2s orbital has buried within it a spherical surface of zero electron probability (a *node*).

The 3s orbital has within it two nodes of zero electron probability.

Node

Node

The different colors of different regions correspond to different algebraic signs of the wave function, analogous to the different phases of a wave.

1s   2s   3s

◀ FIGURE 5.11

Representations of 1s, 2s, and 3s orbitals. Slices through these spherical orbitals are shown on the top and cutaway views on the bottom, with the probability of finding an electron represented by the density of the shading.

Figure It Out

What are the differences between 1s, 2s, and 3s orbitals?

Answer: The size increases from 1s, to 2s, to 3s. Also, the 2s orbital has one node and the 3s orbital has two nodes, or points where the probability of finding an electron is zero.

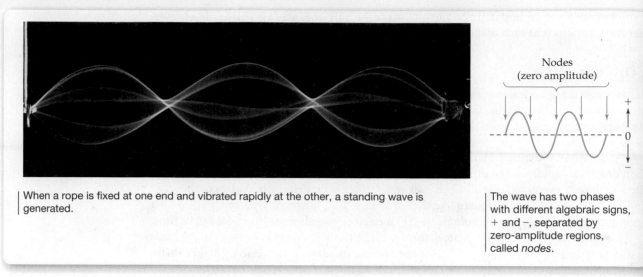

When a rope is fixed at one end and vibrated rapidly at the other, a standing wave is generated.

The wave has two phases with different algebraic signs, + and −, separated by zero-amplitude regions, called *nodes*.

▲ **FIGURE 5.12**
A standing wave in a vibrating rope.

In fact, nodes are an intrinsic property of waves, from moving waves of water in the ocean to the stationary, or standing, wave generated by vibrating a rope or guitar string (**FIGURE 5.12**). A node simply corresponds to the zero-amplitude part of the wave. On either side of the node is a nonzero wave amplitude. Note that a wave has two **phases**—peaks above the zero line and troughs below—corresponding to different algebraic signs, + and −. Similarly, the different regions of 2s and 3s orbitals have different phases, + and −, as indicated in Figure 5.11 by different colors.

p Orbitals

The *p* orbitals are dumbbell-shaped rather than spherical, with their electron distribution concentrated in identical lobes on either side of the nucleus and separated by a planar node cutting through the nucleus. As a result, the probability of finding a *p* electron near the nucleus is zero. The two lobes of a *p* orbital have different phases, as indicated in **FIGURE 5.13** by different colors. We'll see in Chapter 7 that these phases are crucial for bonding because only lobes of the same phase can interact in forming covalent chemical bonds.

▶ **FIGURE 5.13**

Representations of the three 2p orbitals. Each orbital is dumbbell-shaped and oriented in space along one of the three coordinate axes x, y, or z.

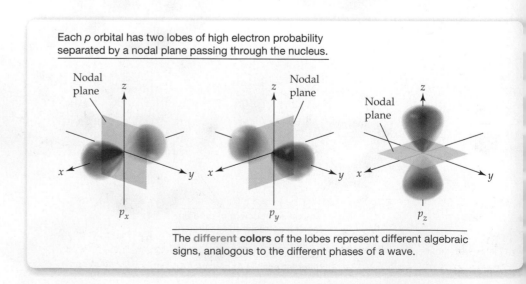

Each *p* orbital has two lobes of high electron probability separated by a nodal plane passing through the nucleus.

Nodal plane

p_x p_y p_z

The **different colors** of the lobes represent different algebraic signs, analogous to the different phases of a wave.

There are three allowable values of m_l when $l = 1$, so each shell beginning with the second has three p orbitals, which are oriented in space at 90° angles to one another along the three coordinate axes x, y, and z. The three p orbitals in the second shell, for example, are designated $2p_x$, $2p_y$, and $2p_z$. As you might expect, p orbitals in the third and higher shells are larger than those in the second shell and extend farther from the nucleus. Their shape is roughly the same, however.

d and f Orbitals

The third and higher shells each contain five d orbitals, which differ from their s and p counterparts because they have two different shapes. Four of the five d orbitals are cloverleaf-shaped and have four lobes of maximum electron probability separated by two nodal planes through the nucleus (**FIGURE 5.14a–d**). The fifth d orbital is similar in shape to a p_z orbital but has an additional donut-shaped region of electron probability centered in the xy plane (Figure 5.14e). In spite of their different shapes, all five d orbitals in a given shell have the same energy. As with p orbitals, alternating lobes of the d orbitals have different phases.

You've probably noticed that both the number of nodal planes through the nucleus and the overall geometric complexity of the orbitals increases with the l quantum number of the subshell: An s orbital has one lobe and no nodal plane through the nucleus; a p orbital has two lobes and one nodal plane; and a d orbital has four lobes and two nodal planes. The seven f orbitals are more complex still, having eight lobes of maximum electron probability separated by three nodal planes through the nucleus. (Figure 5.14f shows one of the seven $4f$ orbitals.) Most of the elements we'll deal with in the following chapters don't use f orbitals in bonding, however, so we won't spend time on them.

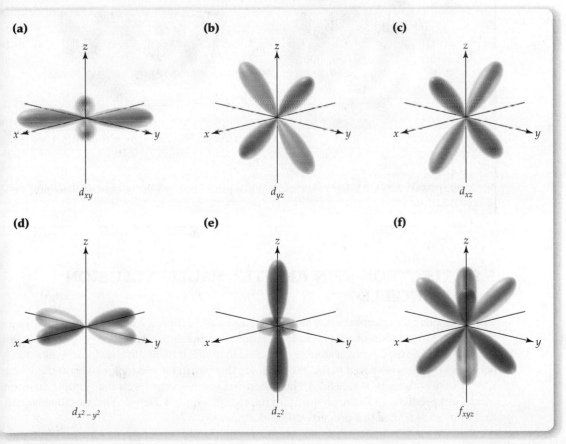

▲ **FIGURE 5.14**

Representations of the five 3d orbitals. Four of the orbitals are shaped like a cloverleaf (**a–d**), and the fifth is shaped like an elongated dumbbell inside a donut (**e**). Also shown is one of the seven $4f$ orbitals (**f**). As with p orbitals in Figure 5.13, the different colors of the lobes reflect different phases.

Conceptual WORKED EXAMPLE 5.7

Assigning Quantum Numbers to an Orbital

Give a possible combination of n and l quantum numbers for the following fourth-shell orbital:

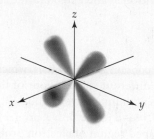

STRATEGY

Since the orbital is in the fourth shell, $n = 4$. Determine the angular momentum quantum number l from the orbital shape. The clover leaf shape represents a d orbital and therefore $l = 2$.

SOLUTION

$$n = 4, l = 2$$

▶ Conceptual **PRACTICE 5.13** Give a possible combination of n and l quantum numbers for the following orbital.

▶ Conceptual **APPLY 5.14** How many nodal planes through the nucleus do you think a g orbital has?

5.9 ▶ ELECTRON SPIN AND THE PAULI EXCLUSION PRINCIPLE

The three quantum numbers n, l, and m_l discussed in Section 5.7 define the energy, shape, and spatial orientation of orbitals, but they don't quite tell the whole story. When the line spectra of many multielectron atoms are studied in detail, it turns out that some lines actually occur as very closely spaced pairs. (You can see this pairing if you look closely at the visible spectrum of sodium in Figure 5.7.) Thus, there are more energy levels than simple quantum mechanics predicts, and a fourth quantum number is required. Denoted m_s, this fourth quantum number is related to a property called *electron spin*.

In some ways, electrons behave as if they were spinning around an axis, somewhat as the Earth spins daily. This spinning charge gives rise to a tiny magnetic field and to a **spin quantum number** (m_s), which can have either of two values, $+1/2$ or $-1/2$ (**FIGURE 5.15**). A spin of $+1/2$ is usually represented by an up arrow ($\uparrow$), and a spin of $-1/2$ by a down arrow ($\downarrow$). Note that the value of m_s is independent of the other three quantum numbers, unlike the values of n, l, and m_l, which are interrelated.

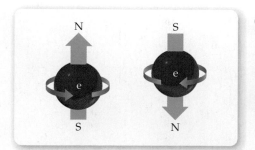

▲ **FIGURE 5.15**

Electron spin. Electrons behave in some ways as if they were tiny charged spheres spinning around an axis. This **spin** (blue arrow) gives rise to a tiny **magnetic field** (green arrow) and to a fourth quantum number, m_s, which can have a value of either $+1/2$ or $-1/2$.

The importance of the spin quantum number comes when electrons occupy specific orbitals in multielectron atoms. According to the **Pauli exclusion principle**, proposed in 1925 by the Austrian physicist Wolfgang Pauli (1900–1958), no two electrons in an atom can have the same four quantum numbers. In other words, the set of four quantum numbers associated with an electron acts as a unique "address" for that electron in an atom, and no two electrons can have the same address.

> **Pauli exclusion principle** No two electrons in an atom can have the same four quantum numbers.

What are the consequences of the Pauli exclusion principle? Electrons that occupy the same orbital have the same three quantum numbers, n, l, and m_l. But if they have the same values for n, l, and m_l, they must have different values for the fourth quantum number, m_s: either $m_s = +1/2$ or $m_s = -1/2$. Thus, an orbital can hold only two electrons, which must have opposite spins.

5.10 ▶ ORBITAL ENERGY LEVELS IN MULTIELECTRON ATOMS

As we said in Section 5.7, the energy level of an orbital in a hydrogen atom, which has only one electron, is determined by its principal quantum number n. Within a shell, all hydrogen orbitals have the same energy, independent of their other quantum numbers. The situation is different in multielectron atoms, however, where the energy level of a given orbital depends not only on the shell but also on the subshell. The s, p, d, and f orbitals within a given shell have slightly different energies in a multielectron atom, as shown previously in Figure 5.10, and there is even some crossover of energies between orbitals in different shells.

The difference in energy between subshells in multielectron atoms results from electron–electron repulsions. In hydrogen, the only electrical interaction is the attraction of the positive nucleus for the negative electron, but in multielectron atoms there are many different interactions. Not only are there the attractions of the nucleus for each electron, there are also the repulsions between every electron and each of its neighbors.

The repulsion of outer-shell electrons by inner-shell electrons is particularly important because the outer-shell electrons are pushed farther away from the nucleus and are thus held less tightly. Part of the attraction of the nucleus for an outer electron is thereby canceled, an effect we describe by saying that the outer electrons are *shielded* from the nucleus by the inner electrons (**FIGURE 5.16**). The nuclear charge actually felt by an electron, called the **effective nuclear charge**, Z_{eff}, is often substantially lower than the actual nuclear charge Z.

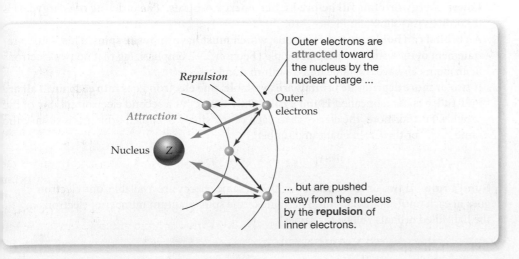

Repulsion

Attraction

Outer electrons are **attracted** toward the nucleus by the nuclear charge ...

Outer electrons

Nucleus Z

... but are pushed away from the nucleus by the **repulsion** of inner electrons.

◀ **FIGURE 5.16**

The origin of electron shielding and Z_{eff}. The outer electrons feel a diminished nuclear attraction because inner electrons shield them from the full charge of the nucleus.

Effective nuclear charge $Z_{eff} = Z_{actual} -$ Electron shielding

How does electron shielding lead to energy differences among orbitals within a shell? The answer is a consequence of the differences in orbital shapes. Compare a 2s orbital with a 2p orbital, for instance. The 2s orbital is spherical and has a large probability density near the nucleus, while the 2p orbitals are dumbbell-shaped and have a node at the nucleus (Section 5.8). An electron in a 2s orbital therefore spends more time closer to the nucleus than an electron in a 2p orbital does and is less shielded. A 2s electron thus feels a higher Z_{eff}, is more tightly held by the nucleus, and is lower in energy than a 2p electron. In the same way, a 3p electron spends more time closer to the nucleus, feels a higher Z_{eff}, and has a lower energy than a 3d electron. More generally, within any given shell, a lower value of the angular-momentum quantum number l corresponds to a higher Z_{eff} and to a lower energy for the electron.

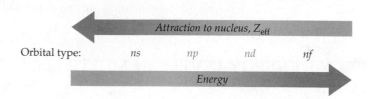

The idea that electrons in different orbitals are shielded differently and feel different values of Z_{eff} is a very useful one to which we'll return on several occasions to explain various chemical phenomena.

5.11 ► ELECTRON CONFIGURATIONS OF MULTIELECTRON ATOMS

All the parts are now in place to provide an electronic description for every element. Knowing the relative energies of the various orbitals, we can predict for each element which orbitals are occupied by electrons—the element's **electron configuration**.

A set of three rules called the **aufbau principle**, from the German word for "building up," guides the filling order of orbitals. In general, each successive electron added to an atom occupies the lowest-energy orbital available. The resultant lowest-energy configuration is called the **ground-state electron configuration** of the atom. Often, several orbitals will have the same energy level—for example, the three p orbitals or the five d orbitals in a given subshell. Orbitals that have the same energy level are said to be **degenerate**.

Rules of the Aufbau Principle:

1. **Lower-energy orbitals fill before higher-energy orbitals.** The ordering of energy levels for orbitals was shown in Figure 5.10.
2. **An orbital can hold only two electrons, which must have opposite spins.** This is just a restatement of the Pauli exclusion principle (Section 5.9), emphasizing that no two electrons in an atom can have the same four quantum numbers.
3. **If two or more degenerate orbitals are available, one electron goes into each until all are half-full,** a statement called **Hund's rule.** Only then does a second electron fill one of the orbitals. Furthermore, the electrons in each of the singly occupied orbitals must have the same value for their spin quantum number.

Hund's rule   If two or more orbitals with the same energy are available, one electron goes in each until all are half full. The value of the spin quantum number of electrons in the half-filled orbitals will be the same.

Hund's rule applies because electrons repel one another and therefore remain as far apart as possible. Electrons will be farther apart and lower in energy if they are in different orbitals describing different spatial regions than if they are in the same orbital occupying the same region. It also turns out that electrons in half-filled orbitals stay farther apart on average if they have the same spin rather than opposite spins.

Electron configurations are normally represented by listing the n quantum number and the s, p, d, or f designation of the occupied orbitals, beginning with the lowest energy one, and with the number of electrons occupying each orbital indicated as a superscript. Let's look at some examples to see how the rules of the aufbau principle are applied.

- **Hydrogen:** Hydrogen has only one electron, which must go into the lowest-energy, $1s$ orbital. Thus, the ground-state electron configuration of hydrogen is $1s^1$.

$$\text{H: } 1s^1$$

- **Helium:** Helium has two electrons, both of which fit into the lowest-energy $1s$ orbital. The two electrons have opposite spins.

$$\text{He: } 1s^2$$

- **Lithium and beryllium:** With the $1s$ orbital full, the third and fourth electrons go into the next available orbital, $2s$.

$$\text{Li: } 1s^2\,2s^1 \qquad \text{Be: } 1s^2\,2s^2$$

- **Boron through neon:** In the six elements from boron through neon, electrons fill the three $2p$ orbitals successively. Because these three $2p$ orbitals have the same energy, they are degenerate and are thus filled according to Hund's rule. In carbon, for instance, the two $2p$ electrons occupy different orbitals, which can be arbitrarily specified as $2p_x$, $2p_y$, or $2p_z$ when writing the electron configuration. The same is true of nitrogen, whose three $2p$ electrons must be in three different orbitals. Per Hund's rule, the electrons in each of the singly occupied carbon and nitrogen $2p$ orbitals must have the same value of the spin quantum number—either $+1/2$ or $-1/2$ —but this is not usually noted in the written electron configuration.

For clarity, we sometimes specify electron configurations using *orbital-filling diagrams*, in which electrons are represented by arrows. The two values of the spin quantum numbers are indicated by having the arrow point either up or down. An up–down pair indicates that an orbital is filled, while a single up (or down) arrow indicates that an orbital is half filled. Note in the diagrams for carbon and nitrogen that the degenerate $2p$ orbitals are half filled rather than filled, according to Hund's rule, and that the electron spin is the same in each.

Electron configuration		Orbital-filling diagram
B: $1s^2\,2s^2\,2p^1$	or	$\underset{1s}{\downarrow\uparrow}\quad\underset{2s}{\downarrow\uparrow}\quad\underset{2p}{\uparrow\ \_\_\ \_\_}$
C: $1s^2\,2s^2\,2p_x{}^1\,2p_y{}^1$	or	$\underset{1s}{\downarrow\uparrow}\quad\underset{2s}{\downarrow\uparrow}\quad\underset{2p}{\uparrow\ \uparrow\ \_\_}$
N: $1s^2\,2s^2\,2p_x{}^1\,2p_y{}^1\,2p_z{}^1$	or	$\underset{1s}{\downarrow\uparrow}\quad\underset{2s}{\downarrow\uparrow}\quad\underset{2p}{\uparrow\ \uparrow\ \uparrow}$

From oxygen through neon, the three $2p$ orbitals are successively filled. For fluorine and neon, it's no longer necessary to distinguish among the different $2p$ orbitals, so we can simply write $2p^5$ and $2p^6$.

O: $\ 1s^2\,2s^2\,2p_x{}^2\,2p_y{}^1\,2p_z{}^1$	or	$\underset{1s}{\downarrow\uparrow}\quad\underset{2s}{\downarrow\uparrow}\quad\underset{2p}{\downarrow\uparrow\ \uparrow\ \uparrow}$
F: $\quad 1s^2\,2s^2\,2p^5$	or	$\underset{1s}{\downarrow\uparrow}\quad\underset{2s}{\downarrow\uparrow}\quad\underset{2p}{\downarrow\uparrow\ \downarrow\uparrow\ \uparrow}$
Ne: $1s^2\,2s^2\,2p^6$	or	$\underset{1s}{\downarrow\uparrow}\quad\underset{2s}{\downarrow\uparrow}\quad\underset{2p}{\downarrow\uparrow\ \downarrow\uparrow\ \downarrow\uparrow}$

- **Sodium and magnesium:** The $3s$ orbital is filled next, giving sodium and magnesium the ground-state electron configurations shown. Note that we often write the configurations in a shorthand version by giving the symbol of the noble gas in the previous row to indicate electrons in filled shells and then specifying only those electrons in partially filled shells.

Neon configuration

Na: $1s^2\,2s^2\,2p^6\;3s^1$ or $[\text{Ne}]\,3s^1$

Mg: $1s^2\,2s^2\,2p^6\;3s^2$ or $[\text{Ne}]\,3s^2$

- **Aluminum through argon:** The $3p$ orbitals are filled according to the same rules used previously for filling the $2p$ orbitals of boron through neon. Rather than explicitly identify which of the degenerate $3p$ orbitals are occupied in Si, P, and S, we'll simplify the writing by giving just the total number of electrons in the subshell. For example, we'll write $3p^2$ for silicon rather than $3p_x^{\,1}\,3p_y^{\,1}$.

Al: $[\text{Ne}]\,3s^2\,3p^1$ **Si:** $[\text{Ne}]\,3s^2\,3p^2$ **P:** $[\text{Ne}]\,3s^2\,3p^3$

S: $[\text{Ne}]\,3s^2\,3p^4$ **Cl:** $[\text{Ne}]\,3s^2\,3p^5$ **Ar:** $[\text{Ne}]\,3s^2\,3p^6$

- **Elements past argon:** Following the filling of the $3p$ subshell in argon, the first crossover in the orbital filling order is encountered. Rather than continue filling the third shell by populating the $3d$ orbitals, the next two electrons in potassium and calcium go into the $4s$ subshell. Only then does filling of the $3d$ subshell occur to give the first transition metal series from scandium through zinc.

K: $[\text{Ar}]\,4s^1$ **Ca:** $[\text{Ar}]\,4s^2$ **Sc:** $[\text{Ar}]\,4s^2\,3d^1$ $\longrightarrow$ **Zn:** $[\text{Ar}]\,4s^2\,3d^{10}$

The experimentally determined ground-state electron configurations of the elements are shown in **FIGURE 5.17**.

5.12 ▶ ANOMALOUS ELECTRON CONFIGURATIONS

The guidelines discussed in the previous section for determining ground-state electron configurations work well but are not completely accurate. A careful look at Figure 5.17 shows that 90 electron configurations are correctly accounted for by the rules but that 21 of the predicted configurations are incorrect.

The reasons for the anomalies often have to do with the unusual stability of both half-filled and fully filled subshells. Chromium, for example, which we would predict to have the configuration $[\text{Ar}]\,4s^2\,3d^4$, actually has the configuration $[\text{Ar}]\,4s^1\,3d^5$. By moving an electron from the $4s$ orbital to an energetically similar $3d$ orbital, chromium trades one filled subshell ($4s^2$) for two half-filled subshells ($4s^1\,3d^5$), thereby allowing the two electrons to be farther apart. In the same way, copper, which we would predict to have the configuration $[\text{Ar}]\,4s^2\,3d^9$, actually has the configuration $[\text{Ar}]\,4s^1\,3d^{10}$. By transferring an electron from the $4s$ orbital to a $3d$ orbital, copper trades one filled subshell ($4s^2$) for a different filled subshell ($3d^{10}$) and gains a half-filled subshell ($4s^1$).

Most of the anomalous electron configurations shown in Figure 5.17 occur in elements with atomic numbers greater than $Z = 40$, where the energy differences between subshells are small. In all cases, the transfer of an electron from one subshell to another lowers the total energy of the atom because of a decrease in electron–electron repulsions.

5.13 ▶ ELECTRON CONFIGURATIONS AND THE PERIODIC TABLE

Why are electron configurations so important, and what do they have to do with the periodic table? The answers emerge when you look closely at Figure 5.17. Focusing only on the electrons in the outermost shell, called the **valence shell**, *all the elements in a given*

1 1A	2 2A	3 3B	4 4B	5 5B	6 6B	7 7B	8 8B	9 8B	10 8B	11 1B	12 2B	13 3A	14 4A	15 5A	16 6A	17 7A	18 8A
1 H $1s^1$																	2 He $1s^2$
3 Li $2s^1$	4 Be $2s^2$											5 B $2s^2 2p^1$	6 C $2s^2 2p^2$	7 N $2s^2 2p^3$	8 O $2s^2 2p^4$	9 F $2s^2 2p^5$	10 Ne $2s^2 2p^6$
11 Na $3s^1$	12 Mg $3s^2$											13 Al $3s^2 3p^1$	14 Si $3s^2 3p^2$	15 P $3s^2 3p^3$	16 S $3s^2 3p^4$	17 Cl $3s^2 3p^5$	18 Ar $3s^2 3p^6$
19 K $4s^1$	20 Ca $4s^2$	21 Sc $4s^2 3d^1$	22 Ti $4s^2 3d^2$	23 V $4s^2 3d^3$	24 Cr $4s^1 3d^5$	25 Mn $4s^2 3d^5$	26 Fe $4s^2 3d^6$	27 Co $4s^2 3d^7$	28 Ni $4s^2 3d^8$	29 Cu $4s^1 3d^{10}$	30 Zn $4s^2 3d^{10}$	31 Ga $4s^2 3d^{10} 4p^1$	32 Ge $4s^2 3d^{10} 4p^2$	33 As $4s^2 3d^{10} 4p^3$	34 Se $4s^2 3d^{10} 4p^4$	35 Br $4s^2 3d^{10} 4p^5$	36 Kr $4s^2 3d^{10} 4p^6$
37 Rb $5s^1$	38 Sr $5s^2$	39 Y $5s^2 4d^1$	40 Zr $5s^2 4d^2$	41 Nb $5s^1 4d^4$	42 Mo $5s^1 4d^5$	43 Tc $5s^2 4d^5$	44 Ru $5s^1 4d^7$	45 Rh $5s^1 4d^8$	46 Pd $4d^{10}$	47 Ag $5s^1 4d^{10}$	48 Cd $5s^2 4d^{10}$	49 In $5s^2 4d^{10} 5p^1$	50 Sn $5s^2 4d^{10} 5p^2$	51 Sb $5s^2 4d^{10} 5p^3$	52 Te $5s^2 4d^{10} 5p^4$	53 I $5s^2 4d^{10} 5p^5$	54 Xe $5s^2 4d^{10} 5p^6$
55 Cs $6s^1$	56 Ba $6s^2$	71 Lu $6s^2 4f^{14} 5d^1$	72 Hf $6s^2 4f^{14} 5d^2$	73 Ta $6s^2 4f^{14} 5d^3$	74 W $6s^2 4f^{14} 5d^4$	75 Re $6s^2 4f^{14} 5d^5$	76 Os $6s^2 4f^{14} 5d^6$	77 Ir $6s^2 4f^{14} 5d^7$	78 Pt $6s^1 4f^{14} 5d^9$	79 Au $6s^1 4f^{14} 5d^{10}$	80 Hg $6s^2 4f^{14} 5d^{10}$	81 Tl $6s^2 4f^{14} 5d^{10} 6p^1$	82 Pb $6s^2 4f^{14} 5d^{10} 6p^2$	83 Bi $6s^2 4f^{14} 5d^{10} 6p^3$	84 Po $6s^2 4f^{14} 5d^{10} 6p^4$	85 At $6s^2 4f^{14} 5d^{10} 6p^5$	86 Rn $6s^2 4f^{14} 5d^{10} 6p^6$
87 Fr $7s^1$	88 Ra $7s^2$	103 Lr $7s^2 5f^{14} 6d^1$	104 Rf $7s^2 5f^{14} 6d^2$	105 Db $7s^2 5f^{14} 6d^3$	106 Sg $7s^2 5f^{14} 6d^4$	107 Bh $7s^2 5f^{14} 6d^5$	108 Hs $7s^2 5f^{14} 6d^6$	109 Mt $7s^2 5f^{14} 6d^7$	110 Ds $7s^1 5f^{14} 6d^9$	111 Rg $7s^1 5f^{14} 6d^{10}$	112 Cn $7s^2 5f^{14} 6d^{10}$	113 —	114 Fl	115 —	116 Lv	117 —	118 —

57 La $6s^2 5d^1$	58 Ce $6s^2 4f^1 5d^1$	59 Pr $6s^2 4f^3$	60 Nd $6s^2 4f^4$	61 Pm $6s^2 4f^5$	62 Sm $6s^2 4f^6$	63 Eu $6s^2 4f^7$	64 Gd $6s^2 4f^7 5d^1$	65 Tb $6s^2 4f^9$	66 Dy $6s^2 4f^{10}$	67 Ho $6s^2 4f^{11}$	68 Er $6s^2 4f^{12}$	69 Tm $6s^2 4f^{13}$	70 Yb $6s^2 4f^{14}$
89 Ac $7s^2 6d^1$	90 Th $7s^2 6d^2$	91 Pa $7s^2 5f^2 6d^1$	92 U $7s^2 5f^3 6d^1$	93 Np $7s^2 5f^4 6d^1$	94 Pu $7s^2 5f^6$	95 Am $7s^2 5f^7$	96 Cm $7s^2 5f^7 6d^1$	97 Bk $7s^2 5f^9$	98 Cf $7s^2 5f^{10}$	99 Es $7s^2 5f^{11}$	100 Fm $7s^2 5f^{12}$	101 Md $7s^2 5f^{13}$	102 No $7s^2 5f^{14}$

▲ **FIGURE 5.17**

Outer-shell, ground-state electron configurations of the elements.

Figure It Out

Identify the elements in period 5 with an anomalous electron configuration.

Answer: Nb, Mo, Ru, Rh, Pd, Ag

TABLE 5.3   Valence-Shell Electron Configurations of Main-Group Elements

Group	Valence-Shell Electron Configuration	
1A	ns^1	(1 total)
2A	ns^2	(2 total)
3A	ns^2np^1	(3 total)
4A	ns^2np^2	(4 total)
5A	ns^2np^3	(5 total)
6A	ns^2np^4	(6 total)
7A	ns^2np^5	(7 total)
8A	ns^2np^6	(8 total)

group of the periodic table have similar valence-shell electron configurations (**TABLE 5.3**). The group 1A elements, for example, all have an s^1 valence-shell configuration; the group 2A elements have an s^2 valence-shell configuration; the group 3A elements have an s^2p^1 valence-shell configuration; and so on across every group of the periodic table (except for the small number of anomalies). Furthermore, because the valence-shell electrons are outermost and least tightly held, they are the most important for determining an element's properties, thus explaining why the elements in a given group of the periodic table have similar chemical behavior.

The periodic table can be divided into four regions, or blocks, of elements according to the orbitals being filled (**FIGURE 5.18**). The group 1A and group 2A elements on the left side of the table are called the **s-block elements** because they result from the filling of an s orbital; the group 3A–8A elements on the right side of the table are the **p-block elements** because they result from the filling of p orbitals; the transition metal **d-block elements** in the middle of the table result from the filling of d orbitals; and the lanthanide / actinide **f-block elements** detached at the bottom of the table result from the filling of f orbitals.

Thinking of the periodic table as outlined in Figure 5.18 provides a useful way to remember the order of orbital filling. Beginning at the top left corner of the periodic table and going across successive rows gives the correct orbital-filling order. The first row of the periodic table, for instance, contains only the two s-block elements H and He, so the first available s orbital (1s) is filled first. The second row begins with two s-block elements (Li and Be) and continues with six p-block elements (B through Ne), so the next available s orbital (2s) and then the first available p orbitals (2p) are filled. Moving similarly across the third row, the 3s and 3p orbitals are filled. The fourth row again starts with two s-block elements (K and Ca) but is then followed by 10 d-block elements (Sc through Zn) and six p-block elements (Ga through Kr). Thus, the order of orbital filling is 4s followed by the first available d orbitals (3d) followed by 4p. Continuing through successive rows of the periodic table gives the entire filling order:

$$1s \rightarrow 2s \rightarrow 2p \rightarrow 3s \rightarrow 3p \rightarrow 4s \rightarrow 3d \rightarrow 4p \rightarrow 5s \rightarrow 4d \rightarrow$$
$$5p \rightarrow 6s \rightarrow 4f \rightarrow 5d \rightarrow 6p \rightarrow 7s \rightarrow 5f \rightarrow 6d \rightarrow 7p$$

▶ **FIGURE 5.18**

Blocks of the periodic table. Each block corresponds to the filling of a different kind of orbital.

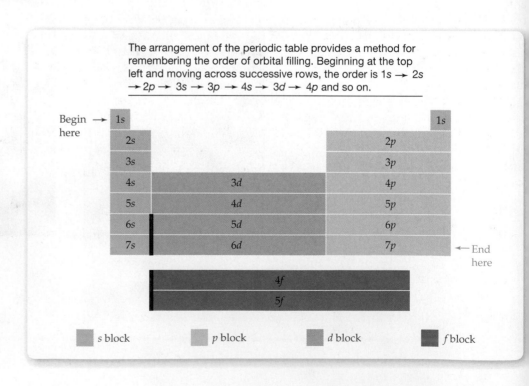

The arrangement of the periodic table provides a method for remembering the order of orbital filling. Beginning at the top left and moving across successive rows, the order is 1s → 2s → 2p → 3s → 3p → 4s → 3d → 4p and so on.

Conceptual WORKED EXAMPLE 5.8

Assigning a Ground-State Electron Configuration to an Atom

Give the ground-state electron configuration of arsenic, $Z = 33$, and draw an orbital-filling diagram, indicating the electrons as up or down arrows.

STRATEGY

Think of the periodic table as having s, p, d, and f blocks of elements, as shown in Figure 5.18. Start with hydrogen at the upper left, and fill orbitals until 33 electrons have been added. Remember that only two electrons can go into an orbital and that each one of a set of degenerate orbitals must be half filled before any one can be completely filled.

SOLUTION

As: $1s^2\, 2s^2\, 2p^6\, 3s^2\, 3p^6\, 4s^2\, 3d^{10}\, 4p^3$ or $[Ar]\, 4s^2\, 3d^{10}\, 4p^3$

An orbital-filling diagram indicates the electrons in each orbital as arrows. Note that the three $4p$ electrons all have the same spin as the orbital fill singly before electrons pair according to Hund's rule.

As: $[Ar]$ ⇅ ⇅ ⇅ ⇅ ⇅ ⇅ ↑ ↑ ↑
       $4s$ $3d$ $4p$

▶ **Conceptual PRACTICE 5.15** Give expected ground-state electron configurations for the following atoms, and draw orbital-filling diagrams.

(a) Ti $(Z = 22)$
(b) Zn $(Z = 30)$
(c) Sn $(Z = 50)$
(d) Pb $(Z = 82)$

▶ **Conceptual APPLY 5.16** Identify the atoms with the following ground-state electron configuration:

(a) $[Kr]$ ⇅   ↑ ↑ ↑ ↑ ↑   _ _ _
          $5s$ $4d$ $5p$

(b) $[Ar]$ ⇅   ⇅ ⇅ ⇅ ↑ ↑   _ _ _
          $4s$ $3d$ $4p$

5.14 ▶ ELECTRON CONFIGURATIONS AND PERIODIC PROPERTIES: ATOMIC RADII

We began this chapter by saying that atomic radius is one of many elemental properties to show periodic behavior. You might wonder, though, how we can talk about a definite size for an atom, having said in Section 5.7 that the electron clouds around atoms have no specific boundaries. What's usually done is to define an atom's radius as being half the distance between the nuclei of two identical atoms when they are bonded together. In Cl_2, for example, the distance between the two chlorine nuclei is 198 pm; in diamond (elemental carbon), the distance between two carbon nuclei is 154 pm. Thus, we say that the atomic radius of chlorine is half the Cl—Cl distance, or 99 pm, and the atomic radius of carbon is half the C—C distance, or 77 pm.

It's possible to check the accuracy of atomic radii by making sure that the assigned values are additive. For instance, since the atomic radius of Cl is 99 pm and the atomic radius of C is 77 pm, the distance between Cl and C nuclei when those two atoms are bonded together ought to be roughly 99 pm + 77 pm, or 176 pm. In fact, the measured distance between chlorine and carbon in chloromethane (CH_3Cl) is 178 pm, remarkably close to the expected value.

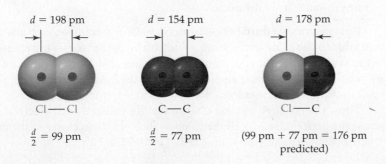

$d = 198$ pm $d = 154$ pm $d = 178$ pm

Cl—Cl   C—C   Cl—C

$\frac{d}{2} = 99$ pm $\frac{d}{2} = 77$ pm   (99 pm + 77 pm = 176 pm predicted)

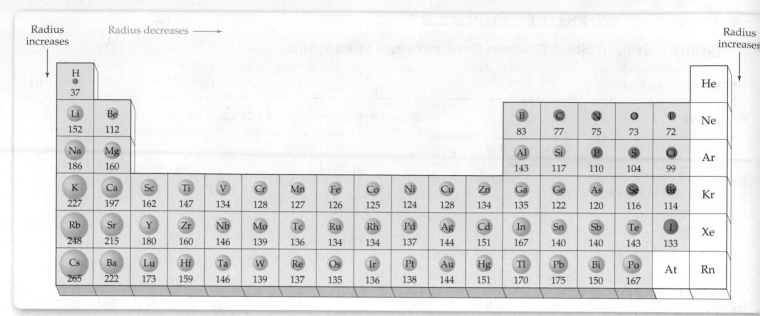

▲ **FIGURE 5.19**

Atomic radii of the elements in picometers.

As shown pictorially in **FIGURE 5.19** and graphically in Figure 5.1 at the beginning of this chapter, a comparison of atomic radius versus atomic number shows a periodic rise-and-fall pattern. Atomic radii increase going down a group of the periodic table (Li < Na < K < Rb < Cs, for instance) but decrease going across a row from left to right (Na > Mg > Al > Si > P > S > Cl, for instance). How can this behavior be explained?

The increase in radius going down a group of the periodic table occurs because successively larger valence-shell orbitals are occupied. In Li, for example, the outermost occupied shell is the second one ($2s^1$); in Na it's the third one ($3s^1$), in K it's the fourth one ($4s^1$); and so on through Rb ($5s^1$), Cs ($6s^1$), and Fr ($7s^1$). Because larger shells are occupied, the atomic radii are also larger.

The decrease in radius from left to right across the periodic table occurs because of an increase in effective nuclear charge caused by the increasing number of protons in the nucleus. As we saw in Section 5.10, Z_{eff}, the effective nuclear charge actually felt by an electron, is lower than the true nuclear charge Z because of shielding by other electrons in the atom. The amount of shielding felt by an electron depends on both the shell and subshell of the other electrons with which it is interacting. As a general rule, a valence-shell electron is:

- Strongly shielded by electrons in inner shells, which are closer to the nucleus.
- Less strongly shielded by other electrons in the same shell, according to the order $s > p > d > f$.
- Only weakly shielded by other electrons in the same subshell, which are at the same distance from the nucleus.

Going across the third period from Na to Cl, for example, each additional electron adds to the same shell (from $3s^1$ for Na to $3s^2\,3p^5$ for Cl). Because electrons in the same shell are at approximately the same distance from the nucleus, they are relatively ineffective at shielding one another. At the same time, though, the nuclear charge Z increases from +11 for Na to +17 for Cl.

Thus, the *effective* nuclear charge for the valence-shell electrons increases across the period, drawing all the valence-shell electrons closer to the nucleus and progressively shrinking the atomic radii (**FIGURE 5.20**).

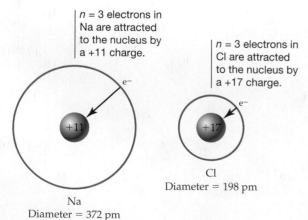

$n = 3$ electrons in Na are attracted to the nucleus by a +11 charge.

$n = 3$ electrons in Cl are attracted to the nucleus by a +17 charge.

Cl
Diameter = 198 pm

Na
Diameter = 372 pm

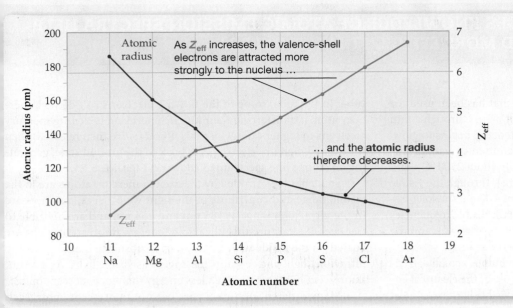

Plots of atomic radius and calculated Z_{eff} for the highest-energy electron versus atomic number.

Figure It Out

What is the relationship between Z_{eff} and atomic radius? Explain how Z_{eff} influences atomic radius.

Answer: As Z_{eff} increases atomic radius decreases. Higher Z_{eff} means the electrons experience a greater attraction to the nucleus, resulting in a smaller atomic radius.

What is true of atomic radius is also true of other atomic properties, whose periodicity can be explained by electron configurations. We'll continue the subject in the next chapter.

WORKED EXAMPLE 5.9

Predicting the Relative Size of Atomic Radii

Which atom in each of the following pairs would you expect to be larger? Explain.
(a) Mg or Ba
(b) W or Hf

STRATEGY

Atomic radius decreases across a period as the effective nuclear charge increases (Z_{eff}), thus pulling electrons to the nucleus with stronger force. Atomic radius increases down a period as a higher-energy valence shell is occupied.

SOLUTION

(a) Ba (valence shell of $n = 6$) is larger than Mg (valence shell of $n = 3$).
(b) Hf (nuclear charge $= +72$) is larger than W (nuclear charge $= +74$).

▶ **PRACTICE 5.17** Which atom in each of the following pairs would you expect to be larger? Explain.

(a) Si or Sn
(b) Os or Lu

▶ **APPLY 5.18** Predict which bond length will be the longest: C—F, C—I, C—Br.

INQUIRY ▶▶▶ HOW DOES KNOWLEDGE OF ATOMIC EMISSION SPECTRA HELP US BUILD MORE EFFICIENT LIGHT BULBS?

In the standard incandescent light bulb that has been used for more than a century, an electrical current passes through a thin tungsten filament, which is thereby heated and begins to glow. The wavelengths and intensity of the light emitted depend on the temperature of the glowing filament—typically about 2500 °C—and cover the range from ultraviolet (200–400 nm), through the visible (400–800 nm), to the infrared (800–2000 nm). The ultraviolet frequencies are blocked by the glass of the bulb, the visible frequencies pass through (the whole point of the light bulb, after all), and the infrared frequencies warm the bulb and its surroundings.

Despite its long history, an incandescent bulb is actually an extremely inefficient device. In fact, only about 5% of the electrical energy consumed by the bulb is converted into visible light, with most of the remaining 95% converted into heat. Thus, many households and businesses are replacing their incandescent light bulbs with modern compact fluorescent bulbs in an effort to use less energy. (Not that fluorescent bulbs are terribly efficient themselves—only about 20% of the energy they consume is converted into light—but that still makes them about four times better than incandescents.)

A fluorescent bulb is, in essence, a variation of the cathode-ray tube described in Section 2.6. The bulb has two main parts, an argon-filled glass tube (either straight or coiled) containing a small amount of mercury vapor, and electronic circuitry that provides a controlled high-voltage current. The current passes through a filament, which heats up and emits a flow of electrons through the tube. In the tube, some of the flowing electrons collide with mercury atoms, transferring their kinetic energy and exciting mercury electrons to higher-energy orbitals. Photons are then released when the excited mercury electrons fall back to the ground state, generating an atomic line spectrum as shown in **FIGURE 5.21**.

Some photons emitted by the excited mercury atoms are in the visible range and contribute to the light we observe, but most are in the ultraviolet range at 185 nm and 254 nm and are invisible to our eyes. To capture this ultraviolet energy, fluorescent bulbs are coated on the inside with a *phosphor*, a substance that absorbs the ultraviolet light and re-emits the energy as visible light. As a result, fluorescent lights waste much less energy than incandescent bulbs.

Many different phosphors are used in fluorescent lights, each emitting its own line spectrum with visible light of various colors. Typically, a so-called triphosphor mixture is used, consisting of several complex metal oxides and rare-earth ions: $Y_2O_3{:}Eu^{3+}$ (red emitting), $CeMgAl_{11}O_{19}{:}Tb^{3+}$ (green emitting), and $BaMgAl_{10}O_{17}{:}Eu^{2+}$ (blue emitting). The final color that results can be tuned as desired by the manufacturer, but typically the three emissions together are distributed fairly evenly over the visible spectrum to provide a color reproduction that our eyes perceive as natural white light (**FIGURE 5.22**).

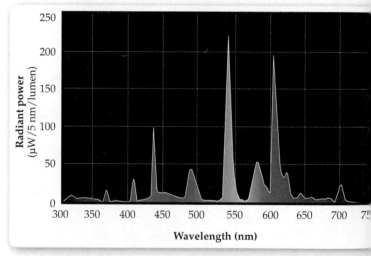

▲ **FIGURE 5.22**

The triphosphor spectrum emitted from a typical fluorescent bulb. The triphosphor spectrum is distributed over the visible spectrum and is perceived by our eyes as white light.

▲ Compact fluorescent bulbs like those shown here are a much more energy-efficient way to light a home than typical incandescent light bulbs.

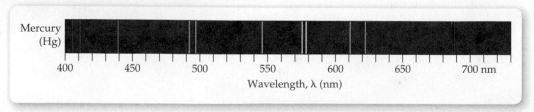

▲ **FIGURE 5.21**

Line emission spectrum for mercury.

PROBLEM 5.19 Mercury vapor is contained inside the fluorescent bulb.

(a) Use the periodic table on the inside front cover of the book to write the electron configuration for the ground state of mercury.

(b) Sketch the orbital filling diagram for the ground state of mercury.

(c) How many unpaired electrons are there in the ground state of mercury?

PROBLEM 5.20 When electricity is used to add energy to the mercury atoms, an electron in a 6s orbital can be "excited" into a 6p orbital.

(a) Write the electron configuration for a mercury atom in the excited state.

(b) Sketch the orbital filling diagram for the excited state of mercury.

(c) How many unpaired electrons are there in the excited state of mercury?

PROBLEM 5.21 Atomic emission spectra arise from electron transitions from higher-energy orbitals to lower-energy orbitals. The blue line at 434.7 nm in the emission spectrum for mercury (Figure 5.21) arises from an electron moving from a 7d to a 6p orbital.

(a) Give the n and l quantum numbers and give the range of possible m_l quantum numbers for the higher energy 7d orbital.

(b) Give the n and l quantum numbers and give the range of possible m_l quantum numbers for the lower energy 6p orbital.

(c) Calculate the energy difference between the 7d and 6p orbital in units of kJ/mol.

PROBLEM 5.22 Three different wavelengths in the line spectrum for mercury and three electronic transitions in mercury are given. Match the three wavelengths with the correct electronic transition between orbitals.

Emission Lines: 185.0 nm, 140.2 nm, 126.8 nm

Electronic Transitions: $8p \rightarrow 6s$, $7p \rightarrow 6s$, $6p \rightarrow 6s$

PROBLEM 5.23

(a) How is the emission spectrum from the triphosphor in the fluorescent bulb different than the emission spectrum from a "white light" source such as the sun or an incandescent bulb?

(b) Why does the fluorescent bulb appear as "white light" to our eyes?

STUDY GUIDE

Section	Concept Summary	Learning Objectives	Test Your Understanding
5.1 ▸ The Nature of Radiant Energy and the Electromagnetic Spectrum	Understanding the nature of atoms and molecules begins with an understanding of light and other kinds of **radiant energy** that make up the **electromagnetic spectrum**. An electromagnetic wave travels through a vacuum at the speed of light (c) and is characterized by its **frequency (ν)**, **wavelength** (λ), and **amplitude**.	**5.1** Label the wavelength, frequency, and amplitude in an electromagnetic wave and understand their meaning. (Figure 5.3)	Problems 5.2, 5.24, 5.32
		5.2 Interconvert between wavelength and frequency of electromagnetic radiation.	Worked Example 5.1; Problems 5.36, 5.37, 5.40, 5.41
5.2 ▸ Particlelike Properties of Radiant Energy: The Photoelectric Effect and Planck's Postulate	Electromagnetic energy exists only in discrete amounts, referred to as a **quantum**. The energy of one quantum is related to the frequency of an electromagnetic wave and is given by **Planck's postulate**. In the **photoelectric effect**, light above a threshold energy is able to remove electrons from a substance. Einstein proposed that light behaves like particles (photons) in his explanation of the photoelectric effect.	**5.3** Calculate the energy of electromagnetic radiation in units of J/photon or kJ/mol, when given the frequency or wavelength.	Worked Example 5.2; Problems 5.38, 5.42, 5.44, 5.118, 5.122
		5.4 Describe the photoelectric effect and explain how it supports the theory of particlelike properties of light. (Figures 5.4 and 5.5)	Problems 5.52, 5.53
		5.5 Calculate the frequency or wavelength of radiation needed to produce the photoelectric effect given the work function of a metal.	Worked Example 5.3; Problems 5.54, 5.55

Section	Concept Summary	Learning Objectives	Test Your Understanding
5.3 ▸ The Interaction of Radiant Energy with Atoms: Line Spectra	Unlike the white light of the Sun, which consists of a nearly continuous distribution of wavelengths, the light emitted by an excited atom consists of only a few discrete wavelengths, a **line spectrum**. The **Balmer–Rydberg equation** is a mathematical model used to calculate the wavelengths in the emission spectrum for hydrogen.	**5.6** Describe the difference between a continuous spectrum and a line spectrum. (Figure 5.7)	Problems 5.56, 5.57
5.4 ▸ The Bohr Model of the Atom: Quantized Energy	Bohr's model of the atom depicts a dense, positively charged nucleus circled by electrons in orbits of different radii. The model is successful in predicting line spectra because it presumes the energy of the orbits is **quantized**. The Bohr model is incorrect, however, in describing fixed electron orbits. Another limitation of the Bohr model is that it is only accurately describes one-electron species.	**5.7** Compare the wavelength and frequency of different electron transitions in the Bohr model of the atom.	Problem 5.25
		5.8 Relate wavelengths calculated using the Balmer–Rydberg equation to energy levels in the Bohr model of the atom.	Worked Example 5.4; Problems 5.46–5.51, 5.114, 5.116
5.5 ▸ Wavelike Properties of Matter: de Broglie's Hypothesis	Just as light behaves in some respects like a stream of small particles (**photons**), electrons and other tiny units of matter behave in some respects like waves. The wavelength of a particle of mass m traveling at a velocity v is given by the **de Broglie equation**.	**5.9** Calculate the wavelength of a moving object using the de Broglie equation.	Worked Example 5.5; Problems 5.58, 5.59, 5.62, 5.63
		5.10 Explain why the wavelength of macroscopic objects is not observed.	Problems 5.60, 5.61
5.6 ▸ The Quantum Mechanical Model of the Atom: Heisenberg's Uncertainty Principle	According to **Heisenberg's uncertainty principle**, we can never know both the position and the velocity of an electron (or of any other object) beyond a certain level of precision.	**5.11** Calculate the uncertainty in the position of the moving object if the velocity is known.	Problems 5.64, 5.65
5.7 ▸ The Quantum Mechanical Model of the Atom: Orbitals and Quantum Numbers	The **quantum mechanical model** proposed in 1926 by Erwin Schrödinger describes an atom by a mathematical equation similar to that used to describe wave motion. The behavior of each electron in an atom is characterized by a **wave function**, or **orbital**, whose square defines the probability of finding the electron in a given volume of space. Each wave function has a set of three parameters called **quantum numbers**. The **principal quantum number** n defines the size of the orbital; the **angular-momentum quantum number** l defines the shape of the orbital; and the **magnetic quantum number** m_l defines the spatial orientation of the orbital.	**5.12** Identify and write valid sets of quantum numbers that describe electrons in different types of orbitals.	Worked Example 5.6; Problems 5.69–5.71, 5.74, 5.75
5.8 ▸ The Shapes of Orbitals	The square of a wave function ψ^2, or orbital, describes the probability of finding the electron within a specific region of space. The shape of that spatial region is defined by the angular-momentum quantum number l, with $l = 0$ called an s orbital, $l = 1$ a p orbital, $l = 2$ a d orbital, and $l = 3$ an f orbital.	**5.13** Identify an orbital based on its shape and describe it using a set of quantum numbers.	Worked Example 5.7; Problems 5.13, 5.26
		5.14 Visualize the nodal planes in different types of orbitals and different shells.	Problems 5.14, 5.72, 5.73
5.9 ▸ Electron Spin and the Pauli Exclusion Principle	A fourth quantum number m_s, describing electron spin, differentiates the two electrons that may occupy a given orbital. The **Pauli exclusion principle** states that no two electrons in an atom can have the same four quantum numbers. The **spin quantum number** m_s specifies the electron spin as either $+1/2$ or $-1/2$. Electrons that occupy the same orbital must have opposing spins.	**5.15** Assign a set of four quantum numbers for electrons in an atom.	Problems 5.80–5.83

Section	Concept Summary	Learning Objectives	Test Your Understanding
5.10 ▶ Orbital Energy Levels in Multielectron Atoms	Within a shell, all hydrogen orbitals have the same energy. In multielectron atoms, the *s*, *p*, *d*, and *f* orbitals within a given shell have slightly different energies. The difference in energy between subshells in multielectron atoms results from the different shapes of the orbitals that lead to differing degrees of electron shielding.	**5.16** Explain how electron shielding gives the order of subshells from lowest to highest in energy (Figure 5.9).	Problems 5.84–5.87, 5.90
		5.17 Predict the order of filling of subshells based upon energy.	Problems 5.92, 5.93
5.11 –5.12 ▶ Electron Configurations of Multielectron Atoms	The **ground-state electron configuration** of a multielectron atom is arrived at by following a series of rules called the **aufbau principle.** 1. The lowest-energy orbitals fill first. 2. Only two electrons go into any one **orbital** (**Pauli exclusion principle**), and these must be of opposite spins. 3. If two or more orbitals are equal in energy (degenerate), each is half filled before any one is completely filled (**Hund's rule**). The aufbau principle usually correctly predicts the electron configuration of elements; however, 21 elements have anomalous configurations due to the unusual stability of both half-filled and fully filled subshells (Figure 5.17).	**5.18** Assign electron configurations to atoms in their ground state.	Worked Example 5.8; Problems 5.94, 5.95, 5.110, 5.111
5.13 ▶ Electron Configurations and the Periodic Table	The periodic table is the most important organizing principle of chemistry. It is successful because elements in each group of the periodic table have similar **valence-shell** electron configurations and therefore have similar properties. The periodic table can be used to predict the order that electrons fill orbitals (Figure 5.18).	**5.19** Draw orbital filling diagrams for the ground state of an atom and determine the number of unpaired electrons.	Worked Example 5.8; Problems 5.15, 5.16, 5.96–5.98
		5.20 Identify atoms from orbital filling diagrams or electron configurations.	Problems 5.29, 5.30, 5.99, 5.120
5.14 ▶ Electron Configurations and Periodic Properties: Atomic Radii	Atomic radii of elements show a periodic rise-and-fall pattern according to the positions of the elements in the table. Atomic radii increase going down a group because *n* increases, and they decrease from left to right across a period because the **effective nuclear charge** (Z_{eff}) increases.	**5.21** Explain the periodic trend in atomic radii.	Problems 5.104, 5.105
		5.22 Predict the relative size of atoms based upon their position in the periodic table.	Worked Example 5.9; Problems 5.106–5.109

KEY TERMS

amplitude   *156*

angular-momentum quantum number (*l*)   *169*

aufbau principle   *176*

Balmer–Rydberg equation   *162*

d-block element   *180*

de Broglie equation   *165*

degenerate   *176*

effective nuclear charge (Z_{eff})   *175*

electromagnetic spectrum   *155*

electron configuration   *176*

f-block element   *180*

frequency (*ν*)   *156*

ground-state electron configuration   *176*

Heisenberg uncertainty principle   *167*

hertz (Hz)   *156*

Hund's rule   *176*

line spectrum   *161*

magnetic quantum number (m_l)   *169*

node   *171*

orbital   *168*

p-block element   *180*

Pauli exclusion principle   *175*

phase   *172*

photon   *158*

photoelectric effect   *158*

Planck's postulate   *158*

principal quantum number (*n*)   *168*

quantum   *159*

quantized   *159*

quantum mechanical model   *167*

quantum number   *168*

radiant energy   *155*

Rutherford's nuclear model   *163*

s-block element   *180*

shell   *168*

spin quantum number (m_s)   *174*

subshell   *169*

valence shell   *178*

wave function   *168*

wavelength (*λ*)   *156*

KEY EQUATIONS

- **The relationship between frequency (ν) and wavelength (λ) (Section 5.1)**

$$\lambda = \frac{c}{\nu} \quad \text{or} \quad \nu = \frac{c}{\lambda}$$

where c is the speed of light 3.00×10^8 m/s.

- **Planck's postulate relating energy (E) and frequency (ν) or wavelength (λ) (Section 5.2)**

$$E = h\nu = \frac{hc}{\lambda}$$

where h is Planck's constant 6.626×10^{-34} J·s.

- **The Balmer–Rydberg equation to account for the line spectrum of hydrogen (Section 5.3)**

$$\frac{1}{\lambda} = R_\infty \left[\frac{1}{m^2} - \frac{1}{n^2} \right] \quad \text{or} \quad \nu = R_\infty \cdot c \left[\frac{1}{m^2} - \frac{1}{n^2} \right]$$

where m and n are integers with $n > m$ and R_∞ is the Rydberg constant.

- **de Broglie's hypothesis relating wavelength (λ) mass (m), and velocity (v) (Section 5.5)**

$$\lambda = \frac{h}{mv}$$

- **Heisenberg's Uncertainty Principle (Section 5.6)**

$$(\Delta x)(\Delta mv) \geq \frac{h}{4\pi}$$

CONCEPTUAL PROBLEMS

Problems 5.1–5.23 appear within the chapter

5.24 Two electromagnetic waves are represented below.

  (a) Which wave has the greater intensity?

  (b) Which wave corresponds to higher-energy radiation?

  (c) Which wave represents yellow light, and which represents infrared radiation?

  (a) 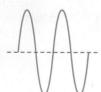 **(b)**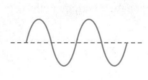

5.25 The following diagram shows the energy levels of the different shells in the hydrogen atom.

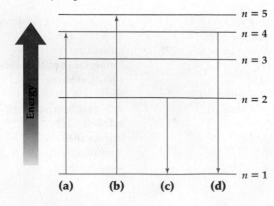

  (a) Which transition corresponds to absorption of light with the longest wavelength?

  (b) Which transition corresponds to emission of light with the shortest wavelength?

5.26 Identify each of the following orbitals, and give n and l quantum numbers for each.

  (a)   **(b)**

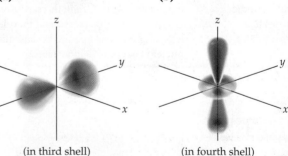

  (in third shell)   (in fourth shell)

5.27 Where on the blank outline of the periodic table do elements that meet the following descriptions appear?

  (a) Elements with the valence-shell ground-state electron configuration $ns^2\, np^5$

  (b) An element with the ground-state electron configuration [Ar] $4s^2\, 3d^{10}\, 4p^5$

  (c) Elements with electrons whose largest principal quantum number is $n = 4$

  (d) Elements that have only one unpaired p electron

  (e) The d-block elements

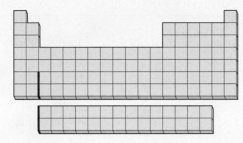

5.28 One of the elements shown on the following periodic table has an anomalous ground-state electron configuration. Which is it—red, blue, or green—and why?

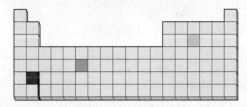

5.29 What atom has the following orbital-filling diagram?

[Ar]

5.30 The following orbital-filling diagram represents an excited state rather than the ground state of an atom. Identify the atom, and give its ground-state electron configuration.

[Ar]

5.31 Which of the following three spheres represents a Ca atom, which an Sr atom, and which a Br atom?

r = 215 pm   r = 114 pm   r = 197 pm

SECTION PROBLEMS

Electromagnetic Energy and Atomic Spectra
(Sections 5.1–5.4)

5.32 Which has the higher frequency, red light or violet light? Which has the longer wavelength? Which has the greater energy?

5.33 Which has the higher frequency, infrared light or ultraviolet light? Which has the longer wavelength? Which has the greater energy?

5.34 The Hubble Space Telescope detects electromagnetic energy in the wavelength range 1.15×10^{-7} m to 2.0×10^{-6} m. What region of the electromagnetic spectrum is found completely within this range? What regions fall partially in this range?

5.35 The Green Bank Telescope in West Virginia—the world's largest steerable radio telescope—detects frequencies from 290 MHz to 90 GHz. What region or regions of the electromagnetic spectrum are found completely or partially within its detection range?

5.36 What is the wavelength in meters of ultraviolet light with $\nu = 5.5 \times 10^{15}\ s^{-1}$?

5.37 What is the frequency of a microwave with $\lambda = 4.33 \times 10^{-3}$ m?

5.38 Calculate the energies of the following waves in kilojoules per mole, and tell which member of each pair has the higher value.

(a) An FM radio wave at 99.5 MHz and an AM radio wave at 1150 kHz

(b) An X ray with $\lambda = 3.44 \times 10^{-9}$ m and a microwave with $\lambda = 6.71 \times 10^{-2}$ m

5.39 The MRI (magnetic resonance imaging) body scanners used in hospitals operate with 400 MHz radio frequency energy. How much energy does this correspond to in kilojoules per mole?

5.40 A certain cellular telephone transmits at a frequency of 825 MHz and receives at a frequency of 875 MHz.

(a) What is the wavelength of the transmitted signal in cm?

(b) What is the wavelength of the received signal in cm?

5.41 Optical fibers allow the fast transmission of vast amounts of data. In one type of fiber, the wavelength of transmitted light is 1.3×10^3 nm.

(a) What is the frequency of the light?

(b) Fiber optic cable is available in 12 km lengths. How long will it take for a signal to travel that distance assuming that the speed of light in the cable is the same as in a vacuum?

5.42 What is the wavelength in meters of photons with the following energies? In what region of the electromagnetic spectrum does each appear?

(a) 90.5 kJ/mol

(b) 8.05×10^{-4} kJ/mol

(c) 1.83×10^3 kJ/mol

5.43 What is the energy of each of the following photons in kilojoules per mole?

(a) $\nu = 5.97 \times 10^{19}\ s^{-1}$

(b) $\nu = 1.26 \times 10^6\ s^{-1}$

(c) $\lambda = 2.57 \times 10^2$ m

5.44 The data encoded on CDs, DVDs, and Blu-ray discs is read by lasers. What is the wavelength in nanometers and the energy in joules of the following lasers?

(a) CD laser, $\nu = 3.85 \times 10^{14}\ s^{-1}$

(b) DVD laser, $\nu = 4.62 \times 10^{14}\ s^{-1}$

(c) Blu-ray laser, $\nu = 7.41 \times 10^{14}\ s^{-1}$

5.45 The semimetal germanium is used as a component in photodetectors, which generate electric current when exposed to light. If a germanium photodetector responds to photons in the range $\lambda = 400 - 1700$ nm, will the following light sources be detected?

(a) a laser with $\nu = 4.35 \times 10^{14}\ s^{-1}$

(b) photons with $E = 43$ kJ/mol

(c) electromagnetic radiation with $\nu = 706$ THz

5.46 According to the equation for the Balmer line spectrum of hydrogen, a value of $n = 3$ gives a red spectral line at 656.3 nm, a value of $n = 4$ gives a green line at 486.1 nm, and a value of $n = 5$ gives a blue line at 434.0 nm. Calculate the energy in kilojoules per mole of the radiation corresponding to each of these spectral lines.

5.47 According to the equation for the Balmer line spectrum of hydrogen, a value of $n = 4$ gives a red spectral line at 486.1 nm. Calculate the energy in kilojoules per mole of the radiation corresponding to this spectral line.

5.48 Calculate the wavelength and energy in kilojoules necessary to completely remove an electron from the first shell ($m = 1$) of a hydrogen atom ($R_\infty = 1.097 \times 10^{-2}\,\text{nm}^{-1}$).

5.49 Calculate the wavelength and energy in kilojoules necessary to completely remove an electron from the second shell ($m = 2$) of a hydrogen atom ($R_\infty = 1.097 \times 10^{-2}\,\text{nm}^{-1}$).

5.50 One series of lines of the hydrogen spectrum is caused by emission of energy accompanying the fall of an electron from outer shells to the fourth shell. The lines can be calculated using the Balmer–Rydberg equation:

$$\frac{1}{\lambda} = R_\infty\left[\frac{1}{m^2} - \frac{1}{n^2}\right]$$

where $m = 4$, $R_\infty = 1.097 \times 10^{-2}\,\text{nm}^{-1}$, and n is an integer greater than 4. Calculate the wavelengths in nanometers and energies in kilojoules per mole of the first two lines in the series. In what region of the electromagnetic spectrum do they fall?

5.51 One series of lines of the hydrogen spectrum is caused by emission of energy accompanying the fall of an electron from outer shells to the third shell. The lines can be calculated using the Balmer–Rydberg equation:

$$\frac{1}{\lambda} = R_\infty\left[\frac{1}{m^2} - \frac{1}{n^2}\right]$$

where $m = 3$, $R_\infty = 1.097 \times 10^{-2}\,\text{nm}^{-1}$, and n is an integer greater than 3. Calculate the wavelengths in nanometers and energies in kilojoules per mole of the first two lines in the series. In what region of the electromagnetic spectrum do they fall?

5.52 The work function of cesium metal is 188 kJ/mol, which corresponds to light with a wavelength of 637 nm. Which of the following will cause the smallest number of electrons to be ejected from cesium?

(a) High-amplitude wave with a wavelength of 500 nm

(b) Low-amplitude wave with a wavelength of 500 nm

(c) High-amplitude wave with a wavelength of 650 nm

(d) Low-amplitude wave with a wavelength of 650 nm

5.53 The work function of calcium metal is kJ/mol, which corresponds to light with a wavelength of 432 nm. Which of the following will cause a largest number of electrons to be ejected from cesium?

(a) High-amplitude wave with a wavelength of 400 nm

(b) Low-amplitude wave with a wavelength of 400 nm

(c) High-amplitude wave with a wavelength of 450 nm

(d) Low-amplitude wave with a wavelength of 450 nm

5.54 The work function of silver metal is 436 kJ/mol. What frequency of light is needed to eject electrons from a sample of silver?

5.55 What is the work function of gold metal in kJ/mol if light with $\lambda = 234$ nm is necessary to eject electrons?

5.56 Spectroscopy is a technique that uses the interaction of radiant energy with matter to identify or quantify a substance in a sample. A deuterium lamp is often used a light source in the ultraviolet region of the spectrum and the emission spectrum is shown. Is this a continuous or line emission spectrum?

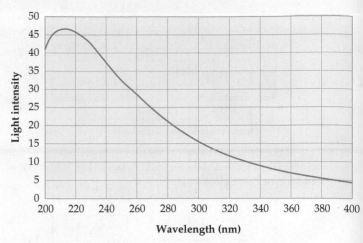

5.57 Sodium-vapor lamps are a common source of lighting. The emission spectrum from this type of lamp is shown. Is this a continuous or line emission spectrum?

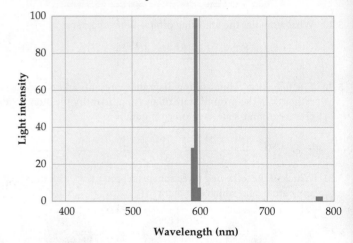

Particles and Waves (Section 5.5)

5.58 Protons and electrons can be given very high energies in particle accelerators. What is the wavelength in meters of an electron (mass $= 9.11 \times 10^{-31}$ kg) that has been accelerated to 5% of the speed of light? In what region of the electromagnetic spectrum is this wavelength?

5.59 What is the wavelength in meters of a proton (mass $= 1.673 \times 10^{-24}$ g) that has been accelerated to 1% of the speed of light? In what region of the electromagnetic spectrum is this wavelength?

5.60 What is the de Broglie wavelength in meters of a baseball weighing 145 g and traveling at 156 km/h? Why do we not observe this wavelength?

5.61 What is the de Broglie wavelength in meters of a mosquito weighing 1.55 mg and flying at 1.38 m/s? Why do we not observe this wavelength?

5.62 At what speed in meters per second must a 145 g baseball be traveling to have a de Broglie wavelength of 0.500 nm?

5.63 What velocity would an electron (mass $= 9.11 \times 10^{-31}$ kg) need for its de Broglie wavelength to be that of red light (750 nm)?

5.64 Use the Heisenberg uncertainty principle to calculate the uncertainty in meters in the position of a honeybee weighing 0.68 g and traveling at a velocity of 0.85 m/s. Assume that the uncertainty in the velocity is 0.1 m/s.

5.65 The mass of a helium atom is 4.0026 amu, and its average velocity at 25 °C is 1.36×10^3 m/s. What is the uncertainty in meters in the position of a helium atom if the uncertainty in its velocity is 1%?

Orbitals and Quantum Mechanics (Sections 5.6–5.9)

5.66 What is the Heisenberg uncertainty principle, and how does it affect our description of atomic structure?

5.67 Why do we have to use an arbitrary value such as 95% to determine the spatial limitations of an orbital?

5.68 What are the four quantum numbers, and what does each specify?

5.69 Tell which of the following combinations of quantum numbers are not allowed. Explain your answers.

(a) $n = 3, l = 0, m_l = -1$

(b) $n = 3, l = 1, m_l = 1$

(c) $n = 4, l = 4, m_l = 0$

5.70 Give the allowable combinations of quantum numbers for each of the following electrons:

(a) A $4s$ electron   (b) A $3p$ electron

(c) A $5f$ electron   (d) A $5d$ electron

5.71 Give the orbital designations of electrons with the following quantum numbers:

(a) $n = 3, l = 0, m_l = 0$

(b) $n = 2, l = 1, m_l = -1$

(c) $n = 4, l = 3, m_l = -2$

(d) $n = 4, l = 2, m_l = 0$

5.72 How many nodal surfaces does a $4s$ orbital have? Draw a cutaway representation of a $4s$ orbital showing the nodes and the regions of maximum electron probability.

5.73 How do the number of nodal planes change as the l quantum number of the subshell changes?

5.74 Which of the following combinations of quantum numbers can refer to an electron in a ground-state cobalt atom $(Z = 27)$?

(a) $n = 3, l = 0, m_l = 2$

(b) $n = 4, l = 2, m_l = -2$

(c) $n = 3, l = 1, m_l = 0$

5.75 Which of the following combinations of quantum numbers can refer to an electron in a ground-state selenium atom $(Z = 34)$?

(a) $n = 3, l = 3, m_l = 2$

(b) $n = 4, l = 2, m_l = -2$

(c) $n = 4, l = 1, m_l = 0$

5.76 What is the maximum number of electrons in an atom whose highest-energy electrons have the principal quantum number $n = 5$?

5.77 What is the maximum number of electrons in an atom whose highest-energy electrons have the principal quantum number $n = 4$ and the angular-momentum quantum number $l = 0$?

5.78 Sodium atoms emit light with a wavelength of 330 nm when an electron moves from a $4p$ orbital to a $3s$ orbital. What is the energy difference between the orbitals in kilojoules per mole?

5.79 Excited rubidium atoms emit red light with $\lambda = 795$ nm. What is the energy difference in kilojoules per mole between orbitals that give rise to this emission?

5.80 Assign a set of four quantum numbers to each electron in carbon.

5.81 Assign a set of four quantum numbers to each electron in oxygen.

5.82 Assign a set of four quantum numbers for the outermost two electrons in Sr.

5.83 Which of the following is a valid set of four quantum numbers for a d electron in Mo?

(a) $n = 4, l = 1, m_l = 0, m_s = +1/2$

(b) $n = 5, l = 2, m_l = 0, m_s = -1/2$

(c) $n = 4, l = 2, m_l = -1, m_s = +1/2$

(d) $n = 5, l = 2, m_l = -3, m_s = +1/2$

Electron Configurations (Sections 5.10–5.13)

5.84 What is meant by the term *effective nuclear charge*, Z_{eff}, and what causes it?

5.85 How does electron shielding in multielectron atoms give rise to energy differences among $3s$, $3p$, and $3d$ orbitals?

5.86 Order the electrons in the following orbitals according to their shielding ability: $4s$, $4d$, $4f$.

5.87 Order the following elements according to increasing Z_{eff}: Ca, Se, Kr, K.

5.88 Why does the number of elements in successive periods of the periodic table increase by the progression 2, 8, 18, 32?

5.89 Which two of the four quantum numbers determine the energy level of an orbital in a multielectron atom?

5.90 Which orbital in each of the following pairs is higher in energy?

(a) $5p$ or $5d$

(b) $4s$ or $3p$

(c) $6s$ or $4d$

5.91 Order the orbitals for a multielectron atom in each of the following lists according to increasing energy:

(a) $4d, 3p, 2p, 5s$

(b) $2s, 4s, 3d, 4p$

(c) $6s, 5p, 3d, 4p$

5.92 According to the aufbau principle, which orbital is filled immediately *after* each of the following in a multielectron atom?

(a) $4s$ (b) $3d$

(c) $5f$ (d) $5p$

5.93 According to the aufbau principle, which orbital is filled immediately *before* each of the following?

(a) $3p$ (b) $4p$

(c) $4f$ (d) $5d$

5.94 Give the expected ground-state electron configurations for the following elements:

(a) Ti   (b) Ru

(c) Sn   (d) Sr

(e) Se

5.95 Give the expected ground-state electron configurations for atoms with the following atomic numbers:

(a) $Z = 55$ (b) $Z = 40$

(c) $Z = 80$ (d) $Z = 62$

5.96 Draw orbital-filling diagrams for the following atoms. Show each electron as an up or down arrow, and use the abbreviation of the preceding noble gas to represent inner-shell electrons.

(a) Rb   (b) W

(c) Ge   (d) Zr

5.97 Draw orbital-filling diagrams for atoms with the following atomic numbers. Show each electron as an up or down arrow, and use the abbreviation of the preceding noble gas to represent inner-shell electrons.

(a) $Z = 25$ (b) $Z = 56$
(c) $Z = 28$ (d) $Z = 47$

5.98 How many unpaired electrons are present in each of the following ground-state atoms?

(a) O   (b) Si
(c) K   (d) As

5.99 Identify the following atoms:

(a) It has the ground-state electron configuration.
    $[Ar] 4s^2 3d^{10} 4p^1$.
(b) It has the ground-state electron configuration $[Kr] 4d^{10}$.

5.100 At what atomic number is the filling of a g orbital likely to begin?

5.101 Assuming that g orbitals fill according to Hund's rule, what is the atomic number of the first element to have a filled g orbital?

5.102 Take a guess. What do you think is a likely ground-state electron configuration for the sodium *ion*, Na^+, formed by loss of an electron from a neutral sodium atom?

5.103 Take a guess. What is a likely ground-state electron configuration for the chloride ion, Cl^-, formed by adding an electron to a neutral chlorine atom?

Electron Configurations and Periodic Properties (Section 5.14)

5.104 Why do atomic radii increase going down a group of the periodic table?

5.105 Why do atomic radii decrease from left to right across a period of the periodic table?

5.106 Order the following atoms according to increasing atomic radius: S, F, O.

5.107 Order the following atoms according to increasing atomic radius: Rb, Cl, As, K

5.108 Which atom in each of the following pairs has a larger radius?

(a) Na or K   (b) V or Ta
(c) V or Zn   (d) Li or Ba

5.109 Which atom in each of the following pairs has a larger radius?

(a) C or Ge   (b) Ni or Pt
(c) Sn or I   (d) Na or Rb

5.110 What is the expected ground-state electron configuration of the recently discovered element with $Z = 116$?

5.111 What is the atomic number and expected ground-state electron configuration of the yet undiscovered element directly below Fr in the periodic table?

CHAPTER PROBLEMS

5.112 Orbital energies in single-electron atoms or ions, such as He^+, can be described with an equation similar to the Balmer–Rydberg equation:

$$\frac{1}{\lambda} = Z^2 R \left[\frac{1}{m^2} - \frac{1}{n^2} \right]$$

where Z is the atomic number. What wavelength of light in nanometers is emitted when the electron in He^+ falls from $n = 3$ to $n = 2$?

5.113 Like He^+, the Li^{2+} ion is a single-electron system (Problem 5.112). What wavelength of light in nanometers must be absorbed to promote the electron in Li^{2+} from $n = 1$ to $n = 4$?

5.114 Use the Balmer equation to calculate the wavelength in nanometers of the spectral line for hydrogen when $n = 6$ and $m = 2$. What is the energy in kilojoules per mole of the radiation corresponding to this line?

5.115 Lines in a certain series of the hydrogen spectrum are caused by emission of energy accompanying the fall of an electron from outer shells to the fifth shell. Use the Balmer–Rydberg equation to calculate the wavelengths in nanometers and energies in kilojoules per mole of the two longest-wavelength lines in the series. In what region of the electromagnetic spectrum do they fall?

5.116 What is the shortest wavelength in nanometers in the series you calculated in Problem 5.115?

5.117 What is the wavelength in meters of photons with the following energies? In what region of the electromagnetic spectrum does each appear?

(a) 142 kJ/mol
(b) 4.55×10^{-2} kJ/mol
(c) 4.81×10^4 kJ/mol

5.118 What is the energy of each of the following photons in kilojoules per mole?

(a) $\nu = 3.79 \times 10^{11}$ s^{-1}
(b) $\nu = 5.45 \times 10^4$ s^{-1}
(c) $\lambda = 4.11 \times 10^{-5}$ m

5.119 The *second* in the SI system is defined as the duration of 9,192,631,770 periods of radiation corresponding to the transition between two energy levels of a cesium-133 atom. What is the energy difference between the two levels in kilojoules per mole?

5.120 Write the symbol, give the ground-state electron configuration, and draw an orbital-filling diagram for each of the following atoms. Use the abbreviation of the preceding noble gas to represent the inner-shell electrons.

(a) The heaviest alkaline earth metal
(b) The lightest transition metal
(c) The heaviest actinide metal
(d) The lightest semimetal
(e) The group 6A element in the fifth period

5.121 Imagine a universe in which the four quantum numbers can have the same possible values as in our universe except that the angular-momentum quantum number l can have integral values of $0, 1, 2, \ldots, n + 1$ (instead of $0, 1, 2, \ldots, n - 1$).

(a) How many elements would be in the first two rows of the periodic table in this universe?
(b) What would be the atomic number of the element in the second row and fifth column?
(c) Draw an orbital-filling diagram for the element with atomic number 12.

5.122 Cesium metal is frequently used in photoelectric cells because the amount of energy necessary to eject electrons from a cesium surface is relatively small—only 206.5 kJ/mol. What wavelength of light in nanometers does this correspond to?

5.123 The laser light used in compact disc players has $\lambda = 780$ nm. In what region of the electromagnetic spectrum does this light appear? What is the energy of this light in kilojoules per mole?

5.124 Draw orbital-filling diagrams for the following atoms. Show each electron as an up or down arrow, and use the abbreviation of the preceding noble gas to represent inner-shell electrons.

   (a) Sr   (b) Cd
   (c) has $Z = 22$ (d) has $Z = 34$

5.125 The atomic radii of Y (180 pm) and La (187 pm) are significantly different, but the radii of Zr (160 pm) and Hf (159 pm) are essentially identical. Explain.

5.126 You're probably familiar with using Scotch Tape for wrapping presents but may not know that it can also generate electromagnetic radiation. When Scotch Tape is unrolled in a vacuum (but not in air), photons with a range of frequencies around $\nu = 2.9 \times 10^{18}$ s^{-1} are emitted in nanosecond bursts.

   (a) What is the wavelength in meters of photons with $\nu = 2.9 \times 10^{18}$ s^{-1}?

   (b) What is the energy in kJ/mol of photons with $\nu = 2.9 \times 10^{18}$ s^{-1}?

   (c) What type of electromagnetic radiation are these photons?

5.127 Hard wintergreen-flavored candies are *triboluminescent*, meaning that they emit flashes of light when crushed. (You can see it for yourself if you look in a mirror while crunching a wintergreen Life Saver in your mouth in a dark room.) The strongest emission is around $\lambda = 450$ nm.

   (a) What is the frequency in s^{-1} of photons with $\lambda = 450$ nm?

   (b) What is the energy in kJ/mol of photons with $\lambda = 450$ nm?

   (c) What is the color of the light with $\lambda = 450$ nm?

5.128 One method for calculating Z_{eff} is to use the equation

$$Z_{eff} = \sqrt{\frac{(E)(n^2)}{1312 \text{ kJ/mol}}}$$

where E is the energy necessary to remove an electron from an atom and n is the principal quantum number of the electron. Use this equation to calculate Z_{eff} values for the highest-energy electrons in potassium ($E = 418.8$ kJ/mol) and krypton ($E = 1350.7$ kJ/mol).

5.129 One watt (W) is equal to 1 J/s. Assuming that 5.0% of the energy output of a 75 W light bulb is visible light and that the average wavelength of the light is 550 nm, how many photons are emitted by the light bulb each second?

5.130 Microwave ovens work by irradiating food with microwave radiation, which is absorbed and converted into heat. Assuming that radiation with $\lambda = 15.0$ cm is used, that all the energy is converted to heat, and that 4.184 J is needed to raise the temperature of 1.00 g of water by 1.00 °C, how many photons are necessary to raise the temperature of a 350 mL cup of water from 20 °C to 95 °C?

5.131 Photochromic sunglasses, which darken when exposed to light, contain a small amount of colorless AgCl embedded in the glass. When irradiated with light, metallic silver atoms are produced and the glass darkens: AgCl $\longrightarrow$ Ag + Cl. Escape of the chlorine atoms is prevented by the rigid structure of the glass, and the reaction therefore reverses as soon as the light is removed. If 310 kJ/mol of energy is required to make the reaction proceed, what wavelength of light is necessary?

5.132 The amount of energy necessary to remove an electron from an atom is a quantity called the *ionization energy*, E_i. This energy can be measured by a technique called *photoelectron spectroscopy*, in which light of wavelength λ is directed at an atom, causing an electron to be ejected. The kinetic energy of the ejected electron (E_k) is measured by determining its velocity, v ($E_k = mv^2/2$), and E_i is then calculated using the conservation of energy principle. That is, the energy of the incident light equals E_i plus E_k. What is the ionization energy of selenium atoms in kilojoules per mole if light with $\lambda = 48.2$ nm produces electrons with a velocity of 2.371×10^6 m/s? The mass, m, of an electron is 9.109×10^{-31} kg.

5.133 X rays with a wavelength of 1.54×10^{-10} m are produced when a copper metal target is bombarded with high-energy electrons that have been accelerated by a voltage difference of 30,000 V. The kinetic energy of the electrons equals the product of the voltage difference and the electronic charge in coulombs, where 1 volt-coulomb = 1 J.

   (a) What is the kinetic energy in joules and the de Broglie wavelength in meters of an electron that has been accelerated by a voltage difference of 30,000 V?

   (b) What is the energy in joules of the X rays emitted by the copper target?

5.134 In the Bohr model of atomic structure, electrons are constrained to orbit a nucleus at specific distances, given by the equation

$$r = \frac{n^2 a_0}{Z}$$

where r is the radius of the orbit, Z is the charge on the nucleus, a_0 is the *Bohr radius* and has a value of 5.292×10^{-11} m, and n is a positive integer ($n = 1, 2, 3, \ldots$) like a principal quantum number. Furthermore, Bohr concluded that the energy level E of an electron in a given orbit is

$$E = \frac{-Ze^2}{2r}$$

where e is the charge on an electron. Derive an equation that will let you calculate the difference ΔE between any two energy levels. What relation does your equation have to the Balmer–Rydberg equation?

5.135 Assume that the rules for quantum numbers are different and that the spin quantum number m_s can have any of three values, $m_s = -1/2, 0, +1/2$, while all other rules remain the same.

   (a) Draw an orbital-filling diagram for the element with $Z = 25$, showing the individual electrons in the outermost subshell as up arrows, down arrows, or 0. How many partially filled orbitals does the element have?

   (b) What is the atomic number of the element in the third column of the fourth row under these new rules? What block does it belong to ($s, p, d,$ or f)?

5.136 Given the subshells $1s, 2s, 2p, 3s, 3p,$ and $3d$, identify those that meet the following descriptions:

   (a) Has $l = 2$

   (b) Can have $m_l = -1$

   (c) Is empty in a nitrogen atom

   (d) Is full in a carbon atom

   (e) Contains the outermost electrons in a beryllium atom

   (f) Can contain two electrons, both with spin $m_s = +1/2$

5.137 A hydrogen atom with an electron in the first shell ($n = 1$) absorbs ultraviolet light with a wavelength of 1.03×10^{-7} m. To what shell does the electron jump?

5.138 A minimum energy of 7.21×10^{-19} J is required to produce the photoelectric effect in chromium metal.

(a) What is the minimum frequency of light needed to remove an electron from chromium?

(b) Light with a wavelength of 2.50×10^{-7} m falls on a piece of chromium in an evacuated glass tube. What is the minimum de Broglie wavelength of the emitted electrons? (Note that the energy of the incident light must be conserved; that is, the photon's energy must equal the sum of the energy needed to eject the electron plus the kinetic energy of the electron.)

MULTICONCEPT PROBLEMS

5.139 A photon produced by an X ray machine has an energy of 4.70×10^{-16} J.

(a) What is the frequency of the photon?

(b) What is the wavelength of radiation of frequency (a)?

(c) What is the velocity of an electron with a de Broglie wavelength equal to (b)?

(d) What is the kinetic energy of an electron traveling at velocity (c)?

5.140 An energetically excited hydrogen atom has its electron in a 5*f* subshell. The electron drops down to the 3*d* subshell, releasing a photon in the process.

(a) Give the *n* and *l* quantum numbers for both subshells, and give the range of possible m_l quantum numbers.

(b) What wavelength of light is emitted by the process?

(c) The hydrogen atom now has a single electron in the 3*d* subshell. What is the energy in kJ/mol required to remove this electron?

5.141 Consider the noble gas xenon.

(a) Write the electron configuration of xenon using the abbreviation of the previous noble gas.

(b) When xenon absorbs 801 kJ/mol of energy, it is excited into a higher-energy state in which the outermost electron has been promoted to the next available subshell. Write the electron configuration for this excited xenon.

(c) The energy required to completely remove the outermost electron from the excited xenon atom is 369 kJ/mol, almost identical to that of cesium (376 kJ/mol). Explain.

Ionic Compounds: Periodic Trends and Bonding Theory

Most ionic compounds, such as table salt, are crystalline solids. Surprisingly, the ionic compound in this reactor is a liquid used as a solvent in chemical processes.

 How has an understanding of ionic compounds led to the production of safer solvents?

The answer to this question can be found in the **INQUIRY** ▶▶▶ on page 214.

CONTENTS

Having described the electronic structure of isolated atoms in Chapter 5, let's now extend that description to atoms in chemical compounds. What is the force that holds atoms together in chemical compounds? Certainly there must be *some* force holding atoms together; otherwise, they would simply fly apart and no chemical compounds could exist. As we saw in Section 2.10, the forces that hold atoms together are called chemical bonds and are of two types: covalent bonds and ionic bonds. As a general rule, ionic bonds form primarily between a metal atom and a nonmetal atom, while covalent bonds form primarily between two nonmetal atoms. Electrostatic forces, the attraction of positive and negative charges, are the fundamental reason that atoms come together to form a bond, but the types of charges in ionic and covalent bonds are different. In this and the next two chapters, we'll look at the nature of chemical bonds and at the energy changes that accompany their formation and breakage. We'll begin in the present chapter with a look at ions and the formation of ionic bonds.

6.1 ▶ ELECTRON CONFIGURATIONS OF IONS

REMEMBER...

The **ground-state electron configuration** of an atom or ion is a description of the atomic orbitals that are occupied in the lowest-energy state of the atom or ion. (Section 5.11)

REMEMBER...

According to the **aufbau principle**, lower-energy orbitals fill before higher-energy ones and an orbital can hold only two electrons, which have opposite spins. (Section 5.11)

We've seen on several occasions, particularly during the discussion of redox reactions in Sections 4.10 and 4.11, that metals (left side of the periodic table) tend to give up electrons in their chemical reactions and form cations. Conversely, halogens and some other nonmetals (right side of the table) tend to accept electrons in their chemical reactions and form anions. What are the **ground-state electron configurations** of the resultant ions?

Ions of main-group elements (groups 1A–7A) lose or gain electrons to achieve a noble gas configuration. Atoms or ions with same electron configuration are **isoelectronic**, thus main group ions are isoelectronic to noble gases. The **aufbau principle** (Section 5.11) applies to the formation of ionic compounds: Electrons given up by a metal in forming a cation come from the highest-energy occupied orbital, while the electrons that are accepted by a nonmetal in forming an anion go into the lowest-energy unoccupied orbital. When a sodium atom ($1s^2 2s^2 2p^6 3s^1$) reacts with a chlorine atom and gives up an electron, for example, the valence-shell $3s$ electron of sodium is lost, giving an Na^+ ion with the noble-gas electron configuration of neon ($1s^2 2s^2 2p^6$). At the same time, when the chlorine atom ($1s^2 2s^2 2p^6 3s^2 3p^5$) accepts an electron from sodium, the electron fills the remaining vacancy in the $3p$ subshell to give a Cl^- ion with the noble-gas electron configuration of argon ($1s^2 2s^2 2p^6 3s^2 3p^6$).

$$\text{Na: } 1s^2\, 2s^2\, 2p^6\, 3s^1 \xrightarrow{-e^-} \text{Na}^+\text{: } 1s^2\, 2s^2\, 2p^6 \text{ or } [\text{Ne}]$$

$$\text{Cl: } 1s^2\, 2s^2\, 2p^6\, 3s^2\, 3p^5 \xrightarrow{+e^-} \text{Cl}^-\text{: } 1s^2\, 2s^2\, 2p^6\, 3s^2\, 3p^6 \text{ or } [\text{Ar}]$$

What is true for sodium is also true for the other elements in group 1A: All form positive ions by losing their valence-shell s electron when they undergo reaction, and all the resultant ions have noble-gas electron configurations. Similarly for the elements in group 2A: All form a doubly positive ion when they react, losing both their valence-shell s electrons. An Mg atom ($1s^2 2s^2 2p^6 3s^2$), for example, goes to an Mg^{2+} ion with the neon configuration $1s^2 2s^2 2p^6$ by loss of its two $3s$ electrons.

$$\text{Group 1A atom: } [\text{Noble gas}]\, ns^1 \xrightarrow{-e^-} \text{Group 1A ion}^+\text{: } [\text{Noble gas}]$$

$$\text{Group 2A atom: } [\text{Noble gas}]\, ns^2 \xrightarrow{-2e^-} \text{Group 2A ion}^{2+}\text{: } [\text{Noble gas}]$$

Just as the group 1A and group 2A metals *lose* the appropriate number of electrons to yield ions with noble-gas configurations, the group 6A and group 7A nonmetals *gain* the appropriate number of electrons when they react with metals. The halogens in group 7A gain one electron to form singly charged anions with noble-gas configurations, and the elements in group 6A gain two electrons to form doubly charged anions with noble-gas configurations. Oxygen ($1s^2 2s^2 2p^4$), for example, becomes the O^{2-} ion with the neon configuration ($1s^2 2s^2 2p^6$) when it reacts with a metal:

$$\text{Group 6A atom: } [\text{Noble gas}]\, ns^2\, np^4 \xrightarrow{+2e^-} \text{Group 6A ion}^{2-}\text{: } [\text{Noble gas}]\, ns^2\, np^6$$

$$\text{Group 7A atom: } [\text{Noble gas}]\, ns^2\, np^5 \xrightarrow{+e^-} \text{Group 7A ion}^-\text{: } [\text{Noble gas}]\, ns^2\, np^6$$

TABLE 6.1 **Some Common Main-Group Ions and Their Noble-Gas Electron Configurations**

Group 1A	Group 2A	Group 3A	Group 6A	Group 7A	Electron Configuration
H^+					[None]
H^-					[He]
Li^+	Be^{2+}				[He]
Na^+	Mg^{2+}	Al^{3+}	O^{2-}	F^-	[Ne]
K^+	Ca^{2+}	$^*Ga^{3+}$	S^{2-}	Cl^-	[Ar]
Rb^+	Sr^{2+}	$^*In^{3+}$	Se^{2-}	Br^-	[Kr]
Cs^+	Ba^{2+}	$^*Tl^{3+}$	Te^{2-}	I^-	[Xe]

*These ions don't have a true noble-gas electron configuration because they have an additional filled d subshell.

The formulas and electron configurations of the most common main-group ions are listed in **TABLE 6.1**.

The situation is a bit different for ion formation from the transition-metal elements than it is for the main-group elements. Transition metals react with nonmetals to form cations by first losing their valence-shell s electrons and then losing one or more d electrons. As a result, all the remaining valence electrons in transition-metal cations occupy d orbitals. Iron, for instance, forms the Fe^{2+} ion by losing its two $4s$ electrons and forms the Fe^{3+} ion by losing two $4s$ electrons and one $3d$ electron:

$$Fe: [Ar]\, 4s^2\, 3d^6 \xrightarrow{-2e^-} Fe^{2+}: [Ar]\, 3d^6$$
$$Fe: [Ar]\, 4s^2\, 3d^6 \xrightarrow{-3e^-} Fe^{3+}: [Ar]\, 3d^5$$

It may seem strange that building up the periodic table adds the $3d$ electrons *after* the $4s$ electrons, whereas ion formation from a transition metal removes the $4s$ electrons *before* the $3d$ electrons. Note, though, that the two processes are not the reverse of one another, so they can't be compared directly. Building up the periodic table adds one electron to the valence shell and also adds one positive charge to the nucleus, but ion formation removes an electron from the valence shell without altering the nucleus.

⟶● WORKED EXAMPLE 6.1

Writing Ground-State Electron Configurations for Ions

Predict the ground-state electron configuration for each of the following ions.
(a) Se^{2-} **(b)** Cs^+ **(c)** Cr^{3+}

STRATEGY

First write the electron configuration for the atom. To account for charge of the ion, add electrons for anions and remove electrons for cations. Remember transition metal cations lose outer shell s electrons before d electrons.

SOLUTION

(a) The valence electron configuration for Se is $[Ar]\, 4s^2\, 4p^4$. Add two electrons to form the Se^{2-} ion:

$$Se: [Ar]\, 4s^2\, 4p^4 \xrightarrow{+2e^-} Se^{2-}: [Ar]\, 4s^2\, 4p^6 \text{ or } [Kr]$$

(b) The electron configuration for Cs is $[Xe]\, 6s^1$. Remove one electron to form the Cs^+ ion:

$$Cs: [Xe]\, 6s^1 \xrightarrow{-e^-} Cs^+: [Kr]\, 5s^2\, 5p^6 \text{ or } [Xe]$$

(c) The electron configuration for Cr is $[Ar]\, 4s^1\, 3d^5$. Three electrons must be removed to form Cr^{3+}. First remove one electron from the 4s orbital and then two electrons from the 3d orbitals to form the Cr^{3+} ion: $Cr: [Ar]\, 4s^1\, 3d^5 \xrightarrow{-3e^-} Cr^{3+}: [Ar]\, 3d^3$

▶ **PRACTICE 6.1** Predict the ground-state electron configuration for each of the following ions:

(a) Ra^{2+} **(b)** Ni^{2+} **(c)** N^{3-}

▶ **APPLY 6.2** Which of the following sets of ions are isoelectronic?

(a) F^-, Cl^-, Br^- **(b)** Ti^{4+}, Ca^{2+}, Cl^-
(c) Na^+, Mg^{2+}, Al^{3+}

6.2 ▶ IONIC RADII

Just as there are systematic differences in the **radii of atoms** (Section 5.14), there are also systematic differences in the radii of ions. As shown in **FIGURE 6.1** for the elements of group 1A and 2A, atoms shrink dramatically when an electron is removed to form a cation. The radius of an Na atom, for example, is 186 pm, but that of an Na^+ cation is 102 pm. Similarly, the radius of an Mg atom is 160 pm and that of an Mg^{2+} cation is 72 pm.

The cation that results when an electron is removed from a neutral atom is smaller than the original atom both because the electron is removed from a large, valence-shell orbital and because there is an increase in the **effective nuclear charge**, Z_{eff}, for the remaining electrons (Section 5.10). On going from a neutral Na atom to a charged Na^+ cation, for example, the electron configuration changes from $1s^2 2s^2 2p^6 3s^1$ to $1s^2 2s^2 2p^6$.

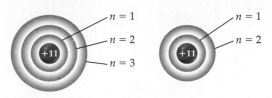

$$Na: 1s^2 2s^2 2p^6 3s^1 \xrightarrow{-e^-} Na^+: 1s^2 2s^2 2p^6 \text{ or } [Ne]$$

The valence shell of the Na *atom* is the *third* shell, but the valence shell of the Na^+ *cation* is the *second* shell. Thus, the Na^+ ion has a smaller valence shell than the Na atom and therefore a smaller size. In addition, the effective nuclear charge felt by the valence-shell electrons is greater in the Na^+ cation than in the neutral atom. The Na atom has 11 protons and 11 electrons, but the Na^+ cation has 11 protons and only 10 electrons. The smaller number of electrons in the cation means that they shield one another to a lesser extent and therefore are pulled in more strongly toward the nucleus.

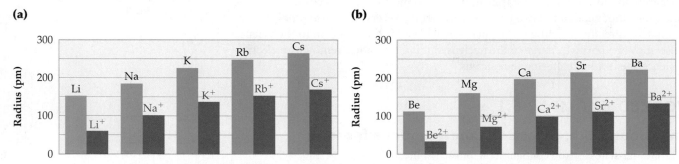

(a)

(b)

Cations are smaller than the corresponding neutral atoms, both because the principal quantum number of the valence-shell electrons is smaller for the cations than it is for the neutral atoms and because Z_{eff} is larger.

▲ **FIGURE 6.1**

Radii of (a) group 1A atoms and their cations; (b) group 2A atoms and their cations.

The same effects felt by the group 1A elements when a single electron is lost are felt by the group 2A elements when two electrons are lost. For example, loss of two valence-shell electrons from an Mg atom $(1s^2\,2s^2\,2p^6\,3s^2)$ gives the Mg^{2+} cation $(1s^2\,2s^2\,2p^6)$. The smaller valence shell of the Mg^{2+} cation and the increase in effective nuclear charge combine to cause a dramatic shrinkage. A similar shrinkage occurs whenever any of the metal atoms on the left-hand two-thirds of the periodic table is converted into a cation.

Just as neutral atoms shrink when converted to cations by loss of one or more electrons, they expand when converted to anions by gain of one or more electrons. As shown in **FIGURE 6.2** for the group 7A elements (halogens), the expansion is dramatic. Chlorine, for example, nearly doubles in radius, from 99 pm for the neutral atom to 184 pm for the chloride anion.

The expansion that occurs when a group 7A atom gains an electron to yield an anion can't be accounted for by a change in the quantum number of the valence shell, because the added electron simply completes an already occupied p subshell. For instance, $[Ne]\,3s^2\,3p^5$ for a Cl atom becomes $[Ne]\,3s^2\,3p^6$ for a Cl^- anion. Thus, the expansion is due entirely to the decrease in effective nuclear charge and the increase in electron–electron repulsions that occurs when an extra electron is added.

$$Cl: [Ne]\,3s^2\,3p^5 \xrightarrow{+e^-} Cl^-: [Ne]\,3s^2\,3p^6$$

◀ **FIGURE 6.2**

Radii of the group 7A atoms (halogens) and their anions.

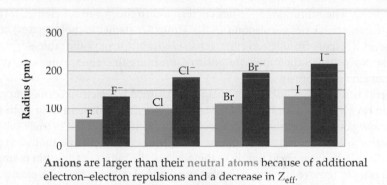

Anions are larger than their neutral atoms because of additional electron–electron repulsions and a decrease in Z_{eff}.

WORKED EXAMPLE 6.2

Predicting Relative Sizes of Ions

Which atom or ion each of the following pairs would you expect to be larger?
(a) S and S^{2-}
(b) Cl^- and I^-
(c) Ni and Ni^{2+}

STRATEGY

For each set, compare the subshell of the valence electrons and the effective nuclear charge (Z_{eff}). Larger size correlates with higher n levels of the valence electrons and smaller Z_{eff}.

SOLUTION

(a) S^{2-} is larger than S because Z_{eff} decreases as the proton-to-electron ratio decreases. The proton-to-electron ratio in S^{2-} $(16p/18e^-)$ is smaller than in S $(16p/16e^-)$. Also, two additional electrons in the $3p$ subshell increase electron–electron repulsions that increase size.

(b) I^- is larger than Cl^-. The valence electrons in I^- are in the $5p$ subshell while the valence electrons in Cl^- are in the $3p$ subshell.

(c) Ni is larger than Ni^{2+} because two electrons are removed from the $4s$ orbital, which is larger than the $3d$ orbital. Z_{eff} also increases for Ni^{2+} because the ratio of protons to electrons increases; Ni $(28p/28e^-)$ and Ni^{2+} $(28p/26e^-)$.

▶ **PRACTICE 6.3** Which atom or ion in each of the following pairs would you expect to be larger?

(a) O or O^{2-}
(b) Fe or Fe^{3+}
(c) H or H^-

▶ **Conceptual APPLY 6.4** Which of the following spheres represents a K^+ ion, which a Ca^{2+} ion, and which a Cl^- ion?

$r = 184$ pm $r = 133$ pm $r = 100$ pm

6.3 ▶ IONIZATION ENERGY

We saw in the previous chapter that the absorption of electromagnetic energy by an atom leads to a change in electron configuration. When energy is added, a valence-shell electron is promoted from a lower-energy orbital to a higher-energy one with a larger principal quantum number n. If enough energy is absorbed, the electron can even be removed completely from the atom, leaving behind a cation. The amount of energy necessary to remove the highest-energy electron from an isolated neutral atom in the gaseous state is called the atom's **ionization energy**, abbreviated E_i. For hydrogen, $E_i = 1312.0$ kJ/mol.

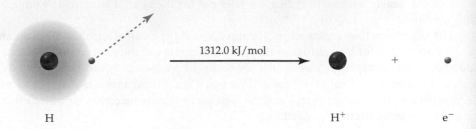

H → 1312.0 kJ/mol → H⁺ + e⁻

As shown by the plot in **FIGURE 6.3**, ionization energies differ widely, from a low of 375.7 kJ/mol for cesium to a high of 2372.3 kJ/mol for helium. Furthermore, the data show a clear periodicity. The minimum E_i values correspond to the group 1A elements (alkali metals), the maximum E_i values correspond to the group 8A elements (noble gases), and a gradual increase in E_i occurs from left to right across a row of the periodic table—from Na to Ar, for example. Note that all the values are positive, meaning that energy must always be added to remove an electron from an atom.

The periodicity evident in Figure 6.3 can be explained by electron configurations. Atoms of the group 8A elements have filled valence subshells, either s for helium or both s and p for the other noble gases. As described in Section 5.14, an electron in a filled valence subshell feels a relatively high Z_{eff} because electrons in the same subshell don't **shield** one another very strongly. As a result, the electrons are held tightly to the nucleus, the radius of the atom is small, and the energy necessary to remove an electron is relatively large. Atoms of group 1A elements, by contrast, have only a single s electron in their valence shell. This single valence electron is shielded from the nucleus by all the inner-shell electrons, called the **core electrons**, resulting in a low Z_{eff}. The valence electron is thus held loosely, and the energy necessary to remove it is relatively small.

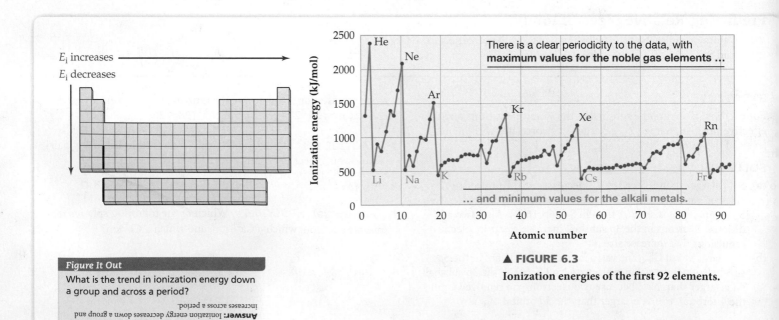

▲ **FIGURE 6.3**

Ionization energies of the first 92 elements.

E_i increases ⟶

E_i decreases

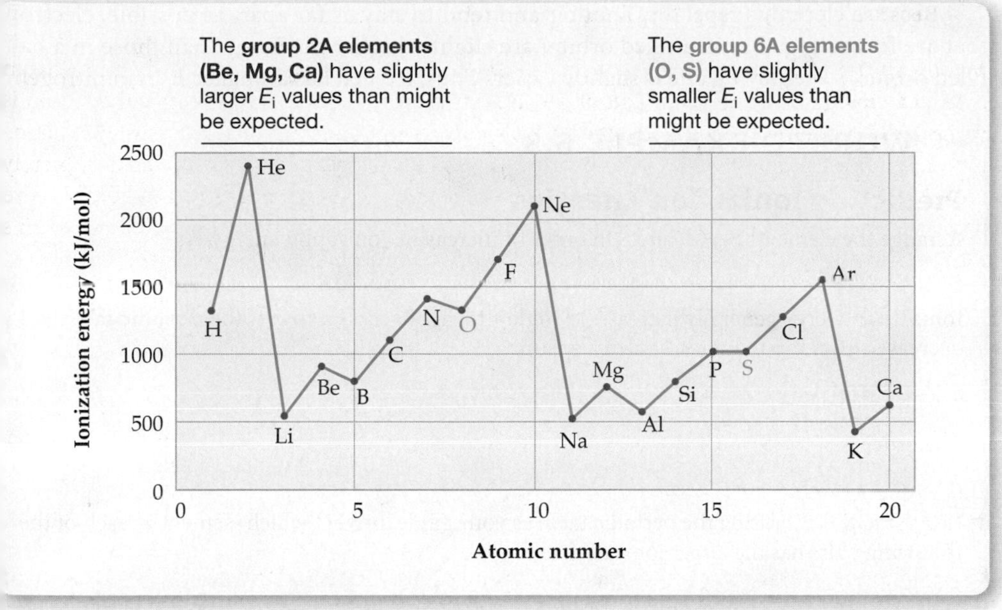

The **group 2A elements (Be, Mg, Ca)** have slightly larger E_i values than might be expected.

The **group 6A elements (O, S)** have slightly smaller E_i values than might be expected.

The plot of ionization energies in Figure 6.3 shows other trends in the data beyond the obvious periodicity. One such trend is that ionization energies gradually decrease going down a group in the periodic table, from He to Rn and from Li to Fr, for instance. As the atomic number increases going down a group, both the principal quantum number of the valence-shell electrons and their average distance from the nucleus also increase. As a result, the valence-shell electrons are less tightly held and E_i is smaller.

Yet another point about the E_i data is that minor irregularities occur across a row of the periodic table. A close look at E_i values of the first 20 elements (**FIGURE 6.4**) shows that the E_i of beryllium is larger than that of its neighbor boron and the E_i of nitrogen is larger than that of its neighbor oxygen. Similarly, magnesium has a larger E_i than aluminum and phosphorus has a slightly larger E_i than sulfur.

The slightly enlarged E_i values for the group 2A elements Be, Mg, and others can be explained by their electron configurations. Compare beryllium with boron, for instance. A $2s$ electron is removed on ionization of beryllium, but a $2p$ electron is removed on ionization of boron:

s electron removed

$$\text{Be } (1s^2\, 2s^2) \longrightarrow \text{Be}^+ (1s^2\, 2s^1) + e^- \qquad E_i = 899.4 \text{ kJ/mol}$$

p electron removed

$$\text{B } (1s^2\, 2s^2\, 2p^1) \longrightarrow \text{B}^+ (1s^2\, 2s^2) + e^- \qquad E_i = 800.6 \text{ kJ/mol}$$

Because a $2s$ electron spends more time closer to the nucleus than a $2p$ electron, it is held more tightly and is harder to remove. Thus, the E_i of beryllium is larger than that of boron. Put another way, the $2p$ electron of boron is shielded somewhat by the $2s$ electrons and is thus more easily removed than a $2s$ electron of beryllium.

The lowered E_i values for atoms of group 6A elements can be explained by their electron configurations as well. Comparing nitrogen with oxygen, for instance, the nitrogen electron is removed from a half-filled orbital, whereas the oxygen electron is removed from a filled orbital:

Half-filled orbital

$$\text{N } (1s^2\, 2s^2\, 2p_x^{\,1}\, 2p_y^{\,1}\, 2p_z^{\,1}) \longrightarrow \text{N}^+ (1s^2\, 2s^2\, 2p_x^{\,1}\, 2p_y^{\,1}) + e^- \qquad E_i = 1402.3 \text{ kJ/mol}$$

Filled orbital

$$\text{O } (1s^2\, 2s^2\, 2p_x^{\,2}\, 2p_y^{\,1}\, 2p_z^{\,1}) \longrightarrow \text{O}^+ (1s^2\, 2s^2\, 2p_x^{\,1}\, 2p_y^{\,1}\, 2p_z^{\,1}) + e^- \qquad E_i = 1313.9 \text{ kJ/mol}$$

Because electrons repel one another and tend to stay as far apart as possible, electron that are forced together in a filled orbital are slightly higher in energy than those in a half filled orbital, so removing one is slightly easier. Thus, oxygen has a smaller E_i than nitrogen.

WORKED EXAMPLE 6.3

Predicting Ionization Energies

Arrange the elements Se, Cl, and S in order of increasing ionization energy.

STRATEGY

Ionization energy generally increases from left to right across a row of the periodic table and decreases from top to bottom down a group.

SOLUTION

The order is Se < S < Cl.

▶ **PRACTICE 6.5** Using the periodic table as your guide, predict which element in each of the following pairs has the larger ionization energy:

(a) K or Br **(b)** S or Te **(c)** Ga or Se **(d)** Ne or Sr

▶ Conceptual **APPLY 6.6** Given the orbital filling diagrams on the left for the valence electrons of elements, rank them from lowest to highest ionization energy.

(a) $\underset{3s}{\uparrow\downarrow}$ $\quad$ $\underset{}{\uparrow}$ $\underset{3p}{\uparrow}$ $\underset{}{\uparrow}$

(b) $\underset{3s}{\uparrow\downarrow}$ $\quad$ $\underset{}{\uparrow\downarrow}$ $\underset{3p}{\uparrow}$ $\underset{}{\uparrow}$

(c) $\underset{5s}{\uparrow\downarrow}$ $\quad$ $\underset{}{}$ $\underset{5p}{}$ $\underset{}{}$

(d) $\underset{2s}{\uparrow\downarrow}$ $\quad$ $\underset{}{\uparrow\downarrow}$ $\underset{2p}{\uparrow\downarrow}$ $\underset{}{\uparrow}$

6.4 ▶ HIGHER IONIZATION ENERGIES

Ionization is not limited to the loss of a single electron from an atom. Two, three, or even more electrons can be lost sequentially from an atom, and the amount of energy associated with each step can be measured.

$$M + \text{Energy} \longrightarrow M^+ + e^- \qquad \text{First ionization energy } (E_{i1})$$
$$M^+ + \text{Energy} \longrightarrow M^{2+} + e^- \qquad \text{Second ionization energy } (E_{i2})$$
$$M^{2+} + \text{Energy} \longrightarrow M^{3+} + e^- \qquad \text{Third ionization energy } (E_{i3})$$
$$\ldots \quad \text{and so forth}$$

Successively larger amounts of energy are required for each ionization step because it is much harder to pull a negatively charged electron away from a positively charged ion than from a neutral atom. Interestingly, though, the energy differences between successive steps vary dramatically from one element to another. Removing the second electron from sodium takes nearly 10 times as much energy as removing the first one (4562 versus 496 kJ/mol), but removing the second electron from magnesium takes only twice as much energy as removing the first one (1451 versus 738 kJ/mol).

Large jumps in successive ionization energies are also found for other elements, as is indicated by the zigzag line in **TABLE 6.2**. Magnesium has a large jump between its second and third ionization energies, aluminum has a large jump between its third and fourth ionization energies, silicon has a large jump between its fourth and fifth ionization energies, and so on.

The large increases in ionization energies highlighted by the zigzag line in Table 6.2 can be understood by examining electron configurations.

Let's first examine sodium. The equations representing the first and second ionization energies in sodium are:

$$Na\,(1s^22s^22p^63s^1) + \text{Energy} \longrightarrow Na^+(1s^22s^22p^6) + e^- \qquad \text{First ionization energy } (E_{i1})$$
$$Na^+(1s^22s^22p^6) + \text{Energy} \longrightarrow Na^{2+}(1s^22s^22p^5) + e^- \qquad \text{Second ionization energy } (E_{i2})$$

In sodium, the large jump in ionization energy between E_{i1} and E_{i2}, can be attributed to the difference in energy required to remove an electron from the 3s subshell and the 2p subshell. The 3s subshell is further from the nucleus and more shielded than the

TABLE 6.2	Higher Ionization Energies (kJ/mol) for Main-Group Third-Row Elements							
Group	1A	2A	3A	4A	5A	6A	7A	8A
E_i Number	Na	Mg	Al	Si	P	S	Cl	Ar
E_{i1}	496	738	578	787	1,012	1,000	1,251	1,520
E_{i2}	4,562	1,451	1,817	1,577	1,903	2,251	2,297	2,665
E_{i3}	6,912	7,733	2,745	3,231	2,912	3,361	3,822	3,931
E_{i4}	9,543	10,540	11,575	4,356	4,956	4,564	5,158	5,770
E_{i5}	13,353	13,630	14,830	16,091	6,273	7,013	6,540	7,238
E_{i6}	16,610	17,995	18,376	19,784	22,233	8,495	9,458	8,781
E_{i7}	20,114	21,703	23,293	23,783	25,397	27,106	11,020	11,995

The zigzag line marks the large jumps in ionization energies.

p subshell, thus the first ionization requires less energy. In other words, Z_{eff} is lower for an electron in a $3s$ orbital compared to a $2p$ orbital.

Now let's examine magnesium, which has a large jump between E_{i2} and E_{i3}. The equations representing the first three ionization energies of magnesium are:

$$Mg\,(1s^2\,2s^2\,2p^6\,3s^2) + \text{Energy} \longrightarrow Mg^+(1s^2\,2s^2\,2p^6\,3s^1) + e^- \quad \text{First ionization energy } (E_{i1})$$

$$Mg^+(1s^2\,2s^2\,2p^6\,3s^1) + \text{Energy} \longrightarrow Mg^{2+}(1s^2\,2s^2\,2p^6) + e^- \quad \text{Second ionization energy } (E_{i2})$$

$$Mg^{2+}(1s^2\,2s^2\,2p^6) + \text{Energy} \longrightarrow Mg^{3+}(1s^2\,2s^2\,2p^5) + e^- \quad \text{Third ionization energy } (E_{i3})$$

In a similar manner to sodium, the large increase in ionization energy arises when the electron must be removed from inner core electrons ($2p$) in the third ionization. It's relatively easier to remove an electron from a *partially* filled valence shell because Z_{eff} is lower, but it's relatively harder to remove an electron from a *filled* valence shell because Z_{eff} is higher. In other words, ions formed by reaction of main-group elements usually have filled s and p subshells (a noble-gas electron configuration), which corresponds to having eight electrons (an *octet*) in the valence shell of an atom or ion. Sodium ([Ne] $3s^1$) loses only one electron easily, magnesium ([Ne] $3s^2$) loses only two electrons easily, aluminum ([Ne] $3s^2\,3p^1$) loses only three electrons easily, and so on across the row.

8 electrons in
outer (2nd) shell

$$Na\,(1s^2\,2s^2\,2p^6\,3s^1) \longrightarrow Na^+(1s^2\,2s^2\,2p^6) + e^-$$

$$Mg\,(1s^2\,2s^2\,2p^6\,3s^2) \longrightarrow Mg^{2+}(1s^2\,2s^2\,2p^6) + 2\,e^-$$

$$Al\,(1s^2\,2s^2\,2p^6\,3s^2\,3p^1) \longrightarrow Al^{3+}(1s^2\,2s^2\,2p^6) + 3\,e^-$$

$$\vdots \qquad\qquad\qquad \vdots$$

$$Cl\,(1s^2\,2s^2\,2p^6\,3s^2\,3p^5) \longrightarrow Cl^{7+}(1s^2\,2s^2\,2p^6) + 7\,e^-$$

WORKED EXAMPLE 6.4

Higher Ionization Energies

Which has the larger fifth ionization energy, Ge or As?

STRATEGY

Look at their positions in the periodic table and write the electron configuration. If the fifth electron must be removed from an inner shell, then that element will have the larger ionization energy.

SOLUTION

Ge: $[Ar]\, 4s^2\, 3d^{10}\, 4p^2$ As: $[Ar]\, 4s^2\, 3d^{10}\, 4p^3$

The group 4A element germanium has four valence-shell electrons and thus has four relatively low ionization energies, whereas the group 5A element arsenic has five valence-shell electrons and has five low ionization energies. Germanium has a larger E_{i5} than arsenic because the fifth electron to be removed in Ge occupies a lower subshell ($n = 3$).

▶ **PRACTICE 6.7**

(a) Which has the larger third ionization energy, Be or N?
(b) Which has the larger fourth ionization energy, Ga or Ge?

▶ **Conceptual** **APPLY 6.8** The figure on the right represents the successive ionization energies of an atom in the third period of the periodic table. Which atom is this most likely to be?

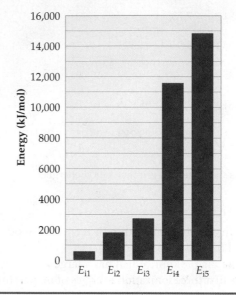

6.5 ▶ ELECTRON AFFINITY

Just as it's possible to measure the energy change on *removing* an electron from an atom to form a *cation*, it's also possible to measure the energy change on *adding* an electron to an atom to form an *anion*. An element's **electron affinity** (E_{ea}) is the energy change that occurs when an electron is added to an isolated atom in the gaseous state.

Ionization energies (Section 6.3) are always positive because energy must always be added to separate a negatively charged electron from the resultant positively charged cation. Electron affinities, however, are generally negative because energy is usually released when a neutral atom adds an additional electron.* We'll see in Chapter 9 that this same convention is used throughout chemistry: A positive energy change means that energy is added, and a negative energy change means that energy is released.

The more negative the E_{ea}, the greater the tendency of the atom to accept an electron and the more stable the anion that results. In contrast, an atom that forms an unstable anion by addition of an electron has, in principle, a positive value of E_{ea}, but no experimental measurement can be made because the process does not take place. All we can say is that the E_{ea} for such an atom is greater than zero. The E_{ea} of hydrogen, for instance, is $-72.8\ kJ/mol$, meaning that energy is released and the H$^-$ anion is stable. The E_{ea} of neon, however, is greater than 0 kJ/mol, meaning that Ne does not add an electron and the Ne$^-$ anion is not stable.

$$H\,(1s^1) + e^- \longrightarrow H^-\,(1s^2) + 72.8\ kJ/mol \qquad E_{ea} = -72.8\ kJ/mol$$
$$Ne\,(1s^2\,2s^2\,2p^6) + e^- + Energy \longrightarrow Ne^-\,(1s^2\,2s^2\,2p^6\,3s^1) \quad E_{ea} > 0\ kJ{>}mol$$

---

*We have defined E_{ea} as the energy *released* when a neutral atom *gains* an electron to form an anion and have given it a negative sign. Some books and reference sources adopt the opposite point of view, defining E_{ea} as the energy *gained* when an anion *loses* an electron to form a neutral atom and giving it a positive value. The two definitions are simply the reverse of one another, so the sign of the energy change is also reversed.

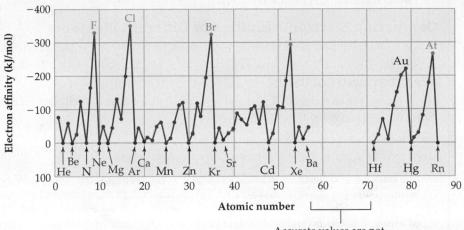

A negative value for E_{ea}, such as those for the group 7A elements (halogens), means that energy is released when an electron adds to an atom.

A value of zero, such as those for the group 2A elements (alkaline earths) and group 8A elements (noble gases), means that energy is absorbed but the exact amount can't be measured.

Accurate values are not known for elements 58–71.

▲ **FIGURE 6.5**

Electron affinities for elements 1–57 and 72–86.

Figure It Out

Which group in the periodic table has the most negative electron affinities? What does a negative electron affinity mean?

Answer: The large negative E_{ea} for the halogens indicates that they form stable anions with a large release of energy.

As with ionization energies, electron affinities show a periodicity that is related to the electron configurations of the elements. The data in **FIGURE 6.5** indicate that group 7A elements have the most negative electron affinities, corresponding to the largest release of energy, while group 2A and group 8A elements have near-zero or positive electron affinities, corresponding to a small release or even an absorption of energy.

The value of an element's electron affinity is due to an interplay of several offsetting factors. Attraction between the additional electron and the nucleus favors a negative E_{ea}, but the increase in electron–electron repulsions that results from addition of the extra electron favors a positive E_{ea}.

Large negative E_{ea}'s are found for the halogens (F, Cl, Br, I) because each of these elements has both a high Z_{eff} and room in its valence shell for an additional electron. *Halide ions*, halogens with a negative charge such as fluoride (F^-), have a noble-gas electron configuration with filled s and p sublevels, and the attraction between the additional electron and the atomic nucleus is high. Positive E_{ea}'s are found for the noble-gas elements (He, Ne, Ar, Kr, Xe), however, because the s and p sublevels in these elements are already full, so the additional electron must go into the next higher shell, where it is shielded from the nucleus and feels a relatively low Z_{eff}. The attraction of the nucleus for the added electron is therefore small and is outweighed by the additional electron–electron repulsions.

A halogen: $Cl\ (\ldots 3s^2 3p^5) + e^- \longrightarrow Cl^-(\ldots 3s^2 3p^6) \qquad E_{ea} = -348.6\ kJ/mol$

A noble gas: $Ar\ (\ldots 3s^2 3p^6) + e^- \longrightarrow Ar^-\ (\ldots 3s^2 3p^6 4s^1) \quad E_{ea} > 0\ kJ/mol$

In looking for other trends in the data of Figure 6.5, the near-zero E_{ea}'s of the alkaline earth metals (Be, Mg, Ca, Sr, Ba) are particularly striking. Atoms of these elements have filled s subshells, which means that the additional electron must go into a p subshell. The higher energy of the p subshell, together with a relatively low Z_{eff} for elements on the left side of the periodic table, means that alkaline earth atoms accept an electron reluctantly and have E_{ea} values near zero.

An alkaline earth metal: $Mg(\ldots 3s^2) + e^- \longrightarrow Mg^-(\ldots 3s^2 3p^1) \quad E_{ea} \approx 0\ kJ/mol$

•—— WORKED EXAMPLE 6.5

Comparing Electron Affinities of Different Elements

Why does nitrogen have a less favorable (more positive) E_{ea} than its neighbors on either side, C and O?

STRATEGY AND SOLUTION

The magnitude of an element's E_{ea} depends on the element's valence-shell electron configuration. The electron configurations of C, N, and O are

Carbon: $1s^2\, 2s^2\, 2p_x^1\, 2p_y^1$ Nitrogen: $1s^2\, 2s^2\, 2p_x^1\, 2p_y^1\, 2p_z^1$

Oxygen: $1s^2\, 2s^2\, 2p_x^2\, 2p_y^1\, 2p_z^1$

Carbon has only two electrons in its $2p$ subshell and can readily accept another in its vacant $2p_z$ orbital. Nitrogen, however, has a half-filled $2p$ subshell, so the additional electron must pair up in a $2p$ orbital where it feels a repulsion from the electron already present. Thus, the E_{ea} of nitrogen is less favorable than that of carbon. Oxygen also must add an electron to an orbital that already has one electron, but the additional stabilizing effect of increased Z_{eff} across the periodic table counteracts the effect of electron repulsion, resulting in a more favorable E_{ea} for O than for N.

▶ **PRACTICE 6.9**   Why does manganese, atomic number 25, have a less favorable E_{ea} than its neighbors on either side?

▶ **Conceptual APPLY 6.10**   Which of the indicated three elements has the least favorable E_{ea}, and which has the most favorable E_{ea}?

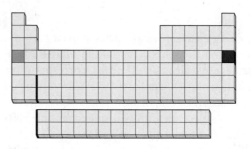

6.6 ▶ THE OCTET RULE

Let's list the important points discussed in the previous four sections and see if we can draw some general conclusions:

- **Group 1A elements have a relatively low E_{i1},** so they tend to lose their ns^1 valence-shell electron easily when they react, thereby adopting the electron configuration of the noble gas in the previous row of the periodic table.

- **Group 2A elements have relatively low E_{i1} and E_{i2},** so they tend to lose both their ns^2 valence-shell electrons easily when they react and adopt a noble-gas electron configuration.

- **Group 7A elements have a relatively large negative E_{ea},** so they tend to gain one electron easily when they react, changing from $ns^2\, np^5$ to $ns^2\, np^6$ and thereby adopting the configuration of the neighboring noble gas in the same row.

- **Group 8A (noble gas) elements are essentially inert** and undergo very few reactions. They neither gain nor lose electrons easily.

All these observations can be gathered into a single statement called the **octet rule**:

> **Octet rule**   Main-group elements tend to undergo reactions that leave them with eight outer-shell electrons. That is, main-group elements react so that they attain a noble-gas electron configuration with filled s and p sublevels in their valence electron shell.

As we'll see in the next chapter, there are exceptions to the octet rule, particularly for elements in the third and lower rows of the periodic table. Nevertheless, the rule is useful for making predictions and for providing insights about chemical bonding.

Why does the octet rule work? What factors determine how many electrons an atom is likely to gain or lose? Clearly, electrons are most likely to be lost if they are held loosely in the first place—that is, if they feel a relatively low effective nuclear charge, Z_{eff}, and have lower ionization energies. Valence-shell electrons in the group 1A, 2A, and 3A metals, for instance, are shielded from the nucleus by core electrons, feel a low Z_{eff}, and are therefore lost relatively easily. Once the next lower noble-gas configuration is reached, though, loss of an additional electron suddenly becomes much more difficult because it must come from an inner shell, where it feels a much higher Z_{eff}.

Conversely, electrons are most likely to be gained if they can be held tightly by a high Z_{eff}. Valence-shell electrons in the group 6A and 7A elements, for example, are poorly shielded, feel high values of Z_{eff}, and aren't lost easily. The high Z_{eff} thus makes possible the gain of one or more additional electrons into vacant valence-shell orbitals. Once the noble-gas configuration is reached, though, lower-energy orbitals are no longer available. An additional electron would have to be placed in a higher-energy orbital, where it would feel only a low Z_{eff}.

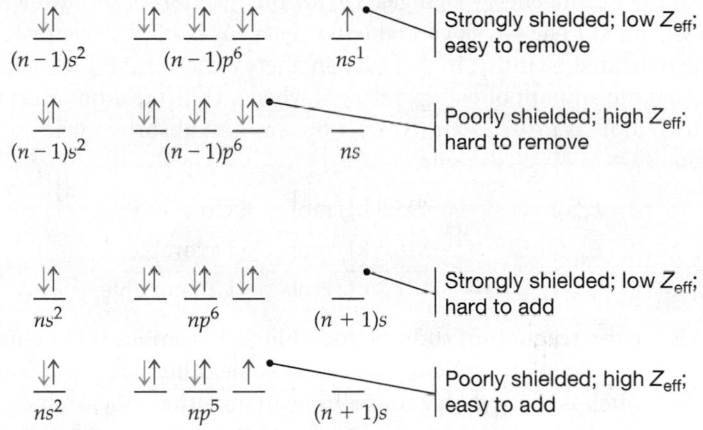

Eight is therefore the magic number for valence-shell electrons. Taking electrons *from* a filled octet is difficult because they are tightly held by a high Z_{eff}; adding more electrons *to* a filled octet is difficult because, with s and p sublevels full, no low-energy orbital is available.

WORKED EXAMPLE 6.6

Chemical Reactions and the Octet Rule

Lithium metal reacts with nitrogen to yield Li_3N. What noble-gas configuration does the nitrogen atom in Li_3N have?

STRATEGY AND SOLUTION

The compound Li_3N contains three Li^+ ions, each formed by loss of a $2s$ electron from lithium metal (group 1A). The nitrogen atom in Li_3N must therefore gain three electrons over the neutral atom, making it triply negative (N^{3-}) and giving it a valence-shell octet with the neon configuration:

$$N: (1s^2\, 2s^2\, 2p^3) \xrightarrow{\ +3e^-\ } N^{3-}:(1s^2\, 2s^2\, 2p^6) \text{ or } [Ne]$$

continued on next page

▶ **PRACTICE 6.11** What noble-gas configurations are the following elements likely to adopt in reactions when they form ions?

(a) Rb **(b)** Ba **(c)** Ga **(d)** F

▶ **APPLY 6.12** What are group 6A elements likely to do when they form ions—gain electrons or lose them? How many? What will be the charge on the ion of a 6A element?

6.7 ▶ IONIC BONDS AND THE FORMATION OF IONIC SOLIDS

An **ionic bond** occurs when one atom transfers an electron to another, creating an electrostatic attraction between positively charged cations and negatively charged anions. A familiar example of an ionic compound is table salt, NaCl, consisting of Na^+ and Cl^- ions. Sodium metal directly reacts with chlorine gas to form ions because sodium gives up an electron relatively easily (that is, has a small positive ionization energy) and chlorine accepts an electron easily (that is, has a large negative electron affinity). In general, an element with a small E_i can transfer an electron to an element with the negative E_{ea}, yielding a cation and an anion. The electron transfer from sodium to chlorine to form ions is:

$$\underset{1s^2\,2s^2\,2p^6\,3s^1}{Na} + \underset{1s^2\,2s^2\,2p^6\,3s^2\,3p^5}{Cl} \longrightarrow \underset{1s^2\,2s^2\,2p^6}{Na^+}\ \underset{1s^2\,2s^2\,2p^6\,3s^2\,3p^6}{Cl^-}$$

What about the overall energy change, ΔE, for the reaction of sodium with chlorine to yield Na^+ and Cl^- ions? (The Greek capital letter delta, Δ, is used to represent a change in the value of the indicated quantity, in this case an energy change ΔE.) It's apparent from E_i and E_{ea} values that the amount of energy released when a chlorine atom accepts an electron ($E_{ea} = -348.6$ kJ/mol) is insufficient to offset the amount absorbed when a sodium atom loses an electron ($E_i = +495.8$ kJ/mol):

E_i for Na	$= +495.8$ kJ/mol	(Unfavorable)
E_{ea} for Cl	$= -348.6$ kJ/mol	(Favorable)
ΔE	$= +147.2$ kJ/mol	(Unfavorable)

REMEMBER...

In an **ionic solid**, oppositely charged ions are attracted to one another by ionic bonds and are packed together in a regular way. (Section 2.11)

The net ΔE for the reaction of sodium and chlorine atoms would be unfavorable by +147.2 kJ/mol, and no reaction would occur, unless some other factors were involved. This additional factor, which is more than enough to overcome the unfavorable energy change of electron transfer, is the large gain in stability due to the electrostatic attractions between product anions and cations in the formation of an **ionic solid**.

The actual reaction of solid sodium metal with gaseous chlorine molecules to form solid sodium chloride occurs all at once rather than in a stepwise manner, but it's easier to make an energy calculation if we imagine a series of hypothetical steps for which exact energy changes can be measured experimentally. There are five steps to take into account to calculate the overall energy change.

Step ① Solid Na metal is first converted into isolated, gaseous Na atoms, a process called *sublimation*. Because energy must be added to disrupt the forces holding atoms together in a solid, the heat of sublimation has a positive value: +107.3 kJ/mol for Na.

$$Na(s) \longrightarrow Na(g)$$
$$+107.3 \text{ kJ/mol}$$

Step ② Gaseous Cl_2 molecules are split into individual Cl atoms. Energy must be added to break molecules apart, and the energy required for bond breaking therefore has a positive value: +243 kJ/mol for Cl_2 (or 122 kJ/mol for $1/2\ Cl_2$). We'll look further into bond dissociation energies in Section 7.2.

$$1/2\ Cl_2(g) \longrightarrow Cl(g)$$
$$+122 \text{ kJ/mol}$$

Step ③ Isolated Na atoms are ionized into Na^+ ions plus electrons. The energy required is the first ionization energy of sodium (E_{i1}) and has a positive value: +495.8 kJ/mol.

$$Na(g) \longrightarrow Na^+(g) + e^-$$
$$+495.8 \text{ kJ/mol}$$

Step ④ Cl^- ions are formed from Cl atoms by addition of an electron. The energy released is the electron affinity of chlorine (E_{ea}) and has a negative value: −348.6 kJ/mol.

$$Cl(g) + e^- \longrightarrow Cl^-(g)$$
$$-348.6 \text{ kJ/mol}$$

Step ⑤ Lastly, solid NaCl is formed from isolated gaseous Na^+ and Cl^- ions. The energy change is a measure of the overall electrostatic interactions between ions in the solid. It is the amount of energy released when isolated ions condense to form a solid, and it has a negative value: −787 kJ/mol for NaCl.

$$Na^+(g) + Cl^-(g) \longrightarrow NaCl(s)$$
$$-787 \text{ kJ/mol}$$

Net reaction:

Net energy change (ΔE):

$$Na(s) + 1/2Cl_2(g) \longrightarrow NaCl(s)$$
$$-411 \text{ kJ/mol}$$

The five hypothetical steps in the reaction between sodium metal and gaseous chlorine are depicted in **FIGURE 6.6** in a pictorial format called a **Born–Haber cycle**, which shows

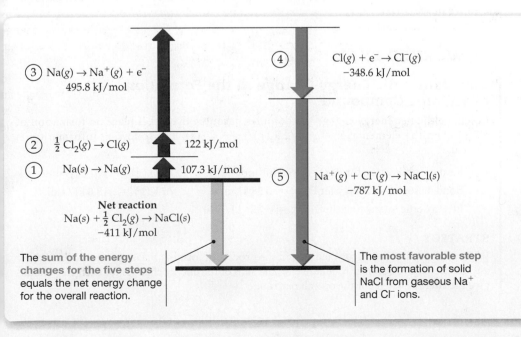

◀ **FIGURE 6.6**

A Born–Haber cycle for the formation of NaCl(s) from Na(s) and $Cl_2(g)$.

Figure It Out

(a) Which step in the Born–Haber cycle releases the largest amount of energy?

(b) Why is this step considered favorable in the formation of ionic compounds from their elements?

Answer: (a) Step 5, the electrostatic attractions between gaseous Na^+ and Cl^- ions when they form solid NaCl. **(b)** If a reaction releases energy, the products are in a lower energy state and are more stable.

③ $Na(g) \rightarrow Na^+(g) + e^-$
495.8 kJ/mol

④ $Cl(g) + e^- \rightarrow Cl^-(g)$
−348.6 kJ/mol

② $\frac{1}{2} Cl_2(g) \rightarrow Cl(g)$ 122 kJ/mol

① $Na(s) \rightarrow Na(g)$ 107.3 kJ/mol

⑤ $Na^+(g) + Cl^-(g) \rightarrow NaCl(s)$
−787 kJ/mol

Net reaction
$Na(s) + \frac{1}{2} Cl_2(g) \rightarrow NaCl(s)$
−411 kJ/mol

The **sum of the energy changes** for the five steps equals the net energy change for the overall reaction.

The **most favorable step** is the formation of solid NaCl from gaseous Na^+ and Cl^- ions.

▶ **FIGURE 6.7**

A Born–Haber cycle for the formation of MgCl₂ from the elements.

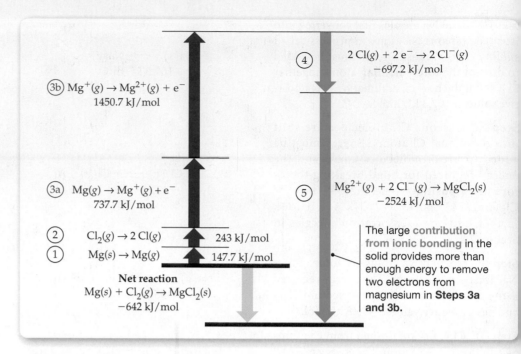

④ $2\,\text{Cl}(g) + 2\,\text{e}^- \rightarrow 2\,\text{Cl}^-(g)$
$-697.2\ \text{kJ/mol}$

③b $\text{Mg}^+(g) \rightarrow \text{Mg}^{2+}(g) + \text{e}^-$
$1450.7\ \text{kJ/mol}$

③a $\text{Mg}(g) \rightarrow \text{Mg}^+(g) + \text{e}^-$
$737.7\ \text{kJ/mol}$

⑤ $\text{Mg}^{2+}(g) + 2\,\text{Cl}^-(g) \rightarrow \text{MgCl}_2(s)$
$-2524\ \text{kJ/mol}$

② $\text{Cl}_2(g) \rightarrow 2\,\text{Cl}(g)$ $243\ \text{kJ/mol}$
① $\text{Mg}(s) \rightarrow \text{Mg}(g)$ $147.7\ \text{kJ/mol}$

Net reaction
$\text{Mg}(s) + \text{Cl}_2(g) \rightarrow \text{MgCl}_2(s)$
$-642\ \text{kJ/mol}$

The large **contribution from ionic bonding** in the solid provides more than enough energy to remove two electrons from magnesium in **Steps 3a and 3b.**

how each step contributes to the overall energy change and how the net process is the sum of the individual steps. As indicated in the diagram, steps 1, 2, and 3 have positive values and absorb energy, while steps 4 and 5 have negative values and release energy. The largest contribution is step 5, which measures the electrostatic forces between ions in the solid product—that is, the strength of the ionic bonding. Were it not for this large amount of stabilization of the solid due to ionic bonding, no reaction would take place.

A similar Born–Haber cycle for the reaction of magnesium with chlorine shows the energy changes involved in the reaction of an alkaline earth element (**FIGURE 6.7**). As in the reaction of sodium and chlorine to form NaCl, there are five contributions to the overall energy change. First, solid magnesium metal must be converted into isolated gaseous magnesium atoms (sublimation). Second, the bond in Cl_2 molecules must be broken to yield chlorine atoms. Third, the magnesium atoms must lose two electrons to form Mg^{2+} ions. Fourth, the chlorine atoms formed in step 2 must accept electrons to form Cl^- ions. Fifth, the gaseous ions must combine to form the ionic solid, MgCl_2. As the Born–Haber cycle indicates, it is the large contribution from ionic bonding that releases enough energy to drive the entire process.

WORKED EXAMPLE 6.7

Calculating the Energy Change in the Formation of an Ionic Compound

Calculate the net energy change in kilojoules per mole that takes place on formation of KF(s) from the elements: $\text{K}(s) + 1/2\,\text{F}_2(g) \longrightarrow \text{KF}(s)$. The following information is needed:

Heat of sublimation for $\text{K}(s) = 89.2\ \text{kJ/mol}$	E_{ea} for $\text{F}(g) = -328\ \text{kJ/mol}$
Bond dissociation energy for $\text{F}_2(g) = 158\ \text{kJ/mol}$	E_{i} for $\text{K}(g) = 418.8\ \text{kJ/mol}$
Electrostatic interactions in $\text{KF}(s) = -821\ \text{kJ/mol}$	

STRATEGY

Write a chemical reaction for each step in the process of forming the ionic solid KF and draw a Born–Haber cycle similar to Figure 6.6. The sum of the energy changes for each step should give the energy change for the overall reaction.

SOLUTION

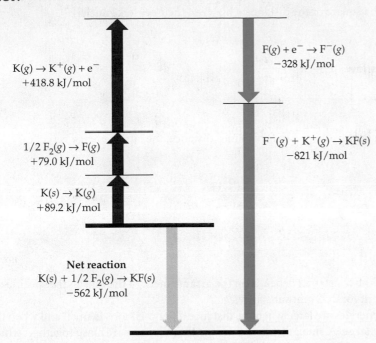

$K(g) \rightarrow K^+(g) + e^-$
$+418.8$ kJ/mol

$1/2\ F_2(g) \rightarrow F(g)$
$+79.0$ kJ/mol

$K(s) \rightarrow K(g)$
$+89.2$ kJ/mol

$F(g) + e^- \rightarrow F^-(g)$
-328 kJ/mol

$F^-(g) + K^+(g) \rightarrow KF(s)$
-821 kJ/mol

Net reaction
$K(s) + 1/2\ F_2(g) \rightarrow KF(s)$
-562 kJ/mol

▶ **PRACTICE 6.13** Calculate the net energy change in kilojoules per mole that takes place on formation of $MgF_2(s)$ from the elements: $Mg(s) + F_2(g) \longrightarrow MgF_2(s)$. The following information is needed:

Heat of sublimation for $Mg(s) = 147.7$ kJ/mol

Bond dissociation energy for $F_2(g) = 158$ kJ/mol

Electrostatic interactions in $MgF_2(s) = -2957$ kJ/mol

E_{ea} for $F(g) = -328$ kJ/mol

E_{i1} for $Mg(g) = 737.7$ kJ/mol

E_{i2} for $Mg(g) = 1450.7$ kJ/mol

▶ **APPLY 6.14** Calculate the energy of electrostatic attractions of $LiCl(s)$. The following information is needed.

Heat of sublimation for $Li(s) = 161$ kJ/mol

Bond dissociation energy for $Cl_2(g) = 243$ kJ/mol

$Li(s) + 1/2\ Cl_2(g) \longrightarrow LiCl(s) = -409$ kJ/mol

E_{ea} for $Cl(g) = -349$ kJ/mol

E_i for $Li(g) = 520$ kJ/mol

6.8 ▶ LATTICE ENERGIES IN IONIC SOLIDS

The measure of the electrostatic interaction energies between ions in a solid—and thus the measure of the strength of the solid's ionic bonds—is called the **lattice energy (U)**. By convention, lattice energy is defined as the amount of energy that must be added to break up an ionic solid into its individual gaseous ions, so it has a positive value. The formation of a solid from ions is the reverse of the breakup, and thus has a negative value; for example, in step 5 in the Born–Haber cycle illustrated in Figure 6.6, the formation of NaCl has a negative value, $-U$:

$NaCl(s) \longrightarrow Na^+(g) + Cl^-(g)$ $U = +787$ kJ/mol   (Energy absorbed)

$Na^+(g) + Cl^-(g) \longrightarrow NaCl(s)$ $-U = -787$ kJ/mol   (Energy released)

What factors affect the magnitude of the lattice energy in an ionic compound? The lattice energy depends on the strength of the ionic bond between the cations and anions in the ionic compound. **Coulomb's law** describes the force (F) that results from the interaction of electric charges and is equal to a constant k times the product of the charges on the ions, z_1 and z_2, divided by the square of the distance d between their centers (nuclei):

Coulomb's law $\quad F = k \times \dfrac{z_1 z_2}{d^2}$

Because energy is equal to force times distance, the lattice energy is

Lattice energy $\quad U = F \times d = k \times \dfrac{z_1 z_2}{d}$

The value of the constant k depends on the arrangement of the ions in the specific compound and is different for different substances.

Lattice energies are large when the distance d between ions is small and when the charges z_1 and z_2 are large. A small distance d means that the ions are close together, which implies that they have small ionic radii. Thus, if z_1 and z_2 are held constant, the largest lattice energies belong to compounds formed from the smallest ions, as listed in **TABLE 6.3**.

Within a series of compounds that have the same anion but different cations, lattice energy increases as the cation becomes smaller. Comparing LiF, NaF, and KF, for example, cation size follows the order $K^+ > Na^+ > Li^+$, so lattice energies follow the order LiF > NaF > KF. Similarly, within a series of compounds that have the same cation but different anions, lattice energy increases as anion size decreases. Comparing LiF, LiCl, LiBr, and LiI, for example, anion size follows the order $I^- > Br^- > Cl^- > F^-$, so lattice energies follow the reverse order LiF > LiCl > LiBr > LiI.

Table 6.3 also shows that compounds of ions with higher charges have larger lattice energies than compounds of ions with lower charges. In comparing NaI, MgI_2, and AlI_3, for example, the order of charges on the cations is $Al^{3+} > Mg^{2+} > Na^+$, and the order of lattice energies is $AlI_3 > MgI_2 > NaI$.

TABLE 6.3 **Lattice Energies of Some Ionic Solids (kJ/mol)**

Cation	Anion				
	F^-	Cl^-	Br^-	I^-	O^{2-}
Li^+	1036	853	807	757	2925
Na^+	923	787	747	704	2695
K^+	821	715	682	649	2360
Be^{2+}	3505	3020	2914	2800	4443
Mg^{2+}	2957	2524	2440	2327	3791
Ca^{2+}	2630	2258	2176	2074	3401

WORKED EXAMPLE 6.8

Lattice Energies

Which has the larger lattice energy, NaCl or CsI?

STRATEGY

The magnitude of a substance's lattice energy is affected both by the charges on its constituent ions and by the sizes of those ions. The higher the charges on the ions and the smaller the sizes of the ions, the larger the lattice energy. In this case, all four ions—Na^+, Cs^+, Cl^-, and I^-—are singly charged, so they differ only in size.

SOLUTION

Because Na^+ is smaller than Cs^+ and Cl^- is smaller than I^-, the distance between ions is smaller in NaCl than in CsI. Thus, NaCl has the larger lattice energy.

▶ **PRACTICE 6.15** Which substance in each of the following pairs has the larger lattice energy?

(a) KCl or RbCl
(b) CaF_2 or BaF_2
(c) CaO or KI

▶ **Conceptual** **APPLY 6.16** One of the following pictures represents NaCl and one represents MgO. Which is which, and which has the larger lattice energy?

(a)

(b)

▲ Crystals of sodium chloride.

INQUIRY ▶▶▶ HOW HAS AN UNDERSTANDING OF IONIC COMPOUNDS LED TO THE PRODUCTION OF SAFER SOLVENTS?

When you think of ionic compounds, you probably think of crystalline, high-melting solids: sodium chloride (mp = 801 °C), magnesium oxide (mp = 2825 °C), lithium carbonate (mp = 732 °C), and so on. It's certainly true that many ionic compounds fit that description, but not all. Some ionic compounds are actually liquid at room temperature. Ionic liquids, in fact, have been known for nearly a century—the first such compound to be discovered was ethylammonium nitrate, $CH_3CH_2NH_3^+ NO_3^-$, with a melting point of just 12 °C (54 °F).

▲ Several hundred different ionic liquids, such as these from the German company BASF, are manufactured and sold commercially for use in a wide variety of chemical processes.

One defining property of a liquid is the ability to flow. To do this, ions in a lattice must move positions relative to one another and overcome the electrostatic forces between them. Therefore, the attractive forces between ions in an ionic liquid are lower than those in an ionic solid, where the position of ions is rigid and fixed. Generally speaking, the ionic liquids used today are salts in which the cation has an irregular shape and in which one or both of the ions are large and bulky so that the charges are dispersed over a large volume. Both factors minimize the crystal lattice energy, thereby making the solid less stable and favoring the liquid. Typical cations are derived from nitrogen-containing organic compounds called *amines*, either tetrabutylammonium ions or *N*-alkylpyridinium ions.

Tetrabutylammonium ion

N-alkylpyridinium ion

Anions are just as varied as the cations, and more than 500 different ionic liquids with different anion/cation combinations are commercially available. Hexafluorophosphate, tetrafluoroborate, alkyl sulfates, trifluoromethanesulfonate, and halides are typical anions.

Hexafluorophosphate   Tetrafluoroborate

Methyl sulfate   Tetrafluoromethanesulfonate   Halide

For many years, ionic liquids were just laboratory curiosities. More recently, though, they have been found to be excellent solvents, particularly for use in green chemistry processes like those described in the Chapter 2 Inquiry. Ionic liquids have many useful properties:

• They dissolve many different types of compounds, giving highly concentrated solutions and thereby minimizing the amount of liquid needed.

- They can be fine-tuned for use in specific reactions by varying cation and anion structures.
- They are nonflammable.
- They are stable at high temperatures.
- They do not evaporate readily.
- They are generally recoverable and can be reused many times.

Among their potential applications, ionic liquids are now being explored as replacements for toxic or flammable organic solvents in many industrial processes, for use as electrolytes in high-temperature batteries, and as solvents for the extraction of heavy organic materials from oil shale. We'll be hearing much more about ionic liquids in the coming years.

PROBLEM 6.17 What structural features do ionic liquids have that prevent them from forming solids easily?

PROBLEM 6.18 Compare the following two ionic liquids: tetraheptylammonium bromide and tetraheptylammonium iodide. The structure of the tetraheptylammonium ion is:

$$\left[\begin{array}{c} CH_2(CH_2)_5CH_3 \\ | \\ H_3C(H_2C)_5H_2C - N - CH_2(CH_2)_5CH_3 \\ | \\ CH_2(CH_2)_5CH_3 \end{array} \right]^+$$

(a) Which picture corresponds to tetraheptylammonium bromide and which to tetraheptylammonium iodide?

(i)   **(ii)**

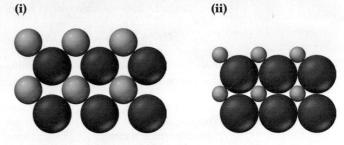

(b) Which ionic liquid has a larger lattice energy?
(c) One ionic liquid has a melting point of 88 °C and the other has a melting point of 39 °C. Match the ionic liquid to its melting point.

PROBLEM 6.19 An ionic liquid consisting of a bulky amine cation and chloride anion has a melting point of 41°C. For a green chemistry solvent application, it is desirable to have a melting point below room temperature (25 °C) so energy is not required to heat the compound. The melting point of the ionic liquid can be altered by replacing chloride with a different anion. The anion possibilities are: F^-, Se^{2-}, O^{2-}, and Br^-.

(a) Write the electron configuration for each of these ions.
(b) Which ions are isoelectronic?
(c) Which ion is the best choice for replacing the chloride ion to make an ionic liquid with a lower melting temperature?

STUDY GUIDE

Section	Concept Summary	Learning Objectives	Test Your Understanding
6.1 ▶ Electron Configurations of Ions	Metallic elements, on the left side of the periodic table, tend to give up electrons to form cations, while the halogens and a few other nonmetallic elements, on the right side of the table, tend to accept electrons to form anions. The electrons given up by a main-group metal in forming a cation come from the highest-energy occupied orbital, while the electrons that are accepted by a nonmetal in forming an anion go into the lowest-energy unoccupied orbital.	**6.1** Write ground-state electron configurations for main group and transition metal ions. **6.2** Determine the number of unpaired electrons in a transition metal ion.	Worked Example 6.1; Problems 6.20, 6.38, 6.40, 6.42 Problems 6.44, 6.45
6.2 ▶ Ionic Radii	Cations have smaller atomic radii than corresponding atoms because removing electrons increases the **effective nuclear charge (Z_{eff})** and in some cases decreases the valence shell. Anions have larger atomic radii than corresponding atoms because gaining electron decreases Z_{eff} and creates additional electron–electron repulsive forces.	**6.3** Predict the relative size of anions, cations, and atoms. **6.4** Predict the relative size of isoelectronic ions.	Worked Example 6.2; Problems 6.21, 6.46, 6.47 Problems 6.48, 6.49
6.3 ▶ Ionization Energy	The amount of energy necessary to remove a valence electron from an isolated neutral atom is called the atom's **ionization energy (E_i)**. Ionization energies are smallest for metallic elements on the left side of the periodic table and largest for nonmetallic elements on the right side. As a result, metals usually give up electrons and act as reducing agents in chemical reactions.	**6.5** Order elements from lowest to highest ionization energy. **6.6** Explain the periodic trend in ionization energy.	Worked Example 6.3; Problems 6.6, 6.52, 6.53 Problem 6.50
6.4 ▶ Higher Ionization Energies	Ionization is not limited to the removal of a single electron from an atom. Two, three, or even more electrons can be removed sequentially from an atom, although larger amounts of energy are required for each successive ionization step. In general, valence-shell electrons are much more easily removed than **core electrons.**	**6.7** Compare successive ionization energies for different elements. **6.8** Identify elements based on values of successive ionization energies.	Worked Example 6.4; Problems 6.23, 6.54, 6.56 Problems 6.8, 6.24, 6.58
6.5 ▶ Electron Affinity	The amount of energy released or absorbed when an electron adds to an isolated neutral atom is called the atom's **electron affinity (E_{ea})**. By convention, a negative E_{ea} corresponds to a release of energy and a positive E_{ea} corresponds to an absorption of energy. Electron affinities are most negative for group 7A elements and most positive for group 2A and 8A elements. As a result, the group 7A elements usually accept electrons and in chemical reactions.	**6.9** Compare the value of electron affinity for different elements. **6.10** Explain the periodic trend in electron affinity.	Worked Example 6.5; Problems 6.62, 6.63, 6.67 Problems 6.64–6.67
6.6 ▶ The Octet Rule	In general, reactions of main-group elements can be described by the **octet rule**, which states that these elements tend to undergo reactions so as to attain a noble-gas electron configuration with filled s and p subshells within their valence shell. Elements on the left side of the periodic table tend to give up electrons until a noble-gas configuration is reached; elements on the right side of the table tend to accept electrons until a noble-gas configuration is reached; and the noble gases themselves are essentially unreactive.	**6.11** Use the octet rule to predict charges on main group ions, electron configurations of main group ions, and formulas for ionic compounds.	Worked Example 6.6; Problems 6.30, 6.68–6.71

Section	Concept Summary	Learning Objectives	Test Your Understanding
6.7 ▸ Ionic Bonds and the Formation of Ionic Solids	Main-group metals in groups 1A and 2A react with nonmetals in groups 5A–7A, during which the metal loses one or more electrons to the nonmetal. The product, such as NaCl, is an ionic solid that consists of metal cations and halide anions electrostatically attracted to one another by **ionic bonds**.	**6.12** Visualize ionic compounds on the molecular level.	Problems 6.25, 6.26, 6.27
		6.13 Draw a Born–Haber cycle and calculate the energy change that occurs when an ionic compound is formed from its elements.	Worked Example 6.7; Problems 6.31, 6.80, 6.82
		6.14 Use the Born–Haber cycle to solve for the energy change associated with one of the steps.	Problems 6.14, 6.94, 6.98
6.8 ▸ Lattice Energies in Ionic Solids	The sum of the interaction energies among all ions in a crystal is called the crystal's **lattice energy (U)**. The higher the charges on the ions and the smaller the sizes of the ions, the larger the lattice energy.	**6.15** Predict the relative magnitude of lattice energy given the formula or molecular representation of an ionic compound.	Worked Example 6.8; Problems 6.28, 6.29, 6.72, 6.73

KEY TERMS

aufbau principle   *196*
Born–Haber cycle   *209*
core electron   *200*
Coulomb's law   *211*

effective nuclear charge (Z_{eff})   *198*
electron affinity (E_{ea})   *204*
ground-state electron
   configuration   *196*

ionic bond   *208*
ionic solid   *208*
ionization energy (E_i)   *200*
isoelectronic   *196*

lattice energy (U)   *211*
octet rule   *207*
shield   *200*

KEY EQUATIONS

- **Coulomb's Law (Section 6.8)**

$$F = k \times \frac{z_1 z_2}{d^2}$$

- **Lattice Energy (U) (Section 6.8)**

$$U = F \times d = k \times \frac{z_1 z_2}{d}$$

CONCEPTUAL PROBLEMS

Problems 6.1–6.19 appear within the chapter.

6.20 Where on the periodic table would you find the element that has an ion with each of the following electron configurations? Identify each ion.

(a) 3+ ion: $1s^2 2s^2 2p^6$ (b) 3+ ion: $[Ar] 3d^3$

(c) 2+ ion: $[Kr] 5s^2 4d^{10}$ (d) 1+ ion: $[Kr] 4d^{10}$

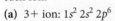

6.21 Which of the following spheres is likely to represent a metal atom, and which a nonmetal atom? Which sphere in the products represents a cation, and which an anion?

6.22 Circle the approximate part or parts of the periodic table where the following elements appear:

(a) Elements with the smallest values of E_{i1}

(b) Elements with the largest atomic radii

(c) Elements with the most negative values of E_{ea}

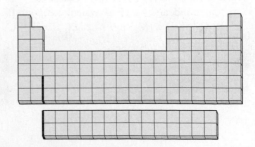

6.23 Order the indicated three elements according to the ease with which each is likely to lose its third electron.

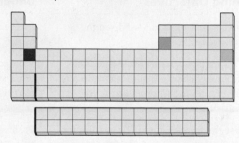

6.24 This figure represents the successive ionization energy of an atom in the third period of the periodic table. Which atom is this most likely to be?

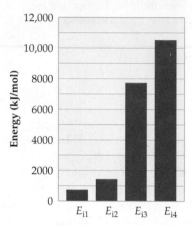

6.25 In the following drawings, red spheres represent cations and blue spheres represent anions. Match each of the drawings (**a**)–(**d**) with the following ionic compounds:

(**i**) $Ca_3(PO_4)_2$ (**ii**) Li_2CO_3 (**iii**) $FeCl_2$ (**iv**) $MgSO_4$

(**a**)   (**b**)

(**c**)   (**d**)

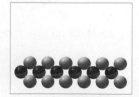 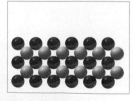

6.26 Which of the following drawings is more likely to represent an ionic compound, and which a covalent compound?

(**a**)   (**b**)

6.27 Each of the pictures (**a**)–(**d**) represents one of the following substances at 25 °C: sodium, chlorine, iodine, sodium chloride. Which picture corresponds to which substance?

(**a**)   (**b**)   (**c**)   (**d**)

  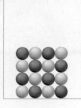 

6.28 Which of the following alkali metal halides has the largest lattice energy, and which has the smallest lattice energy? Explain.

(**a**)   (**b**)   (**c**)

6.29 Which of the following alkali metal halides has the larger lattice energy, and which the smaller lattice energy? Explain.

(**a**)   (**b**)

6.30 Three binary compounds are represented on the following drawing: red with red, blue with blue, and green with green. Give a likely formula for each compound.

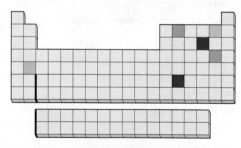

6.31 Given the following values for the formation of $LiCl(s)$ from its elements, draw a Born–Haber cycle similar to that shown in Figure 6.6.

E_{ea} for $Cl(g) = -348.6$ kJ/mol
Heat of sublimation for $Li(s) = +159.4$ kJ/mol
E_{i1} for $Li(g) = +520$ kJ/mol
Bond dissociation energy for $Cl_2(g) = +243$ kJ/mol
Lattice energy for $LiCl(s) = +853$ kJ/mol

SECTION PROBLEMS

Ions, Ionization Energy, and Electron Affinity (Sections 6.1–6.5)

6.32 What is the difference between a covalent bond and an ionic bond?

6.33 Characterize bonds between the two atoms as covalent or ionic.
- (a) Cl and I
- (b) Rb and F
- (c) Na and S
- (d) N and S

6.34 What is the difference between a molecule and an ion?

6.35 Label the following species as molecules or ions.
- (a) NO_3^-
- (b) NH_4^+
- (c) NO_2
- (d) $CH_3CO_2^-$

6.36 How many protons and electrons are in each of the following ions?
- (a) Be^{2+}
- (b) Rb^+
- (c) Se^{2-}
- (d) Au^{3+}

6.37 What is the identity of the element X in the following ions?
- (a) X^{2+}, a cation that has 36 electrons
- (b) X^-, an anion that has 36 electrons

6.38 What are the likely ground-state electron configurations of the following cations?
- (a) La^{3+}
- (b) Ag^+
- (c) Sn^{2+}

6.39 What are the likely ground-state electron configurations of the following anions?
- (a) Se^{2-}
- (b) N^{3-}

6.40 What is the electron configuration of Ca^{2+}? What is the electron configuration of Ti^{2+}?

6.41 Identify the element whose 2+ ion has the ground-state electron configuration $[Ar] 3d^{10}$.

6.42 What doubly positive ion has the following ground-state electron configuration? $1s^2 2s^2 2p^6$

6.43 What tripositive ion has the electron configuration $[Kr] 4d^3$? What neutral atom has the electron configuration $[Kr] 5s^2 4d^2$?

6.44 There are two elements in the transition-metal series Sc through Zn that have four unpaired electrons in their 2+ ions. Identify them.

6.45 Which element in the transition-metal series Sc through Zn has five unpaired electrons in its +3 ion?

6.46 Which atom or ion in the following pairs would you expect to be larger?
- (a) S or S^{2-}
- (b) Ca or Ca^{2+}
- (c) O^- or O^{2-}

6.47 Which atom or ion in the following pairs would you expect to be larger?
- (a) Rb or Rb^+
- (b) N or N^{3-}
- (c) Cr^{3+} or Cr^{6+}

6.48 Order the following ions from smallest to largest. Sr^{2+}, Se^{2-}, Br^-, Rb^+

6.49 Order the following ions from smallest to largest. Mg^{2+}, O^{2-}, F^-, Na^+

6.50 Which group of elements in the periodic table has the largest E_{i1}, and which group has the smallest? Explain.

6.51 Which element in the periodic table has the smallest ionization energy? Which has the largest?

6.52 Which element in each of the following sets has the smallest first ionization energy, and which has the largest?
- (a) Li, Ba, K
- (b) B, Be, Cl
- (c) Ca, C, Cl

6.53 Order the elements in each set from the smallest to largest first ionization energy.
- (a) Na, I, P
- (b) P, Sr, Mg
- (c) Ca, Cs, Se

6.54 (a) Which has the smaller second ionization energy, K or Ca?
(b) Which has the larger third ionization energy, Ga or Ca?

6.55 (a) Which has the smaller fourth ionization energy, Sn or Sb?
(b) Which has the larger sixth ionization energy, Se or Br?

6.56 Three atoms have the following electron configurations:
- (a) $1s^2 2s^2 2p^6 3s^2 3p^3$
- (b) $1s^2 2s^2 2p^6 3s^2 3p^6$
- (c) $1s^2 2s^2 2p^6 3s^2 3p^6 4s^2$

Which of the three has the largest E_{i2}? Which has the smallest E_{i7}?

6.57 Three atoms have the following electron configurations:
- (a) $1s^2 2s^2 2p^6 3s^2 3p^1$
- (b) $1s^2 2s^2 2p^6 3s^2 3p^5$
- (c) $1s^2 2s^2 2p^6 3s^2 3p^6 4s^1$

Which of the three has the largest E_{i1}? Which has the smallest E_{i4}?

6.58 The first four ionization energies in kJ/mol of a certain second-row element are 801, 2427, 3660, and 25,025. What is the likely identity of the element?

6.59 The first four ionization energies in kJ/mol of a certain second-row element are 900, 1757, 14,849, and 21,007. What is the likely identity of the element?

6.60 What is the relationship between the electron affinity of a singly charged cation such as Na^+ and the ionization energy of the neutral atom?

6.61 What is the relationship between the ionization energy of a singly charged anion such as Cl^- and the electron affinity of the neutral atom?

6.62 Which has the more negative electron affinity, Na^+ or Na? Na^+ or Cl?

6.63 Which has the more negative electron affinity, Br or Br^-?

6.64 Why is energy usually released when an electron is added to a neutral atom but absorbed when an electron is removed from a neutral atom?

6.65 Why does ionization energy increase regularly across the periodic table from group 1A to group 8A, whereas electron affinity increases irregularly from group 1A to group 7A and then falls dramatically for group 8A?

6.66 No element has a negative second electron affinity. That is, the process $A^-(g) + e^- \longrightarrow A^{2-}(g)$ is unfavorable for every element. Suggest a reason.

6.67 Why does phosphorus have a less-negative electron affinity than its neighbors silicon and sulfur?

Octet Rule, Ionic Bonds, and Lattice Energy (Sections 6.6–6.8)

6.68 What noble-gas configurations and charge are the following elements likely to attain in reactions in which they form ions?
- (a) N
- (b) Ca
- (c) S
- (d) Br

6.69 Each of the following pairs of elements will react to form a binary ionic compound. Write the formula of each compound formed, and give its name.

(a) Magnesium and chlorine

(b) Calcium and oxygen

(c) Lithium and nitrogen

(d) Aluminum and oxygen

6.70 Element X reacts with element Y to give a product containing X^{3+} ions and Y^{2-} ions.

(a) Is element X likely to be a metal or a nonmetal? Explain.

(b) Is element Y likely to be a metal or a nonmetal? Explain.

(c) What is the formula of the product?

(d) In what groups of the periodic table are elements X and Y likely to be found?

6.71 Element X reacts with element Y to give a product containing X^{2+} ions and Y^- ions.

(a) Is element X likely to be a metal or a nonmetal? Explain.

(b) Is element Y likely to be a metal or a nonmetal? Explain.

(c) What is the formula of the product?

(d) In what groups of the periodic table are elements X and Y likely to be found?

6.72 Order the following compounds according to their expected lattice energies: LiCl, KCl, KBr, MgCl$_2$.

6.73 Order the following compounds according to their expected lattice energies: AlBr$_3$, MgBr$_2$, LiBr, CaO.

6.74 Calculate the energy change in kilojoules per mole when lithium atoms lose an electron to bromine atoms to form isolated Li^+ and Br^- ions. [The E_i for Li(g) is 520 kJ/mol; the E_{ea} for Br(g) is -325 kJ/mol.]

6.75 Cesium has the smallest ionization energy of all elements (376 kJ/mol), and chlorine has the most negative electron affinity (-349 kJ/mol). Will a cesium atom transfer an electron to a chlorine atom to form isolated $Cs^+(g)$ and $Cl^-(g)$ ions? Explain.

6.76 Find the lattice energy of LiBr(s) in Table 6.3, and calculate the energy change in kilojoules per mole for the formation of solid LiBr from the elements. [The sublimation energy for Li(s) is $+159.4$ kJ/mol, the bond dissociation energy of Br$_2$(g) is $+224$ kJ/mol, and the energy necessary to convert Br$_2$(l) to Br$_2$(g) is 30.9 kJ/mol.]

6.77 Look up the lattice energies in Table 6.3, and calculate the energy change in kilojoules per mole for the formation of the following substances from their elements:

(a) LiF(s) [The sublimation energy for Li(s) is $+159.4$ kJ/mol, the E_i for Li(g) is 520 kJ/mol, the E_{ea} for F(g) is -328 kJ/mol, and the bond dissociation energy of F$_2$(g) is $+158$ kJ/mol.]

(b) CaF$_2$(s) [The sublimation energy for Ca(s) is $+178.2$ kJ/mol, $E_{i1} = +589.8$ kJ/mol, and $E_{i2} = +1145$ kJ/mol.]

6.78 Born–Haber cycles, such as those shown in Figures 6.6 and 6.7, are called *cycles* because they form closed loops. If any five of the six energy changes in the cycle are known, the value of the sixth can be calculated. Use the following five values to calculate the lattice energy in kilojoules per mole for sodium hydride, NaH(s):

E_{ea} for H(g) = -72.8 kJ/mol

E_{i1} for Na(g) = $+495.8$ kJ/mol

Heat of sublimation for Na(s) = $+107.3$ kJ/mol

Bond dissociation energy for H$_2$(g) = $+435.9$ kJ/mol

Net energy change for the formation of NaH$_2$(s) from its elements = -60 kJ/mol

6.79 Calculate a lattice energy for CaH$_2$(s) in kilojoules per mole using the following information:

E_{ea} for H(g) = -72.8 kJ/mol

E_{i1} for Ca(g) = $+589.8$ kJ/mol

E_{i2} for Ca(g) = $+1145$ kJ/mol

Heat of sublimation for Ca(s) = $+178.2$ kJ/mol

Bond dissociation energy for H$_2$(g) = $+435.9$ kJ/mol

Net energy change for the formation of CaH$_2$(s) from its elements = -186.2 kJ/mol

6.80 Calculate the overall energy change in kilojoules per mole for the formation of CsF(s) from its elements using the following data:

E_{ea} for F(g) = -328 kJ/mol

E_{i1} for Cs(g) = $+375.7$ kJ/mol

E_{i2} for Cs(g) = $+2422$ kJ/mol

Heat of sublimation for Cs(s) = $+76.1$ kJ/mol

Bond dissociation energy for F$_2$(g) = $+158$ kJ/mol

Lattice energy for CsF(s) = $+740$ kJ/mol

6.81 The estimated lattice energy for CsF$_2$(s) is $+2347$ kJ/mol. Use the data given in Problem 6.80 to calculate an overall energy change in kilojoules per mole for the formation of CsF$_2$(s) from its elements. Does the overall reaction absorb energy or release it? In light of your answer to Problem 6.80, which compound is more likely to form in the reaction of cesium with fluorine, CsF or CsF$_2$?

6.82 Calculate the overall energy change in kilojoules per mole for the formation of CaCl(s) from the elements. The following data are needed:

E_{ea} for Cl(g) = -348.6 kJ/mol

E_{i1} for Ca(g) = $+589.8$ kJ/mol

E_{i2} for Ca(g) = $+1145$ kJ/mol

Heat of sublimation for Ca(s) = $+178.2$ kJ/mol

Bond dissociation energy for Cl$_2$(g) = $+243$ kJ/mol

Lattice energy for CaCl$_2$(s) = $+2258$ kJ/mol

Lattice energy for CaCl(s) = $+717$ kJ/mol (estimated)

6.83 Use the data in Problem 6.82 to calculate an overall energy change for the formation of CaCl$_2$(s) from the elements. Which is more likely to form, CaCl or CaCl$_2$?

6.84 Use the data and the result in Problem 6.78 to draw a Born–Haber cycle for the formation of NaH(s) from its elements.

6.85 Use the data and the result in Problem 6.77(a) to draw a Born–Haber cycle for the formation of LiF(s) from its elements.

CHAPTER PROBLEMS

6.86 Cu$^+$ has an ionic radius of 77 pm, but Cu^{2+} has an ionic radius of 73 pm. Explain.

6.87 The following ions all have the same number of electrons: Ti^{4+}, Sc^{3+}, Ca^{2+}, S^{2-}. Order them according to their expected sizes, and explain your answer.

6.88 Calculate overall energy changes in kilojoules per mole for the formation of MgF(s) and MgF$_2$(s) from their elements. In light of your answers, which compound is more likely to form in the reaction of magnesium with fluorine, MgF or MgF$_2$? The following data are needed:

E_{ea} for F(g) = -328 kJ/mol

E_{i1} for Mg(g) = $+737.7$ kJ/mol

E_{i2} for Mg(g) = $+1450.7$ kJ/mol

Heat of sublimation for Mg(s) = $+147.7$ kJ/mol

Bond dissociation energy for $F_2(g) = +158$ kJ/mol

Lattice energy for $MgF_2(s) = +2952$ kJ/mol

Lattice energy for $MgF(s) = 930$ kJ/mol (estimated)

6.89 Draw Born–Haber cycles for the formation of both MgF and MgF_2 (Problem 6.88).

6.90 We saw in Section 6.7 that the reaction of solid sodium with gaseous chlorine to yield solid sodium chloride (Na^+Cl^-) is favorable by 411 kJ/mol. Calculate the energy change for the alternative reaction that yields chlorine sodide (Cl^+Na^-), and then explain why sodium chloride formation is preferred.

$$2\,Na(s) + Cl_2(g) \longrightarrow 2\,Cl^+Na^-(s)$$

Assume that the lattice energy for Cl^+Na^- is the same as that for Na^+Cl^-. The following data are needed in addition to that found in Section 6.7:

$$E_{ea} \text{ for } Na(g) = -52.9 \text{ kJ/mol}$$
$$E_{i1} \text{ for } Cl(g) = +1251 \text{ kJ/mol}$$

6.91 Draw a Born–Haber cycle for the reaction of sodium with chlorine to yield chlorine sodide (Problem 6.90).

6.92 Many early chemists noted a diagonal relationship among elements in the periodic table, whereby a given element is sometimes more similar to the element below and to the right than it is to the element directly below. Lithium is more similar to magnesium than to sodium, for example, and boron is more similar to silicon than to aluminum. Use your knowledge about the periodic trends of such properties as atomic radii and Z_{eff} to explain the existence of diagonal relationships.

6.93 Heating elemental cesium and platinum together for two days at 973 K gives a dark red ionic compound that is 57.67% Cs and 42.33% Pt.

(a) What is the empirical formula of the compound?

(b) What are the charge and electron configuration of the cesium ion?

(c) What are the charge and electron configuration of the platinum ion?

6.94 Use the following information plus the data given in Tables 6.2 and 6.3 to calculate the second electron affinity, E_{ea2}, of oxygen. Is the O^{2-} ion stable in the gas phase? Why is it stable in solid MgO?

Heat of sublimation for $Mg(s) = +147.7$ kJ/mol

Bond dissociation energy for $O_2(g) = +498.4$ kJ/mol

E_{ea1} for $O(g) = -141.0$ kJ/mol

Net energy change for formation of $MgO(s)$ from its elements $= -601.7$ kJ/mol

6.95 (a) Which element from each set has the largest atomic radius? Explain.

   (i) Ba, Ti, Ra, Li   (ii) F, Al, In, As

(b) Which element from each set has the smallest ionization energy? Explain.

   (i) Tl, Po, Se, Ga   (ii) Cs, Ga, Bi, Se

6.96 (a) Which of the elements Be, N, O, and F has the most negative electron affinity? Explain.

(b) Which of the ions Se^{2-}, F^-, O^{2-}, and Rb^+ has the largest radius? Explain.

6.97 Given the following information, construct a Born–Haber cycle to calculate the lattice energy of $CaC_2(s)$.

Net energy change for the formation of $CaC_2(s) = -60$ kJ/mol

Heat of sublimation for $Ca(s) = +178$ kJ/mol

E_{i1} for $Ca(g) = +590$ kJ/mol

E_{i2} for $Ca(g) = +1145$ kJ/mol

Heat of sublimation for $C(s) = +717$ kJ/mol

Bond dissociation energy for $C_2(g) = +614$ kJ/mol

E_{ea1} for $C_2(g) = -315$ kJ/mol

E_{ea2} for $C_2(g) = +410$ kJ/mol

6.98 Given the following information, construct a Born–Haber cycle to calculate the lattice energy of $CrCl_2I(s)$:

Net energy change for the formation of $CrCl_2I(s) = -420$ kJ/mol

Bond dissociation energy for $Cl_2(g) = +243$ kJ/mol

Bond dissociation energy for $I_2(s) = +151$ kJ/mol

Heat of sublimation for $I_2(s) = +62$ kJ/mol

Heat of sublimation for $Cr(s) = +397$ kJ/mol

E_{i1} for $Cr(g) = 652$ kJ/mol

E_{i2} for $Cr(g) = 1588$ kJ/mol

E_{i3} for $Cr(g) = 2882$ kJ/mol

E_{ea} for $Cl(g) = -349$ kJ/mol

E_{ea} for $I(g) = -295$ kJ/mol

MULTICONCEPT PROBLEMS

6.99 Consider the electronic structure of the element bismuth.

(a) The first ionization energy of bismuth is $E_{i1} = +703$ kJ/mol. What is the longest possible wavelength of light that could ionize an atom of bismuth?

(b) Write the electron configurations of neutral Bi and the Bi^+ cation.

(c) What are the n and l quantum numbers of the electron removed when Bi is ionized to Bi^+?

(d) Would you expect element 115 to have an ionization energy greater than, equal to, or less than that of bismuth? Explain.

6.100 Iron is commonly found as Fe, Fe^{2+}, and Fe^{3+}.

(a) Write electron configurations for each of the three.

(b) What are the n and l quantum numbers of the electron removed on going from Fe^{2+} to Fe^{3+}?

(c) The third ionization energy of Fe is $E_{i3} = +2952$ kJ/mol. What is the longest wavelength of light that could ionize $Fe^{2+}(g)$ to $Fe^{3+}(g)$?

(d) The third ionization energy of Ru is less than the third ionization energy of Fe. Explain.

6.101 The ionization energy of an atom can be measured by photoelectron spectroscopy, in which light of wavelength λ is directed at an atom, causing an electron to be ejected. The kinetic energy of the ejected electron (E_K) is measured by determining its velocity, v, since $E_K = 1/2\,mv^2$. The E_i is then calculated using the relationship that the energy of the incident light equals the sum of E_i plus E_K.

(a) What is the ionization energy of rubidium atoms in kilojoules per mole if light with $\lambda = 58.4$ nm produces electrons with a velocity of 2.450×10^6 m/s? (The mass of an electron is 9.109×10^{-31} kg.)

(b) What is the ionization energy of potassium in kilojoules per mole if light with $\lambda = 142$ nm produces electrons with a velocity of 1.240×10^6 m/s?

Covalent Bonding and Electron-Dot Structures

O₂N—C... structure (parathion)

Parathion is an organophosphate insecticide used to control pests in many crops such as corn, wheat, and cotton. The toxicity of organophosphate compounds is influenced by the properties of chemical bonds discussed in this chapter.

? How do we make organophosphate insecticides less toxic to humans?

The answer to this question can be found in the **INQUIRY** ▸▸▸ on page 250.

CONTENTS

The chapter-opening image shows the chemical structure of the insecticide para-thion. In this *compound*, all the bonds are between two nonmetal elements such as C—C, C—H, C—O, and P—S. We saw in the previous chapter that an ionic bond between a metal and a reactive nonmetal is typically formed by the transfer of electrons be-tween atoms. The metal loses one or more electrons and becomes a cation, while the non-metal atom gains one or more electrons and becomes an anion. The oppositely charged ions are held together by electrostatic attractions called *ionic bonds*.

How, though, do bonds form between atoms of the same or similar elements? Simply put, the answer is that the bonds in such compounds are formed by the *sharing* of electrons between atoms rather than by the transfer of electrons from one atom to another. As we saw in Section 2.10, a bond formed by the sharing of electrons is called a *covalent bond*, and the unit of matter held together by one or more covalent bonds is called a *molecule*. We'll explore the nature of covalent bonding in this chapter.

7.1 ▶ COVALENT BONDING IN MOLECULES

The covalent bonding model involves a *sharing* of electrons between two atoms, in contrast, to the ionic bond in which a *transfer* of electrons occurs. To see how the formation of a cova-lent bond between atoms can be described, let's look at the H—H bond in the H_2 molecule as the simplest example. When two hydrogen atoms come close together, electrostatic interac-tions begin to develop between them. The two positively charged nuclei repel each other, and the two negatively charged electrons repel each other, but each nucleus attracts both electrons (**FIGURE 7.1**). If the attractive forces are stronger than the repulsive forces, a covalent bond is formed, with the two atoms held together and the two shared electrons occupying the region between the nuclei.

In essence, the shared electrons act as a kind of "glue" to bind the two atoms into an H_2 molecule. Both nuclei are simultaneously attracted to the same electrons and are therefore held together, much as two tug-of-war teams pulling on the same rope are held together.

The magnitudes of the various attractive and repulsive forces between nuclei and elec-trons in a covalent bond depend on how close the atoms are. If the hydrogen atoms are too far apart, the attractive forces are small and no bond exists. If the hydrogen atoms are too close together, the repulsive interaction between the nuclei becomes so strong that it pushes the atoms apart. Thus, there is an optimum distance between nuclei called the **bond length** where net attractive forces are maximized and the H—H molecule is most stable. In the H_2 molecule, the bond length is 74 pm. On a graph of energy versus internu-clear distance, the bond length is the H—H distance in the minimum energy, most stable arrangement (**FIGURE 7.2**).

Every bond in every molecule has its own specific *bond length*, the distance between nu-clei of two bonded atoms. Not surprisingly, though, bonds between the same pairs of atoms usually have similar lengths. For example, carbon–carbon single bonds usually have lengths in the range 152.0–153.5 pm regardless of the exact structure of the molecule. Note in the

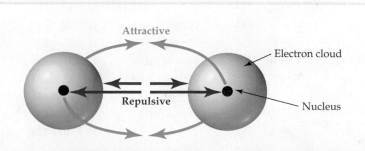

The nucleus–electron **attractions** are greater than the nucleus–nucleus and electron–electron **repulsions**, resulting in a net attractive force that binds the atoms together.

◀ **FIGURE 7.1**

A covalent H—H bond. The bond is the net result of attractive and repulsive electrostatic forces.

▶ **FIGURE 7.2**

A graph of potential energy versus internuclear distance for the H_2 molecule.

Figure It Out

Why is 74 pm the bond length of H_2?

Answer: At an internuclear distance of 74 pm, the energy is lowest because attractive forces are maximized. The lowest energy state is the most stable.

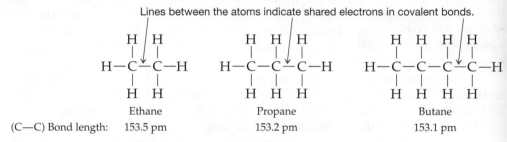

If the atoms are **too close**, strong repulsions occur.

If the atoms are too far apart, attractions are weak and no bonding occurs.

HH (too close)

H ······ H (too far)

H–H

When the atoms are optimally separated, energy is at a minimum.

Energy -436 kJ/mol

Bond length (74 pm)

Internuclear distance ⟶

following examples that covalent bonds are indicated by **lines between atoms**, as described in Section 2.10.

Lines between the atoms indicate shared electrons in covalent bonds.

Ethane	Propane	Butane

(C—C) Bond length: 153.5 pm   153.2 pm   153.1 pm

Because similar bonds have similar lengths, it's possible to construct a table of average values to compare different kinds of bonds (**TABLE 7.1**). Keep in mind, though, that the actual value in a specific molecule might vary by $\pm 10\%$ from the average.

Every covalent bond has its own characteristic length that leads to maximum stability and that is roughly predictable from the knowledge of atomic radii (Section 5.14). For example, because the atomic radius of hydrogen is 37 pm and the atomic radius of chlorine is 99 pm, the H—Cl bond length in a hydrogen chloride molecule should be approximately 37 pm + 99 pm = 136 pm. The actual value is 127 pm. Note that in the series of diatomic halogen molecules that bond length increases down the group as the atomic radius of the element increases.

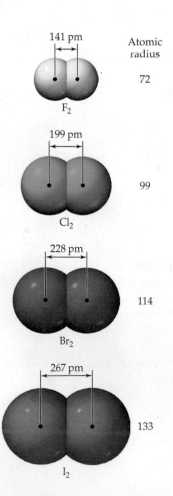

141 pm

Atomic radius

F_2 72

199 pm

Cl_2 99

228 pm

Br_2 114

267 pm

I_2 133

TABLE 7.1   Average Bond Lengths (pm)

H—H	74^a	C—H	110	N—H	98	O—F	130	I—I	267^a
H—C	110	C—C	154	N—C	147	O—Cl	165	S—F	168
H—F	92^a	C—F	141	N—F	134	O—Br	180	S—Cl	203
H—Cl	127^a	C—Cl	176	N—Cl	169	O—I	199	S—Br	218
H—Br	142^a	C—Br	191	N—Br	184	O—N	136	S—S	208
H—I	161^a	C—I	176	N—N	140	O—O	132		
H—N	98	C—N	147	N—O	136	F—F	141^a		
H—O	94	C—O	143	O—H	94	Cl—Cl	199^a		
H—S	132	C—S	181	O—C	143	Br—Br	228^a		

Multiple covalent bonds[b]

C=C	134	C≡C	120	C=O	121	O=O	121^a	N≡N	113^a

[a]Exact value.

[b]We'll discuss multiple covalent bonds in Section 7.5.

7.2 ▶ STRENGTHS OF COVALENT BONDS

Look again at Figure 7.2, the graph of energy versus internuclear distance for the H_2 molecule, and note how the H_2 molecule is lower in energy than two separate hydrogen atoms. When pairs of hydrogen atoms bond together, they form lower-energy H_2 molecules and release 436 kJ/mol. In other words, 436 kJ must be *added* to split 1 mol of H_2 molecules apart into 2 mol of hydrogen atoms.

The amount of energy that must be supplied to break a chemical bond in an isolated molecule in the gaseous state—and thus the amount of energy released when the bond forms—is called the **bond dissociation energy (D)**. Bond dissociation energies are always positive because energy must always be supplied to break a bond. Conversely, the amount of energy released on forming a bond always has a negative value.

Similar to bond length, every bond has its own specific bond dissociation energy and bonds between the same pairs of atoms usually have similar D values. For example, carbon–carbon single bonds usually have D values of approximately 350–380 kJ/mol regardless of the exact structure of the molecule.

Ethane $\quad$ Propane $\quad$ Butane
$D = 377$ kJ/mol $\quad$ $D = 370$ kJ/mol $\quad$ $D = 372$ kJ/mol

The average bond dissociation energies for specific types of bonds are listed in **TABLE 7.2**. These also vary by ±10% depending on the actual structure of molecule. Bond dissociation energies cover a wide range, from a low of 151 kJ/mol for the I—I bond to a high of 570 kJ/mol for the H—F bond. As a rule of thumb, though, most of the bonds commonly encountered in naturally occurring molecules (C—H, C—C, C—O) have values in the range of 350–400 kJ/mol.

Periodic properties and bond lengths can be used to explain why some bonds are stronger than others. Let's examine the bond lengths and values of D for a series of hydrogen halide bonds:

Bond	Bond Dissociation Energy (D) (kJ/mol)	Bond Length (pm)
H—F	570	92
H—Cl	432	127
H—Br	366	142
H—I	298	161

As we proceed down through the period of halogens, the value of D becomes smaller, meaning the bond is weaker. As the atomic radius of the halogen increases, the shared electrons are farther away and more shielded from the positively charged nucleus, leading to a longer and weaker bond. Although there are exceptions, *shorter bonds are typically stronger*. For example, the F—F bond is predicted to be stronger than the Cl—Cl bond because fluorine atoms are smaller, but in fact the bond dissociation energy for F—F is 159 kJ/mol

TABLE 7.2   Average Bond Dissociation Energies, *D* (kJ/mol)

H—H	436[a]	C—H	410	N—H	390	O—F	180	I—I	151[a]
H—C	410	C—C	350	N—C	300	O—Cl	200	S—F	310
H—F	570[a]	C—F	450	N—F	270	O—Br	210	S—Cl	250
H—Cl	432[a]	C—Cl	330	N—Cl	200	O—I	220	S—Br	210
H—Br	366[a]	C—Br	270	N—Br	240	O—N	200	S—S	225
H—I	298[a]	C—I	240	N—N	240	O—O	180		
H—N	390	C—N	300	N—O	200	F—F	159[a]		
H—O	460	C—O	350	O—H	460	Cl—Cl	243[a]		
H—S	340	C—S	260	O—C	350	Br—Br	193[a]		

Multiple covalent bonds[b]

C=C	728	C≡C	965	C=O	732	O=O	498[a]	N≡N	945[a]

[a]Exact value.

[b]We'll discuss multiple covalent bonds in Section 7.5.

compared to 243 kJ/mol for Cl—Cl. Chapter 8 will describe two theories for covalent bonding, Molecular Orbital Theory and Valence Bond Theory, used to explain observed bond properties; including length and strength. We will see that certain models work well in explaining and predicting some properties while other models are better suited for different properties.

The correlation between bond length and strength also holds true for multiple bonds. In speaking of molecules with multiple bonds, we often use the term **bond order** to refer to the number of electron pairs shared between atoms. Thus, the F—F bond in the F_2 molecule has a bond order of 1, the O=O bond in the O_2 molecule has a bond order of 2, and the N≡N bond in the N_2 molecule has a bond order of 3.

Multiple bonds are both shorter and stronger than their corresponding single-bond counterparts because there are more shared electrons holding the atoms together. Compare, for example, the O=O double bond in O_2 with the O—O single bond in H_2O_2 (hydrogen peroxide), and compare the N≡N triple bond in N_2 with the N—N single bond in N_2H_4 (hydrazine):

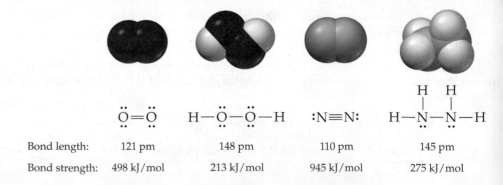

Bond length:	121 pm	148 pm	110 pm	145 pm
Bond strength:	498 kJ/mol	213 kJ/mol	945 kJ/mol	275 kJ/mol

7.3 ▶ POLAR COVALENT BONDS: ELECTRONEGATIVITY

We've given the impression up to this point that a given bond is either purely ionic, with electrons completely transferred, or purely covalent, with electrons shared equally. In fact, though, ionic and covalent bonds represent only the two extremes of a continuous range of possibilities. Between these two extremes are the large majority of bonds in which the bonding electrons are shared unequally between two atoms but are not completely transferred, called **polar covalent bonds** (**FIGURE 7.3**). The lowercase Greek letter delta (δ) is used to

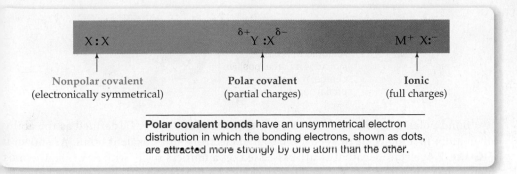

The bonding continuum from nonpolar covalent to ionic. The symbol δ (Greek delta) means partial charge, either partial positive ($\delta+$) or partial negative ($\delta-$).

denote the resultant partial charges on the atoms, either partial positive ($\delta+$) for the atom that has a smaller share of the bonding electrons or partial negative ($\delta-$) for the atom that has a larger share.

The extent of electron transfer in a compound is most easily visualized with what are called *electrostatic potential maps*, which use color to portray the calculated electron distribution in an isolated, gas-phase molecule. Yellow–green represents a neutral, nonpolar atom; blue represents a deficiency of electrons on an atom (partial positive charge); and red represents a surplus of electrons on an atom (partial negative charge). As examples of different points along the bonding spectrum, let's look at the three substances Cl_2, HCl, and NaCl.

- **Cl_2—Nonpolar Covalent** The bond in a chlorine molecule is nonpolar covalent, with the bonding electrons attracted equally to the two identical chlorine atoms. A similar situation exists in all such molecules that contain a covalent bond between two identical atoms. Neither chlorine atom has a partial positive or partial negative charge as shown by their identical yellow–green coloration in an electrostatic potential map.

The two bonding electrons, shown here as dots, are symmetrically distributed between the two Cl atoms. | Cl:Cl | **A nonpolar covalent bond.** Yellow-green represents a neutral atom.

- **HCl—Polar Covalent** The bond in a hydrogen chloride molecule is polar covalent. The chlorine atom attracts the bonding electron pair more strongly than hydrogen does, resulting in an unsymmetrical distribution of electrons. Chlorine thus has a partial negative charge (orange in the electrostatic potential map), and hydrogen has a partial positive charge (blue in the electrostatic potential map). Experimentally, the H—Cl bond has been found to be about 83% covalent and 17% ionic.

$$^{\delta+}H — Cl^{\delta-}$$

$$[H\ :Cl]$$

A polar covalent bond. The two bonding electrons (dots) are attracted more strongly by Cl than by H.

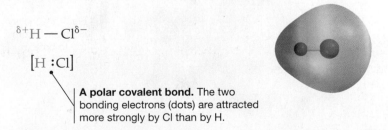

- **NaCl—Ionic** The bond in solid sodium chloride is a largely ionic one between Na^+ and Cl^-. In spite of what we've said previously, though, experiments show that the NaCl bond is only about 80% ionic and that the electron transferred from Na to Cl still spends some of its time near sodium. Thus, the electron-poor sodium atom is blue in an electrostatic potential map, while the electron-rich chlorine is red.

$Na^+ Cl^-$ **An ionic bond.** Blue indicates a partial positive charge; red indicates a partial negative charge.

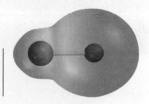

Bond polarity is due to differences in **electronegativity (EN)**, defined as the ability of an atom in a molecule to attract the shared electrons in a covalent bond. As shown in **FIGURE 7.4**, electronegativities are expressed on a unitless scale, with fluorine, the most highly electronegative element, assigned a value of 4.0. Metallic elements on the left of the periodic table attract electrons only weakly and are the least electronegative elements. Halogens and other reactive nonmetals in the upper right of the table attract electrons strongly and are the most electronegative. Figure 7.4 also shows that electronegativity generally decreases down the periodic table within a group.

The polarity of a bond is another important factor influencing bond dissociation energy. The American chemist, Linus Pauling, developed the electronegativity scale shown in Figure 7.4 by examining extra bond strength due to bond polarity. For example, the HF bond dissociation energy would be expected to be approximately the average of the bond dissociation energies of H_2 (436 kJ/mol) and F_2 (159 kJ/mol); or about 297 kJ/mol. However, the HF bond strength is much higher (570 kJ/mol). Pauling explained this discrepancy through bond polarity. The H_2 bond is nonpolar while the HF bond is highly polar, resulting in a partial negative charge on fluorine and a partial positive charge on hydrogen. The attraction between these partial charges *increases* the energy required to break the bond. In general, *increased bond polarity leads to increased bond strength.* Using the bond dissociation energies of many compounds, Pauling arrived at a scale of relative electronegativity values.

Another way to calculate electronegativity is by taking the average of the absolute value of **electron affinity** (E_{ea}, Section 6.5) and **ionization energy** (E_i, Section 6.3) and setting up a scale with fluorine assigned a value of 4.0. *Electron affinity* is a measure of the tendency of an isolated atom to gain an electron, and *ionization energy* is a measure of the tendency of an isolated atom to lose an electron. Both E_{ea} and E_i are related to the ability of an atom to attract shared electrons in a chemical bond.

REMEMBER...

Electron affinity (E_{ea}) is defined as the energy change that occurs when an electron is added to an isolated gaseous atom. (Section 6.5)
Ionization energy (E_i) in contrast, is the amount of energy needed to remove the highest-energy electron from an isolated neutral atom in the gaseous state. (Section 6.3)

H 2.1																		**He**
Li 1.0	**Be** 1.5											**B** 2.0	**C** 2.5	**N** 3.0	**O** 3.5	**F** 4.0	**Ne**
Na 0.9	**Mg** 1.2											**Al** 1.5	**Si** 1.8	**P** 2.1	**S** 2.5	**Cl** 3.0	**Ar**
K 0.8	**Ca** 1.0	**Sc** 1.3	**Ti** 1.5	**V** 1.6	**Cr** 1.6	**Mn** 1.5	**Fe** 1.8	**Co** 1.9	**Ni** 1.9	**Cu** 1.9	**Zn** 1.6	**Ga** 1.6	**Ge** 1.8	**As** 2.0	**Se** 2.4	**Br** 2.8	**Kr**
Rb 0.8	**Sr** 1.0	**Y** 1.2	**Zr** 1.4	**Nb** 1.6	**Mo** 1.8	**Tc** 1.9	**Ru** 2.2	**Rh** 2.2	**Pd** 2.2	**Ag** 1.9	**Cd** 1.7	**In** 1.7	**Sn** 1.8	**Sb** 1.9	**Te** 2.1	**I** 2.5	**Xe**
Cs 0.7	**Ba** 0.9	**Lu** 1.1	**Hf** 1.3	**Ta** 1.5	**W** 1.7	**Re** 1.9	**Os** 2.2	**Ir** 2.2	**Pt** 2.2	**Au** 2.4	**Hg** 1.9	**Tl** 1.8	**Pb** 1.9	**Bi** 1.9	**Po** 2.0	**At** 2.1	**Rn**

Electronegativity increases from left to right.

Most electronegative

Electronegativity decreases from top to bottom.

▲ **FIGURE 7.4**

Electronegativity values and trends in the periodic table.

Figure It Out

How does the electronegativity change down a group of elements and across a period of elements? Can you propose an explanation for the trend?

Answer: Electronegativity increases across a period because Z_{eff} increases as more protons are added to the nucleus. Electronegativity decreases down a group because Z_{eff} decreases as the positive nucleus is shielded by increasing numbers of core electrons. In general, electronegativity is inversely related to atomic size.

How can we use electronegativity to predict bond polarity? The following guidelines generally apply:

- Bonds between atoms with the same or similar electronegativity are usually nonpolar covalent.
- Bonds between atoms whose electronegativities differ by more than 2 units are largely ionic.
- Bonds between atoms whose electronegativities differ by less than 2 units are usually polar covalent.

WORKED EXAMPLE 7.1

Classifying Bond Type

Classify the carbon–chlorine bond in chloroform, $CHCl_3$, and the sodium–chlorine bond in sodium chloride as nonpolar covalent, polar covalent, or ionic.

STRATEGY

Calculate the electronegativity difference between elements using values in Figure 7.4 and follow the classification guidelines.

SOLUTION

We can be reasonably sure that a $C—Cl$ bond in chloroform, $CHCl_3$, is polar covalent, while an Na^+Cl^- bond in sodium chloride is largely ionic.

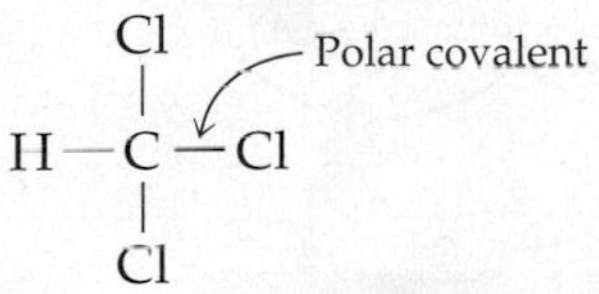

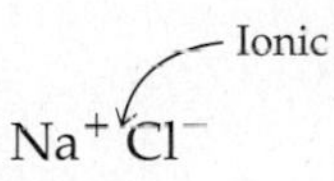

Chlorine:	EN = 3.0	Chlorine:	EN = 3.0
Carbon:	EN = 2.5	Sodium:	EN = 0.9
Difference = 0.5		Difference = 2.1	

PRACTICE 7.1 Use the electronegativity values in Figure 7.4 to predict whether the bonds in the following compounds are polar covalent or ionic:

(a) $SiCl_4$ **(b)** CsBr   **(c)** $FeBr_3$ **(d)** CH_4

Conceptual APPLY 7.2 An electrostatic potential map of water is shown below. Which atom, H or O, is positively polarized (electron-poor) and which is negatively polarized (electron-rich)? Is this polarity pattern consistent with the electronegativity values of O and H given in Figure 7.4?

Water

7.4 ▶ A COMPARISON OF IONIC AND COVALENT COMPOUNDS

Look at the comparison between NaCl and HCl in **TABLE 7.3** to get an idea of the difference between ionic and covalent compounds. Sodium chloride, an ionic compound, is a white solid with a melting point of 801 °C and a boiling point of 1465 °C. Hydrogen chloride, a covalent compound, is a colorless gas with a melting point of −115 °C and a boiling point of −84.9 °C.

TABLE 7.3	Some Physical Properties of NaCl and HCl	
Property	NaCl	HCl
Formula mass	58.44 amu	36.46 amu
Physical appearance	White solid	Colorless gas
Type of bond	Ionic	Covalent
Melting point	801 °C	−115 °C
Boiling point	1465 °C	−84.9 °C

What accounts for such large differences in properties between ionic compounds and covalent compounds?

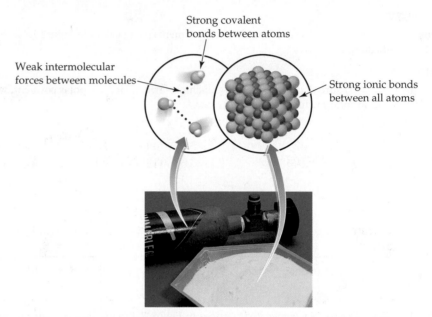

Strong covalent bonds between atoms

Weak intermolecular forces between molecules

Strong ionic bonds between all atoms

▲ Sodium chloride, an ionic compound, is a white, crystalline solid that melts at 801 °C. Hydrogen chloride, a molecular compound, is a gas at room temperature.

REMEMBER...

Lattice energy (*U*) is the amount of energy that must be supplied to break an ionic solid into its individual gaseous ions and is thus a measure of the strength of the crystal's ionic bonds. (Section 6.8)

Ionic compounds are high-melting solids because there are strong ionic bonds between each atom in a lattice structure. As discussed previously in Section 2.11, a visible sample of sodium chloride consists not of NaCl molecules but a vast three-dimensional network of ions in which each Na$^+$ cation is attracted to many surrounding Cl$^-$ anions and each Cl$^-$ ion is attracted to many surrounding Na$^+$ ions. To melt or boil sodium chloride, all the ions must be separated from one another. This means that every ionic attraction in the entire crystal—the **lattice energy (*U*)**—must be overcome, a process that requires a large amount of energy.

Covalent compounds, by contrast, are low-melting solids, liquids, or even gases. A sample of a covalent compound, such as hydrogen chloride, consists of discrete HCl molecules. In order to melt or boil a covalent compound, forces between molecules must be broken. In covalent compounds, bonds exist *within* molecules but not *between* molecules. Attractive forces between molecules do exist and are called **intermolecular forces**. *Intermolecular forces are weaker than ionic or covalent bonds* and therefore relatively little energy is required to overcome them to melt or boil the substance. We'll look at the nature of intermolecular forces in Chapter 8.

7.5 ▶ ELECTRON-DOT STRUCTURES: THE OCTET RULE

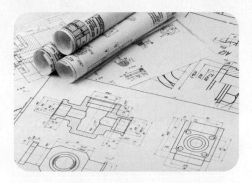

One way to picture the sharing of electrons between atoms in covalent or polar covalent bonds is to use **electron-dot structures**, or *Lewis structures*, named after G. N. Lewis of the University of California at Berkeley. An electron-dot structure represents an atom's valence electrons by dots and indicates by the placement of the dots how the valence electrons are distributed in a molecule. The central idea is that when atoms form a bond they share electrons to achieve a *complete valence shell* or *noble gas electron configuration*.

The electron-dot model, like most models, is a simplified picture of bonding that has been developed because it is useful in making predictions. Just as a blueprint serves as a model for a building, the Lewis structures can be thought of as a "blueprint" containing a set of rules for combining atoms to make molecules. Some key uses of the electron-dot model for bonding are in predicting molecular formulas, reactivity, and shape.

An electron-dot structure for an H_2 molecule is written showing a pair of dots between the hydrogen atoms, indicating that the hydrogens share the pair of electrons in a covalent bond:

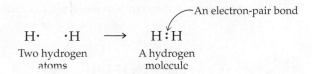

By sharing two electrons in a covalent bond, each hydrogen effectively has one electron pair and the stable, $1s^2$ electron configuration of the noble gas helium.

(b)

▲ Lewis structures can be thought of as a "blueprint" that provides a set of guidelines for combining atoms to make molecules and predicting their three-dimensional structure.

Atoms other than hydrogen also form covalent bonds by sharing electron pairs, and the electron-dot structures of the resultant molecules are drawn by assigning the correct number of valence electrons to each atom. Group 3A atoms, such as boron, have three valence electrons; group 4A atoms, such as carbon, have four valence electrons; and so on across the periodic table. The group 7A element fluorine has seven valence electrons, and an electron-dot structure for the F_2 molecule shows how a covalent bond can form:

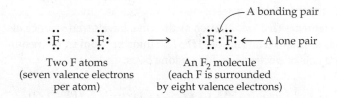

Six of the seven valence electrons in a fluorine atom are already paired in three filled atomic orbitals and thus are not shared in bonding. The seventh fluorine valence electron, however, is unpaired and can be shared in a covalent bond with another fluorine. Each atom in the resultant F_2 molecule thereby gains a noble-gas configuration with eight valence-shell electrons and thus obeys the **octet rule**, discussed in Section 6.6. The three pairs of nonbonding electrons on each fluorine atom are called **lone pairs**, or *nonbonding pairs*, and the shared electrons are called a *bonding pair*.

REMEMBER...

According to the **octet rule**, main-group elements tend to undergo reactions that give them a noble-gas electron configuration with filled *s* and *p* subshells in their valence electron shell. (Section 6.6)

The tendency of main-group atoms to fill their *s* and *p* subshells and thereby achieve a noble-gas configuration when they form bonds is an important guiding principle that makes it possible to predict the formulas and electron-dot structures of a great many molecules. As a general rule, a main-group atom shares as many of its valence-shell electrons as possible, either until it has no more to share or until it reaches an octet configuration. The following guidelines apply:

- **Group 3A elements**, such as boron, have three valence electrons and can therefore form three electron-pair bonds in neutral molecules such as borane, BH_3. The boron atom in the resultant molecule has only three bonding pairs of electrons, however, and can't reach an electron octet. (The bonding situation in BH_3 is actually more complicated than suggested here; we'll deal with it in Section 22.6.)

$$\cdot \overset{\cdot}{B} \cdot \;+\; 3\,H\cdot \;\longrightarrow\; H \overset{\overset{H}{\cdot\cdot}}{\underset{}{B}} H$$

Only six electrons around boron

Borane

- **Group 4A elements**, such as carbon, have four valence electrons and form four bonds, as in methane, CH_4. The carbon atom in the resultant molecule has four bonding pairs of electrons.

$$\cdot \overset{\cdot}{\underset{\cdot}{C}} \cdot \;+\; 4\,H\cdot \;\longrightarrow\; H \overset{\overset{H}{\cdot\cdot}}{\underset{\underset{H}{\cdot\cdot}}{C}} H$$

Methane

- **Group 5A elements**, such as nitrogen, have five valence electrons and form three bonds, as in ammonia, NH_3. The nitrogen atom in the resultant molecule has three bonding pairs of electrons and one lone pair.

$$\cdot \overset{\cdot}{\underset{\cdot\cdot}{N}} \cdot \;+\; 3\,H\cdot \;\longrightarrow\; H \overset{\overset{H}{\cdot\cdot}}{\underset{\cdot\cdot}{N}} H$$

Ammonia

- **Group 6A elements**, such as oxygen, have six valence electrons and form two bonds, as in water, H_2O. The oxygen atom in the resultant molecule has two bonding pairs of electrons and two lone pairs.

$$\cdot \overset{\cdot\cdot}{\underset{\cdot\cdot}{O}} \cdot \;+\; 2\,H\cdot \;\longrightarrow\; H \overset{\cdot\cdot}{\underset{\cdot\cdot}{O}} H$$

Water

- **Group 7A elements** (halogens), such as fluorine, have seven valence electrons and form one bond, as in hydrogen fluoride, HF. The fluorine atom in the resultant molecule has one bonding pair of electrons and three lone pairs.

$$:\overset{\cdot\cdot}{\underset{\cdot\cdot}{F}} \cdot \;+\; H\cdot \;\longrightarrow\; H \overset{\cdot\cdot}{\underset{\cdot\cdot}{F}}:$$

Hydrogen
fluoride

- **Group 8A elements** (noble gases), such as neon, rarely form covalent bonds because they already have valence-shell octets.

$$:\overset{\cdot\cdot}{\underset{\cdot\cdot}{Ne}}:$$ Does not form covalent bonds

These conclusions are summarized in **TABLE 7.4**.

TABLE 7.4 Covalent Bonding for Second-Row Elements

Group	Number of Valence Electrons	Number of Bonds	Example
3A	3	3	BH_3
4A	4	4	CH_4
5A	5	3	NH_3
6A	6	2	H_2O
7A	7	1	HF
8A	8	0	Ne

Not all covalent bonds contain just one shared electron pair, or **single bond**, like those just discussed. In molecules such as O_2, N_2, and many others, the atoms share more than one pair of electrons, leading to the formation of *multiple* covalent bonds. The oxygen atoms in the O_2 molecule, for example, reach valence-shell octets by sharing two pairs, or four electrons, in a **double bond**. Similarly, the nitrogen atoms in the N_2 molecule share three pairs, or six electrons, in a **triple bond**. (Although the O_2 molecule does have a double bond, the following electron-dot structure is incorrect in some respects, as we'll see in Section 8.8.)

$$\cdot \ddot{\text{O}} \cdot \; + \; \cdot \ddot{\text{O}} \cdot \; \longrightarrow \; \ddot{\text{O}} :: \ddot{\text{O}}$$

Two electron pairs form the double bond, O=O

$$: \dot{\text{N}} \cdot \; + \; \cdot \dot{\text{N}} : \; \longrightarrow \; : \text{N} ::: \text{N} :$$

Three electron pairs form the triple bond, N≡N

One final point about covalent bonds involves the origin of the bonding electrons. Although most covalent bonds form when two atoms each contribute one electron, bonds can also form when one atom donates both electrons (a lone pair) to another atom that has a vacant valence orbital. The ammonium ion (NH_4^+), for instance, forms when the two lone-pair electrons from the nitrogen atom of ammonia, $:NH_3$, bond to H^+. Such bonds are sometimes called **coordinate covalent bonds**, which form when one atom donates an electron pair to an empty orbital on another atom.

An ordinary covalent bond—each atom donates one electron.

$$\text{H} \cdot \; + \; \cdot \text{H} \; \longrightarrow \; \text{H} : \text{H}$$

A coordinate covalent bond—the nitrogen atom donates both electrons.

$$\text{H}^+ \; + \; \begin{array}{c} \text{H} \\ : \ddot{\text{N}} : \text{H} \\ \text{H} \end{array} \; \longrightarrow \; \left[\begin{array}{c} \text{H} \\ \text{H} : \ddot{\text{N}} : \text{H} \\ \text{H} \end{array} \right]^+$$

Note that the nitrogen atom in the ammonium ion (NH_4^+) has more than the usual number of bonds—four instead of three—but that it still has an octet of valence electrons. Nitrogen, oxygen, phosphorus, and sulfur form coordinate covalent bonds frequently.

WORKED EXAMPLE 7.2

Drawing an Electron-Dot Structure

Draw an electron-dot structure for phosphine, PH_3.

STRATEGY

The number of covalent bonds formed by a main-group element depends on the element's group number. Phosphorus, a group 5A element, has five valence electrons and can achieve a valence-shell octet by forming three bonds and leaving one lone pair. Each hydrogen supplies one electron.

SOLUTION

$$H\overset{\displaystyle H}{\underset{..}{:\!\ddot{P}\!:}}H \qquad \text{Phosphine}$$

▶ **PRACTICE 7.3**   Draw electron-dot structures for the following molecules:

(a) H_2S, hydrogen sulfide, a poisonous gas produced by rotten eggs
(b) $CHCl_3$, chloroform, an anesthetic

▶ **APPLY 7.4**   Use the octet rule to predict the molecular formula of compounds that form between the elements

(a) oxygen and fluorine.
(b) silicon and chlorine.

7.6 ▶ PROCEDURE FOR DRAWING ELECTRON-DOT STRUCTURES

In drawing structures, we'll follow the usual convention of indicating a two-electron covalent bond by a line. Similarly, we'll use two lines between atoms to represent four shared electrons (two pairs) in a double bond, and three lines to represent six shared electrons (three pairs) in a triple bond. Arranging individual electrons when drawing electron-dot structures can be tedious, especially in large molecules or in those with multiple bonds. Instead, a general method of drawing electron-dot structures is outlined below.

Drawing Electron-Dot Structures

Step 1.   Find the total number of valence electrons in the molecule or ion. Find the sum of the valence electrons for all atoms and add one additional electron for each negative charge in an anion, and subtract one electron for each positive charge in a cation. In SF_4, for example, the total is 34 (6 from sulfur and 7 from each of 4 fluorines). In OH^-, the total is 8 (6 from oxygen, 1 from hydrogen, and 1 for the negative charge). In NH_4^+, the total is 8 (5 from nitrogen, and 1 from each of 4 hydrogens, minus 1 for the positive charge).

$$SF_4 \qquad\qquad OH^- \qquad\qquad NH_4^+$$

$$:\!\dot{\underset{..}{S}}\!\cdot \quad 4\!:\!\ddot{\underset{..}{F}}\!\cdot \qquad :\!\ddot{\underset{..}{O}}\!\cdot \ \ H\!\cdot \qquad :\!\dot{\underset{.}{N}}\!\cdot \ \ 4\,H\!\cdot$$

$$6e^- + (4 \times 7e^-) \qquad 6e^- + 1e^- + 1e^- \qquad 5e^- + (4 \times 1e^-) - 1e^-$$
$$= 34e^- \qquad\qquad = 8e^- \qquad\qquad = 8e^-$$

Step 2.   Decide what the connections are between atoms, and draw lines to represent the bonds. Often, you'll be told the connections; other times you'll have to guess. Guidelines for connecting atoms in an electron-dot structure are:

- The central atom is typically the one with the lowest electronegativity (except H).
- Hydrogen and the halogens usually form only one bond.
- Elements in the second row usually form the number of bonds given in Table 7.4.
- *(Octet Rule Exception)* Elements in the third row and lower often have an **expanded octet**, as they form more bonds than predicted by the octet rule. Atoms of these

1 1A																	18 8A
H	2 2A											13 3A	14 4A	15 5A	16 6A	17 7A	He
Li	Be											B	C	N	O	F	Ne
Na	Mg	3 3B	4 4B	5 5B	6 6B	7 7B	8	9 8B	10	11 1B	12 2B	Al	Si	P	S	Cl	Ar
K	Ca	Sc	Ti	V	Cr	Mn	Fe	Co	Ni	Cu	Zn	Ga	Ge	As	Se	Br	Kr
Rb	Sr	Y	Zr	Nb	Mo	Tc	Ru	Rh	Pd	Ag	Cd	In	Sn	Sb	Te	I	Xe
Cs	Ba	Lu	Hf	Ta	W	Re	Os	Ir	Pt	Au	Hg	Tl	Pb	Bi	Po	At	Rn
Fr	Ra	Lr	Rf	Db	Sg	Bh	Hs	Mt	Ds	Rg	Cn						

Atoms of these elements, all of which are in the third row or lower, are larger than their second-row counterparts and can therefore accommodate more bonded atoms.

▲ **FIGURE 7.5**

The octet rule occasionally fails for the main-group elements shown in blue.

Figure It Out

Which of the following elements can have an expanded octet when they are the central atom in an electron-dot structure? Cl, O, S, I, C

Answer: Cl, S, I.

elements are larger than their second row counterparts, can accommodate more than 4 atoms around them, and therefore form more than four bonds (**FIGURE 7.5**). The second-row element nitrogen, for instance, bonds to only three chlorine atoms in forming NCl_3 and thus obeys the octet rule, while the third-row element phosphorus bonds to five chlorine atoms in forming PCl_5 and thus does not follow the octet rule.

• *(Octet Rule Exception)* Elements in Group 3A, such as B and Al, are frequently **electron deficient**, meaning they are surrounded by less than eight electrons. The electron-dot structure for BH_3 in Section 7.5 is a representative example; B has only three bonds, or a total of six electrons.

If, for example, you were asked to predict the connections in SF_4, a good guess would be that each fluorine forms one bond to sulfur, which occurs as the central atom.

$$\begin{array}{ccc} F & & F \\ & \diagdown \ \diagup & \\ & S & \\ & \diagup \ \diagdown & \\ F & & F \end{array}$$ Sulfur tetrafluoride, SF_4

Step 3. Subtract the number of valence electrons used for bonding from the total number calculated in Step 1 to find the number that remain. Assign as many of these remaining electrons as necessary to the terminal atoms (other than hydrogen) so that each has an octet. In SF_4, 8 of the 34 total valence electrons are used in covalent bonding, leaving $34 - 8 = 24$. Twenty-four of the 26 are assigned to the four terminal fluorine atoms to reach an octet configuration for each:

$$\begin{array}{ccc} :\ddot{F} & & \ddot{F}: \\ & \diagdown \ \diagup & \\ & S & \\ & \diagup \ \diagdown & \\ :\ddot{F} & & \ddot{F}: \end{array}$$ $8 + 24 = 32$ electrons distributed

Step 4. If unassigned electrons remain after Step 3, place them on the central atom. In SF_4, 32 of the 34 electrons have been assigned, leaving the final 2 to be placed on the central S atom. Sulfur is in the third row and can therefore expand its octet using *d* orbitals.

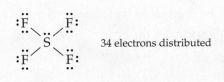

34 electrons distributed

Step 5. Form multiple bonds to fulfill an incomplete octet on the central atom. If no unassigned electrons remain after Step 3 but the central atom does not yet have an octet, use one or more lone pairs of electrons from a neighboring atom to form a multiple bond (either double or triple). Oxygen, carbon, nitrogen, and sulfur often form multiple bonds. Worked Example 7.4 shows how to deal with such a case.

Drawing Electron-Dot Structures for Molecules with One Central Atom

Worked Examples 7.3 and 7.4 demonstrate how to draw electron-dot structures for molecules with one central atom.

WORKED EXAMPLE 7.3

Drawing Electron-Dot Structures for Molecules with One Central Atom and Single Bonds

Draw an electron-dot structure for OF_2.

STRATEGY

Follow the five steps outlined at the start of this section under "Drawing Electron-Dot Structures."

SOLUTION

Step 1. First count the total number of valence electrons. Oxygen has 6 and each fluorine has 7 for a total of 20.

Step 2. Next, decide how atoms may be connected and draw lines to indicate bonds. Since oxygen is less electronegative than fluorine and halogens typically only form one bond, place oxygen in the center.

$$F—O—F$$

Step 3. Subtract the number of valence electrons used for bonding (4 total, 2 from each bond) from the total number calculated in Step 1 (20) to find the number that remain $(20 - 4 = 16)$. Assign as many of these remaining electrons as necessary to the terminal atoms so that each has an octet. Each fluorine atom requires three lone pairs to satisfy the octet rule for a total of 12 nonbonding electrons.

$$:\ddot{F}—O—\ddot{F}: \quad 4 + 12 = 16 \text{ electrons distributed}$$

Step 4. If unassigned electrons remain after Step 3, place them on the central atom.

$$:\ddot{F}—\ddot{O}—\ddot{F}: \quad 20 \text{ electrons distributed}$$

Step 5. If the central atom does not have an octet, make multiple bonds. This step is not necessary because the central atom oxygen has a complete octet.

CHECK

Electron-dot structures can always be checked by evaluating two criteria. (1) Does the final structure have the same number of electrons as the total number of valence electrons in molecule? (2) Is the octet rule satisfied for each atom? The structure for OF_2 contains 20 electrons and each atom has an octet; therefore, a valid electron-dot structure has been drawn.

▶ **PRACTICE 7.5** Draw an electron-dot structure for the following molecules or ions:

(a) CH_2F_2 (b) $AlCl_3$ (c) ClO_4^- (d) PCl_5 (e) $XeOF_4$ (f) NH_4^+

▶ **APPLY 7.6** Identify the correct electron-dot structure for $POCl_3$. Explain what is wrong with the incorrect ones.

(a)   (b)   (c)

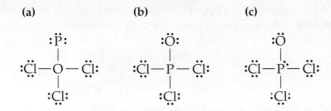

●— WORKED EXAMPLE 7.4

Drawing Electron-Dot Structures for Molecules with One Central Atom and Multiple Bonds

Draw an electron-dot structure for formaldehyde, CH_2O, a compound used in manufacturing the adhesives for making plywood and particle board.

STRATEGY

Follow the five steps outlined in the list above entitled "Drawing Electron-Dot Structures."

SOLUTION

Step 1. Count the total number of valence electrons. Carbon has 4, each hydrogen has 1, and the oxygen has 6, for a total of 12.

Step 2. Decide on the probable connections between atoms, and draw a line to indicate each bond. In the case of formaldehyde, the less electronegative atom (carbon) is the central atom, and both hydrogens and the oxygen are bonded to carbon:

$$\begin{array}{c} O \\ | \\ H-C-H \end{array}$$

Steps 3 and 4. Six of the 12 valence electrons are used for bonds, leaving 6 for assignment to the terminal oxygen atom. There are no more electrons to add to the central carbon atom.

$$\begin{array}{c} :\ddot{O}: \\ | \\ H-C-H \end{array} \quad \text{12 electrons distributed}$$

↖—Only six electrons around carbon

Step 5. At this point, all the valence electrons are assigned, but the central carbon atom still does not have an octet. To achieve an octet, we move two of the oxygen electrons from a lone pair into a bonding pair, generating a carbon–oxygen double bond that satisfies the octet rule for both oxygen and carbon.

$$\begin{array}{c} :O: \\ \| \\ H-C-H \end{array}$$

Formaldehyde, CH_2O

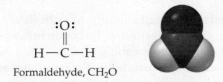

CHECK

The structure for CH_2O contains the correct number of valence electrons (12) and each atom has an octet.

continued on next page

▶ **PRACTICE 7.7** Draw an electron-dot structure for the following molecules or ions:

(a) CO   (b) HCN   (c) CO_3^{2-} (d) CO_2

▶ **APPLY 7.8** Identify the correct electron-dot structure(s) for phosphate, PO_4^{3-}. Explain what is wrong with the incorrect ones. Is it reasonable to have more than one correct electron-dot structure? Explain.

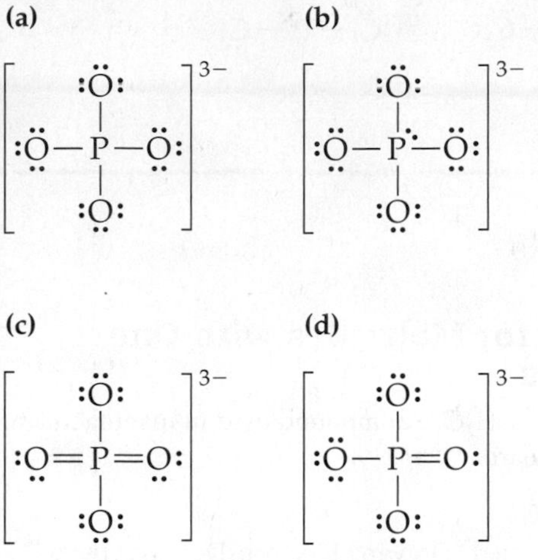

(a)   (b)

(c)   (d)

7.7 ▶ DRAWING ELECTRON-DOT STRUCTURES FOR RADICALS

Except for hydrogen, nearly all atoms in almost all the 60 million known chemical compounds follow the octet rule. As a result, most compounds have an even number of electrons, which are paired up in filling orbitals. A very few substances, called **radicals** (or sometimes *free radicals*), have an odd number of electrons, meaning that at least one of their electrons must be unpaired in a half-filled orbital.

Because of their unpaired electron, radicals are usually very reactive, rapidly undergoing some kind of reaction that allows them to pair their electrons and form a more stable product. This pairing of electrons is why isolated halogen atoms dimerize to form halogen molecules, and nitrogen dioxide (NO_2) dimerizes to form dinitrogen tetroxide (N_2O_4).

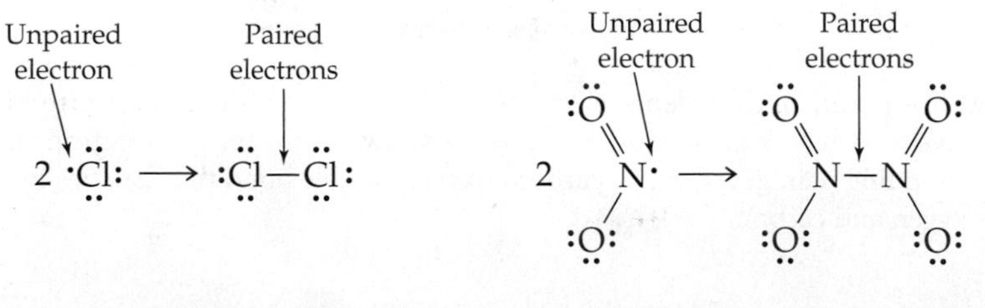

Despite their high reactivity, small amounts of radicals are involved in many common processes, including the combustion of fuels, the industrial preparation of polyethylene and other polymers, and the ongoing destruction of the Earth's ozone layer in the upper atmosphere by chlorofluorocarbons. Perhaps more surprisingly, radicals are also present in the human body, where they both mediate important life processes and act as causal factors for many diseases, including cancer, emphysema, stroke, and diabetes. Even the changes that take place as a result of normal aging are thought to involve radicals.

For example, nitric oxide (NO) is a radical emitted from combustion engines that initi-ates a set of reactions that contribute to atmospheric ozone pollution. However, NO serves several useful purposes in the body and the Nobel Prize in Physiology and Medicine in 1998 was awarded to scientists who discovered that nitric oxide is an important signaling molecule in the cardiovascular system. Nitric oxide, also known as the endothelium-derived relaxing factor (EDRF), is released by cells lining the interior surface of blood vessels. The release of NO causes the muscles surrounding vessels to relax, thereby dilating blood vessels and increasing blood flow. Hair growth, pulmonary hypertension, and angina are all treated by drugs that affect NO release.

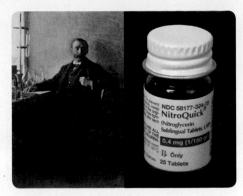

▲ Alfred Nobel used his fortune from the invention of dynamite, a formulation with the nitroglycerin molecule, to institute the highly prestigious Nobel prizes. He found it ironic that later in life his doctor prescribed nitroglycerin, which releases nitric oxide, to treat his heart condition.

──• WORKED EXAMPLE 7.5

Drawing an Electron-Dot Structure for a Radical

Draw an electron-dot structure for nitric oxide, NO.

STRATEGY

Follow the five steps outlined in the list entitled "Drawing Electron-Dot Structures" (Section 7.6). In a radical, the *least electronegative* atom will have an odd number of electrons and an incomplete octet.

SOLUTION

Step 1. Count the total number of valence electrons. Nitrogen has 5 and oxygen has 6, for a total of 11. An odd number of electrons indicate the presence of a radical.

Step 2. Determine the connectivity of atoms. Nitrogen and oxygen are bonded together as a simple diatomic molecule.

Steps 3 and 4. Subtract the number of valence electrons used for bonding (2) from the total number calculated in Step 1 (11) to find the number that remain (9). Assign as many of these remaining electrons as necessary to the terminal atoms. Give the more electronegative atom, oxygen, an octet and place any remaining electrons on nitrogen.

$$\cdot \ddot{N}-\ddot{O}:$$ 11 electrons distributed

Step 5. All the valence electrons are assigned, but nitrogen is not close to satisfying its octet with only five electrons in the structure. We therefore move two of the oxygen electrons from a lone pair into a bonding pair, generating a double bond. The octet rule is satisfied for oxygen as it is the more electronegative atom and nitrogen is now closer to an octet. The least electronegative atom will never have an octet since a radical has an odd number of electrons.

$$\cdot \ddot{N}=\ddot{O}$$ 11 electrons distributed

CHECK

The structure for NO contains 11 electrons and the most electronegative atom (O) has a complete octet.

▶ **PRACTICE 7.9** Draw an electron-dot structure for the following radicals:

(a) ClO_2 (involved in depletion of stratospheric ozone)
(b) Hydroxyl, OH (important in atmospheric chemistry, aging, and the development of disease)
(c) Ethyl, C_2H_5 (involved in synthesis of polyethylene polymer)

▶ **APPLY 7.10** Which oxygen species do you predict to be most reactive? O_3, O_2, O_2^-.

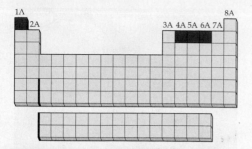

7.8 ▶ ELECTRON-DOT STRUCTURES OF COMPOUNDS CONTAINING ONLY HYDROGEN AND SECOND-ROW ELEMENTS

Many of the naturally occurring compounds on which life is based—proteins, fats, carbohydrates, and numerous others—contain only hydrogen and one or more of the second-row elements carbon, nitrogen, and oxygen. Molecules that contain carbon atoms bonded together in a chain are called **organic compounds**. Electron-dot structures are relatively easy to draw for such compounds because the octet rule is almost always followed and the number of bonds formed by each element is predictable (Table 7.4).

For molecules that contain second-row atoms in addition to hydrogen, the second-row atoms are bonded to one another in a central core, with hydrogens on the periphery. In ethane (C_2H_6), for instance, two carbon atoms, each of which forms four bonds, combine with six hydrogens, each of which forms one bond. Joining the two carbon atoms and adding the appropriate number of hydrogens to each yields only one possible structure:

$$\left.\begin{array}{c} 6\ H\cdot \\[1em] 2\ \cdot\overset{\displaystyle\cdot}{C}\cdot \end{array}\right\} \longrightarrow \quad \begin{array}{c} H\ H \\ H:\overset{\displaystyle..}{\underset{\displaystyle..}{C}}:\overset{\displaystyle..}{\underset{\displaystyle..}{C}}:H \\ H\ H \end{array} \quad \text{or} \quad \begin{array}{c} H\quad H \\ |\quad\ | \\ H-C-C-H \\ |\quad\ | \\ H\quad H \end{array}$$

Ethane, C_2H_6

For larger molecules that contain numerous second-row atoms, there is usually more than one possible electron-dot structure. In such cases, some additional knowledge about the order of connections among atoms is necessary before a structure can be drawn.

● WORKED EXAMPLE 7.6

Drawing an Electron-Dot Structure for Molecules with More than One Central Atom

Draw an electron-dot structure for acetylene, C_2H_2, commonly used as a fuel for torches and as a starting material in chemical synthesis.

STRATEGY

Follow the five steps for drawing electron-dot structures outlined in Section 7.6. For molecules with more than one central atom, the second-row atoms are bonded to one another in a central core.

SOLUTION

Step 1. Count the total number of valence electrons in the molecule. Each carbon has 4 and each hydrogen has 1, for a total of 10.

Step 2. Determine the connectivity of atoms. Connect the second row atoms (C) in a chain.

$$H-C-C-H$$

Steps 3 and 4. Subtract the number of valence electrons used for bonding (6) from the total number calculated in step 1 (10) to find the number that remain (4). Assign as many of these remaining electrons as necessary to the terminal atoms, except hydrogen. Since four electrons remain and all terminal atoms are hydrogen, place the electrons on the central carbon atoms.

$$H-\overset{\displaystyle..}{C}-\overset{\displaystyle..}{C}-H$$

Step 5. All the valence electrons are assigned, but the two central carbon atoms do not have an octet. We can move electron pairs on the two carbon atoms to form a triple bond between them, thus satisfying the octet rule for carbon. The valence shell for hydrogen is also filled with two electrons.

$$H-C\equiv C-H$$

CHECK

The structure for C_2H_2 contains the correct number of valence electrons (10), and each atom has a complete valence shell; carbon with 8 and hydrogen with 2.

▶ **PRACTICE 7.11** Draw an electron-dot structure for the following molecules or ions:

(a) Methylamine, CH_5N
(b) Ethylene, C_2H_4
(c) Propane, C_3H_8
(d) Hydrogen Peroxide, H_2O_2
(e) Hydrazine, N_2H_4

▶ **APPLY 7.12** There are two molecules with the formula C_2H_6O. Draw electron-dot structures for both. (Hint: The connection of atoms is different in the two structures.)

Conceptual WORKED EXAMPLE 7.7

Identifying Multiple Bonds and Lone Pairs in Organic Molecules

Below is a representation of histidine, an amino acid constituent of proteins. Only the connections between atoms are shown; multiple bonds are not indicated. Give the chemical formula of histidine, and complete the structure by showing where the multiple bonds and lone pairs are located (red = O, gray = C, blue = N, ivory = H).

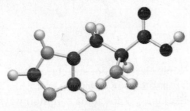

Histidine

STRATEGY

Count the atoms of each element to find the formula. Then look at each atom in the structure to find what is needed for completing its octet. Table 7.4 indicates the number of bonds formed by elements in different groups of the periodic table. Each carbon (gray) should have four bonds, each oxygen (red) should have two bonds, and each nitrogen (blue) should have three bonds. Complete the octet for each atom by adding lone pairs; carbon already has a complete octet with four bonds, nitrogen needs one lone pair of electrons, and oxygen needs two lone pairs of electrons.

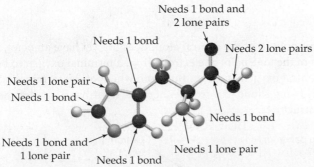

Needs 1 bond and 2 lone pairs

Needs 1 bond

Needs 2 lone pairs

Needs 1 lone pair

Needs 1 bond

Needs 1 bond

Needs 1 bond and 1 lone pair

Needs 1 lone pair

Needs 1 bond

continued on next page

SOLUTION

Histidine has the formula $C_6H_9N_3O_2$.

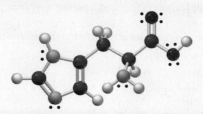

CHECK

It would be time consuming to count all electrons in this structure and compare the number to total valence electrons. The best way to check a structure for a large molecule is to ensure that each atom (except H) has an octet.

▶ **PRACTICE 7.13** The following structure is a representation of cytosine, a constituent of the DNA found in all living cells. Only the connections between atoms are shown; multiple bonds are not indicated. Give the formula of cytosine, and complete the structure by showing where the multiple bonds and lone pairs are located (red = O, gray = C, blue = N, ivory = H).

Cytosine

▶ **APPLY 7.14** Draw two possible electron-dot structures for the molecules with the following formulas. You may connect the atoms in different ways to draw the structures. (Hint: Look at the structures in Worked Example 7.7 for some common ways that second row atoms are connected.)

(a) C_6H_7N **(b)** $C_2H_5O_2N$

7.9 ▶ ELECTRON-DOT STRUCTURES AND RESONANCE

The stepwise procedure given in Section 7.6 for drawing electron-dot structures sometimes leads to an interesting problem. Look at ozone, O_3, for instance. Step 1 indicates the molecule has 18 valence electrons, and following Steps 2–4 we draw the following structure:

$$:\ddot{O}-\ddot{O}-\ddot{O}:$$

We find at this point that the central atom does not yet have an octet, and we therefore have to move one of the lone pairs of electrons from a terminal oxygen to become a bonding pair, giving the central oxygen an octet. But from which of the terminal oxygens should we take a lone pair? Moving a lone pair from either the "right-hand" or "left-hand" oxygen produces acceptable structures:

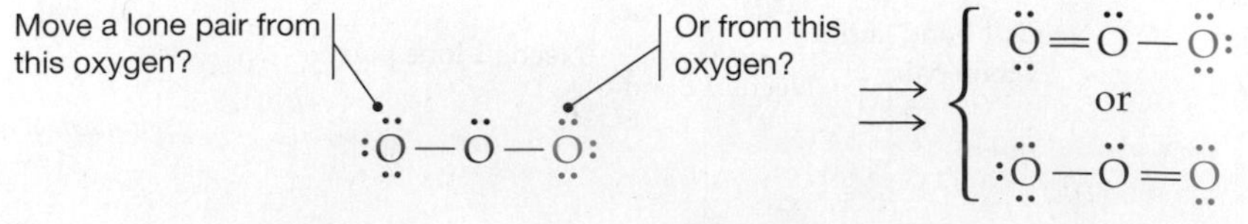

Move a lone pair from this oxygen?   Or from this oxygen?

$$:\ddot{O}-\ddot{O}-\ddot{O}: \longrightarrow \begin{cases} \ddot{O}=\ddot{O}-\ddot{O}: \\ \text{or} \\ :\ddot{O}-\ddot{O}=\ddot{O} \end{cases}$$

Which of the two structures for O_3 is correct? In fact, neither is correct by itself. This bonding description is at odds with experimental evidence about the actual structure of ozone. If this were an accurate depiction, then one oxygen–oxygen bond should be shorter than the other and the double bond should show a region of higher electron density. Measurements show that both oxygen–oxygen bond lengths in the ozone molecule are equivalent. The electrostatic potential map also indicates that both bonds are equivalent in electron density. A double bond would show increased electron density, or red color between the central and terminal oxygen atoms. The symmetrical distribution of electron density indicates that the electrons from the double bond are **delocalized**, or spread out over the two oxygen–oxygen bonds.

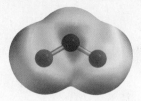

▲ Electrostatic potential map for ozone showing two equivalent oxygen-oxygen bonds.

Bond lengths and bond strengths for ozone can be measured experimentally and are compared with average single (O—O) and double (O=O) bonds from other molecules. The characteristics of the oxygen–oxygen bond in ozone are between those of a single and double bond. The two bonds in ozone have an identical length of 128 pm and we describe ozone as having a bond order of 1.5, midway between pure single bonds and pure double bonds.

	(O=O) in O_2	Average (O—O)	Oxygen–Oxygen Bond in Ozone
Bond Length (pm)	121	132	128
Bond Strength (kJ/mol)	498	146	374

Noting that the electron-dot model is at odds with the actual structure of ozone, we underscore that Lewis theory is just a model for bonding that chemists use to make predictions about how atoms combine. Lewis theory is extremely useful, but keep in mind that all models have limitations and they may not be a true representation of reality.

How does the "hybrid" nature of the bonds in ozone arise? The concept of **resonance theory**, developed by Linus Pauling in the 1930s, describes many molecules by drawing two or more electron-dot structures and considering the actual molecule to be a composite of these two structures. Curved arrows are a bookkeeping method showing how electrons can be rearranged to create a different electron-dot structure. Note that electrons are not really flowing through the molecule; the curved arrows are used to keep track of electron changes in dot structures. The tail indicates which electrons will move and the arrowhead indicates where they will now reside. Electrons can move into a new bond or onto an atom as a lone pair. When drawing resonance structures do not move atoms, only electrons! Curved arrows show how the two contributing electron-dot structures for the ozone molecule are interconverted.

$$:\ddot{O}-\ddot{O}=\ddot{O} \longleftrightarrow \ddot{O}=\ddot{O}-\ddot{O}:$$

Resonance structures are indicated when two (or more) alternative electron-dot structures are connected by a double-headed *resonance arrow* (↔). A straight, double-headed arrow always indicates resonance; it is never used for any other purpose. Two resonance structures can be drawn for ozone and the *average* of these two structures is called a **resonance hybrid**. Note that the two resonance forms differ only in the placement of valence shell electrons (both bonding and nonbonding). The total number of valence electrons is the same in both structures, the connections between atoms remain the same, and the relative positions of the atoms remain the same.

This double-headed arrow means that structures on either side are contributors to the resonance hybrid.

A solid and dashed line indicates a bond with properties between those of a single and double bond.

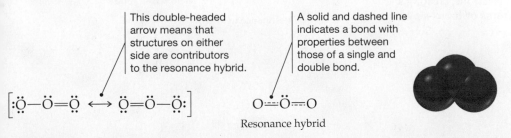

$$\left[:\ddot{O}-\ddot{O}=\ddot{O} \longleftrightarrow \ddot{O}=\ddot{O}-\ddot{O}:\right] \qquad O=\!\!=\ddot{O}=\!\!=O$$

Resonance hybrid

Recognizing that single electron-dot structures can't be written for all molecules tells us that the electron-dot model oversimplifies how bonds form and how electrons are distributed within a molecule. A more accurate way of describing electron distributions called *molecular orbital theory* will be presented in Chapter 8. Chemists nevertheless make routine use of electron-dot and resonance structures because they are a simple and powerful tool for predicting molecular structure and reactive sites in molecules.

WORKED EXAMPLE 7.8

Drawing Resonance Structures

The nitrate ion, NO_3^-, has three equivalent nitrogen–oxygen bonds, and its electronic structure is a resonance hybrid of three electron-dot structures.

(a) Draw a valid electron-dot structure for nitrate.

(b) Use curved arrows to convert the structure drawn in part (a) into two other resonance structures.

(c) Draw all three resonance structures and a resonance hybrid.

(a) STRATEGY AND SOLUTION

Begin as you would for drawing any electron-dot structure. There are 24 valence electrons in the nitrate ion: 5 from nitrogen, 6 from each of 3 oxygens, and 1 for the negative charge. The three equivalent oxygens are all bonded to nitrogen, the less electronegative central atom:

6 of 24 valence electrons assigned

Distributing the remaining 18 valence electrons among the three terminal oxygen atoms completes the octet of each oxygen but leaves nitrogen with only 6 electrons.

To give nitrogen an octet, one of the oxygen atoms must use a lone pair to form a nitrogen–oxygen double bond.

(b) STRATEGY AND SOLUTION

Use curved arrows to change the location of electrons, creating a new electron-dot structure. Do not move atoms or break single bonds when drawing resonance structures!

(c) STRATEGY AND SOLUTION

Show all the valid electron-dot structures connected by the resonance arrow(s). The resonance hybrid is considered to be an average of all three structures and is drawn with three equivalent nitrogen–oxygen bonds. The bonds are represented with a solid line and a dashed line to indicate that they are between a single and a double bond.

SOLUTION

Resonance hybrid

CHECK

All structures have the correct number of valence electrons and satisfy the octet rule.

▶ **PRACTICE 7.15** Called "laughing gas," nitrous oxide (N_2O) is sometimes used by dentists as an anesthetic.

(a) Given the connections N—N—O draw an electron-dot structure.

(b) Use curved arrows to show how the structure in part (a) can be converted into a resonance structure.

▶ **APPLY 7.16** Draw as many resonance structures as possible for each of the following molecules or ions, giving all atoms (except H) octets. Used curved arrows to show how one structure can be converted into another.

(a) SO_2 (b) CO_3^{2-} (c) BF_3

● **CONCEPTUAL** **WORKED EXAMPLE 7.9**

Drawing Resonance Structures in Organic Compounds

A structural formula for histidine (Worked Example 7.7) is shown. Evaluate whether the following arrows result in a valid electron-dot structure for histidine. If so, draw the resonance structure.

(a)

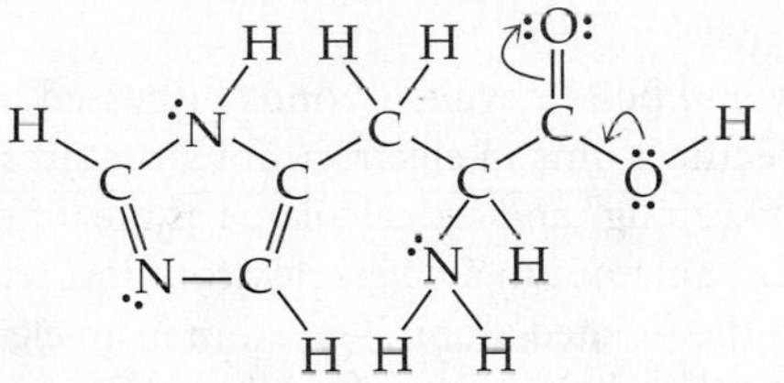

(b)

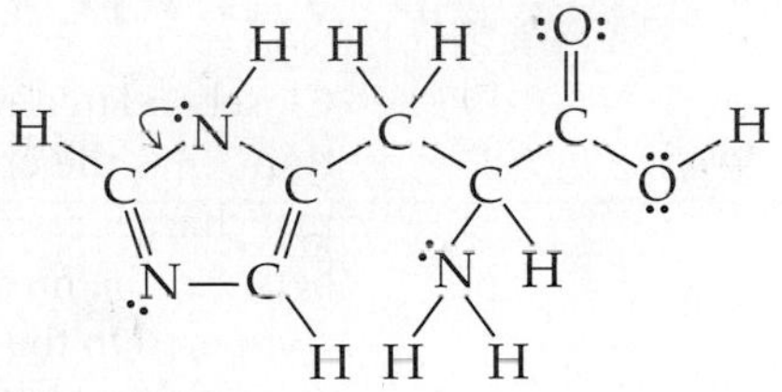

STRATEGY

Rearrange electrons as indicated by the curved arrows and check to make sure that each atom satisfies the octet rule. An invalid electron-dot structure results when the octet rule is exceeded.

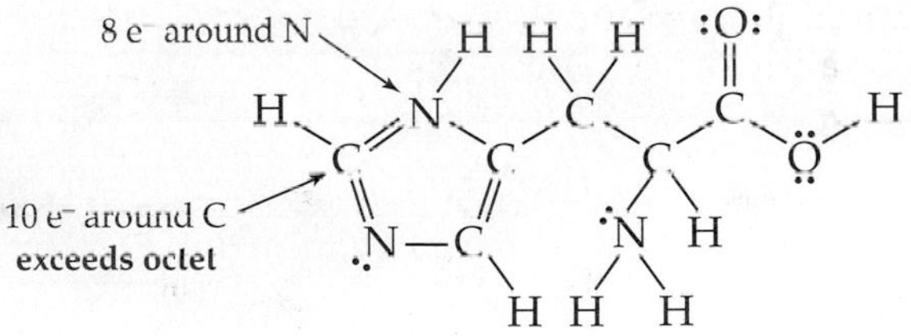

SOLUTION

(a) The octet rule is satisfied for each atom. The resonance structure is seen below.

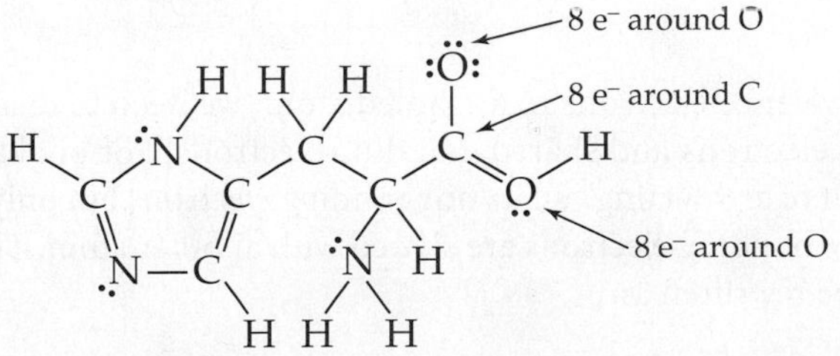

(b) The arrow results in an invalid electron-dot structure because C has five bonds (10 e^-) and therefore exceeds an octet. A resonance structure cannot be drawn from the arrow indicated.

▶ **PRACTICE 7.17** Evaluate whether the following arrows result in a valid electron-dot structure for histidine. If so, draw the resonance structure.

(a)

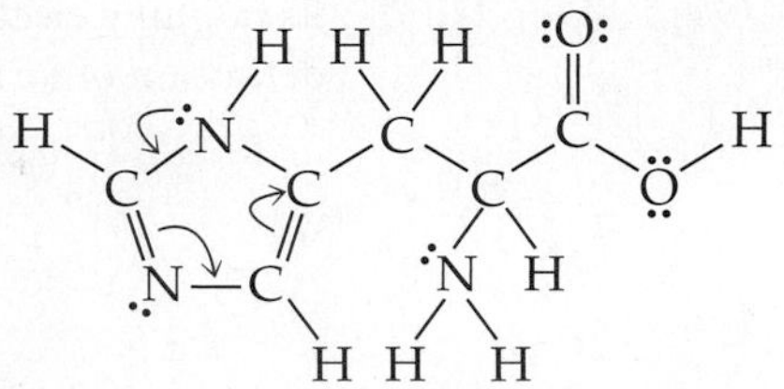

continued on next page

(b)

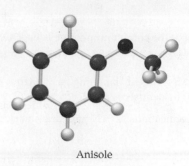

Anisole

▶**APPLY 7.18** The following structure shows the connections between atoms in anisole, a compound used in perfumery. (red = O, gray = C, ivory = H)

(a) Draw a structural formula for anisole showing positions of multiple bonds and lone pairs.

(b) Use curved arrows to show how to convert between the original structure and two additional resonance structures.

7.10 ▶ FORMAL CHARGES

Closely related to the ideas of electronegativity and polar covalent bonds discussed in Section 7.3 is the concept of **formal charge** on specific atoms in electron-dot structures. Formal charges result from a kind of electron "bookkeeping" and are calculated as the difference between the number of valence electrons around an atom in a given electron-dot structure compared to the number of valence electrons in the isolated atom. If an atom in an electron-dot structure has a different number of electrons than its number of valence electrons, then the atom has either gained or lost electrons and thus has a formal charge. If the atom in a molecule has more electrons than the isolated atom, it has a negative formal charge; if it has fewer electrons, it has a positive formal charge.

$$\text{Formal charge} = \left(\begin{array}{c}\text{Number of}\\\text{valence electrons}\\\text{in free atom}\end{array}\right) - \left(\begin{array}{c}\text{Number of}\\\text{valence electrons}\\\text{in bonded atom}\end{array}\right)$$

In counting the number of valence electrons in a bonded atom, we need to distinguish between unshared, nonbonding electrons and shared, bonding electrons. For bookkeeping purposes, an atom can be thought of as "owning" all its nonbonding electrons but only half of its bonding electrons, because the bonding electrons are shared with another atom. Thus, the definition of formal charge can be rewritten as:

$$\text{Formal charge} = \left(\begin{array}{c}\text{Number of}\\\text{valence electrons}\\\text{in free atom}\end{array}\right) - \frac{1}{2}\left(\begin{array}{c}\text{Number of}\\\text{bonding}\\\text{electrons}\end{array}\right) - \left(\begin{array}{c}\text{Number of}\\\text{nonbonding}\\\text{electrons}\end{array}\right)$$

In the ammonium ion (NH_4^+), for instance, each of the four equivalent hydrogen atoms has 2 valence electrons in its covalent bond to nitrogen and the nitrogen atom has 8 valence electrons, 2 from each of its four N—H bonds:

$$\left[\begin{array}{c} H \\ H:\ddot{N}:H \\ \ddot{H} \end{array} \right]^+ \quad \begin{array}{l} \text{Ammonium ion} \\ \text{8 valence electrons around nitrogen} \\ \text{2 valence electrons around each hydrogen} \end{array}$$

For bookkeeping purposes, each hydrogen atom owns half of its 2 shared bonding electrons, or 1, while the nitrogen atom owns half of its 8 shared bonding electrons, or 4. Because an isolated hydrogen atom has 1 electron and the hydrogens in the ammonium ion each still own 1 electron, they have neither gained nor lost electrons and thus have no formal charge. An isolated nitrogen atom, however, has 5 valence electrons, while the nitrogen atom in NH_4^+ owns only 4 and thus has a formal charge of +1. The sum of the formal charges on all the atoms (+1 in this example) must equal the overall charge on the ion.

$$\left[\begin{array}{c} H \\ H:\ddot{N}:H \\ \ddot{H} \end{array} \right]^+$$

For hydrogen: Isolated hydrogen valence electrons   1
Bound hydrogen bonding electrons   2
Bound hydrogen nonbonding electrons   0

$$\text{Formal charge} - 1 - \frac{1}{2}(2) - 0 = 0$$

For nitrogen: Isolated nitrogen valence electrons   5
Bound nitrogen bonding electrons   8
Bound nitrogen nonbonding electrons   0

$$\text{Formal charge} = 5 - \frac{1}{2}(8) - 0 = +1$$

The value of formal charge calculations comes from their application to the resonance structures described in the previous section. It often happens that the resonance structures of a given substance are not equivalent. One of the structures may be "better" than the others, meaning that it approximates the actual electronic structure of the substance more closely. The resonance hybrid in such cases is thus weighted more strongly toward the more favorable structure.

When evaluating the relative importance of different resonance structures, three criteria are frequently employed.

- Smaller formal charges (either positive or negative) are preferable to larger ones. Zero is preferred over −1, but −1 is preferred over −2.
- Negative formal charges should reside on more electronegative atoms.
- Like charges should not be on adjacent atoms.

We can use these criteria to evaluate the resonance structures for N_2O. Formal charges have been assigned to each atom in the structure:

Preferred electron-dot structure because formal charges are minimized and negative formal charge is on the most electronegative element.

$$:N{\equiv}\overset{+}{N}{-}\overset{-}{\ddot{O}}: \quad \longleftrightarrow \quad \overset{-}{\ddot{N}}{=}\overset{+}{N}{=}\overset{}{\ddot{O}} \quad \longleftrightarrow \quad \overset{-2}{:\ddot{N}}{-}\overset{+}{N}{\equiv}\overset{+}{O}:$$

Notice that when the formal charges assigned in each resonance structure are added together they give the overall charge on N_2O, which is zero. The first and second structures are preferred over the third because the magnitude of the formal charges is lower. The first structure is "best" because the negative charge resides on the more electronegative atom oxygen. The best structure based on formal charge will make the largest contribution to the resonance hybrid. In this case, in N_2O the nitrogen–oxygen bond will have more single-bond character while the nitrogen–nitrogen bond will have more triple bond character.

● WORKED EXAMPLE 7.10

Calculating Formal Charges

Calculate the formal charge on each atom in the following electron-dot structure for SO_2:

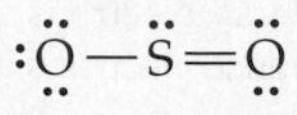

$$:\overset{..}{\underset{..}{O}}-\overset{..}{S}=\overset{..}{\underset{..}{O}}$$

STRATEGY

Use the formula for calculating formal charge. Find the number of valence electrons on each atom (its periodic group number). Then subtract half the number of the atom's bonding electrons and all of its nonbonding electrons.

SOLUTION

For sulfur:	Isolated sulfur valence electrons	6
	Sulfur bonding electrons	6
	Sulfur nonbonding electrons	2
	Formal charge $= 6 - \frac{1}{2}(6) - 2 = +1$	

For singly bonded oxygen:	Isolated oxygen valence electrons	6
	Oxygen bonding electrons	2
	Oxygen nonbonding electrons	6
	Formal charge $= 6 - \frac{1}{2}(2) - 6 = -1$	

For doubly bonded oxygen:	Isolated oxygen valence electrons	6
	Oxygen bonding electrons	4
	Oxygen nonbonding electrons	4
	Formal charge $= 6 - \frac{1}{2}(4) - 4 = 0$	

The sulfur atom of SO_2 has a formal charge of +1, and the singly bonded oxygen atom has a formal charge of −1. We might therefore write the structure for SO_2 as

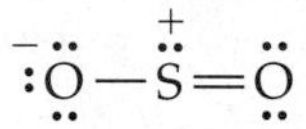

$$:\overset{..}{\underset{..}{O}}{}^{-}-\overset{+}{\overset{..}{S}}=\overset{..}{\underset{..}{O}}$$

CHECK

The sum of the formal charges must equal the overall charge on the molecule or ion. In this case the sum of formal charges is zero which equals the charge on the neutral SO_2 molecule.

▶ **PRACTICE 7.19**   Calculate the formal charge on each atom in the following electron-dot structures:

(a) Cyanate ion: $\left[:\overset{..}{\underset{..}{N}}=C=\overset{..}{\underset{..}{O}}\right]^{-}$ **(b)** Ozone: $:\overset{..}{\underset{..}{O}}-\overset{..}{\underset{..}{O}}=\overset{..}{\underset{..}{O}}$

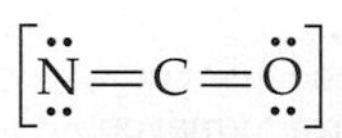

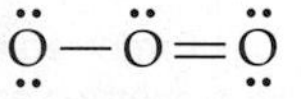

▶ **APPLY 7.20**   Start with the electron-dot structure for the cyanate ion shown in Problem 7.19a.

(a) Use curved arrows to convert the original structure into two additional electron-dot structures.

(b) Calculate the formal charge on each atom in all three resonance structures and decide which makes the largest contribution to the resonance hybrid.

(c) Which bond do you predict to be the shortest (carbon–nitrogen or carbon–oxygen)?

—● WORKED EXAMPLE 7.11

Calculating Formal Charges in Organic Compounds

Calculate formal charges on the C, O, and N atoms in the two resonance structures for acetamide, a compound related to proteins. Decide which structure makes a larger contribution to the resonance hybrid.

STRATEGY

Use the formula for calculating formal charge. Evaluate the importance of the resonance structures using the three criteria listed in this section; lowest formal charge, negative formal charge on electronegative atoms, and separation of like charges.

SOLUTION

The structure without formal charges makes a larger contribution to the resonance hybrid because energy is required to separate + and − charges. Thus, the actual electronic structure of acetamide is closer to that of the more favorable, lower energy structure.

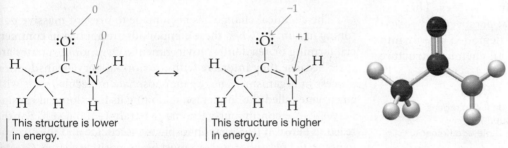

| This structure is lower in energy. | This structure is higher in energy. |

Acetamide

▶ **PRACTICE 7.21** Calculate formal charges on the C and O atoms in two resonance structures for acetic acid, the main component of vinegar. Decide which structure makes a larger contribution to the resonance hybrid.

▶ **APPLY 7.22** Three resonance structures for anisole (Problem 7.18) are shown. Calculate formal charges on C and O atoms and decide which structure makes the largest contribution to the resonance hybrid.

INQUIRY ▸▸▸ HOW DO WE MAKE ORGANOPHOSPHATE INSECTICIDES LESS TOXIC TO HUMANS?

▲ Organophosphate insecticides are used on many crops throughout the world.

O rganophosphates are a class of chemical compounds that have been widely used as insecticides. They are beneficial from an environmental perspective because they are nonpersistent, meaning they are degraded rapidly into harmless, water-soluble products. A general chemical structure for an organophosphate is

$$
\begin{array}{c}
\text{O} \\
\parallel \\
\text{X} - \text{P} - \text{OR}_1 \\
\mid \\
\text{OR}_2
\end{array}
\quad \bullet
$$

R is a general designation for an organic (carbon containing) group.

R_1 and R_2 could be the same or different.

X represents a halogen, SR, or OR group.

This class of insecticides must be highly potent due to their short lifetime and changes in their chemical structure can be used to "tune" toxicity. In fact, nerve agents are highly toxic organophosphate compounds in which substituent groups X and R have been changed. In this Inquiry, we will examine how bond polarity influences the potency of these compounds and how substituting atoms of different electronegativity can make these compounds safer for humans to work with.

Examine the chemical structure of parathion and disulfoton, two organophosphate insecticides used to control a variety of pests that attack m any field and vegetable crops. The double bond between phosphorus and oxygen in the general structure of an organophosphate had been replaced with a double bond to sulfur.

Parathion

Disulfoton

The chemical change has been made to prevent massive poisoning in humans when these compounds are applied in commercial farming or residential environments. Organophosphates are toxic because they interfere with the normal nerve transduction process in organisms. The organophosphate molecule reacts with an enzyme called cholinesterase, inhibiting its function and causing nerves to fire continuously leading to paralysis and death. For the reaction between the organophosphate insecticide and the enzyme to occur, the phosphorus atom must bear a positive charge. Greater positive charge leads to increased rate of reaction and increased toxicity of the insecticide.

FIGURE 7.6 shows electrostatic potential maps and atomic charges of an organophosphate with a phosphorus–oxygen double bond ($P{=}O$) and a phosphorus–sulfur ($P{=}S$) double bond. Since oxygen is more electronegative than sulfur it has greater ability to pull shared electrons toward itself. Therefore, the molecule with the ($P{=}O$) bond is more toxic due to the greater positive charge on phosphorus ($+1.80$). In the less toxic form ($P{=}S$), the positive charge on phosphorus is lower ($+1.30$). The electrostatic potential map reflects the higher negative charge on oxygen as indicated by the strong red color. The sulfur atom is larger and less electronegative and the lower negative charge is depicted with a combination of yellow and red.

Organophosphates with a $P{=}S$ bond are still highly potent insecticides due to differences in biochemistry between insects and

▶ **FIGURE 7.6**

Molecular modeling of organophosphate compounds. In organophosphate insecticides, the positive charge on the phosphorus atom can be altered by binding it to elements with varying electronegativity. Oxygen is more electronegative than sulfur, thus the P=O is more polar than the P=S bond and the positive charge on phosphorus is larger.

$$S$$
$$\|$$
$$H_3CO-P-OCH_3$$
$$|$$
$$OCH_3$$

Less toxic insecticide

S = −0.6
P = +1.3

$$O$$
$$\|$$
$$H_3CO-P-OCH_3$$
$$|$$
$$OCH_3$$

More toxic insecticide

O − −0.8
P = +1.8

mammals. Once inside the insect, the sulfur atom is rapidly converted back into oxygen by oxidative enzymes, thus creating a potent neurotoxin. Animals have much lower level of these enzymes so the conversion of sulfur to oxygen does not occur to a significant extent. Therefore, when these compounds enter mammals they remain in their less toxic form, but are converted to the highly toxic form in insects. Although the P=S bond makes the organophosphate less toxic, care must still be taken in applying these insecticides; several cases of human poisoning and even death have occurred.

PROBLEM 7.23 Use Figure 7.4 to calculate the difference in electronegativity between phosphorus and oxygen in the P=O bond and between phosphorus and sulfur in the P=S bond. Are these bonds classified as nonpolar, polar covalent, or ionic?

PROBLEM 7.24 Consider the substituted phenyldiethylphosphate insecticide with the following structure:

Which chemical group at position X would result in greatest toxicity?

(a) I, Br, Cl

(b) CF_3, CHF_2, CH_2F, CH_3

PROBLEM 7.25 In organophosphate compounds, phosphorus has an expanded octet. Why can phosphorus accommodate more than eight electrons in its electron-dot structure?

PROBLEM 7.26 The following structure is a representation of the organophosphate insecticide diazinon. Only the connections between atoms are shown; multiple bonds are not indicated. Give the formula of diazinon, and complete the structure by showing where the multiple bonds and lone pairs are located (red = O, gray = C, yellow = sulfur, blue = N, ivory = H, purple = P). (Hint: There is a double bond between phosphorus and sulfur.)

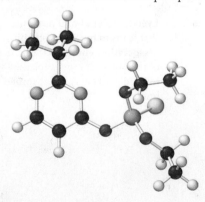

continued on next page

continued from previous page

PROBLEM 7.27 The electron-dot structure for the nerve agent, Sarin is shown. Calculate the formal charges on P and the atoms bonded to it; F, C, and two O's.

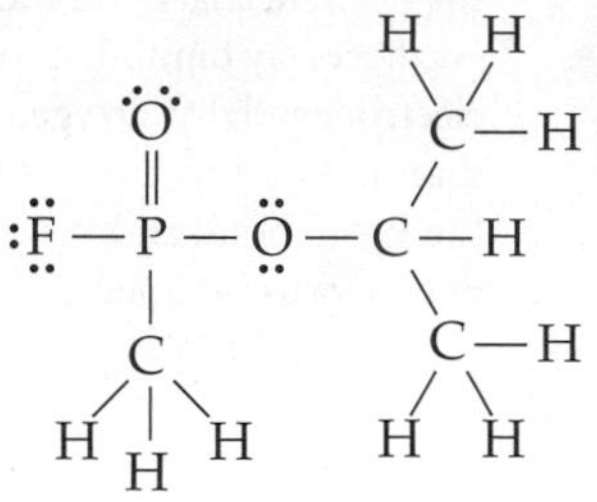

(a)

(b)

PROBLEM 7.28 Draw the new electron-dot structures indicated by the curved arrows in the Sarin molecule. Decide if the new structure is valid and if so, calculate formal charges on F, P, and both oxygen atoms. Compare formal charges to those calculated in Problem 7.27. Use formal charge to evaluate which structure is the major contributor to the resonance hybrid.

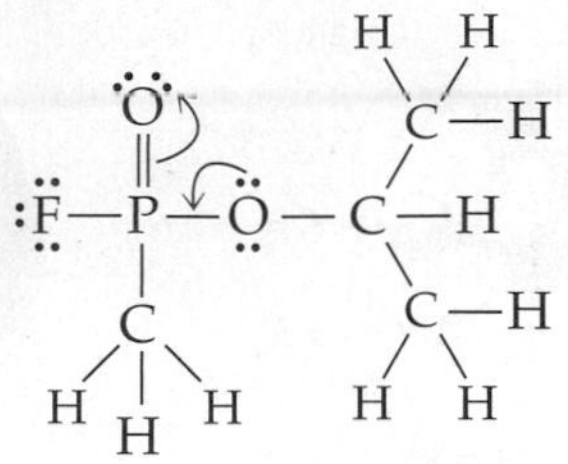

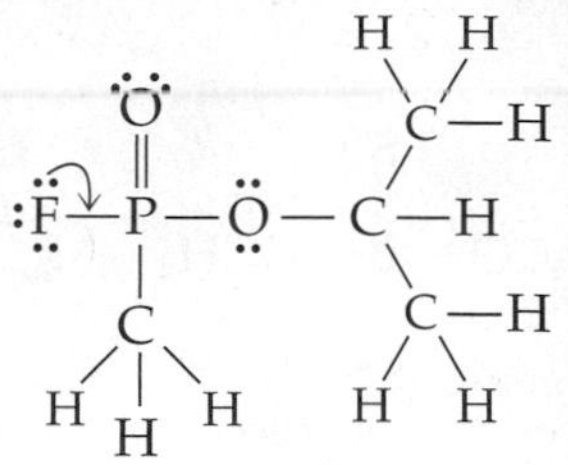

(c)

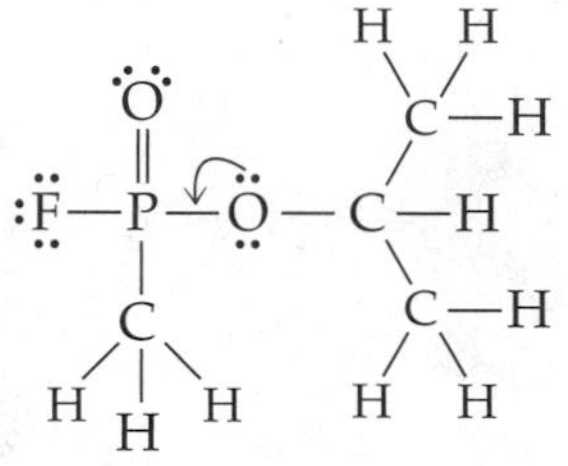

(d)

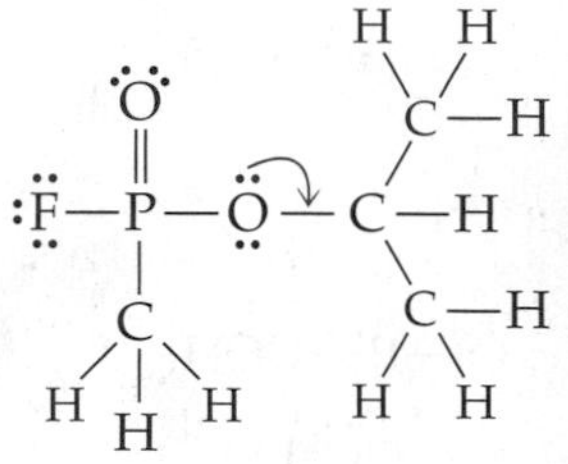

STUDY GUIDE

Section	Concept Summary	Learning Objectives	Test Your Understanding
7.1 ▸ Covalent Bonding in Molecules	A **covalent bond** results from the sharing of electrons between atoms. The force joining atoms is the electrostatic attraction of the negatively charged electrons in each atom to the positively charged nuclei of the two atoms.	7.1 Describe the difference between an ionic and covalent bond.	Problem 7.38
		7.2 Describe the changes in energy that occur as two nuclei approach to form a covalent bond. (Figure 7.2)	Problems 7.29, 7.30, 7.39
		7.3 Name a covalent compound given the chemical formula.	Problems 7.60, 7.61
7.2 ▸ Strengths of Covalent Bonds	Every covalent bond has a specific **bond length** that leads to optimum stability and a specific **bond dissociation energy (D)** that describes the strength of the bond. Energy is released when a bond is formed, and energy is absorbed when a bond is broken.	7.4 Predict trends in bond length and bond dissociation energy based on bond order and atomic size.	Problems 7.54–7.59

Section	Concept Summary	Learning Objectives	Test Your Understanding
7.3 ▸ Polar Covalent Bonds: Electronegativity	In a bond between dissimilar atoms, such as that in HCl, one atom often attracts the bonding electrons more strongly than the other, giving rise to a **polar covalent bond**. Bond polarity is due to differences in **electronegativity (EN)**, the ability of an atom in a molecule to attract shared electrons. Electronegativity increases from left to right across a row and generally decreases from top to bottom in a group of the periodic table.	**7.5** Rank elements by increasing value of electronegativity.	Problems 7.42, 7.43
		7.6 Classify bonds as nonpolar covalent, polar covalent, or ionic.	Worked Example 7.1; Problems 7.48, 7.49, 7.50, 7.51
		7.7 Visualize regions of high and low electron density in a polar covalent bond.	Problems 7.31, 7.32
		7.8 Predict trends in bond dissociation energy based upon both atomic size and polarity of the bond.	Problems 7.56, 7.57
7.4 ▸ A Comparison of Ionic and Covalent Compounds	Ionic compounds are solids with high melting points because strong ionic bonds exist between all atoms in a crystal. In general, covalent compounds have lower melting points because they are molecules held together by intermolecular forces that are much weaker than actual bonds.	**7.9** Explain the different physical properties of ionic and covalent compounds.	Problems 7.33, 7.62, 7.63
7.5 ▸ Electron-Dot Structures and the Octet Rule	As a general rule, a main-group atom shares as many of its valence-shell electrons as possible, either until it has no more to share or until it reaches an octet. Atoms in the third and lower rows of the periodic table can accommodate more than the number of bonds predicted by the octet rule.	**7.10** Draw an electron dot structure by using valence electrons to give all atoms (except H) an octet.	Worked Example 7.2; Problem 7.3
7.6 ▸ Procedure for Drawing Electron-Dot Structures	An **electron-dot structure** represents an atom's valence electrons by dots and shows the two electrons in a **single bond** as a pair of dots shared between atoms. Similarly, a **double bond** is represented as four dots or two lines between atoms, and a **triple bond** is represented as six dots or three lines between atoms.	**7.11** Use the five-step procedure for drawing electron-dot structures for all molecules including those with, expanded octets, and those containing multiple bonds.	Worked Examples 7.3, 7.4; Problems 7.66-7.71
7.7 ▸ Drawing Electron-Dot Structures for Radicals	**Radicals** are highly reactive species that contain an odd number of electrons. The least electronegative atom does not have a complete octet in the electron-dot structure for a radical.	**7.12** Draw electron-dot structures for radicals.	Worked Example 7.5; Problems 7.9, 7.10
7.8 ▸ Electron-Dot Structures of Compounds Containing Only Hydrogen and Second-Row Elements	Many biological and organic compounds contain only second row elements and hydrogen. Drawing electron-dot structures is straightforward as they follow the pattern of bonding outlined in Table 7.4.	**7.13** Draw electron-dot structures for molecules with more than one central atom.	Worked Examples 7.6 and 7.7; Problems 7.34–7.36, 7.74–7.77
7.9 ▸ Electron-Dot Structures and Resonance	Some molecules with multiple bonds can be represented by more than one electron-dot structure. In such cases, no single structure is adequate by itself. The actual electronic structure of the molecule is a **resonance hybrid** of the different individual structures.	**7.14** Draw resonance structures and use curved arrows to depict how one structure can be converted to another.	Worked Examples 7.8, 7.9; Problems, 7.78–7.83, 7.94
7.10 ▸ Formal Charge	**Formal charge** denotes whether an atom has more or less electrons in an electron-dot structure compared to its number of valence electrons. Formal charges are used to assess the relative contribution of individual resonance structures to the resonance hybrid.	**7.15** Calculate the formal charge on atoms in an electron-dot structure.	Worked Example 7.10; Problems 7.84–7.87, 7.90
		7.16 Use the formal charge to evaluate the contribution of different resonance structures to the resonance hybrid.	Worked Example 7.11; Problems 7.88, 7.89, 7.92, 7.93

KEY TERMS

bond dissociation energy (D) *225*	delocalized *243*	formal charge *246*	polar covalent bond *226*
bond length *223*	double bond *233*	lattice energy (U) *230*	radicals *238*
bond order *226*	electron deficient *235*	lines between atoms *224*	resonance hybrid *243*
bonding pair *231*	electron-dot structure *231*	lone pair *231*	resonance theory *243*
coordinate covalent bond *233*	electronegativity (EN) *228*	octet rule *231*	single bond *233*
	expanded octet *234*	organic compounds *240*	triple bond *233*

KEY EQUATIONS

• **Formal charge (Section 7.10)**

$$\text{Formal charge} = \begin{pmatrix} \text{Number of} \\ \text{valence electrons} \\ \text{in free atom} \end{pmatrix} - \frac{1}{2}\begin{pmatrix} \text{Number of} \\ \text{bonding} \\ \text{electrons} \end{pmatrix} - \begin{pmatrix} \text{Number of} \\ \text{nonbonding} \\ \text{electrons} \end{pmatrix}$$

CONCEPTUAL PROBLEMS

Problems 7.1–7.28 appear within the chapter.

7.29 The following diagram shows the potential energy of two atoms as a function of internuclear distance. Match the descriptions with the indicated letter on the plot.

(a) Repulsive forces are high between the two atoms.

(b) The two atoms neither exert attractive nor repulsive forces on one another.

(c) The attractive forces between atoms are maximized resulting in the lowest energy state.

(d) Attractive forces between atoms are present, but are not at maximum strength.

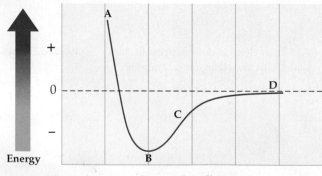

7.30 The following diagram shows the potential energy of two atoms as a function of internuclear distance. Which bond is the strongest? Which bond is the longest?

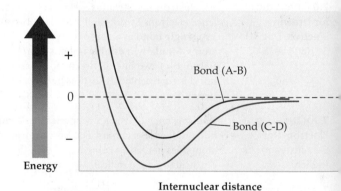

7.31 Two electrostatic potential maps are shown, one of methyllithium (CH_3Li) and the other of chloromethane (CH_3Cl). Based on their polarity patterns, which do you think is which?

(a)   **(b)**

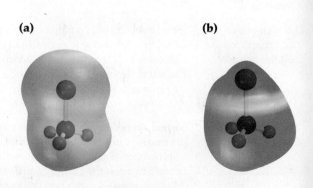

7.32 Electrostatic potential maps of acetaldehyde (C_2H_4O), ethane (C_2H_6), ethanol, C_2H_6O, and fluorethane (C_2H_5F) are shown. Which do you think is which?

(a)

(b)

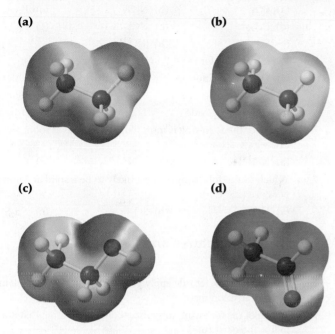

(c)

(d)

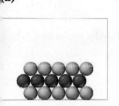

7.33 Which of the following drawings is most likely to represent an ionic compound and which a covalent compound?

(a)

(b)

(c)

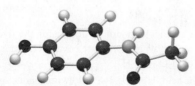

7.34 The following ball-and-stick molecular model is a representation of acetaminophen, the active ingredient in such over-the-counter headache remedies as Tylenol (red = O, gray = C, blue = N, ivory = H). Give the formula of acetaminophen and indicate the positions of the double bonds and lone pairs.

Acetaminophen

7.35 The following ball-and-stick molecular model is a representation of thalidomide, a drug that causes birth defects when taken by expectant mothers but is valuable for its use against leprosy. The lines indicate only the connections between atoms, not whether the bonds are single, double, or triple (red = O, gray = C, blue = N, ivory = H). Give the formula of thalidomide, and indicate the positions of multiple bonds and lone pairs.

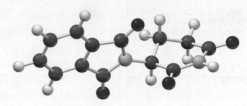

Thalidomide

7.36 Sinapaldehyde, a compound present in the toasted wood used for aging wine, has the following connections among atoms. Complete the electron-dot structure for sinapaldehyde showing lone pairs and identifying multiple bonds.

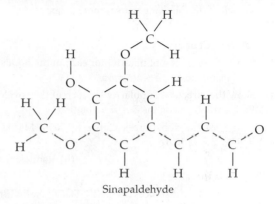

Sinapaldehyde

7.37 Vitamin C (ascorbic acid) has the following connections among atoms. Complete the electron-dot structure for Vitamin C, showing lone pairs and identifying multiple bonds.

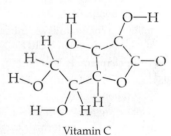

Vitamin C

SECTION PROBLEMS

Strengths of Covalent Bonds and Electronegativity (Sections 7.1–7.4)

7.38 Match the following descriptions with the type of bond (ionic, nonpolar covalent, covalent).

(a) One or more electrons are transferred from a metal to a nonmetal atom.

(b) Electrons are shared equally between two atoms.

(c) Electrons are shared unequally by two atoms.

7.39 Why do two atoms come together to form a covalent bond?

(a) Attractive forces between the positively charged nuclei and the electrons in both atoms occur when the atoms are close together.

(b) Repulsive forces between protons in the nuclei of the two atoms are minimized when the atoms are close together.

(c) Repulsive forces between electrons in the two atoms are minimized when the atoms are close together.

7.40 What general trends in electronegativity occur in the periodic table?

7.41 Predict the electronegativity of the undiscovered element with $Z = 119$.

7.42 Order the following elements according to increasing electronegativity: Li, Br, Pb, K, Mg, C.

7.43 Order the following elements according to decreasing electronegativity: C, Ca, Cs, Cl, Cu.

7.44 Which of the following substances are largely ionic, and which are covalent?

(a) HF   (b) HI   (c) $PdCl_2$

(d) BBr_3 (e) NaOH   (f) CH_3Li

7.45 Use the electronegativity data in Figure 7.4 to predict which bond in each of the following pairs is more polar:

(a) C—H or C—Cl   (b) Si—Li or Si—Cl

(c) N—Cl or N—Mg

7.46 Show the direction of polarity for each of the bonds in Problem 7.45, using the $\delta+/\delta-$ notation.

7.47 Show the direction of polarity for each of the covalent bonds in Problem 7.44, using the $\delta+/\delta-$ notation.

7.48 Which of the substances $CdBr_2$, P_4, BrF_3, MgO, NF_3, $BaCl_2$, $POCl_3$, and LiBr are:

(a) largely ionic?   (b) nonpolar covalent?

(c) polar covalent?

7.49 Which of the substances S_8, $CaCl_2$, $SOCl_2$, NaF, CBr_4, BrCl, LiF, and AsH_3 are:

(a) largely ionic?   (b) nonpolar covalent?

(c) polar covalent?

7.50 Order the following compounds according to the increasing ionic character of their bonds: CCl_4, $BaCl_2$, $TiCl_3$, ClO_2.

7.51 Order the following compounds according to the increasing ionic character of their bonds: NH_3, NCl_3, Na_3N, NO_2.

7.52 Using only the elements P, Br, and Mg, give formulas for the following:

(a) an ionic compound

(b) a molecular compound with polar covalent bonds that obeys the octet rule and has no formal charges

7.53 Using only the elements Ca, Cl, Si, give formulas for the following:

(a) an ionic compound

(b) a molecular compound with polar covalent bonds that obeys the octet rule and has no formal charges

7.54 Which compound do you expect to have the stronger N—N bond, N_2H_2 or N_2H_4? Explain.

7.55 Which compound do you expect to have the stronger N—O bond, NO or NO_2? Explain.

7.56 Explain the difference in the bond dissociation energies for the following bonds: (C—F, 450 kJ/mol), (N—F, 270 kJ/mol), (O—F, 180 kJ/mol), (F—F, 159 kJ/mol)

7.57 Explain the difference in the bond dissociation energies for the following bonds: (C—F, 450 kJ/mol), (C—Cl, 330 kJ/mol), (C—Br, 270 kJ/mol), (C—I, 240 kJ/mol)

7.58 Predict which of the following bonds should be strongest: N—H, O—H, S—H

7.59 Predict which of the following bonds should be weakest Cl—Cl, Br—Br, I—I

7.60 Name the following molecular compounds:

(a) PCl_3 (b) N_2O_3 (c) P_4O_7

(d) BrF_3 (e) NCl_3 (f) P_4O_6

(g) S_2F_2 (h) SeO_2

7.61 Write formulas for the following molecular compounds:

(a) Disulfur dichloride   (b) Iodine monochloride

(c) Nitrogen triiodide   (d) Dichlorine monoxide

(e) Chlorine trioxide   (f) Tetrasulfur tetranitride

7.62 Which of the following is most likely to be a gas at room temperature (25 °C)?

(a) NH_3 (b) K_3N (c) Ca_3N_2

7.63 Which of the following is most likely to be a solid at room temperature (25 °C)?

(a) H_2S (b) SO_2 (c) Na_2S

Electron-Dot Structures and Resonance (Sections 7.5–7.9)

7.64 Why does the octet rule apply primarily to main-group elements not to transition metals?

7.65 Which of the following substances contains an atom that does not follow the octet rule?

(a) $AlCl_3$ (b) PCl_3

(c) PCl_5 (d) $SiCl_4$

7.66 Draw electron-dot structures for the following molecules or ions

(a) CBr_4 (b) NCl_3 (c) C_2H_5Cl

(d) BF_4^- (e) O_2^{2-} (f) NO^+

7.67 Draw electron-dot structures for the following molecules, which contain atoms from the third row or lower:

(a) $SbCl_3$ (b) KrF_2 (c) ClO_2

(d) PF_5 (e) H_3PO_4 (f) $SeCl_2$

7.68 Identify the correct electron-dot structure for XeF_5^+.

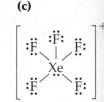

7.69 Draw an electron-dot structure for the hydronium ion, H_3O^+ and show how a coordinate covalent bond is formed by the reaction of H_2O with H^+.

7.70 Oxalic acid, $H_2C_2O_4$, is a mildly poisonous substance found in the leaves of rhubarb, spinach, and many other plants. (You'd have to eat about 15 lbs. or so of spinach leaves to ingest a lethal amount.) If oxalic acid has a C—C single bond and no C—H bond, draw its electron-dot structure showing lone pairs and identifying any multiple bonds.

7.71 Draw an electron-dot structure for carbon disulfide, CS_2, showing lone pairs and identifying any multiple bonds.

7.72 Identify the third-row elements, X, that form the following ions:

(a)
$$\left[\begin{array}{c} :\ddot{C}l: \\ | \\ :\ddot{C}l - X - \ddot{C}l: \\ | \\ :\ddot{C}l: \end{array} \right]^{-}$$

(b)
$$\left[\begin{array}{c} H \\ | \\ H - X - H \\ | \\ H \end{array} \right]^{+}$$

7.73 Identify the fourth-row elements, X, that form the following compounds:

(a)
$$\ddot{O} = \ddot{X} - \ddot{O}:$$

(b)
$$:\ddot{F} \diagdown \diagup \ddot{F}: \\ \cdot \ddot{X} \cdot$$

7.74 Write electron-dot structures for molecules with the following connections, showing lone pairs and identifying any multiple bonds:

(a)
$$\begin{array}{ccc} O & & H \\ | & & | \\ Cl - C - O - C - H \\ & & | \\ & & H \end{array}$$

(b)
$$\begin{array}{c} H \\ | \\ H - C - C C - H \\ | \\ H \end{array}$$

7.75 Write electron-dot structures for molecules with the following connections, showing lone pairs and identifying any multiple bonds:

(a)
$$\begin{array}{cc} O & H \\ | & | \\ H - C - N - H \end{array}$$

(b)
$$\begin{array}{c} H \\ | \\ H - C - C - N - O \\ | \\ H \end{array}$$

7.76 Methylphenidate ($C_{14}H_{19}NO_2$), marketed as Ritalin, is often prescribed for attention deficit disorder. Complete the following electron-dot structure, showing the position of any multiple bonds and lone pairs:

Methylphenidate
(Ritalin)

7.77 Pregabalin ($C_8H_{17}NO_2$), marketed as Lyrica, is an anticonvulsant drug prescribed for treatment of seizures. Complete the following electron-dot structure, showing the position of any multiple bonds and lone pairs:

Pregabalin
(Lyrica)

7.78 Draw as many resonance structures as you can that obey the octet rule for each of the following molecules or ions. Use curved arrows to depict the conversion of one structure into another.
  (a) HN_3 (b) SO_3 (c) SCN^-

7.79 Draw as many resonance structures as you can for the following nitrogen containing compounds. Not all will obey the octet rule. Use curved arrows to depict the conversion of one structure into another.
  (a) N_2O
  (b) NO
  (c) NO_2
  (d) N_2O_3 ($ONNO_2$)

7.80 Which of the following pairs of structures represent resonance forms, and which do not?

(a) $H - C \equiv N - \ddot{O}:$ and $H - C = \ddot{N} - \ddot{O}:$

(b)
$$:\ddot{O} - \overset{\overset{\textstyle :O:}{\|}}{S} - \ddot{O}: \quad \text{and} \quad :\ddot{O} - \overset{\overset{\textstyle :\ddot{O}:}{|}}{S} - \ddot{O}:$$

(c)
$$\left[\begin{array}{cc} :\ddot{O}: & H \\ \diagdown & \diagup \\ C = C \\ \diagup & \diagdown \\ H & H \end{array} \right]^{-} \text{and} \left[\begin{array}{cc} :O: & H \\ \diagdown & \diagup \\ C - C: \\ \diagup & \diagdown \\ H & H \end{array} \right]^{-}$$

(d)
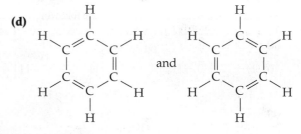

and

7.81 Which of the following pairs of structures represent resonance forms, and which do not?

(a)
$$\begin{array}{cc} :\ddot{F}: & :\ddot{F}: \\ \diagdown & \diagdown \\ B - \ddot{F}: \text{ and } B = \ddot{F} \\ \diagup & \diagup \\ :\ddot{F}: & :\ddot{F}: \end{array}$$

(b)

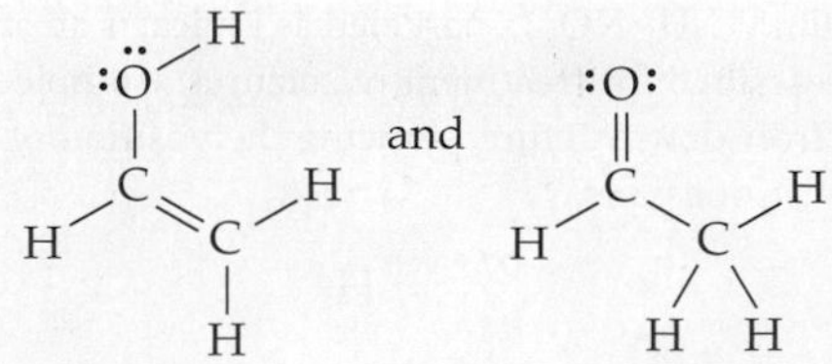

and

(c)

and

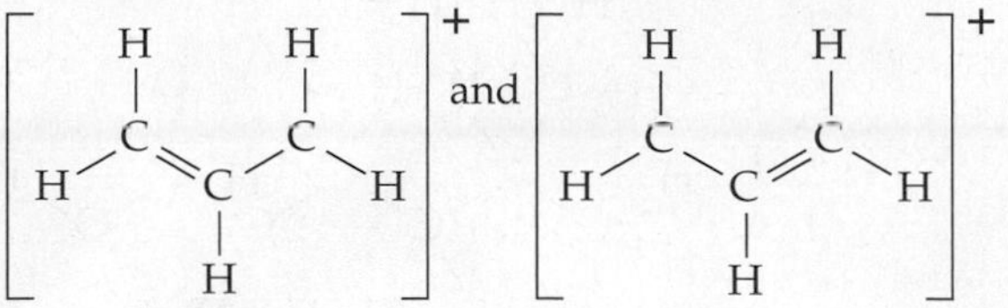

7.82 Draw as many resonance structures as you can that obey the octet rule for phenol (C_6H_6O). Use curved arrows to depict the conversion of one structure into another.

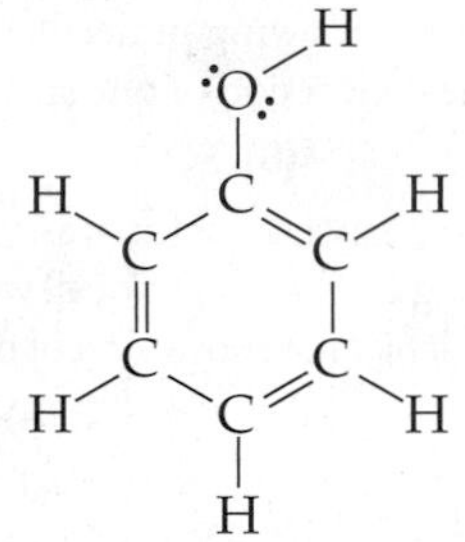

7.83 Draw as many resonance structures as you can that obey the octet rule for C_2H_5N. Use curved arrows to depict the conversion of one structure into another.

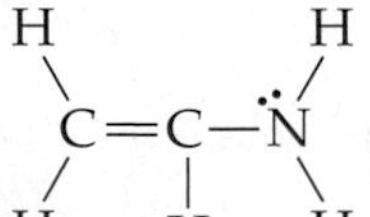

Formal Charges (Section 7.10)

7.84 Draw an electron-dot structure for carbon monoxide, CO, and assign formal charges to both atoms.

7.85 Assign formal charges to the atoms in the following structures:

(a)

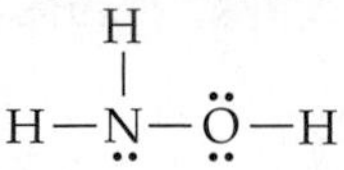

(b)

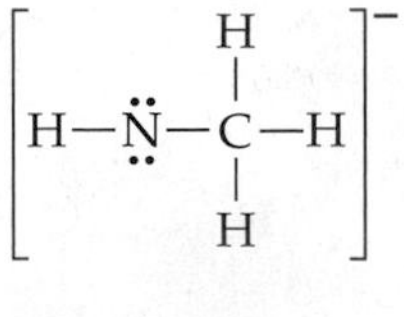

(c)

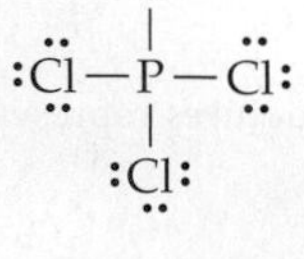

7.86 Assign formal charges to the atoms in the following resonance forms of ClO_2^-:

7.87 Assign formal charges to the atoms in the following resonance forms of H_2SO_3:

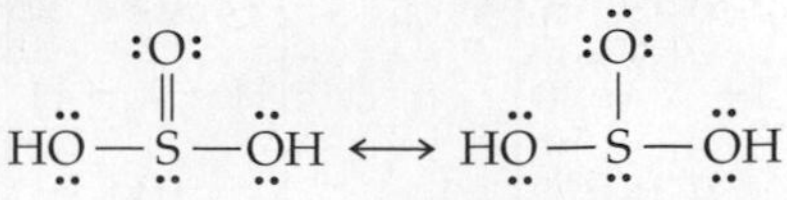

7.88 Assign formal charges to the atoms in the following structures. Which of the two do you think is the more important contributor to the resonance hybrid?

(a)   **(b)**

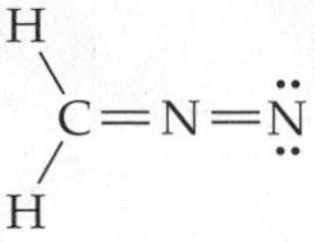

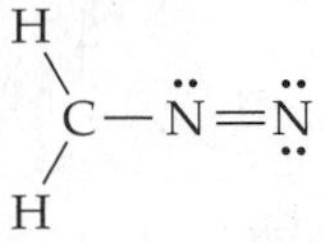

7.89 Calculate formal charges for the C and O atoms in the following two resonance structures. Which structure do you think is the more important contributor to the resonance hybrid? Explain.

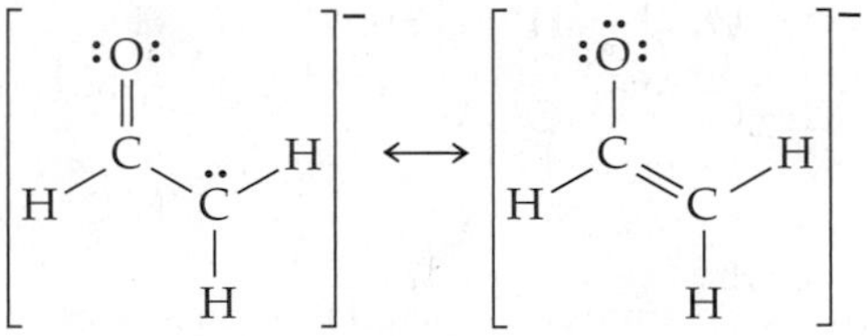

7.90 Draw two electron-dot resonance structures that obey the octet rule for trichloronitromethane, CCl_3NO_2, and show the formal charge on N and O in both structures. (Carbon is connected to the chlorines and to nitrogen; nitrogen is also connected to both oxygens.)

7.91 Draw two electron-dot resonance structures that obey the octet rule for nitrosyl chloride, NOCl (nitrogen is the central atom). Show formal charges, if present, and predict which of the two structures is a larger contributor to the resonance hybrid.

7.92 Draw the resonance structure indicated by the curved arrows. Assign formal charges and evaluate which of the two structures is a larger contributor to the resonance hybrid.

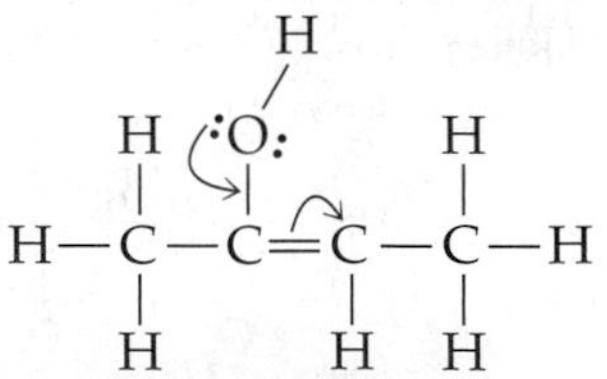

7.93 Draw the resonance structure indicated by the curved arrows. Assign formal charges and evaluate which of the two structures is a larger contributor to the resonance hybrid.

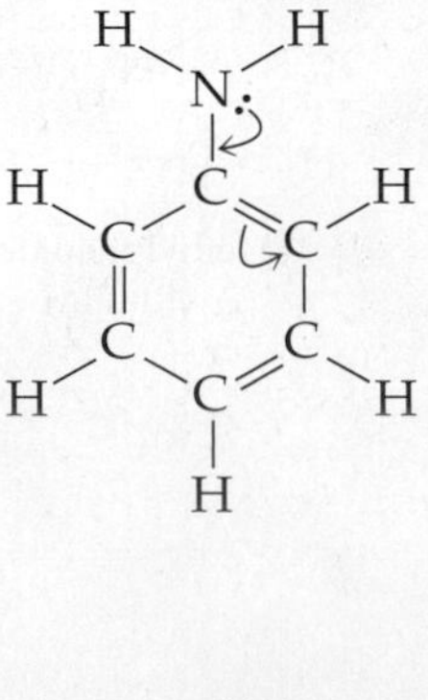

CHAPTER PROBLEMS

7.94 Benzene has the following structural formula.

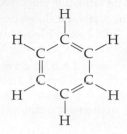

(a) Use curved arrows to show how to convert the original structure into a resonance structure.

(b) Which statement best describes the carbon–carbon bonds in benzene?

   i) Three carbon–carbon bonds are longer and weaker than the other three carbon–carbon bonds.

   ii) All six carbon–carbon bonds are identical and their length and strength is between a double and single bond.

   iii) The length of carbon–carbon double bond switches back and forth between the length of a double and a single bond.

7.95 Boron trifluoride reacts with dimethyl ether to form a compound with a coordinate covalent bond. Assign formal charges to the B and O atoms in both the reactants and products.

$$F-B \begin{matrix} F \\ | \\ \\ | \\ F \end{matrix} + \quad :O: \begin{matrix} CH_3 \\ | \\ \\ | \\ CH_3 \end{matrix} \longrightarrow F-B \begin{matrix} F \\ | \\ \\ | \\ F \end{matrix} -O: \begin{matrix} CH_3 \\ \\ \\ CH_3 \end{matrix}$$

Boron   Dimethyl
trifluoride   ether

7.96 Draw an electron-dot structure for each of the following substances.

(a) F_3S-S-F (b) $CH_3-C=C-CO_2^-$

7.97 Thiofulminic acid, $H-C\equiv N-S$ has recently been detected at very low temperatures.

(a) Draw an electron-dot structure for thiofulminic acid, and assign formal charges.

(b) A related compound with the same formula and the connection $H-N-C-S$ is also known. Draw an electron-dot structure for this related compound and assign formal charges.

(c) Which of the two molecules is likely to be more stable? Explain.

7.98 Draw three resonance structures for sulfur tetroxide, SO_4, whose connections are shown below. (This is a neutral molecule; it is not a sulfate ion.) Assign formal charges to the atoms in each structure.

$$\begin{matrix} O \\ | \\ O \\ | \\ O-S-O \end{matrix}$$

Sulfur tetroxide

7.99 Draw two resonance structures for methyl isocyanate, CH_3NCO, a toxic gas that was responsible for the deaths of at least 3000 people when it was accidentally released into the atmosphere in December 1984 in Bhopal, India. Assign formal charges to the atoms in each resonance structure.

7.100 Write an electron-dot structure for chloral hydrate, also known in old detective novels as "knockout drops."

$$\begin{matrix} Cl & O-H \\ | & | \\ Cl-C-C-O-H \\ | & | \\ Cl & H \end{matrix} \quad \text{Chloral hydrate}$$

7.101 The overall energy change during a chemical reaction can be calculated from the knowledge of bond dissociation energies using the following relationship:

$$\text{Energy change} = D\,(\text{Bonds broken}) - D\,(\text{Bonds formed})$$

Use the data in Table 7.2 to calculate an energy change for the reaction of methane with chlorine.

$$CH_4(g) + Cl_2(g) \longrightarrow CH_3Cl(g) + HCl(g)$$

7.102 The following molecular model is that of aspartame, $C_{14}H_{18}N_2O_5$, known commercially as NutraSweet. Only the connections between atoms are shown; multiple bonds are not indicated. Complete the structure by indicating the positions of the multiple bonds and lone pairs.

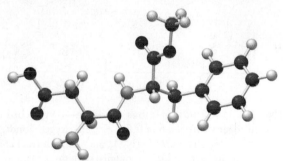

Aspartame

7.103 Ibuprofen ($C_{13}H_{18}O_2$), marketed under such a brand name as Advil and Motrin, is a drug sold over the counter for treatment of pain and inflammation. Complete the structure of ibuprofen by adding hydrogen atoms and lone pairs where needed.

Ibuprofen

7.104 Some mothballs used when storing clothes are made of naphthalene ($C_{10}H_8$), which has the following incomplete structure.

Naphthalene

(a) Add double bonds where needed to draw a complete electron-dot structure.

(b) Starting from this structure, use curved arrows to indicate how a new resonance structure can be drawn.

7.105 Four different structures **(a)**, **(b)**, **(c)**, and **(d)** can be drawn for compounds named dibromobenzene, but only three different compounds actually exist. Explain.

(a)

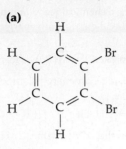

(b)

(c)

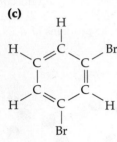

(d)

7.106 The N_2O_5 molecule has nitrogen–oxygen bonds, but no nitrogen–nitrogen bonds nor oxygen–oxygen bonds. Draw eight resonance structures for N_2O_5, and assign formal charges to the atoms in each. Which resonance structures make the more important contributions to the resonance hybrid?

7.107 In the cyanate ion, OCN^-, carbon is the central atom.

(a) Draw as many resonance structures as you can for OCN^-, and assign formal charges to the atoms in each.

(b) Which resonance structure makes the greatest contribution to the resonance hybrid? Which makes the least contribution? Explain.

MULTICONCEPT PROBLEMS

7.108 Sulfur reacts with chlorine to give a product that contains 47.5% by mass sulfur and 52.5% by mass chlorine and has no formal charges on any of its atoms. Draw the electron-dot structure of the product.

7.109 Sulfur reacts with ammonia to give a product **A** that contains 69.6% by mass sulfur and 30.4% by mass nitrogen and has a molar mass of 184.3 g.

(a) What is the formula of product **A**?

(b) The S and N atoms in the product **A** alternate around a ring with half of the atoms having formal charges. Draw two possible electron-dot structures for **A**.

(c) When compound **A** is heated with metallic silver at 250 °C a new product **B** is formed. Product **B** has the same percent composition as **A** but has a molar mass of 92.2 g. Draw two possible electron-dot structures for **B** which, like **A**, also has a ring structure.

7.110 The neutral OH molecule has been implicated in certain ozone destroying processes that take place in the upper atmosphere.

(a) Draw electron-dot structures for the OH molecule and the OH^- ion.

(b) Electron affinity can be defined for molecules just as it is defined for single atoms. Assuming that the electron added to OH is localized in a single atomic orbital on one atom, identify which atom is accepting the electron, and give the n and quantum numbers of the atomic orbital.

(c) The electron affinity of OH is similar to but slightly more negative than that of O atoms. Explain.

7.111 Suppose that the Pauli exclusion principle were somehow changed to allow three electrons per orbital rather than two.

(a) Instead of an octet, how many outer-shell electrons would be needed for a noble-gas electron configuration?

(b) How many electrons would be shared in a covalent bond?

(c) Give the electron configuration, and draw an electron-dot structure for element X with Z = 12.

(d) Draw an electron-dot structure for the molecule X_2.

7.112 The dichromate ion, $Cr_2O_7^{2-}$, has neither Cr—Cr nor O—O bonds.

(a) Taking both $4s$ and $3d$ electrons into account, draw an electron-dot structure that minimizes the formal charges on the atoms.

(b) How many outer-shell electrons does each Cr atom have in your electron-dot structure?

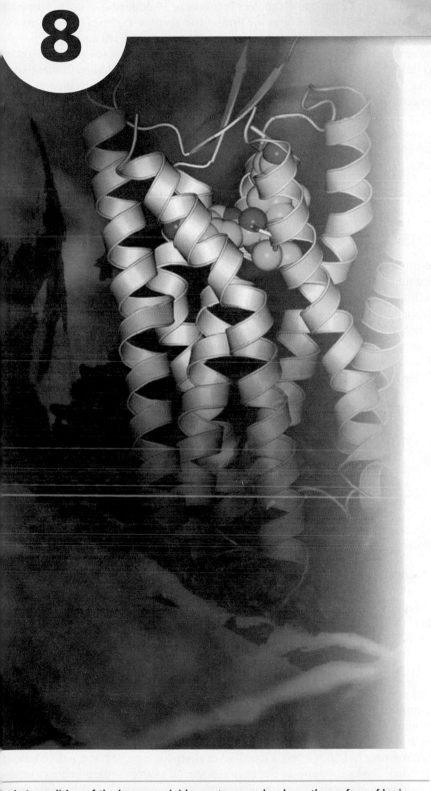

Covalent Compounds: Bonding Theories and Molecular Structure

Artist's rendition of the human opioid receptor, a molecule on the surface of brain cells that binds to opioids and is centrally involved in pleasure, pain, addiction, and depression. Binding of small molecules to biological receptors is governed by shape and intermolecular forces, two properties of molecules that we'll discuss in this chapter.

 Why do different drugs have different physiological responses?

The answer to this question can be found in the **INQUIRY** ▶▶▶ on page 299.

CONTENTS

What is the importance of molecular structure? The exact three-dimensional shape and charge distribution of a molecule is crucial in understanding its propertie and behavior. The Inquiry at the end of the chapter explores the importanc of molecular structure in biological processes such as the sense of smell and functioning o drugs. The electron-dot structures described in the previous chapter provide a straightfor ward way to indicate the connections between atoms in a molecule; that is, to identify th locations of the covalent bonds. However, more information is needed for a detailed picture o molecular structure. For one thing, electron-dot structures are two-dimensional, while mol ecules are three-dimensional. Furthermore, electron-dot structures do not explain the natur of covalent bonds in quantum mechanical terms useful for describing electron probability.

In this chapter, we will build upon the concepts of electron-dot structures and pola covalent bonds to describe shapes of molecules and interactions between them. More com plex models of chemical bonding including valence bond theory and molecular orbital theor will be used to explain and predict properties of molecules including structure and reactivity

8.1 ▶ MOLECULAR SHAPES: THE VSEPR MODEL

Often, three-dimensional shape plays a crucial part in determining a molecule's chemistr and biological function. Look at the following ball-and-stick models of water, ammonia, and methane. Each of these molecules—and every other molecule as well—has a specific three dimensional shape. Below the ball-and-stick model is the structural model that uses soli wedges and dashed lines to represent three-dimensional shape. Solid lines are assumed to b in the plane of the paper, dashed lines recede behind the plane of the paper away from th viewer, and heavy, wedged lines protrude out of the paper toward the viewer. It is a usefu skill to be able to draw molecules using this type of three-dimensional representation.

Water, H_2O Ammonia, NH_3 Methane, CH_4

Like so many other properties, the shape of a molecule is determined by the electroni structure of its atoms. That shape can often be predicted using what is called the **valence-shel electron-pair repulsion (VSEPR) model**. Electrons in bonds and in lone pairs can be thought o as "charge clouds" that repel one another and stay as far apart as possible, thus causing molecule to assume specific shapes. There are only two steps to remember in applying the VSEPR model:

Applying the VSEPR Model

Step 1. Write an electron-dot structure for the molecule, as described in Section 7.6, an count the number of electron charge clouds surrounding the atom of interest. A charg cloud is simply a group of electrons, either in a bond or in a lone pair, that occupy a region o space around an atom. The following bonds and electron groups represent *one* charge cloud a single electron (in a radical species), a lone pair of electrons, a single bond, a double bond and a triple bond. The total number of charge clouds around an atom is the number of region of space that contain electrons. For example, an atom with four single bonds has four charg clouds. An atom with a lone pair, a single bond, and a double bond has three charge clouds.

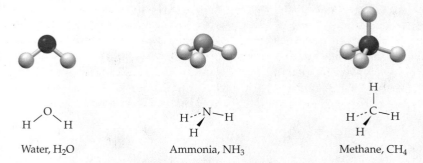

4 single bonds = 4 charge clouds

1 double bond
1 lone pair of e⁻ } = 3 charge clouds
1 single bond

Step 2. Predict the geometric arrangement of charge clouds around each atom by assuming that the clouds are oriented in space as far away from one another as possible. How they achieve this orientation depends on their number. Let's look at the possibilities.

Two Charge Clouds

When there are only two charge clouds on an atom, as occurs on the carbon atoms of CO_2 (two double bonds) and HCN (one single bond and one triple bond), the clouds are farthest apart when they point in opposite directions. Thus, CO_2 and HCN are linear molecules with **bond angles** of 180°.

A CO_2 molecule is linear, with a bond angle of 180°.

An HCN molecule is linear, with a bond angle of 180°.

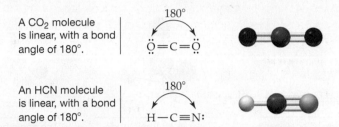

Three Charge Clouds

When there are three charge clouds on an atom, as occurs on the carbon atom of formaldehyde (two single bonds and one double bond) and the sulfur atom of SO_2 (one single bond, one double bond, and one lone pair), the clouds are farthest apart when they lie in the same plane and point to the corners of an equilateral triangle. Thus, a formaldehyde molecule has a *trigonal planar shape*, with H—C—H and H—C=O bond angles near 120°. Similarly, an SO_2 molecule has a trigonal planar arrangement of its three charge clouds on sulfur, but one corner of the triangle is occupied by a lone pair and two corners by oxygen atoms. The molecule therefore has a bent shape, with an O—S=O bond angle of approximately 120° rather than 180°. Note that the number of charge clouds determines the bond angles, but the actual shape of the molecule is given by positions of atoms only and lone pairs of electrons are not "seen." For consistency, we'll use the word *shape* to refer to the overall arrangement of atoms in a molecule, not to the geometric arrangement of charge clouds around a specific atom.

A formaldehyde molecule is trigonal planar, with bond angles of roughly 120°.

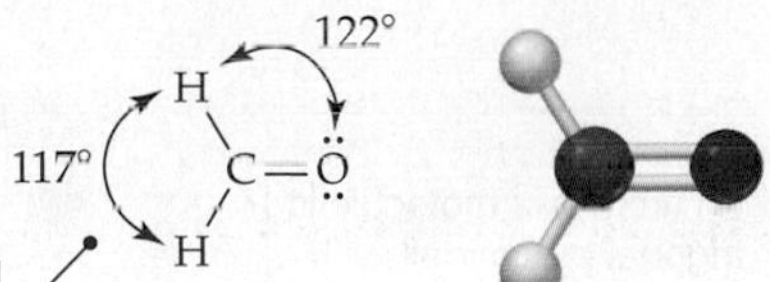

Top view

Side view

An SO_2 molecule is bent, with a bond angle of approximately 120°.

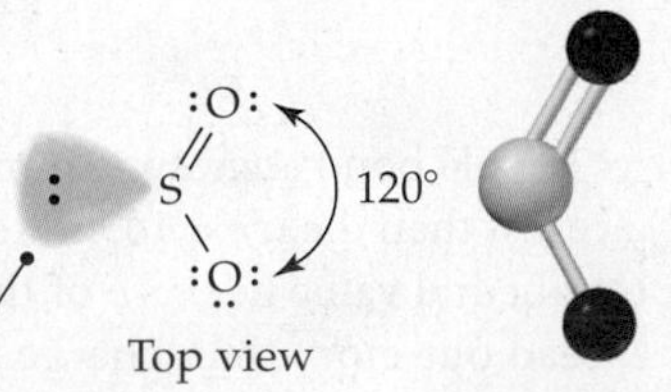

Top view

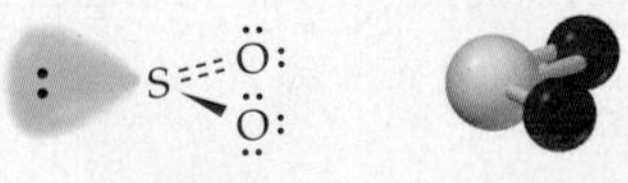

Side view

Four Charge Clouds

When there are four charge clouds on an atom, as occurs on the central atoms in CH_4 (fou single bonds), NH_3 (three single bonds and one lone pair), and H_2O (two single bonds an two lone pairs), the clouds are farthest apart if they extend toward the corners of a regula tetrahedron. As illustrated in **FIGURE 8.1**, a regular tetrahedron is a geometric solid whos four identical faces are equilateral triangles. The central atom lies in the center of the tetrahe dron, the charge clouds point toward the four corners, and the angle between two lines draw from the center to any two corners is 109.5°.

▶ **FIGURE 8.1**

The tetrahedral geometry of an atom with four charge clouds.

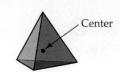

Center

A regular tetrahedron

The atom is located in the **center** of a regular tetrahedron.

The four charge clouds point to the **four corners** of the tetrahedron.

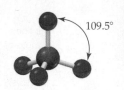

109.5°

A tetrahedral molecule

The angle between any two bonds is 109.5°.

Because valence electron octets are so common, particularly for second-row elements the atoms in a great many molecules have geometries based on the tetrahedron. Methane for example, has a tetrahedral shape, with H—C—H bond angles of 109.5°. In NH_3, th nitrogen atom has a tetrahedral arrangement of its four charge clouds, but one corner of the tetrahedron is occupied by a lone pair, resulting in a *trigonal pyramidal* shape for the mol ecule. Similarly, H_2O has two corners of the tetrahedron occupied by lone pairs and thus ha a *bent* shape.

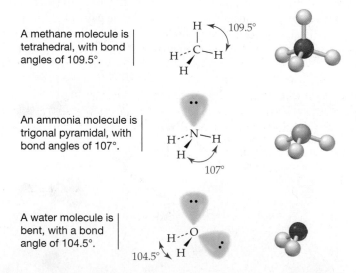

A methane molecule is tetrahedral, with bond angles of 109.5°.

An ammonia molecule is trigonal pyramidal, with bond angles of 107°.

A water molecule is bent, with a bond angle of 104.5°.

Note that the H—N—H bond angles in ammonia (107°) and the H—O—H bond an gle in water (104.5°) are less than the ideal 109.5° tetrahedral value. The angles are diminished somewhat from the tetrahedral value because of the presence of lone pairs. Charge clouds o lone-pair electrons spread out more than charge clouds of bonding electrons because the aren't confined to the space between two nuclei. As a result, the somewhat enlarged lone-pai charge clouds tend to compress the bond angles in the rest of the molecule.

Five Charge Clouds

Five charge clouds, such as are found on the central atoms in PCl_5, SF_4, ClF_3, and I_3^-, ar oriented toward the corners of a geometric figure called a *trigonal bipyramid*. Three cloud

...e in a plane and point toward the corners of an equilateral triangle, the fourth cloud points ...irectly up, and the fifth cloud points down:

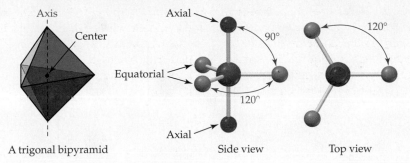

A trigonal bipyramid   Side view   Top view

Trigonal bipyramidal geometry differs from the linear, trigonal planar, and tetrahedral ...eometries discussed previously because it has two kinds of positions—three *equatorial* posi-...ions (around the "equator" of the bipyramid) and two *axial* positions (along the "axis" of ...he bipyramid). The three equatorial positions are at angles of 120° to one another and at an ...ngle of 90° to the axial positions. The two axial positions are at angles of 180° to each other ...nd at an angle of 90° to the equatorial positions.

Different substances containing a trigonal bipyramidal arrangement of charge clouds on ...n atom adopt different shapes, depending on whether the five charge clouds contain bond ...ng or nonbonding electrons. Phosphorus pentachloride, for instance, has all five positions ...round phosphorus occupied by chlorine atoms and thus has a trigonal bipyramidal shape:

A PCl_5 molecule is trigonal bipyramidal.

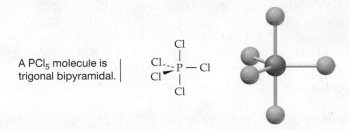

The sulfur atom in SF_4 is bonded to four other atoms and has one nonbonding electron ...one pair. Because an electron lone pair spreads out and occupies more space than a bonding ...air, the nonbonding electrons in SF_4 occupy an equatorial position where they are close to (90° ...way from) only two charge clouds. Were they instead to occupy an axial position, they would ...e close to three charge clouds. As a result, SF_4 has a shape often described as that of a seesaw. ...he two axial bonds form the board, and the two equatorial bonds form the legs of the seesaw.

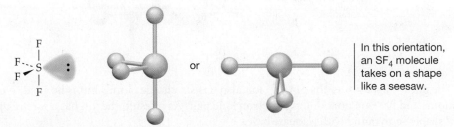

In this orientation, an SF_4 molecule takes on a shape like a seesaw.

The chlorine atom in ClF_3 is bonded to three other atoms and has two nonbonding elec-...ron lone pairs. Both lone pairs occupy equatorial positions, resulting in a T shape for the ...lF_3 molecule.

A ClF_3 molecule is T-shaped.

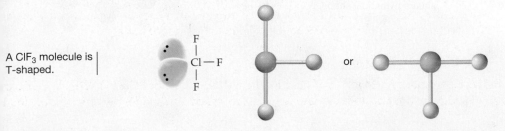

The central iodine atom in the I_3^- ion is bonded to two other atoms and has three lone pairs. All three lone pairs occupy equatorial positions, resulting in a linear shape for I_3^-.

An I_3^- ion is linear.

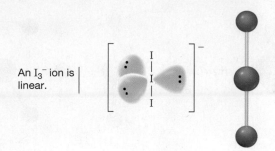

Six Charge Clouds

Six charge clouds around an atom orient toward the six corners of a regular octahedron, a geometric solid whose eight faces are equilateral triangles. All six positions are equivalent, and the angle between any two adjacent positions is 90°.

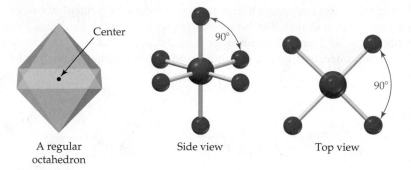

A regular octahedron   Side view   Top view

As was true in the case of five charge clouds, different shapes are possible for molecules having atoms with six charge clouds, depending on whether the clouds are of bonding or nonbonding electrons. In sulfur hexafluoride, for instance, all six positions around sulfur are occupied by fluorine atoms:

An SF_6 molecule is octahedral.

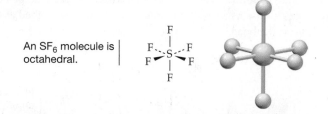

The antimony atom in the $SbCl_5^{2-}$ ion also has six charge clouds but is bonded to only five atoms and has one nonbonding electron lone pair. As a result, the ion has a *square pyramidal* shape—a pyramid with a square base:

An $SbCl_5^{2-}$ ion has a square pyramidal shape.

The xenon atom in XeF_4 is bonded to four atoms and has two lone pairs. The lone pairs orient as far away from each other as possible to minimize electronic repulsions, giving the molecule a *square planar* shape:

An XeF_4 molecule has a square planar shape.

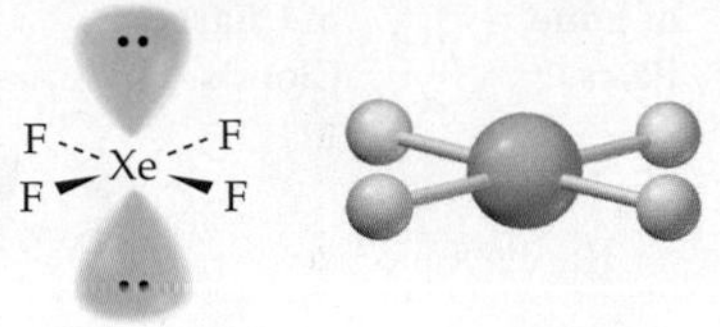

All the geometries for two to six charge clouds are summarized in **TABLE 8.1**.

WORKED EXAMPLE 8.1

Using the VSEPR Model to Predict Molecular Shape

Predict the shape of BrF_5.

STRATEGY

First, draw an electron-dot structure for BrF_5 to determine that the central bromine atom has six charge clouds (five bonds and one lone pair). Then predict how the six charge clouds are arranged.

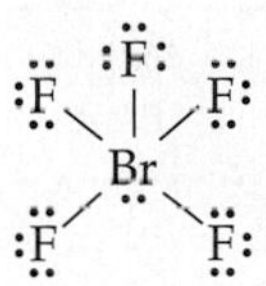

Bromine pentafluoride

SOLUTION

Six charge clouds imply an octahedral arrangement. The five attached atoms and one lone pair give BrF_5 a square pyramidal shape:

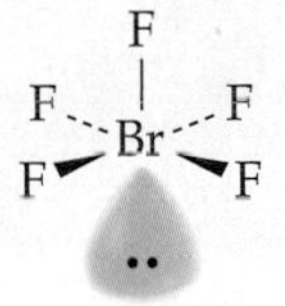

▶ **Conceptual APPLY 8.2**   What is the number and geometric arrangement of charge clouds around the central atom in each of the following molecular models? What is the shape of the molecule?

(a)   **(b)**

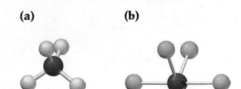

▶ **PRACTICE 8.1**   Predict the shapes of the following molecules or ions:

(a) O_3 (b) H_3O^+ (c) XeF_2 (d) PF_6^-
(e) $XeOF_4$ (f) AlH_4^- (g) BF_4^- (h) $SiCl_4$
(i) ICl_4^- (j) $AlCl_3$

TABLE 8.1		Geometry Around Atoms with 2, 3, 4, 5, and 6 Charge Clouds		
Number of Bonds	Number of Lone Pairs	Number of Charge Clouds	Geometry and Shape	Example
2	0	2	Linear	$O{=}C{=}O$
3	0	3	Trigonal planar	$H{\cdots}C{=}O$, H
2	1		Bent	$O{\cdots}S$, O
4	0	4	Tetrahedral	$H{\cdots}C{-}H$ with H top and bottom
3	1		Trigonal pyramidal	$H{\cdots}N{-}H$, H
2	2		Bent	$H{\cdots}O$, H
5	0	5	Trigonal bipyramidal	$Cl{\cdots}P{-}Cl$ with Cl top and bottom
4	1		Seesaw	$F{\cdots}S$ with F top and bottom
3	2		T-shaped	$Cl{-}F$ with F top and bottom
2	3		Linear	$[I{\cdots}I{\cdots}I]^{-}$
6	0	6	Octahedral	$F{\cdots}S{\cdots}F$ with F around
5	1		Square pyramidal	$[Cl{\cdots}Sb{\cdots}Cl]^{2-}$
4	2		Square planar	$F{\cdots}Xe{\cdots}F$

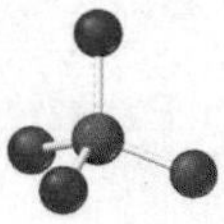

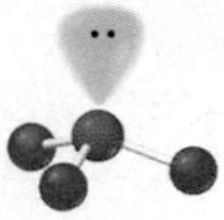

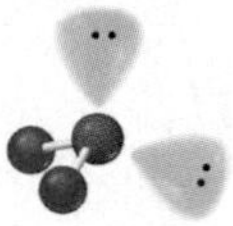

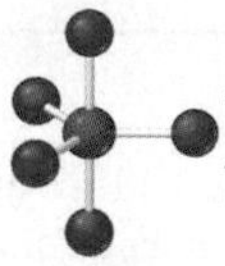

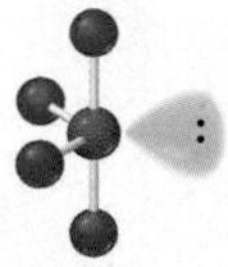

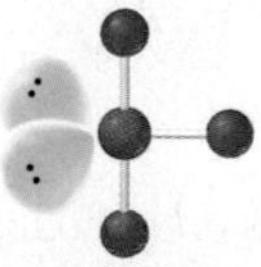

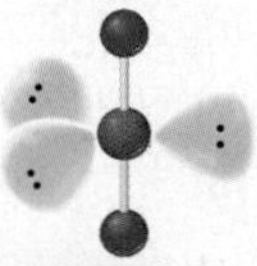

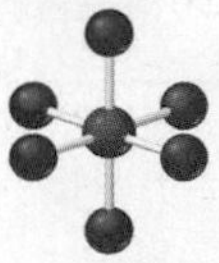

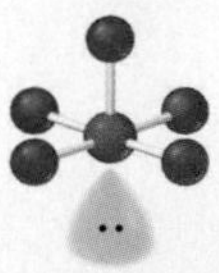

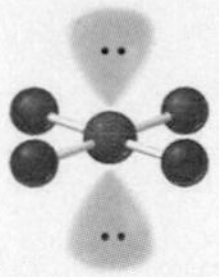

Shapes of Larger Molecules

The geometries around individual atoms in larger molecules can also be predicted from the rules summarized in Table 8.1. For example, each of the two carbon atoms in ethylene ($H_2C=CH_2$) has three charge clouds, giving rise to trigonal planar geometry for each carbon. The molecule as a whole has a planar shape, with $H-C-C$ and $H-C-H$ bond angles of approximately 120°.

Each carbon atom in ethylene has trigonal planar geometry. As a result, the entire molecule is planar, with bond angles of 120°.

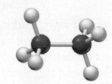

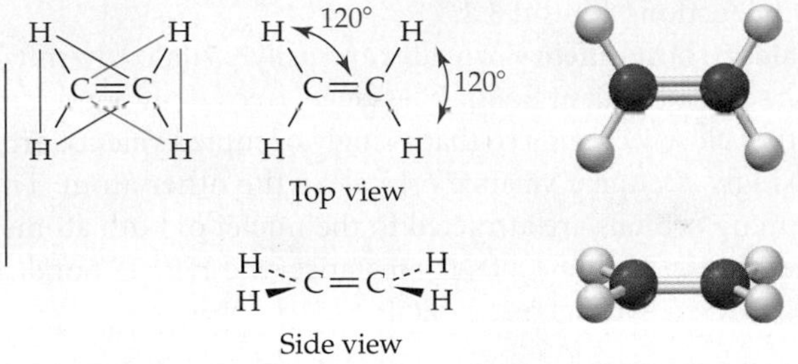

Top view

Side view

Carbon atoms bonded to four other atoms are each at the center of a tetrahedron. As shown below for ethane, H_3C-CH_3, the two tetrahedrons are joined so that the central carbon atom of one is a corner atom of the other.

Each carbon atom in ethane has tetrahedral geometry, with bond angles of 109.5°.

Conceptual WORKED EXAMPLE 8.2

Predicting Shape in Molecules with More Than One Central Atom

Pyruvic acid is a key substance in the metabolism of both carbohydrates and several amino acids. Describe the geometry around each of the central atoms in pyruvic acid, and draw the overall shape of the molecule.

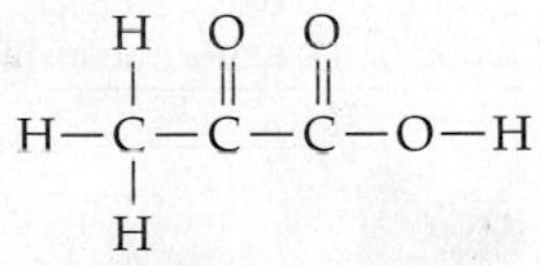

$$
\begin{array}{ccc}
H & O & O \\
| & \| & \| \\
H-C-C-C-O-H \\
| \\
H
\end{array}
$$

Pyruvic acid

STRATEGY

Use the VSEPR model to predict the geometric arrangement of charge clouds around each of the three carbon atoms. Draw the overall molecule showing the bond angles predicted by VSEPR and using dashed lines and wedges to represent three-dimensional shape.

SOLUTION

The carbon on the left has four charge clouds (four single bonds) and tetrahedral geometry with 109.5° bond angles. Both the carbon in the middle and on the right have three charge clouds (two single bonds and one double bond) and trigonal planar geometry with 120° bond angles.

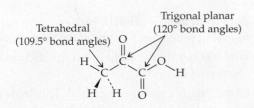

Tetrahedral
(109.5° bond angles)

Trigonal planar
(120° bond angles)

▶ **PRACTICE 8.3** Acetic acid, CH_3CO_2H, is the main organic constituent of vinegar. Draw an electron-dot structure for acetic acid and give the bond angles around each carbon atom. Draw a three-dimensional representation of the molecule. (The two carbons are connected by a single bond, and both oxygens are connected to the same carbon.)

▶ **APPLY 8.4** Benzene, C_6H_6, is a cyclic molecule in which all six carbon atoms are joined in a ring, with each carbon also bonded to one hydrogen. Draw an electron-dot structure for benzene and give the bond angles around each carbon. Draw a three-dimensional representation of the molecule.

8.2 ▶ VALENCE BOND THEORY

The VSEPR model discussed in the previous section provides a simple way to predict molecular shapes, but it says nothing about the electronic nature of covalent bonds. To describe bonding, two models, called *valence bond theory* and *molecular orbital theory* have been developed. We'll look first at valence bond theory, followed by molecular orbital theory in Sections 8.7 and 8.8.

Valence bond theory provides an easily visualized orbital picture of how electron pairs are shared in a covalent bond. In essence, a covalent bond results when two atoms approach each other closely enough so that a singly occupied valence orbital on one atom spatially *overlaps* a singly occupied valence orbital on the other atom. The now-paired electrons in the overlapping orbitals are attracted to the nuclei of both atoms and thus bond the two atoms together. In the H_2 molecule, for instance, the H—H bond results from the overlap of two singly occupied hydrogen 1s orbitals.

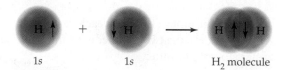

Recall that atomic orbitals arise from the **Schrödinger wave equation** and that the two lobes of a *p* atomic orbital have different **phases**, as represented by different colors. In the valence bond model, the two overlapping lobes must be of the same phase, and the strength of the covalent bond that forms depends on the amount of orbital overlap: the greater the overlap, the stronger the bond. This, in turn, means that bonds formed by overlap of other than *s* orbitals have a directionality to them. In the F_2 molecule, for instance, each fluorine atom has the ground state electron configuration $[He] 2s^2 2p_x^2 2p_y^2 2p_z^1$ and the F—F bond results from the overlap of two singly occupied 2p orbitals. The two 2p orbitals must point directly at each other for optimum overlap to occur, and the F—F bond forms along the orbital axis. Such bonds that result from head-on orbital overlap are called **sigma (σ) bonds**.

Bonds form between two lobes of the same phase.

A **sigma (σ) bond** forms from head-on orbital overlap.

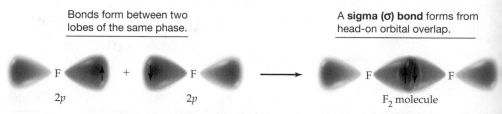

In HCl, the covalent bond involves overlap of a hydrogen 1s orbital with a chlorine 3p orbital and forms along the *p*-orbital axis:

The key ideas of valence bond theory can be summarized as follows:

Principles of Valence Bond Theory

- Covalent bonds are formed by overlap of atomic orbitals, each of which contains one electron of opposite spin. The two overlapping lobes must be of the same phase.
- Each of the bonded atoms maintains its own atomic orbitals, but the electron pair in the overlapping orbitals is shared by both atoms.
- The greater the amount of orbital overlap, the stronger the bond. This leads to a directional character for the bond when other than *s* orbitals are involved.

8.3 ▶ HYBRIDIZATION AND sp^3 HYBRID ORBITALS

How does valence bond theory describe the electronic structure of complex polyatomic molecules, and how does it account for the observed geometries around atoms in molecules? Let's look first at a simple tetrahedral molecule such as methane, CH_4. There are several problems to be dealt with.

Carbon has the ground-state electron configuration $[He]2s^2\ 2p_x^1\ 2p_y^1$. It thus has four valence electrons, two of which are paired in a 2s orbital and two of which are unpaired in different 2p orbitals that we'll arbitrarily designate as $2p_x$ and $2p_y$. But how can carbon form four bonds if two of its valence electrons are already paired and only two unpaired electrons remain for sharing? The answer is that an electron must be promoted from the lower-energy 2s orbital to the vacant, higher-energy $2p_z$ orbital, giving an *excited-state configuration* $[He]\ 2s^1\ 2p_x^1\ 2p_y^1\ 2p_z^1$ that has *four* unpaired electrons and can thus form four bonds.

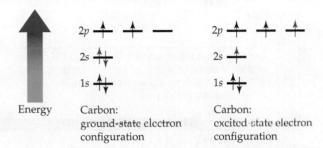

Energy

Carbon:
ground-state electron
configuration

Carbon:
excited-state electron
configuration

A second problem is more difficult to resolve: If excited-state carbon uses two kinds of orbitals for bonding, 2s and 2p, how can it form four *equivalent* bonds? Furthermore, if the three 2p orbitals in carbon are at angles of 90° to one another, and if the 2s orbital has no directionality, how can carbon form bonds with angles of 109.5° directed to the corners of a regular tetrahedron? The answers to these questions were provided in 1931 by Linus Pauling, who introduced the idea of *hybrid orbitals.*

Pauling showed how the quantum mechanical wave functions for s and p atomic orbitals derived from the Schrödinger wave equation can be mathematically combined to form a new set of equivalent wave functions called **hybrid atomic orbitals.** When one s orbital combines with three p orbitals, as occurs in an excited-state carbon atom, four equivalent hybrid orbitals, called **sp^3 hybrid orbitals**, result. (The superscript 3 in the name sp^3 tells how many p atomic orbitals are combined to construct the hybrid orbitals, not how many electrons occupy the orbital.)

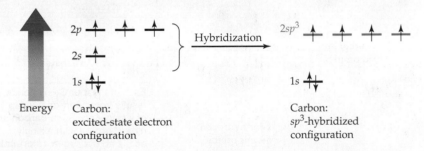

Energy

Carbon:
excited-state electron
configuration

Carbon:
sp^3-hybridized
configuration

Each of the four equivalent sp^3 hybrid orbitals has two lobes of different phase like an atomic p orbital (Section 5.8) but one of the lobes is larger than the other, giving the orbital a directionality. The four large lobes are oriented toward the four corners of a tetrahedron at angles of 109.5°, as shown in **FIGURE 8.2**. For consistency in the use of colors, we'll routinely show the different phases of orbitals in red and blue and will show the large lobes of the resultant hybrid orbitals in green.

The shared electrons in a covalent bond made with a spatially directed hybrid orbital spend most of their time in the region between the two bonded nuclei. As a result, covalent bonds made with sp^3 hybrid orbitals are often strong ones. In fact, the energy released on forming the four strong C—H bonds in CH_4 more than compensates for the energy required to form the excited state of carbon. **FIGURE 8.3** shows how the four C—H sigma bonds in methane can form by head-on overlap of carbon sp^3 hybrid orbitals with hydrogen 1s orbitals.

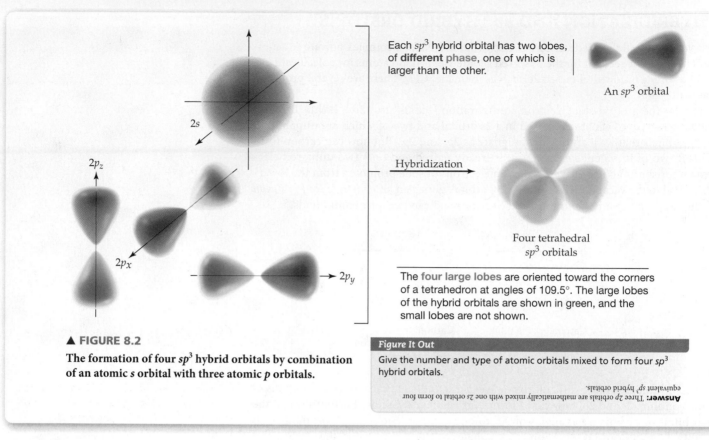

Each sp^3 hybrid orbital has two lobes, of **different phase**, one of which is larger than the other.

An sp^3 orbital

Hybridization →

Four tetrahedral sp^3 orbitals

The **four large lobes** are oriented toward the corners of a tetrahedron at angles of 109.5°. The large lobes of the hybrid orbitals are shown in green, and the small lobes are not shown.

▲ **FIGURE 8.2**

The formation of four sp^3 hybrid orbitals by combination of an atomic s orbital with three atomic p orbitals.

Figure It Out

Give the number and type of atomic orbitals mixed to form four sp^3 hybrid orbitals.

Answer: Three $2p$ orbitals are mathematically mixed with one $2s$ orbital to form four equivalent sp^3 hybrid orbitals.

▶ **FIGURE 8.3**

The bonding in methane (CH_4).

Figure It Out

What types of orbitals overlap to form a C—H bond in methane?

Answer: Singly occupied carbon sp^3 hybrid orbital overlaps with a singly occupied hydrogen $1s$ orbital.

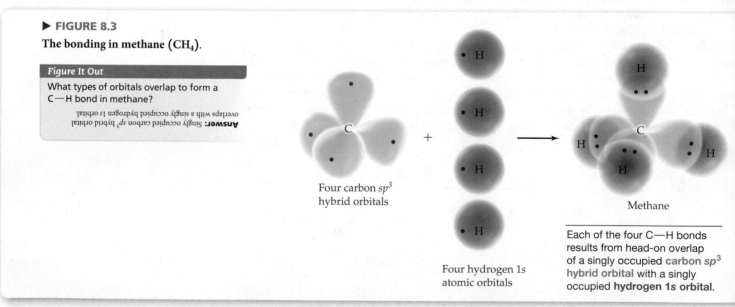

Four carbon sp^3 hybrid orbitals

Four hydrogen $1s$ atomic orbitals

Methane

Each of the four C—H bonds results from head-on overlap of a singly occupied **carbon sp^3 hybrid orbital** with a singly occupied **hydrogen $1s$ orbital**.

The same kind of sp^3 hybridization that describes the bonds to carbon in the tetrahedral methane molecule can also be used to describe bonds to nitrogen in the trigonal pyramidal ammonia molecule, to oxygen in the bent water molecule, and to all other atoms that the VSEPR model predicts to have a tetrahedral arrangement of four charge clouds.

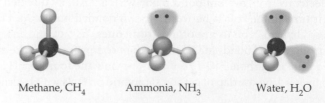

Methane, CH_4 Ammonia, NH_3 Water, H_2O

WORKED EXAMPLE 8.3

Identifying Orbital Overlap in Tetrahedral Geometry

Identify the orbitals that overlap to form the P—Cl bond in PCl_3.

STRATEGY

Write the electron-dot structure for PCl_3 and determine the geometry and hybrid orbitals around the central atom. Determine which orbitals on the terminal atoms are singly occupied and overlap them with the singly occupied hybrid orbitals on the central atom to form a bond.

SOLUTION

The electron-dot structure for PCl_3 has three single bonds and one lone pair of electrons. There are four charge clouds in a tetrahedral arrangement which corresponds to sp^3 hybrid orbitals. The shape of PCl_3 is trigonal pyramidal because there are three bonds and one lone pair.

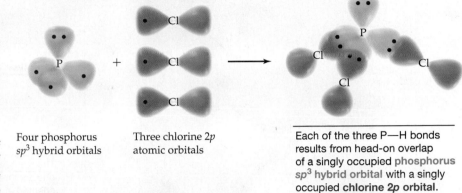

Four phosphorus sp^3 hybrid orbitals

Three chlorine $2p$ atomic orbitals

Each of the three P—H bonds results from head-on overlap of a singly occupied **phosphorus** sp^3 **hybrid orbital** with a singly occupied **chlorine 2p orbital**.

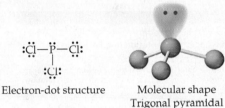

Electron-dot structure

Molecular shape
Trigonal pyramidal

▶ **PRACTICE 8.5** Identify the orbitals that overlap to form the C—Cl bonds and C—H bonds in CH_2Cl_2.

▶ **APPLY 8.6** Describe the bonding in propane, C_3H_8, a fuel often used to heat rural homes and campers. Give the kinds of orbitals on each atom that overlap to form the C—C and C—H bonds.

8.4 ▶ OTHER KINDS OF HYBRID ORBITALS

Other geometries shown in Table 8.1 can also be accounted for by specific kinds of orbital hybridization, although the situation becomes more complex for atoms with five and six charge clouds. Let's look at each.

sp^2 Hybridization

Atoms with three charge clouds undergo hybridization by combination of one atomic s orbital with two p orbitals, resulting in three sp^2 **hybrid orbitals**. These three sp^2 hybrids lie in a plane and are oriented toward the corners of an equilateral triangle at angles of 120° to one another. One p orbital remains unchanged and is oriented at a 90° angle to the plane of the sp^2 hybrids, as shown in **FIGURE 8.4**.

The presence of the unhybridized p orbital on an sp^2-hybridized atom has some interesting consequences. Look, for example, at ethylene, $H_2C\!=\!CH_2$, a colorless gas used as starting material for the industrial preparation of polyethylene. Each carbon atom in ethylene has three charge clouds and is sp^2-hybridized. The two sp^2-hybridized carbon atoms approach each other with sp^2 orbitals aligned head-on to form a sigma (σ) bond. The unhybridized p orbitals on the carbons likewise approach each other and form a bond, but in a parallel, sideways manner rather than head-on. Such a sideways bond, in which the shared electrons occupy regions above and below a line connecting the nuclei rather than directly between the nuclei, is called a **pi (π) bond** (**FIGURE 8.5**). A double bond, such as the C$=$C bond in ethylene, consists of one sigma bond and one pi bond. In addition, four C—H bonds form in ethylene by overlap of the remaining four sp^2 orbitals with hydrogen $1s$ orbitals.

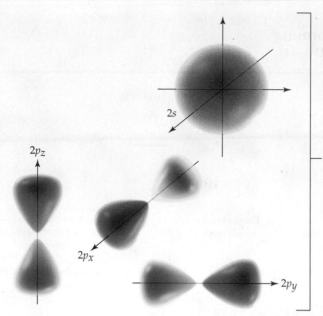

An sp^2 hybrid orbital has two lobes of **different** phase, one of which is larger than the other.

An sp^2 orbital

Hybridization

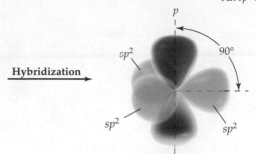

The hybrid orbitals lie in a plane at angles of 120° to one another, and one unhybridized p orbital remains, oriented at a 90° angle to the sp^2 hybrids. The **large lobes** of the hybrid orbitals are shown in green, and the small lobes are not shown.

▲ **FIGURE 8.4**

The formation of sp^2 hybrid orbitals by combination of one s orbital and two p orbitals.

Figure It Out

Give the number and type of atomic orbitals mixed to form three sp^2 hybrid orbitals. Which atomic orbital is not mixed and what is its orientation relative to the sp^2 hybrid orbitals?

Answer: One $2s$ orbital and two $2p$ orbitals are mathematically mixed to form three sp^2 hybrid orbitals. One $2p$ atomic orbital remains which is oriented 90° to the sp^2 hybrid orbitals.

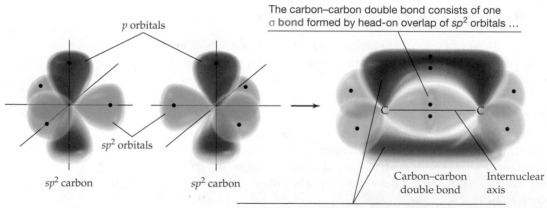

p orbitals

sp^2 orbitals

sp^2 carbon sp^2 carbon

The carbon–carbon double bond consists of one σ **bond** formed by head-on overlap of sp^2 orbitals ...

Carbon–carbon   Internuclear
double bond   axis

... and one π bond formed by sideways overlap of p orbitals. The π bond has two regions of orbital overlap—**one above** and **one below** the internuclear axis.

▲ **FIGURE 8.5**

The structure of a carbon–carbon double bond.

Figure It Out

Which orbitals overlap to form the σ bond, and which overlap to form the π bond between carbon atoms in ethylene?

Answer: The σ bond is formed by overlap of C(sp^2) with C(sp^2), and the π bond is formed by overlap of C($2p$) with C($2p$).

The π bond has two regions of orbital overlap, one above and one below a line drawn between the nuclei, known as the *internuclear axis*. Both regions are part of the same bond, and the two shared electrons are spread over both regions. As always, the p lobes must be of the same phase for overlap leading to bond formation. The net result of both σ and π overlap is the sharing of four electrons and the formation of a carbon–carbon double bond.

WORKED EXAMPLE 8.4

Identifying Hybridization and Orbital Overlap in Single and Double Bonds

Propylene, C_3H_6, a starting material used to manufacture polypropylene polymers, has its three carbon atoms connected in a row. Draw the overall shape of the molecule, and indicate what kinds of orbitals on each atom overlap to form the carbon–carbon and carbon–hydrogen bonds.

STRATEGY

Draw an electron-dot structure of the molecule and count the number of charge clouds around each carbon atom. Use the number of charge clouds and the VSEPR model to predict geometry and hybridization.

SOLUTION

A molecule with the formula C_3H_6 and its carbons connected in a row does not have enough hydrogen atoms to give each carbon four single bonds. Thus, propylene must contain at least one carbon–carbon double bond. In the structure below, the carbon atom on the left has four charge clouds (four single bonds) and tetrahedral geometry with sp^3 hybrid orbitals. The other carbon atoms have three charge clouds (two single bonds and one double bond) and trigonal planar geometry with sp^2 hybrid orbitals. The carbon on the left has bond angles of 109.5° and the other carbon atoms have bond angles of 120°.

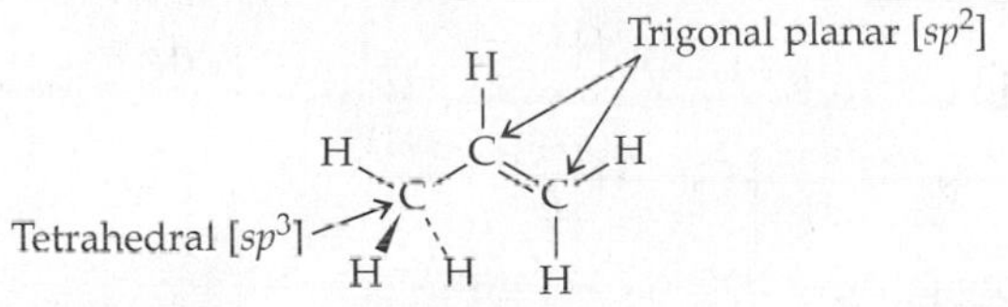

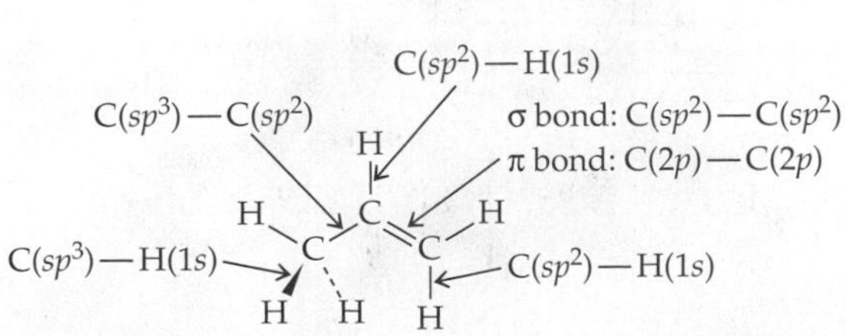

The carbon–carbon single bond is formed by an overlap of a sp^3 orbital and a sp^2 orbital. The carbon–carbon double bond consists of a σ bond formed from head-on overlap of two sp^2 hybrid orbitals and a π bond formed from sideways overlap of two p orbitals. Each carbon–hydrogen bond is formed from overlap of a hydrogen 1s orbital with the hybrid orbital on the carbon to which it is attached.

▶ **PRACTICE 8.7** Describe the hybridization of the carbon atom in formaldehyde, $H_2C=O$, and make a rough sketch of the molecule showing the orbitals involved in bonding.

▶ **APPLY 8.8** Describe the hybridization of each carbon atom in lactic acid, a metabolite formed in tired muscles. Tell what kinds of orbitals on each atom overlap to form the carbon–carbon, carbon–oxygen, and carbon–hydrogen bonds. Draw the overall shape of the molecule.

Propylene

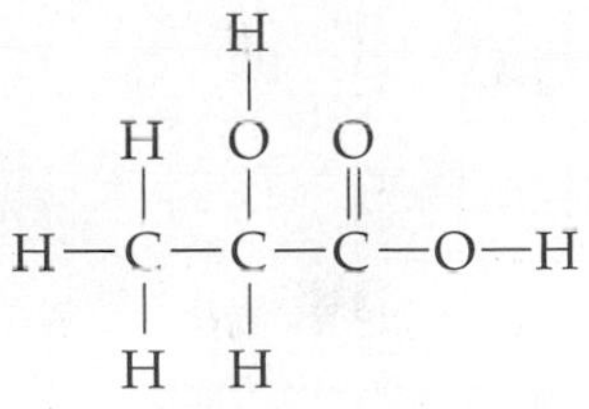

Lactic acid

sp Hybridization

Atoms with two charge clouds undergo hybridization by combination of one atomic *s* orbital with one *p* orbital, resulting in two *sp* **hybrid orbitals** that are oriented 180° from each other. Since only one *p* orbital is involved when an atom undergoes *sp* hybridization, the other two *p* orbitals are unchanged and are oriented at 90° angles to the *sp* hybrids, as shown in **FIGURE 8.6**.

▶ **FIGURE 8.6**

sp **Hybridization.**

The combination of one *s* and one *p* orbital gives **two *sp* hybrid orbitals** oriented 180° apart.

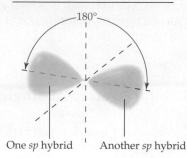

One *sp* hybrid   Another *sp* hybrid

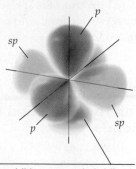

In addition, two unhybridized **p orbitals** remain, oriented at 90° angles to the *sp* hybrids.

One of the simplest examples of *sp* hybridization occurs in acetylene, H—C≡C—H, a colorless gas used in welding. Both carbon atoms in the acetylene molecule have linear geometry and are *sp*-hybridized. When the two *sp*-hybridized carbon atoms approach each other with their *sp* orbitals aligned head-on for σ bonding, the unhybridized *p* orbitals on each carbon are aligned for π bonding. Two *p* orbitals are aligned in an up/down position, and two are aligned in an in/out position. Thus, there are two mutually perpendicular π bonds that form by sideways overlap of *p* orbitals, along with one σ bond that forms by head-on overlap of the *sp* orbitals. The net result is the sharing of six electrons and formation of a triple bond (**FIGURE 8.7**). In addition, two C—H bonds form in acetylene by overlap of the remaining two *sp* orbitals with hydrogen 1*s* orbitals.

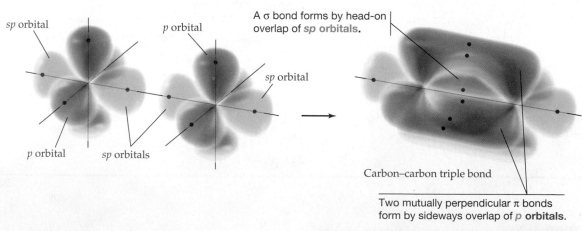

sp orbital

p orbital

p orbital   *sp* orbitals

sp orbital

A σ bond forms by head-on overlap of ***sp* orbitals.**

Carbon–carbon triple bond

Two mutually perpendicular π bonds form by sideways overlap of **p orbitals.**

▲ **FIGURE 8.7**

Formation of a triple bond by two *sp*-hybridized atoms.

● WORKED EXAMPLE 8.5

Identifying Hybridization and Orbital Overlap in Single and Double Bonds

Describe the hybridization of the carbon atoms in allene, $H_2C=C=CH_2$, and make a rough sketch of the molecule showing orbital overlap in bond formation.

STRATEGY

Draw an electron-dot structure to find the number of charge clouds on each atom.

Two charge clouds

Three charge clouds —→ H   H ←— Three charge clouds

C=C=C

H   H

Then predict the geometry around each atom using the VSEPR model (Table 8.1).

SOLUTION

Because the central carbon atom in allene has two charge clouds (two double bonds), it has a linear geometry and is sp-hybridized. Because the two terminal carbon atoms have three charge clouds each (one double bond and two single bonds), they have trigonal planar geometry and are sp^2-hybridized. The central carbon uses its sp orbitals to form two σ bonds at 180° angles and uses its two unhybridized p orbitals to form π bonds, one to each of the terminal carbons. The unhybridized p orbitals are oriented at 90° from each other and are designated at $2p_x$ and $2p_y$ to show this spatial arrangement. Each terminal carbon atom uses an sp^2 orbital for σ bonding to carbon, a p orbital for π bonding, and its two remaining sp^2 orbitals for C—H bonds.

σ bond: C(sp)—C(sp^2)   σ bond: C(sp)—C(sp^2)
π bond: C($2p_x$)—C($2p_x$)   π bond: C($2p_y$)—C($2p_y$)

H   H

C=C=C

H   H

C(sp^2)—H($1s$)

Note that the mutually perpendicular arrangement of the two π bonds results in a similar perpendicular arrangement of the two CH_2 groups.

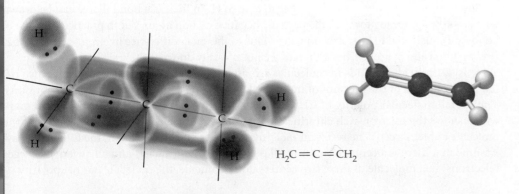

$H_2C=C=CH_2$

▶ **PRACTICE 8.9**   Describe the hybridization of the carbon atom in carbon dioxide, and make a rough sketch of the molecule showing its hybrid orbitals and π bonds.

▶ **APPLY 8.10**   Describe the hybridization of the carbon atom in the hydrogen cyanide molecule, H—C≡N, and make a rough sketch to show the hybrid orbitals it uses for bonding.

TABLE 8.2 Hybrid Orbitals and Their Geometry		
Number of Charge Clouds	Arrangement of Charge Clouds	Hybridization
2	Linear	sp
3	Trigonal planar	sp^2
4	Tetrahedral	sp^3

Atoms with Five and Six Charge Clouds

Main-group atoms with five or six charge clouds, such as the phosphorus in PCl_5 and the sulfur in SF_6, were at one time thought to undergo hybridization by combination of five and six atomic orbitals, respectively. Because a given shell has a total of only four s and p orbitals, however, the need to use five or six orbitals implies that d orbitals are involved. As we'll see in Section 20.11, hybridization involving d orbitals is indeed involved for many compounds of transition metals. Recent quantum mechanical calculations indicate, however, that main-group compounds do not use d orbitals in hybridization but instead use a more complex bonding pattern that is not easily explained by valence bond theory.

A summary of the three common kinds of hybridization for main-group elements and the geometry that each corresponds to is given in **TABLE 8.2**.

8.5 ▶ POLAR COVALENT BONDS AND DIPOLE MOMENTS

Molecular polarity is an important characteristic that influences both chemical and physical properties of molecules. To understand the notion of molecular polarity, it's first necessary to develop the ideas of *bond dipoles* and *dipole moments*. We saw in Section 7.3 that **polar covalent bonds** form between atoms of different **electronegativity**. In chloromethane (CH_3Cl), chlorine is more electronegative than carbon so the chlorine atom attracts the electrons in the C—Cl bond more strongly than carbon does. The C—Cl bond is therefore polarized so that the chlorine atom is slightly electron-rich ($\delta-$) and the carbon atom is slightly electron-poor ($\delta+$).

Because the polar C—Cl bond in chloromethane has a positive end and a negative end—we describe it as being a bond **dipole**, and we often represent the dipole using an arrow with a cross at one end ($\mapsto$) to indicate the direction of electron displacement. The point of the arrow represents the negative end of the dipole ($\delta-$), and the crossed end (which looks like a plus sign) represents the positive end ($\delta+$). The C—H bond in chloromethane also has a small bond dipole because carbon is slightly more electronegative than H. In theory, a bond dipole can be drawn with the arrow pointing towards the partially negative carbon atom, but in practice the C—H bonds are considered to be nearly nonpolar. Due to the small difference in electronegativity between carbon and hydrogen, bond dipoles are not typically drawn for C—H bonds.

Just as individual bonds in molecules are often polar, molecules as a whole are also often polar because of the net sum of individual bond polarities and lone-pair contributions. The overall molecular polarity of CH_3Cl is clearly visible in an **electrostatic potential map**, which shows the electron-rich chlorine atom as red and the electron-poor remainder of the molecule as blue-green. The resultant *molecular dipole* can be looked at in the following way: Assume that there is a center of mass of all positive charges (nuclei) and all negative charges (electrons) in a molecule. If these two centers don't coincide, the molecule has net polarity.

Chlorine is at the negative end of the bond dipole.

Carbon is at the positive end of the bond dipole.

Chloromethane, CH_3Cl

Net

The measure of net molecular polarity is a quantity called the **dipole moment (μ)**, (Greek nu), which is defined as the magnitude of the charge Q at either end of the molecular dipole times the distance r between the charges: $\mu = Q \times r$. Dipole moments are expressed in *debyes* (D), where $1\ D = 3.336 \times 10^{-30}$ coulomb meters (C·m) in SI units. To calibrate your thinking, the charge on an electron is 1.60×10^{-19} C. Thus, if a proton and an electron were separated by 100 pm (a bit less than the length of a typical covalent bond), then the dipole moment would be 1.60×10^{-29} C·m, or 4.80 D:

$$\mu = Q \times r$$

$$\mu = (1.60 \times 10^{-19}\,C)(100 \times 10^{-12}\,m)\left(\frac{1\,D}{3.336 \times 10^{-30}C \cdot m}\right) = 4.80\ D$$

It's relatively easy to measure dipole moments experimentally, and values for some common substances are given in **TABLE 8.3**. Once the dipole moment is known, it's then possible to get an idea of the amount of charge separation in a molecule. In chloromethane, for example, the experimentally measured dipole moment is $\mu = 1.90$ D. If we assume that the contributions of the nonpolar C—H bonds are small, then most of the chloromethane dipole moment is due to the C—Cl bond. Since the C—Cl bond distance is 179 pm, we can calculate that the dipole moment of chloromethane would be $1.79 \times 4.80\ D = 8.59$ D if the C—Cl bond were ionic—that is, if a full negative charge on chlorine were separated from a full positive charge on carbon by a distance of 179 pm. But because the measured dipole moment of chloromethane is only 1.90 D, we can conclude that the C—Cl bond is only about $(1.90/8.59)(100\%) = 22\%$ ionic. Thus, the chlorine atom in chloromethane has an excess of about 0.2 electron, and the carbon atom has a deficiency of about 0.2 electron.

Perhaps it's not surprising that by far the largest dipole moment listed in Table 8.3 belongs to the ionic compound NaCl, which exists in the gas phase as a pair of Na$^+$ and Cl$^-$ ions held together by a strong ionic bond. The molecular compounds water and ammonia also have substantial dipole moments because both oxygen and nitrogen are electronegative relative to hydrogen and because both O and N have lone pairs of electrons that make substantial contributions to net molecular polarity:

TABLE 8.3 Dipole Moments of Some Common Compounds	
Compound	**Dipole Moment (D)**
NaCl[a]	9.0
CH$_3$Cl	1.90
H$_2$O	1.85
NH$_3$	1.47
HCl	1.11
CO$_2$	0
CCl$_4$	0

[a]Measured in the gas phase.

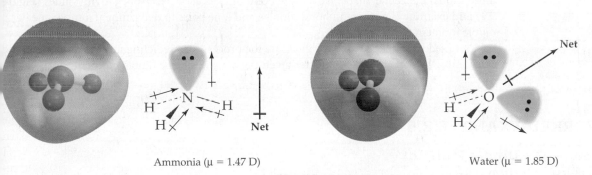

Ammonia (μ = 1.47 D)   Water (μ = 1.85 D)

In contrast with water and ammonia, carbon dioxide and tetrachloromethane (CCl$_4$) have zero dipole moments. Molecules of both substances contain *individual* polar covalent bonds, but because of the symmetry of their structures, the individual bond polarities exactly cancel.

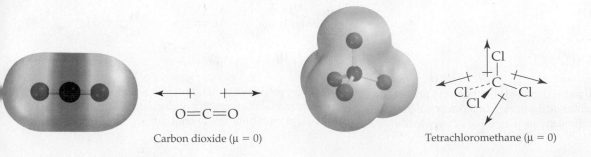

Carbon dioxide (μ = 0)   Tetrachloromethane (μ = 0)

▶ **FIGURE 8.8**

Strategy for predicting molecular polarity.

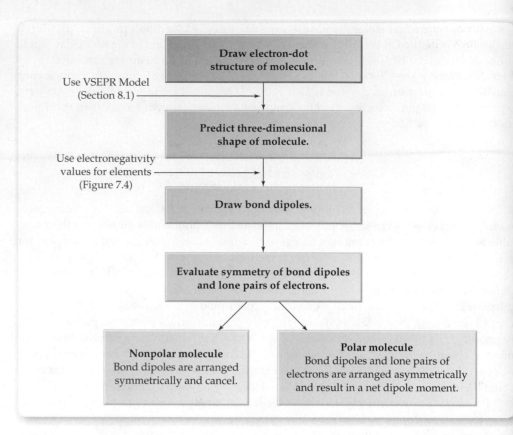

FIGURE 8.8 outlines the strategy for predicting molecular polarity. The first step is to draw an electron-dot structure and use the VSEPR model (Section 8.1) to determine three-dimensional molecular shape. Next draw bond dipoles using electronegativity values (Figure 7.4) and examine symmetry in bond dipoles and lone pairs to determine if a molecule has a dipole moment.

Worked Examples 8.6 and 8.7 illustrate the process for predicting molecular polarity and calculating the percent ionic character given the magnitude of a dipole moment.

Conceptual WORKED EXAMPLE 8.6

Predicting the Presence of a Dipole Moment

Would you expect vinyl chloride (H_2C=$CHCl$), the starting material used for preparation of poly(vinyl chloride) polymer, to have a dipole moment? If so, indicate the direction.

STRATEGY AND SOLUTION

Follow the procedure outlined in Figure 8.8 for determining molecular polarity.

The electron-dot structure for vinyl chloride is:

The VSEPR model described in Section 8.1 is then used to predict the molecular shape of vinyl chloride. Each carbon has trigonal planar geometry because both carbon atoms have three charge clouds.

The molecule as a whole is planar because the unhybridized carbon $2p$ orbitals in the pi part of the C=C double bond overlap in a sideways manner:

Vinyl chloride

Top view   Side view

Then, assign polarities to the individual bonds according to the differences in electronegativity of the bonded atoms, and make a reasonable guess about the overall polarity that would result by evaluating symmetry in bond dipoles and summing the individual contributions. Only the C—Cl bond has a substantial polarity, giving the molecule a net polarity in the direction of the chlorine atom:

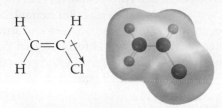

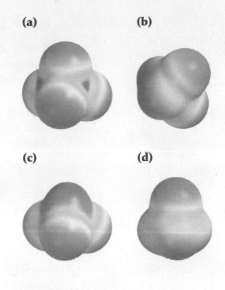

(a)　　　**(b)**

(c)　　　**(d)**

▶ **Conceptual PRACTICE 8.11**　Tell which of the following compounds is likely to have a dipole moment, and show the direction of each.

(a) SF_6
(c) SO_3

(b) $H_2C{=}CF_2$
(d) BrF_3

▶ **Conceptual APPLY 8.12**　Match the following electrostatic potential maps with their corresponding molecule: CH_3F, CHF_3, CF_4, CH_2F_2.

─● WORKED EXAMPLE 8.7

Calculating Percent Ionic Character from a Dipole Moment

The dipole moment of HCl is 1.11 D, and the distance between atoms is 127 pm. What is the percent ionic character of the HCl bond?

STRATEGY

If HCl were 100% ionic, a negative charge (Cl^-) would be separated from a positive charge (H^+) by 127 pm. Calculate the expected dipole moment, and compare that calculated value to the actual value.

SOLUTION

The calculated dipole moment is

$$\mu = Q \times r$$

$$\mu = (1.60 \times 10^{-19}C)(127 \times 10^{-12}m)\left(\frac{1\,D}{3.336 \times 10^{-30}C \cdot m}\right) = 6.09\,D$$

The observed dipole moment of 1.11 D for HCl implies that the H—Cl bond is only about 18% ionic:

$$\frac{1.11\,D}{6.09\,D} \times 100\% = 18.2\%$$

▶ **PRACTICE 8.13**　The dipole moment of HF is $\mu = 1.83$ D, and the bond length is 92 pm. Calculate the percent ionic character of the H—F bond. Is HF more ionic or less ionic than HCl (Worked Example 8.7)?

▶ **APPLY 8.14**　Without performing any calculations, predict which bond has a greater percent ionic character, H—Br or H—I. Check your prediction using the following data: the dipole moment of HBr is $\mu = 0.82$ D, and the bond length is 142 pm; the dipole moment of HI is $\mu = 0.38$ D, and the bond length is 161 pm.

8.6 ▶ INTERMOLECULAR FORCES

Now that we know a bit about molecular polarities, let's see how they give rise to some of the forces that occur between molecules, called **intermolecular forces**. These forces are different than covalent bonds between atoms *within* a molecule. For example, in H_2O there are two covalent bonds between hydrogen and oxygen, but there are also attractive forces between two H_2O molecules that we refer to as intermolecular forces. (Strictly speaking, the term *intermolecular* refers only to molecular substances, but we'll use it generally to refer to interactions among all kinds of particles, including molecules, ions, and atoms.)

Intermolecular forces influence many important macroscopic properties of matter such as solubility, melting point, and boiling point. They also play a key role in stabilizing the shapes and interactions of biomolecules, as we'll see at the end of this section. To understand how the strength of substance's intermolecular forces influences its melting and boiling point we must first visualize the difference between a gas, liquid, and solid at the molecular level. **FIGURE 8.9** shows that the particles in a gas are free to move around at random and fill the volume of a container because the attractive forces (intermolecular forces) between particles are weak. Liquids and solids are distinguished from gases by the presence of substantial attractive forces between particles. In liquids, these attractive forces are strong enough to hold the particles in close contact while still allowing them to move freely past one another. In solids, the forces are so strong that they hold the particles rigidly in place and prevent their movement.

Temperature influences the phase of matter because at higher temperatures particles have greater kinetic energy and can overcome the intermolecular forces holding them together. We are familiar with a sample of H_2O that exists either as solid ice, liquid water, or gaseous vapor, depending on its temperature. Nitrogen (N_2) offers another example; nitrogen is a gas at higher temperatures but becomes a liquid at low temperature (**FIGURE 8.10**).

Intermolecular forces as a whole are usually called **van der Waals forces** after the Dutch scientist Johannes van der Waals (1837–1923). These forces are of several different types, including *dipole–dipole forces*, *London dispersion forces*, and *hydrogen bonds*. In addition, *ion-dipole forces* exist between ions and molecules. All these intermolecular forces are electrostatic in origin and result from the mutual attraction of unlike charges or the mutual repulsion of like charges. If the particles are ions, then full charges are present and the ion–ion attractions are so strong (energies on the order of 500–1000 kJ/mol) that they give rise to what we call **ionic bonds**. If the particles are neutral, then only partial charges are present, but even so, the attractive forces can be substantial.

REMEMBER...

Ionic bonds generally form between the cation of a metal and the anion of a reactive nonmetal. (Section 6.7)

▶ **FIGURE 8.9**

A molecular comparison of gases, liquids, and solids.

Figure It Out

Which phase of matter has the strongest intermolecular forces?

Answer: Solid

In **gases**, the particles feel little attraction for one another and are free to move about randomly.

In **liquids**, the particles are held close together by attractive forces but are free to move around one another.

In **solids**, the particles are held in an ordered arrangement.

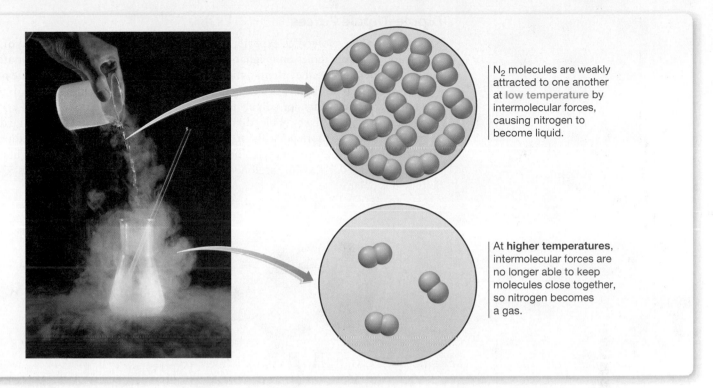

N₂ molecules are weakly attracted to one another at **low temperature** by intermolecular forces, causing nitrogen to become liquid.

At **higher temperatures**, intermolecular forces are no longer able to keep molecules close together, so nitrogen becomes a gas.

▲ **FIGURE 8.10**
Intermolecular forces.

Ion–Dipole Forces

We said in the previous section that a molecule has a net polarity and an overall dipole moment if the sum of its individual bond dipoles is nonzero. One side of the molecule has a net excess of electrons and a partial negative charge ($\delta-$), while the other side has a net deficiency of electrons and a partial positive charge ($\delta+$). An **ion–dipole force** is the result of electrical interactions between an ion and the partial charges on a polar molecule (**FIGURE 8.11**).

The favored orientation of a polar molecule in the presence of ions is one where the positive end of the molecular dipole is near an anion and the negative end of the dipole is near a cation. The magnitude of the interaction energy E depends on the charge on the ion z, the strength of the dipole as measured by its dipole moment μ, and the inverse square of the distance r from the ion to the dipole: $E \propto z\mu/r^2$. Ion–dipole forces are highly variable in strength and are particularly important in aqueous solutions of ionic substances such as NaCl, in which polar water molecules surround the ions. We'll explore the **formation of solutions** in more detail in Chapter 12.

LOOKING AHEAD...

In Section 12.2, we'll look at the energy changes accompanying the **formation of solutions** and at how those energy changes are affected by different kinds of intermolecular forces.

◀ **FIGURE 8.11**
Ion–dipole forces.

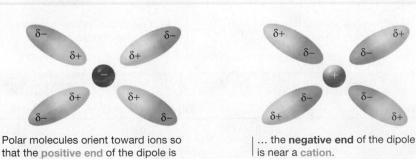

Polar molecules orient toward ions so that the **positive end** of the dipole is near an **anion** and ...

... the **negative end** of the dipole is near a **cation**.

Dipole–Dipole Forces

Neutral but polar molecules experience **dipole–dipole forces** as the result of electrical interactions among dipoles on neighboring molecules. The forces can be either attractive or repulsive, depending on the orientation of the molecules (**FIGURE 8.12**), and the net force in a large collection of molecules is a summation of many individual interactions of both types. The forces are weaker than ion–dipole forces, with energies on the order of 3–4 kJ/mol.

The strength of a given dipole–dipole interaction depends on the sizes of the dipole moments involved. The more polar the substance, the greater the strength of its dipole–dipole interactions. Butane, for instance, is a nonpolar molecule with a molecular weight of 58 and a boiling point of −0.5 °C, while acetone has the same molecular weight yet boils 57 °C higher because it is polar.

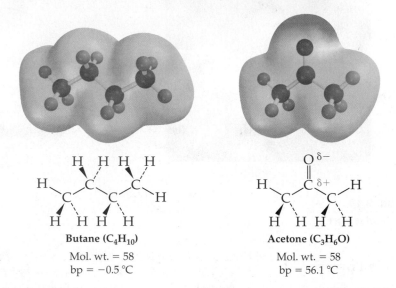

Butane (C_4H_{10})
Mol. wt. = 58
bp = −0.5 °C

Acetone (C_3H_6O)
Mol. wt. = 58
bp = 56.1 °C

TABLE 8.4 lists several substances with similar molecular weights but different dipole moments and indicates that there is a rough correlation between dipole moment and boiling point. The larger the dipole moment, the stronger the intermolecular forces and the greater the amount of heat that must be added to overcome those forces. Thus, substances with higher dipole moments generally have higher boiling points.

▶ **FIGURE 8.12**
Dipole–dipole forces.

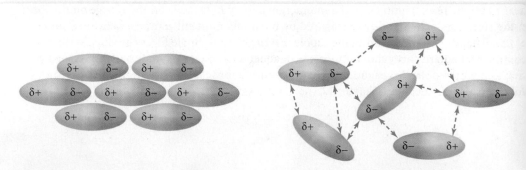

Polar molecules **attract** one another when they orient with unlike charges close together, but …

… they **repel** one another when they orient with like charges together.

TABLE 8.4	**Comparison of Molecular Weights, Dipole Moments, and Boiling Points**		
Substance	Mol. Wt.	Dipole Moment (D)	bp (K)
$CH_3CH_2CH_3$	44.10	0.08	231
CH_3OCH_3	46.07	1.30	248
CH_3CN	41.05	3.93	355

London Dispersion Forces

The causes of intermolecular forces among charged and polar particles are relatively easy to understand, but it's less obvious how attractive forces arise among nonpolar molecules or among the individual atoms of a noble gas. Benzene (C_6H_6), for instance, has zero dipole moment and therefore experiences no dipole–dipole forces. Nevertheless, there must be *some* intermolecular forces present among benzene molecules because the substance is a liquid rather than a gas at room temperature, with a melting point of 5.5 °C and a boiling point of 80.1 °C.

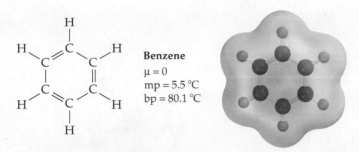

Benzene
$\mu = 0$
mp = 5.5 °C
bp = 80.1 °C

All atoms and molecules, regardless of structure, experience **London dispersion forces**, which result from the motion of electrons. Take even a simple nonpolar molecule like Br_2, for instance. Averaged over time, the distribution of electrons throughout the molecule is symmetrical, but at any given instant there may be more electrons at one end of the molecule than at the other, giving the molecule a short-lived dipole moment. This instantaneous dipole on one molecule can affect the electron distributions in neighboring molecules and *induce* temporary dipoles in those neighbors (**FIGURE 8.13**). As a result, weak attractive forces develop and Br_2 is a liquid at room temperature rather than a gas.

Dispersion forces are generally small, with energies in the range $1 - 10 \, \text{kJ/mol}$, and their exact magnitude depends on the ease with which a molecule's electron cloud can be distorted by a nearby electric field, a property referred to as **polarizability**. A smaller molecule or lighter atom is less polarizable and has smaller dispersion forces because it has only a few, tightly held electrons. A larger molecule or heavier atom, however, is more polarizable and has larger dispersion forces because it has many electrons, some of which are less tightly held and are farther from the nucleus. Among the halogens, for instance, the F_2 molecule is small and less polarizable, while I_2 is larger and more polarizable. As a result, F_2 has smaller dispersion forces and is a gas at room temperature, while I_2 has larger dispersion forces and is a solid (**TABLE 8.5**).

Shape is also important in determining the magnitude of the dispersion forces affecting a molecule. More spread-out shapes, which maximize molecular surface area, allow greater contact between molecules and give rise to higher dispersion forces than do more compact shapes, which minimize molecular contact. Pentane, for example, boils at 309.2 K, whereas 2,2-dimethylpropane boils at 282.6 K. Both substances have the same molecular formula,

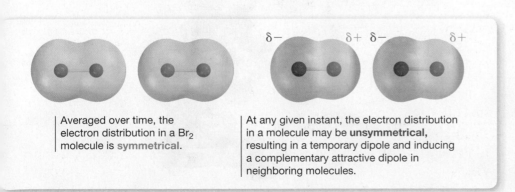

$\delta-$	$\delta+$	$\delta-$	$\delta+$

Averaged over time, the electron distribution in a Br_2 molecule is **symmetrical**.

At any given instant, the electron distribution in a molecule may be **unsymmetrical**, resulting in a temporary dipole and inducing a complementary attractive dipole in neighboring molecules.

◀ **FIGURE 8.13**
London dispersion forces.

TABLE 8.5	Melting Points and Boiling Points of the Halogens	
Halogen	mp (K)	bp (K)
F_2	53.5	85.0
Cl_2	171.6	239.1
Br_2	265.9	331.9
I_2	386.8	457.5

▶ **FIGURE 8.14**

The effect of molecular shape on London dispersion forces.

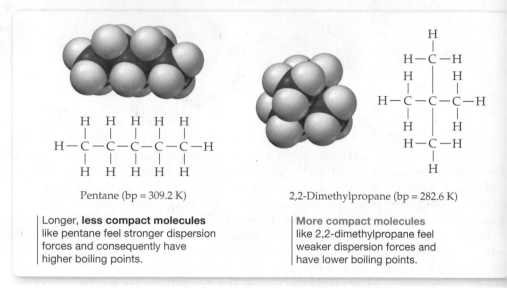

Pentane (bp = 309.2 K)

2,2-Dimethylpropane (bp = 282.6 K)

| Longer, **less compact molecules** like pentane feel stronger dispersion forces and consequently have higher boiling points. | **More compact molecules** like 2,2-dimethylpropane feel weaker dispersion forces and have lower boiling points. |

C_5H_{12}, but pentane is longer and somewhat spread out, whereas 2,2-dimethylpropane is mor spherical and compact (**FIGURE 8.14**).

Hydrogen Bonds

In many ways, *hydrogen bonds* are responsible for life on Earth. They cause water to be liquid rather than a gas at ordinary temperatures, and they are the primary intermolecula force that holds huge biomolecules in the shapes needed to play their essential roles in bio chemistry. **Deoxyribonucleic acid (DNA)**, a molecule that stores our genetic code, contain two enormously long molecular strands that are coiled around each other and held togethe by hydrogen bonds.

LOOKING AHEAD...

In Section 23.11, we'll examine the molecular composition of **DNA** and the role of hydrogen bonding in structure and replication.

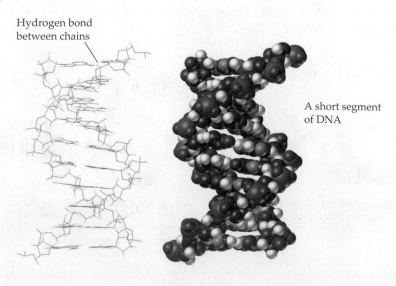

Hydrogen bond between chains

A short segment of DNA

A **hydrogen bond** is an attractive interaction between a hydrogen atom bonded to a very electronegative atom (O, N, or F) and an electron-rich region elsewhere in the same molecule or in a different molecule. Most often, this electron-rich region is an unshared electron pair on another electronegative atom. For example, hydrogen bonds occur in both water and ammonia:

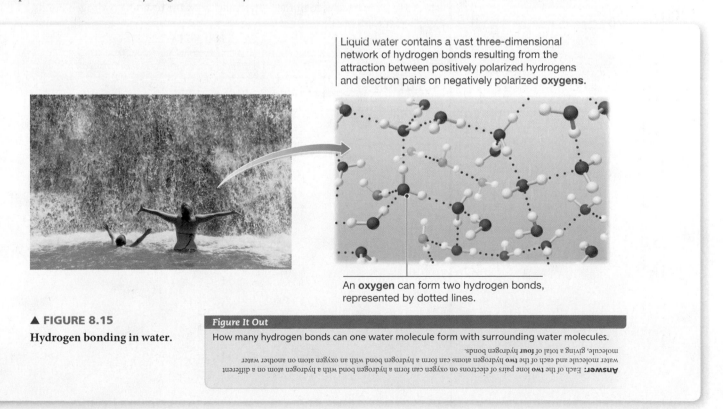

Water   Ammonia

Hydrogen bonds arise because O—H, N—H, and F—H bonds are highly polar, with a partial positive charge on the hydrogen and a partial negative charge on the electronegative atom. In addition, the hydrogen atom has no core electrons to shield its nucleus and it has a small size so it can be approached closely by other molecules. As a result, the dipole–dipole attraction between the hydrogen and an unshared electron pair on a nearby atom is unusually strong, giving rise to a hydrogen bond. Water, in particular, is able to form a vast three-dimensional network of hydrogen bonds because each H_2O molecule has two hydrogens and two electron pairs (**FIGURE 8.15**).

Hydrogen bonds can be quite strong, with energies up to 40 kJ/mol. To see one effect of hydrogen bonding, look at **TABLE 8.6**, which plots the boiling points of the covalent binary hydrides for the group 4A–7A elements. As you might expect, the boiling points generally increase with molecular weight down a group of the periodic table as a result of increased London dispersion forces—for example, $CH_4 < SiH_4 < GeH_4 < SnH_4$. Three substances, however, are clearly anomalous: NH_3, H_2O, and HF. All three have higher boiling points than might be expected because of the hydrogen bonds they contain.

Liquid water contains a vast three-dimensional network of hydrogen bonds resulting from the attraction between positively polarized hydrogens and electron pairs on negatively polarized **oxygens**.

An **oxygen** can form two hydrogen bonds, represented by dotted lines.

▲ **FIGURE 8.15**

Hydrogen bonding in water.

Figure It Out

How many hydrogen bonds can one water molecule form with surrounding water molecules.

Answer: Each of the two lone pairs of electrons on oxygen can form a hydrogen bond with a hydrogen atom on a different water molecule and each of the **two** hydrogen atoms can form a hydrogen bond with an oxygen atom on another water molecule, giving a total of **four** hydrogen bonds.

TABLE 8.6 Boiling Points of the Covalent Binary Hydrides of Groups 4A, 5A, 6A, and 7A

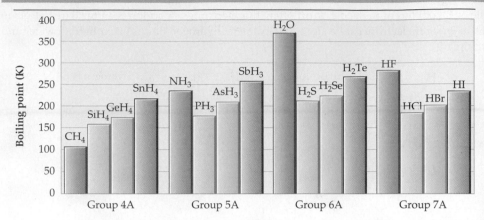

The boiling points generally increase with increasing molecular weight down a group of the periodic table, but the hydrides of nitrogen (**NH₃**), oxygen (**H₂O**), and fluorine (**HF**) have abnormally high boiling points because these molecules form hydrogen bonds.

Conceptual WORKED EXAMPLE 8.8

Drawing Hydrogen Bonds

Methanol (CH_3OH) is a liquid that can be used in fuel cells. Draw all the hydrogen bonds that can form when one methanol molecule is surrounded by other methanol molecules in a liquid sample.

STRATEGY

Hydrogen bonding occurs as a result of a strong dipole–dipole interaction and follows the pattern shown below. X represents one of the highly electronegative atoms F, O, or N. Align two CH_3OH molecules to fit the pattern and draw the resulting hydrogen bonds.

$$\overset{\delta+}{X-H}\cdots\cdots\overset{\delta-}{:X}$$

SOLUTION

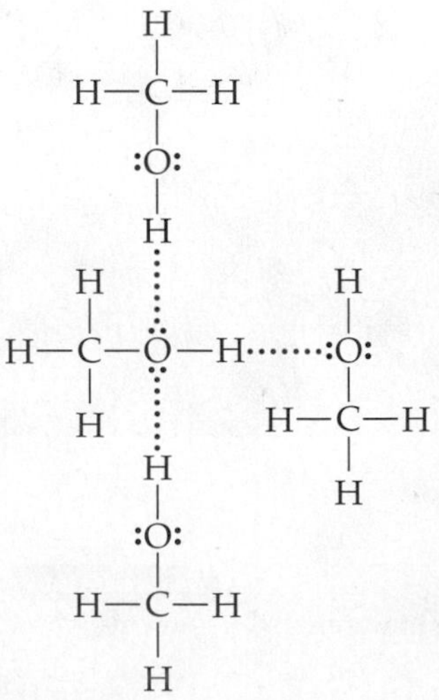

Conceptual PRACTICE 8.15 A liquid sample contains formaldehyde (CH_2O) dissolved in water. Which of the following illustrations depicts hydrogen bonding that can occur in this solution? Select all the correct answers. (Hint: Consider if hydrogen bonds will form between two formaldehyde molecules, between two water molecules, or between a water molecule and a formaldehyde molecule.)

(a)

(b)

(c)

(d)

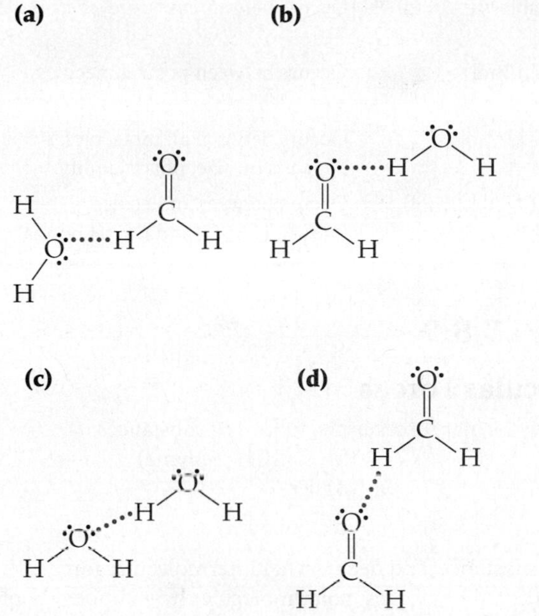

Conceptual APPLY 8.16 Hydrogen bonding between pairs of guanine (G)–cytosine (C) molecules and adenine (A)–thymine (T) molecules holds two DNA strands together. In the diagram below, these pairs of molecules are aligned as they exist in two DNA strands. The red squiggly line represents where the molecule is attached to the sugar–phosphate backbone of DNA.

(a) Draw the hydrogen bonds that occur in the G–C pair and A–T pair.
(b) Which region of DNA would have the higher melting point; regions high in A–T pairs or regions high in G–C pairs?

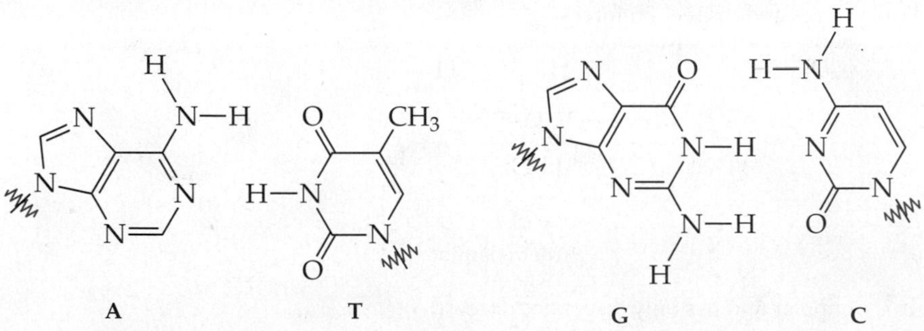

TABLE 8.7 compares the various kinds of intermolecular forces.

TABLE 8.7 **A Comparison of Intermolecular Forces**

Force	Strength	Characteristics
Ion–dipole	Highly variable (10–70 kJ/mol)	Occurs between ions and polar molecules
Dipole–dipole	Weak (3–4 kJ/mol)	Occurs between polar molecules
London dispersion	Weak (1–10 kJ/mol)	Occurs between all molecules; strength depends on size, polarizability
Hydrogen bond	Moderate (10–40 kJ/mol)	Occurs between molecules with O—H, N—H, and F—H bonds

● WORKED EXAMPLE 8.9

Identifying Intermolecular Forces

Identify the likely kinds of intermolecular forces in the following substances:
(a) HCl
(b) CH_3CH_3 (ethane)
(c) CH_3NH_2 (methylamine)
(d) Kr

STRATEGY

Determine the structure of each substance, and decide what intermolecular forces are present. Remember: All molecules have dispersion forces; polar molecules have dipole–dipole forces; and molecules with O—H, N—H, or F—H bonds have hydrogen bonds.

SOLUTION

(a) HCl is a polar molecule but can't form hydrogen bonds. It has dipole–dipole forces and dispersion forces.
(b) CH_3CH_3 is a nonpolar molecule and has only dispersion forces.

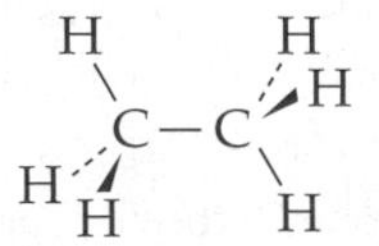

Ethane

(c) CH_3NH_2 is a polar molecule that can form hydrogen bonds. In addition, it has dipole-dipole forces and dispersion forces.

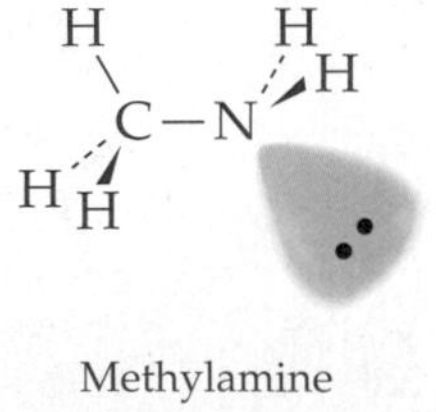

Methylamine

(d) Kr is nonpolar and has only dispersion forces.

▶ **PRACTICE 8.17** Of the substances Ar, Cl_2, CCl_4, CH_3F, and HNO_3, which has:

(a) Dipole–dipole forces?
(b) Hydrogen-bonding?
(c) The smallest dispersion forces?
(d) The largest dispersion forces?

▶ **APPLY 8.18** Consider the kinds of intermolecular forces present in the following compounds, and rank the substances in likely order of increasing boiling point H_2S(mol. wt. = 34), CH_3OH(mol. wt. = 32), C_2H_6 (mol. wt. = 30), Ar (mol. wt. = 40).

8.7 ► MOLECULAR ORBITAL THEORY: THE HYDROGEN MOLECULE

The valence bond model that describes covalent bonding through orbital overlap is easy to visualize and leads to a satisfactory description for most molecules. It does, however, have some problems. Perhaps the most serious flaw in the valence bond model is that it sometimes leads to an incorrect electronic description. For this reason, another bonding description called **molecular orbital (MO) theory** is often used. The molecular orbital model is more complex and less easily visualized than the valence bond model, particularly for larger molecules, but it sometimes gives a more satisfactory accounting of chemical and physical properties.

To introduce some of the basic ideas of molecular orbital theory, let's look again at orbitals. The concept of an orbital derives from the quantum mechanical wave equation, in which the square of the wave function gives the probability of finding an electron within a given region of space. The kinds of orbitals that we've been concerned with up to this point are called *atomic orbitals* because they are characteristic of individual atoms. Atomic orbitals on the same atom can combine to form hybrids, and atomic orbitals on different atoms can overlap to form covalent bonds, but the orbitals and the electrons in them remain localized on specific atoms.

Atomic orbital

A wave function whose square gives the probability of finding an electron within a given region of space *in an atom.*

Molecular orbital theory takes a different approach to bonding by considering the molecule as a whole rather than concentrating on individual atoms. A **molecular orbital** is to a molecule what an atomic orbital is to an atom.

Molecular orbital

A wave function whose square gives the probability of finding an electron within a given region of space in a molecule.

Like atomic orbitals, molecular orbitals have specific energy levels and specific shapes, and they can be occupied by a maximum of two electrons with opposite spins. The energy and shape of a molecular orbital depend on the size and complexity of the molecule and can thus be fairly complicated, but the fundamental analogy between atomic and molecular orbitals remains. Let's look at the molecular orbital description of the simple diatomic molecule H_2 to see some general features of MO theory.

Imagine what might happen when two isolated hydrogen atoms approach each other and begin to interact. The $1s$ orbitals begin to blend together, and the electrons spread out over both atoms. Molecular orbital theory says that there are two ways for the orbital interaction to occur—an additive way and a subtractive way. The additive interaction leads to formation of a molecular orbital that is roughly egg-shaped, whereas the subtractive interaction leads to formation of a molecular orbital that contains a **node** between atoms (FIGURE 8.17).

The additive combination, denoted σ, is lower in energy than the two isolated $1s$ orbitals and is called a **bonding molecular orbital** because any electrons it contains spend most of their time in the region between the two nuclei, bonding the atoms together. The subtractive combination, denoted σ^* and spoken as "sigma star," is higher in energy than the two isolated $1s$ orbitals and is called an **antibonding molecular orbital**. Any electrons it contains can't occupy the central region between the nuclei and can't contribute to bonding.

REMEMBER...

A **node** is a surface of zero electron probability separating regions of nonzero probability within an orbital. (Section 5.8)

The additive combination of atomic 1s orbitals forms a lower-energy, **bonding molecular orbital**, σ.

The subtractive combination of atomic 1s orbitals forms a higher-energy, **antibonding molecular orbital**, σ*, that has a **node** between the nuclei.

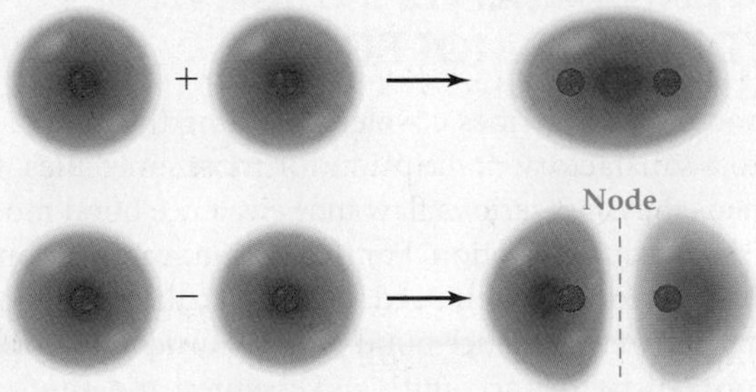

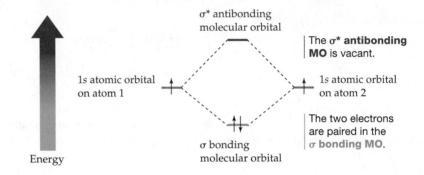

Node

▲ **FIGURE 8.17**

Formation of molecular orbitals in the H₂ molecule.

Figure It Out

Why are some molecular orbitals referred to as bonding and others as antibonding?

Answer: A bond corresponds to a higher electron density between two atoms. A bonding orbital increases electron density between atoms while an antibonding orbital decreases electron density between atoms due to the node.

▶ **FIGURE 8.18**

Energy levels of molecular orbitals for the H₂ molecule.

Figure It Out

How does the energy of the σ and σ* molecular orbitals in H₂ compare to the energy of the 1s orbital in an isolated hydrogen atom?

Answer: The σ orbital is lower in energy than a hydrogen 1s orbital, and the σ* orbital is higher in energy than a hydrogen 1s orbital.

σ* antibonding molecular orbital

1s atomic orbital on atom 1

1s atomic orbital on atom 2

The σ* **antibonding MO** is vacant.

The two electrons are paired in the σ **bonding MO**.

σ bonding molecular orbital

Energy

Diagrams such as the one shown in **FIGURE 8.18** are used to show the energy relationships of the various orbitals. The two isolated H atomic orbitals are shown on either side, and the two H₂ molecular orbitals are shown in the middle. Each of the starting hydrogen atomic orbitals has one electron, which pair up and occupy the lower-energy bonding MO after covalent bond formation.

Similar MO diagrams can be drawn, and predictions about stability can be made, for diatomic species such as H₂⁻. For example, we might imagine constructing the H₂⁻ ion by bringing together a neutral H atom with one electron and an H⁻ anion with two electrons. Since the resultant H₂⁻ ion has three electrons, two of them will occupy the lower-energy bonding σ MO and one will occupy the higher-energy antibonding σ* MO as shown in **FIGURE 8.19.** Two electrons are lowered in energy while only one electron is raised in energy, so a net gain in stability results. We therefore predict (and find experimentally) that the H₂⁻ ion is stable.

Bond orders—the number of electron pairs shared between atoms—can be calculated from MO diagrams by subtracting the number of antibonding electrons from the number of bonding electrons and dividing the difference by 2:

$$\text{Bond order} = \frac{\left(\begin{array}{c}\text{Number of}\\\text{bonding electrons}\end{array}\right) - \left(\begin{array}{c}\text{Number of}\\\text{antibonding electrons}\end{array}\right)}{2}$$

The H₂ molecule, for instance, has a bond order of 1 because it has two bonding electrons and no antibonding electrons. In the same way, the H₂⁻ ion has a bond order of ½.

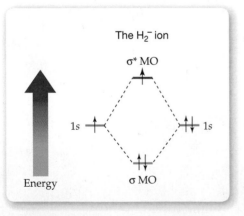

The H₂⁻ ion

σ* MO

1s   1s

Energy   σ MO

▲ **FIGURE 8.19**

Energy levels of molecular orbitals for the stable H₂⁻ ion.

REMEMBER...

Bond order is the number of electron pairs shared between atoms. Higher bond orders correspond to shorter and stronger bonds. (Section 7.2)

The key ideas of the molecular orbital theory of bonding can be summarized as follows:

Key Ideas of Molecular Orbital Theory

- Molecular orbitals are to molecules what atomic orbitals are to atoms. A molecular orbital describes a region of space in a molecule where electrons are most likely to be found, and it has a specific size, shape, and energy level.
- Molecular orbitals are formed by combining atomic orbitals on different atoms. The number of molecular orbitals formed is the same as the number of atomic orbitals combined.
- Molecular orbitals that are lower in energy than the starting atomic orbitals are bonding, and MOs that are higher in energy than the starting atomic orbitals are antibonding.
- Electrons occupy molecular orbitals beginning with the MO of lowest energy. A maximum of two electrons can occupy each orbital, and their spins are paired.
- Bond order can be calculated by subtracting the number of electrons in antibonding MOs from the number in bonding MOs and dividing the difference by 2.

—● WORKED EXAMPLE 8.10

Constructing an MO Diagram for a First Row Diatomic Molecule

Construct an MO diagram for the He_2 molecule. What is its bond order? Is this ion likely to be stable?

STRATEGY

Count the total number of valence electrons in the He_2 molecule and fill orbitals in the MO diagram starting with the lowest energy. Calculate the bond order according to the equation in this section. Bond orders greater than zero represent stable species; the higher the bond order, the stronger the bond.

SOLUTION

The hypothetical He_2 molecule has four electrons, two of which occupy the lower-energy bonding orbital, and two of which occupy the higher-energy antibonding orbital. Since the decrease in energy for the two bonding electrons is counteracted by the increase in energy for the two antibonding electrons, the He_2 molecule has no net bonding energy and is not stable. The hypothetical He_2 molecule has two bonding and two antibonding electrons and a bond order of $(2 - 2)/2 = 0$, which accounts for the instability of He_2.

The He_2 molecule

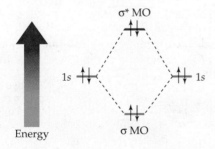

PRACTICE 8.19 Construct an MO diagram for the He_2^+ ion. What is its bond order? Is this ion likely to be stable?

APPLY 8.20 According to MO theory, is He_2^{2+} predicted to have a stronger or weaker bond than He_2^+?

8.8 ▶ MOLECULAR ORBITAL THEORY: OTHER DIATOMIC MOLECULES

Having looked at bonding in the H_2 molecule, let's move up a level in complexity by looking at the bonding in several second-row diatomic molecules—N_2, O_2, and F_2. The valence bond model developed in Section 8.2 predicts that the nitrogen atoms in N_2 are triply bonded and have one lone pair each, that the oxygen atoms in O_2 are doubly bonded and have two lone pairs each, and that the fluorine atoms in F_2 are singly bonded and have three lone pairs each.

Valence bond theory predicts:

:N≡N: Ö=Ö   :F—F:

1 σ bond   1 σ bond   1 σ bond
and 2 π bonds   and 1 π bond

Unfortunately, this simple valence bond picture can't be right because it predicts that the electrons in all three molecules are *spin-paired*. In other words, the electron-dot structures indicate that the occupied atomic orbitals in all three molecules contain two electrons each. It can be demonstrated experimentally, however, that the O_2 molecule has two electrons that are not spin-paired and that these electrons therefore must be in different, singly occupied orbitals.

Experimental evidence for the electronic structure of O_2 rests on the observation that substances with unpaired electrons are attracted by magnetic fields and are thus said to be **paramagnetic**. The more unpaired electrons a substance has, the stronger the paramagnetic attraction. Substances whose electrons are all spin-paired, by contrast, are weakly repelled by magnetic fields and are said to be **diamagnetic**. Both N_2 and F_2 are diamagnetic, just as predicted by their electron-dot structures, but O_2 is paramagnetic. When liquid O_2 is poured over the poles of a strong magnet, the O_2 sticks to the poles, as shown in **FIGURE 8.20**.

Why is O_2 paramagnetic? Electron-dot structures and valence bond theory fail to answer this question, but MO theory explains the experimental results nicely. In a molecular orbital description of N_2, O_2, and F_2, two atoms come together and their valence-shell atomic orbitals interact to form molecular orbitals. Four orbital interactions occur, leading to the formation of four bonding MOs and four antibonding MOs, whose relative energies are shown in **FIGURE 8.21**. (Note that the relative energies of the σ_{2p} and π_{2p} orbitals in N_2 are different from those in O_2 and F_2.)

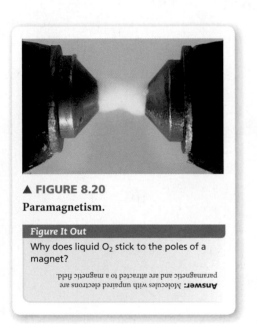

▲ **FIGURE 8.20**

Paramagnetism.

Figure It Out

Why does liquid O_2 stick to the poles of a magnet?

Answer: Molecules with unpaired electrons are paramagnetic and are attracted to a magnetic field.

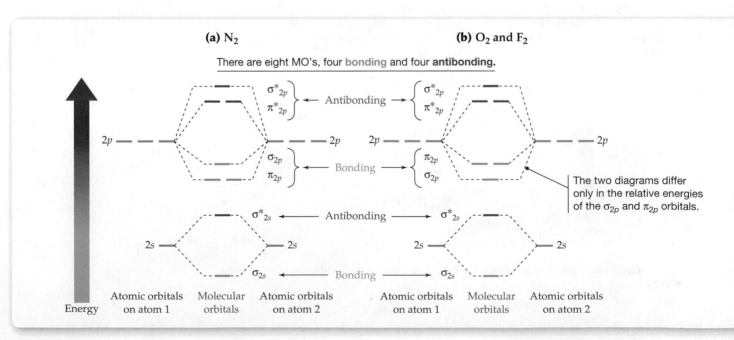

(a) N_2 **(b)** O_2 and F_2

There are eight MO's, four **bonding** and four **antibonding**.

The two diagrams differ only in the relative energies of the σ_{2p} and π_{2p} orbitals.

Energy | Atomic orbitals on atom 1 | Molecular orbitals | Atomic orbitals on atom 2 | Atomic orbitals on atom 1 | Molecular orbitals | Atomic orbitals on atom 2

▲ **FIGURE 8.21**

Energy levels of molecular orbitals for (a) N_2 and (b) O_2 and F_2.

The diagrams in Figure 8.21 show the following orbital interactions:

- The 2s orbitals interact to give σ_{2s} and σ_{2s}^* MOs.
- The two 2p orbitals that lie on the internuclear axis interact head-on to give σ_{2p} and σ_{2p}^* MOs.
- The two remaining pairs of 2p orbitals that are perpendicular to the internuclear axis interact in a sideways manner to give two **degenerate** π_{2p} and two degenerate π_{2p}^* MOs oriented 90° apart.

We should also point out that MO diagrams like those in Figure 8.21 are usually obtained from mathematical calculations and can't necessarily be predicted. MO theory is therefore less easy to visualize and understand on an intuitive level than valence bond theory. The shapes of the σ_{2p}, σ_{2p}^*, π_{2p}, and π_{2p}^* MOs are shown in **FIGURE 8.22**.

When appropriate numbers of valence electrons are added to occupy the molecular orbitals, the results shown in **FIGURE 8.23** are obtained. Both N_2 and F_2 have all their electrons spin-paired, but O_2 has two unpaired electrons in the degenerate π_{2p}^* orbitals. Both N_2 and F_2 are therefore diamagnetic, whereas O_2 is paramagnetic.

REMEMBER...

Degenerate orbitals are those that have the same energy. (Section 5.11)

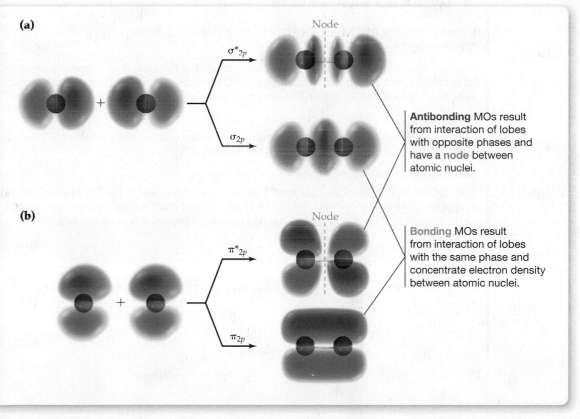

(a)

σ_{2p}^*

σ_{2p}

Node

Antibonding MOs result from interaction of lobes with opposite phases and have a **node** between atomic nuclei.

(b)

π_{2p}^*

π_{2p}

Node

Bonding MOs result from interaction of lobes with the same phase and concentrate electron density between atomic nuclei.

▲ **FIGURE 8.22**

Formation of (a) σ_{2p} and σ_{2p}^* MOs by head-on interaction of two p atomic orbitals, and (b) π_{2p} and π_{2p}^* MOs by sideways interaction.

▶ **FIGURE 8.23**
Energy levels of molecular orbitals for the second-row diatomic molecules (a) N_2, (b) O_2, and (c) F_2.

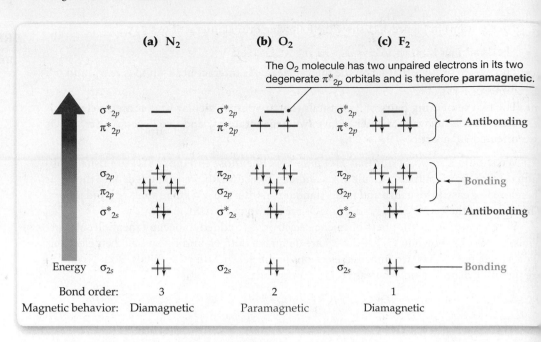

WORKED EXAMPLE 8.11

Constructing an MO Diagram for a Second Row Diatomic Molecule

Nitrogen monoxide (NO), typically called nitric oxide, is a diatomic molecule with a molecular orbital diagram similar to that of N_2 (Figure 8.21a). Show the MO diagram for nitric oxide, and predict its bond order. Is NO diamagnetic or paramagnetic?

STRATEGY

Count the number of valence electrons in NO, and assign them to the available MOs shown in Figure 8.21a, beginning with the lowest energy orbital.

SOLUTION

Nitric oxide has 5 valence electrons from N and 6 from O for a total of 11. The 11 electrons are assigned to MOs in Figure 8.21a. NO has 8 bonding and 3 antibonding electrons and a bond order of $(8 - 3)/2 = 2.5$. Because it has one unpaired electron, it is paramagnetic.

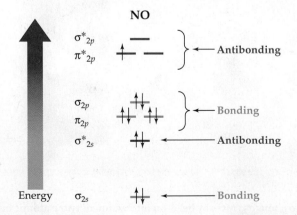

▶ **PRACTICE 8.21** The B_2 and C_2 molecules have MO diagrams similar to that of N_2 in Figure 8.21a. What MOs are occupied in B_2 and C_2, and what is the bond order in each? Would you expect either of these substances to be paramagnetic?

▶ **APPLY 8.22** Determine the bond order for the species: O_2^{2-}, O_2^-, O_2, O_2^+, O_2^{2+}. List them in order of increasing bond energy and increasing bond length.

8.9 ▶ COMBINING VALENCE BOND THEORY AND MOLECULAR ORBITAL THEORY

Whenever two different theories explain the same concept, the question comes up: Which theory is better? The question isn't easy to answer, though, because it depends on what is meant by "better." Valence bond theory is better because of its simplicity and ease of visualization, but MO theory is better because of its accuracy. Best of all, though, is a joint use of the two theories that combines the strengths of both.

Valence bond theory has two main problems: (1) It incorrectly predicts aspects of electronic structure such as magnetic properties and energy levels. In the O_2 molecule, all electrons are paired in the electron-dot structure even though experimental data indicates that it is paramagnetic (has unpaired electrons). (2) Valence bond theory cannot adequately represent bonding in molecules with a single electron-dot structure (such as O_3). No single structure is adequate; the concept of a resonance hybrid, a blend of multiple electron-dot structures (Section 7.9) is not well described by valence bond theory. The first problem occurs infrequently, but the second is much more common. To better deal with resonance, chemists often use a combination of bonding theories in which the σ bonds in a given molecule are described by valence bond theory and π bonds in the same molecule are described by MO theory.

Take O_3, for instance. Valence bond theory says that ozone is a resonance hybrid of two equivalent structures, both of which have two O—O σ bonds and one O=O π bond (Section 7.9). One structure has a lone pair of electrons in the p orbital on the left-hand oxygen atom and a π bond to the right-hand oxygen. The other structure has a lone pair of electrons in the p orbital on the right-hand oxygen and a π bond to the left-hand oxygen. The actual structure of O_3 is an average of the two resonance forms in which four electrons occupy the entire region encompassed by the overlapping set of three p orbitals. The only difference between the resonance structures is in the placement of p electrons. The atoms themselves are in the same positions in both, and the geometries are the same in both (**FIGURE 8.24**).

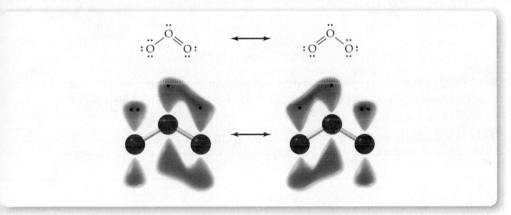

◀ **FIGURE 8.24**
The structure of ozone.

Valence bond theory thus gives a good description of the ozone O—O σ bonds, whose electrons are localized between specific pairs of atoms, but a poor description of the π bonds among p atomic orbitals, whose four electrons are spread out, or *delocalized*, over the molecule. Yet this is exactly what MO theory does best—describe bonds in which electrons are delocalized throughout a molecule. Thus, a combination of valence bond theory and MO theory is used. The σ bonds are best described in valence bond terminology as being localized between pairs of atoms, and the π electrons are best described by MO theory as being delocalized over the entire molecule. The lowest energy π molecular orbital best represents the resonance hybrid, which is a blend of the two resonance structures of ozone.

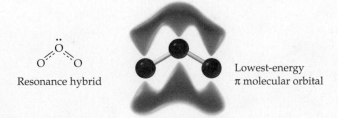

Resonance hybrid

Lowest-energy
π molecular orbital

> ● **WORKED EXAMPLE 8.12**

Combining Valence Bond and MO Theory

Acetamide (C_2H_5NO), an organic compound that has been detected in interstellar space, has the following connections among atoms and can be drawn using two nonequivalent electron-dot resonance structures. Draw both, and sketch a π molecular orbital showing how the π electrons are delocalized over the oxygen and nitrogen atoms.

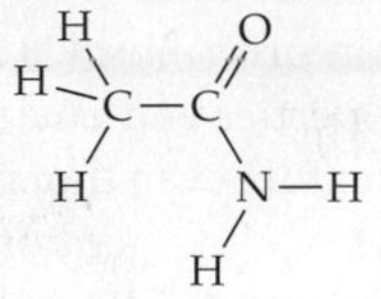

Acetamide

STRATEGY

Draw the two electron-dot structures in the usual way (Section 7.9), including all nonbonding electrons. The two structures differ in that one has a carbon-oxygen double bond and a lone pair on nitrogen, while the other has a carbon–nitrogen double bond and an additional lone pair on oxygen. Thus, a π molecular orbital corresponding to the O—C—N resonance includes all three atoms.

SOLUTION

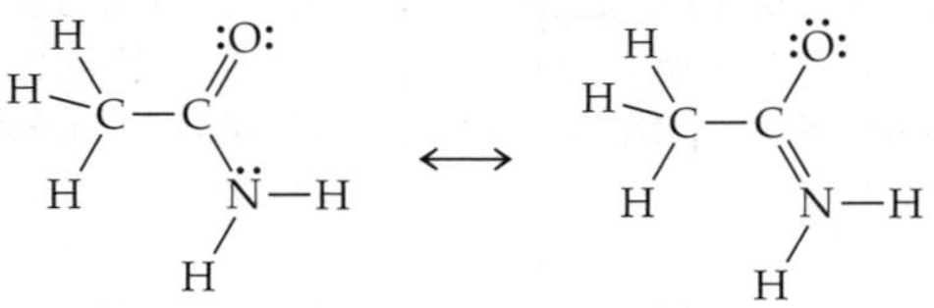

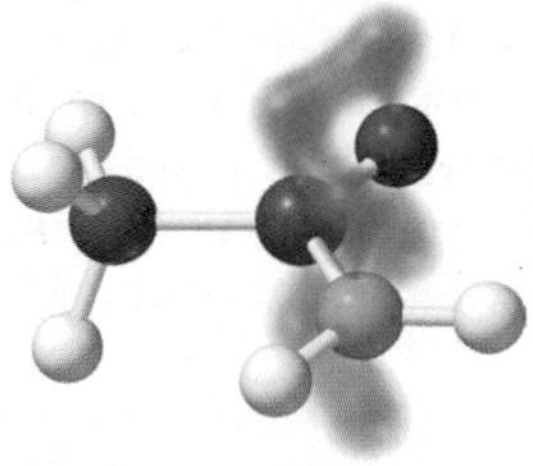

▶ **PRACTICE 8.23** Draw two resonance structures for the formate ion, HCO_2^-, and sketch a π molecular orbital showing how the π electrons are delocalized over both oxygen atoms.

▶ **APPLY 8.24** Draw two resonance structures for the benzene molecule, C_6H_6, and sketch a π molecular orbital showing how the π electrons are delocalized over the carbon atoms.

INQUIRY ▶▶▶ WHY DO DIFFERENT DRUGS HAVE DIFFERENT PHYSIOLOGICAL RESPONSES?

Have you ever wondered why medicines have specific physiological effects in your body? Acetaminophen and morphine relieve pain, ephedrine and caffeine stimulate the central nervous system, and penicillin fights bacterial infections. The answer is that a small molecule (or a part of a large molecule) binds to a very specific receptor site on a biological molecule and either promotes or inhibits a response. Nearly every process that happens in your body from responses to medication, to nerve impulses, to your sense of taste and smell occur because of molecular recognition between a receptor and specific target molecule. This recognition depends on both the molecule's shape and its charge distribution and is often described as a lock-and-key fit.

One way to visualize the specificity of molecular shape is to think about *handedness*. Why does a right-handed glove fit only on your right hand and not on your left hand? Why do the threads on a light bulb twist only in one direction so that you have to turn the bulb clockwise to screw it in? Gloves and the light bulb threads fit specific objects not just because of their shape but because of their handedness. Handedness can be observed with a mirror. When a right-handed glove is held up to a mirror, the reflected image looks like a left-handed glove. (Try it.) When the light bulb with clockwise threads is reflected in a mirror, the threads in the mirror image appear to twist in a counterclockwise direction.

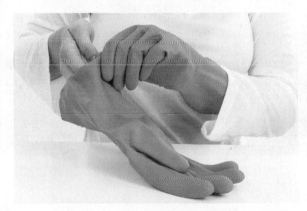

▲ A right hand fits only into a right-handed glove with a complementary shape, not into a left-handed glove.

Molecules, too, can have shapes that give them a handedness and can thus exist in mirror-image forms, one right-handed and one left-handed. Take, for example, a molecule with a tetrahedral central atom and four different substituent groups, designated as Compound 1. This molecule is handed because in the mirror image, the substituent groups have a different geometric arrangement in space. Notice in

Compound 1 that the groups A, B, C, and D are in clockwise arrangement, while in the mirror image the same groups are arranged counter clockwise. The mirror image cannot be rotated to position its substituent groups in the same locations as in Compound 1. As a result, Compound 1 cannot be superimposed on its mirror image.

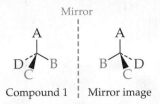

Compound 1 ¦ Mirror image

Objects, like hands or Compound 1, that are not superimposable on their mirror images are called **chiral**. Objects that are superimposable on their mirror image are **achiral**. The most common source of molecular chirality (or handedness) is the presence of a carbon atom with four different substituent groups called a **chiral center** (or chiral carbon).

A specific binding site in a large biological molecule such as an enzyme can distinguish between right- and left-handed forms of a chiral molecule such as Compound 1 and its mirror image. Consider, for example, the binding sites on the receptor shown in **FIGURE 8.16a**, which match up with and therefore bind to specific substituent groups in Compound 1 (Figure 8.16b). The substituent groups in the mirror image of Compound 1, however, do not line up with the binding sites on the receptor. If the mirror image molecule is rotated so that substituent B aligns with its receptor site, the groups C and D do not align and binding does not occur (Figure 8.16c).

The main classes of molecules found in living organisms: carbohydrates (sugars), proteins, fats, and nucleic acids are chiral, and usually only one of the two possible mirror-image forms occurs naturally in a given organism. The other form can often be made in the laboratory but does not occur naturally. The biological consequences of molecular shape can be dramatic. Look at the structures of dextromethorphan and levomethorphan, for instance. Both of these molecules bind to opioid receptors like the one depicted in the chapter opening image. (The Latin prefixes *dextro-* and *levo-* mean

(a)

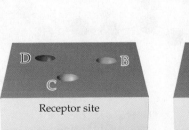

Receptor site

(b)

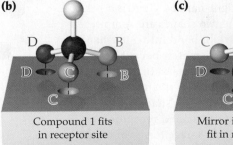

Compound 1 fits in receptor site

(c)

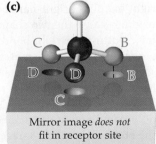

Mirror image *does not* fit in receptor site

▲ **FIGURE 8.16**

A schematic diagram of a receptor site capable of interacting with Compound 1 but not its mirror image.

continued on next page

continued from previous page

"right" and "left," respectively.) Dextromethorphan is a common cough suppressant found in many over-the-counter cold medicines, but its mirror image, levomethorphan, is a powerful narcotic pain reliever similar in its effects to morphine. The two substances are chemically identical except for their handedness, yet their biological properties are very different

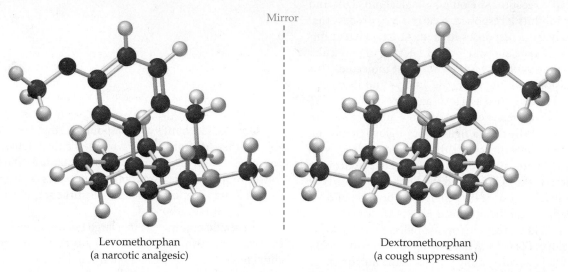

Levomethorphan
(a narcotic analgesic)

Dextromethorphan
(a cough suppressant)

▲ The black spheres in these structures represent carbon atoms, the ivory spheres represent hydrogen, the red spheres represent oxygen, and the blue spheres represent nitrogen.

As another example of the effects of shape and chirality, look at the substance called *carvone*. Left-handed carvone occurs in mint plants and has the characteristic odor of spearmint, while right-handed carvone occurs in several herbs and has the odor of caraway seeds. Again, the two structures are the same except for their handedness, yet they have entirely different odors.

Why do different mirror-image forms of molecules have different biological properties? The answer goes back to the question about why a right-handed glove fits only the right hand. A right hand in a right-handed glove is a perfect match because the two shapes are complementary. Putting the right hand into a left-handed glove produces a mismatch because the two shapes are *not* complementary. In the same way, handed molecules such as dextromethorphan and carvone have specific shapes that match only complementary-shaped receptor sites in the body. The mirror-image forms of the molecules can't fit into the receptor sites and thus don't elicit the same biological response.

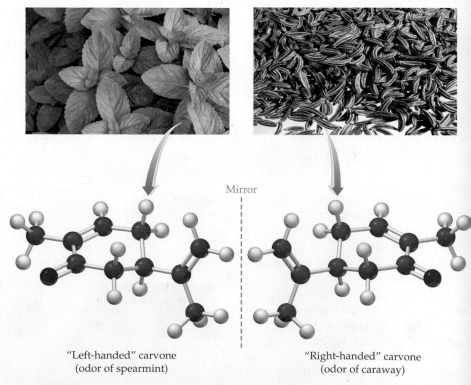

"Left-handed" carvone
(odor of spearmint)

"Right-handed" carvone
(odor of caraway)

▲ Two plants both produce carvone, but the mint leaves yield the "left-handed" form, while the caraway seeds yield the "right-handed" form.

Precise molecular shape is critically important to every living organism. Almost every chemical interaction in living systems is governed by complementarity between handed molecules and their glove-like receptors. Although shape is one important criteria in binding, molecules must also be held in place in the receptor site. Intermolecular forces such as electrostatic attractions, hydrogen bonds, dipole–dipole interactions, and dispersion forces are used to "glue" molecules to one another. Scientists can experimentally determine the three-dimensional atomic structure of a molecule such as the opioid receptor that binds dextromethorphan and levomethorphan along with dozens of other legal and illegal drugs. The detailed atomic shape paves the way for the design of safer and more effective opioid drugs.

PROBLEM 8.25 Which of these objects are chiral (exhibit handedness)?

(a) basketball   (b) foot
(c) golf club   (d) coffee mug
(e) seashell with a helical twist

PROBLEM 8.26 One of the following two molecules is chiral and can exist in two mirror-image forms; the other does not. Which is which? Why?

(a)   (b)

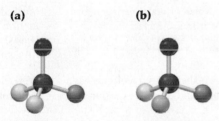

PROBLEM 8.27 Identify which of the following molecules are chiral and can exist in two mirror-image forms.

(a) Lactic acid: formed during muscle contraction.

$$HOOC\overset{H}{\underset{H_3C}{\overset{|}{-}C-}}OH$$

(b) Acetic acid: the main component of vinegar.

$$HOOC\overset{H}{\underset{H}{\overset{|}{-}C-}}H$$

(c) Propanoic acid: a preservative that inhibits the growth of mold.

$$HOOC\overset{CH_3}{\underset{H}{\overset{|}{-}C-}}H$$

PROBLEM 8.28 Asparagine is a naturally occurring amino acid that exists in two mirror-image forms. One form tastes bitter while the other form tastes sweet.

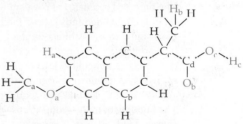

(a) Indicate which of the carbon atoms (labeled a–d) is the chiral center.
(b) How many σ bonds and how many π bonds does the asparagine molecule contain?
(c) Indicate the hybridization of carbon atoms (labeled a–d).
(d) Circle the atoms that can participate in hydrogen bonding with the asparagine receptor site.

PROBLEM 8.29 Naproxen, the active ingredient in Aleve, is a nonsteroidal anti-inflammatory drug. The connections between atoms are shown.

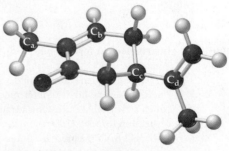

(a) Complete the electron-dot structure by filling in multiple bonds and lone pairs of electrons.
(b) Indicate which of the carbon atoms (labeled a–d) is the chiral center.
(c) How many σ bonds and how many π bonds does the Naproxen molecule contain?
(d) Indicate the hybridization of carbon atoms labeled (a–d).
(e) Indicate which of the labeled hydrogen and oxygen atoms can participate in hydrogen bonding with a receptor site.

PROBLEM 8.30 The left- and right-handed carvone molecules fit different smell receptors due to differences in shape.

"Left-handed" carvone
(odor of spearmint)

(a) Indicate which of the carbon atoms (labeled a–d) is the chiral center.
(b) Indicate the hybridization of the carbon atoms labeled (a–d).
(c) Can the carvone molecule participate in hydrogen bonding with its receptor site?

STUDY GUIDE

Section	Concept Summary	Learning Objectives	Test Your Understanding
8.1 ▶ Molecular Shapes: The VSEPR Model	Molecular shape can often be predicted using the **valence-shell electron-pair repulsion (VSEPR) model**, which treats the electrons around atoms as charge clouds that repel one another and therefore orient themselves as far away from one another as possible. Atoms with two charge clouds adopt a linear arrangement of the clouds, atoms with three charge clouds adopt a trigonal planar arrangement, and atoms with four charge clouds adopt a tetrahedral arrangement. Similarly, atoms with five charge clouds are trigonal bipyramidal and atoms with six charge clouds are octahedral.	**8.1** Use the VSEPR model to predict geometry from the total number of charge clouds and lone pairs of electrons around an atom.	Problems, 8.31–8.33, 8.42–8.45
		8.2 Use the VSEPR model to predict bond angles and overall shape of a molecule or ion with one central atom.	Worked Example 8.1; Problems 8.46–8.53, 8.105, 8.118
		8.3 Use the VSEPR model to predict bond angles and overall shape of a molecule with more than one central atom.	Worked Example 8.2; Problems 8.54–8.59, 8.117
8.2 ▶ Valence Bond Theory	According to **valence bond theory**, covalent bond formation occurs by the overlap of two singly occupied atomic orbital lobes of the same phase, either head-on along the internuclear axis to form a σ **bond** or sideways above and below the internuclear axis to form a π **bond**.	**8.4** Describe the difference between a sigma and a pi bond.	Problems 8.60, 8.61
8.3–8.4 ▶ Hybrid Orbitals	The observed geometry of covalent bonding in main-group compounds is described by assuming that s and p atomic orbitals combine to generate hybrid orbitals, which are strongly oriented in specific directions: sp **hybrid orbitals** have linear geometry, sp^2 **hybrid orbitals** have trigonal planar geometry, and sp^3 **hybrid orbitals** have tetrahedral geometry. The bonding of main-group atoms with five and six charge clouds is more complex.	**8.5** Determine the type of hybrid orbitals based upon the number of charge clouds around an atom.	Problems 8.62, 8.63
		8.6 Write an electron-dot structure for a molecule and determine hybridization and bond angles on nonterminal atoms.	Problems 8.35–8.37, 8.64–8.67, 8.100–8.103, 8.108, 8.111, 8.112
		8.7 Identify which orbitals overlap to form σ and π bonds in molecules.	Worked Examples 8.3, 8.4, 8.5; Problems 8.5–8.10, 8.68, 8.69, 8.109, 8.110
8.5 ▶ Polar Covalent Bonds and Dipole Moments	The difference in electronegativity between elements in a bond can results in a polar covalent bond. If the bond dipoles and lone pairs of electrons occur in an asymmetric arrangement around the center of the molecule, the molecule has a net polarity, a property measured by the **dipole moment**.	**8.8** Predict whether a given molecule has a dipole moment and draw its direction.	Worked Example 8.6; Problems 8.76, 8.77, 8.80–8.83, 8.98
		8.9 Interpret electrostatic potential maps of molecules.	Problems 8.12, 8.38, 8.40
		8.10 Calculate the percent ionic character in a bond.	Worked Example 8.7; Problems 8.13, 8.14, 8.78, 8.79
8.6 ▶ Intermolecular Forces	**Intermolecular forces**, collectively known as van der Waals forces, are the attractions responsible for holding particles together in the solid and liquid phases. There are several kinds of intermolecular forces, all which arise from electrostatic attractions. **London dispersion forces** are characteristic of all molecules and result from the presence of temporary dipole moments caused by momentarily unsymmetrical electron distributions. **Dipole–dipole forces** occur between two polar molecules. A **hydrogen bond** is the attraction between a positively polarized hydrogen bonded to O, N, or F and a lone pair of electrons on an O, N, or F atom. In addition, **ion–dipole forces** occur between an ion and a polar molecule.	**8.11** Identify the types of intermolecular forces experienced by a molecule.	Worked Example 8.9; Problems 8.17, 8.72, 8.73, 8.75
		8.12 Relate the strength of intermolecular forces to physical properties such as melting point and boiling point.	Problems 8.18, 8.41, 8.74, 8.99
		8.13 Sketch the hydrogen bonding that occurs between two molecules.	Worked Example 8.8; Problems 8.15, 8.16, 8.84–8.87

Section	Concept Summary	Learning Objectives	Test Your Understanding
8.7–8.8 ▶ Molecular Orbital Theory	**Molecular orbital theory** sometimes gives a more accurate picture of electronic structure than the valence bond model. A **molecular orbital** is a wave function whose square gives the probability of finding an electron in a given region of space in a molecule. Combination of two atomic orbitals gives two molecular orbitals, a **bonding MO** that is lower in energy than the starting atomic orbitals and an **antibonding MO** that is higher in energy than the starting atomic orbitals. Molecular orbital theory is particularly useful for describing delocalized π bonding in molecules.	**8.14** Interpret the molecular orbital diagram for a first row diatomic molecule or ion. **8.15** Interpret the molecular orbital diagram for a second row diatomic molecule or ion. Calculate the bond order and predict magnetic properties.	Worked Example 8.10; Problems 8.25, 8.26 Worked Example 8.11; Problems 8.27, 8.28, 8.90–8.95, 8.114–8.116
8.9 ▶ Combining Valence Bond Theory and Molecular Orbital Theory	Molecules that exhibit resonance are often described by a combination of valence bond theory and molecular orbital theory. Valence bond theory is a simple model that works well for describing σ bonds, but often inaccurately predicts properties such as magnetism. Molecular orbital theory is a more sophisticated model for describing electron delocalization in π bonds and more accurately predicts properties.	**8.16** Draw orbital overlap diagrams for molecules and describe the use of both valence bond theory and molecular orbital theory.	Worked Example, 8.12; Problems 8.29, 8.30, 8.96, 8.97

KEY TERMS

achiral *299*
antibonding molecular orbital *291*
bond angle *263*
bond order *292*
bonding molecular orbital *291*
chiral *299*
chiral center *299*
diamagnetic *294*

dipole *278*
dipole–dipole force *284*
dipole moment (μ) *302*
electronegativity *278*
electrostatic potential map *278*
hybrid atomic orbital *271*
hydrogen bond *287*
intermolecular force *282*
ion–dipole force *283*

London dispersion force *285*
molecular orbital *291*
molecular orbital (MO) theory *291*
paramagnetic *294*
pi (π) bond *273*
polarizability *285*
sigma (σ) bond *270*
sp hybrid orbital *275*

sp^2 hybrid orbital *273*
sp^3 hybrid orbital *271*
valence bond theory *270*
valence-shell electron-pair repulsion (VSEPR) model *262*
van der Waals forces *282*

CONCEPTUAL PROBLEMS

Problems 8.1–8.30 appear within the chapter.

8.31 What is the geometry around the central atom in each of the following molecular models?

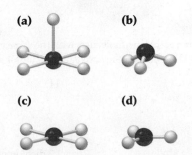

8.32 What is the geometry around the central atom in each of the following molecular models? (There may be a "hidden" atom directly behind a visible atom in some cases.)

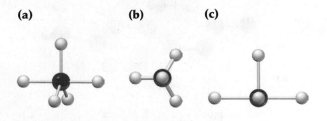

8.33 Three of the following molecular models have a tetrahedral central atom, and one does not. Which is the odd one? (There may be a "hidden" atom directly behind a visible atom in some cases.)

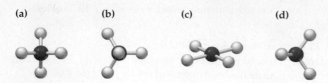

(a)   **(b)**   **(c)**   **(d)**

8.34 Identify each of the following sets of hybrid orbitals:

(a)   **(b)**   **(c)**

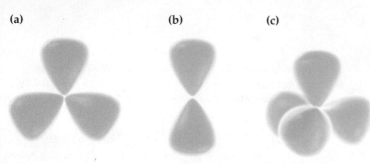

8.35 The VSEPR model is a simple predictive tool that is usually, but not always, correct. Take urea, for instance, a waste product excreted in animal urine:

Urea

What hybridization would you expect for the C and N atoms in urea according to the VSEPR model, and what approximate values would you expect for the various bond angles? What are the actual hybridizations and bond angles based on the molecular model shown? (Red = O, gray = C, blue = N, ivory = H).

8.36 The following ball-and-stick molecular model is a representation of acetaminophen, the active ingredient in such over-the-counter headache remedies as Tylenol (red = O, gray = C, blue = N, ivory = H):

(a) What is the formula of acetaminophen?

(b) Indicate the positions of the multiple bonds in acetaminophen.

(c) What is the geometry around each carbon?

(d) What is the hybridization of each carbon?

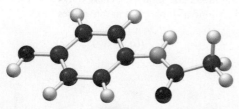

Acetaminophen

8.37 The following ball-and-stick molecular model is a representation of thalidomide, a drug that causes birth defects when taken by expectant mothers but is valuable for its use against leprosy. The lines indicate only the connections between

atoms, not whether the bonds are single, double, or triple (red = O, gray = C, blue = N, ivory = H).

(a) What is the formula of thalidomide?

(b) Indicate the positions of the multiple bonds in thalidomide.

(c) What is the geometry around each carbon?

(d) What is the hybridization of each carbon?

Thalidomide

8.38 Ethyl acetate, $CH_3CO_2CH_2CH_3$, is commonly used as a solvent and nail-polish remover. Look at the following electrostatic potential map of ethyl acetate, and explain the observed polarity.

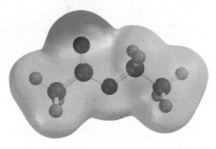

8.39 The dipole moment of methanol is $\mu = 1.70$ D. Use arrows to indicate the direction in which electrons are displaced.

8.40 Methylamine, CH_3NH_2, is responsible for the odor of rotting fish. Look at the following electrostatic potential map of methylamine and explain the observed polarity.

8.41 Two dichloroethylene molecules with the same chemical formula ($C_2H_2Cl_2$), but different arrangements of atoms are shown.

cis 1,2 dichloroethylene   *trans* 1,2 dichloroethylene

(a) Match each form of dichloroethylene to its electrostatic potential map.

(i)   **(ii)**

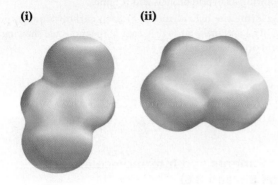

(b) Which form of dichloroethylene has a dipole moment of 2.39 D and which has dipole moment of 0.00 D?

(c) Which form of dichloroethylene has the highest boiling point?

SECTION PROBLEMS

The VSEPR Model (Section 8.1)

8.42 What geometric arrangement of charge clouds do you expect for atoms that have the following number of charge clouds?
(a) 3   (b) 5   (c) 2   (d) 6

8.43 What shape do you expect for molecules that meet the following descriptions?
(a) A central atom with two lone pairs and three bonds to other atoms
(b) A central atom with two lone pairs and two bonds to other atoms
(c) A central atom with two lone pairs and four bonds to other atoms

8.44 How many charge clouds are there around the central atom in molecules that have the following geometry?
(a) Tetrahedral   (b) Octahedral
(c) Bent   (d) Linear
(e) Square pyramidal   (f) Trigonal pyramidal

8.45 How many charge clouds are there around the central atom in molecules that have the following geometry?
(a) Seesaw   (b) Square planar
(c) Trigonal bipyramidal   (d) T-shaped
(e) Trigonal planar   (f) Linear

8.46 What shape do you expect for each of the following molecules?
(a) H_2Se (b) $TiCl_4$
(c) O_3 (d) GaH_3

8.47 What shape do you expect for each of the following molecules?
(a) XeO_4 (b) SO_2Cl_2
(c) OsO_4 (d) SeO_2

8.48 What shape do you expect for each of the following molecules or ions?
(a) SbF_5 (b) IF_4^+
(c) SeO_3^{2-} (d) CrO_4^{2-}

8.49 Predict the shape of each of the following ions:
(a) NO_3^- (b) NO_2^+ (c) NO_2^-

8.50 What shape do you expect for each of the following anions?
(a) PO_4^{3-} (b) MnO_4^- (c) SO_4^{2-}
(d) SO_3^{2-} (e) ClO_4^- (f) SCN^-

8.51 What shape do you expect for each of the following cations?
(a) XeF_3^+ (b) SF_3^+
(c) ClF_2^+ (d) CH_3^+

8.52 What bond angles do you expect for each of the following?
(a) The F—S—F angle in SF_2
(b) The H—N—N angle in N_2H_2
(c) The F—Kr—F angle in KrF_4
(d) The Cl—N—O angle in NOCl

8.53 What bond angles do you expect for each of the following?
(a) The Cl—P—Cl angle in PCl_6^-
(b) The Cl—I—Cl angle in ICl_2^-
(c) The O—S—O angle in SO_4^{2-}
(d) The O—B—O angle in BO_3^{3-}

8.54 Acrylonitrile is used as the starting material for manufacturing acrylic fibers. Predict values for all bond angles in acrylonitrile.

$$H_2C{=}C{-}C{\equiv}N\textbf{:}$$

with H attached above the second C. Acrylonitrile

8.55 Predict values for all bond angles in dimethyl sulfoxide, a powerful solvent used in veterinary medicine to treat inflammation.

$$H_3C{-}S{-}CH_3$$

with :O: double-bonded above S, and lone pairs on S. Dimethyl sulfoxide

8.56 Oceanographers study the mixing of water masses by releasing tracer molecules at a site and then detecting their presence at other places. The molecule trifluoromethylsulfur pentafluoride is one such tracer. Draw an electron-dot structure for CF_3SF_5, and predict the bond angles around both carbon and sulfur.

8.57 A potential replacement for the chlorofluorocarbon refrigerants that harm the Earth's protective ozone layer is a compound called E143a, or trifluoromethyl methyl ether, F_3COCH_3. Draw an electron-dot structure for F_3COCH_3, and predict the geometry around both the carbons and the oxygen.

8.58 Explain why cyclohexane, a substance that contains a six-membered ring of carbon atoms, is not flat but instead has a puckered, nonplanar shape. Predict the values of the C—C—C bond angles.

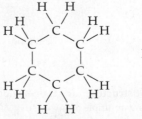

Cyclohexane

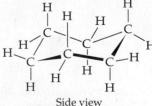

Side view

8.59 Like cyclohexane (Problem 8.58), benzene also contains a six-membered ring of carbon atoms, but it is flat rather than puckered. Explain, and predict the values of the C—C—C bond angles.

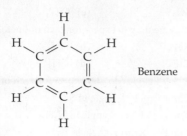

Benzene

Valence Bond Theory and Hybridization (Sections 8.2–8.4)

8.60 What is the difference in spatial distribution between electrons in a π bond and electrons in a σ bond?

8.61 The average C—C bond dissociation energy (D) is 350 kJ/mol and the average C=C bond dissociation energy is 728 kJ/mol. Based on these values, which is stronger: a σ or a π bond?

8.62 What hybridization do you expect for atoms that have the following numbers of charge clouds?
  (a) 2   (b) 3   (c) 4

8.63 What spatial arrangement of charge clouds corresponds to each of the following kinds of hybridization?
  (a) sp^3 (b) sp^2 (c) sp

8.64 What hybridization would you expect for the indicated atom in each of the following molecules?
  (a) H_2C=O   (b) BH_3
  (c) CH_3SH (d) H_2C=NH

8.65 What hybridization would you expect for the indicated atom in each of the following ions?
  (a) BH_4^- (b) HCO_2^-
  (c) CH_3^+ (d) CH_3^-

8.66 Oxaloacetic acid is an intermediate involved in the citric acid cycle of food metabolism. What are the hybridizations of the various carbon atoms in oxaloacetic acid, and what are the approximate values of the various bond angles?

$$H-O-\overset{\overset{\textstyle O}{\|}}{C}-\overset{\overset{\textstyle O}{\|}}{C}-\overset{\overset{\textstyle H}{|}}{\underset{\underset{\textstyle H}{|}}{C}}-\overset{\overset{\textstyle O}{\|}}{C}-O-H \qquad \text{Oxaloacetic acid}$$

8.67 The atoms in the amino acid glycine are connected as shown:

$$H-\overset{\overset{\textstyle H}{|}}{\underset{\underset{\textstyle H}{|}}{N}}-\overset{\overset{\textstyle H}{|}}{\underset{\underset{\textstyle H}{|}}{C}}-\overset{\overset{\textstyle O}{\|}}{C}-O-H \qquad \text{Glycine}$$

  (a) Draw an electron-dot structure for glycine, showing lone pairs and identifying any multiple bonds.
  (b) Predict approximate values for the H—C—H, O—C—O, and H—N—H bond angles.
  (c) Which hybrid orbitals are used by the C and N atoms?

8.68 Describe the hybridization of the carbon atom in the poisonous gas phosgene, Cl_2CO, and make a rough sketch of the molecule showing its hybrid orbitals and π bonds.

8.69 Describe the hybridization of each carbon atom in propyne (C_3H_4) and make a rough sketch of the molecule showing its hybrid orbitals and π bonds.

$$H-C\equiv C-\overset{\overset{\textstyle H}{|}}{\underset{\underset{\textstyle H}{|}}{C}}-H$$

Dipole Moments and Intermolecular Forces (Sections 8.5 and 8.6)

8.70 Why don't all molecules with polar covalent bonds have dipole moments?

8.71 What is the difference between London dispersion forces and dipole–dipole forces?

8.72 What are the most important kinds of intermolecular forces present in each of the following substances?
  (a) Chloroform, $CHCl_3$
  (b) Oxygen, O_2
  (c) Polyethylene, C_nH_{2n+2}
  (d) Methanol, CH_3OH

8.73 Of the substances Xe, CH_3Cl, HF, which has:
  (a) The smallest dipole–dipole forces?
  (b) The largest hydrogen bond forces?
  (c) The largest dispersion forces?

8.74 Methanol (CH_3OH; bp = 65 °C) boils nearly 230 °C higher than methane (CH_4; bp = −164 °C), but 1-decanol ($C_{10}H_{21}OH$; bp = 231 °C) boils only 57 °C higher than decane ($C_{10}H_{22}$; bp = 174 °C). Explain.

Methanol

1-Decanol

8.75 Which substance in each of the following pairs would you expect to have larger dispersion forces?
  (a) Ethane, C_2H_6, or octane, C_8H_{18}
  (b) HCl or HI
  (c) H_2O or H_2Se

8.76 Which of the following substances would you expect to have a nonzero dipole moment? Explain, and show the direction of each.
  (a) Cl_2O
  (b) XeF_4
  (c) Chloroethane, CH_3CH_2Cl
  (d) BF_3

8.77 Which of the following substances would you expect to have a nonzero dipole moment? Explain, and show the direction of each.

(a) NF_3 (b) CH_3NH_2
(c) XeF_2 (d) PCl_5

8.78 The dipole moment of BrCl is 0.518 D and the distance between atoms is 213.9 pm. What is the percent ionic character of the BrCl bond?

8.79 The dipole moment of ClF is 0.887 D and the distance between atoms is 162.8 pm. What is the percent ionic character of the ClF bond?

8.80 Why is the dipole moment of SO_2 1.63 D, but that of CO_2 is zero?

8.81 Draw three-dimensional structures of PCl_3 and PCl_5, and then explain why one of the molecules has a dipole moment and one does not.

8.82 The class of ions $PtX_4{}^{2-}$, where X is a halogen, has a square planar geometry.

(a) Draw a structure for a $PtBr_2Cl_2{}^{2-}$ ion that has no dipole moment.

(b) Draw a structure for a $PtBr_2Cl_2{}^{2-}$ ion that has a dipole moment.

8.83 Of the two compounds SiF_4 and SF_4, which is polar and which is nonpolar?

8.84 Draw a picture showing how hydrogen bonding takes place between two ammonia molecules.

8.85 1,3-Propanediol can form *intra*molecular as well as *inter*molecular hydrogen bonds. Draw a structure of 1,3-propanediol showing an intramolecular hydrogen bond.

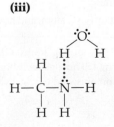

1,3-Propanediol

8.86 A liquid sample contains methylamine (CH_3NH_2) dissolved in water. Which of the following illustrations depicts the hydrogen bonding that occurs between methylamine and water?

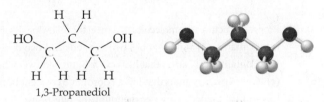

(i)   (ii)

(iii)   (iv)

8.87 Dimethyl ether has the following structure.

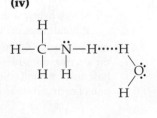

(a) Does hydrogen bonding occur in a pure sample of dimethyl ether? If so, sketch the hydrogen bonds.

(b) Which of the following illustrations depicts the hydrogen bonding that occurs between dimethyl ether and water?

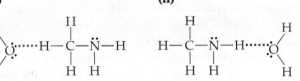

(i)   (ii)

Molecular Orbital Theory (Sections 8.7–8.9)

8.88 What is the difference in spatial distribution between electrons in a bonding MO and electrons in an antibonding MO?

8.89 For a given type of MO, use a σ_{2s} as an example, is the bonding or antibonding orbital higher in energy? Explain.

8.90 Use the MO energy diagram in Figure 8.23b to describe the bonding in $O_2{}^+$, O_2, and $O_2{}^-$. Which of the three is likely to be stable? What is the bond order of each? Which contain unpaired electrons?

8.91 Use the MO energy diagram in Figure 8.23a to describe the bonding in $N_2{}^+$, N_2, and $N_2{}^-$. Which of the three is likely to be stable? What is the bond order of each? Which contain unpaired electrons?

8.92 The C_2 molecule can be represented by an MO diagram similar to that in Figure 8.23a.

(a) What is the bond order of C_2?

(b) To increase the bond order of C_2, should you add or remove an electron?

(c) Give the charge and the bond order of the new species made in part (b).

8.93 Look at the molecular orbital diagram for O_2 in Figure 8.23 and answer the following questions:

(a) What is the bond order of O_2?

(b) To increase the bond order of O_2, should you add or remove an electron?

(c) Give the charge and the bond order of the new species made in part (b).

8.94 Look at the MO diagrams of corresponding neutral diatomic species in Figure 8.23, and predict whether each of the following ions is diamagnetic or paramagnetic. Diagrams for Li_2 and C_2 are similar to N_2; Cl_2 is similar to F_2.

(a) $C_2{}^{2-}$ (b) $C_2{}^{2+}$ (c) $F_2{}^-$
(d) Cl_2 (e) $Li_2{}^+$

8.95 Look at the MO diagrams of corresponding neutral diatomic species in Figure 8.23, and predict whether each of the following ions is diamagnetic or paramagnetic. MO diagrams for Li_2 and C_2 are similar to N_2; Cl_2 is similar to F_2.

(a) $O_2{}^{2+}$ (b) $N_2{}^{2+}$ (c) $C_2{}^+$
(d) $F_2{}^{2+}$ (e) $Cl_2{}^+$

8.96 Make a sketch showing the location and geometry of the *p* orbitals in the allyl cation. Describe the bonding in this cation using a localized valence bond model for σ bonding and a delocalized MO model for π bonding.

$$H_2C=\overset{\overset{\displaystyle H}{|}}{C}-CH_2{}^+ \quad \text{Allyl cation}$$

8.97 Make a sketch showing the location and geometry of the *p* orbitals in the nitrite ion, $NO_2{}^-$. Describe the bonding in this ion using a localized valence bond model for σ bonding and a delocalized MO model for π bonding.

CHAPTER PROBLEMS

8.98 Fluorine is more electronegative than chlorine, yet fluoromethane (CH_3F; $\mu = 1.86$ D) has a smaller dipole moment than chloromethane (CH_3Cl; $\mu = 1.90$ D). Explain.

8.99 For each of the following substances, identify the intermolecular force or forces that predominate. Using your knowledge of the relative strengths of the various forces, rank the substances in order of their normal boiling points.

$$Al_2O_3, F_2, H_2O, Br_2, ICl, NaCl$$

8.100 Bupropion, marketed as Wellbutrin, is a heavily prescribed medication used in the treatment of depression. Complete the following electron-dot structure for Bupropion by adding lone pairs of electrons.

Bupropion

   (a) How many σ and how many π bonds are in the molecule?
   (b) Give the hybridization of each carbon atom in the molecule.
   (c) Give the bond angles of each carbon atom.
   (d) Give the hybridization of the N atom in the molecule.

8.101 Efavirenz, marketed as Sustivam is a medication used in the treatment of human immunodeficiency virus, HIV. Complete the following electron-dot structure for Efavirenz by adding lone pairs of electrons.

Sustiva

   (a) How many σ and how many π bonds are in the molecule?
   (b) Give the hybridization of each carbon atom in the molecule.
   (c) Give the hybridization of the N atom in the molecule.
   (d) Give the hybridization of the oxygen atoms in the molecule.

8.102 What is the hybridization of the B and N atoms in borazine, what are the values of the B—N—B and N—B—N bond angles, and what is the overall shape of the molecule?

Borazine

8.103 Benzyne, C_6H_4, is a highly energetic and reactive molecule. What hybridization do you expect for the two triply bonded carbon atoms? What are the "theoretical" values for the C—C≡C bond angles? Why do you suppose benzyne is so reactive?

Benzyne

8.104 Propose structures for molecules that meet the following descriptions:
   (a) Contains a C atom that has two π bonds and two σ bonds
   (b) Contains an N atom that has one π bond and two σ bonds
   (c) Contains an S atom that has a coordinate covalent bond

8.105 Use VSEPR theory to answer the following questions:
   (a) Which molecule, BF_3 or PF_3, has the smaller F—X—F angles?
   (b) Which ion, PCl_4^+ or ICl_2^-, has the smaller Cl—X—Cl angles?
   (c) Which ion, CCl_3^- or PCl_6^-, has the smaller Cl—X—Cl angles?

8.106 In the cyanate ion, OCN^-, carbon is the central atom.
   (a) Draw as many resonance structures as you can for OCN^-, and assign formal charges to the atoms in each.
   (b) Which resonance structure makes the greatest contribution to the resonance hybrid? Which makes the least contribution? Explain.
   (c) Is OCN^- linear or bent? Explain.
   (d) Which hybrid orbitals are used by the C atom, and how many π bonds does the C atom form?

8.107 Aspirin has the following connections among atoms. Complete the electron-dot structure for aspirin, tell how many σ bonds and how many π bonds the molecule contains, and tell the hybridization of each carbon atom.

Aspirin

8.108 The cation $[H-C-N-Xe-F]^+$ is entirely linear. Draw an electron-dot structure consistent with that geometry, and tell the hybridization of the C and N atoms.

8.109 Acrylonitrile (C_3H_3N) is a molecule that is polymerized to make carpets and fabrics. The connections between atoms are shown.

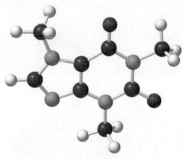

    (a) Complete the electron-dot structure for acrylonitrile by adding multiple bonds and lone pairs of electrons.

    (b) Tell how many σ bonds and how many π bonds the molecule contains.

    (c) Tell the hybridization of each carbon atom.

    (d) Identify the shortest bond in the molecule.

8.110 The odor of cinnamon oil is due to cinnamaldehyde, C_9H_8O. What is the hybridization of each carbon atom in cinnamaldehyde? How many σ and how many π bonds does cinnamaldehyde have?

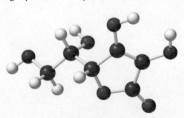

Cinnamaldehyde

8.111 The following molecular model is a representation of caffeine. Identify the position(s) of multiple bonds in caffeine, and tell the hybridization of each carbon atom (red = O, blue = N, gray = C, ivory = H).

Caffeine

8.112 The following molecular model is a representation of ascorbic acid, or vitamin C. Identify the position(s) of multiple bonds in ascorbic acid, and tell the hybridization of each carbon atom (red = O, gray = C, ivory = H).

Asorbic acid

8.113 Draw a molecular orbital energy diagram for Li_2. What is the bond order? Is the molecule likely to be stable? Explain.

8.114 Calcium carbide, CaC_2, reacts with water to produce acetylene, C_2H_2, and is sometimes used as a convenient source of that substance. Use the MO energy diagram in Figure 8.23a to describe the bonding in the carbide anion, C_2^{2-}. What is its bond order?

8.115 At high temperatures, sulfur vapor is predominantly in the form of $S_2(g)$ molecules.

    (a) Assuming that the molecular orbitals for third-row diatomic molecules are analogous to those for second-row molecules, construct an MO diagram for the valence orbitals of $S_2(g)$.

    (b) Is S_2 likely to be paramagnetic or diamagnetic?

    (c) What is the bond order of $S_2(g)$?

    (d) When two electrons are added to S_2, the disulfide ion S_2^{2-} is formed. Is the bond length in S_2^{2-} likely to be shorter or longer than the bond length in S_2? Explain.

8.116 Carbon monoxide is produced by incomplete combustion of fossil fuels.

    (a) Give the electron configuration for the valence molecular orbitals of CO. The orbitals have the same energy order as those of the N_2 molecule.

    (b) Do you expect CO to be paramagnetic or diamagnetic?

    (c) What is the bond order of CO? Does this match the bond order predicted by the electron-dot structure?

    (d) CO can react with OH^- to form the formate ion, HCO_2^-. Draw an electron-dot structure for the formate ion, and give any resonance structures if appropriate.

8.117 Draw an electron-dot structure for each of the following substances, and predict the molecular geometry of every nonterminal atom.

    (a) F_3S-S-F **(b)** $CH_3-C\equiv C-CO_2^-$

8.118 The ion I_5^- is shaped like a big "V." Draw an electron-dot structure consistent with this overall geometry.

MULTICONCEPT PROBLEMS

8.119 The dichromate ion, $Cr_2O_7^{2-}$, has neither Cr—Cr nor O—O bonds.

    (a) Taking both $4s$ and $3d$ electrons into account, draw an electron-dot structure that minimizes the formal charges on the atoms.

    (b) How many outer-shell electrons does each Cr atom have in your electron-dot structure? What is the likely geometry around the Cr atoms?

8.120 Just as individual bonds in a molecule are often polar, molecules as a whole are also often polar because of the net sum of individual bond polarities. There are three possible structures for substances with the formula $C_2H_2Cl_2$, two of which are polar overall and one of which is not.

    (a) Draw the three possible structures for $C_2H_2Cl_2$, predict an overall shape for each, and explain how they differ.

    (b) Which of the three structures is nonpolar, and which two are polar? Explain.

    (c) Two of the three structures can be interconverted by a process called cis–trans isomerization, in which rotation around the central carbon–carbon bond takes place when the molecules are irradiated with ultraviolet light. If light with a wavelength of approximately 200 nm is required for isomerization, how much energy in kJ/mol is involved?

(d) Sketch the orbitals involved in the central carbon–carbon bond, and explain why so much energy is necessary for bond rotation to occur.

8.121 Cyclooctatetraene dianion, $C_8H_8^{2-}$, is an organic ion with the structure shown. Considering only the π bonds and not the σ bonds, cyclooctatetraene dianion can be described by the following energy diagram of its π molecular orbitals:

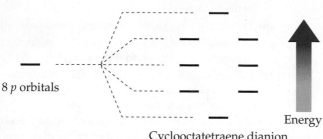

Cyclooctatetraene dianion

Cyclooctatetraene dianion
pi (π) molecular orbitals

(a) What is the hybridization of the 8 carbon atoms?

(b) Three of the π molecular orbitals are bonding, three are antibonding, and two are *nonbonding*, meaning that they have the same energy level as isolated p orbitals. Which is which?

(c) Complete the MO energy diagram by assigning the appropriate numbers of p electrons to the various molecular orbitals, indicating the electrons using up/down arrows ($\uparrow\downarrow$).

(d) Based on your MO energy diagram, is the dianion paramagnetic or diamagnetic?

CHAPTER

9

Thermochemistry: Chemical Energy

Vegetable oil from the bright yellow rapeseed plant is a leading candidate for large-scale production of biodiesel fuel.

 How is the energy content of new fuels determined?

The answer to this question can be found in the **INQUIRY** ▸▸▸ on page 344.

CONTENTS

Why do chemical reactions occur? Stated simply, the answer involves *stability*. For a reaction to take place spontaneously, the final products of the reaction must be more stable than the starting reactants.

But what is "stability," and what does it mean to say that one substance is more stable than another? The key factor in determining the stability of a substance is energy. Less stable substances have higher energy and are generally converted into more stable substances with lower energy. But what, in turn, does it mean for a substance to have higher or lower energy? We'll explore some different forms of energy in this chapter and look at the subject of **thermochemistry**, the absorption or release of heat energy that accompanies chemical reactions.

9.1 ▶ ENERGY AND ITS CONSERVATION

The word *energy*, though familiar to everyone, is surprisingly hard to define in simple, nontechnical terms. A good working definition, however, is to say that **energy** is the capacity to do work or supply heat. The water falling over a dam, for instance, contains energy that can be used to turn a turbine and generate electricity. A tank of propane gas contains energy that, when released in the chemical process of combustion, can heat a house or barbecue a hamburger.

Energy is classified as either *kinetic* or *potential*. **Kinetic energy (E_K)** is the energy of motion:

Kinetic energy $E_K = \dfrac{1}{2}mv^2$ where m = mass and v = velocity

The derived SI unit for energy ($kg \cdot m^2/s^2$) follows from the expression for kinetic energy, $E_K = (1/2)mv^2$, and is given the name **joule (J)**.

$$1\ (kg \cdot m^2/s^2) = 1\ J$$

Potential energy (E_P), by contrast, is stored energy—perhaps stored in an object because of its height or in a molecule because of reactions it can undergo. The water sitting in a reservoir behind a dam contains potential energy because of its height above the stream at the bottom of the dam. When the water is allowed to fall, its potential energy is converted to kinetic energy. Propane and other substances used as fuels contain potential energy because they can undergo a reaction with oxygen—a *combustion reaction*—that releases heat.

Let's pursue the relationship between potential energy and kinetic energy a bit further. According to the **conservation of energy law**, energy can be neither created nor destroyed; it can only be converted from one form into another.

Conservation of energy law   Energy cannot be created or destroyed; it can only be converted from one form into another.

To take an example, think about a hydroelectric dam. The water sitting motionless in the reservoir behind the dam has potential energy because of its height above the outlet stream, but it has no kinetic energy because it isn't moving ($v = 0$). When the water falls through the penstocks of the dam, however, its height and potential energy decrease while its velocity and kinetic energy increase. The moving water then spins the turbine of a generator, converting its kinetic energy into electrical energy (**FIGURE 9.1**).

The conversion of the kinetic energy in falling water into electricity illustrates several other important points about energy. One is that energy has many forms. Thermal energy, for example, seems different from the kinetic energy of falling water, yet is really quite similar.

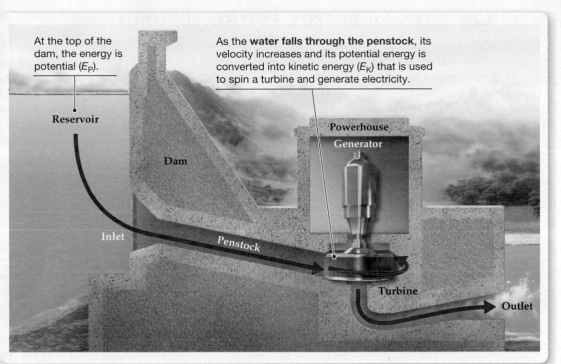

At the top of the dam, the energy is potential (E_P).

Reservoir

As the **water falls through the penstock**, its velocity increases and its potential energy is converted into kinetic energy (E_K) that is used to spin a turbine and generate electricity.

Dam

Powerhouse

Generator

Inlet

Penstock

Turbine

Outlet

◀ **FIGURE 9.1**

Conservation of energy. The total amount of energy contained by the water in the reservoir is constant.

Thermal energy is just the kinetic energy of molecular motion, which we measure by finding the **temperature** of an object. An object has a low temperature and we perceive it as cold if its atoms or molecules are moving slowly. Conversely, an object has a high temperature and we perceive it as hot if its atoms or molecules are moving rapidly and are colliding forcefully with a thermometer or other measuring device.

Heat, in turn, is the amount of thermal energy transferred from one object to another as the result of a temperature difference between the two. Rapidly moving molecules in a hotter object collide with more slowly moving molecules in a colder object, transferring kinetic energy and causing the slower moving molecules to speed up.

Chemical energy is another kind of energy that seems different from that of the water in a reservoir, yet again is really quite similar. Chemical energy is a kind of potential energy in which chemical bonds act as the storage medium. Just as water releases its potential energy when it falls to a more stable position, chemicals can release their potential energy in the form of heat or light when they undergo reactions and form more stable products. We'll explore this topic shortly.

A second point illustrated by falling water involves the conservation of energy law. To keep track of all the energy involved, it's necessary to take into account the entire chain of events that ensue from the falling water: the sound of the crashing water, the heating of the rocks at the bottom of the dam, the driving of turbines and electrical generators, the transmission of electrical power, the appliances powered by the electricity, and so on. Carrying the process to its logical extreme, it's necessary to take the entire universe into account when keeping track of all the energy in the water because the energy lost in one form *always* shows up elsewhere in another form. So important is the conservation of energy law that it's also known as the **first law of thermodynamics**.

First law of thermodynamics   Energy cannot be created or destroyed; it can only be converted from one form into another.

9.2 ▶ INTERNAL ENERGY AND STATE FUNCTIONS

When keeping track of the energy changes in a chemical reaction, it's often helpful to think of the reaction as being isolated from the world around it. The substances we focus on in an experiment—the starting reactants and the final products—are collectively called the **system**, while everything else—the reaction flask, the solvent, the room, the building, and so on—is called the **surroundings**. If the system could be truly isolated from its surroundings so that no energy transfer could occur between them, then the total **internal energy** (E) of the system, defined as the sum of all the kinetic and potential energies for every molecule or ion in the system, would be conserved and remain constant throughout the reaction. In fact, this assertion is just a restatement of the first law of thermodynamics:

> **First law of thermodynamics (restated)**   The total internal energy E of an isolated system is constant.

In practice, of course, it's not possible to truly isolate a chemical reaction from its surroundings. In any real situation, the chemicals are in physical contact with the walls of a flask or container, and the container itself is in contact with the surrounding air or laboratory bench. What's important, however, is not that the system be isolated but that we be able to measure accurately any energy that enters the system from the surroundings or leaves the system and flows to the surroundings (**FIGURE 9.2**). That is, we must be able to measure any *change* in the internal energy of the system, ΔE. The energy change ΔE represents the difference in internal energy between the final state of the system after reaction and the initial state of the system before reaction:

$$\Delta E = E_{final} - E_{initial}$$

By convention, energy changes are measured from the point of view of the system. Any energy that flows *from* the system *to* the surroundings has a negative sign because the system has lost it (that is, E_{final} is smaller than $E_{initial}$). Any energy that flows *to* the system *from* the surroundings has a positive sign because the system has gained it (E_{final} is larger than $E_{initial}$). If, for instance, we were to burn 1.00 mol of methane in the presence of 2.00 mol of oxygen, 802 kJ would be released as heat and transferred from the system to the surroundings. The system has 802 kJ less energy, so $\Delta E = -802$ kJ. This energy flow can be detected and measured by placing the reaction vessel in a water bath and noting the temperature rise of the bath during the reaction.

$$CH_4(g) + 2\,O_2(g) \longrightarrow CO_2(g) + 2\,H_2O(g) + 802\text{ kJ energy} \quad \Delta E = -802\text{ kJ}$$

▶ **FIGURE 9.2**

Energy changes in a chemical reaction.

Figure It Out

What is the sign of ΔE_{system} for a chemical reaction that causes a decrease in the temperature of the surroundings?

Answer: ΔE_{system} is positive because the chemical reaction (system) has absorbed heat from the surroundings.

The **system** is the mixture of reactants, and the *surroundings* are the flask, the solvent, the room, and the rest of the universe.

Surroundings

System

The energy change is the difference between final and initial states.

$$\Delta E = E_{final} - E_{initial}$$

Energy flowing out of the system to the surroundings has a negative sign because $E_{final} < E_{initial}$.

Energy flowing into the system from the surroundings has a positive sign because $E_{final} > E_{initial}$.

The methane combustion experiment tells us that the products of the reaction, $CO_2(g)$ and $2 H_2O(g)$, have 802 kJ less internal energy than the reactants, $CH_4(g)$ and $2 O_2(g)$, even though we don't know the exact values at the beginning ($E_{initial}$) and end (E_{final}) of the reaction. Note that the value $\Delta E = -802$ kJ for the reaction refers to the energy released when reactants are converted to products *in the molar amounts represented by coefficients in the balanced equation*. That is, 802 kJ is released when 1 mol of gaseous methane reacts with 2 mol of gaseous oxygen to give 1 mol of gaseous carbon dioxide and 2 mol of gaseous water vapor (**FIGURE 9.3**).

The internal energy of a system depends on many things: chemical identity, sample size, temperature, pressure, physical state (gas, liquid, or solid), and so forth. What the internal energy does not depend on is the system's past history. It doesn't matter what the system's temperature or physical state was an hour ago, and it doesn't matter how the chemicals were obtained. All that matters is the present condition of the system. Thus, internal energy is said to be a **state function**, one whose value depends only on the present state of the system. Pressure, volume, and temperature are other examples of state functions, but work and heat are not.

> **State function** A function or property whose value depends only on the present state, or condition, of the system, not on the path used to arrive at that state.

We can illustrate the idea of a state function by imagining a cross-country trip, say from the Artichoke Capitol of the World (Castroville, California), to the Hub of the Universe (Boston, Massachusetts). You are the system, and your position is a state function because how you got to wherever you are is irrelevant. Because your position is a state function, the *change* in your position after you complete your travel (Castroville and Boston are about 2720 miles apart) is independent of the path you take, whether through North Dakota or Louisiana (**FIGURE 9.4**).

The cross-country trip shown in Figure 9.4 illustrates an important point about state functions: their reversibility. Imagine that, after traveling from Castroville to Boston, you turn around and go back. Because your final position is now identical to your initial position, the change in your position is zero. The overall change in any state function is zero when the system returns to its original condition. For a nonstate function, however, the overall change is not zero when the system returns to its original condition. Any work you do in making the trip is not recovered when you return to your initial position, and any money or time you spend does not reappear.

Reactants
1 mol CH_4 (g) + 2 mol O_2 (g)

$\Delta E = -802$ kJ

Products
1 mol CO_2 (g) + 2 mol H_2O (g)

▲ **FIGURE 9.3**
Energy change in the combustion of one mole of methane (CH_4).

▲ Natural gas (methane, CH_4) burns and releases energy as heat to its surroundings.

State function | Nonstate functions
Position | Work expended
 | Money spent

◀ **FIGURE 9.4**

State functions. Because your position is a state function, the change in your position on going from Castroville, California, to Boston, Massachusetts, is independent of the path you take.

Figure It Out

Which of the following is a state function? The speed of the car on the highway or the distance of the route from Castroville to Boston.

Answer: The speed of the car is a state function because it does not depend on the pathway taken to achieve it. The car is traveling at 60 mph whether it has sped up from 30 mph or slowed down from 75 mph. The distance of the trip is a nonstate function because it depends on the pathway taken.

Measurements of the caloric content of food are based on the principle that energy changes in chemical reactions are state functions. The energy content of biological molecules such as carbohydrates and fats can be measured by burning them in oxygen in a calorimetry experiment as described in Section 9.7. The energy change is the same as when these compounds are "burned" for energy in your body because both processes have an identical overall chemical reaction.

$$C_6H_{12}O_6(s) + 6\,O_2(g) \xrightarrow{\text{combustion or metabolism}} 6\,CO_2(g) + 6\,H_2O(l) \quad \Delta E = -2810 \text{ kJ}$$

The same amount of energy is released whether glucose is directly burned or whether it is metabolized to carbon dioxide and water in a series of reactions in your body because state functions do not depend on the pathway taken from the initial to final state. Energy content of foods is typically reported in units of **Calories (Cal)** per gram, and the energy change for the combustion of glucose can be converted to the unit given on food labels as follows:

$$\frac{2810 \text{ kJ}}{1 \text{ mol } C_6H_{12}O_6} \times \frac{1000 \text{ J}}{1 \text{ kJ}} \times \frac{1 \text{ cal}}{4.184 \text{ J}} \times \frac{1 \text{ Cal}}{1000 \text{ cal}} \times \frac{1 \text{ mol } C_6H_{12}O_6}{180.0 \text{ g } C_6H_{12}O_6} = 3.73 \text{ Cal/g}$$

9.3 ▶ EXPANSION WORK

Just as energy comes in many forms, so too does *work*. In physics, **work (w)** is defined as the force (F) that produces the movement of an object times the distance moved (d):

$$\text{Work} = \text{Force} \times \text{Distance}$$
$$w = F \times d$$

When you run up stairs, for instance, your leg muscles provide a force sufficient to overcome gravity and lift you higher. When you swim, you provide a force sufficient to push water out of the way and pull yourself forward.

The most common type of work encountered in chemical systems is the *expansion work* (also called *pressure–volume*, or *PV, work*) done as the result of a volume change in the system. In the combustion reaction of propane (C_3H_8) with oxygen, for instance, the balanced equation says that 7 mol of products come from 6 mol of reactants:

$$\underbrace{C_3H_8(g) + 5\,O_2(g)}_{\text{6 mol of gas}} \longrightarrow \underbrace{3\,CO_2(g) + 4\,H_2O(g)}_{\text{7 mol of gas}}$$

If the reaction takes place inside a container outfitted with a movable piston, the greater volume of gas in the product will force the piston outward against the pressure of the atmosphere (P), moving air molecules aside and thereby doing work (**FIGURE 9.5**).

▲ This hiker going uphill is doing a lot of work to overcome gravity.

▶ **FIGURE 9.5**

Expansion work in a chemical reaction.

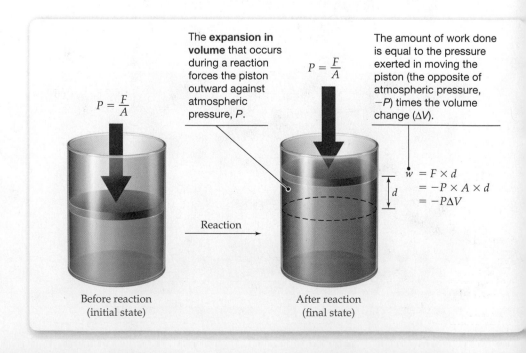

The **expansion in volume** that occurs during a reaction forces the piston outward against atmospheric pressure, P.

$$P = \frac{F}{A}$$

$$P = \frac{F}{A}$$

The amount of work done is equal to the pressure exerted in moving the piston (the opposite of atmospheric pressure, $-P$) times the volume change (ΔV).

$$w = F \times d$$
$$= -P \times A \times d$$
$$= -P\Delta V$$

Before reaction
(initial state)

After reaction
(final state)

Reaction

A short calculation gives the exact amount of work done during the expansion. We know from physics that force (F) is area (A) times pressure (P). In Figure 9.5, the force that the expanding gas exerts is the area of the piston times the pressure that the gas exerts against the piston. This pressure is equal in magnitude but opposite in sign to the external atmospheric pressure that opposes the movement, so it has the value $-P$.

$$F = -P \times A \quad \text{where } P \text{ is the external atmospheric pressure}$$

If the piston is pushed out a distance d, then the amount of work done is equal to force times distance, or pressure times area times distance:

$$w = F \times d = -P \times A \times d$$

This equation can be simplified by noticing that the area of the piston times the distance the piston moves is just the volume change in the system: $\Delta V = A \times d$. Thus, the amount of work done is equal to the pressure the gas exerts against the piston times the volume change, hence the name PV work:

A negative value   A positive value

$$w = -P\Delta V \qquad \text{Work done during expansion}$$

What about the sign of the work done during the expansion? Because the work is done by the system to move air molecules aside as the piston rises, work energy must be leaving the system. Thus, the negative sign of the work in the preceding equation is consistent with the convention previously established for ΔE (Section 9.2), whereby we always adopt the point of view of the system. Any energy that flows out of the system has a negative sign because the system has lost it ($E_{final} < E_{initial}$).

If the pressure is given in the unit atmospheres (atm) and the volume change is given in liters, then the amount of work done has the derived unit liter atmosphere ($L \cdot atm$), where 1 atm $= 101 \times 10^3 \, kg/(m \cdot s^2)$. Thus, 1 L $\cdot$ atm $= 101$ J:

$$1 \, L \cdot atm = (1 \, L)\left(\frac{10^{-3} \, m^3}{1 \, L}\right)\left(101 \times 10^3 \, \frac{kg}{m \cdot s^2}\right) = 101 \, \frac{kg \cdot m^2}{s^2} = 101 \, J$$

When a reaction takes place with a contraction in volume rather than an expansion, the ΔV term has a negative sign and the work has a positive sign. This is again consistent with adopting the point of view of the system because the system has now gained work energy ($E_{final} > E_{initial}$). An example is the industrial synthesis of ammonia by reaction of hydrogen with nitrogen. Four moles of gaseous reactants yield only 2 mol of gaseous products, so the volume of the system contracts and work is gained by the system.

$$\underbrace{3 \, H_2(g) + N_2(g)}_{\text{4 mol of gas}} \longrightarrow \underbrace{2 \, NH_3(g)}_{\text{2 mol of gas}}$$

A positive value   A negative value

$$w = -P\Delta V \qquad \text{Work gained during contraction}$$

If there is no volume change, then $\Delta V = 0$ and there is no work. Such is the case for the combustion of methane, where 3 mol of gaseous reactants give 3 mol of gaseous products:

$$CH_4(g) + 2 \, O_2(g) \longrightarrow CO_2(g) + 2 \, H_2O(g)$$

—• WORKED EXAMPLE 9.1

Calculating the Amount of PV Work

Calculate the work in kilojoules done during a reaction in which the volume expands from 12.0 L to 14.5 L against an external pressure of 5.0 atm.

IDENTIFY

Known	Unknown
Pressure (P) = 5.0 atm	Work (w)
Initial and Final Volume (V)	

continued on next page

STRATEGY

Expansion work done during a chemical reaction is calculated with the formula $w = -P\Delta V$, where P is the external pressure opposing the change in volume. In this instance, $P = 5.0$ atm and $\Delta V = (14.5 - 12.0)\, L = 2.5\, L$.

SOLUTION

$$w = -(5.0\text{ atm})(2.5\text{ L}) = -12.5\text{ L} \cdot \text{atm}$$

$$(-12.5\text{ L} \cdot \text{atm})\left(101\,\frac{\text{J}}{\text{L} \cdot \text{atm}}\right) = -1.3 \times 10^3\text{ J} = -1.3\text{ kJ}$$

CHECK

It is always useful to check the sign of energy change to ensure it makes sense. Changes in energy and work are expressed from the perspective of the system. In this case, the system has done work on the surroundings by expanding the volume, and thus the sign of w should be negative.

▶ **PRACTICE 9.1** Calculate the work (kJ) done during a synthesis of ammonia in which the volume contracts from 8.6 L to 4.3 L at a constant external pressure of 44 atm. In which direction does the work energy flow? What is the sign of the energy change?

▶ **Conceptual APPLY 9.2** How much work is done in kilojoules, and in which direction, as a result of the following reaction?

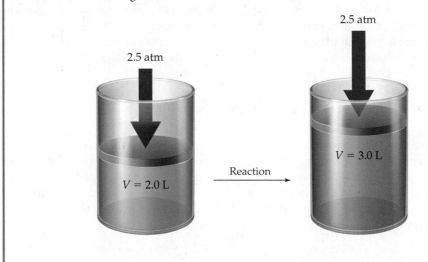

9.4 ▶ ENERGY AND ENTHALPY

We've seen up to this point that a system can exchange energy with its surroundings either by transferring heat or by doing work. The burning of gasoline (C_8H_{18}) in your car's engine heats up the surroundings causing the product gases CO_2 and H_2O to expand and do work by pushing the piston (**FIGURE 9.6**). Using the symbol q to represent transferred heat and the formula $w = -P\Delta V$, we can represent the total change in internal energy of a system, ΔE as follows:

Change in the internal energy of a system (ΔE) $\quad \Delta E = q + w = q - P\Delta V$

where q has a positive sign if the system gains heat and a negative sign if the system loses heat

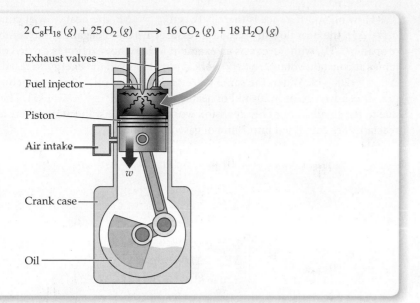

$$2\,C_8H_{18}\,(g) + 25\,O_2\,(g) \longrightarrow 16\,CO_2\,(g) + 18\,H_2O\,(g)$$

Exhaust valves

Fuel injector

Piston

Air intake

w

Crank case

Oil

◀ **FIGURE 9.6**

Energy is transferred as heat and work when gasoline (C_8H_{18}) burns in a car's engine.

Rearranging the equation to solve for q gives the amount of heat transferred:

$$q = \Delta E + P\Delta V$$

Let's look at two ways in which a chemical reaction might be carried out. On the one hand, a reaction might be carried out in a closed container with a constant volume, so that $\Delta V = 0$. In such a case, no PV work can be done so the energy change in the system is due entirely to heat transfer. We indicate this heat transfer at constant volume by the symbol q_v.

$$q_v = \Delta E \quad \text{At constant volume; } \Delta V = 0$$

Alternatively, a reaction might be carried out in an open flask or other apparatus that keeps the pressure constant and allows the volume of the system to change freely. In such a case, $\Delta V \neq 0$ and the energy change in the system is due to both heat transfer and PV work. We indicate the heat transfer at constant pressure by the symbol q_p:

$$q_p = \Delta E + P\Delta V \qquad \text{At constant pressure}$$

Because reactions carried out at constant pressure in open containers are so common in chemistry, the heat change q_p for such a process is given a special symbol and is called the **heat of reaction**, or **enthalpy change (ΔH)**. The **enthalpy (H)** of a system is the name given to the quantity $E + PV$.

▲ Chemical reactions are often carried out in open vessels at constant atmospheric pressure.

Enthalpy change (ΔH) $\quad q_p = \underset{\text{Enthalpy change}}{\Delta E + P\Delta V = \Delta H}$

Note that only the enthalpy *change* during a reaction is important. As with internal energy, enthalpy is a state function whose value depends only on the current state of the system, not on the path taken to arrive at that state. Thus, we don't need to know the exact value of the system's enthalpy before and after a reaction. We need to know only the difference between final and initial states:

$$\Delta H = H_{\text{final}} - H_{\text{initial}}$$
$$= H_{\text{products}} - H_{\text{reactants}}$$

How big a difference is there between $q_v = \Delta E$, the heat flow at constant volume, and $q_p = \Delta H$, the heat flow at constant pressure? Let's look again at the combustion reaction of propane, C_3H_8, with oxygen as an example. When the reaction is carried out in a closed container at constant volume, no PV work is possible so all the energy released is released as heat $\Delta E = -2046$ kJ. When the same reaction is carried out in an open container at constant pressure, however, only 2044 kJ of heat is released ($\Delta H = -2044$ kJ). The difference, 2 kJ, is due to the small amount of expansion work done against the atmosphere as 6 mol of gaseous reactants are converted into 7 mol of gaseous products.

$$C_3H_8(g) + 5\,O_2(g) \longrightarrow 3\,CO_2(g) + 4\,H_2O(g) \qquad \Delta E = -2046\,\text{kJ}$$
$$\Delta H = -2044\,\text{kJ}$$
$$P\Delta V = +2\,\text{kJ}$$

That is:

$$\left[\begin{array}{l} q_v = q_p + w \\ \Delta E = \Delta H - P\Delta V \\ -2046\,\text{kJ} = -2044\,\text{kJ} - (+2\,\text{kJ}) \end{array}\right]$$

What is true of the propane + oxygen reaction is also true of most other reactions: The difference between ΔH and ΔE is usually small, so the two quantities are nearly equal. Of course, if no volume change occurs and no work is done, such as in the combustion of methane in which 3 mol of gaseous reactants give 3 mol of gaseous products, then ΔH and ΔE are the same:

$$CH_4(g) + 2\,O_2(g) \longrightarrow CO_2(g) + 2\,H_2O(g) \qquad \Delta E = \Delta H = -802\,\text{kJ}$$

Although the amount of work is small compared to heat in most chemical reactions such as the combustion of propane, a significant amount of work can be obtained by engineering systems that convert heat into work. In the example of a car's engine, most of the work done on the pistons comes from the expansion of the product gases as a result of their temperature increase from the heat transfer of the reaction.

●─ WORKED EXAMPLE 9.2

Calculating Internal Energy Change (ΔE) for a Reaction

The reaction of nitrogen with hydrogen to make ammonia has $\Delta H = -92.2$ kJ. What is the value of ΔE in kilojoules if the reaction is carried out at a constant pressure of 40.0 atm and the volume change is -1.12 L?

$$N_2(g) + 3\,H_2(g) \longrightarrow 2\,NH_3(g) \qquad \Delta H = -92.2\,\text{kJ}$$

IDENTIFY

Known	Unknown
Change in enthalpy ($\Delta H = -92.2$ kJ)	Change in internal energy (ΔE)
Pressure ($P = 40.0$ atm)	
Volume Change ($\Delta V = -1.12$ L)	

STRATEGY

We are given an enthalpy change ΔH, a volume change ΔV, and a pressure P and asked to find an energy change ΔE. Rearrange the equation $\Delta H = \Delta E + P\Delta V$ to the form $\Delta E = \Delta H - P\Delta V$ and substitute the appropriate values for ΔH, P, and ΔV.

SOLUTION

$$\Delta E = \Delta H - P\Delta V$$
$$\text{where} \quad \Delta H = -92.2\,\text{kJ}$$
$$P\Delta V = (40.0\,\text{atm})(-1.12\,\text{L}) = -44.8\,\text{L}\cdot\text{atm}$$
$$= (-44.8\,\text{L}\cdot\text{atm})\left(101\,\frac{\text{J}}{\text{L}\cdot\text{atm}}\right) = -4520\,\text{J} = -4.52$$
$$\Delta E = (-92.2\,\text{kJ}) - (-4.52\,\text{kJ}) = -87.7\,\text{kJ}$$

CHECK

The sign of ΔE is similar in size and magnitude ΔH, which is to be expected because energy transfer as work is usually small compared to heat.

▶ **PRACTICE 9.3** The reaction between hydrogen and oxygen to yield water vapor has $\Delta H = -484$ kJ. How much PV work is done, and what is the value of ΔE in kilojoules for the reaction of 2.00 mol of H_2 with 1.00 mol of O_2 at atmospheric pressure if the volume change is -24.4 L?

$$2\,H_2(g) + O_2(g) \longrightarrow 2\,H_2O(g) \quad \Delta H° = -484\text{ kJ}$$

▶ **Conceptual APPLY 9.4** The following reaction has $\Delta E = -186$ kJ/mol.

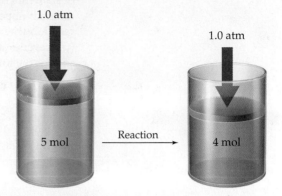

(a) Is the sign of $P\Delta V$ positive or negative? Explain.
(b) What is the sign and approximate magnitude of ΔH? Explain.

9.5 ▶ THERMOCHEMICAL EQUATIONS AND THE THERMODYNAMIC STANDARD STATE

A **thermochemical equation** gives a balanced chemical equation along with the value of the enthalpy change (ΔH), the amount of heat released or absorbed when reactants are converted to products. In the combustion of propane the thermochemical equation is:

$$C_3H_8(g) + 5\,O_2(g) \longrightarrow 3\,CO_2(g) + 4\,H_2O(g) \quad \Delta H = -2044\text{ kJ}$$

To ensure that all measurements are reported in the same way so that different reactions can be compared, a set of conditions called the **thermodynamic standard state** has been defined.

> **Thermodynamic standard state** Most stable form of a substance at 1 atm pressure* and at a specified temperature, usually 25 °C; 1 M concentration for all substances in solution.

Measurements made under these standard conditions are indicated by addition of the superscript ° to the symbol of the quantity reported. Thus, an enthalpy change measured under standard conditions is called a **standard enthalpy of reaction** and is indicated by the symbol $\Delta H°$. The reaction of propane with oxygen in the thermodynamic standard state is written as:

$$C_3H_8(g) + 5\,O_2(g) \longrightarrow 3\,CO_2(g) + 4\,H_2O(g) \quad \Delta H° = -2044\text{ kJ (25 °C, 1 atm)}$$

A thermochemical equation specifies the amount of each substance, and therefore the equation for combustion of propane above means that the reaction of 1 mol of propane gas with 5 mol of oxygen gas to give 3 mol of CO_2 gas and 4 mol of water vapor releases 2044 kJ. The amount of heat released in a specific reaction, however, depends on the amounts of reactants. Thus, reaction of 2.000 mol of propane with 10.00 mol of O_2 releases 2.000×2044 kJ $= 4088$ kJ.

$$2\,C_3H_8(g) + 10\,O_2(g) \longrightarrow 6\,CO_2(g) + 8\,H_2O(g) \quad \Delta H° = -4088\text{ kJ}$$

---

*The standard pressure, listed here and in most other books as 1 atmosphere (atm), has been redefined to be 1 bar, which is equal to 0.986 923 atm. The difference is small, however.

It should also be emphasized that $\Delta H°$ values refer to the reaction going in the direction written. For the *reverse reaction, the sign of* $\Delta H°$ *must be changed*. Because of the reversibility of state functions (Section 9.2), the enthalpy change for a reverse reaction is equal in magnitude but opposite in sign to that for the corresponding forward reaction. Now consider the reaction where gaseous carbon dioxide and water vapor react to form propane and oxygen (the reverse of the combustion of propane). Heat must be absorbed for the reaction to occur and for the formation of one mole of propane the value of $\Delta H° = +2044$ kJ:

$$3\,CO_2(g) + 4\,H_2O(g) \longrightarrow C_3H_8(g) + 5\,O_2(g) \quad \Delta H° = +2044 \text{ kJ}$$
$$C_3H_8(g) + 5\,O_2(g) \longrightarrow 3\,CO_2(g) + 4\,H_2O(g) \quad \Delta H° = -2044 \text{ kJ}$$

Note that the physical states of reactants and products must be specified as solid (s), liquid (l), gaseous (g), or aqueous (aq) when enthalpy changes are reported. The enthalpy change for the reaction of propane with oxygen is $\Delta H° = -2044$ kJ if water is produced as a gas but $\Delta H° = -2220$ kJ if water is produced as a liquid.

$$C_3H_8(g) + 5\,O_2(g) \longrightarrow 3\,CO_2(g) + 4\,H_2O(g) \quad \Delta H° = -2044 \text{ kJ}$$
$$C_3H_8(g) + 5\,O_2(g) \longrightarrow 3\,CO_2(g) + 4\,H_2O(l) \quad \Delta H° = -2220 \text{ kJ}$$

The difference of 176 kJ between the values of $\Delta H°$ for the two reactions arises because the conversion of liquid water to gaseous water requires energy. If liquid water is produced $\Delta H°$ is larger (more negative), but if gaseous water is produced, $\Delta H°$ is smaller (less negative) because 44.0 kJ/mol is needed for the vaporization.

$$H_2O(l) \longrightarrow H_2O(g) \quad \Delta H° = 44.0 \text{ kJ}$$
$$\text{or} \qquad 4\,H_2O(l) \longrightarrow 4\,H_2O(g) \quad \Delta H° = 176 \text{ kJ}$$

Worked Example 9.3 illustrates how to calculate the amount of heat transfer for a given reaction given the amounts of reactants and direction of reaction.

WORKED EXAMPLE 9.3

Calculating the Amount of Heat Released in a Reaction

How much heat in kilojoules is evolved when 5.00 g of propane reacts with excess O_2?

$$C_3H_8(g) + 5\,O_2(g) \longrightarrow 3\,CO_2(g) + 4\,H_2O(g) \quad \Delta H° = -2044 \text{ kJ}$$

IDENTIFY

Known	Unknown
Amount of propane (5.00 g)	Heat transfer (kJ)
Balanced thermochemical equation	

STRATEGY

According to the balanced equation, 2044 kJ of heat is evolved from the reaction of 1 mol of propane. To find out how much heat is evolved from the reaction of 5.00 g of propane, we have to convert from grams to moles and then use the thermochemical equation to relate quantity to heat.

SOLUTION

The molar mass of propane (C_3H_8) is 44.09 g/mol, so 5.00 g of C_3H_8 equals 0.113 mol:

$$5.00 \text{ g } C_3H_8 \times \frac{1 \text{ mol } C_3H_8}{44.09 \text{ g } C_3H_8} = 0.113 \text{ mol } C_3H_8$$

Because 1 mol of C_3H_8 releases 2044 kJ of heat, 0.113 mol of C_3H_8 releases 231 kJ of heat:

$$0.113 \text{ mol } C_3H_8 \times \frac{2044 \text{ kJ}}{1 \text{ mol } C_3H_8} = 231 \text{ kJ}$$

CHECK

Since the molar mass of C_3H_8 is about 44 g/mol, 5 g of C_3H_8 is roughly 0.1 mol. Therefore, the heat evolved is roughly 1/10 of the amount released for 1 mol of C_3H_8, and the estimate yields an answer of 204 kJ, which is reasonably close to 231 kJ.

▶ **PRACTICE 9.5**   Use the following thermochemical equation to calculate how much heat in kilojoules is evolved or absorbed in each of the following processes.

$$2\,H_2(g) + O_2(g) \longrightarrow 2\,H_2O(l) \qquad \Delta H° = -571.6\text{ kJ}$$

(a) 10.00 g of gaseous hydrogen burns in excess oxygen.
(b) 5.500 mol of liquid water are converted to hydrogen and oxygen gas.

▶ **APPLY 9.6**   Approximately, 1.8×10^6 kJ of energy is required to heat an average home in the Midwest during the month of January. How much propane (kg) must be burned to provide 1.8×10^6 kJ of energy?

$$C_3H_8(g) + 5\,O_2(g) \longrightarrow 3\,CO_2(g) + 4\,H_2O(g) \quad \Delta H° = -2044\text{ kJ}$$

9.6 ▶ ENTHALPIES OF CHEMICAL AND PHYSICAL CHANGES

Almost every change in a system involves either a gain or a loss of enthalpy. The change can be either physical, such as the melting of a solid to a liquid, or chemical, such as the burning of propane. Let's look at examples of both kinds.

Enthalpies of Chemical Change

We saw in Section 9.4 that an enthalpy change is often called a *heat of reaction* because it is a measure of the heat flow into or out of a system at constant pressure. If the products of a reaction have more enthalpy than the reactants, then heat has flowed into the system from the surroundings and ΔH has a positive sign. Reactions in which the system gains heat are said to be **endothermic** (*endo* means "within," so heat flows in). The reaction of 1 mol of barium hydroxide octahydrate* with ammonium chloride, for example, absorbs 80.3 kJ from the surroundings ($\Delta H° = +80.3$ kJ). The surroundings, having lost heat, become cold—so cold, in fact, that the temperature drops below freezing (**FIGURE 9.7**).

$$Ba(OH)_2 \cdot 8\,H_2O(s) + 2\,NH_4Cl(s) \longrightarrow$$
$$BaCl_2(aq) + 2\,NH_3(aq) + 10\,H_2O(l) \quad \Delta H° = +80.3\text{ kJ}$$

If the products of a reaction have less enthalpy than the reactants, then heat has flowed out of the system to the surroundings and ΔH has a negative sign. Reactions that lose heat to the surroundings are said to be **exothermic** (*exo* means "out," so heat flows out). The so-called thermite reaction of aluminum with iron(III) oxide, for instance, releases so much heat ($\Delta H° = -852$ kJ), and the surroundings get so hot, that the reaction is used in construction work to weld iron.

$$2\,Al(s) + Fe_2O_3(s) \longrightarrow 2\,Fe(s) + Al_2O_3(s) \quad \Delta H° = -852\text{ kJ}$$

As noted previously, the value of $\Delta H°$ given for an equation assumes that the equation is balanced to represent the numbers of moles of reactants and products, that all substances are in their standard states, and that the physical state of each substance is as specified.

▲ A balloon filled with hydrogen burns in air to release energy to the surroundings.

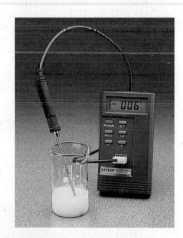

▲ **FIGURE 9.7**

The endothermic reaction of barium hydroxide octahydrate with ammonium chloride. The reaction draws so much heat from the surroundings that the temperature falls below 0 °C.

Figure It Out

Identify the system and the surroundings in the photo of the endothermic reaction.

Answer: The system is the reaction.
$$Ba(OH)_2 \cdot 8\,H_2O(s) + 2\,NH_4Cl(s) \longrightarrow$$
$$BaCl_2(aq) + 2\,NH_3(aq) + 10\,H_2O(l)$$
The surroundings are the solution, beaker, tabletop, temperature probe, and air in the room.

---

*Barium hydroxide octahydrate, $Ba(OH)_2 \cdot 8\,H_2O$, is a crystalline compound that contains eight water molecules clustered around the barium ion.

Enthalpies of Physical Change

What would happen if you started with a block of ice at a low temperature, say −10 °C, and slowly increased its enthalpy by adding heat? The initial input of heat would cause the temperature of the ice to rise until it reached 0 °C. Additional heat would then cause the ice to melt without raising its temperature as the added energy is expended in overcoming the forces that hold H_2O molecules together in the ice crystal. The amount of heat necessary to melt a substance without changing its temperature is called the *enthalpy of fusion*, or *heat of fusion* (ΔH_{fusion}). For H_2O, ΔH_{fusion} = 6.01 kJ/mol at 0 °C and the process is endothermic because heat is absorbed. The thermochemical equation representing the heat of fusion of water is:

$$H_2O(s) \longrightarrow H_2O(l) \quad \Delta H_{fusion} = +6.01 \text{ kJ/mol}$$

Once the ice has melted, further input of heat raises the temperature of the liquid water until it reaches 100 °C, and adding still more heat then causes the water to boil. Once again, energy is necessary to overcome the forces holding molecules together in the liquid, so the temperature does not rise again until all the liquid has been converted into vapor. The amount of heat required to vaporize a substance without changing its temperature is called the *enthalpy of vaporization*, or *heat of vaporization* (ΔH_{vap}). For H_2O, ΔH_{vap} = 40.7 kJ/mol at 100 °C, and the process is endothermic. The thermochemical equation representing the heat of vaporization of water is:

$$H_2O(l) \longrightarrow H_2O(g) \quad \Delta H_{vap} = +40.7 \text{ kJ/mol}$$

Another kind of physical change in addition to melting and boiling is **sublimation**, the direct conversion of a solid to a vapor without going through a liquid state. Solid CO_2 (dry ice), for example, changes directly from solid to vapor at atmospheric pressure without first melting to a liquid. Since enthalpy is a state function, the enthalpy change on going from solid to vapor must be constant regardless of the path taken. Thus, at a given temperature, a substance's *enthalpy of sublimation*, or *heat of sublimation* (ΔH_{subl}), equals the sum of the heat of fusion and the heat of vaporization (**FIGURE 9.8**).

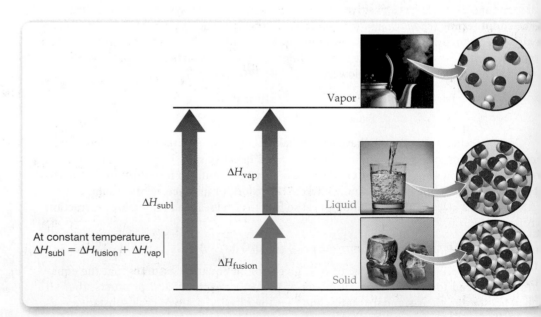

▲ **FIGURE 9.8**

Enthalpy as a state function. Because enthalpy is a state function, the value of the enthalpy change from solid to vapor does not depend on the path taken between the two states.

WORKED EXAMPLE 9.4

Classifying Endo- and Exothermic Processes

Indicate the direction of heat transfer between the system and the surroundings, classify the following processes as endo- or exothermic, and give the sign of ΔH°.
(a) Water freezing on the surface of a lake in the winter.
(b) Sweat evaporating from your skin.
(c) Dissolving the salt magnesium sulfate in water to produce a hand warmer.

STRATEGY

Write an equation for each process and evaluate whether heat is absorbed or released based on observed temperature changes.

SOLUTION

(a) It is difficult to observe the temperature change in the surroundings as water freezes. However, it is intuitive that heat must be absorbed from the surroundings to melt solid ice into liquid water. In the reverse process of freezing, $H_2O(l) \longrightarrow H_2O(s)$, the system releases heat to the surroundings. Therefore, the process is exothermic and ΔH°. is negative.

(b) When sweat evaporates, liquid water is turned into gaseous water, $H_2O(l) \longrightarrow H_2O(g)$, and the system absorbs heat from the surroundings. Sweating cools you down because the phase change absorbs heat from your body (part of the surroundings). Therefore, the sign ΔH°. is positive and the process is endothermic.

(c) In a hand warmer, the temperature of the solution increases because the system, $MgSO_4(s) \longrightarrow Mg^{2+}(aq) + SO_4^{2-}(aq)$, transfers heat to the surroundings. Therefore, the process is exothermic and ΔH°. is negative.

▶ **PRACTICE 9.7** Indicate the direction of heat transfer between the system and the surroundings, classify the following processes as endo- or exothermic, and give the sign of ΔH°.

(a) Steam from the shower condensing on the bathroom mirror.
(b) Burning charcoal on the grill.
(c) Dry ice, CO_2 (s), subliming.

▶ **APPLY 9.8** Instant hot packs and cold packs contain a solid salt in a capsule that is broken and dissolved in water to produce a temperature change. Which of the following salts would result in greatest decrease in temperature when dissolved in 100.0 mL of water in a cold pack? Use the following thermochemical data to answer the question.

$LiF(s) \longrightarrow Li^+(aq) + F^-(aq)$		$\Delta H^\circ = +5.5\,kJ$
$LiCl(s) \longrightarrow Li^+(aq) + Cl^-(aq)$		$\Delta H^\circ = -37.1\,kJ$
$BaSO_4(s) \longrightarrow Ba^{2+}(aq) + SO_4^{2-}(aq)$		$\Delta H^\circ = +26.3\,kJ$
$CaCl_2(s) \longrightarrow Ca^{2+}(aq) + 2Cl^-(aq)$		$\Delta H^\circ = -81.8\,kJ$

(a) 10.0 g of LiF
(b) 10.0 g of $BaSO_4$
(c) 10.0 g of LiCl
(d) 10.0 g of $CaCl_2$

▲ Instant cold packs are used to treat athletic injuries.

9.7 ▶ CALORIMETRY AND HEAT CAPACITY

The amount of heat transferred during a reaction can be measured with a device called a *calorimeter*, shown schematically in **FIGURE 9.9**. At its simplest, a calorimeter is just an insulated vessel with a stirrer, a thermometer, and a loose-fitting lid to keep the contents at atmospheric pressure. The reaction is carried out inside the vessel, and the heat evolved or absorbed is calculated from the temperature change. Because the pressure inside the calorimeter is constant (atmospheric pressure), the temperature measurement makes it possible to calculate the enthalpy change ΔH during a reaction. This type of calorimeter is used to measure the enthalpy change for the reaction of a salt dissolving in water such as the reaction that occurs in hot packs and cold packs.

A somewhat more complicated device called a *bomb calorimeter* is used to measure the heat released during a combustion reaction, or burning of a flammable substance. (More generally, a *combustion* reaction is any reaction that produces a flame.) The sample is placed in

▶ **FIGURE 9.9**

A calorimeter for measuring the heat flow in a reaction at constant pressure (ΔH).

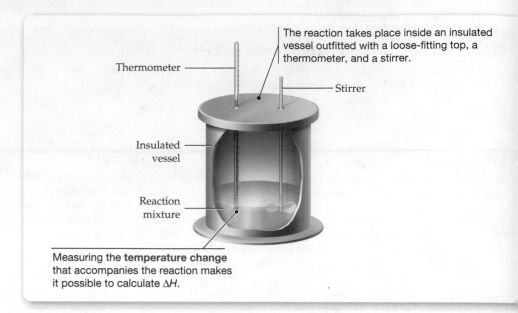

Thermometer

The reaction takes place inside an insulated vessel outfitted with a loose-fitting top, a thermometer, and a stirrer.

Stirrer

Insulated vessel

Reaction mixture

Measuring the **temperature change** that accompanies the reaction makes it possible to calculate ΔH.

a small cup and sealed in an oxygen atmosphere inside a steel bomb that is itself placed in a insulated, water-filled container (**FIGURE 9.10**). The reactants are ignited electrically, and th evolved heat is calculated from the temperature change of the surrounding water. Since th reaction takes place at constant volume rather than constant pressure, the measurement pro vides a value for ΔE rather than ΔH. The caloric content of food, such as a potato chip, ca be measured by burning it in a bomb calorimeter.

How can the temperature change inside a calorimeter be used to calculate ΔH (or ΔE for a reaction? When a calorimeter and its contents absorb a given amount of heat, the tem perature rise that results depends on the calorimeter's *heat capacity*. **Heat capacity (C) i** the amount of heat required to raise the temperature of an object or substance by a give amount, a relationship that can be expressed by the equation

$$C = \frac{q}{\Delta T}$$

where q is the quantity of heat transferred and ΔT is the temperature change that result ($\Delta T = T_{final} - T_{initial}$).

The greater the heat capacity, the greater the amount of heat needed to produce a give temperature change. A bathtub full of water, for instance, has a greater heat capacity than

▶ **FIGURE 9.10**

A bomb calorimeter for measuring the heat evolved at constant volume (ΔE) in a combustion reaction.

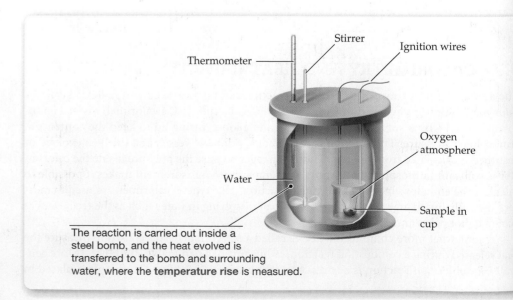

Stirrer

Ignition wires

Thermometer

Oxygen atmosphere

Water

Sample in cup

The reaction is carried out inside a steel bomb, and the heat evolved is transferred to the bomb and surrounding water, where the **temperature rise** is measured.

TABLE 9.1   Specific Heats and Molar Heat Capacities for Some Common Substances at 25 °C

Substance	Specific Heat (c) $J/(g \cdot °C)$	Molar Heat Capacity (C_m) $J/(mol \cdot °C)$
Air (dry)	1.01	29.1
Aluminum	0.897	24.2
Copper	0.385	24.4
Gold	0.129	25.4
Iron	0.449	25.1
Mercury	0.140	28.0
NaCl	0.859	50.2
Water(s)[a]	2.03	36.6
Water(l)	4.179	75.3

[a]At −11°C

coffee cup full, and it therefore takes far more heat to warm the tubful than the cupful. The exact amount of heat absorbed is equal to the heat capacity times the temperature rise:

$$q = C \times \Delta T$$

Heat capacity is an **extensive property**, so its value depends on both the size of an object and its composition. To compare different substances, it's useful to define a quantity called the *specific heat capacity*, or simply **specific heat** (c), the amount of heat necessary to raise the temperature of 1 g of a substance by 1 °C. The amount of heat necessary to raise the temperature of a given object is the specific heat times the mass of the object times the rise in temperature:

$$q = \text{Specific heat } (c) \times \text{Mass of substance } (m) \times \Delta T$$

Worked Example 9.5 shows how specific heats (c) and heat capacities (C) are determined.

Closely related to specific heat is the **molar heat capacity** (C_m), defined as the amount of heat necessary to raise the temperature of 1 mol of a substance by 1 °C. The amount of heat necessary to raise the temperature of a given number of moles of a substance is thus

$$q = C_m \times \text{Moles of substance} \times \Delta T$$

Values of specific heats and molar heat capacities for some common substances are given in **TABLE 9.1**. The values are temperature dependent, so the temperatures at which the measurements are taken must be specified.

As indicated in Table 9.1, the specific heat of liquid water is considerably higher than that of most other substances, so a large transfer of heat is necessary to either cool or warm a given amount of water. One consequence is that large lakes or other bodies of water tend to moderate the air temperature in surrounding areas. Another consequence is that the human body, which is about 60% water, is able to maintain a relatively steady internal temperature under changing outside conditions.

> **REMEMBER...**
>
> **Extensive properties**, such as length and volume, have values that depend on the sample size. Intensive properties, such as temperature and melting point, have values that do not depend on the amount of the sample. (Section 2.3)

▲ Lake Chelan in the North Cascades of Washington State is the third deepest freshwater lake in the United States at 1486 ft. Such large masses of water moderate the temperature of the surroundings because of their high heat capacity.

WORKED EXAMPLE 9.5

Calculating Specific Heat (C) and Molar Heat Capacity (C_M)

What is the specific heat of silicon in $J/(g \cdot °C)$ if it takes 192 J to raise the temperature of a 45.0 g block by 6.0 °C? What is the molar heat capacity of silicon in $J/(mol \cdot °C)$?

IDENTIFY

Known	Unknown
Heat transfer ($q = 192$ J)	Specific heat (c)
Mass of Si ($m = 45.0$ g)	Molar heat capacity (C_m)
Change in temperature ($\Delta T = 6.0$ °C)	

continued on next page

STRATEGY

To find a specific heat of a substance, calculate the amount of energy necessary to raise the temperature of 1 g of the substance by 1 °C. To calculate molar heat capacity, use the molar mass of silicon to convert from grams to moles in specific heat capacity.

SOLUTION

$$\text{Specific heat }(c)\text{ of Si} = \frac{192 \text{ J}}{(45.0 \text{ g})(6.0 \text{ °C})} = 0.71 \text{ J/(g} \cdot \text{°C)}$$

$$\text{Molar heat capacity}(C_m)\text{: } \frac{0.71 \text{ J}}{(\text{g} \cdot \text{°C})} \times \frac{28.1 \text{ g Si}}{1 \text{ mol Si}} = \frac{20.0 \text{ J}}{(\text{mol} \cdot \text{°C})}$$

CHECK

Table 9.1 lists specific and molar heat capacities for several substances. The values calculated are within the range given in the table and are therefore reasonable numbers.

▶ **PRACTICE 9.9** What is the specific heat of lead in J/(g · °C) if it takes 97.2 J to raise the temperature of a 75.0 g block by 10.0 °C? What is the molar heat capacity of lead in J/(mol · °C)?

▶ **APPLY 9.10** Calculate the heat capacity (C) of a bomb calorimeter if 1650 J raised the temperature of the entire calorimeter by 2.00 °C.

WORKED EXAMPLE 9.6

Calculating ΔH in a Calorimetry Experiment

Aqueous silver ion reacts with aqueous chloride ion to yield a white precipitate of solid silver chloride:

$$\text{Ag}^+(aq) + \text{Cl}^-(aq) \longrightarrow \text{AgCl}(s)$$

When 10.0 mL of 1.00 M $AgNO_3$ solution is added to 10.0 mL of 1.00 M NaCl solution at 25.0 °C in a calorimeter, a white precipitate of AgCl forms and the temperature of the aqueous mixture increases to 32.6 °C. Assuming that the specific heat of the aqueous mixture is 4.18 J/(g · °C), that the density of the mixture is 1.00 g/mL, and that the calorimeter itself absorbs a negligible amount of heat, calculate ΔH in kilojoules/mol AgCl for the reaction.

IDENTIFY

Known	Unknown
Volume of $AgNO_3$ solution (10.0 mL)	Enthalpy change for reaction (ΔH)
Concentration of $AgNO_3$ solution (1.00 M)	
Volume of NaCl solution (10.0 mL)	
Concentration of NaCl solution (1.00 M)	
Initial temperature (25.0 °C)	
Final temperature (32.6 °C)	
Specific heat of mixture (c = 4.18 J/(g·°C))	
Density of mixture (d = 1.00 g/mL)	

STRATEGY

Because the temperature rises during the reaction, heat must be released and ΔH must be negative. The amount of heat evolved during the reaction is equal to the amount of heat absorbed by the mixture:

Heat evolved = Specific heat (c) × Mass of mixture (m) × Temperature change (ΔT)

Dividing the heat evolved by the number of moles gives the enthalpy change ΔH.

SOLUTION

$$\text{Specific heat }(c) = 4.18 \text{ J/(g} \cdot \text{°C)}$$

$$\text{Mass }(m) = (20.0 \text{ mL})\left(1.00\frac{\text{g}}{\text{mL}}\right) = 20.0 \text{ g}$$

$$\text{Temperature change }(\Delta T) = 32.6 \text{ °C} - 25.0 \text{ °C} = 7.6 \text{ °C}$$

$$\text{Heat evolved} = \left(4.18\frac{\text{J}}{\text{g} \cdot \text{°C}}\right)(20.0 \text{ g})(7.6 \text{ °C}) = 6.4 \times 10^2 \text{J}$$

According to the balanced equation, the number of moles of AgCl produced equals the number of moles of Ag^+ (or Cl^-) reacted:

$$\text{Moles of } Ag^+ = (10.0 \text{ mL})\left(\frac{1.00 \text{ mol } Ag^+}{1000 \text{ mL}}\right) = 1.00 \times 10^{-2} \text{ mol } Ag^+$$

$$\text{Moles of AgCl} = 1.00 \times 10^{-2} \text{ mol AgCl}$$

$$\text{Heat evolved per mole of AgCl} = \frac{6.4 \times 10^2 \text{ J}}{1.00 \times 10^{-2} \text{ mol AgCl}} = 64 \text{ kJ/mol AgCl}$$

Therefore, $\Delta H = -64$ kJ (negative because heat is released by the reaction to warm the solution)

CHECK

Combustion reactions release large amounts of heat and typically have ΔH values ranging from -300 to -2000 kJ/mol. The heat released when a salt dissolves has significantly less energy than combustion so $\Delta H = -64$ kJ/mol is a reasonable answer.

▶ **PRACTICE 9.11** When 25.0 mL of 1.0 M H_2SO_4 is added to 50.0 mL of 1.0 M NaOH at 25.0 °C in a calorimeter, the temperature of the aqueous solution increases to 33.9 °C. Assuming that the specific heat of the solution is 4.18 J/(g · °C), that its density is 1.00 g/mL, and that the calorimeter itself absorbs a negligible amount of heat, calculate ΔH in kilojoules/mol H_2SO_4 for the reaction.

$$H_2SO_4(aq) + 2 NaOH(aq) \longrightarrow 2 H_2O(l) + Na_2SO_4(aq)$$

▶ **APPLY 9.12** When 2.00 g of glucose, $C_6H_{12}O_6$, is burned in a bomb calorimeter, the temperature of the water and calorimeter rises by 18.6 °C. Assuming that the bath contains 250.0 g of water and that the heat capacity for the calorimeter is 623 J/°C, calculate combustion energies (ΔE) for glucose in both kJ/g and Cal/g.

▲ The reaction of aqueous $AgNO_3$ with aqueous NaCl to yield solid AgCl is an exothermic process.

9.8 ▶ HESS'S LAW

Now that we've discussed in general terms the energy changes that occur during chemical reactions, let's look at a specific example in detail. In particular, let's look at the *Haber process*, the industrial method by which more than 120 million metric tons of ammonia is produced each year worldwide, primarily for use as fertilizer (1 metric ton = 1000 kg). The reaction of hydrogen with nitrogen to make ammonia is exothermic, with $\Delta H° = -92.2$ kJ.

$$3 H_2(g) + N_2(g) \longrightarrow 2 NH_3(g) \quad \Delta H° = -92.2 \text{ kJ}$$

If we dig into the details of the reaction, we find that it's not as simple as it looks. In fact, the overall reaction occurs in a series of steps, with hydrazine (N_2H_4) produced at an intermediate stage:

$$2 H_2(g) + N_2(g) \longrightarrow \underset{\text{Hydrazine}}{N_2H_4(g)} \xrightarrow{H_2} \underset{\text{Ammonia}}{2 NH_3(g)}$$

The enthalpy change for the conversion of hydrazine to ammonia can be measured as $\Delta H° = -187.6$ kJ, but if we wanted to measure $\Delta H°$ for the formation of hydrazine from hydrogen and nitrogen, we would have difficulty because the reaction doesn't go cleanly. Some of the hydrazine is converted into ammonia and some of the starting nitrogen remains.

Fortunately, there's a way around the difficulty—a way that makes it possible to measure an energy change indirectly when a direct measurement can't be made. The trick is to realize that, because enthalpy is a state function, ΔH is the same no matter what path is taken between two states. Thus, the sum of the enthalpy changes for the individual steps in a sequence must equal the enthalpy change for the overall reaction, a statement known as **Hess's law**:

Hess's law The overall enthalpy change for a reaction is equal to the sum of the enthalpy changes for the individual steps in the reaction.

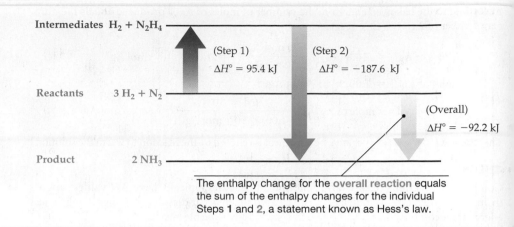

▶ **FIGURE 9.11**

Enthalpy changes for steps in the synthesis of ammonia from nitrogen and hydrogen. If $\Delta H°$ values for step 2 and for the overall reaction are known, then ΔH for step 1 can be calculated.

Figure It Out

In a different reaction that occurs by two steps, what is $\Delta H°$ for step 1 if $\Delta H°$ for step 2 is +25.0 kJ and $\Delta H°$ for the overall reaction is −25.0 kJ?

Answer: −50.0 kJ

The enthalpy change for the **overall reaction** equals the sum of the enthalpy changes for the individual Steps **1** and **2**, a statement known as Hess's law.

Reactants and products in the individual steps can be added and subtracted like algebraic quantities in determining the overall equation. In the synthesis of ammonia, for example, the sum of steps 1 and 2 is equal to the overall reaction. Thus, the sum of the enthalpy changes for steps 1 and 2 is equal to the enthalpy change for the overall reaction. With this knowledge, we can calculate the enthalpy change for step 1. **FIGURE 9.11** shows the situation pictorially, and Worked Example 9.7 gives an example of Hess's law calculations.

Step 1. $\quad 2\,H_2(g) + N_2(g) \longrightarrow N_2H_4(g) \qquad \Delta H°_1 = ?$

Step 2. $\quad N_2H_4(g) + H_2(g) \longrightarrow 2\,NH_3(g) \qquad \Delta H°_2 = -187.6\ \text{kJ}$

Overall reaction $\quad 3\,H_2(g) + N_2(g) \longrightarrow 2\,NH_3(g) \qquad \Delta H°_{\text{reaction}} = -92.2\ \text{kJ}$

Since $\quad \Delta H°_1 + \Delta H°_2 = \Delta H°_{\text{reaction}}$

then $\quad \Delta H°_1 = \Delta H°_{\text{reaction}} - \Delta H°_2$

$$= (-92.2\ \text{kJ}) - (-187.6\ \text{kJ}) = +95.4\ \text{kJ}$$

WORKED EXAMPLE 9.7

Using Hess's Law to Calculate $\Delta H°$

Methane, the main constituent of natural gas, burns in oxygen to yield carbon dioxide and water:

$$CH_4(g) + 2\,O_2(g) \longrightarrow CO_2(g) + 2\,H_2O(l)$$

Use the following information to calculate $\Delta H°$ in kilojoules for the combustion of methane:

$$CH_4(g) + O_2(g) \longrightarrow CH_2O(g) + H_2O(g) \quad \Delta H° = -275.6\ \text{kJ}$$
$$CH_2O(g) + O_2(g) \longrightarrow CO_2(g) + H_2O(g) \quad \Delta H° = -526.7\ \text{kJ}$$
$$H_2O(l) \longrightarrow H_2O(g) \quad \Delta H° = 44.0\ \text{kJ}$$

STRATEGY

It often takes some trial and error, but the idea is to combine the individual reactions so that their sum is the desired reaction. The important points are that:

- All the reactants [$CH_4(g)$ and $O_2(g)$] must appear on the left.
- All the products [$CO_2(g)$ and $H_2O(l)$] must appear on the right.

- All intermediate products [$CO_2(g)$ and $H_2O(g)$] must occur on *both* the left and the right so that they cancel.
- A reaction written in the reverse of the direction given [$H_2O(g) \longrightarrow H_2O(l)$] must have the sign of its $\Delta H°$ reversed (Section 9.5).
- If a reaction is multiplied by a coefficient [$H_2O(g) \longrightarrow H_2O(l)$ is multiplied by 2], then $\Delta H°$ for the reaction must be multiplied by that same coefficient (Section 9.5).

SOLUTION

$$CH_4(g) + O_2(g) \longrightarrow \cancel{CH_2O(g)} + \cancel{H_2O(g)} \qquad \Delta H° = -275.6 \text{ kJ}$$
$$\cancel{CH_2O(g)} + O_2(g) \longrightarrow CO_2(g) + \cancel{H_2O(g)} \qquad \Delta H° = -526.7 \text{ kJ}$$
$$2\,[\cancel{H_2O(g)} \longrightarrow H_2O(l)] \quad 2\,[\Delta H° = -44.0 \text{ kJ}] = -88.0 \text{ kJ}$$

$$\overline{CH_4(g) + 2\,O_2(g) \longrightarrow CO_2(g) + 2\,H_2O(l) \qquad \Delta H° = -890.3 \text{ kJ}}$$

▶ **PRACTICE 9.13**   *Water gas* is the name for the mixture of CO and H_2 prepared by reaction of steam with carbon at 1000 °C:

$$C(s) + H_2O(g) \longrightarrow CO(g) + H_2(g)$$
"Water gas"

The hydrogen is then purified and used as a starting material for preparing ammonia. Use the following information to calculate $\Delta H°$ in kilojoules for the water–gas reaction:

$$C(s) + O_2(g) \longrightarrow CO_2(g) \qquad \Delta H° = -393.5 \text{ kJ}$$
$$2\,CO(g) + O_2(g) \longrightarrow 2\,CO_2(g) \qquad \Delta H° = -566.0 \text{ kJ}$$
$$2\,H_2(g) + O_2(g) \longrightarrow 2\,H_2O(g) \qquad \Delta H° = -483.6 \text{ kJ}$$

▶ **Conceptual APPLY 9.14**   The reaction of A with B to give D proceeds in two steps and can be represented by the following Hess's law diagram.

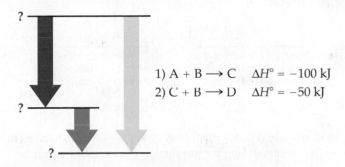

1) A + B ⟶ C $\Delta H° = -100$ kJ
2) C + B ⟶ D $\Delta H° = -50$ kJ

(a) What is the equation and $\Delta H°$ for the net reaction?
(b) Which arrow on the diagram corresponds to which step, and which arrow corresponds to the net reaction?
(c) The diagram shows three energy levels. Which substances are represented by each energy level?

9.9 ▶ STANDARD HEATS OF FORMATION

Where do the $\Delta H°$ values we've been using in previous sections come from? There are so many chemical reactions—hundreds of millions are known—that it's impossible to measure $\Delta H°$ for all of them. A more efficient method is to measure **standard heats of formation** ($\Delta H°_f$) for different substances and use those values to calculate $\Delta H°$ for any reaction that involves those substances.

> **Standard heat of formation**   The enthalpy change ($\Delta H°_f$) for the formation of 1 mol of a substance in its standard state from its constituent elements in their standard states.

Note several points about this definition. First, the "reaction" to form a substance from its constituent elements can be (and often is) hypothetical. We can't combine carbon and hydrogen in the laboratory to make methane, for instance, yet the heat of formation for methane is $\Delta H°_f = -74.8$ kJ/mol, which corresponds to the standard enthalpy change for the hypothetical reaction

$$C(s) + 2 H_2(g) \longrightarrow CH_4(g) \qquad \Delta H° = -74.8 \text{ kJ}$$

Second, each substance in the reaction must be in its most stable, standard form at 1 atm pressure and the specified temperature (usually 25 °C). Carbon, for instance, is most stable as solid graphite rather than as diamond under these conditions, and hydrogen is most stable as gaseous H_2 molecules rather than as H atoms. **TABLE 9.2** gives standard heats of formation for some common substances, and Appendix B gives a more detailed list.

No elements are listed in Table 9.2 because, by definition, the most stable form of any element, such as $Ar(g)$ or $Au(s)$, in its standard state has $\Delta H°_f = 0$ kJ. The enthalpy change for the formation of an element from itself is zero. Defining $\Delta H°_f$ as zero for all element thus establishes a thermochemical "sea level," or reference point, from which other enthalpy changes are measured.

How can standard heats of formation be used for thermochemical calculations? The standard enthalpy change for any chemical reaction is found by subtracting the sum of the heats of formation of all reactants from the sum of the heats of formation of all products, with each heat of formation multiplied by the coefficient of that substance in the balanced equation.

$$\Delta H°_{reaction} = \Delta H°_f(\text{Products}) - \Delta H°_f(\text{Reactants})$$

To find $\Delta H°$ for the reaction

$$a\,A + b\,B + \cdots \longrightarrow c\,C + d\,D + \cdots$$

| Subtract the sum of the heats of formation for these reactants . . . | . . . from the sum of the heats of formation for these products. |

$$\Delta H°_{reaction} = [c\,\Delta H°_f(\text{C}) + d\,\Delta H°_f(\text{D}) + \cdots] - [a\,\Delta H°_f(\text{A}) + b\,\Delta H°_f(\text{B}) + \cdots]$$

As an example, let's calculate $\Delta H°$ for the fermentation of glucose to make ethyl alcohol (ethanol), the reaction that occurs during the production of alcoholic beverages:

$$C_6H_{12}O_6(s) \longrightarrow 2 C_2H_5OH(l) + 2 CO_2(g) \qquad \Delta H° = ?$$

Using the data in Table 9.2 gives the following answer:

$$\begin{aligned}\Delta H° &= [2\,\Delta H°_f(\text{Ethanol}) + 2\,\Delta H°_f(\text{CO}_2)] - [\Delta H°_f(\text{Glucose})] \\ &= (2 \text{ mol})(-277.7 \text{ kJ/mol}) + (2 \text{ mol})(-393.5 \text{ kJ/mol}) - (1 \text{ mol})(-1273.3 \text{ kJ/mol}) \\ &= -69.1 \text{ kJ}\end{aligned}$$

The fermentation reaction is exothermic by 69.1 kJ.

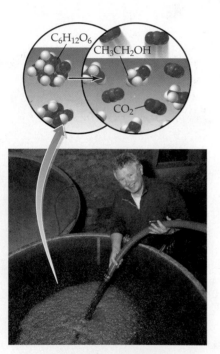

▲ Fermentation of the sugar from grapes yields the ethyl alcohol in wine.

TABLE 9.2	Standard Heats of Formation for Some Common Substances at 25 °C				
Substance	Formula	$\Delta H°_f$ (kJ/mol)	Substance	Formula	$\Delta H°_f$ (kJ/mol)
Acetylene	$C_2H_2(g)$	227.4	Hydrogen chloride	$HCl(g)$	−92.3
Ammonia	$NH_3(g)$	−46.1	Iron(III) oxide	$Fe_2O_3(s)$	−824.2
Carbon dioxide	$CO_2(g)$	−393.5	Magnesium carbonate	$MgCO_3(s)$	−1095.8
Carbon monoxide	$CO(g)$	−110.5	Methane	$CH_4(g)$	−74.8
Ethanol	$C_2H_5OH(l)$	−277.7	Nitric oxide	$NO(g)$	91.3
Ethylene	$C_2H_4(g)$	52.3	Water (g)	$H_2O(g)$	−241.8
Glucose	$C_6H_{12}O_6(s)$	1273.3	Water (l)	$H_2O(l)$	−285.8

Why does this calculation "work"? It works because enthalpy is a state function and the calculation is really just an application of Hess's law. That is, the sum of the individual equations corresponding to the heat of formation for each substance in the reaction equals the enthalpy change for the overall reaction:

(1)	$C_6H_{12}O_6(s) \longrightarrow 6\,\cancel{C}(s) + 6\,\cancel{H_2}(g) + 3\,\cancel{O_2}(g)$	$-\Delta H°_f =$	$+1273.3\text{ kJ}$
(2)	$2\,[2\,\cancel{C}(s) + 3\,\cancel{H_2}(g) + 1/2\,\cancel{O_2}(g) \longrightarrow C_2H_5OH(l)]$	$2\,[\Delta H°_f = -277.7\text{ kJ}] =$	-555.4 kJ
(3)	$2\,[\cancel{C}(s) + \cancel{O_2}(g) \longrightarrow CO_2(g)]$	$2\,[\Delta H°_f = -393.5\text{ kJ}] =$	-787.0 kJ
(Net)	$C_6H_{12}O_6(s) \longrightarrow 2\,C_2H_5OH(l) + 2\,CO_2(g)$	$\Delta H° =$	-69.1 kJ

Note that reaction (1) represents the formation of glucose from its elements written in reverse, so the sign of $\Delta H°_f$ is reversed. Note also that reactions (2) and (3), which represent the formation of ethyl alcohol and carbon dioxide, respectively, are multiplied by 2 to arrive at the balanced equation for the overall reaction.

● WORKED EXAMPLE 9.8

Using Standard Heats of Formation to Calculate ΔH°

Oxyacetylene welding torches burn acetylene gas, $C_2H_2(g)$. Use the information in Table 9.2 to calculate $\Delta H°$ in kilojoules for the combustion reaction of acetylene to yield $CO_2(g)$ and $H_2O(g)$.

STRATEGY

Write the balanced equation and look up the appropriate heats of formation for each reactant and product in Table 9.2. Carry out the calculation using the general formula:

$$\Delta H°_{reaction} = \Delta H°_f\,(Products) - \Delta H°_f\,(Reactants)$$

Multiply each $\Delta H°_f$ by the coefficient given in the balanced equation and remember that $\Delta H°_f(O_2) = 0\text{ kJ/mol}$ because O_2 is in its elemental form.

SOLUTION

The balanced equation is

$$2\,C_2H_2(g) + 5\,O_2(g) \longrightarrow 4\,CO_2(g) + 2\,H_2O(g)$$

The necessary heats of formation are

$$\Delta H°_f\,[C_2H_2(g)] = 227.4\text{ kJ/mol} \qquad \Delta H°_f\,[H_2O(g)] = -241.8\text{ kJ/mol}$$
$$\Delta H°_f\,[CO_2(g)] = -393.5\text{ kJ/mol}$$

The standard enthalpy change for the reaction is

$$\Delta H° = [4\,\Delta H°_f\,(CO_2)] + 2\,\Delta H°_f\,(H_2O)] - [2\,\Delta H°_f\,(C_2H_2)]$$
$$= (4\text{ mol})(-393.5\text{ kJ/mol}) + (2\text{ mol})(-241.8\text{ kJ/mol}) - (2\text{ mol})(227.4\text{ kJ/mol})$$
$$= -2512.4\text{ kJ}$$

▶ **PRACTICE 9.15** Use the information in Table 9.2 to calculate $\Delta H°$ in kilojoules for the reaction of ammonia with O_2 to yield nitric oxide (NO) and $H_2O(g)$, a step in the Ostwald process for the commercial production of nitric acid.

▶ **APPLY 9.16** The thermochemical equation for the combustion of octane (C_8H_{18}) in your car's engine is: $2\,C_8H_{18}(l) + 25\,O_2(g) \longrightarrow 8\,CO_2(g) + 9\,H_2O(l) \quad \Delta H° = -5220\text{ kJ}$

Use the information in Table 9.2 to calculate $\Delta H°_f\,[C_8H_{18}(l)]$.

9.10 ▶ BOND DISSOCIATION ENERGIES

The procedure described in the previous section for determining heats of reaction from heats of formation is extremely useful, but it still presents a problem. To use the method, it's necessary to know $\Delta H°_f$ for every substance in a reaction. This implies, in turn, that vast numbers of measurements are needed because there are over 40 million known chemical compounds. In practice, though, only a few thousand $\Delta H°_f$ values have been determined.

For those reactions where insufficient $\Delta H°_f$ data are available to allow an exact calculation of $\Delta H°$, it's often possible to estimate $\Delta H°$ by using the average **bond dissociation energies (D)** discussed previously in Section 7.2. Although we didn't identify it as such at the time, a bond dissociation energy is really just a standard enthalpy change for the corresponding bond-breaking reaction.

REMEMBER...

Bond dissociation energy (D) is the amount of energy that must be supplied to break a chemical bond in an isolated molecule in the gaseous state and is thus the amount of energy released when the bond forms. (Section 7.2)

For the reaction $X - Y \longrightarrow X + Y$ $\Delta H° = D =$ Bond dissociation energy

When we say, for example, that the bond dissociation energy of Cl_2 is 243 kJ/mol, we mean that the standard enthalpy change for the reaction $Cl_2(g) \longrightarrow 2\,Cl(g)$ is $\Delta H° = 243$ kJ. Bond dissociation energies are always positive because energy must always be put into bonds to break them.

Applying Hess's law, we can calculate an approximate enthalpy change for any reaction by subtracting the sum of the bond dissociation energies in the products from the sum of the bond dissociation energies in the reactants:

$$\Delta H° = D(\text{Reactant bonds}) - D(\text{Product bonds})$$

In the reaction of H_2 with Cl_2 to yield HCl, for example, the reactants have one $Cl-Cl$ bond and one $H-H$ bond, while the product has two $H-Cl$ bonds.

$$H_2(g) + Cl_2(g) \longrightarrow 2\,HCl(g)$$

According to the data in Table 7.2, the bond dissociation energy of Cl_2 is 243 kJ/mol, that of H_2 is 436 kJ/mol, and that of HCl is 432 kJ/mol. We can thus calculate an approximate standard enthalpy change for the reaction.

$$\begin{aligned}
\Delta H° &= D(\text{Reactant bonds}) - D(\text{Product bonds}) \\
&= (D_{Cl-Cl} + D_{H-H}) - (2\,D_{H-Cl}) \\
&= [(1\,\text{mol})(243\,\text{kJ/mol}) + (1\,\text{mol})(436\,\text{kJ/mol})] - (2\,\text{mol})(432\,\text{kJ/mol}) \\
&= -185\,\text{kJ}
\end{aligned}$$

The reaction is exothermic by approximately 185 kJ.

● WORKED EXAMPLE 9.9

Using Bond Dissociation Energies to Calculate $\Delta H°$

Use the data in Table 7.2 to find an approximate $\Delta H°$ in kilojoules for the industrial synthesis of chloroform by reaction of methane with Cl_2.

$$CH_4(g) + 3\,Cl_2(g) \longrightarrow CHCl_3(g) + 3\,HCl(g)$$

STRATEGY

Identify all the bonds in the reactants and products, and look up the appropriate bond dissociation energies in Table 7.2. Then subtract the sum of the bond dissociation energies in the products from the sum of the bond dissociation energies in the reactants to find the enthalpy change for the reaction.

SOLUTION

The reactants have four $C-H$ bonds and three $Cl-Cl$ bonds; the products have one $C-H$ bond, three $C-Cl$ bonds, and three $H-Cl$ bonds. The bond dissociation energies from Table 7.2 are:

$C-H$	$D = 410\,\text{kJ/mol}$	$Cl-Cl$	$D = 243\,\text{kJ/mol}$
$C-Cl$	$D = 330\,\text{kJ/mol}$	$H-Cl$	$D = 432\,\text{kJ/mol}$

Subtracting the product bond dissociation energies from the reactant bond dissociation energies gives the enthalpy change for the reaction:

$$\Delta H° = [3\,D_{Cl-Cl} + 4\,D_{C-H}] - [D_{C-H} + 3\,D_{H-Cl} + 3\,D_{C-Cl}]$$
$$= [(3\,mol)(243\,kJ/mol) + (4\,mol)(410\,kJ/mol)] - [(1\,mol)(410\,kJ/mol)$$
$$+ (3\,mol)(432\,kJ/mol) + (3\,mol)(330\,kJ/mol)]$$
$$= -327\,kJ$$

The reaction is exothermic by approximately 330 kJ.

▶ **PRACTICE 9.17**   Use the data in Table 7.2 to calculate an approximate $\Delta H°$ in kilojoules for the industrial synthesis of ethyl alcohol from ethylene: $C_2H_4(g) + H_2O(g) \longrightarrow C_2H_5OH(g)$.

Ethyl alcohol

▶ **APPLY 9.18**   Benzene (C_6H_6) has two resonance structures meaning each carbon–carbon bond is equivalent, with a bond strength between a single C—C bond and double C=C.

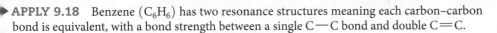

Estimate the carbon–carbon bond strength in benzene given:

$$2\,C_6H_6(g) + 15\,O_2(g) \longrightarrow 12\,CO_2(g) + 6\,H_2O(g) \quad \Delta H° = -6339\,kJ$$

Bond dissociation data can be found in Table 7.2. (The strength of the O=O bond in O_2 is 498 kJ/mol, and that of a C=O bond in CO_2 is 804 kJ/mol.)

9.11 ▶ FOSSIL FUELS, FUEL EFFICIENCY, AND HEATS OF COMBUSTION

Surely the most familiar of all exothermic reactions is the one that takes place every time you turn up a thermostat, drive a car, or light a match: the burning of a carbon-based fuel by reaction with oxygen to yield H_2O, CO_2, and heat. The amount of energy released on burning a substance is called its **heat of combustion** ($\Delta H°_c$), or *combustion enthalpy*, $\Delta H°_c$, and is simply the standard enthalpy change for the reaction of 1 mol of the substance with oxygen. Hydrogen, for instance, has $\Delta H°_c = -285.8\,kJ/mol$, and methane has $\Delta H°_c = -890.3\,kJ/mol$. Note that the H_2O product in giving heats of combustion is $H_2O(l)$ rather than $H_2O(g)$.

$$H_2(g) + 1/2\,O_2(g) \longrightarrow H_2O(l) \qquad\qquad \Delta H°_c = -285.8\,kJ/mol$$
$$CH_4(g) + 2\,O_2(g) \longrightarrow CO_2(g) + 2\,H_2O(l) \quad \Delta H°_c = -890.3\,kJ/mol$$

To compare the efficiency of different fuels, it's more useful to calculate combustion enthalpies per gram or per milliliter of substance rather than per mole (**TABLE 9.3**). For applications where weight is important, as in rocket engines, hydrogen is ideal because its combustion enthalpy per gram is the highest of any known fuel. For applications where volume is important, as in automobiles, a mixture of hydrocarbons—compounds of carbon and hydrogen—such as those in gasoline is most efficient because hydrocarbon combustion enthalpies per milliliter are relatively high. Octane and toluene are representative examples.

TABLE 9.3 **Thermochemical Properties of Some Fuels**

Fuel	Combustion Enthalpy		
	kJ/mol	kJ/g	kJ/mL
Hydrogen, $H_2(l)$	−285.8	−141.8	−9.9[a]
Ethanol, $C_2H_5OH(l)$	−1366.8	−29.7	−23.4
Graphite, $C(s)$	−393.5	−32.8	−73.8
Methane, $CH_4(g)$	−890.3	−55.5	−30.8[a]
Methanol, $CH_3OH(l)$	−725.9	−22.7	−17.9
Octane, $C_8H_{18}(l)$	−5470	−47.9	−33.6
Toluene, $C_7H_8(l)$	−3910	−42.3	−36.7

[a]Calculated for the compressed liquid at 0 °C

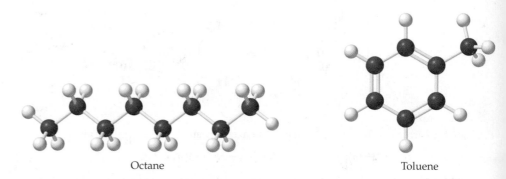

Octane   Toluene

▲ Much coal lies near the surface of the Earth and is obtained by strip mining.

With the exception of hydrogen, all common fuels are organic compounds, whose energy is derived ultimately from the sun through the photosynthesis of carbohydrates in green plants. Although the details are complex, the net result of the photosynthesis reaction is the conversion of carbon dioxide and water into glucose, $C_6H_{12}O_6$, plus oxygen. Glucose, once formed, is converted into cellulose and starch, which in turn act as structural materials for plants and as food sources for animals. The conversion is highly endothermic and therefore requires a large input of solar energy. It has been estimated that the total annual amount of solar energy absorbed by the Earth's vegetation is approximately 10^{19} kJ, an amount sufficient to synthesize 5×10^{14} kg of glucose per year.

$$6\,CO_2(g) + 6\,H_2O(l) \longrightarrow C_6H_{12}O_6(s) + 6\,O_2(g) \qquad \Delta H° = 2803 \text{ kJ}$$

The fossil fuels we use most—coal, petroleum, and natural gas—are derived from the decayed remains of organisms from previous geologic eras. Both coal and petroleum are enormously complex mixtures of compounds. Coal is primarily of vegetable origin, and many of the compounds it contains are structurally similar to graphite (pure carbon). Petroleum is a viscous liquid mixture of hydrocarbons that are primarily of marine origin, and natural gas is primarily methane, CH_4.

Coal is burned just as it comes from the mine, but petroleum must be *refined* before use. Refining begins with *distillation*, the separation of crude liquid oil into fractions on the basis of their boiling points (bp). So-called straight-run gasoline (bp 30 − 200 °C) consists of compounds with 5–11 carbon atoms per molecule; kerosene (bp 175 − 300 °C) contains compounds in the $C_{11} − C_{14}$ range; gas oil (bp 275–400 °C) contains $C_{14} − C_{25}$ substances, and lubricating oils contain whatever remaining compounds will distill. Left over is a tarry residue of asphalt (**FIGURE 9.12**).

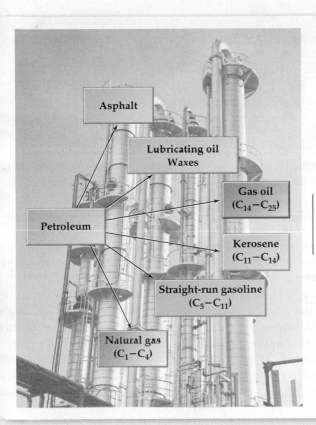

◀ **FIGURE 9.12**

The products of petroleum refining are separated based upon differences in boiling point.

Figure It Out

Which type of intermolecular force (Section 8.6) causes hydrocarbons with different numbers of carbon atoms to have different boiling points?

Answer: London dispersion forces

The different fractions are grouped according to the number of carbon atoms their molecules contain.

As the world's petroleum deposits become more scarce, other sources of energy will have to be found to replace them. Hydrogen, although it burns cleanly and is relatively nonpolluting, has numerous drawbacks: low availability, difficulty in transport and storage, and low combustion enthalpy per milliliter, to name a few. Ethanol and methanol look like potential choices for alternative fuels because both can be produced relatively cheaply and have reasonable combustion enthalpies. At present, ethanol is produced largely by fermentation of corn or cane sugar, but methods are being developed to produce it from waste wood by the breakdown of cellulose to glucose and subsequent fermentation. Methanol is produced directly from natural gas in a two-step process:

$$CH_4(g) + H_2O(g) \longrightarrow CO(g) + 3 H_2(g)$$
$$CO(g) + 2 H_2(g) \longrightarrow CH_3OH(l)$$

9.12 ▶ AN INTRODUCTION TO ENTROPY

We said in the introduction to this chapter that chemical reactions (and physical processes) occur when the final state is more stable than the initial state. Because less stable substances generally have higher internal energy and are converted into more stable substances with lower internal energy, energy is generally released in chemical reactions. At the same time, though, we've seen that some reactions and processes occur even though they absorb rather than release energy. The endothermic reaction of barium hydroxide octahydrate with ammonium chloride shown previously in Figure 9.7, for example, absorbs 80.3 kJ of heat ($\Delta H° = +80.3$ kJ) and leaves the surroundings so cold that the temperature drops below 0 °C.

$$Ba(OH)_2 \cdot 8 H_2O(s) + 2 NH_4Cl(s) \longrightarrow BaCl_2(aq) + 2 NH_3(aq) + 10 H_2O(l)$$
$$\Delta H° = +80.3 \text{ kJ}$$

An example of a physical process that takes place spontaneously yet absorbs energy takes place every time an ice cube melts. At a temperature of 0 °C, ice spontaneously absorbs heat from the surroundings to turn from solid into liquid water.

Before exploring the situation further, it's important to understand what the word *spontaneous* means in chemistry, for it's not quite the same as in everyday language. In chemistry, a **spontaneous process** is one that, once started, proceeds on its own without a continuous external influence. The change need not happen quickly, like a spring uncoiling or a sled going downhill. It can also happen slowly, like the gradual rusting away of an iron bridge or abandoned car. A *nonspontaneous* process, by contrast, takes place only in the presence of a continuous external influence. Energy must be continuously expended to recoil a spring or to push a sled uphill. When the external influence stops, the process also stops.

Note that the reaction of barium hydroxide octahydrate with ammonium chloride and the melting ice cube absorb heat yet still take place spontaneously. What's going on? There must be some other factor in addition to energy that determines whether a reaction or process will occur. We'll take only a brief look at this additional factor now and return for a more in-depth study in Chapter 17 on **thermodynamics**.

What do the reaction of barium hydroxide octahydrate and the melting of an ice cube have in common that allows the two processes to take place spontaneously even though they absorb heat? The common feature of these and all other processes that absorb heat yet occur spontaneously is an increase in the amount of molecular randomness of the system. The eight water molecules rigidly held in the $Ba(OH)_2 \cdot 8 H_2O$ crystal break loose and become free to move about randomly in the aqueous liquid product. Similarly, the rigidly held H_2O molecules in the ice lose their crystalline ordering and move around freely in liquid water.

The amount of molecular randomness in a system is called the system's **entropy** (S). Entropy has the units J/K (not kJ/K) and is a quantity that can be determined for pure substances. The larger the value of S, the greater the molecular randomness of the particles in the system. Gases, for example, have more randomness and higher entropy than liquids, and liquids have more randomness and higher entropy than solids (**FIGURE 9.13**).

A change in entropy is represented as $\Delta S = S_{final} - S_{initial}$. When randomness increases, as it does when barium hydroxide octahydrate reacts or ice melts, ΔS has a positive value because $S_{final} > S_{initial}$. The reaction of $Ba(OH)_2 \cdot 8 H_2O(s)$ with $NH_4Cl(s)$ has $\Delta S° = +428 \text{ J/K}$, and the melting of ice has $\Delta S° = +22.0 \text{ J/(K} \cdot \text{mol)}$. When randomness decreases, ΔS is negative because $S_{final} < S_{initial}$. The freezing of water, for example, has $\Delta S° = -22.0 \text{ J/(K} \cdot \text{mol)}$. (As with $\Delta H°$, the superscript ° is used in $\Delta S°$ to refer to the standard entropy change in a reaction where products and reactants are in their standard states.)

Thus, two factors determine the spontaneity of a chemical or physical change in a system: a release or absorption of heat (ΔH) and an increase or decrease in molecular randomness (ΔS). To decide whether a process is spontaneous, both enthalpy and entropy changes must be taken into account:

LOOKING AHEAD...

A consolidated study of **thermodynamics**, including a more detailed discussion of entropy and free energy, is provided in Chapter 17.

▲ Sledding downhill is a spontaneous process that, once started, continues on its own. Dragging the sled back uphill is a nonspontaneous process that requires a continuous input of energy.

Spontaneous process:

Favored by decrease in H (negative ΔH)

Favored by increase in S (positive ΔS)

Nonspontaneous process:

Favored by increase in H (positive ΔH)

Favored by decrease in S (negative ΔS)

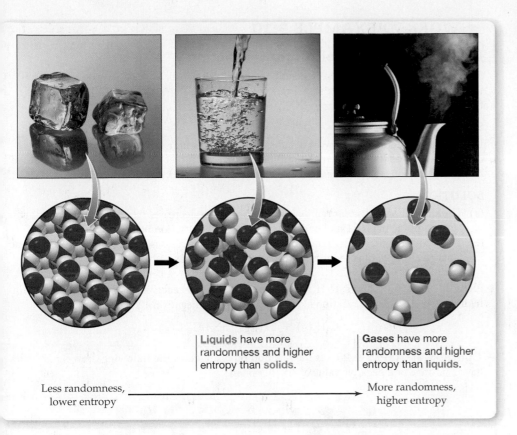

Liquids have more randomness and higher entropy than **solids**.

Gases have more randomness and higher entropy than **liquids**.

Less randomness, lower entropy

More randomness, higher entropy

Note that the two factors don't have to operate in the same direction. Thus, it's possible for a process to be *disfavored* by enthalpy (endothermic, positive ΔH) yet still be spontaneous because it is strongly *favored* by entropy (positive ΔS). The melting of ice $[\Delta H° = +6.01 \text{ kJ/mol}; \Delta S° = +22.0 \text{ J/(K} \cdot \text{mol)}]$ is just such a process, as is the reaction of barium hydroxide octahydrate with ammonium chloride $(\Delta H° = +80.3 \text{ kJ}; \Delta S° = +428 \text{ J/K})$. In the latter case, 3 mol of solid reactants produce 10 mol of liquid water, 2 mol of dissolved ammonia, and 3 mol of dissolved ions (1 mol of Ba^{2+} and 2 mol of Cl^-), with a consequent large increase in molecular randomness:

$$Ba(OH)_2 \cdot 8 \, H_2O(s) + 2 \, NH_4Cl(s) \longrightarrow BaCl_2(aq) + 2 \, NH_3(aq) + 10 \, H_2O(l)$$

3 mol solid reactants

3 mol dissolved ions

2 mol dissolved molecules

10 mol liquid water molecules

$$\Delta H° = +80.3 \text{ kJ} \longleftarrow \text{Unfavorable}$$

$$\Delta S° = +428 \text{ J/K} \longleftarrow \text{Favorable}$$

Conversely, it's also possible for a process to be favored by enthalpy (exothermic, negative ΔH) yet be nonspontaneous because it is strongly disfavored by entropy (negative ΔS). The conversion of liquid water to ice is nonspontaneous above 0 °C, for example, because the process is disfavored by entropy $[\Delta S° = -22.0 \text{ J/(K} \cdot \text{mol)}]$ even though it is favored by enthalpy $(\Delta H° = -6.01 \text{ kJ/mol})$.

WORKED EXAMPLE 9.10

Predicting the Sign of ΔS for a Reaction

Predict whether $\Delta S°$ is likely to be positive or negative for each of the following reactions:

(a) $H_2C{=}CH_2(g) + Br_2(g) \longrightarrow BrCH_2CH_2Br(l)$
(b) $2\,C_2H_6(g) + 7\,O_2(g) \longrightarrow 4\,CO_2(g) + 6\,H_2O(g)$

STRATEGY

Look at each reaction, and try to decide whether molecular randomness increases or decreases. Reactions that increase the number of gaseous molecules generally have a positive ΔS, while reactions that decrease the number of gaseous molecules have a negative ΔS.

SOLUTION

(a) The amount of molecular randomness in the system decreases when 2 mol of gaseous reactants combine to give 1 mol of liquid product, so the reaction has a negative $\Delta S°$.

(b) The amount of molecular randomness in the system increases when 9 mol of gaseous reactants give 10 mol of gaseous products, so the reaction has a positive $\Delta S°$.

▶ **PRACTICE 9.19** Ethane, C_2H_6, can be prepared by the reaction of acetylene, C_2H_2, with hydrogen. Is $\Delta S°$ for the reaction likely to be positive or negative? Explain.

$$C_2H_2(g) + 2\,H_2(g) \longrightarrow C_2H_6(g)$$

▶ **Conceptual APPLY 9.20** Is the reaction represented in the following drawing likely to have a positive or a negative value of $\Delta S°$? Explain.

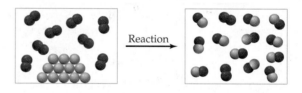

9.13 ▶ AN INTRODUCTION TO FREE ENERGY

How do we weigh the relative contributions of enthalpy changes (ΔH) and entropy changes (ΔS) to the overall spontaneity of a process? To take both factors into account when deciding the spontaneity of a chemical reaction or other process, we define a quantity called the **Gibbs free-energy change** (ΔG), which is related to ΔH and ΔS by the equation $\Delta G = \Delta H - T\Delta S$.

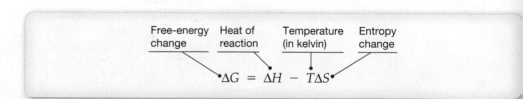

The value of the free-energy change ΔG determines whether a chemical or physical process will occur spontaneously. If ΔG has a negative value, free energy is released and the process is spontaneous. If ΔG has a value of 0, the process is neither spontaneous nor nonspontaneous but is instead at an equilibrium. And if ΔG has a positive value, free energy is absorbed and the process is nonspontaneous.

$\Delta G < 0$ Process is spontaneous

$\Delta G = 0$ Process is at equilibrium—neither spontaneous nor nonspontaneous

$\Delta G > 0$ Process is nonspontaneous

Because the $T\Delta S$ term in the free-energy equation is temperature dependent, we deduce that some processes might be either spontaneous or nonspontaneous depending on the temperature. At low temperature, for instance, an unfavorable (positive) ΔH term might be larger than a favorable (positive) $T\Delta S$ term, but at higher temperature, the $T\Delta S$ term might be larger. Thus, an endothermic process that is nonspontaneous at low temperature can become spontaneous at higher temperature. This, in fact, is exactly what happens in the ice–water transition. At a temperature below 0 °C, the melting of ice is nonspontaneous because the unfavorable ΔH term outweighs the favorable $T\Delta S$ term. At a temperature above 0 °C, however, the melting of ice is spontaneous because the favorable $T\Delta S$ term outweighs the unfavorable ΔH term (**FIGURE 9.14**). At exactly 0 °C, the two terms are balanced.

$$\Delta G° = \Delta H° - T\Delta S°$$

At -10 °C (263 K): $\Delta G° = 6.01 \dfrac{\text{kJ}}{\text{mol}} - (263 \text{ K})\left(0.0220 \dfrac{\text{kJ}}{\text{K} \cdot \text{mol}}\right) = +0.22 \text{ kJ/mol (nonspontaneous)}$

At 0 °C (273 K): $\Delta G° = 6.01 \dfrac{\text{kJ}}{\text{mol}} - (273 \text{ K})\left(0.0220 \dfrac{\text{kJ}}{\text{K} \cdot \text{mol}}\right) = 0.00 \text{ kJ/mol (equilibrium)}$

At $+10$ °C (283 K): $\Delta G° = 6.01 \dfrac{\text{kJ}}{\text{mol}} - (283 \text{ K})\left(0.0220 \dfrac{\text{kJ}}{\text{K} \cdot \text{mol}}\right) = -0.22 \text{ kJ/mol (spontaneous)}$

An example of a chemical reaction in which temperature controls spontaneity is that of carbon with water to yield carbon monoxide and hydrogen. The reaction has an unfavorable ΔH term (positive) but a favorable $T\Delta S$ term (positive) because randomness increases when a solid and 1 mol of gas are converted into 2 mol of gas:

$$C(s) + H_2O(g) \longrightarrow CO(g) + H_2(g) \quad \Delta H° = +131 \text{ kJ} \quad \textbf{Unfavorable}$$
$$\Delta S° = +134 \text{ J/K} \quad \textbf{Favorable}$$

No reaction occurs if carbon and water are mixed at room temperature because the unfavorable ΔH term outweighs the favorable $T\Delta S$ term. At approximately 978 K (705 °C), however, the reaction becomes spontaneous because the favorable $T\Delta S$ term becomes larger than the unfavorable ΔH term. Below 978 K, ΔG has a positive value; at 978 K, $\Delta G = 0$; and above 978 K, ΔG has a negative value. (The calculation is not exact because values of ΔH and ΔS themselves vary somewhat with temperature.)

◀ **FIGURE 9.14**

Melting and freezing. The melting of ice is disfavored by enthalpy ($\Delta H > 0$) but favored by entropy ($\Delta S > 0$). The freezing of water is favored by enthalpy ($\Delta H < 0$) but disfavored by entropy ($\Delta S < 0$).

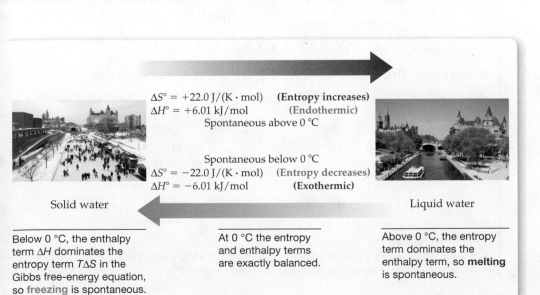

$\Delta S° = +22.0 \text{ J/(K} \cdot \text{mol)}$ **(Entropy increases)**
$\Delta H° = +6.01 \text{ kJ/mol}$ **(Endothermic)**
Spontaneous above 0 °C

Spontaneous below 0 °C
$\Delta S° = -22.0 \text{ J/(K} \cdot \text{mol)}$ **(Entropy decreases)**
$\Delta H° = -6.01 \text{ kJ/mol}$ **(Exothermic)**

Solid water

Liquid water

Below 0 °C, the enthalpy term ΔH dominates the entropy term $T\Delta S$ in the Gibbs free-energy equation, so **freezing** is spontaneous.

At 0 °C the entropy and enthalpy terms are exactly balanced.

Above 0 °C, the entropy term dominates the enthalpy term, so **melting** is spontaneous.

$$\Delta G° = \Delta H° - T\Delta S°$$

At 695 °C (968 K): $\Delta G° = 131 \text{ kJ} - (968 \text{ K})\left(0.134 \dfrac{\text{kJ}}{\text{K}}\right) = +1 \text{ kJ (nonspontaneous)}$

At 705 °C (978 K): $\Delta G° = 131 \text{ kJ} - (978 \text{ K})\left(0.134 \dfrac{\text{kJ}}{\text{K}}\right) = 0 \text{ kJ (equilibrium)}$

At 715 °C (988 K): $\Delta G° = 131 \text{ kJ} - (988 \text{ K})\left(0.134 \dfrac{\text{kJ}}{\text{K}}\right) = -1 \text{ kJ (spontaneous)}$

The reaction of carbon with water is, in fact, the first step of an industrial process for manufacturing methanol (CH_3OH). As supplies of natural gas and oil diminish, this reaction may become important for the manufacture of synthetic fuels.

A process is at equilibrium when it is balanced between spontaneous and nonspontaneous—that is, when $\Delta G = 0$ and it is energetically unfavorable to go either from reactants to products or from products to reactants. Thus, at the equilibrium point, we can set up the equation

$$\Delta G = \Delta H - T\Delta S = 0 \text{ At equilibrium}$$

Solving this equation for T gives

$$T = \frac{\Delta H}{\Delta S}$$

which makes it possible to calculate the temperature at which a changeover in behavior between spontaneous and nonspontaneous occurs. Using the known values of $\Delta H°$ and $\Delta S°$ for the melting of ice, for instance, we find that the point at which liquid water and solid ice are in equilibrium is

$$T = \frac{\Delta H°}{\Delta S°} = \frac{6.01 \text{ kJ}}{0.0220 \dfrac{\text{kJ}}{\text{K}}} = 273 \text{ K} = 0 °C$$

Not surprisingly, the ice–water equilibrium point is 273 K, or 0 °C, the melting point of ice.

In the same way, the temperature at which the reaction of carbon with water changes between spontaneous and nonspontaneous is 978 K, or 705 °C:

$$T = \frac{\Delta H°}{\Delta S°} = \frac{131 \text{ kJ}}{0.134 \dfrac{\text{kJ}}{\text{K}}} = 978 \text{ K}$$

This section and the preceding one serve only as an introduction to entropy and free energy. We'll return in Chapter 17 for a more in-depth look at these two important topics.

Conceptual WORKED EXAMPLE 9.11

Predicting the Signs of ΔH, ΔS, and ΔG for a Reaction

What are the signs of ΔH, ΔS, and ΔG for the following nonspontaneous transformation?

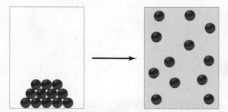

STRATEGY

First, decide what kind of process is represented in the drawing. Then decide whether the process increases or decreases the entropy of the system and whether it is exothermic or endothermic.

SOLUTION

The drawing shows ordered particles in a solid subliming to give a gas. Formation of a gas from a solid increases molecular randomness, so ΔS is positive. Furthermore, because we're told that the process is nonspontaneous, ΔG is also positive. Because the process is favored by ΔS (positive) yet still nonspontaneous, ΔH must be unfavorable (positive). This makes sense, because conversion of a solid to a liquid or gas requires energy and is always endothermic.

▶ **Conceptual PRACTICE 9.21** What are the signs of ΔH, ΔS, and ΔG for the following spontaneous reaction?

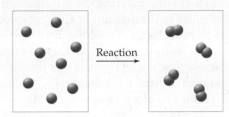

▶ **Conceptual APPLY 9.22** The following reaction is exothermic:

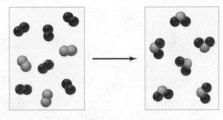

(a) Write a balanced equation for the reaction. (Red spheres = A; Green spheres = B)
(b) What are the signs of ΔH and ΔS for the reaction?
(c) Is the reaction likely to be spontaneous at low temperatures only, at high temperatures only, or at all temperatures? Explain.

WORKED EXAMPLE 9.12

Using the Free-Energy Equation to Calculate Equilibrium Temperature

Lime (CaO) is produced by heating limestone ($CaCO_3$) to drive off CO_2 gas, a reaction used to make Portland cement. Is the reaction spontaneous under standard conditions at 25 °C? Calculate the temperature at which the reaction becomes spontaneous.

$$CaCO_3(s) \longrightarrow CaO(s) + CO_2(g) \quad \Delta H° = 179.2 \text{ kJ}; \quad \Delta S° = 160.0 \text{ J/K}$$

STRATEGY

The spontaneity of the reaction at a given temperature can be found by determining whether ΔG is positive or negative at that temperature. The changeover point between spontaneous and nonspontaneous can be found by setting $\Delta G = 0$ and solving for T.

SOLUTION

At 25 °C (298 K), we have

$$\Delta G = \Delta H - T\Delta S = 179.2 \text{ kJ} - (298 \text{ K})\left(0.1600 \frac{\text{kJ}}{\text{K}}\right) = +131.5 \text{ kJ}$$

Because ΔG is positive at this temperature, the reaction is nonspontaneous. The changeover point between spontaneous and nonspontaneous is approximately

$$T = \frac{\Delta H}{\Delta S} = \frac{179.2 \text{ kJ}}{0.1600 \dfrac{\text{kJ}}{\text{K}}} = 1120 \text{ K}$$

The reaction becomes spontaneous above approximately 1120 K (847 °C).

▶ **PRACTICE 9.23** Is the Haber process for the industrial synthesis of ammonia spontaneous or nonspontaneous under standard conditions at 25 °C? At what temperature (°C) does the changeover occur?

$$N_2(g) + 3 H_2(g) \longrightarrow 2 NH_3(g) \quad \Delta H° = -92.2 \text{ kJ}; \Delta S° = -199 \text{ J/K}$$

▶ **APPLY 9.24** A certain reaction has $\Delta H° = +75.0$ kJ; $\Delta S° = +231$ J/K. At what temperature is the reaction at equilibrium? Should the temperature be increased or decreased to make the reaction spontaneous?

INQUIRY ▶▶▶ HOW IS THE ENERGY CONTENT OF NEW FUELS DETERMINED?

The petroleum era began in August 1859, when the world's first oil well was drilled near Titusville, Pennsylvania. Since that time, approximately 1.5×10^{12} barrels of petroleum have been used throughout the world, primarily as fuel for automobiles (1 barrel = 42 gallons).

No one really knows how much petroleum remains on Earth. Current world consumption is approximately 3.1×10^{10} barrels per year, and currently known recoverable reserves are estimated at 1.4×10^{12} barrels. Thus, the world's known petroleum reserves will be exhausted in approximately 45 years at the current rate of consumption. Additional petroleum reserves will surely be found, but consumption is also increasing, making any prediction of the amount of time remaining highly inaccurate. Only two things are certain: The amount of petroleum remaining *is* finite, and we *will* run out at some point, whenever that might be. Thus, alternative energy sources are needed.

▲ Fermentation of cane sugar is one method for producing ethanol for use as an alternative fuel.

Of the various alternative energy sources now being explored, *biofuels*—fuels derived from recently living organisms such as trees, corn, sugar cane, and rapeseed—look promising because they are renewable and because they are more nearly *carbon neutral* than fossil fuels, meaning that the amount of CO_2 released to the environment during the manufacture and burning of a biofuel is similar to the amount of CO_2 removed from the environment by photosynthesis during the plant's growth. Note that the phrase *carbon neutral* doesn't mean that biofuels don't release CO_2 when burned; they often release just as much CO_2 as any other fuel. The Inquiry in Chapter 3 compares CO_2 emissions of different fuels using principles of stoichiometry.

The two biofuels receiving the most attention at present are ethanol and biodiesel. Ethanol, sometimes called bioethanol to make it sound more attractive, is simply ethyl alcohol, the same substance found in alcoholic drinks and produced in the same way by yeast-catalyzed fermentation of carbohydrate.

Glucose
($C_6H_{12}O_6$)

Yeast enzymes →

Ethanol
(C_2H_5OH)

The only difference between beverage ethanol and fuel ethanol is the source of the sugar. Beverage ethanol comes primarily from fermentation of sugar in grapes (for wine) or grains (for distilled liquors), while fuel ethanol comes primarily from fermentation of cane sugar or corn. Much current work is being done, however, on developing economical methods of converting cheap cellulose-based agricultural and logging wastes into fermentable sugars.

Biodiesel consists primarily of organic compounds called *long-chain methyl esters*, which are produced by reaction of common vegetable oils with methyl alcohol in the presence of an acid catalyst. Any vegetable oil can be used, but rapeseed oil and soybean oil are the most common. (Canola oil, often used for cooking, is a specific cultivar of generic rapeseed.) Once formed, the biodiesel is typically mixed in up to 30% concentration with petroleum-based diesel fuel for use in automobiles and trucks.

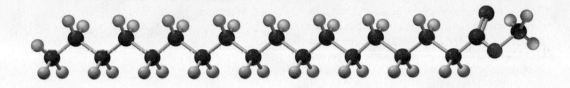

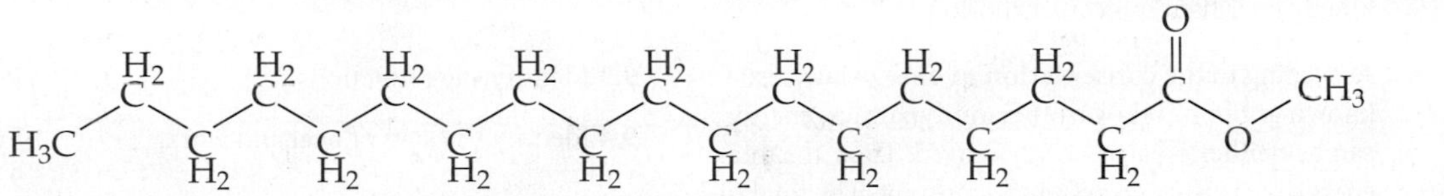

A typical long-chain methyl ester in biodiesel

PROBLEM 9.25

(a) Balance the equation for the combustion of liquid ethanol (C_2H_6O).

$$C_2H_6O(l) + O_2(g) \longrightarrow CO_2(g) + H_2O(l)$$

(b) Use the information in Table 9.2 and $[\Delta H°_f$ for C_2H_6O $(l) = -277.7$ kJ/mol] to calculate the combustion enthalpy ($\Delta H°_c$) in kilojoules per mole of ethanol.

(c) Calculate the combustion enthalpy ($\Delta H°_c$) for ethanol in units of kJ/g.

(d) The density of ethanol at 25 °C is 0.789 g/mL. Calculate the combustion enthalpy ($\Delta H°_c$) in units of kJ/mL.

PROBLEM 9.26 Bond dissociation energies can be used to estimate the enthalpy change for reaction if the reactants and products are in the gas phase. In the engine of a car, liquid fuel is converted to the vapor phase before combustion. Calculate an approximate heat of combustion for gaseous ethanol in kilojoules per mole ethanol by using the bond dissociation energies in Table 7.2. (The strength of the O=O bond is 498 kJ/mol, and that of a C=O bond in CO_2 is 804 kJ/mol.) The structural formula for ethanol is:

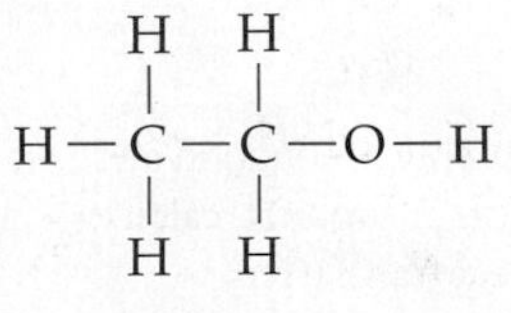

$$
\begin{array}{c}
\quad H \quad H \\
\quad | \quad\; | \\
H-C-C-O-H \\
\quad | \quad\; | \\
\quad H \quad H
\end{array}
$$

PROBLEM 9.27

(a) Calculate the PV work in kilojoules done during the combustion of 1 mol of ethanol in which the volume contracts from 73.5 L to 49.0 L at a constant external pressure of 1 atm. In which direction does the work energy flow? (1 L · atm = 101 J.)

(b) If $\Delta H° = -1367$ kJ/mol, calculate the change in internal energy (ΔE) for the combustion of 1 mol of ethanol.

PROBLEM 9.28

(a) When 0.350 g of biodiesel ($C_{19}H_{38}O_2$) is burned in a bomb calorimeter, the temperature of both the water and the calorimeter rise in temperature by 6.85 °C. Assuming that the bath contains 300.0 g of water and that the heat capacity for the calorimeter is 675 J/°C, calculate combustion energy(ΔE) for biodiesel in units of kJ/g

(b) Calculate combustion energy (ΔE) for biodiesel in units of kJ/mol.

(c) If the density of biodiesel is 0.880 g/mL, calculate combustion energy (ΔE) for biodiesel in units of kJ/mL.

PROBLEM 9.29 A 12.0-gallon fuel tank can hold 35.8 kg of ethanol and 39.9 kg of biodiesel. Use the following thermochemical equations to calculate the total amount of heat (MJ) released by burning one tank of each type of fuel. Compare your answer to a 12.0-gallon tank of octane (main component of gasoline) which burns to release 1530 MJ.

$$C_2H_6O(l) + 3\,O_2(g) \longrightarrow 2\,CO_2(g) + 3\,H_2O(l) \quad \Delta H° = -1367\ \text{kJ}$$
$$C_{19}H_{38}O_2(l) + 27.5\,O_2(g) \longrightarrow 19\,CO_2(g) + 19\,H_2O(l) \quad \Delta H° = -11{,}236\ \text{kJ}$$

STUDY GUIDE

Section	Concept Summary	Learning Objectives	Test Your Understanding
9.1 ▶ Energy and Its Conservation	Energy is either *kinetic* or *potential*. **Kinetic energy** (E_K) is the energy of motion. Its value depends on both the mass m and velocity v of an object according to the equation $E_K = (1/2)mv^2$. **Potential energy** (E_P) is the energy stored in an object because of its position or in a chemical substance because of its composition. **Heat** is the thermal energy transferred between two objects as the result of a temperature difference, whereas **temperature** is a measure of the kinetic energy of molecular motion.	**9.1** Calculate the kinetic energy of an object in motion. **9.2** Convert between common units for energy.	Problems 9.42, 9.43 Problems 9.48, 9.49
9.2 ▶ Internal Energy and State Functions	According to the **conservation of energy law**, also known as the **first law of thermodynamics**, energy can be neither created nor destroyed. Thus, the total energy of an isolated system is constant. The total **internal energy (E)** of a system—the sum of all kinetic and potential energies for each particle in the system—is a **state function** because its value depends only on the present condition of the system, not on how that condition was reached.	**9.3** Identify state functions. **9.4** Identify the sign of heat and work.	Problems 9.44, 9.45 Problems 9.30, 9.31, 9.32
9.3 ▶ Expansion Work	**Work (w)** is defined as the distance moved times the force that produces the motion. In chemistry, most work is expansion work (*PV* work) done as the result of a volume change during a reaction when air molecules are pushed aside. The amount of work done by an expanding gas is given by the equation $w = -P\Delta V$, where P is the pressure against which the system must push and ΔV is the change in volume of the system.	**9.5** Calculate *PV* work.	Worked Example 9.1; Problems 9.33, 9.50, 9.51, 9.56, 9.130, 9.131
9.4 ▶ Energy and Enthalpy	The total internal energy change that takes place during a reaction is the sum of the heat transferred (q) and the work done ($-P\Delta V$). The equation $\Delta E = q + (-P\Delta V)$ or $q = \Delta E + P\Delta V = \Delta H$ where ΔH is the **enthalpy change** of the system, is a fundamental equation of thermochemistry. In general, the $P\Delta V$ term is much smaller than the ΔE term, so that the total internal energy change of a reacting system is approximately equal to ΔH, also called the **heat of reaction**.	**9.6** Calculate the internal energy change (ΔE) for a reaction.	Worked Example 9.2; Problems 9.57, 9.58, 9.60, 9.61, 9.132
9.5 and 9.6 ▶ Thermochemical Equations and Enthalpies of Chemical and Physical Changes	**Thermochemical equations** show a balanced chemical equation along with the value of the enthalpy change (ΔH) for the reaction. The amount of heat transferred depends on the quantity of reactants and products, and the thermochemical equation specifies this relationship for chemical and physical changes. Reactions that have a negative ΔH are **exothermic** because heat is lost by the system, and reactions that have a positive ΔH are **endothermic** because heat is absorbed by the system.	**9.7** Given a thermochemical equation and the amount of reactant or product, calculate the amount of heat transferred. **9.8** Classify endo- and exothermic reactions.	Worked Example 9.3; Problems 9.64–9.69 Worked Example 9.4; Problems 9.62, 9.63, 9.70, 9.71

Section	Concept Summary	Learning Objectives	Test Your Understanding
9.7 ▸ Calorimetry and Heat Capacity	**Heat capacity** (C), is a value that specifies how much heat is required to raise the temperature of an object by 1 °C. Similarly, **specific heat** (c) and **molar heat capacity** (C_m) are values that specify the amount of heat required to raise the temperature of a given amount of a substance by 1 °C. A calorimeter is a device used to determine ΔH of a reaction by surrounding the reaction with water and measuring the temperature change of the water.	**9.9** Calculate heat capacities, temperature changes, or heat transfer using the equation for heat capacity (C), specific heat (c), or molar heat capacity (C_m).

9.10 Calculate enthalpy changes in a calorimetry experiment. | Worked Example 9.5; Problems 9.74–9.79, 9.144, 9.146

Worked Example 9.6; Problems 9.80–9.83, 9.147 |
| **9.8 ▸** Hess's Law | Because enthalpy is a state function, ΔH is the same regardless of the path taken between reactants and products. Thus, the sum of the enthalpy changes for the individual steps in a reaction is equal to the overall enthalpy change for the entire reaction, a relationship known as **Hess's law**. Using this law, it is possible to calculate overall enthalpy changes for individual steps that can't be measured directly. | **9.11** Use Hess's Law to find ΔH for an overall reaction, given reaction steps and their ΔH values. | Worked Example 9.7; Problems 9.34, 9.35, 9.85–9.87 |
| **9.9 ▸** Standard Heats of Formation | Hess's law also makes it possible to calculate the enthalpy change of any reaction if the standard heats of formation ($\Delta H°_f$) are known for the reactants and products. The **standard heat of formation** ($\Delta H°_f$) is the enthalpy change for the hypothetical formation of 1 mol of a substance in its **thermodynamic standard state** from the most stable forms of the constituent elements in their standard states (1 atm pressure and a specified temperature, usually 25 °C). | **9.12** Identify standard states of elements.

9.13 Write standard enthalpy of formation reactions ($\Delta H°_f$) for compounds from their elements.

9.14 Use values of ($\Delta H°_f$) for elements and compounds to calculate $\Delta H°$ for a reaction. | Problems 9.90, 9.91

Problems 9.92, 9.93

Worked Example 9.8; Problems 9.94–9.105, 9.140 |
| **9.10 ▸** Bond Dissociation Energies | Breaking a bond is an endothermic process and forming a bond is an exothermic process, which is the underlying reason for enthalpy changes of chemical reactions. Bond dissociation energies (D) can be used to estimate $\Delta H°$ for a gas-phase reaction by adding up all the bonds broken in the reactants and subtracting all the bonds formed in the products. | **9.15** Use bond dissociation energies to estimate $\Delta H°$ for a reaction. | Worked Example 9.9; Problems 9.17, 9.18, 9.106–9.109 |
| **9.11 ▸** Fossil Fuels, Fuel Efficiency, and Heats of Combustion | The combustion enthalpy ($\Delta H°_c$) is the enthalpy change associated with reacting one mole of a substance with oxygen. Most common fuels are organic compounds and $\Delta H°_c$ values are reported on a per-mass or per-volume basis because these criteria are important in vehicle design. | **9.16** Calculate $\Delta H°_c$ for various fuels using thermochemical principles such as Hess's Law, calorimetry, and/or bond dissociation enthalpies. | Problems 9.110, 9.111 |
| **9.12 and 9.13 ▸** An Introduction to Entropy and Free Energy | In addition to enthalpy, **entropy** (S)—a measure of the amount of molecular randomness in a system—is important in determining whether a process will occur spontaneously. Together, changes in enthalpy and entropy define a quantity called the **Gibbs free-energy change** (ΔG) according to the equation $\Delta G = \Delta H - T\Delta S$. If ΔG is negative, the reaction is a **spontaneous process**; if ΔG is positive, the reaction is nonspontaneous. | **9.17** Predict the sign of the entropy change (ΔS) given the chemical equation or a molecular diagram.

9.18 Using the relationship between Gibbs free energy (ΔG) and spontaneity, predict the sign of $\Delta G, \Delta H, \Delta S$.

9.19 Use the Gibbs free energy equation to calculate an equilibrium temperature. | Worked Example 9.10; Problems 9.24, 9.25, 9.116, 9.117

Worked Example 9.11; Problems 9.26, 9.27, 9.36–9.39, 9.118–9.123

Worked Example 9.12; Problems 9.124–9.129, 9.136, 9.137 |

KEY TERMS

bond dissociation
   energies (*D*)   334
conservation of energy law   312
endothermic   323
enthalpy (*H*)   319
enthalpy change (Δ*H*)   319
entropy (*S*)   338
exothermic   323
first law of
   thermodynamics   313

Gibbs free-energy
   change (Δ*G*)   340
heat   313
heat capacity (*C*)   326
heat of combustion (Δ*H*°$_c$)   335
heat of reaction (Δ*H*)   319
Hess's law   329
internal energy (*E*)   314
joule (J)   312
kinetic energy (*E*$_K$)   312

molar heat capacity (*C*$_m$)   327
potential energy (*E*$_P$)   312
specific heat (*c*)   327
spontaneous process   338
standard enthalpy of
   reaction (Δ*H*°)   321
standard heats of
   formation (Δ*H*°$_f$)   331
state function   315
sublimation   324

system   314
surroundings   314
temperature   313
thermochemical equation   321
thermochemistry   312
thermodynamic standard
   state   321
work (*w*)   316

KEY EQUATIONS

- **Kinetic Energy (Section 9.1)**

$$E_K = \frac{1}{2}mv^2 \quad \text{where } m = \text{mass and } v = \text{velocity}$$

- **Work (Section 9.3)**

Work $(w) = \text{Force } (F) \times \text{Distance } (d) = -P\Delta V$

- **Internal Energy (Section 9.4)**

$\Delta E = q + w = q - P\Delta V$

- **Enthalpy (Section 9.4)**

$\Delta H = q_p = \Delta E + P\Delta V$ where q_p = heat at constant pressure, P = pressure

$\qquad\qquad \Delta E$ = internal energy change, and ΔV = volume change

- **Heat Capacity (Section 9.7)**

$$C = \frac{q}{\Delta T}$$

- **Heat Transfer (Section 9.7)**

$q = C \times \Delta T$

$q = \text{Specific heat } (c) \times \text{Mass of substance } (m) \times \Delta T$

$q = C_m \times \text{Moles of substance} \times \Delta T$

- **Heat of Reaction (Section 9.9)**

For the reaction $a\,\text{A} + b\,\text{B} + \ldots \longrightarrow c\,\text{C} + d\,\text{D} + \ldots$

$\Delta H°_{\text{reaction}} = [c\,\Delta H°_f(\text{C}) + d\,\Delta H°_f(\text{D}) + \ldots\,] - [a\,\Delta H°_f(\text{A}) + b\,\Delta H°_f(\text{B}) + \ldots\,]$

- **Free Energy Change (Section 9.13)**

$\Delta G = \Delta H - T\Delta S$ where ΔH = enthalpy change, T = temperature, ΔS = entropy change

CONCEPTUAL PROBLEMS

Problems 9.1–9.29 appear within the chapter.

9.30 A piece of dry ice (solid CO_2) is placed inside a balloon and the balloon is tied shut. Over time, the carbon dioxide sublimes, causing the balloon to increase in volume. Give the sign of the enthalpy change and the sign of work for the sublimation of CO_2.

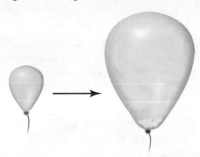

9.31 Imagine a reaction that results in a change in both volume and temperature:

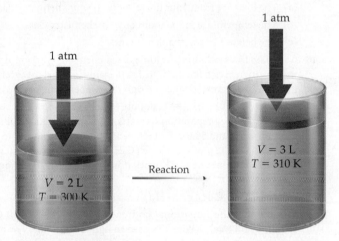

(a) Has any work been done? If so, is its sign positive or negative?

(b) Has there been an enthalpy change? If so, what is the sign of ΔH? Is the reaction exothermic or endothermic?

9.32 Redraw the following diagram to represent the situation (a) when work has been gained by the system and (b) when work has been lost by the system:

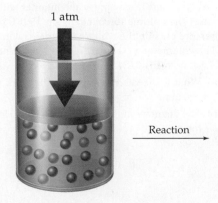

9.33 A reaction is carried out in a cylinder fitted with a movable piston. The starting volume is $V = 5.00$ L, and the apparatus is held at constant temperature and pressure. Assuming that $\Delta H = -35.0$ kJ and $\Delta E = -34.8$ kJ, redraw the piston to show its position after reaction. Does V increase, decrease, or remain the same?

9.34 The reaction of A with B to give D proceeds in two steps:

1) A + B $\longrightarrow$ C $\Delta H° = -20$ kJ
2) C + B $\longrightarrow$ D $\Delta H° = +50$ kJ
3) A + 2 B $\longrightarrow$ D $\Delta H° = ?$

(a) Which Hess's Law diagram represents the reaction steps and the overall reaction?

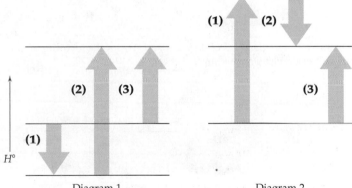

Diagram 1   Diagram 2

(b) What is the value of $\Delta H°$ for the overall reaction

A + 2 B $\longrightarrow$ D $\Delta H° = ?$

9.35 Acetylene, C_2H_2, reacts with H_2 in two steps to yield ethane, CH_3CH_3:

(1)   HC≡CH + H_2 $\longrightarrow$ H_2C=CH_2 $\Delta H° = -175.1$ kJ
(2) H_2C=CH_2 + H_2 $\longrightarrow$ CH_3CH_3 $\Delta H° = -136.3$ kJ
Net   HC≡CH + 2 H_2 $\longrightarrow$ CH_3CH_3 $\Delta H° = -311.4$ kJ

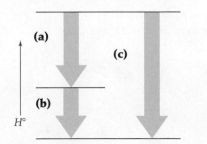

Hess's law diagram

Which arrow (**a**, **b**, **c**) in the Hess's law diagram corresponds to which step, and which arrow corresponds to the net reaction? Where are the reactants located on the diagram, and where are the products located?

9.36 The following reaction is exothermic:

(a) Write a balanced equation for the reaction (red spheres represent A atoms and ivory spheres represent B atoms).

(b) What are the signs (+ or −) of ΔH and ΔS for the reaction?

(c) Is the reaction likely to be spontaneous at lower temperatures only, at higher temperatures only, or at all temperatures?

9.37 The following drawing portrays a reaction of the type A ⟶ B + C, where the different colored spheres represent different molecular structures. Assume that the reaction has $\Delta H° = +55$ kJ. Is the reaction likely to be spontaneous at all temperatures, nonspontaneous at all temperatures, or spontaneous at some but nonspontaneous at others? Explain.

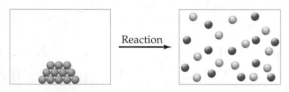

9.38 What are the signs of ΔH, ΔS, and ΔG for the following spontaneous change? Explain.

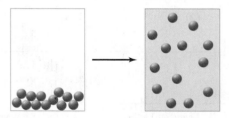

9.39 The following reaction of A_3 molecules is spontaneous:

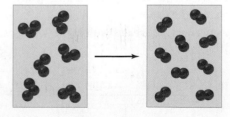

(a) Write a balanced equation for the reaction.

(b) What are the signs of ΔH, ΔS, and ΔG for the reaction? Explain.

Heat, Work, and Energy (Sections 9.1–9.3)

9.40 What is the difference between heat and temperature? Between work and energy? Between kinetic energy and potential energy?

9.41 What is internal energy?

9.42 Which has more kinetic energy, a 1400-kg car moving at 115 km/h or a 12,000-kg truck moving at 38 km/h?

9.43 Assume that the kinetic energy of a 1400-kg car moving at 115 km/h (Problem 9.42) could be converted entirely into heat. What amount of water could be heated from 20 °C to 50 °C by the car's energy? 4.184 J are required to heat 1 g of water by 1 °C.

9.44 Which of the following are state functions, and which are not?

(a) The temperature of an ice cube

(b) The volume of an aerosol can

(c) The amount of time required for Paula Radcliffe to run her world-record marathon: 2:15:25

9.45 Which of the following are state functions, and which are not?

(a) The distance from your dorm room to your chemistry class.

(b) The temperature in the room of your chemistry class.

(c) The balance in your bank account.

9.46 Calculate the work done in joules by a chemical reaction if the volume increases from 3.2 L to 3.4 L against a constant external pressure of 3.6 atm. What is the sign of the energy change?

9.47 The addition of H_2 to C=C double bonds is an important reaction used in the preparation of margarine from vegetable oils. If 50.0 mL of H_2 and 50.0 mL of ethylene (C_2H_4) are allowed to react at 1.5 atm, the product ethane (C_2H_6) has a volume of 50.0 mL. Calculate the amount of PV work done, and tell the direction of the energy flow.

$$C_2H_4(g) + H_2(g) \longrightarrow C_2H_6(g)$$

9.48 Assume that the nutritional content of an apple—say 50 Cal (1 Cal = 1000 cal)—could be used to light a light bulb. For how many minutes would there be light from each of the following?

(a) a 100-watt incandescent bulb ((1 W = 1 J/s)

(b) a 23-watt compact fluorescent bulb, which provides a similar amount of light

9.49 A double cheeseburger has a caloric content of 440 Cal (1 Cal = 1000 cal). If this energy could be used to operate a television set, for how many hours would the following sets run?

(a) a 275-watt 46-in. plasma TV (1 W = 1 J/s)

(b) a 175-watt 46-in. LCD TV

9.50 A reaction inside a cylindrical container with a movable piston causes the volume to change from 12.0 L to 18.0 L while the pressure outside the container remains constant at 0.975 atm. (The volume of a cylinder is $V = \pi r^2 h$, where h is the height. 1 L·atm = 101.325 J.)

(a) What is the value in joules of the work w done during the reaction?

(b) The diameter of the piston is 17.0 cm. How far does the piston move?

9.51 At a constant pressure of 0.905 atm, a chemical reaction takes place in a cylindrical container with a movable piston having a diameter of 40.0 cm. During the reaction, the height of the piston drops by 65.0 cm. (The volume of a cylinder is $V = \pi r^2 h$, where h is the height; $1\ L \cdot atm = J$.)

(a) What is the change in volume in liters during the reaction?

(b) What is the value in joules of the work w done during the reaction?

Energy and Enthalpy (Section 9.4)

9.52 What is the difference between the internal energy change ΔE and the enthalpy change ΔH? Which of the two is measured at constant pressure, and which at constant volume?

9.53 What is the sign of ΔH for an exothermic reaction? For an endothermic reaction?

9.54 Under what circumstances are ΔE and ΔH essentially equal?

9.55 Which of the following has the highest enthalpy content, and which the lowest at a given temperature: $H_2O(s)$, $H_2O(l)$, or $H_2O(g)$? Explain.

9.56 The explosion of 2.00 mol of solid trinitrotoluene (TNT; $C_7H_5N_3O_6$) with a volume of approximately 274 mL produces gases with a volume of 448 L at room temperature and 1.0 atm pressure. How much PV work in kilojoules is done during the explosion?

Trinitrotoluene

$$2\ C_7H_5N_3O_6(s) \longrightarrow 12\ CO(g) + 5\ H_2(g) + 3\ N_2(g) + 2\ C(s)$$

9.57 The reaction between hydrogen and oxygen to yield water vapor has $\Delta H° = -484$ kJ. How much PV work is done, and what is the value of ΔE in kilojoules for the reaction of 0.50 mol of H_2 with 0.25 mol of O_2 at atmospheric pressure if the volume change is -5.6 L?

$$2\ H_2(g) + O_2(g) \longrightarrow 2\ H_2O(g) \quad \Delta H° = -484\ kJ$$

9.58 The enthalpy change for the reaction of 50.0 mL of ethylene with 50.0 mL of H_2 at 1.5 atm pressure (Problem 9.47) is $\Delta H = -0.31$ kJ. What is the value of ΔE?

9.59 Assume that a particular reaction evolves 244 kJ of heat and that 35 kJ of PV work is gained by the system. What are the values of ΔE and ΔH for the system? For the surroundings?

9.60 What is the enthalpy change (ΔH) for a reaction at a constant pressure of 1.00 atm if the internal energy change (ΔE) is 44.0 kJ and the volume increase is 14.0 L? ($1\ L \cdot atm = 101.325\ J$.)

9.61 A reaction takes place at a constant pressure of 1.10 atm with an internal energy change (ΔE) of 71.5 kJ and a volume decrease of 13.6 L. What is the enthalpy change (ΔH) for the reaction? ($1\ L \cdot atm = 101.325\ J$.)

Thermochemical Equations for Chemical and Physical Changes (Sections 9.5 and 9.6)

9.62 Indicate the direction of heat transfer between the system and the surroundings, classify the following processes as endo- or exothermic, and give the sign of $\Delta H°$.

(a) Evaporating rubbing alcohol from your skin.

(b) Solidifying molten gold into a gold bar.

(c) Melting solid gallium metal with the heat from your hand.

9.63 Indicate the direction of heat transfer between the system and the surroundings, classify the following processes as endo- or exothermic, and give the sign of $\Delta H°$.

(a) $N_2(g) + O_2(g) \longrightarrow 2\ NO(g)$ $\Delta H° = +182.6$ kJ

(b) $2\ H_2O(g) \longrightarrow 2\ H_2(g) + O_2(g)$ $\Delta H° = +483.6$ kJ

(c) $H_2(g) + Cl_2(g) \longrightarrow 2HCl(g)$ $\Delta H° = -184.6$ kJ

9.64 The familiar "ether" used as an anesthetic agent is diethyl ether, $C_4H_{10}O$. Its heat of vaporization is $+26.5$ kJ/mol at its boiling point. How much energy in kilojoules is required to convert 100 mL of diethyl ether at its boiling point from liquid to vapor if its density is 0.7138 g/mL?

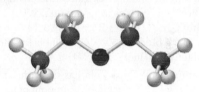

Diethyl ether

9.65 How much energy in kilojoules is required to convert 100 mL of water at its boiling point from liquid to vapor, and how does this compare with the result calculated in Problem 9.64 for diethyl ether? $[\Delta H_{vap}(H_2O) = +40.7$ kJ/mol$]$

9.66 Aluminum metal reacts with chlorine with a spectacular display of sparks:

$$2\ Al(s) + 3\ Cl_2(g) \longrightarrow 2\ AlCl_3(s) \quad \Delta H° = -1408.4\ kJ$$

How much heat in kilojoules is released on reaction of 5.00 g of Al?

9.67 How much heat in kilojoules is evolved or absorbed in the reaction of 1.00 g of Na with H_2O? Is the reaction exothermic or endothermic?

$$2\ Na(s) + 2\ H_2O(l) \longrightarrow 2\ NaOH(aq) + H_2(g)$$
$$\Delta H° = -368.4\ kJ$$

9.68 How much heat in kilojoules is evolved or absorbed in each of the following reactions?

(a) Burning of 15.5 g of propane:

$$C_3H_8(g) + 5\ O_2(g) \longrightarrow 3\ CO_2(g) + 4\ H_2O(l)$$
$$\Delta H° = -2220\ kJ$$

(b) Reaction of 4.88 g of barium hydroxide octahydrate with ammonium chloride:

$$Ba(OH)_2 \cdot 8\ H_2O(s) + 2\ NH_4Cl(s) \longrightarrow$$
$$BaCl_2(aq) + 2\ NH_3(aq) + 10\ H_2O(l) \quad \Delta H° = +80.3\ kJ$$

9.69 Nitromethane (CH_3NO_2), sometimes used as a fuel in drag racers, burns according to the following equation. How much heat is released by burning 100.0 g of nitromethane?

$$4\ CH_3NO_2(l) + 7\ O_2(g) \longrightarrow 4\ CO_2(g) + 6\ H_2O(g) + 4\ NO_2(g)$$
$$\Delta H° = -2441.6\ kJ$$

9.70 How much heat in kilojoules is evolved or absorbed in the reaction of 2.50 g of Fe_2O_3 with enough carbon monoxide to produce iron metal? Is the process exothermic or endothermic?

$$Fe_2O_3(s) + 3\,CO(g) \longrightarrow 2\,Fe(s) + 3\,CO_2(g)$$
$$\Delta H° = -24.8\ kJ$$

9.71 How much heat in kilojoules is evolved or absorbed in the reaction of 233.0 g of calcium oxide with enough carbon to produce calcium carbide? Is the process exothermic or endothermic?

$$CaO(s) + 3\,C(s) \longrightarrow CaC_2(s) + CO(g)$$
$$\Delta H° = 464.6\ kJ$$

Calorimetry and Heat Capacity (Section 9.7)

9.72 What is the difference between heat capacity and specific heat?

9.73 Does a measurement carried out in a bomb calorimeter give a value for ΔH or ΔE? Explain.

9.74 Sodium metal is sometimes used as a cooling agent in heat-exchange units because of its relatively high molar heat capacity of 28.2 J/(mol·°C). What is the specific heat and molar heat capacity of sodium in J/(g·°C)?

9.75 Titanium metal is used as a structural material in many high-tech applications, such as in jet engines. What is the specific heat of titanium in J/(g·°C) if it takes 89.7 J to raise the temperature of a 33.0 g block by 5.20 °C? What is the molar heat capacity of titanium in J/(mol·°C)?

9.76 Assuming that Coca Cola has the same specific heat as water [4.18 J/(g·°C)], calculate the amount of heat in kilojoules transferred when one can (about 350 g) is cooled from 25 °C to 3 °C.

9.77 Calculate the amount of heat required to raise the temperature of 250.0 g (approximately 1 cup) of hot chocolate from 25.0 °C to 80.0 °C. Assume hot chocolate has the same specific heat as water [4.18 J/(g·°C)].

9.78 Instant cold packs used to treat athletic injuries contain solid NH_4NO_3 and a pouch of water. When the pack is squeezed, the pouch breaks and the solid dissolves, lowering the temperature because of the endothermic reaction

$$NH_4NO_3(s) \xrightarrow{H_2O} NH_4NO_3(aq) \quad \Delta H = +25.7\ kJ$$

What is the final temperature in a squeezed cold pack that contains 50.0 g of NH_4NO_3 dissolved in 125 mL of water? Assume a specific heat of 4.18 J/(g·°C) for the solution, an initial temperature of 25.0 °C, and no heat transfer between the cold pack and the environment.

9.79 Instant hot packs contain a solid and a pouch of water. When the pack is squeezed, the pouch breaks and the solid dissolves, increasing the temperature because of the exothermic reaction. The following reaction is used to make a hot pack:

$$LiCl(s) \xrightarrow{H_2O} Li^+(aq) + Cl^-(aq) \quad \Delta H = -36.9\ kJ$$

What is the final temperature in a squeezed hot pack that contains 25.0 g of LiCl dissolved in 125 mL of water? Assume a specific heat of 4.18 J/(g·°C) for the solution, an initial temperature of 25.0 °C, and no heat transfer between the hot pack and the environment.

9.80 When 1.045 g of CaO is added to 50.0 mL of water at 25.0 °C in a calorimeter, the temperature of the water increases to 32.3 °C. Assuming that the specific heat of the solution is 4.18 J/(g·°C) and that the calorimeter itself absorbs a negligible amount of heat, calculate ΔH in kilojoules/mol $Ca(OH)_2$ for the reaction

$$CaO(s) + H_2O(l) \longrightarrow Ca(OH)_2(aq)$$

9.81 When a solution containing 8.00 g of NaOH in 50.0 g of water at 25.0 °C is added to a solution of 8.00 g of HCl in 250.0 g of water

at 25.0 °C in a calorimeter, the temperature of the solution increases to 33.5 °C. Assuming that the specific heat of the solution is 4.18 J/(g·°C) and that the calorimeter itself absorbs a negligible amount of heat, calculate ΔH in kilojoules/mol for the reaction

$$NaOH(aq) + HCl(aq) \longrightarrow NaCl(aq) + H_2O(l)$$

When the experiment is repeated using a solution of 10.00 g of HCl in 248.0 g of water, the same temperature increase is observed. Explain.

9.82 When 0.187 g of benzene, C_6H_6, is burned in a bomb calorimeter, the temperature of both the water and the calorimeter rise by 4.53 °C. Assuming that the bath contains 250.0 g of water and that the heat capacity for the calorimeter is 525 J/°C, calculate combustion energies (ΔE) for benzene in units of kJ/g and kJ/mol.

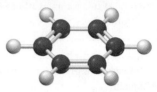

Benzene

9.83 When 0.500 g of ethanol, C_2H_6O, is burned in a bomb calorimeter, the temperature of both the water and the calorimeter rise by 9.15 °C. Assuming that the bath contains 250.0 g of water and that the heat capacity for the calorimeter is 575 J/°C, calculate combustion energies (ΔE) for ethanol in units of kJ/g and kJ/mol.

Hess's Law and Heats of Formation (Sections 9.8 and 9.9)

9.84 How is the standard state of an element defined? What is Hess's law, and why does it "work"?

9.85 The following steps occur in the reaction of ethyl alcohol (CH_3CH_2OH) with oxygen to yield acetic acid (CH_3CO_2H). Show that equations 1 and 2 sum to give the net equation and calculate $\Delta H°$ for the net equation.

(1) $CH_3CH_2OH(l) + 1/2\,O_2(g) \longrightarrow$
$$CH_3CHO(g) + H_2O(l) \quad \Delta H° = -174.2\ k$$

(2) $CH_3CHO(g) + 1/2\,O_2(g) \longrightarrow$
$$CH_3CO_2H(l) \quad \Delta H° = -318.4\ k$$

(Net) $CH_3CH_2OH(l) + O_2(g) \longrightarrow$
$$CH_3CO_2H(l) + H_2O(l) \quad \Delta H° =$$

9.86 The industrial degreasing solvent methylene chloride, CH_2Cl_2, is prepared from methane by reaction with chlorine:

$$CH_4(g) + 2\,Cl_2(g) \longrightarrow CH_2Cl_2(g) + 2\,HCl(g)$$

Use the following data to calculate $\Delta H°$ in kilojoules for the reaction:

$$CH_4(g) + Cl_2(g) \longrightarrow CH_3Cl(g) + HCl(g) \quad \Delta H° = -98.3\ kJ$$
$$CH_3Cl(g) + Cl_2(g) \longrightarrow CH_2Cl_2(g) + HCl(g) \quad \Delta H° = -104$$

Methylene chloride

9.87 Hess's law can be used to calculate reaction enthalpies for hypothetical processes that can't be carried out in the laboratory. Set up a Hess's law cycle that will let you calculate $\Delta H°$ for the conversion of methane to ethylene:

$$2\,CH_4(g) \longrightarrow C_2H_4(g) + 2\,H_2(g)$$

You can use the following information:

$$2\,C_2H_6(g) + 7\,O_2(g) \longrightarrow 4\,CO_2(g) + 6\,H_2O(l)$$
$$\Delta H° = -3120.8\ kJ$$

$$CH_4(g) + 2\,O_2(g) \longrightarrow CO_2(g) + 2\,H_2O(l)$$
$$\Delta H° = -890.3\ kJ$$

$$C_2H_4(g) + H_2(g) \longrightarrow C_2H_6(g) \qquad \Delta H° = -136.3\ kJ$$
$$H_2O(l) \quad \Delta H°_f = -285.8\ kJ/mol$$

9.88 What is a compound's standard heat of formation?

9.89 How is the standard state of an element defined? Why do elements always have $\Delta H°_f = 0$?

9.90 What phase of matter is associated with the standard states of the following elements and compounds?

(a) Cl_2 (b) Hg (c) CO_2 (d) Ga

9.91 What is the phase of the standard states of the following elements and compounds?

(a) NH_3 (b) Fe (c) N_2 (d) Br_2

9.92 Write balanced equations for the formation of the following compounds from their elements:

(a) iron(III) oxide

(b) sucrose (table sugar, $C_{12}H_{22}O_{11}$)

(c) uranium hexafluoride (a solid at 25 °C)

9.93 Write balanced equations for the formation of the following compounds from their elements:

(a) ethanol (C_2H_6O)

(b) sodium sulfate

(c) dichloromethane (a liquid, CH_2Cl_2)

9.94 Sulfuric acid (H_2SO_4), the most widely produced chemical in the world, is made by a two-step oxidation of sulfur to sulfur trioxide, SO_3, followed by reaction with water. Calculate $\Delta H°_f$ for SO_3 in kJ/mol, given the following data:

$$S(s) + O_2(g) \longrightarrow SO_2(g) \qquad \Delta H° = -296.8\ kJ$$
$$SO_2(g) + 1/2\,O_2(g) \longrightarrow SO_3(g) \quad \Delta H° = -98.9\ kJ$$

9.95 Calculate $\Delta H°_f$ in kJ/mol for benzene, C_6H_6, from the following data:

$$2\,C_6H_6(l) + 15\,O_2(g) \longrightarrow 12\,CO_2(g) + 6\,H_2O(l)$$
$$\Delta H° = -6534\ kJ$$

$$\Delta H°_f\,(CO_2) = -393.5\ kJ/mol$$
$$\Delta H°_f\,(H_2O) = -285.8\ kJ/mol$$

9.96 The standard enthalpy change for the reaction of $SO_3(g)$ with $H_2O(l)$ to yield $H_2SO_4(aq)$ is $\Delta H° = -227.8\ kJ$. Use the information in Problem 9.94 to calculate $\Delta H°_f$ for $H_2SO_4(aq)$ in kJ/mol. [For $H_2O(l)$, $\Delta H°_f = -285.8\ kJ/mol$.]

9.97 Acetic acid (CH_3CO_2H), whose aqueous solutions are known as vinegar, is prepared by reaction of ethyl alcohol (CH_3CH_2OH) with oxygen:

$$CH_3CH_2OH(l) + O_2(g) \longrightarrow CH_3CO_2H(l) + H_2O(l)$$

Use the following data to calculate $\Delta H°$ in kilojoules for the reaction:

$$\Delta H°_f\,[CH_3CH_2OH(l)] = -277.7\ kJ/mol$$
$$\Delta H°_f\,[CH_3CO_2H(l)] = -484.5\ kJ/mol$$
$$\Delta H°_f\,[H_2O(l)] = -285.8\ kJ/mol$$

9.98 Styrene (C_8H_8), the precursor of polystyrene polymers, has a standard heat of combustion of $-4395\ kJ/mol$. Write a balanced equation for the combustion reaction, and calculate $\Delta H°_f$ for styrene in kJ/mol.

$$\Delta H°_f\,[CO_2(g)] = -393.5\ kJ/mol;$$
$$\Delta H°_f\,[H_2O(l)] = -285.8\ kJ/mol$$

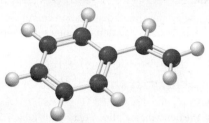

Styrene

9.99 Methyl *tert*-butyl ether (MTBE), $C_5H_{12}O$, a gasoline additive used to boost octane ratings, has $\Delta H°_f = -313.6\ kJ/mol$. Write a balanced equation for its combustion reaction, and calculate its standard heat of combustion in kilojoules.

9.100 Methyl *tert*-butyl ether is prepared by reaction of methanol(l)($\Delta H°_f = -239.2\ kJ/mol$) with 2-methyl-propene($g$), according to the equation

$$CH_3-\overset{\displaystyle CH_3}{\underset{}{C}}=CH_2 + CH_3OH \longrightarrow$$

2-Methylpropene

$$CH_3-\overset{\displaystyle CH_3}{\underset{\displaystyle CH_3}{C}}-O-CH_3 \quad \Delta H° = -57.5\ kJ$$

Methyl *tert*-butyl ether

Calculate $\Delta H°_f$ in kJ/mol for 2-methylpropene.

9.101 One possible use for the cooking fat left over after making french fries is to burn it as fuel. Write a balanced equation, and use the following data to calculate the amount of energy released in kJ/mL from the combustion of cooking fat:

$$\text{Formula} = C_{51}H_{88}O_6 \quad \Delta H°_f = -1310\ kJ/mol$$
$$\text{Density} = 0.94\ g/ml$$

9.102 Given the standard heats of formation shown in Appendix B, what is $\Delta H°$ in kilojoules for the reaction $CaCO_3(s) \longrightarrow CaO(s) + CO_2(g)$?

9.103 Given the standard heats of formation shown in Appendix B, what is $\Delta H°$ in kilojoules for the reaction

$$3\,N_2O_4(g) + 2\,H_2O(l) \longrightarrow 4\,HNO_3(aq) + 2\,NO(g).$$

9.104 Calculate $\Delta H°$ in kilojoules for the synthesis of lime (CaO) from limestone ($CaCO_3$), the key step in the manufacture of cement.

$$CaCO_3(s) \longrightarrow CaO(s) + CO_2(g) \quad \Delta H°_f\,[CaCO_3(s)] = -1207.6\ kJ/mol$$
$$\Delta H°_f\,[CaO(s)] = -634.9\ kJ/mol$$
$$\Delta H°_f\,[CO_2(g)] = -393.5\ kJ/mol$$

9.105 Use the information in Table 9.2 to calculate $\Delta H°$ in kilojoules for the photosynthesis of glucose ($C_6H_{12}O_6$) and O_2 from CO_2 and liquid H_2O, a reaction carried out by all green plants.

Bond Dissociation Energies (Section 9.10)

9.106 Use the bond dissociation energies in Table 7.2 to calculate an approximate $\Delta H°$ in kilojoules for the reaction of ethylene with hydrogen to yield ethane.

$$H_2C{=}CH_2(g) + H_2(g) \longrightarrow CH_3CH_3(g)$$

9.107 Use the data in Table 7.2 to calculate an approximate $\Delta H°$ in kilojoules for the synthesis of hydrazine from ammonia: $2\,NH_3(g) + Cl_2(g) \longrightarrow N_2H_4(g) + 2\,HCl(g)$.

9.108 Use the average bond dissociation energies in Table 7.2 to calculate approximate reaction enthalpies in kilojoules for the following processes:

(a) $2\,CH_4(g) \longrightarrow C_2H_6(g) + H_2(g)$

(b) $C_2H_6(g) + F_2(g) \longrightarrow C_2H_5F(g) + HF(g)$

(c) $N_2(g) + 3\,H_2(g) \longrightarrow 2\,NH_3(g)$

9.109 Use the bond dissociation energies in Table 7.2 to calculate an approximate $\Delta H°$ in kilojoules for the high-temperature industrial synthesis of isopropyl alcohol (rubbing alcohol) by reaction of water vapor with propene.

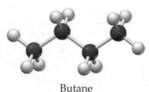

$$\underset{\text{Propene}}{CH_3CH{=}CH_2} + H_2O \longrightarrow \underset{\text{Isopropyl alcohol}}{CH_3\overset{\displaystyle OH}{\overset{\displaystyle |}{C}}HCH_3}$$

Fossil Fuels and Heats of Combustion (Section 9.11)

9.110 Liquid butane (C_4H_{10}), the fuel used in many disposable lighters, has $\Delta H°_f = -147.5$ kJ/mol and a density of 0.579 g/mL. Write a balanced equation for the combustion of butane, and use Hess's law to calculate the enthalpy of combustion in kJ/mol, kJ/g, and kJ/mL.

Butane

9.111 Calculate an approximate heat of combustion for ethane (C_2H_6) in kJ/mol and kJ/g by using the bond dissociation energies in Table 7.2. (The strength of the $O{=}O$ bond is 498 kJ/mol, and that of a $C{=}O$ bond in CO_2 is 804 kJ/mol.)

Free Energy and Entropy (Sections 9.12 and 9.13)

9.112 What does entropy measure?

9.113 What are the two terms that make up the free-energy change for a reaction, ΔG, and which of the two is usually more important?

9.114 How is it possible for a reaction to be spontaneous yet endothermic?

9.115 Is it possible for a reaction to be nonspontaneous yet exothermic? Explain.

9.116 Tell whether the entropy changes for the following processes are likely to be positive or negative:

(a) The fizzing of a newly opened can of soda

(b) The growth of a plant from seed

9.117 Tell whether the entropy changes, ΔS, for the following processes are likely to be positive or negative:

(a) The conversion of liquid water to water vapor at 100 °C

(b) The freezing of liquid water to ice at 0 °C

(c) The eroding of a mountain by a glacier

9.118 Tell whether the free-energy changes, ΔG, for the processes listed in Problem 9.117 are likely to be positive, negative, or zero.

9.119 When a bottle of perfume is opened, odorous molecules mix with air and slowly diffuse throughout the entire room. Is ΔG for the diffusion process positive, negative, or zero? What about ΔH and ΔS for the diffusion?

9.120 One of the steps in the cracking of petroleum into gasoline involves the thermal breakdown of large hydrocarbon molecules into smaller ones. For example, the following reaction might occur:

$$C_{11}H_{24} \longrightarrow C_4H_{10} + C_4H_8 + C_3H_6$$

Is ΔS for this reaction likely to be positive or negative? Explain.

9.121 The commercial production of 1,2-dichloroethane, a solvent used in dry cleaning, involves the reaction of ethylene with chlorine:

$$C_2H_4(g) + Cl_2(g) \longrightarrow C_2H_4Cl_2(l)$$

Is ΔS for this reaction likely to be positive or negative? Explain.

9.122 Tell whether reactions with the following values of ΔH and ΔS are spontaneous or nonspontaneous and whether they are exothermic or endothermic:

(a) $\Delta H = -48$ kJ; $\Delta S = +135$ J/K at 400 K

(b) $\Delta H = -48$ kJ; $\Delta S = -135$ J/K at 400 K

(c) $\Delta H = +48$ kJ; $\Delta S = +135$ J/K at 400 K

(d) $\Delta H = +48$ kJ; $\Delta S = -135$ J/K at 400 K

9.123 Tell whether reactions with the following values of ΔH and ΔS are spontaneous or nonspontaneous and whether they are exothermic or endothermic:

(a) $\Delta H = -128$ kJ; $\Delta S = 35$ J/K at 500 K

(b) $\Delta H = +67$ kJ; $\Delta S = -140$ J/K at 250 K

(c) $\Delta H = +75$ kJ; $\Delta S = 95$ J/K at 800 K

9.124 Suppose that a reaction has $\Delta H = -33$ kJ and $\Delta S = -58$ J/K. At what temperature will it change from spontaneous to nonspontaneous?

9.125 Suppose that a reaction has $\Delta H = +41$ kJ and $\Delta S = -27$ J/K. At what temperature, if any, will it change between spontaneous and nonspontaneous?

9.126 Which of the reactions (a)–(d) in Problem 9.122 are spontaneous at all temperatures, which are nonspontaneous at all temperatures, and which have an equilibrium temperature?

9.127 Vinyl chloride ($H_2C{=}CHCl$), the starting material used in the industrial preparation of poly(vinyl chloride), is prepared by a two-step process that begins with the reaction of Cl_2 with ethylene to yield 1,2-dichloroethane:

$$Cl_2(g) + H_2C{=}CH_2(g) \longrightarrow ClCH_2CH_2Cl(l)$$
$$\Delta H° = -217.5 \text{ kJ}$$
$$\Delta S° = -233.9 \text{ J/K}$$

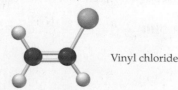

Vinyl chloride

(a) Tell whether the reaction is favored by entropy, by enthalpy, by both, or by neither, and then calculate $\Delta G°$ at 298 K.

(b) Tell whether the reaction has an equilibrium temperature between spontaneous and nonspontaneous. If yes, calculate the equilibrium temperature.

9.128 Ethyl alcohol has $\Delta H_{fusion} = 5.02$ kJ/mol and melts at -114.1 °C. What is the value of ΔS_{fusion} for ethyl alcohol?

9.129 Chloroform has $\Delta H_{vaporization} = 29.2$ kJ/mol and boils at 61.2 °C. What is the value of $\Delta S_{vaporization}$ for chloroform?

CHAPTER PROBLEMS

9.130 When a sample of a hydrocarbon fuel is ignited and burned in oxygen, the internal energy decreases by 7.20 kJ. If 5670 J of heat were transferred to the surroundings, what is the sign and magnitude of work? If the reaction took place in an environment with a pressure of 1 atm, what was the volume change?

9.131 Used in welding metals, the reaction of acetylene with oxygen has $\Delta H° = -1256.2$ kJ:

$$C_2H_2(g) + 5/2\,O_2(g) \longrightarrow H_2O(g) + 2\,CO_2(g)$$
$$\Delta H° = -1256.2 \text{ kJ}$$

How much PV work is done in kilojoules and what is the value of ΔE in kilojoules for the reaction of 6.50 g of acetylene at atmospheric pressure if the volume change is -2.80 L?

9.132 Ethyl chloride (C_2H_5Cl), a substance used as a topical anesthetic, is prepared by reaction of ethylene with hydrogen chloride:

$$C_2H_4(g) + HCl(g) \longrightarrow C_2H_5Cl(g) \quad \Delta H° = -72.3 \text{ kJ}$$

Ethyl chloride

How much PV work is done in kilojoules, and what is the value of ΔE in kilojoules if 89.5 g of ethylene and 125 g of HCl are allowed to react at atmospheric pressure and the volume change is -71.5 L?

9.133 When 1.50 g of magnesium metal is allowed to react with 200 mL of 6.00 M aqueous HCl, the temperature rises from 25.0 °C to 42.9 °C. Calculate ΔH in kilojoules for the reaction, assuming that the heat capacity of the calorimeter is 776 J/°C, that the specific heat of the final solution is the same as that of water $[(4.18 \text{ J}/(g \cdot °C)]$, and that the density of the solution is 1.00 g/mL.

9.134 Use the data in Appendix B to find standard enthalpies of reaction in kilojoules for the following processes:

(a) $C(s) + CO_2(g) \longrightarrow 2\,CO(g)$

(b) $2\,H_2O_2(aq) \longrightarrow 2\,H_2O(l) + O_2(g)$

(c) $Fe_2O_3(s) + 3\,CO(g) \longrightarrow 2\,Fe(s) + 3\,CO_2(g)$

9.135 Find $\Delta H°$ in kilojoules for the reaction of nitric oxide with oxygen, $2\,NO(g) + O_2(g) \longrightarrow N_2O_4(g)$, given the following data:

$$N_2O_4(g) \longrightarrow 2\,NO_2(g) \qquad \Delta H° = 55.3 \text{ kJ}$$
$$NO(g) + 1/2\,O_2(g) \longrightarrow NO_2(g) \quad \Delta H° = -58.1 \text{ kJ}$$

9.136 The boiling point of a substance is defined as the temperature at which liquid and vapor coexist in equilibrium. Use the heat of vaporization ($\Delta H_{vap} = 30.91$ kJ/mol) and the entropy of

vaporization $[\Delta S_{vap} = 93.2 \text{ J}/(K \cdot mol)]$ to calculate the boiling point (°C) of liquid bromine.

9.137 What is the melting point of benzene in kelvin if $\Delta H_{fusion} = 9.95$ kJ/mol and $\Delta S_{fusion} = 35.7 \text{ J}/(K \cdot mol)$?

9.138 Metallic mercury is obtained by heating the mineral cinnabar (HgS) in air:

$$HgS(s) + O_2(g) \longrightarrow Hg(l) + SO_2(g)$$

(a) Use the data in Appendix B to calculate $\Delta H°$ in kilojoules for the reaction.

(b) The entropy change for the reaction is $\Delta S° = +36.7$ J/K. Is the reaction spontaneous at 25 °C?

(c) Under what conditions, if any, is the reaction nonspontaneous? Explain.

9.139 Methanol (CH_3OH) is made industrially in two steps from CO and H_2. It is so cheap to make that it is being considered for use as a precursor to hydrocarbon fuels, such as methane (CH_4):

Step 1. $\ CO(g) + 2\,H_2(g) \longrightarrow CH_3OH(l)$
$$\Delta S° = -332 \text{ J/K}$$

Step 2. $\ CH_3OH(l) \longrightarrow CH_4(g) + 1/2\,O_2(g)$
$$\Delta S° = 162 \text{ J/K}$$

(a) Calculate $\Delta H°$ in kilojoules for step 1.

(b) Calculate $\Delta G°$ in kilojoules for step 1.

(c) Is step 1 spontaneous at 298 K?

(d) Which term is more important, $\Delta H°$ or $\Delta S°$?

(e) In what temperature range is step 1 spontaneous?

(f) Calculate $\Delta H°$ for step 2.

(g) Calculate $\Delta G°$ for step 2.

(h) Is step 2 spontaneous at 298 K?

(i) Which term is more important, $\Delta H°$ or $\Delta S°$?

(j) In what temperature range is step 2 spontaneous?

(k) Calculate an overall $\Delta G°$, $\Delta H°$, and $\Delta S°$ for the formation of CH_4 from CO and H_2.

(l) Is the overall reaction spontaneous at 298 K?

(m) If you were designing a production facility, would you plan on carrying out the reactions in separate steps or together? Explain.

9.140 Isooctane, C_8H_{18}, is the component of gasoline from which the term *octane rating* derives.

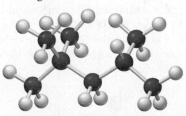

Isooctane

(a) Write a balanced equation for the combustion of isooctane (l) with O_2 to yield $CO_2(g)$ and $H_2O(l)$.

(b) The standard molar heat of combustion for isooctane(l) is -5461 kJ/mol. Calculate $\Delta H°_f$ for isooctane (l).

9.141 We said in Section 9.1 that the potential energy of water at the top of a dam or waterfall is converted into heat when the water dashes against rocks at the bottom. The potential energy of the water at the top is equal to $E_p = mgh$, where m is the mass of the water, g is the acceleration of the falling water due to gravity

(g = 9.81 m/s^2), and h is the height of the water. Assuming that all the energy is converted to heat, calculate the temperature rise of the water in degrees Celsius after falling over California's Yosemite Falls, a distance of 739 m. The specific heat of water is 4.18 J/(g · K).

9.142 For a process to be spontaneous, the total entropy of the system and its surroundings must increase; that is

$$\Delta S_{total} = \Delta S_{system} + \Delta S_{surr} > 0 \quad \text{For a spontaneous process}$$

Furthermore, the entropy change in the surroundings, ΔS_{surr}, is related to the enthalpy change for the process by the equation $\Delta S_{surr} = -\Delta H/T$.

(a) Since both ΔG and ΔS_{total} offer criteria for spontaneity, they must be related. Derive a relationship between them.

(b) What is the value of ΔS_{surr} for the photosynthesis of glucose from CO_2 at 298 K?

$$6\,CO_2(g) + 6\,H_2O(l) \longrightarrow C_6H_{12}O_6(s) + 6\,O_2(g)$$
$$\Delta G° = 2879\ \text{kJ}$$
$$\Delta S° = -262\ \text{J/K}$$

9.143 Set up a Hess's law cycle, and use the following information to calculate $\Delta H°_f$ for aqueous nitric acid, $HNO_3(aq)$. You will need to use fractional coefficients for some equations.

$$3\,NO_2(g) + H_2O(l) \longrightarrow 2\,HNO_3(aq) + NO(g)$$
$$\Delta H° = -137.3\ \text{kJ}$$
$$2\,NO(g) + O_2(g) \longrightarrow 2\,NO_2(g) \quad \Delta H° = -116.2\ \text{kJ}$$
$$4\,NH_3(g) + 5\,O_2(g) \longrightarrow 4\,NO(g) + 6\,H_2O(l)$$
$$\Delta H° = -1165.2\ \text{kJ}$$
$$NH_3(g) \quad \Delta H°_f = -46.1\ \text{kJ/mol}$$
$$H_2O(l) \quad \Delta H°_f = -285.8\ \text{kJ/mol}$$

9.144 A 110.0 g piece of molybdenum metal is heated to 100.0 °C and placed in a calorimeter that contains 150.0 g of water at 24.6 °C. The system reaches equilibrium at a final temperature of 28.0 °C. Calculate the specific heat of molybdenum metal in J/(g · °C). The specific heat of water is 4.184 J/(g · °C).

9.145 Given 400.0 g of hot tea at 80.0 °C, what mass of ice at 0 °C must be added to obtain iced tea at 10.0 °C? The specific heat of the tea is 4.18 J/(g · °C), and ΔH_{fusion} for ice is +6.01 kJ/mol.

9.146 Citric acid has three dissociable hydrogens. When 5.00 mL of 0.64 M citric acid and 45.00 mL of 0.77 M NaOH are mixed at an initial temperature of 26.0 °C, the temperature rises to 27.9 °C as the citric acid is neutralized. The combined mixture has a mass of 51.6 g and a specific heat of 4.0 J/(g · °C). Assuming that no heat is transferred to the surroundings, calculate the enthalpy change for the reaction of 1.00 mol of citric acid in kJ. Is the reaction exothermic or endothermic?

Citric acid

9.147 Assume that 100.0 mL of 0.200 M CsOH and 50.0 mL of 0.4[...] M HCl are mixed in a calorimeter. The solutions start out [...] 22.50 °C, and the final temperature after reaction is 24.28 °C. T[...] densities of the solutions are all 1.00 g/mL, and the specific he[...] of the mixture is 4.2 J/(g · °C). What is the enthalpy change f[...] the neutralization reaction of 1.00 mol of CsOH in kJ?

9.148 Imagine that you dissolve 10.0 g of a mixture of NaNO$_3$ and K[...] in 100.0 g of water and find that the temperature rises by 2.22 °[...] Using the following data, calculate the mass of each compound [...] the original mixture. Assume that the specific heat of the soluti[...] is 4.18 J/(g · °C).

$$NaNO_3(s) \longrightarrow NaNO_3(aq) \quad \Delta H = +20.4\ \text{kJ/mol}$$
$$KF(s) \longrightarrow KF(aq) \qquad \Delta H = -17.7\ \text{kJ/mol}$$

9.149 Consider the reaction: $4\,CO(g) + 2\,NO_2(g) \longrightarrow 4\,CO_2(g)$ [...] N$_2$(g). Using the following information, determine $\Delta H°$ for t[...] reaction at 25 °C.

$NO(g)$	$\Delta H°_f = +91.3\ \text{kJ/mo}$
$CO_2(g)$	$\Delta H°_f = -393.5\ \text{kJ/m}$
$2\,NO(g) + O_2(g) \longrightarrow 2\,NO_2(g)$	$\Delta H° = -116.2\ \text{kJ}$
$2\,CO(g) + O_2(g) \longrightarrow 2\,CO_2(g)$	$\Delta H° = -566.0\ \text{kJ}$

MULTICONCEPT PROBLEMS

9.150 The reaction $S_8(g) \longrightarrow 4\,S_2(g)$ has $\Delta H° = +237$ kJ

(a) The S$_8$ molecule has eight sulfur atoms arranged in a rin[...] What is the hybridization and geometry around each sulf[...] atom in S$_8$?

(b) The average S—S bond dissociation energy is 225 kJ/m[...] Using the value of $\Delta H°$ given above, what is the S=S doub[...] bond energy in S$_2(g)$?

(c) Assuming that the bonding in S$_2$ is similar to the bonding [...] O$_2$, give a molecular orbital description of the bonding in [...] Is S$_2$ likely to be paramagnetic or diamagnetic?

9.151 Phosgene, $COCl_2(g)$, is a toxic gas used as an agent of warfare [...] World War I.

(a) Draw an electron-dot structure for phosgene.

(b) Using the table of bond dissociation energies (Table 7.2) a[...] the value $\Delta H°_f = 716.7$ kJ/mol for C(g), estimate $\Delta H°_f$ f[...] $COCl_2(g)$ at 25 °C. Compare your answer to the actual ΔH[...] given in Appendix B, and explain why your calculation [...] only an estimate.

9.152 Acid spills are often neutralized with sodium carbonate or s[...] dium hydrogen carbonate. For neutralization of acetic acid, t[...] unbalanced equations are

(1) $CH_3CO_2H(l) + Na_2CO_3(s) \longrightarrow$
$$CH_3CO_2Na(aq) + CO_2(g) + H_2O($$

(2) $CH_3CO_2H(l) + NaHCO_3(s) \longrightarrow$
$$CH_3CO_2Na(aq) + CO_2(g) + H_2O($$

(a) Balance both equations.

(b) How many kilograms of each substance is need[...] to neutralize a 1.000 gallon spill of pure acetic ac[...] (density = 1.049 g/mL)?

(c) How much heat in kilojoules is absorbed or liberated in ea[...] reaction? See Appendix B for standard heats of formatio[...] $\Delta H°_f = -726.1$ kJ/mol for $CH_3CO_2\,Na(aq)$.

9.153 (a) Write a balanced equation for the reaction of potassium metal with water.

(b) Use the data in Appendix B to calculate $\Delta H°$ for the reaction of potassium metal with water.

(c) Assume that a chunk of potassium weighing 7.55 g is dropped into 400.0 g of water at 25.0 °C. What is the final temperature of the water if all the heat released is used to warm the water?

(d) What is the molarity of the KOH solution prepared in part (c), and how many milliliters of 0.554 M H_2SO_4 are required to neutralize it?

9.154 Hydrazine, a component of rocket fuel, undergoes combustion to yield N_2 and H_2O:

$$N_2H_4(l) + O_2(g) \longrightarrow N_2(g) + 2 H_2O(l)$$

(a) Draw an electron-dot structure for hydrazine, predict the geometry about each nitrogen atom, and tell the hybridization of each nitrogen.

(b) Use the following information to set up a Hess's law cycle, and then calculate $\Delta H°$ for the combustion reaction. You will need to use fractional coefficients for some equations.

$$2 NH_3(g) + 3 N_2O(g) \longrightarrow 4 N_2(g) + 3 H_2O(l)$$
$$\Delta H° = -1011.2 \text{ kJ}$$

$$N_2O(g) + 3 H_2(g) \longrightarrow N_2H_4(l) + H_2O(l)$$
$$\Delta H° = -317.2 \text{ kJ}$$

$$4 NH_3(g) + O_2(g) \longrightarrow 2 N_2H_4(l) + 2 H_2O(l)$$
$$\Delta H° = -286.0 \text{ kJ}$$

$$H_2O(l) \quad \Delta H°_f = -285.8 \text{ kJ/mol}$$

(c) How much heat is released on combustion of 100.0 g of hydrazine?

9.155 Reaction of gaseous fluorine with compound X yields a single product Y, whose mass percent composition is 61.7% F and 38.3% Cl.

(a) What is a probable molecular formula for product Y, and what is a probable formula for X?

(b) Draw an electron-dot structure for Y, and predict the geometry around the central atom.

(c) Calculate $\Delta H°$ for the synthesis of Y using the following information:

$$2 ClF(g) + O_2(g) \longrightarrow Cl_2O(g) + OF_2(g)$$
$$\Delta H° = +205.4 \text{ kJ}$$

$$2 ClF_3(l) + 2 O_2(g) \longrightarrow Cl_2O(g) + 3 OF_2(g)$$
$$\Delta H° = +532.8 \text{ kJ}$$

$$OF_2(g) \quad \Delta H°_f = +24.5 \text{ kJ/mol}$$

(d) How much heat in kilojoules is released or absorbed in the reaction of 25.0 g of X with a stoichiometric amount of F_2, assuming 87.5% yield for the reaction?

CHAPTER

10

Gases: Their Properties and Behavior

The Earth's atmosphere is not only beautiful to behold but serves many critical functions. We need oxygen to survive and plants require carbon dioxide and nitrogen to grow. Even small changes in atmospheric composition from human activities can greatly impact the quality of the air we breathe and the delicate thermal balance that controls our climate.

 Which gases are greenhouse gases?

The answer to this question can be found in the **INQUIRY** ▶▶▶ on page 392.

CONTENTS

A quick look around tells you that matter takes many forms. Most of the things around you are *solids*, substances whose constituent atoms, molecules, or ions are held rigidly together in a definite way, giving the solid a definite volume and shape. Other substances are *liquids*, whose constituent atoms or molecules are held together less strongly, giving the liquid a definite volume but a changeable and indefinite shape. Still other substances are *gases*, whose constituent atoms or molecules have little attraction for one another and are therefore free to move about in whatever volume is available.

Although gases are few in number—only about a hundred substances are gases at room temperature and atmospheric pressure—experimental studies of their properties were enormously important in the historical development of atomic theories. We'll look briefly at this historical development in the present chapter, and we'll see how the behavior of gases can be described.

0.1 ▶ GASES AND GAS PRESSURE

We live surrounded by a blanket of air—the mixture of gases that make up the Earth's atmosphere. As shown in TABLE 10.1, nitrogen and oxygen account for more than 99% by volume of dry air. The remaining 1% is largely argon, with trace amounts of several other substances also present. Carbon dioxide, about which there is so much current concern because of its relationship to global warming, is present in air only to the extent of about 0.040%, or 400 parts per million (ppm). Although small, this value has risen in the past 160 years from an estimated 290 ppm in 1850, as the burning of fossil fuels and the deforestation of tropical rain forests have increased.

Air is typical of gases in many respects, and its behavior illustrates several important points about gases. For instance, gas mixtures are always *homogeneous*, meaning that they are uniform in composition. Unlike liquids, which often fail to mix with one another and which may separate into distinct layers—oil and water, for example—gases always mix completely. Furthermore, gases are *compressible*. When pressure is applied, the volume of a gas contracts proportionately. Solids and liquids, however, are nearly incompressible, and even the application of great pressure changes their volume only slightly.

Homogeneous mixing and compressibility of gases both occur because the constituent particles—whether atoms or molecules—are far apart (FIGURE 10.1). Mixing occurs because individual gas particles have little interaction with their neighbors so the chemical identity of those neighbors is irrelevant. In solids and liquids, by contrast, the constituent particles are packed closely together, where they are affected by various attractive and repulsive forces that can inhibit their mixing. Compressibility is possible in gases because less than 0.1% of the volume of a typical gas is taken up by the particles themselves under normal circumstances; the remaining 99.9% is empty space. By contrast, approximately 70% of a solid's or liquid's volume is taken up by the particles.

One of the most obvious characteristics of gases is that they exert a measurable *pressure* on the walls of their container (Figure 10.1). We're all familiar with inflating a balloon or

TABLE 10.1	Composition of Dry Air at Sea Level	
Constituent	% Volume	% Mass
N_2	78.08	75.52
O_2	20.95	23.14
Ar	0.93	1.29
CO_2	0.040	0.060
Ne	1.82×10^{-3}	1.27×10^{-3}
He	5.24×10^{-4}	7.24×10^{-5}
CH_4	1.7×10^{-4}	9.4×10^{-5}
Kr	1.14×10^{-4}	3.3×10^{-4}

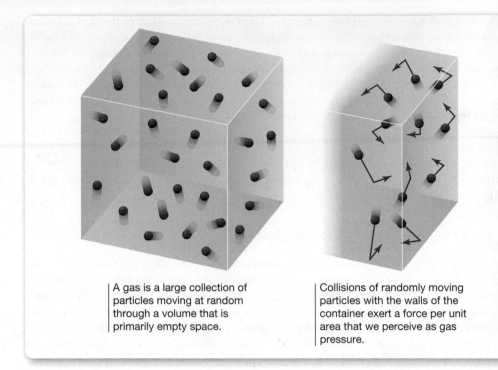

A gas is a large collection of particles moving at random through a volume that is primarily empty space.

Collisions of randomly moving particles with the walls of the container exert a force per unit area that we perceive as gas pressure.

pumping up a bicycle tire and feeling the hardness that results from the pressure inside. I scientific terms, **pressure (P)** is defined as a force (F) exerted per unit area (A). Force, in tur is defined as mass (m) times acceleration (a), which, on Earth, is usually the acceleration d to gravity, $a = 9.81 \text{ m/s}^2$.

$$\text{Pressure } (P) = \frac{F}{A} = \frac{m \times a}{A}$$

The SI unit for force is the **newton (N)**, where $1 \text{ N} = 1 \ (\text{kg} \cdot \text{m})/\text{s}^2$, and the SI unit for pre sure is the **pascal (Pa)**, where $1 \text{ Pa} = 1 \text{ N/m}^2 = 1 \text{ kg}/(\text{m} \cdot \text{s}^2)$. Expressed in more famili units, a pascal is actually a very small amount—the pressure exerted by a mass of 10.2 m resting on an area of 1.00 cm².

$$P = \frac{m \times a}{A} = \frac{(10.2 \text{ mg})\left(\dfrac{1 \text{ kg}}{10^6 \text{ mg}}\right)\left(9.81 \dfrac{\text{m}}{\text{s}^2}\right)}{(1.00 \text{ cm}^2)\left(\dfrac{1 \text{ m}}{10^2 \text{ cm}}\right)^2} = \frac{1.00 \times 10^{-4} \dfrac{\text{kg} \cdot \text{m}}{\text{s}^2}}{1.00 \times 10^{-4} \text{ m}^2}$$

$$= 1.00 \frac{\text{kg}}{\text{m} \cdot \text{s}^2} = 1.00 \text{ Pa}$$

In rough terms, a penny sitting on the tip of your finger exerts a pressure of about 250 Pa. Ju as the air in a tire and a penny on your finger exert pressure, the mass of air in the atmosphe pressing down on the Earth's surface exerts what we call *atmospheric pressure*. In fact, a 1 r column of air extending from the Earth's surface through the upper atmosphere has a mass about 10,300 kg, producing an atmospheric pressure of approximately 101,000 Pa, or 101 kI (**FIGURE 10.2**).

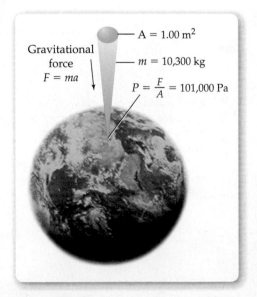

Gravitational force
$F = ma$

A = 1.00 m²
m = 10,300 kg
$P = \dfrac{F}{A} = 101{,}000 \text{ Pa}$

▲ **FIGURE 10.2**

Atmospheric pressure. A column of air 1 m² in cross-sectional area extending from the Earth's surface through the upper atmosphere has a mass of about 10,300 kg, producing an atmospheric pressure of approximately 101,000 Pa.

$$P = \frac{m \times a}{A} = \frac{10{,}300 \text{ kg} \times 9.81 \dfrac{\text{m}}{\text{s}^2}}{1.00 \text{ m}^2} = 101{,}000 \text{ Pa} = 101 \text{ kPa}$$

Pressure in everyday objects such as tires and basketballs is measured in units of poun per square inch (lb/in.² or psi). Atmospheric pressure can be expressed in lb/in.² by convertii the mass (10,300 kg) and the area (1 m²) of a column of the atmosphere from SI units to units of pounds and inches (Figure 10.2).

$$\frac{10{,}300 \text{ kg}}{1 \text{ m}^2} \times \frac{2.21 \text{ lb}}{1 \text{ kg}} \times \left(\frac{1 \text{ m}}{10^2 \text{ cm}}\right)^2 \times \left(\frac{2.54 \text{ cm}}{1 \text{ in.}}\right)^2 = 14.7 \frac{\text{lb}}{\text{in.}^2}$$

We don't feel the atmosphere pushing down on us because of an equivalent force within our body pushing outward. However, the force due to atmospheric pressure can be demonstrated by attaching a metal can to a vacuum pump. When the pump is turned on, the pressure inside the can decreases and atmospheric pressure is strong enough to crush the can (**FIGURE 10.3**)!

As is frequently the case with SI units, which must serve many disciplines, the pascal is an inconvenient size for most chemical measurements. Thus, the alternative pressure units *millimeter of mercury (mm Hg), atmosphere (atm)*, and *bar* are more often used.

The **millimeter of mercury**, also called a *torr* after the seventeenth-century Italian scientist Evangelista Torricelli (1608–1647), is based on atmospheric pressure measurements using a mercury *barometer*. As shown in **FIGURE 10.4**, a barometer consists of a long, thin tube that is sealed at one end, filled with mercury, and then inverted into a dish of mercury. Some mercury runs from the tube into the dish until the downward pressure of mercury due to the pull of gravity inside the column is exactly balanced by the outside atmospheric pressure, which presses on the mercury in the dish and pushes it up the column. The height of the mercury column varies slightly from day to day depending on the altitude and weather conditions, but atmospheric pressure at sea level is defined as exactly 760 mm Hg.

Knowing the density of mercury ($1.359\,51 \times 10^4 \text{ kg/m}^3$ at 0 °C) and the acceleration due to gravity ($9.806\,65 \text{ m/s}^2$), it's possible to calculate the pressure exerted by the column of mercury 760 mm (0.760 m) in height. Thus, 1 standard **atmosphere (atm)** of pressure (1 atm) is now defined as exactly 101,325 Pa:

$$P = (0.760 \text{ m})\left(1.359\,51 \times 10^4 \frac{\text{kg}}{\text{m}^3}\right)\left(9.806\,65 \frac{\text{m}}{\text{s}^2}\right) = 101{,}325 \text{ Pa}$$

$$1 \text{ atm} = 760 \text{ mm Hg} = 101{,}325 \text{ Pa}$$

Although not strictly an SI unit, the **bar** is quickly gaining popularity as a unit of pressure because it is a convenient power of 10 of the SI unit pascal and because it differs from 1 atm by only about 1%:

$$1 \text{ bar} = 100{,}000 \text{ Pa} = 100 \text{ kPa} = 0.986\,923 \text{ atm}$$

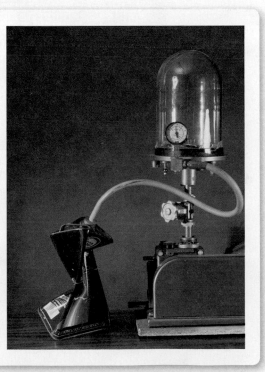

◀ **FIGURE 10.3**

Effect of atmospheric pressure on an evacuated can. (**a**) A vacuum pump is connected to a can. (**b**) Once the vacuum is turned on pressure inside the can decreases and atmospheric pressure crushes the can.

▶ **FIGURE 10.4**

A mercury barometer. The barometer measures atmospheric pressure by determining the height of a mercury column supported in a sealed glass tube.

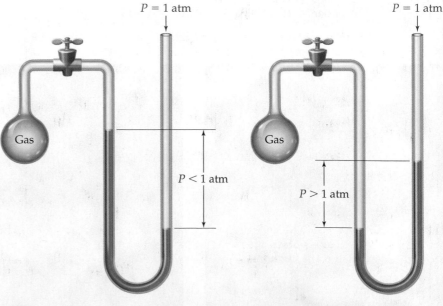

The empty space above the mercury in the sealed end of the tube is a vacuum.

The downward pressure of the mercury in the column is exactly balanced by the outside **atmospheric pressure** that presses down on the mercury in the dish and pushes it up the column.

Atmospheric pressure

760 mm

Mercury-filled dish

TABLE 10.2 Conversions between Common Units of Pressure

	1 atm
Pa (1 N/m²)	*1.013 25 × 10⁵
kPa	*101.325
bar	*1.013 25
mm Hg	*760
lb/in.² or psi	14.7

*The following conversions are exact and do not limit the number of significant figures in a calculation.

TABLE 10.2 summarizes different units of pressure and the conversion factor between atmospheres and the specified unit.

Gas pressure inside a container is often measured using an open-end **manometer**, a simple instrument similar in principle to the mercury barometer. As shown in **FIGURE 10.5**, an open-end manometer consists of a U-tube filled with mercury, with one end connected to a gas-filled container and the other end open to the atmosphere. The difference between the pressure of the gas in the container and the pressure of the atmosphere is equal to the difference between the heights of the mercury levels in the two arms of the U-tube. If the gas pressure

▶ **FIGURE 10.5**

Open-end manometers for measuring pressure in a gas-filled bulb.

Figure It Out

Describe the mercury level in the arm open to the bulb and the arm open to the atmosphere when the gas pressure equals 1 atm.

Answer: The mercury levels in both arms are identical.

$P = 1$ atm

$P = 1$ atm

Gas

Gas

$P < 1$ atm

$P > 1$ atm

(a) The mercury level is higher in the arm open to the bulb because the pressure in the bulb is lower than atmospheric.

(b) The mercury level is higher in the arm open to the atmosphere because the pressure in the bulb is higher than atmospheric.

inside the container is less than atmospheric, the mercury level is higher in the arm connected to the container (Figure 10.5a). If the gas pressure inside the container is greater than atmospheric, the mercury level is higher in the arm open to the atmosphere (Figure 10.5b).

WORKED EXAMPLE 10.1

Converting Between Different Units of Pressure

Typical atmospheric pressure on top of Mt. Everest, whose official altitude is 8848 m, is 265 mm Hg. Convert this value to pascals, atmospheres, and bars.

STRATEGY

Use the conversion factors 101,325 Pa/760 mm Hg, 1 atm/760 mm Hg, and 1 bar/10^5 Pa to carry out the necessary calculations.

SOLUTION

$$(265 \text{ mm Hg})\left(\frac{101,325 \text{ Pa}}{760 \text{ mm Hg}}\right) = 3.53 \times 10^4 \text{ Pa}$$

$$(265 \text{ mm Hg})\left(\frac{1 \text{ atm}}{760 \text{ mm Hg}}\right) = 0.349 \text{ atm}$$

$$(3.53 \times 10^4 \text{ Pa})\left(\frac{1 \text{ bar}}{10^5 \text{ Pa}}\right) = 0.353 \text{ bar}$$

CHECK

One atmosphere equals 760 mm Hg pressure. Since 265 mm Hg is about one-third of 760 mm Hg, the air pressure on Mt. Everest is about one-third of standard atmospheric pressure—approximately 30,000 Pa, 0.3 atm, or 0.3 bar.

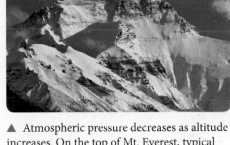

▲ Atmospheric pressure decreases as altitude increases. On the top of Mt. Everest, typical atmospheric pressure is 265 mm Hg.

▶ **PRACTICE 10.1** Barometric pressure changes are used by meteorologists to forecast short-term changes in the weather and are reported in units of inches of mercury. If the local barometric pressure is 28.48 inches of mercury, calculate the pressure in units of atm, bar, and Pa. (1 inch = 2.54 cm exactly)

▶ **APPLY 10.2**

(a) If a barometer were filled with liquid water instead of mercury (Figure 10.4), what would be the height (m) of the column of water if the atmospheric pressure were 1 atm? In other words, express the pressure of 1 atm in units of meters of water instead of millimeters of mercury. (The density of mercury is 13.6 g/mL and the density of water is 1.00 g/mL.)

(b) Why is Hg more commonly used in a barometer than water?

Conceptual WORKED EXAMPLE 10.2

Using an Open-End Manometer to Measure Gas Pressure

What is the pressure of the gas inside the apparatus shown in mm Hg if the outside pressure is 750 mm Hg?

STRATEGY

The gas pressure in the bulb equals the difference between the outside pressure and the manometer reading. The pressure of the gas in the bulb is higher than atmospheric pressure because the liquid level is higher in the arm open to the atmosphere.

SOLUTION

$$P_{\text{gas}} = 750 \text{ mm Hg} + \left(25 \text{ cm Hg} \times \frac{10 \text{ mm}}{1 \text{ cm}}\right) = 1000 \text{ mm Hg} = 1.0 \times 10^3 \text{ mm Hg}$$

continued on next page

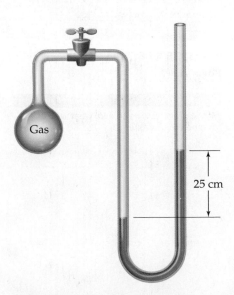

▲ Open-end manometer for Worked Example 10.2.

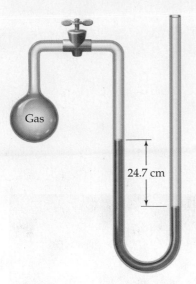

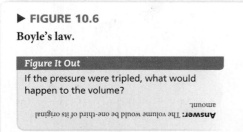

▲ Open-end manometer for Problem 10.3.

▶ **Conceptual PRACTICE 10.3** What is the pressure of the gas inside the apparatus in mm Hg if the outside pressure is 0.975 atm?

▶ **Conceptual APPLY 10.4** Assume that you are using an open-end manometer filled with mineral oil rather than mercury. The level of mineral oil in the arm connected to the bulb is 237 mm higher than the level in the arm connected to the atmosphere and atmospheric pressure is 746 mm Hg.

(a) Draw a picture of the manometer similar to Figure 10.5.
(b) What is the gas pressure in the bulb in mm of Hg? (The density of mercury is 13.6 g/mL, and the density of mineral oil is 0.822 g/mL.)

10.2 ▶ THE GAS LAWS

Unlike solids and liquids, different gases show remarkably similar physical behavior regardless of their chemical makeup. Helium and fluorine, for example, are vastly different in their chemical properties yet are almost identical in much of their physical behavior. Numerous observations made in the late 1600s showed that the properties of any gas can be defined by four variables: pressure (P), temperature (T), volume (V), and amount, or number of moles (n). The specific relationships among these four variables are called the **gas laws**, and a gas whose behavior follows the laws exactly is called an **ideal gas**.

Boyle's Law: The Relationship between Gas Volume and Pressure

Imagine that you have a sample of gas inside a cylinder with a movable piston at one end (**FIGURE 10.6**). What would happen if you were to increase the pressure on the gas by pushing down on the piston? Experience probably tells you that the volume of gas in the cylinder would decrease as you increase the pressure. According to **Boyle's law**, the volume of a fixed amount of gas at a constant temperature varies inversely with its pressure. If the gas pressure is doubled, the volume is halved; if the pressure is halved, the gas volume doubles.

> **Boyle's law** $V \propto 1/P$ or $PV = k$ **at constant n and T** The volume of an ideal gas varies inversely with pressure. That is, P times V is constant when n and T are kept constant. (The symbol $\propto$ means "is proportional to," and k denotes a constant.)

▶ **FIGURE 10.6**

Boyle's law.

Figure It Out

If the pressure were tripled, what would happen to the volume?

Answer: The volume would be one-third of its original amount.

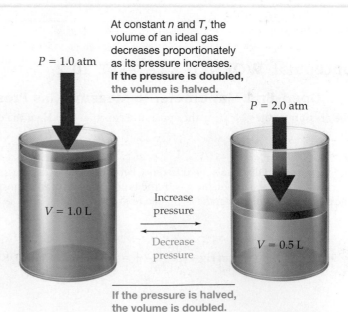

At constant n and T, the volume of an ideal gas decreases proportionately as its pressure increases. **If the pressure is doubled, the volume is halved.**

$P = 1.0$ atm

$P = 2.0$ atm

$V = 1.0$ L

Increase pressure

Decrease pressure

$V = 0.5$ L

If the pressure is halved, the volume is doubled.

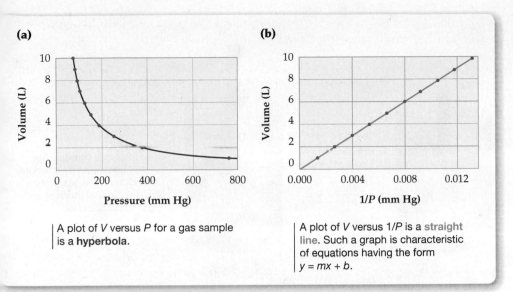

(a)

A plot of *V* versus *P* for a gas sample is a **hyperbola**.

(b)

A plot of *V* versus 1/*P* is a straight line. Such a graph is characteristic of equations having the form $y = mx + b$.

The validity of Boyle's law can be demonstrated by making a simple series of pressure–volume measurements on a gas sample (**TABLE 10.3**) and plotting them as in **FIGURE 10.7**. When *V* is plotted versus *P*, the result is a curve in the form of a hyperbola (Figure 10.7a). When *V* is plotted versus 1/*P*, however, the result is a straight line (Figure 10.7b). Such graphical behavior is characteristic of mathematical equations of the form $y = mx + b$. In this case, $y = V$, $m =$ the slope of the line (the constant *k* in the present instance), $x = 1/P$, and $b =$ the *y*-intercept (a constant; 0 in the present instance). (See Appendix A.3 for a review of linear equations.)

$$V = k\left(\frac{1}{P}\right) + 0 \quad (\text{or } PV = k)$$
$$\begin{array}{cccc} \uparrow & \uparrow\uparrow & & \uparrow \\ y & = m\ x & + & b \end{array}$$

Charles's Law: The Relationship between Gas Volume and Temperature

Imagine again that you have a gas sample inside a cylinder with a movable piston at one end (**FIGURE 10.8**). What would happen if you were to raise the temperature of the sample while letting the piston move freely to keep the pressure constant? Experience tells you that the piston would move up because the volume of the gas in the cylinder would expand. According to **Charles's law**, the volume of a fixed amount of an ideal gas at a constant pressure varies directly with its absolute temperature. If the gas temperature in kelvins is doubled, the volume is doubled; if the gas temperature is halved, the volume is halved.

> **Charles's law** $V \propto T$ or $V/T = k$ **at constant *n* and *P***   The volume of an ideal gas varies directly with absolute temperature. That is, *V* divided by *T* is constant when *n* and *P* are held constant.

The validity of Charles's law can be demonstrated by making a series of temperature–volume measurements on a gas sample, giving the results listed in **TABLE 10.4**. Like Boyle's law, Charles's law takes the mathematical form $y = mx + b$, where $y = V$, $m =$ the slope

TABLE 10.3   Pressure–Volume Measurements on a Gas Sample at Constant *n*, *T*

Pressure (mm Hg)	Volume (L)
760	1
380	2
253	3
190	4
152	5
127	6
109	7
95	8
84	9
76	10

TABLE 10.4   Temperature–Volume Measurements on a Gas Sample at Constant *n* and *P*

Temperature (K)	Volume (L)
123	0.45
173	0.63
223	0.82
273	1.00
323	1.18
373	1.37

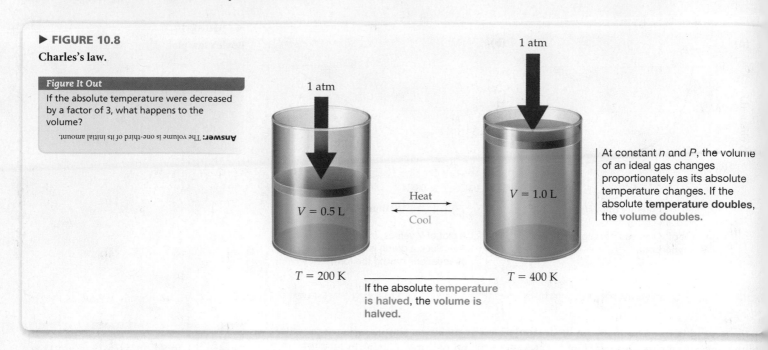

▶ **FIGURE 10.8**

Charles's law.

Figure It Out

If the absolute temperature were decreased by a factor of 3, what happens to the volume?

Answer: The volume is one-third of its initial amount.

1 atm

1 atm

$V = 0.5$ L

Heat

Cool

$V = 1.0$ L

At constant n and P, the volume of an ideal gas changes proportionately as its absolute temperature changes. If the absolute **temperature doubles**, the **volume doubles**.

$T = 200$ K

If the absolute **temperature** is halved, the **volume** is halved.

$T = 400$ K

of the line (the constant k in the present instance), $x = T$, and $b =$ the y-intercept (0 in the present instance). A plot of V versus T is therefore a straight line whose slope is the constant k (**FIGURE 10.9**).

$$V = kT + 0 \quad \left(\text{or } \frac{V}{T} = k\right)$$
$$\uparrow \quad \uparrow\uparrow \quad \uparrow$$
$$y = mx + b$$

The plots of volume versus temperature in Figure 10.9 demonstrate an interesting point. When temperature is plotted on the Celsius scale, the straight line can be extrapolated to $V = 0$ at $T = -273.15$ (Figure 10.9a). But because matter can't have a negative volume, this extrapolation suggests that -273.15 must be the lowest possible temperature, or *absolute zero* on the Kelvin scale (Figure 10.9b). In fact, the approximate value of absolute zero was first determined using this simple method.

▶ **FIGURE 10.9**

Charles's law plot.

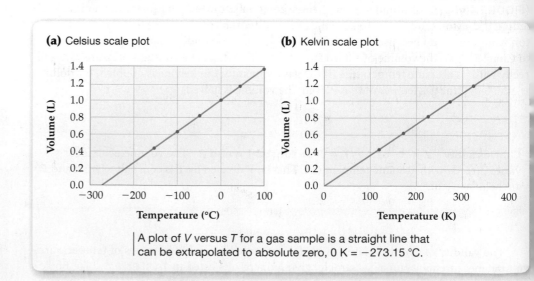

(a) Celsius scale plot

(b) Kelvin scale plot

Volume (L)

Temperature (°C)

Volume (L)

Temperature (K)

A plot of V versus T for a gas sample is a straight line that can be extrapolated to absolute zero, 0 K = -273.15 °C.

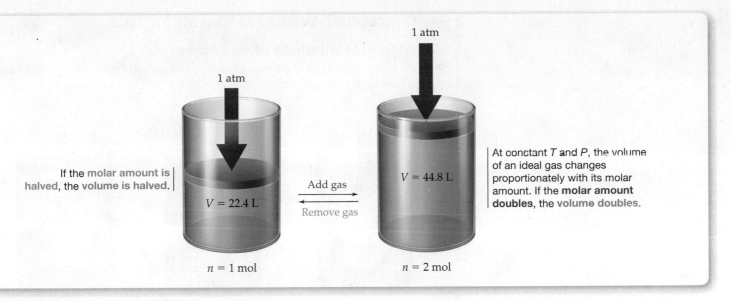

If the molar amount is halved, the volume is halved.

$V = 22.4$ L

$n = 1$ mol

Add gas

Remove gas

$V = 44.8$ L

At constant T and P, the volume of an ideal gas changes proportionately with its molar amount. If the **molar amount doubles**, the **volume doubles**.

$n = 2$ mol

▲ FIGURE 10.10

Avogadro's law.

Avogadro's Law: The Relationship between Volume and Amount

Imagine that you have two gas samples inside cylinders with movable pistons (**FIGURE 10.10**). One cylinder contains 1 mol of a gas and the other contains 2 mol of gas at the same temperature and pressure as the first. Common sense says that the gas in the second cylinder will have twice the volume of the gas in the first cylinder because there is twice as much of it. According to **Avogadro's law**, the volume of an ideal gas at a fixed pressure and temperature depends only on its molar amount. If the amount of the gas is doubled, the gas volume is doubled; if the amount is halved, the volume is halved.

> **Avogadro's law** $V \propto n$ or $V/n = k$ **at constant T and P** The volume of an ideal gas varies directly with its molar amount. That is, V divided by n is constant when T and P are held constant.

Put another way, Avogadro's law also says that equal volumes of different gases at the same temperature and pressure contain the same molar amounts. A 1 L container of oxygen contains the same number of moles as a 1 L container of helium, fluorine, argon, or any other gas at the same T and P. Furthermore, 1 mol of an ideal gas occupies a volume, called the **standard molar volume**, of 22.414 L at 0 °C and exactly 1 atm pressure. For comparison, the standard molar volume is nearly identical to the volume of three basketballs.

● Conceptual WORKED EXAMPLE 10.3

Visual Representations of Gas Laws

Show the approximate level of the movable piston in drawings **(a)** and **(b)** after the indicated changes have been made to the initial gas sample.

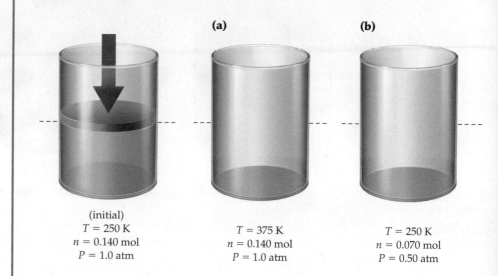

(initial)
$T = 250$ K
$n = 0.140$ mol
$P = 1.0$ atm

(a)
$T = 375$ K
$n = 0.140$ mol
$P = 1.0$ atm

(b)
$T = 250$ K
$n = 0.070$ mol
$P = 0.50$ atm

STRATEGY

Identify which of the variables P, n, and T have changed, and calculate the effect of each change on the volume according to the appropriate gas law.

SOLUTION

(a) The temperature T has increased by a factor of $375/250 = 1.5$, while the molar amount n and the pressure P are unchanged. Charles's law states that $V \propto T$, therefore the volume will increase by a factor of 1.5.

(b) The temperature T is unchanged, while both the molar amount n and the pressure P are halved. Avogadro's laws states that $V \propto n$, therefore halving the molar amount will halve the volume. Similarly, since $V \propto 1/P$ (Boyle's law), halving the pressure will double the volume. The two changes cancel, so the volume is unchanged.

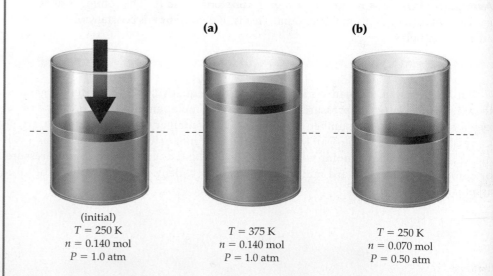

(initial)
$T = 250$ K
$n = 0.140$ mol
$P = 1.0$ atm

(a)
$T = 375$ K
$n = 0.140$ mol
$P = 1.0$ atm

(b)
$T = 250$ K
$n = 0.070$ mol
$P = 0.50$ atm

▶ **Conceptual PRACTICE 10.5** Show the approximate level of the movable piston in drawings **(a)** and **(b)** after the indicated changes have been made to the initial gas sample at a constant pressure of 1.0 atm.

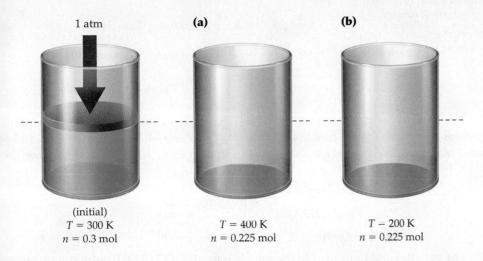

(a) **(b)**

1 atm

(initial)
$T = 300$ K
$n = 0.3$ mol

$T = 400$ K
$n = 0.225$ mol

$T = 200$ K
$n = 0.225$ mol

Conceptual APPLY 10.6   Show the approximate level of the movable piston in drawings (a) and (b) after the indicated changes have been made to the initial gas sample.

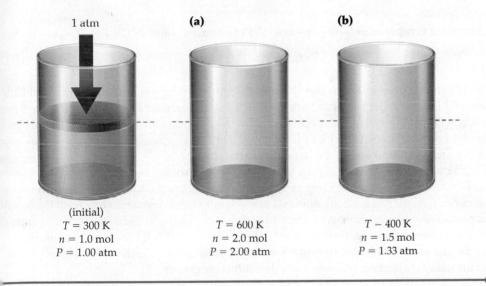

(a) **(b)**

1 atm

(initial)
$T = 300$ K
$n = 1.0$ mol
$P = 1.00$ atm

$T = 600$ K
$n = 2.0$ mol
$P = 2.00$ atm

$T = 400$ K
$n = 1.5$ mol
$P = 1.33$ atm

10.3 ▶ THE IDEAL GAS LAW

All three gas laws discussed in the previous section can be combined into a single statement called the **ideal gas law**, which describes how the volume of a gas is affected by changes in pressure, temperature, and amount. When the values of any three of the variables P, V, T, and n are known, the value of the fourth can be calculated using the ideal gas law. The proportionality constant R in the equation is called the **gas constant (R)** and has the same value for all gases.

Ideal gas law $V = \dfrac{nRT}{P}$ or $PV = nRT$

The ideal gas law can be rearranged in different ways to take the form of Boyle's law, Charles's law, or Avogadro's law.

Boyle's law: $PV = nRT = k$ (When n and T are constant)

Charles's law: $\dfrac{V}{T} = \dfrac{nR}{P} = k$ (When n and P are constant)

Avogadro's law: $\dfrac{V}{n} = \dfrac{RT}{P} = k$ (When T and P are constant)

The value of the gas constant R can be calculated from knowledge of the standard molar volume of a gas. Since 1 mol of a gas occupies a volume of 22.414 L at 0 °C (273.15 K) and 1 atm pressure, the gas constant R is equal to 0.082 058 (L·atm)/(K·mol), or 8.3145 J/(K·mol) in SI units:

$$R = \frac{P \cdot V}{n \cdot T} = \frac{(1\ \text{atm})(22.414\ \text{L})}{(1\ \text{mol})(273.15\ \text{K})} = 0.082\ 058 \frac{\text{L} \cdot \text{atm}}{\text{K} \cdot \text{mol}}$$

$$= 8.3145\ \text{J}/(\text{K} \cdot \text{mol})\qquad (\text{When } P \text{ is in pascals and } V \text{ is in cubic meters })$$

The specific conditions used in the calculation—0 °C (273.15 K) and 1 atm pressure—are said to represent **standard temperature and pressure**, abbreviated **STP**. These standard conditions are generally used when reporting measurements on gases. Note that the standard temperature for gas measurements (0 °C, or 273.15 K) is different from that usually assumed for thermodynamic measurements (25 °C, or 298.15 K; Section 9.5).

> **Standard temperature and pressure (STP) for gases** $T = 0\,°\text{C}$ $P = 1\,\text{atm}$

We should also point out that the standard pressure for gas measurements, still listed here and in most other books as 1 atm (101,325 Pa), has actually been redefined to be 1 bar, or 100,000 Pa. This new standard pressure is now 0.986 923 atm, making the newly defined standard molar volume 22.711 L rather than 22.414 L. Like most other books, we'll continue for the present using 1 atm as the standard pressure.

The name *ideal* gas law implies that there must be some gases whose behavior is *non-ideal*. In fact, there is no such thing as an ideal gas that obeys the equation perfectly under all circumstances. All real gases are nonideal to some extent and deviate slightly from the behavior predicted by the gas laws. As **TABLE 10.5** shows, for example, the actual molar volume of a real gas often differs slightly from the 22.414 L ideal value. Under most conditions, though, the deviations from ideal behavior are so slight as to make little difference. We'll discuss circumstances in Section 10.8 where the deviations are greater.

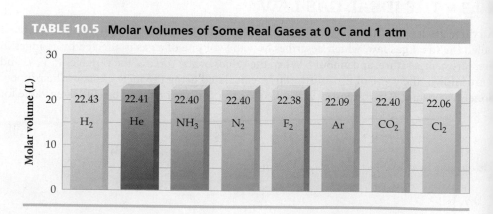

TABLE 10.5   Molar Volumes of Some Real Gases at 0 °C and 1 atm

Gas	Molar volume (L)
H_2	22.43
He	22.41
NH_3	22.40
N_2	22.40
F_2	22.38
Ar	22.09
CO_2	22.40
Cl_2	22.06

WORKED EXAMPLE 10.4

Using the Ideal Gas Law to Solve for an Unknown Variable

How many moles of gas (air) are in the lungs of an average adult with a lung capacity of 3.8 L? Assume that the lungs are at 1.00 atm pressure and a normal body temperature of 37 °C.

IDENTIFY

Known	Unknown
Volume ($V = 3.8$ L)	Moles of gas (n)
Pressure ($P = 1.00$ atm)	
Temperature ($T = 37$ °C)	

STRATEGY

This problem asks for a value of n when V, P, and T are given. Rearrange the ideal gas law to the form $n = PV/RT$, convert the temperature from degrees Celsius to kelvin, and substitute the given values of P, V, and T into the equation.

SOLUTION

$$n = \frac{PV}{RT} = \frac{(1.00 \text{ atm})(3.8 \text{ L})}{\left(0.082\,06\,\dfrac{\text{L}\cdot\text{atm}}{\text{K}\cdot\text{mol}}\right)(310 \text{ K})} = 0.15 \text{ mol}$$

The lungs of an average adult hold 0.15 mol of air.

CHECK

A lung volume of 4 L is about one-sixth of 22.4 L, the standard molar volume of an ideal gas. Thus, the lungs have a capacity of about one-sixth mol, or 0.17 mol.

▶ **PRACTICE 10.7** How many moles of methane gas, CH_4, are in a storage tank with a volume of 1.000×10^5 L at STP? How many grams?

▶ **APPLY 10.8** An aerosol spray can with a volume of 350 mL contains 3.2 g of propane gas (C_3H_8) as propellant. What is the pressure in atmospheres of gas in the can at 20 °C?

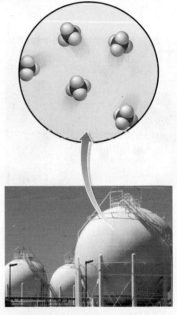

▲ How many moles of methane are in these tanks?

WORKED EXAMPLE 10.5

Using the Ideal Gas Law When Variables Change

In a typical automobile engine, the mixture of gasoline and air in a cylinder is compressed from 1.0 atm to 9.5 atm prior to ignition. If the uncompressed volume of the cylinder is 410 mL, what is the volume in milliliters when the mixture is fully compressed?

IDENTIFY

Known	Unknown
Initial pressure ($P_i = 1.0$ atm)	Final volume (V_f)
Final pressure ($P_f = 9.5$ atm)	
Initial volume ($V_i = 410$ mL)	

STRATEGY

Rearrange the ideal gas law so that variables that have changing values are on one side of the equation and constants are on the other side of the equation.

SOLUTION

In the ideal gas law, pressure and volume change and therefore remain on the left side of the equation, while n, R, and T remain constant.

$$\underbrace{PV}_{\substack{\text{Variables} \\ \text{that change}}} = \underbrace{nRT}_{\text{Constants}} = k$$

continued on next page

Because pressure times volume is a constant (Boyle's law), the equation $P_iV_i = P_fV_f$ can be used to solve for the final volume:

$$V_f = \frac{P_iV_i}{P_f} = \frac{(1.0\text{ atm})(410\text{ mL})}{(9.5\text{ atm})} = 43\text{ mL}$$

CHECK

Because the pressure in the cylinder increases about 10-fold, the volume must decrease about 10-fold according to Boyle's law, from approximately 400 mL to 40 mL.

▶ **PRACTICE 10.9**   What final temperature (°C) is required for the pressure inside an automobile tire to increase from 2.15 atm at 0 °C to 2.37 atm, assuming the volume remains constant?

▶ **APPLY 10.10**   A weather balloon has a volume of 45.0 L when released under conditions of 745 mm Hg and 25.0 °C. What is the volume of the balloon at an altitude of 10,000 m where the pressure is 178 mm Hg and the temperature is 225 K?

10.4 ▶ STOICHIOMETRIC RELATIONSHIPS WITH GASES

Many chemical reactions, including some of the most important processes in the chemical industry, involve gases. Approximately 130 million metric tons of ammonia, for instance, is manufactured each year worldwide by the reaction of hydrogen with nitrogen according to the equation $3\,H_2(g) + N_2(g) \longrightarrow 2\,NH_3(g)$. Thus, it's necessary to be able to calculate amounts of gaseous reactants just as it's necessary to calculate amounts of solids, liquids, and solutions.

Most gas calculations are just applications of the ideal gas law in which three of the variables P, V, T, and n are known and the fourth variable must be calculated. The reaction used in the deployment of automobile air bags, for instance, is the high-temperature decomposition of sodium azide, NaN_3, to produce N_2 gas. (The sodium is then removed by a subsequent reaction.) Worked Example 10.6 shows how to calculate the volume of a gaseous product given the amount reactant.

▲ Automobile air bags are inflated with N_2 gas produced by decomposition of sodium azide.

FO4305OZ02

┌─● **WORKED EXAMPLE 10.6**

Calculating the Volume of Gas Produced in a Chemical Reaction

How many liters of N_2 gas at 1.15 atm and 30 °C are produced by decomposition of 45.0 g of NaN_3?

$$2\,NaN_3(s) \longrightarrow 2\,Na(s) + 3\,N_2(g)$$

IDENTIFY

Known	Unknown
Pressure ($P = 1.15$ atm)	Volume of N_2 (V)
Temperature ($T = 30$ °C)	
Mass of NaN_3 (45.0 g)	

STRATEGY

Use stoichiometric relationships to find the number of moles of N_2 produced from 45.0 g of NaN_3 and then use the ideal gas law to find the volume of N_2.

SOLUTION

To find n, the number of moles of N_2 gas produced, we first need to find how many moles of NaN_3 are in 45.0 g:

$$\text{Molar mass of } NaN_3 = 65.0\text{ g/mol}$$

$$\text{Moles of } NaN_3 = (45.0\text{ g } NaN_3)\left(\frac{1\text{ mol } NaN_3}{65.0\text{ g } NaN_3}\right) = 0.692\text{ mol } NaN_3$$

Next, find how many moles of N_2 are produced in the decomposition reaction. According to the balanced equation, 2 mol of NaN_3 yields 3 mol of N_2, so 0.692 mol of NaN_3 yields 1.04 mol of N_2:

$$\text{Moles of } N_2 = (0.692 \text{ mol } NaN_3)\left(\frac{3 \text{ mol } N_2}{2 \text{ mol } NaN_3}\right) = 1.04 \text{ mol } N_2$$

Finally, use the ideal gas law to calculate the volume of N_2. Remember to use the Kelvin temperature (303 K) rather than the Celsius temperature (30 °C) in the calculation.

$$V = \frac{nRT}{P} = \frac{(1.04 \text{ mol } N_2)\left(0.08206 \dfrac{L \cdot atm}{K \cdot mol}\right)(303 \text{ K})}{1.15 \text{ atm}} = 22.5 \text{ L}$$

▲ Carbonate-bearing rocks like limestone ($CaCO_3$) react with dilute acids such as HCl to produce bubbles of carbon dioxide.

▸ **PRACTICE 10.11** Carbonate-bearing rocks like limestone ($CaCO_3$) react with dilute acids such as HCl to produce carbon dioxide, according to the equation

$$CaCO_3(s) + 2 \text{ HCl}(aq) \longrightarrow CaCl_2(aq) + CO_2(g) + H_2O(l)$$

How many grams of CO_2 are formed by complete reaction of 33.7 g of limestone? What is the volume in liters of this CO_2 at STP?

▸ **APPLY 10.12** Approximately 83% of all ammonia produced is used as fertilizer for crops. Ammonia is synthesized from hydrogen and nitrogen gas according to the equation

$$3 \text{ H}_2(g) + N_2(g) \longrightarrow 2 \text{ NH}_3(g)$$

What volume of hydrogen and nitrogen gas is needed to synthesize 500.0 L of ammonia at STP?

Other applications of the ideal gas law make it possible to calculate such properties as density and molar mass. Densities are calculated by weighing a known volume of a gas at a known temperature and pressure, as shown in **FIGURE 10.11**. Gas density changes dramatically with temperature and pressure so values for these variables must be specified. Gas density values are commonly reported at STP for consistency. The ideal gas law can be used to convert a density measured at any temperature and pressure to its value at STP. For example, if a sample of ammonia gas weighs 0.672 g and occupies a 1.000 L bulb at 25 °C and 733.4 mm Hg pressure, the density at STP can be calculated as follows. The density of any substance is mass divided by volume. For the ammonia sample, the mass is 0.672 g but the volume of the gas is given under nonstandard conditions and must first be converted to STP. Because the amount

◀ **FIGURE 10.11**

Determining the density of an unknown gas.

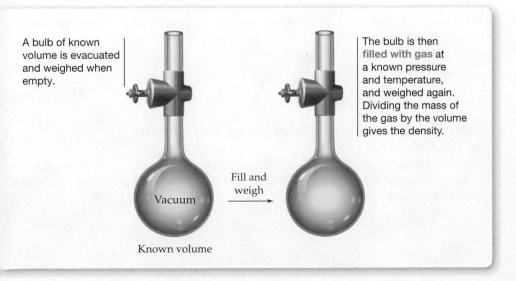

A bulb of known volume is evacuated and weighed when empty.

The bulb is then filled with gas at a known pressure and temperature, and weighed again. Dividing the mass of the gas by the volume gives the density.

Vacuum

Fill and weigh

Known volume

of sample n is constant, we can set the quantity PV/RT measured under nonstandard conditions equal to PV/RT at STP and then solve for V at STP.

$$n = \left(\frac{PV}{RT}\right)_{\text{measured}} = \left(\frac{PV}{RT}\right)_{\text{STP}} \quad \text{or} \quad V_{\text{STP}} = \left(\frac{PV}{RT}\right)_{\text{measured}} \left(\frac{RT}{P}\right)_{\text{STP}}$$

$$V_{\text{STP}} = \left(\frac{733.4 \text{ mm Hg} \times 1.000 \text{ L}}{298 \text{ K}}\right)\left(\frac{273 \text{ K}}{760 \text{ mm Hg}}\right) = 0.884 \text{ L}$$

The amount of gas in the 1.000 L bulb under the measured nonstandard conditions would have a volume of only 0.884 L at STP. Dividing the given mass by this volume gives the density of ammonia at STP:

$$\text{Density} = \frac{\text{Mass}}{\text{Volume}} = \frac{0.672 \text{ g}}{0.884 \text{ L}} = 0.760 \text{ g/L}$$

An equation relating gas density to molar mass can be found by rearranging the ideal gas law. Since density equals mass divided by volume, a term for mass must be added to the ideal gas law. Multiplying both sides of the equation by molar mass (M) incorporates mass into the equation because moles multiplied by molar mass equals mass as shown.

$$PV(M) = \underbrace{n(M)}_{\text{mass } (m)}RT$$

The equation can be rearranged to solve for density, which is defined as mass divided by volume:

$$PV(M) = mRT$$

$$\frac{P\cancel{V}(M)}{\cancel{V}} = \frac{mRT}{V} \quad \text{dividing both sides of the equation by } (V) \text{ yields:}$$

$$PM = \left(\frac{m}{V}\right)RT$$

$$\frac{PM}{RT} = \frac{\left(\frac{m}{V}\right)\cancel{RT}}{\cancel{RT}} \quad \text{dividing both sides of the equation by } (RT) \text{ yields:}$$

Gas density $\quad d = \dfrac{m}{V} = \dfrac{PM}{RT}$

If the density of an unknown gas is measured under conditions of known pressure and temperature, the molar mass of the gas can be calculated. Worked Example 10.7 shows how to identify an unknown gas from a density measurement.

WORKED EXAMPLE 10.7

Identifying an Unknown by Using Gas Density to Find Molar Mass

An unknown gas found bubbling up in a swamp is collected, placed in a glass bulb, and found to have a density of 0.714 g/L at STP. What is the molar mass of the gas? What is a possible identity of the gas?

IDENTIFY

Known	Unknown
Density ($d = 0.714\,\text{g/L}$)	Molar Mass (M)
Temperature ($T = 0\,°C$)	
Pressure ($P = 1\,\text{atm}$)	

STRATEGY

Since d, T, and P are known, the equation for gas density can be rearranged to solve for molar mass. Once molar mass is known the identity of the gas can be suggested.

SOLUTION

Remember to use the Kelvin temperature (273 K) rather than the Celsius temperature ($0\,°C$) in the gas density equation.

$$d = \frac{P(M)}{RT} \quad \text{or} \quad M = \frac{dRT}{P}$$

$$M = \frac{(0.714\text{g/L})\left(0.08206\,\dfrac{\text{L} \cdot \text{atm}}{\text{mol} \cdot \text{K}}\right)(273K)}{(1\,\text{atm})} = 16.0\,\text{g/mol}$$

Thus, the molar mass of the unknown gas (actually methane, CH_4) is 16.0 g/mol.

▶ **PRACTICE 10.13** A foul-smelling gas produced by the reaction of HCl with Na_2S was collected, and a 1.00 L sample was found to have a mass of 1.52 g at STP. What is the molar mass of the gas? What is its likely formula and name?

▶ **APPLY 10.14** The image shows carbon dioxide gas generated by adding dry ice, CO_2 (s), to water. The carbon dioxide flows downward because it is denser than air.

(a) What is the density in g/L of carbon dioxide at 1 atm and 25 °C?
(b) Why is carbon dioxide denser than air?

10.5 ▶ MIXTURES OF GASES: PARTIAL PRESSURE AND DALTON'S LAW

Just as the gas laws apply to all pure gases, regardless of chemical identity, they also apply to *mixtures* of gases, such as air. The pressure, volume, temperature, and amount of a gas mixture are all related by the ideal gas law.

What is responsible for the pressure in a gas mixture? Because the pressure of a pure gas at constant temperature and volume is proportional to its amount ($P = nRT/V$), the pressure contribution from each individual gas in a mixture is also proportional to its amount in the mixture. In other words, the total pressure exerted by a mixture of gases in a container at constant V and T is equal to the sum of the pressures of each individual gas in the container, a statement known as **Dalton's law of partial pressures**.

Dalton's law of partial pressures $P_{total} = P_1 + P_2 + P_3 + \ldots$ at constant V and T, where $P_1, P_2, \ldots$ refer to the pressures each individual gas would have if it were alone.

The individual pressure contributions of the various gases in the mixture, P_1, P_2, and so forth, are called partial pressures and refer to the pressure each individual gas would exert if it were alone in the container. That is,

$$P_1 = n_1\left(\frac{RT}{V}\right) \qquad P_2 = n_2\left(\frac{RT}{V}\right) \qquad P_3 = n_3\left(\frac{RT}{V}\right) \ldots \text{ and so forth}$$

But because all the gases in the mixture have the same temperature and volume, we can rewrite Dalton's law to indicate that the total pressure depends only on the total molar amount of gas present and not on the chemical identities of the individual gases:

$$P_{total} = (n_1 + n_2 + n_3 + \ldots)\left(\frac{RT}{V}\right)$$

The concentration of any individual component in a gas mixture is usually expressed as a **mole fraction** (X), which is defined simply as the number of moles of the component divided by the total number of moles in the mixture:

Mole fraction $(X) = \dfrac{\text{Moles of component}}{\text{Total moles in mixture}}$

The mole fraction of component 1, for example, is

$$X_1 = \frac{n_1}{n_1 + n_2 + n_3 + \ldots} = \frac{n_1}{n_{total}}$$

But because $n = PV/RT$, we can also write

$$X_1 = \frac{P_1\left(\dfrac{V}{RT}\right)}{P_{total}\left(\dfrac{V}{RT}\right)} = \frac{P_1}{P_{total}}$$

which can be rearranged to solve for P_1.
The partial pressure of component 1 in a gas mixture is:

Partial pressure $\quad P_1 = X_1 \cdot P_{total}$

This equation says that the partial pressure exerted by each component in a gas mixture is equal to the mole fraction of that component times the total pressure. In air, for example, the mole fractions of N_2, O_2, Ar, and CO_2 are 0.7808, 0.2095, 0.0093, and 0.000 39, respectively (Table 10.1), and the total pressure of the air is the sum of the individual partial pressures:

$$P_{air} = P_{N_2} + P_{O_2} + P_{Ar} + P_{CO_2} + \ldots$$

Thus, at a total air pressure of 1 atm (760 mm Hg), the partial pressures of the individual components are

$$
\begin{aligned}
P_{N_2} &= 0.7808 \times 1.00\ \text{atm} &&= 0.7808\ \text{atm} \\
P_{O_2} &= 0.2095 \times 1.00\ \text{atm} &&= 0.2095\ \text{atm} \\
P_{Ar} &= 0.0093 \times 1.00\ \text{atm} &&= 0.0093\ \text{atm} \\
P_{CO_2} &= 0.0004 \times 1.00\ \text{atm} &&= 0.0004\ \text{atm} \\
P_{air} &= P_{N_2} + P_{O_2} + P_{Ar} + P_{CO_2} &&= 1.0000\ \text{atm}
\end{aligned}
$$

There are numerous practical applications of Dalton's law, ranging from the use of anesthetic agents in hospital operating rooms, where partial pressures of both oxygen and anesthetic in the patient's lungs must be constantly monitored, to the composition of diving gases used for underwater exploration. Worked Example 10.8 gives an illustration.

WORKED EXAMPLE 10.8

Calculating Partial Pressure

A 1.50 L steel container at 90 °C contains 5.50 g of H_2, 7.31 g of N_2, and 2.42 g of NH_3. What is the partial pressure of each gas and the total pressure in the container?

IDENTIFY

Known	Unknown
Mass of each gas	Partial pressure (P_{H_2}, P_{N_2}, P_{NH_3})
Volume ($V = 1.50$ L)	Total pressure (P_{tot})
Temperature ($T = 90$ °C)	

STRATEGY

Convert the mass of each gas into moles using molar mass. Find the partial pressure of each gas using the general formula $P_1 = n_1(RT/V)$. Find the total pressure of the gas mixture by summing the partial pressure of each gas.

SOLUTION

Moles of each gas:

$$5.50 \text{ g } H_2 \times \frac{1 \text{ mol } H_2}{2.0 \text{ g } H_2} = 2.75 \text{ mol } H_2 \qquad 7.31 \text{ g } N_2 \times \frac{1 \text{ mol } N_2}{28.0 \text{ g } N_2} = 0.261 \text{ mol } N_2$$

$$2.42 \text{ g } NH_3 \times \frac{1 \text{ mol } NH_3}{17.0 \text{ g } NH_3} = 0.142 \text{ mol } NH_3$$

Partial pressure of each gas:

$$P_{H_2} = \frac{(2.75 \text{ mol } H_2)\left(0.08206\dfrac{\text{L} \cdot \text{atm}}{\text{K} \cdot \text{mol}}\right)(363 \text{ K})}{1.50 \text{ L}} = 54.6 \text{ atm}$$

$$P_{N_2} = \frac{(0.261 \text{ mol } N_2)\left(0.08206\dfrac{\text{L} \cdot \text{atm}}{\text{K} \cdot \text{mol}}\right)(363 \text{ K})}{1.50 \text{ L}} = 5.18 \text{ atm}$$

$$P_{NH_3} = \frac{(0.142 \text{ mol } NH_3)\left(0.08206\dfrac{\text{L} \cdot \text{atm}}{\text{K} \cdot \text{mol}}\right)(363 \text{ K})}{1.50 \text{ L}} = 2.81 \text{ atm}$$

The total pressure is the sum of the partial pressure of each gas:

$$P_{tot} = P_{H_2} + P_{N_2} + P_{NH_3} = 54.6 \text{ atm} + 5.18 \text{ atm} + 2.81 \text{ atm} = 62.6 \text{ atm}$$

PRACTICE 10.15 Nitrox is a gas mixture used by scuba divers to prevent nitrogen narcosis, a loss of mental and physical function, caused by increased levels of dissolved nitrogen in the blood. The mole fraction of O_2 is 0.36, and the mole fraction of N_2 is 0.64 in a 10.0 L tank with a pressure of 50 atm at 25 °C.

(a) Calculate the partial pressure of O_2 and N_2.
(b) Calculate the number of moles of O_2 and N_2.

APPLY 10.16 At an underwater depth of 250 ft, the pressure is 8.38 atm. What should the mole fraction of oxygen in the diving gas be for the partial pressure of oxygen in the gas to be 0.21 atm, the same as in air at 1.0 atm?

▲ The partial pressure of oxygen in the scuba tanks must be the same underwater as in air at atmospheric pressure.

10.6 ▶ THE KINETIC–MOLECULAR THEORY OF GASES

Thus far, we've concentrated on just describing the behavior of gases rather than on under standing the reasons for that behavior. Actually, the reasons are straightforward and wer explained more than a century ago using a model called the **kinetic–molecular theory**. Th kinetic–molecular theory is based on the following assumptions:

1. A gas consists of tiny particles, either atoms or molecules, moving about at random.
2. The volume of the particles themselves is negligible compared with the total volume of th gas. Most of the volume of a gas is empty space.
3. The gas particles act independently of one another; there are no attractive or repulsiv forces between particles.
4. Collisions of the gas particles, either with other particles or with the walls of a con tainer, are elastic. That is, they bounce off the walls at the same speed and therefore th same energy they hit with, so that the total kinetic energy of the gas particles is constant constant T.
5. The average kinetic energy of the gas particles is proportional to the Kelvin temperature the sample.

Beginning with these assumptions, it's possible not only to understand the behavior gases but also to derive quantitatively the ideal gas law (though we'll not do so here). Fo example, look at how the individual gas laws follow from the five postulates of kinetic–mo lecular theory:

- **Boyle's law ($P \propto 1/V$):** Gas pressure is a measure of the number and forcefulness collisions between gas particles and the walls of their container. The smaller the volum at constant n and T, the smaller the distance between the particles and the greater th frequency of collisions. Thus, pressure increases as volume decreases (**FIGURE 10.12a**).

- **Charles's law ($V \propto T$):** Temperature is a measure of the average kinetic energy of th gas particles. The higher the temperature at constant n and P, the faster the gas particle move. A greater volume is required to avoid increasing the number of collisions with th walls of the container in order to maintain constant pressure. Thus, volume increases a temperature increases (**FIGURE 10.12b**).

- **Avogadro's law ($V \propto n$):** The more particles there are in a gas sample, the more volum the particles need at constant P and T to avoid increasing the number collisions with th walls of the container in order to maintain constant pressure. Thus, volume increases a amount increases (**FIGURE 10.12c**).

- **Dalton's law ($P_{total} = P_1 + P_2 + \dots$):** Because gas particles are far apart and ac independently of one another, the chemical identity of the particles is irrelevant. Tot pressure of a fixed volume of gas depends only on the temperature T and the total numbe of moles of gas n. The pressure exerted by a specific kind of particle thus depends on th mole fraction of that kind of particle in the mixture, not on the identity of the particl (**FIGURE 10.12d**).

One of the more important conclusions from kinetic–molecular theory comes from assumption 5—the relationship between temperature and E_K, the kinetic energy of molecula motion. Although we won't do so in this book, it can be shown that the total kinetic energy of a mole of gas particles equals $3RT/2$ and that the average kinetic energy per particle is thu $3RT/2N_A$, where N_A is Avogadro's number. Knowing this relationship makes it possible t calculate the average speed u of a gas particle at a given temperature. To take a helium atom a room temperature (298 K), for instance, we can write

$$E_K = \frac{3\,RT}{2\,N_A} = \frac{1}{2}mu^2$$

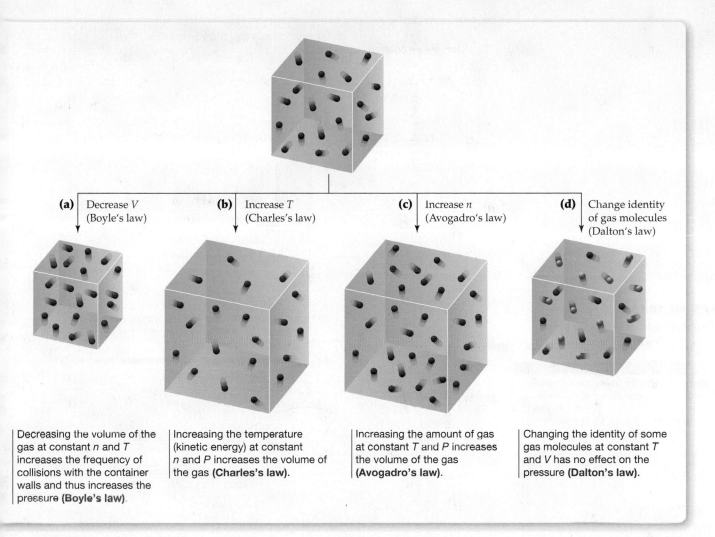

(a) Decrease V (Boyle's law)

(b) Increase T (Charles's law)

(c) Increase n (Avogadro's law)

(d) Change identity of gas molecules (Dalton's law)

Decreasing the volume of the gas at constant n and T increases the frequency of collisions with the container walls and thus increases the pressure **(Boyle's law)**.

Increasing the temperature (kinetic energy) at constant n and P increases the volume of the gas **(Charles's law)**.

Increasing the amount of gas at constant T and P increases the volume of the gas **(Avogadro's law)**.

Changing the identity of some gas molecules at constant T and V has no effect on the pressure **(Dalton's law)**.

▲ **FIGURE 10.12**

kinetic–molecular view of the gas laws.

which can be rearranged to give

Average speed of a gas particle (U).

$$u^2 = \frac{3\,RT}{mN_A}$$

$$\text{or} \quad u = \sqrt{\frac{3RT}{mN_A}} = \sqrt{\frac{3RT}{M}} \quad \text{where } M \text{ is the molar mass}$$

ubstituting appropriate values for $R\,[8.314\,\text{J}/(\text{K}\cdot\text{mol})]$ and for M, the molar mass of he-
um $(4.00\times10^{-3}\,\text{kg/mol})$, we have

$$u = \sqrt{\frac{(3)\left(8.314\,\dfrac{\text{J}}{\text{K}\cdot\text{mol}}\right)(298\,\text{K})}{4.00\times10^{-3}\,\dfrac{\text{kg}}{\text{mol}}}} = \sqrt{1.86\times10^{6}\,\dfrac{\text{J}}{\text{kg}}}$$

$$= \sqrt{1.86\times10^{6}\,\dfrac{\text{kg}\cdot\text{m}^2}{\text{s}^2}} = 1.36\times10^{3}\,\text{m/s}$$

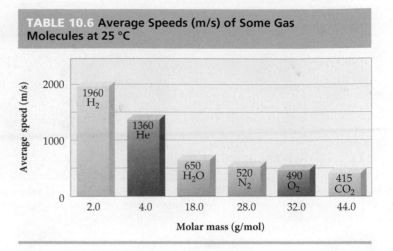

TABLE 10.6 Average Speeds (m/s) of Some Gas Molecules at 25 °C

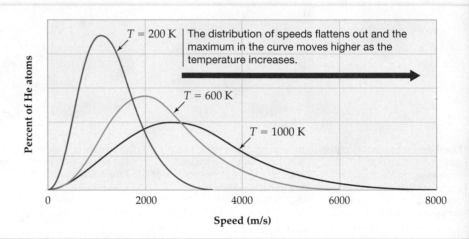

▶ **FIGURE 10.13**

The distribution of speeds for helium atoms at different temperatures.

Figure It Out

What happens to the average speed of gas particles as temperature increases? What happens to the shape of the curve for the distribution of speeds of gas particles as the temperature increases?

Answer: The average speed of gas particles increases as temperature increases. The shape of the curve becomes more broad meaning there is a wider range of speeds at higher temperature.

Thus, the average speed of a helium atom at room temperature is more than 1.3 km/s, about 3000 mi/h! Average speeds of some other molecules at 25 °C are given in TABLE 10.6 The heavier the molecule, the slower the average speed.

Just because the average speed of helium atoms at 298 K is 1.36 km/s doesn't mean tha all helium atoms are moving at that speed or that a given atom will travel from Maine California in one hour. As shown in **FIGURE 10.13**, there is a broad distribution of speed among particles in a gas, a distribution that flattens out and moves higher as the temperatur increases. This means that there is a greater range in molecular speed at higher temperature Furthermore, an individual gas particle is likely to travel only a very short distance before collides with another particle and bounces off in a different direction. Thus, the actual pat followed by a gas particle is a random zigzag.

For helium at room temperature and 1 atm pressure, the average distance between col lisions, called the *mean free path*, is only about 2×10^{-7} m, or 1000 atomic diameters, an there are approximately 10^{10} collisions per second. For a larger O_2 molecule, the mean fre path is about 6×10^{-8} m.

10.7 ▶ GAS DIFFUSION AND EFFUSION: GRAHAM'S LAW

The constant motion and high speeds of gas particles have some important practical conse quences. One such consequence is that gases mix rapidly when they come in contact. Tak the stopper off a bottle of perfume, for instance, and the odor will spread rapidly through room as perfume molecules mix with the molecules in the air. This mixing of different mol ecules by random molecular motion with frequent collisions is called **diffusion**. A simila process in which gas molecules escape without collisions through a tiny hole into a vacuum called **effusion** (**FIGURE 10.14**).

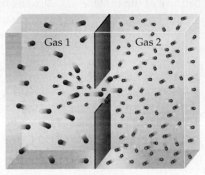 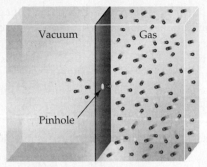

| **Diffusion** is the mixing of gas molecules by random motion under conditions where molecular collisions occur. | **Effusion** is the escape of a gas through a pinhole into a vacuum without molecular collisions. |

According to **Graham's law**, formulated in the mid-1800s by the Scottish chemist Thomas Graham (1805–1869), the rate of effusion of a gas is inversely proportional to the square root of its mass. In other words, the lighter the molecule, the more rapidly it effuses.

Graham's law   Rate of effusion $\propto \dfrac{1}{\sqrt{m}}$

The rate of effusion of a gas is inversely proportional to the square root of its mass, **m**.

In comparing two gases at the same temperature and pressure, we can set up an equation showing that the ratio of the effusion rates of the two gases is inversely proportional to the ratio of the square roots of their masses:

$$\frac{\text{Rate}_1}{\text{Rate}_2} = \frac{\sqrt{m_2}}{\sqrt{m_1}} = \sqrt{\frac{m_2}{m_1}}$$

The inverse relationship between the rate of effusion and the square root of the mass follows directly from the connection between temperature and kinetic energy described in the previous section. Because temperature is a measure of average kinetic energy and is independent of the gas's chemical identity, different gases at the same temperature have the same average kinetic energy:

$$\text{Since} \quad \frac{1}{2}mu^2 = \frac{3\,RT}{2\,N_A} \quad \text{for any gas}$$

$$\text{then} \quad \left(\frac{1}{2}mu^2\right)_{\text{gas 1}} = \left(\frac{1}{2}mu^2\right)_{\text{gas 2}} \quad \text{at the same } T$$

Canceling the factor of 1/2 from both sides and rearranging, we find that the average speeds of the molecules in two gases vary as the inverse ratio of the square roots of their masses:

$$\text{Since} \quad \left(\frac{1}{2}mu^2\right)_{\text{gas 1}} = \left(\frac{1}{2}mu^2\right)_{\text{gas 2}}$$

$$\text{then} \quad (mu^2)_{\text{gas 1}} = (mu^2)_{\text{gas 2}} \quad \text{and} \quad \frac{(u_{\text{gas 1}})^2}{(u_{\text{gas 2}})^2} = \frac{m_2}{m_1}$$

$$\text{so} \quad \frac{u_{\text{gas 1}}}{u_{\text{gas 2}}} = \frac{\sqrt{m_2}}{\sqrt{m_1}} = \sqrt{\frac{m_2}{m_1}}$$

If, as seems reasonable, the rate of effusion of a gas is proportional to the average speed of the gas molecules, then Graham's law results.

Diffusion is more complex than effusion because of the molecular collisions that occur, but Graham's law usually works as a good approximation. One of the most important practical consequences is that mixtures of gases can be separated into their pure components by taking advantage of the different rates of diffusion of the components. For example, naturally occurring uranium is a mixture of isotopes, primarily ^{235}U (0.72%) and ^{238}U (99.28%). In uranium enrichment plants that purify the fissionable uranium-235 used for fuel in nuclear reactors, elemental uranium is converted into volatile uranium hexafluoride (bp 56 °C), and the UF_6 gas is allowed to diffuse from one chamber to another through a permeable membrane. The $^{235}UF_6$ and $^{238}UF_6$ molecules diffuse through the membrane at slightly different rates according to the square root of the ratio of their masses:

$$\text{For } ^{235}UF_6, m = 349.03$$
$$\text{For } ^{238}UF_6, m = 352.04$$
$$\text{so} \quad \frac{\text{Rate of } ^{235}UF_6 \text{ diffusion}}{\text{Rate of } ^{238}UF_6 \text{ diffusion}} = \sqrt{\frac{352.04}{349.03}} = 1.0043$$

The UF_6 gas that passes through the membrane is thus very slightly enriched in the lighter, faster-moving isotope. After repeating the process many thousands of times, a separation of isotopes can be achieved. Approximately 30% of the Western world's nuclear fuel supply—some 5000 tons per year—is produced by this gas diffusion method, although the percentage is dropping because better methods are now available.

▲ Much of the uranium-235 used as a fuel in nuclear reactors is obtained by gas diffusion of UF_6 in cylinders like these.

●━ WORKED EXAMPLE 10.9

Using Graham's Law to Calculate Diffusion Rates

Assume that you have a sample of hydrogen gas containing H_2, HD, and D_2 that you want to separate into pure components. What are the relative rates of diffusion of the three molecules according to Graham's law? $(H = {}^1H)$ and $(D = {}^2H)$

STRATEGY

First, find the masses of the three molecules: for H_2, $m = 2.016$; for HD, $m = 3.022$; for D_2, $m = 4.028$. Then apply Graham's law to different pairs of gas molecules.

SOLUTION

The relative rate of diffusion of each gas relative to the heaviest gas, D_2, can be calculated as shown.

Comparing HD with D_2, we have

$$\frac{\text{Rate of HD diffusion}}{\text{Rate of } D_2 \text{ diffusion}} = \sqrt{\frac{\text{mass of } D_2}{\text{mass of HD}}} = \sqrt{\frac{4.028}{3.022}} = 1.155$$

Comparing H_2 with D_2, we have

$$\frac{\text{Rate of } H_2 \text{ diffusion}}{\text{Rate of } D_2 \text{ diffusion}} = \sqrt{\frac{\text{mass of } D_2}{\text{mass of } H_2}} = \sqrt{\frac{4.028}{2.016}} = 1.414$$

Thus, the relative rates of diffusion are H_2 (1.414) > HD (1.155) > D_2 (1.000).

CHECK

The answer makes sense because the lower the mass of the gas, the higher the relative diffusion rate.

▶ **PRACTICE 10.17** Which gas in each of the following pairs diffuses more rapidly, and what are the relative rates of diffusion?

(a) Kr and O_2 **(b)** N_2 and acetylene (C_2H_2)

▶ **APPLY 10.18** An unknown gas is found to diffuse through a porous membrane 1.414 times faster than SO_2. What is the molecular weight of the gas? What is the likely identity of the gas?

| **At lower pressure**, the volume of the gas particles is negligible compared to the total volume. | **At higher pressure**, the volume of the gas particles is more significant compared to the total volume. As a result, the volume of a real gas at high pressure is somewhat larger than the ideal value. |

0.8 ▶ THE BEHAVIOR OF REAL GASES

ere we expand on a point made earlier: The behavior of a real gas is often slightly different om that of an ideal gas. For instance, kinetic–molecular theory assumes that the volume of ιe gas particles themselves is negligible compared with the total gas volume. The assumption valid at STP, where the volume taken up by molecules of a typical gas is less than 0.1% of ιe total volume, but the assumption is not valid at 500 atm and 0 °C, where the volume of the ιolecules is about 20% of the total volume (**FIGURE 10.15**). As a result, the volume of a real ιs at high pressure is larger than predicted by the ideal gas law.

A second issue arising with real gases is the assumption that there are no attractive forces etween particles. At lower pressures, this assumption is reasonable because the gas particles ιe so far apart. At higher pressures, however, the particles are much closer together and the ιtractive forces between them become more important. In general, intermolecular attrac ιons become significant at a distance of about 10 molecular diameters and increase rapidly ; the distance diminishes (**FIGURE 10.16**). The result is to draw the molecules of real gases ιgether slightly, decreasing the volume at a given pressure (or decreasing the pressure for a ιven volume).

Note that the effect of molecular volume—to increase V—is opposite that of intermo ιcular attractions—to decrease V. The two factors therefore tend to cancel at intermediate ιessures, but the effect of molecular volume dominates above about 350 atm.

Both deviations in the behavior of real gases can be dealt with mathematically by a modi ιcation of the ideal gas law called the **van der Waals equation**, which uses two correction ιctors, called a and b. The increase in V, caused by the volume of the individual gas particles, corrected by subtracting the amount nb from the observed volume, where n is the number ɛ moles of gas. The decrease in V at constant P (or, equivalently, the decrease in P at constant), caused by the effect of attractive forces between gas particles, is corrected by adding an ιmount an^2/V^2 to the pressure.

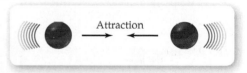

▲ **FIGURE 10.16**

Molecules attract one another at distances up to about 10 molecular diameters. The result is a decrease in the actual volume of most real gases when compared with ideal gases at pressures up to 300 atm.

van der Waals equation

Correction for
intermolecular attractions ↘

Correction for
molecular volume ↙

$$\left(P + \frac{an^2}{V^2}\right)(V - nb) = nRT$$

$$\text{or } P = \frac{nRT}{V - nb} - \frac{an^2}{V^2}$$

10.9 ▶ THE EARTH'S ATMOSPHERE AND AIR POLLUTION

The mantle of gases surrounding the Earth is far from the uniform mixture you might ex pect. Although atmospheric pressure decreases in a regular way at higher altitudes, the pr file of temperature versus altitude is much more complex (**FIGURE 10.17**). Four regions the atmosphere have been defined based on this temperature curve. The temperature in th *troposphere*, the region nearest the Earth's surface, decreases regularly up to about 12 km a titude, where it reaches a minimum value, and then increases in the *stratosphere*, up to abo 50 km. Above the stratosphere, in the *mesosphere* (50–85 km), the temperature again de creases but then again increases in the *thermosphere* (above 85 km). To give you a feeling fo these altitudes, passenger jets normally fly near the top of the troposphere at altitudes of to 12 km, and the world altitude record for jet aircraft is 37.65 km—roughly in the middle the stratosphere.

The most well-known environmental problems relating to the atmosphere are air poll tion, acid rain, climate change, and ozone depletion. In this chapter we'll explore the che istry of air pollution and climate change. Ozone depletion will be considered in Chapter 1 which covers principles of chemical kinetics that are central to the issue. Acid rain will l addressed in the chapter on acids and bases (Chapter 15).

Not surprisingly, it's the layer nearest the Earth's surface—the troposphere—that is th most easily disturbed by human activities. Air pollution has appeared in the last two centuri as an unwanted by-product of industrialized societies. Its causes are relatively straightfo ward; its control is difficult. Pollutants can be grouped into two main categories. **Prima pollutants** are those that enter the environment directly from a source such as vehicle industrial emissions. **Secondary pollutants** are formed by the chemical reaction of a prima pollutant and are not directly emitted from a source.

The United States Environmental Protection Agency has set air quality standards f six principal pollutants called "criteria" pollutants. The standards for the criteria pollutant shown in **TABLE 10.7**, are maximum concentrations that are safe for public health and th environment based upon current knowledge of their effects. Both toxicity and exposure tin are considered in setting the standards. More toxic pollutants have standards set at lowe concentrations although higher concentrations of toxic substances are acceptable for limite

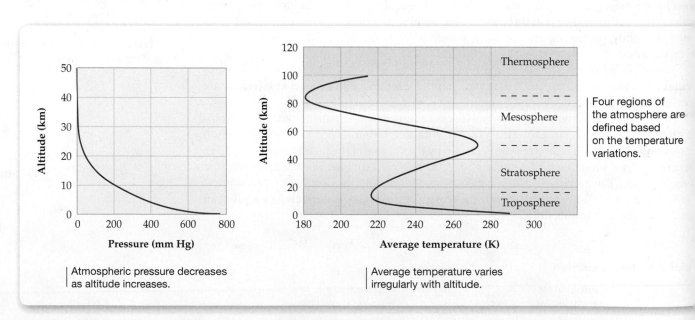

| Atmospheric pressure decreases as altitude increases. | Average temperature varies irregularly with altitude. |

▲ **FIGURE 10.17**
Variations of atmospheric pressure and average temperature with altitude.

mounts of time. For example, a concentration of 9 ppm of carbon monoxide is safe over a period of 8 hours, while exposure to 35 ppm is safe for only 1 hour. As new research emerges about health effects, the standards are modified. In 2008, the 8-hour standard for ozone was lowered from 80 to 75 ppb and further lowering of the standard to 65 ppb is currently under discussion.

The units of measure for the standards (parts per million and parts per billion by volume) are similar to the unit of percent volume of atmospheric components given in Table 10.1 at the start of the chapter. These quantities are all found by multiplying the mole fraction of a gas (X) by a given factor: 100 for percentage, 10^6 for ppm, and 10^9 for ppb. The mole fraction can be used to calculate volume percent because Avogadro's law states that $V \propto n$). These quantities can be calculated for a given gas (a) as follows:

$$\text{Percent by volume} = (X_a) \times 100$$
$$\text{Parts per million (ppm) by volume} = (X_a) \times 10^6$$
$$\text{Parts per billion (ppm) by volume} = (X_a) \times 10^9$$

The concentration units of ppm and ppb represent extremely low levels. One way to grasp the magnitude of these units is to relate them to the timescale. One ppm corresponds to the first second in a total of 10^6 seconds or 11.6 days. If a pollutant has a concentration of 1 ppm, this represents one molecule of pollutant for every 10^6 (one million) air molecules. One ppb represents an even smaller concentration and corresponds to the first second in total of 10^9 seconds or 31.7 years. The 8-hour standard for ozone of 75 ppb therefore represents 75 ozone molecules in a total of 10^9 (one billion) air molecules. Worked Example 10.10 demonstrates how to calculate different concentration units for pollutants.

—• WORKED EXAMPLE 10.10

Calculating Concentration Units for Pollutants

A home carbon monoxide (CO) meter gives a reading of 155 ppm.
(a) Express the concentration of CO as a percentage.
(b) Find the number of CO molecules inhaled in one breath if an average breath of air contains 3.3×10^{-2} total moles of gas.

IDENTIFY

Known	Unknown
Concentration of CO (155 ppm)	Concentration in units of percent
Moles of gas (3.3×10^{-2})	Number of CO molecules

STRATEGY

(a) Use the formula for calculating ppm to find the mole fraction of CO. Once the mole fraction is found it can be substituted into the formula for percent volume.
(b) To find the number of moles of CO, multiply the mole fraction by the total number of moles of air. Convert moles of CO to molecules using Avogadro's number.

SOLUTION

(a) $155 \text{ ppm} = (X_{CO}) \times 10^6$ $X_{CO} = \dfrac{155}{1 \times 10^6} = 1.55 \times 10^{-4}$

$\text{Percent by volume} = (X_{CO}) \times 100 = (1.55 \times 10^{-4}) \times 100 = 0.0155\%$

(b) $3.32 \times 10^{-2} \text{ mol air} \times \dfrac{(1.55 \times 10^{-4} \text{ mol CO})}{1 \text{ mol air}} \times \dfrac{(6.02 \times 10^{23} \text{ molecules CO})}{1 \text{ mol CO}}$

$= 3.00 \times 10^{18} \text{ molecules CO}$

continued on next page

TABLE 10.7 National Ambient Air Quality Standard for Criteria Pollutants, 2014.

Units of measure for the standards are parts per million (ppm) by volume, parts per billion (ppb) by volume, and micrograms per cubic meter of air ($\mu g/m^3$)

Pollutant	Standard
Carbon monoxide (CO)	
1-hour	35 ppm
8-hour	9 ppm
Lead (Pb)	
3-month average	$0.15 \mu g/m^3$
Nitrogen dioxide (NO$_2$)	
1-hour	100 ppb
Annual	53 ppb
Ozone (O$_3$)	
8-hour	75 ppb
Sulfur dioxide (SO$_2$)	
1-hour	75 ppb
Particle pollution	
Particle diameter less than 2.5μm (PM$_{2.5}$)	
Annual	$12 \mu g/m^3$
24-hour	$35 \mu g/m^3$
Particle diameter between 2.5 and 10μm (PM$_{10}$)	
24-hour	$150 \mu g/m^3$

CHECK

The low value for volume percent is reasonable because units of ppm are very small. The multiplication factor between percentage (10^2) and ppm (10^6) differ by a factor of 10^4. Therefore, a simple way to convert between ppm and percent is to move the decimal four places to the left to arrive at 0.0155%. The number of molecules of CO is very large even though the concentration is small. If one breath contains ~10^{-2} moles of air total, this corresponds to ~10^{21} molecules. There are a large number of total molecules in each breath so we would also expect a rather large number of CO molecules.

▶**PRACTICE 10.19**   The local news in an urban area reports an ozone concentration of 110 ppb on a hot summer day. What is the concentration of ozone in units of ppm and volume percent? How many ozone molecules are inhaled in one breath that contains 3.3×10^{-2} moles of air?

▶**APPLY 10.20**   A hot summer day in Denver has an ozone concentration of 95 ppb. Calculate the partial pressure of ozone if the total pressure in the "mile high city" is 0.79 atm.

Primary Pollutants

Carbon Monoxide   Carbon monoxide (CO) is a colorless, odorless gas emitted from combustion processes. So far in this book, we have written the products of hydrocarbon combustion reactions as CO_2 and H_2O. The balanced combustion reaction for natural gas (CH_4) is:

$$CH_4(g) + 2\,O_2(g) \longrightarrow CO_2(g) + 2\,H_2O(l)$$

In reality, combustion reactions produce a mixture of CO_2 and CO and the proportion of CO increases under conditions of limited oxygen. The balanced combustion reaction of natural gas when CO is the product is:

$$CH_4(g) + \frac{3}{2}\,O_2(g) \longrightarrow CO(g) + 2\,H_2O(l)$$

Notice that the coefficient for O_2 (3/2) in the reaction forming CO is smaller than the coefficient for O_2 (2) when CO_2 is product. In the United States, particularly in urban areas, the majority of CO emissions to ambient air come from the combustion of gasoline in vehicles. If an engine is running efficiently, the amount of CO emitted is less. Many states require vehicle emissions tests that directly measure CO in the tailpipe exhaust and if the vehicle fails to meet the standard (1.20% in several states) then the engine must be serviced so that it complies.

Carbon monoxide is a poisonous gas because it enters the bloodstream through the lungs and forms *carboxyhemoglobin*, a compound that inhibits the blood's capacity to carry oxygen to organs and tissues. CO levels above the 8-hour standard of 9 ppm and 1-hour standard of 35 ppm can cause headaches, confusion, and even death. It is extremely important to have a CO monitor in your home because a malfunctioning furnace can emit fatal levels of CO.

Lead   Lead (Pb) is a metal found naturally in the environment and in man-made products. Historically, the major source of lead emissions was leaded gasoline in fuel for vehicles. Tetraethyl lead, $(CH_3CH_2)_4Pb$, was an additive to gasoline that allowed engine compression to be increased, improving vehicle performance and fuel economy. Even small amounts of lead are extremely toxic and once taken into the body accumulates in the bones. Depending on the level of exposure, lead can adversely affect kidney function, neurological development, and the immune, reproductive, and cardiovascular systems. Infants and young children are especially sensitive to even very low levels of lead. Blood concentrations as low as 10 $\mu g/dL$

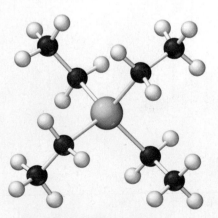

▲ Tetraethyl lead $(CH_3CH_2)_4Pb$ was added to gasoline to improve engine performance but was phased out of use in the 1970s due to the toxic effects of lead.

use learning deficits and reduced IQ levels, impaired hearing, and stunted growth in children.

Lead additives to fuel were phased out in the United States starting in the early 1970s; s a result, lead emissions from the transportation sector have decreased dramatically. By 090 many other countries had banned lead additives or significantly reduced the lead concentration in leaded gasoline (from about 1 g/L to 0.1 g/L). However, leaded gasoline is still sed in many parts of Africa, Asia, and South America. In the United States today, major ources of lead emissions to the air are ore and metal processing and some grades of aviaon gasoline.

Particulate Pollution Particulate matter (PM) is composed of extremely small solid articles and liquid droplets that can be made up of a number of different components ncluding salts of sulfates and nitrates, organic molecules, soot, metal, and dust. The particles hat are of concern for human health are too small to see and have diameters less than 0 μm. For reference a human hair has a diameter of about 70 μm. The PM_{10} classification orresponds to "inhalable coarse particles" (such as those found near roadways and industry), hich have diameters between 2.5 and 10 μm. $PM_{2.5}$ are particles less than 2.5 μm in diameter nd are referred to as "fine particles" (such as those found in smoke and haze).

Both classifications of particles are small enough to pass through the throat and nose and nter the lungs. Health effects include premature death in people with heart or lung disease, rregular heartbeat, aggravated asthma, and decreased lung function. While sensitive groups, ich as asthmatics or the elderly, are most affected, healthy individuals may experience temprary symptoms such as irritation of the airways, coughing, or difficulty breathing. Particue pollution also affects the environment by decreasing visibility (haze) in many parts of the nited States, including treasured national parks. Particulates can also be transported over ng distances and can damage forests, crops, and aquatic ecosystems by increasing acidity nd depleting nutrients.

▲ The difference between a clear and a hazy day from pollution, as seen from Dickey Ridge, Shenandoah National Park. New regulations have decreased the level of haze and improved visibility in some national parks.

Sulfur Dioxide Sulfur dioxide (SO_2) is a highly reactive gas that is a major contributor o particulate pollution and **acid rain**. The largest sources of SO_2 emissions are from e combustion of fossil fuels at power plants and other industrial facilities. While coal onsists mostly of carbon and hydrogen, it contains between 0.5% and 4% sulfur, which is acorporated either into its organic components or as inorganic mineral impurities such s pyrite (FeS_2). When coal is burned, sulfur is converted in gaseous SO_2, which is emitted irectly to the atmosphere unless it is removed by a process called **scrubbing**. A smaller ource of SO_2 emissions is smelting which extracts metals from their ores.

SO_2 is very irritating to tissues and damaging to vegetation. Exposure to $SO_2(g)$ has dverse respiratory effects including constriction of the airways that results in coughing, heezing, and shortness of breath. Scientific studies show a connection between short-term xposure and increased visits to emergency departments for respiratory illnesses.

LOOKING AHEAD...

The causes of **acid rain** and the **scrubbing** process used to remove sulfur from coal will be described in the *Inquiry* in Chapter 15.

Secondary Pollutants

Nitrogen Dioxide, Ozone, and Photochemical Smog Photochemical smog is a azy, brownish layer lying over many cities. Nitrogen dioxide (NO_2) and ozone (O_3) are econdary pollutants in photochemical smog that are harmful to human health, vegetation, nd materials such as rubber. Ground-level ozone is a powerful oxidant that irritates and aflames airways causing coughing, a burning sensation, wheezing, and shortness of breath. It reatly exacerbates the effects of asthma and other lung diseases. It is a serious concern even r healthy individuals who exercise or work outdoors because it can permanently scar lung ssue.

Photochemical smog is a worldwide problem and in the United States both urban and ural areas often issue "ozone alerts" or "ozone action days," especially during hot, summer nonths when ozone reaches unhealthy levels.

▲ The photochemical smog over many cities is the end result of pollution from vehicles and industry.

(a) Ozone degrades rubber.

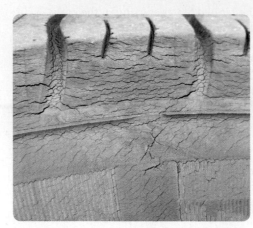

(b) Brown spots indicate ozone damage on this tulip poplar leaf.

▲ Effects of ozone on materials and vegetation.

Let's look at the mechanism for ozone formation to understand why photochemical sm[e] typically occurs in areas with lots of sunshine or in warm summer months. The process b[e] gins with the emission of a primary pollutant, nitric oxide (NO), from combustion reaction[s]. At the high temperature in a car's engine, nitrogen and oxygen react to form a small amou[nt] of NO.

$$N_2(g) + O_2(g) \xrightarrow{\text{heat}} 2\,NO(g)$$

The NO is further oxidized by reaction with oxygen to yield nitrogen dioxide, NO[] which splits into NO plus free oxygen atoms (O) in the presence of sunlight (symbo[l] ized by $h\nu$). Highly reactive oxygen atoms then combine with oxygen molecules to ma[ke] ozone (O_3).

$$NO_2(g) + h\nu\ (\lambda < 400\text{nm}) \longrightarrow NO(g) + O(g)$$
$$O(g) + O_2(g) \longrightarrow O_3(g)$$

These two reactions alone would not necessarily produce unhealthy levels of ozone b[e] cause the amount of ozone depends on the initial amount of NO_2. An additional reactio[n] that regenerates NO_2, however, can initiate a cyclic process in which one molecule of N[O] forms many molecules of ozone. Volatile organic compounds (VOCs) participate in rea[c] tions that regenerate NO_2 from NO. *Volatile* means that the substance readily evaporat[es] from the liquid to the gaseous phase. VOCs are emitted as unburned hydrocarbons in fu[el] exhaust, industrial processes, and even by some natural sources such as trees. You hav[e] directly observed the emission of VOCs when you fuel your car with gasoline and smell th[e] fumes in the air. Thus, reaction 2 above can be modified to show that oxygen-containin[g] VOCs initiate a complex series of reactions to turn NO back into NO_2 which can produ[ce] more ozone. The cycle continues as long as there is nitric oxide from combustion, sunligh[t] and VOCs.

$$\overset{\text{VOCs}}{\overset{\curvearrowleft}{NO_2(g) + h\nu\ (\lambda < 400\ \text{nm}) \longrightarrow NO(g) + O(g)}}$$

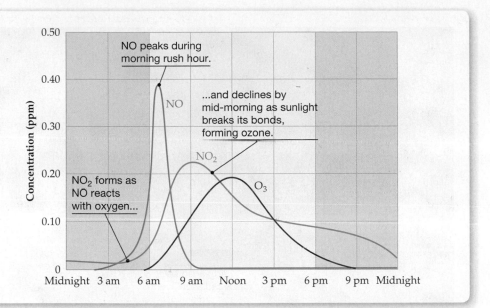

Time profile for pollutants in a photochemical smog event.

he mechanism for ozone formation explains the time profile of pollutants in an urban hotochemical smog event (**FIGURE 10.18**). Measurements of NO are typically high in the orning as a consequence of emissions from the morning commute. By midmorning NO as been oxidized to produce appreciable amounts of NO_2. By noon NO_2 begins to decline as unlight breaks the nitrogen–oxygen bond, forming O atoms which react with O_2 to form O_3. 3 reaches a peak concentration in the afternoon during periods of intense sunlight.

As scientists discover the chemistry behind pollution, solutions can be developed to re- ace the problems. The good news is that air quality in the United States has improved over e last quarter century. **TABLE 10.8** shows that the national average of nearly every pollutant as decreased and certain pollutants, such as Pb and CO have declined dramatically. New chnology, such as catalytic converters in cars (described in Section 13.13), sulfur scrubbing om power plant emissions, and oxygenated fuels, along with state and government regula- ons have made the air much safer to breathe today than in the 1970s and 1980s. But many r quality issues remain to be addressed. Ozone and particulate matter frequently contribute unhealthy levels of air quality in many regions of the United States. While national aver- es for SO_2, Pb, and CO have decreased significantly, high levels persist in industrial areas, articularly in the Midwest and East where electricity is primarily generated from coal. In- rnationally, severe air quality problems exist in major urban areas such as Beijing, China; Iexico City, Mexico; and Paris, France.

TABLE 10.8 Trends in United States Air Quality Based on Measurements from Hundreds of Sites

Pollutant	Percent Change (1980–2012)
Carbon monoxide (CO)	−83%
Lead (Pb)	−91%
Nitrogen dioxide (NO_2)	−60%
Ozone (O_3)	−25%
Sulfur dioxide (SO_2)	−78%
Particulate Pollution	
$PM_{2.5}$	−33% (2000–2012)
PM_{10}	−39% (1990–2012)

0.10 ▶ THE GREENHOUSE EFFECT

he **greenhouse effect** refers to the absorption of infrared (IR) radiation, also known as eat radiation, by gases in the atmosphere which causes an increase in planetary temper- ture. **Greenhouse gases** are gases that absorb infrared (IR) radiation. The greenhouse ffect is a naturally occurring phenomenon that is critical in regulating climate. The average mperature of our planet would be only about 0 °F if the amount of solar radiation reach- g the Earth was the only factor controlling climate. Most water would be frozen and the lanet would not be suitable for maintaining life! The presence of greenhouse gases such as ater, carbon dioxide, and methane increases the Earth's average global temperature to early 60 °F.

▶ **FIGURE 10.19**

The greenhouse effect.

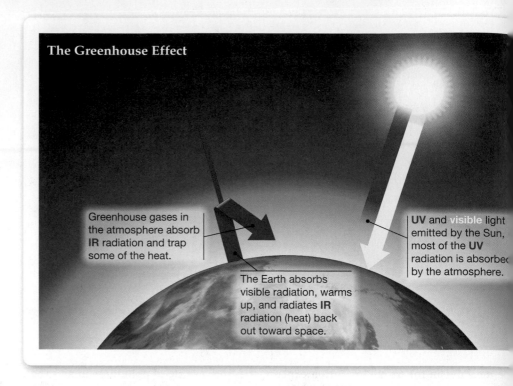

The Greenhouse Effect

Greenhouse gases in the atmosphere absorb **IR** radiation and trap some of the heat.

The Earth absorbs visible radiation, warms up, and radiates **IR** radiation (heat) back out toward space.

UV and **visible** light emitted by the Sun, most of the **UV** radiation is absorbed by the atmosphere.

The principle of the greenhouse effect is illustrated in **FIGURE 10.19**. The Sun, Earth' primary energy source, emits radiation most strongly in the ultraviolet (UV) and visible r gions of the electromagnetic spectrum (purple and yellow arrow). Most high-energy, bi logically damaging UV radiation (purple arrow) is absorbed by ozone and oxygen in t stratosphere and therefore does not reach ground level. Consequently, most of the radiatic reaching Earth is visible light (yellow arrow), which is absorbed by the Earth's surface (ve etation, rocks, water, concrete), causing it to heat up. The warm surface then re-emits he or IR radiation toward space (red arrow). You can sense the radiation of IR from Earth the heat you feel well after dark from a black asphalt road that warmed up during the da The IR radiation emitted from the Earth either escapes into space or is absorbed by green house gases. The greenhouse gases, in turn, reradiate infrared energy and some of it return to Earth resulting in an increase in the temperature of the planet. The greenhouse effect so named because the glass windows in a greenhouse act in a similar manner to the Earth atmosphere. Visible light transmitted through the glass and absorbed by surfaces causes t inside to warm. Infrared radiation emitted by warm interior surfaces is absorbed by the glas preventing some of the heat from escaping. Soon the greenhouse interior is much warm than the temperature outside.

    FIGURE 10.20 illustrates differences in the nature of the electromagnetic radiatic emitted by the Sun and Earth. Incoming solar radiation has a maximum intensity at 483 n which is in the visible region of the electromagnetic spectrum (blue light), while the Earth emission has a maximum intensity near 10,000 nm in the IR. Why do the Earth and Sun em different wavelengths of light? The answer is that the sun is a lot hotter than the Earth. Bo the Sun and Earth emit light as a *black-body*, meaning that a continuous spectrum is emitte from a hot object with wavelength dependent on temperature. A familiar example of blacl body radiation is the reddish-orange glow of a hot electric burner. The peak wavelength 483 nm in the sun's emission spectrum corresponds to a very hot temperature, 6000 K, whi the peak wavelength in the Earth's emission spectrum near 10,000 nm corresponds to a ten perature of 288 K (15 °C or 59 °F).

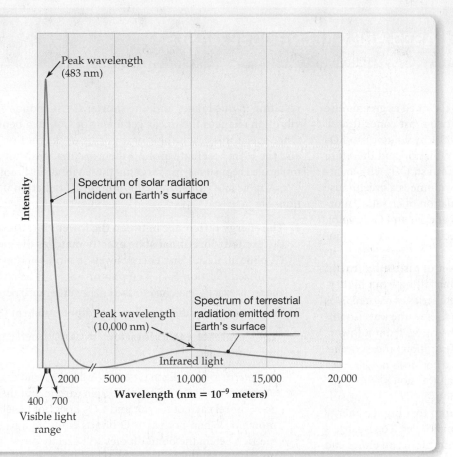

◀ **FIGURE 10.20**
Spectrum of solar radiation incident on the
Earth's surface and spectrum of radiation
emitted from Earth.

▲ A hot electric burner emits black body radiation.

INQUIRY ▶▶▶ WHICH GASES ARE GREENHOUSE GASES?

Why are some atmospheric gases classified as greenhouse gases and others are not? The two most concentrated gases in the atmosphere N_2 (~80% by volume) and O_2 (~20% by volume) do not absorb infrared radiation and therefore are not classified as greenhouse gases. In contrast, CO_2 with an atmospheric concentration of only 0.04% by volume is a greenhouse gas that plays an important role in the regulation of climate. In order to explain why CO_2 is a greenhouse gas and N_2 and O_2 are not, we must understand what occurs on a molecular level when a photon of IR radiation is absorbed.

Let's begin by revisiting the interaction of visible light with atoms. Recall from Section 5.3 that electrons falling from higher-energy to lower-energy orbitals emit a discrete series of colored lines called atomic line spectra. Conversely, when these same wavelengths of light are *absorbed* by an atom, electrons move from a lower-energy to higher-energy orbitals. The same absorption process occurs in molecules, but in this case electrons change positions not between atomic orbitals, but between molecular orbitals (Section 8.7).

Absorbed IR radiation, in the greenhouse effect, is not sufficiently energetic to cause electrons to jump to a higher energy orbital but does increase bond vibrations in molecules. To visualize a bond vibration, imagine two balls representing atoms on either end of a spring (**FIGURE 10.21**). The spring can stretch and retract. The absorption of IR radiation causes a molecule to reach an excited vibrational state in which stretching moves atoms further apart.

In a bond vibration the atoms first move outward along the axis of the imaginary spring.	Once the vibration has reached its maximum amplitude the atoms return to their original position.

▲ **FIGURE 10.21**

Visualization of a bond vibration in a diatomic molecule.

Let's examine some bond vibrations that occur in the greenhouse gas CO_2. CO_2 has two double bonds around the central carbon atom resulting in linear geometry. However, these bonds are not stationary and vibrate in three different ways, as shown below.

	Symmetric stretch	Asymmetric stretch	Bend "scissoring"
Movement of atoms in bond vibration.	O=C=O	O=C=O	O=C=O
Position of atoms after bond vibration.	O=C=O	O=C=O	O—C—O

In the symmetric stretch, both C=O bonds lengthen and in the asymmetric stretch both oxygen atoms move in the same direction

resulting in one longer and one shorter C=O bond. The bending vibration changes the linear geometry of CO_2 to a bent geometry. Once the atoms reach their new positions the motion is reversed and the atoms oscillate around their original position. Every CO_2 molecule is constantly undergoing these vibrational motions.

A molecule will absorb a photon of IR radiation if two conditions are met:

1. **The energy difference between the lower vibrational state and the excited vibrational state exactly matches the energy of the IR photon**. Recall that energy levels in atoms and molecules are quantized (Section 5.4). The photon can only be absorbed if its energy is exactly matched to the energy difference between states.

2. **The vibration results in a change in dipole moment** (Section 8.5).

Let's evaluate each vibration in CO_2 to determine if the requirement is met.

Symmetric stretch: Each C=O bond has a bond dipole because oxygen is more electronegative than carbon, but the dipoles exactly cancel each other out and CO_2 does not have a net dipole moment. When both C=O bonds elongate during the symmetric stretch, the bond dipoles still exactly cancel and there is no change in dipole moment. Thus the symmetric stretch vibration does not absorb IR radiation.

Asymmetric stretch: In the asymmetric stretch, both oxygen atoms carry a partial negative charge and they move in the same direction during the vibration. This results in shifting in negative charge to one side of the CO_2 molecule creating a net dipole moment. Therefore, the asymmetric stretch absorbs IR radiation.

The asymmetric stretch produces an alternating dipole moment.

Bending (scissoring) vibration: The bending vibration changes CO_2 from a linear geometry with no dipole moment to a bent geometry with a dipole moment. Both bond dipoles point toward the partial negative charges on the oxygen atoms, giving the end of CO_2 with two oxygen atoms

The bending vibration produces an alternating dipole moment.

a partial negative charge and the end with the carbon atom a partial positive charge. Thus the bending vibration creates a change in dipole moment and absorbs IR radiation.

REMEMBER...

The measure of net molecular polarity is a quantity called the **dipole moment,** μ (Greek mu), which is defined as the magnitude of the charge Q at either end of the molecular dipole times the distance r between the charges: $\mu = Q \times r$. (Section 8.5)

FIGURE 10.22 shows the infrared absorption spectrum for O₂. The vertical axis is the fraction of light transmitted by the sample, and the negative peaks indicate absorption of IR radiation. Notice that both the asymmetric stretch and the bending vibration have absorption peaks. The wavelength (frequency) for the symmetric stretch is indicated, but no absorption peak is present because this vibration does not cause a net change in dipole moment.

Molecules that absorb IR radiation are greenhouse gases, and water is one of the most important examples. The observation that clear nights are often cooler than cloudy ones is a familiar example of the greenhouse effect; water vapor in clouds strongly absorbs IR. Although water vapor is the largest contributor to the natural greenhouse effect, the amount in the atmosphere is mainly controlled by air temperature and not by emissions from human activities. The greenhouse gas emissions from human activities that are of greatest concern are carbon dioxide (CO_2), nitrous oxide (N_2O), methane (CH_4), and halogen-containing gases. Carbon dioxide is added to the atmosphere primarily from burning fossil fuels, but industrial processes and decaying organic matter also contribute significant amounts. Methane is emitted from fossil fuel mining and use, landfills, agricultural practices, livestock cultivation, and termites. Natural microbial activity in soil and oceans and the use of fertilizer in agricultural practices are sources of nitrous oxide. Halogenated gases, such as hydrofluorocarbons ($CHClF_2$) or sulfur hexafluoride (SF_6), are synthetic greenhouse gases emitted from a variety of industrial applications such as refrigeration and air conditioning.

Global-warming potential (GWP) is a relative measure of how much heat a greenhouse gas traps in the atmosphere on a per mass basis and carbon dioxide is set to a reference value of 1. A GWP is calculated over a specific time interval, commonly 20, 100, or 500 years, and is influenced by how strongly a gas absorbs IR and how long it remains in the atmosphere. **TABLE 10.9** gives GWP values for the major greenhouse gases associated with human activities. Table 10.9 also includes radiative forcing values which describe the

TABLE 10.9 Global-Warming Potentials and Radiative Forcing of Greenhouse Gases

Greenhouse Gas	Global Warming Potential (100-year value)	Atmospheric Concentration	Radiative Forcing (W/m^2)
CO_2	1	400 ppm	1.82
CH_4	21	1.8 ppm	0.48
N_2O	310	325 ppb	0.17
CFC-12	4600	0.52 ppb	0.17
SF_6	22,800	0.007 ppb	0.004
Halogenated gases (total)			0.360

(Intergovernmental Panel on Climate Change   Climate Change 2013: The Physical Science Basis)

relative contribution of each gas to the greenhouse effect. Higher positive values for radiative forcing indicate larger warming effects. Radiative forcing is defined as the net change in the energy balance of the Earth and is expressed in units of watts per square meter (W/m^2). Although CO_2 has the lowest GWP of the greenhouse gases listed in Table 10.9, it makes the greatest contribution to the greenhouse effect because it has a higher concentration than the other gases.

PROBLEM 10.21 Explain why nitrogen and oxygen are not greenhouse gases.

PROBLEM 10.22 The water molecule has similar bond vibrations to carbon dioxide. Decide whether the symmetric, asymmetric, and bending vibrations in water will result in the absorption of IR radiation.

PROBLEM 10.23 Bond vibrations for the symmetric and asymmetric stretch in methane are illustrated below. Decide whether each vibration will result in the absorption of IR radiation. Arrows indicate the movement of atoms during the vibration.

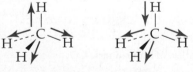

Symmetric stretch   Asymmetric stretch

PROBLEM 10.24 How many times larger is carbon dioxide's contribution to the greenhouse effect than methane? (Use radiative forcing values in Table 10.9 to compare the two gases.)

PROBLEM 10.25 N_2O has a GWP value of 310 and CO_2 has a GWP value 1, but CO_2 makes a greater contribution to the greenhouse effect. Explain.

PROBLEM 10.26 Although human activities do not have significant influence on the concentration of water vapor in the atmosphere, the surface warming caused by other greenhouse gases could increase rates of evaporation. If more water vapor entered the atmosphere, what would be the effect on climate based on the greenhouse effect?

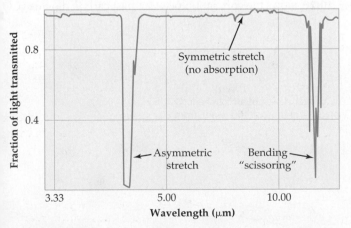

▲ **FIGURE 10.22**

Infrared absorption spectrum of carbon dioxide.

10.11 ▶ CLIMATE CHANGE

Global warming refers to the idea that increasing concentrations of greenhouse gases w̄ upset the delicate thermal balance of incoming and outgoing radiation on Earth. Rising leve of greenhouse gases will absorb more IR radiation and cause an enhanced warming effec The term global warming has been replaced with **climate change** because there will not a uniform rise in temperature at all locations. Instead, different areas will warm by varyii degrees and other areas may even experience cooling. Climate science is complex because is a global issue and greenhouse gas concentration is only one of many factors that influen climate. Cloud cover, particulate matter, solar energy, and changing surface reflectivity d to melting polar ice caps or deforestation are just a few variables that influence global and r gional temperature. In this section, we will describe how concentrations of greenhouse gas have changed over time and their impact on climate both today and in the future.

Since the Industrial Revolution began in the late 1700s, people have added a significai amount of greenhouse gases to the atmosphere by burning fossil fuels for energy, cuttii down forests, and producing industrial goods such as cement or metals. Although natur mechanisms remove greenhouse gases from the atmosphere, concentrations are rising b cause the rate of addition exceeds the rate of loss. For example, careful measurements sho that concentrations of atmospheric carbon dioxide have been rising in the last 160 years fro: an estimated 290 parts per million (ppm) in 1850 to 400 ppm in 2014 (**FIGURE 10.23**).

It is also useful to examine greenhouse gas levels on a longer timescale to compare o current situation to other time periods. Samples from polar ice caps are used to create histor cal records because the depth of the ice core sample can be correlated with time. Greenhou gases trapped in air bubbles in the ice can be analyzed when the ice is melted. Measuremen show that current global atmospheric concentrations of CO_2 and CH_4 are unprecedente compared with their levels in the past 650,000 years (**FIGURE 10.24**). Concentrations these greenhouse gases have fluctuated over time, but dramatic increases are evident in th last century. Note that CO_2 levels never exceeded 300 ppm in the long-term historical recor but burning of fossil fuels has caused the level to rise to 400 ppm. Methane concentratio did not exceed 0.8 ppm (800 ppb) in the historical record, but have recently risen to 1.8 pp (1,800 ppb). Similarly, N_2O levels have not exceeded 280 ppb over the past 100,000 years bi have increased to a concentration of 325 ppb. Most of the halogenated gases are man-mad and their atmospheric concentration began rising as they were used in industrial processe over the last few decades.

Ever since Earth formed, its climate has undergone dramatic shifts, from "ice ages" wit high glacial coverage to relatively warm periods. Factors such as surface reflectivity, airborr dust, variations in the Earth's orbit, and solar intensity combined with greenhouse gas le els contributed to these periodic fluctuations in climate. Ice core samples can also be used measure a long-term historical record of past climate. Analysis of the ratio of hydrogen isc topes ($^2H{:}^1H$) in H_2O in ice core samples helps scientists estimate long-term shifts in averag global temperature. **FIGURE 10.25** shows the correlation between past carbon dioxide (CO

▶ **FIGURE 10.23**

Annual concentration of atmospheric CO_2 since 1850.

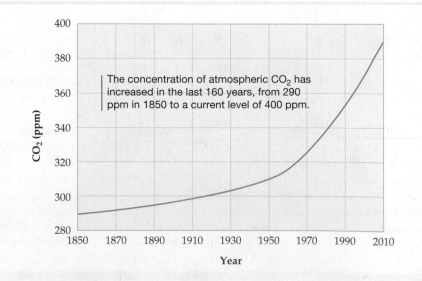

The concentration of atmospheric CO_2 has increased in the last 160 years, from 290 ppm in 1850 to a current level of 400 ppm.

(a) CO_2

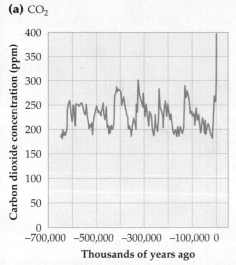

(b) CH_4

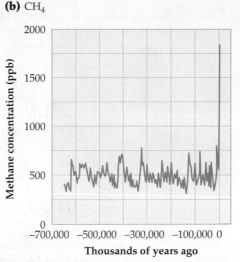

Source: U.S. Environmental Protection Agency, Climate Change, Greenhouse Gases.

◀ **FIGURE 10.24**

Long-term historical record of greenhouse gases obtained from ice core and atmospheric measurements.

Figure It Out

How do current levels of CO_2 and CH_4 in the atmosphere compare to levels over the past 650,000 years?

Answer: The historical record shows that both CO_2 and CH_4 levels have risen dramatically in the past few centuries, and the current level is significantly higher than any time in the past 650,000 years.

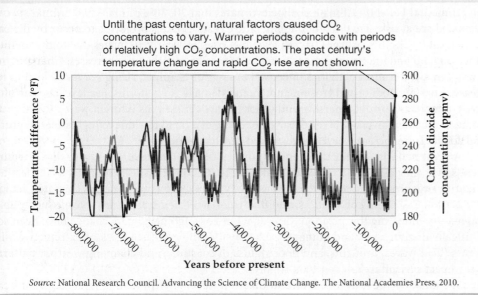

Until the past century, natural factors caused CO_2 concentrations to vary. Warmer periods coincide with periods of relatively high CO_2 concentrations. The past century's temperature change and rapid CO_2 rise are not shown.

Source: National Research Council. Advancing the Science of Climate Change. The National Academies Press, 2010.

◀ **FIGURE 10.25**

Correlation of CO_2 levels and past global temperatures measured in ice cores.

Figure It Out

Does this figure show that an increase in CO_2 concentration causes an increase in temperature?

Answer: No, it does not prove that the temperature changed in direct response to changes in CO_2 levels. It is possible that the opposite occurred: a temperature change could have caused changes in CO_2 levels. The factors are related in a complex way, but the data clearly shows that higher temperatures correspond with higher CO_2 levels.

oncentrations (top) and Antarctic temperature (bottom) over the past 800,000 years. Warmer eriods coincide with higher CO_2 concentrations.

Most atmospheric scientists believe the climate has changed rapidly in recent years due o human-induced changes to the atmosphere. **FIGURE 10.26** illustrates the change in global urface temperature (°C) from 1880 to present relative to the average temperature during the me period 1951–1980. Roughly two-thirds of the warming occurred since 1975. While the mperature increase may seem small (approximately 1 °C from 1990 to present), it is impor- nt to realize that warming is occurring at a much more rapid pace than past climate shifts, hich occurred over thousands of years. Climate models predict that average global tempera- ures may increase by 3.5–4 °C (~6−7 °F) by 2100, depending on levels of future greenhouse as emissions and that warming will continue at an accelerated pace.

Scientists track the effects of climate change using indicators that show trends in envi- onmental conditions, such as the extent of sea ice, changes in sea level, changes in biodiver- ty, and ocean acidity. The Arctic is warming at a rate much faster than the rest of the Earth nd sea ice is melting at an astonishing rate. The area of Arctic sea ice in September 2012 was .3 million square miles (an area five times the size of Texas) less than the historical 1979– 000 average (**FIGURE 10.27**). Climate is an important influence on ecosystems and the rate f temperature change may simply be too fast for many species to adapt in order to survive. pecies that are especially climate sensitive such as those dependent on sea ice (polar bears nd seals) or those adapted to mountain or cold water environments (pikas and salmon) may

▶ **FIGURE 10.26**

Change in global surface temperature (°C) relative to 1951–1980 average temperatures. The dotted green line is the annual mean and the solid red line is the 5-year mean.

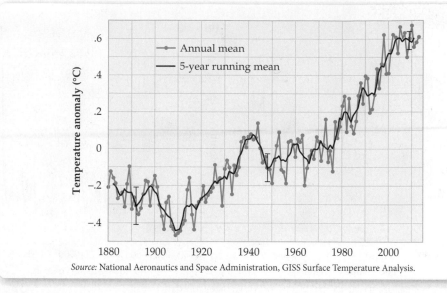

Source: National Aeronautics and Space Administration, GISS Surface Temperature Analysis.

▲ Climate change may be the leading factor decreasing the populations of the American pika.

suffer significant declines in population or in extreme cases extinction. A study by the Intergovernmental Panel on Climate change estimates that 20–30% of plant and animal species evaluated are at risk of extinction if temperature reaches levels projected to occur by the end of the century. In the past century, sea level has risen by nearly 10 inches due to the melting of ice on land and the thermal expansion of water as its temperature increases. Changing sea levels can affect human activities in coastal areas. For example, rising sea levels lead to increased coastal flooding and erosion and greater damage from storms. Sea level rise can alter ecosystems, transforming marshes and other freshwater systems into salt water. Coral reefs, home to much biodiversity in the oceans, are dying at a rapid rate due to higher temperatures and acidity from increased levels of dissolved carbon dioxide.

A great majority of climate scientists (over 97% in a recent survey) agree that warming trends over the past century are very likely due to human activities, and most of the leading scientific organizations worldwide, such as the Intergovernmental Panel on Climate Change, The National Academy of Sciences, and the American Chemical Society, have issued public statements endorsing this position. Consequences of ongoing climate change have been scientifically documented, and humans must consider the increasing risk of extreme weather events (heat waves, droughts, and floods), loss of coastline, and changing weather patterns that impact agriculture and our food supply.

We must decide how to respond to this long-term environmental issue. Solutions can be complex, costly, and difficult to implement, but several options are being researched and tested. These include alternative energy sources (such as solar, wind, and nuclear), carbon sequestration—the trapping and storing of carbon dioxide—and conservation measures (increasing efficiency or decreasing use). Many scientists and policy makers believe that it is more economical to develop a sustainable energy system rather than dealing with the costs associated with the negative consequences of climate change.

▶ **FIGURE 10.27**

Sea ice extent in the Arctic. September 2012 had the lowest sea ice extent on record, 49% below the 1979–2000 average for that month shown by the yellow line.

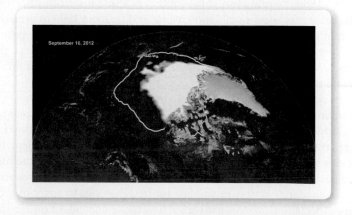

STUDY GUIDE

Section	Concept Summary	Learning Objectives	Test Your Understanding
10.1 ▶ Gases and Gas Pressure	A gas is a collection of atoms or molecules moving independently through a volume that is largely empty space. Collisions of randomly moving particles with the walls of their container exert a force per unit area that we perceive as **pressure (P)**. The SI unit for pressure is the **pascal (Pa)**, but the **atmosphere (atm)**, the **millimeter of mercury**, and the **bar** are more commonly used.	10.1 Convert between different units of pressure.	Worked Example 10.1; Problems 10.38, 10.42, 10.43
		10.2 Describe how a barometer and manometer measure pressure.	Worked Example 10.2; Problems 10.27, 10.28, 10.29, 10.44–10.46
10.2 ▶ The Gas Laws	The physical condition of any gas is defined by four variables: pressure (P), temperature (T), volume (V), and molar amount (n). The specific relationships among these variables are called the **gas laws**: Boyle's law, Charles's law, and Avogadro's law.	10.3 Use the individual gas laws to calculate pressure, volume, molar amount, or temperature for a gas sample when conditions change.	Worked Example 10.3; Problems 10.31, 10.36, 10.50
10.3 ▶ The Ideal Gas Law	The three individual gas laws can be combined into a single **ideal gas law**, $PV = nRT$. If any three of the four variables P, V, T, and n are known, the fourth can be calculated. The constant R in the equation is called the **gas constant (R)** and has the same value for all gases. At **standard temperature and pressure (STP**; 1 atm and $0\ ^{\circ}C$), the **standard molar volume** of an ideal gas is 22.414 L.	10.4 Use the ideal gas law to calculate pressure, volume, molar amount, or temperature for a gas sample.	Worked Examples 10.4, 10.5; Problems 10.52–10.61
10.4 ▶ Stoichiometric Relationships with Gases	The ideal gas law can be used to calculate the number of moles in a gas sample. The number of moles in gaseous reactant or product can be related to other substances in a chemical reaction using stoichiometry. Gas density can be used to identify unknown gases and depends on pressure, molar mass, and temperature.	10.5 Calculate volumes of gases in chemical reactions.	Worked Example 10.6; Problems 10.72–10.77, 10.156
		10.6 Calculate the density or molar mass of a gas using the formula for gas density.	Worked Example 10.7; Problems 10.68–10.71
10.5 ▶ Mixtures of Gases: Partial Pressure and Dalton's Law	The gas laws apply to mixtures of gases as well as to pure gases. According to **Dalton's law of partial pressures**, the total pressure exerted by a mixture of gases in a container is equal to the sum of the pressures each individual gas would exert alone.	10.7 Calculate the partial pressure, mole fraction, or amount of each gas in a mixture.	Worked Example 10.8; Problems 10.34, 10.80–10.88, 10.147, 10.148, 10.151, 10.159
10.6 ▶ The Kinetic–Molecular Theory of Gases	The behavior of gases can be accounted for using a model called the **kinetic–molecular theory**, a group of five postulates: 1. A gas consists of tiny particles moving at random. 2. The volume of the gas particles is negligible compared with the total volume. 3. There are no forces between particles, either attractive or repulsive. 4. Collisions of gas particles are elastic. 5. The average kinetic energy of gas particles is proportional to their absolute temperature.	10.8 Use the assumptions of kinetic–molecular theory to predict gas behavior. 10.9 Calculate the average molecular speed of a gas particle at a given temperature.	Problems 10.30, 10.32, 10.90, 10.146 Problems 10.94–10.99

Section	Concept Summary	Learning Objectives	Test Your Understanding
10.7 ▶ Gas Diffusion and Effusion: Graham's Law	The connection between temperature and kinetic energy obtained from the kinetic–molecular theory makes it possible to calculate the average speed of a gas particle at any temperature. An important practical consequence of this relationship is **Graham's law**, which states that the rate of a gas's **effusion**, or spontaneous passage through a pinhole in a membrane, depends inversely on the square root of the gas's mass.	**10.10** Visualize the processes of effusion and diffusion.	Problems 10.35, 10.37
		10.11 Use Graham's Law to estimate relative rates of diffusion for two gases.	Worked Example 10.9; Problems 10.100–10.103, 10.143
10.8 ▶ The Behavior of Real Gases	Real gases differ in their behavior from that predicted by the ideal gas law, particularly at high pressure, where gas particles are forced close together and intermolecular attractions become significant. The deviations from ideal behavior can be dealt with mathematically by the **van der Waals equation**.	**10.12** Understand the conditions when gases deviate the most from ideal behavior.	Problems 10.106, 10.107
		10.13 Use the van der Waals equation to calculate the properties of real gases.	Problems 10.108, 10.109
10.9 ▶ The Earth's Atmosphere and Pollution	The Earth's atmosphere is divided into layers based upon changes in temperature. Air pollution occurs in the lowest layer of the atmosphere called the troposphere. Pollutants that are emitted directly from sources are called **primary pollutants** and include lead, particulate matter, nitric oxide, sulfur dioxide, and carbon monoxide. **Secondary pollutants** are formed from reactions of primary pollutants and include ozone and nitrogen dioxide.	**10.14** Convert between different units used to express the concentration of pollutants.	Worked Example 10.10; Problems 10.113, 10.116–10.118
		10.15 Use the gas laws, Dalton's Law, and stoichiometry to calculate amounts of pollutant gases in the atmosphere.	Problems 10.111, 10.120, 10.121
		10.16 Identify the components and causes of photochemical smog.	Problems 10.123–10.129
10.10 and 10.11 ▶ The Greenhouse Effect and Climate Change	The **greenhouse effect** is the trapping of heat emitted from the Earth by gases that absorb infrared radiation (**greenhouse gases**). **Climate change** is occurring as a result of rising levels of greenhouse gases from human activities. Measurements and models show that certain regions such as the Arctic and land masses will warm more than other regions, such as the ocean.	**10.17** Explain the principle of the greenhouse effect.	Problems 10.130–10.132
		10.18 Describe the trends in greenhouse gas concentrations over time and measured and predicted effects of climate change.	Problems 10.134–10.137

KEY TERMS

atmosphere (atm) *361*
Avogadro's law *367*
bar *361*
Boyle's law *364*
Charles's law *365*
climate change *394*
Dalton's law of partial pressures *375*

diffusion *380*
effusion *380*
gas constant (*R*) *369*
gas laws *364*
global warming *394*
Graham's law *381*
greenhouse effect *389*
greenhouse gas *389*

ideal gas *364*
ideal gas law *369*
kinetic–molecular theory *378*
manometer *362*
millimeter of mercury (mm Hg) *361*
mole fraction (*X*) *376*
newton (N) *360*

pascal (Pa) *360*
primary pollutant *384*
pressure (*P*) *360*
secondary pollutant *384*
standard molar volume *367*
standard temperature and pressure (STP) *370*
van der Waals equation *383*

KEY EQUATIONS

Boyle's law (Section 10.2)

$V \propto 1/P$ or $PV = k$ at constant n and T

Charles's law (Section 10.2)

$V \propto T$ or $V/T = k$ at constant n and P

Avogadro's law (Section 10.2)

$V \propto n$ or $V/n = k$ at constant T and P

Ideal gas law (Section 10.3)

$V = \dfrac{nRT}{P}$ or $PV = nRT$

Standard temperature and pressure (STP) for gases (Section 10.3)

$T = 0\,°C$ $P = 1\,atm$

Gas density (Section 10.4)

$d = \dfrac{m}{V} = \dfrac{PM}{RT}$

Dalton's law of partial pressures (Section 10.5)

$P_{total} = P_1 + P_2 + P_3 + \ldots$

where $P_1, P_2, \ldots$ are the pressures each individual gas would have if it were alone.

Mole fraction (Section 10.6)

$(X) = \dfrac{\text{Moles of component}}{\text{Total moles in mixture}}$

Partial pressure (Section 10.6)

$P_1 = X_1 \cdot P_{total}$

Average speed of gas particle at temperature T (Section 10.6)

$u = \sqrt{\dfrac{3RT}{mN_A}} = \sqrt{\dfrac{3RT}{M}}$ where M is the molar mass.

Graham's law (Section 10.7)

Rate of effusion $\propto \dfrac{1}{\sqrt{m}}$

van der Waals equation (Section 10.8)

$\left(P + \dfrac{an^2}{V^2}\right)(V - nb) = nRT$ or $P = \dfrac{nRT}{V - nb} - \dfrac{an^2}{V^2}$

where a and b are correction factors to the ideal gas law.

CONCEPTUAL PROBLEMS

Problems 10.1–10.26 appear within the chapter.

10.27 A glass tube has one end in a dish of mercury and the other end closed by a stopcock. The distance from the surface of the mercury to the bottom of the stopcock is 850 mm. The apparatus is at 25 °C, and the mercury level in the tube is the same as that in the dish.

(1) **(2)** **(3)**

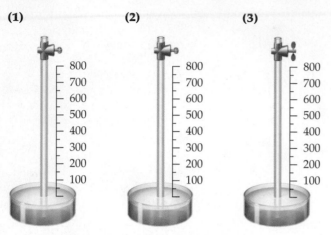

(a) Show on drawing **(1)** what the approximate level of mercury in the tube will be when the temperature of the entire apparatus is lowered from +25 °C to −25 °C.

(b) Show on drawing **(2)** what the approximate level of mercury in the tube will be when a vacuum pump is connected to the top of the tube, the stopcock is opened, the tube is evacuated, the stopcock is closed, and the pump is removed.

(c) Show on drawing **(3)** what the approximate level of mercury in the tube will be when the stopcock in drawing **(2)** is opened.

10.28 The apparatus shown is called a *closed-end* manometer because the arm not connected to the gas sample is closed to the atmosphere and is under vacuum. Explain how you can read the gas pressure in the bulb.

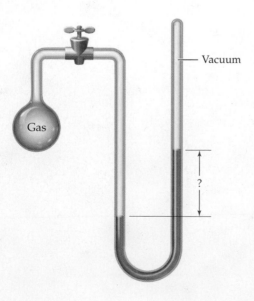

10.29 Redraw the following open-end manometer to show what would look like when stopcock A is opened.

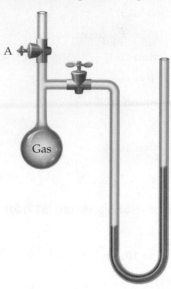

10.30 A 1:1 mixture of helium (red) and argon (blue) at 300 K portrayed below on the left. Draw the same mixture when th temperature is lowered to 150 K.

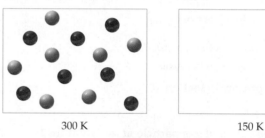

300 K   150 K

10.31 Assume that you have a sample of gas in a cylinder with movable piston, as shown in the following drawing:

Redraw the apparatus to show what the sample will look like after **(a)** the temperature is increased from 300 K to 450 K at con stant pressure, **(b)** the pressure is increased from 1 atm to 2 atm at constant temperature, and **(c)** the temperature is decrease from 300 K to 200 K and the pressure is decreased from 3 atm to 2 atm.

10.32 Assume that you have a sample of gas at 350 K in a sealed con tainer, as represented in **(a)**. Which of the drawings **(b)**–**(d**

represents the gas after the temperature is lowered from 350 K to 150 K? The boiling point of the gas is 90 K.

(a)

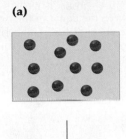

10.34 The following drawing represents a container holding a mixture of four gases, red, blue, green, and black. If the total pressure inside the container is 420 mm Hg, what is the partial pressure of each individual component?

(c) **(d)**

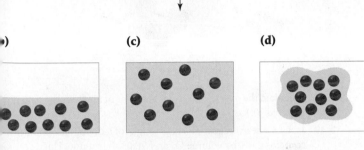

10.35 Three bulbs, two of which contain different gases and one of which is empty, are connected as shown in the following drawing. Redraw the apparatus to represent the gases after the stopcocks are opened and the system is allowed to come to equilibrium.

0.33 Assume that you have a mixture of He (atomic weight = 4) and Xe (atomic weight = 131) at 300 K. Which of the drawings best represents the mixture (blue = He; red = Xe)?

(a) **(b)** **(c)**

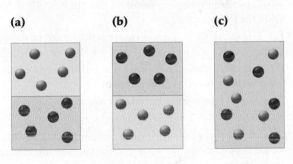

10.36 Show the approximate level of the movable piston in drawings **(a)**, **(b)**, and **(c)** after the indicated changes have been made to the gas.

(a) **(b)** **(c)**

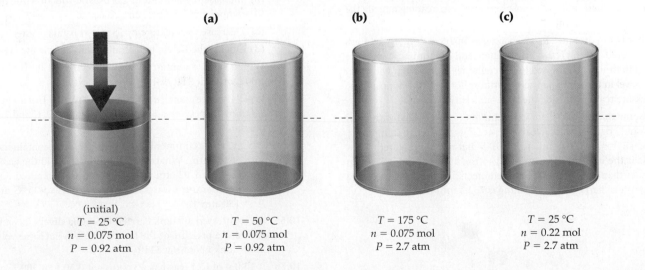

(initial)
$T = 25\,°C$
$n = 0.075$ mol
$P = 0.92$ atm

$T = 50\,°C$
$n = 0.075$ mol
$P = 0.92$ atm

$T = 175\,°C$
$n = 0.075$ mol
$P = 2.7$ atm

$T = 25\,°C$
$n = 0.22$ mol
$P = 2.7$ atm

10.37 Effusion of a 1:1 mixture of two gases through a small pinhole produces the results shown below.

(a) Which gas molecules—yellow or blue—have a higher average speed?

(b) If the yellow molecules have a molecular weight of 25, what is the molecular weight of the blue molecules?

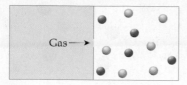

SECTION PROBLEMS

Gases and Gas Pressure (Section 10.1)

10.38 Yet another common measure of pressure is the unit pounds per square inch (psi). How many pounds per square inch correspond to 1.00 atm? To 1.00 mm Hg?

10.39 If the density of water is 1.00 g/mL and the density of mercury is 13.6 g/mL, how high a column of water in meters can be supported by standard atmospheric pressure? By 1 bar?

10.40 What is temperature a measure of?

10.41 Why are gases so much more compressible than solids or liquids?

10.42 Atmospheric pressure at the top of Pikes Peak in Colorado is approximately 480 mm Hg. Convert this value to atmospheres and to pascals.

10.43 Carry out the following conversions:

(a) 352 torr to kPa

(b) 0.255 atm to mm Hg

(c) 0.0382 mm Hg to Pa

10.44 What is the pressure in millimeters of mercury inside a container of gas connected to a mercury-filled open-end manometer of the sort shown in Figure 10.5 when the level in the arm connected to the container is 17.6 cm lower than the level in the arm open to the atmosphere and the atmospheric pressure reading outside the apparatus is 754.3 mm Hg?

10.45 What is the pressure in atmospheres inside a container of gas connected to a mercury-filled open-end manometer when the level in the arm connected to the container is 28.3 cm higher than the level in the arm open to the atmosphere and the atmospheric pressure reading outside the apparatus is 1.021 atm?

10.46 Assume that you have an open-end manometer filled with ethyl alcohol (density = 0.7893 g/mL at 20 °C) rather than mercury (density = 13.546 g/mL at 20°C). What is the pressure in pascals if the level in the arm open to the atmosphere is 55.1 cm higher than the level in the arm connected to the gas sample and the atmospheric pressure reading is 752.3 mm Hg?

Ethyl alcohol

10.47 Assume that you have an open-end manometer filled with chloroform (density = 1.4832 g/mL at 20 °C) rather than mercury (density = 13.546 g/mL at 20 °C). What is the difference in height between the liquid in the two arms if the pressure the arm connected to the gas sample is 0.788 atm and the atmospheric pressure reading is 0.849 atm? In which arm is the chloroform level higher?

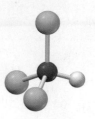

Chloroform

10.48 Calculate the average molecular weight of air from the data given in Table 10.1.

10.49 What is the average molecular weight of a diving-gas mixture that contains 2.0% by volume O_2 and 98.0% by volume He?

The Gas Laws (Sections 10.2 and 10.3)

10.50 Assume that you have a cylinder with a movable piston. What would happen to the gas pressure inside the cylinder if you were to do the following?

(a) Triple the Kelvin temperature while holding the volume constant

(b) Reduce the amount of gas by one-third while holding the temperature and volume constant

(c) Decrease the volume by 45% at constant T

(d) Halve the Kelvin temperature and triple the volume

10.51 Assume that you have a cylinder with a movable piston. What would happen to the gas volume of the cylinder if you were to do the following?

(a) Halve the Kelvin temperature while holding the pressure constant

(b) Increase the amount of gas by one-fourth while holding the temperature and pressure constant

(c) Decrease the pressure by 75% at constant T

(d) Double the Kelvin temperature and double the pressure

10.52 Which sample contains more molecules: 1.00 L of O_2 at STP, 1.00 L of air at STP, or 1.00 L of H_2 at STP?

10.53 Which sample contains more molecules: 2.50 L of air at 50 °C and 750 mm Hg pressure or 2.16 L of CO_2 at −10 °C and 765 mm Hg pressure?

10.54 Oxygen gas is commonly sold in 49.0 L steel containers at a pressure of 150 atm. What volume in liters would the gas occupy at a pressure of 1.02 atm if its temperature remained unchanged? If its temperature was raised from 20 °C to 35 °C at constant $P = 150$ atm?

10.55 A compressed air tank carried by scuba divers has a volume of 8.0 L and a pressure of 140 atm at 20 °C. What is the volume of air in the tank in liters at STP?

10.56 If 15.0 g of CO_2 gas has a volume of 0.30 L at 300 K, what is its pressure in millimeters of mercury?

10.57 If 2.00 g of N_2 gas has a volume of 0.40 L and a pressure of 6.0 atm, what is its Kelvin temperature?

10.58 The matter in interstellar space consists almost entirely of hydrogen atoms at a temperature of 100 K and a density of approximately 1 atom/cm^3. What is the gas pressure in millimeters of mercury?

10.59 Methane gas, CH_4, is sold in a 43.8 L cylinder containing 5.54 kg. What is the pressure inside the cylinder in kilopascals at 20 °C?

10.60 Many laboratory gases are sold in steel cylinders with a volume of 43.8 L. What mass in grams of argon is inside a cylinder whose pressure is 17,180 kPa at 20 °C?

10.61 A small cylinder of helium gas used for filling balloons has a volume of 2.30 L and a pressure of 13,800 kPa at 25 °C. How many balloons can you fill if each one has a volume of 1.5 L and a pressure of 1.25 atm at 25 °C?

Gas Stoichiometry (Section 10.4)

10.62 Which sample contains more molecules, 15.0 L of steam (gaseous H_2O) at 123.0 °C and 0.93 atm pressure or a 10.5 g ice cube at −5.0 °C?

10.63 Which sample contains more molecules, 3.14 L of Ar at 85.0 °C and 1111 mm Hg pressure or 11.07 g of Cl_2?

10.64 Imagine that you have two identical flasks, one containing hydrogen at STP and the other containing oxygen at STP. How can you tell which is which without opening them?

10.65 Imagine that you have two identical flasks, one containing chlorine gas and the other containing argon at the same temperature and pressure. How can you tell which is which without opening them?

10.66 What is the total mass in grams of oxygen in a room measuring 4.0 m by 5.0 m by 2.5 m? Assume that the gas is at STP and that air contains 20.95% oxygen by volume.

10.67 The average oxygen content of arterial blood is approximately 0.25 g of O_2 per liter. Assuming a body temperature of 37 °C, how many moles of oxygen are transported by each liter of arterial blood? How many milliliters?

10.68 One mole of an ideal gas has a volume of 22.414 L at STP. Assuming ideal behavior, what are the densities of the following gases in g/L at STP?

(a) CH_4 (b) CO_2 (c) O_2

10.69 What is the density in g/L of a gas mixture that contains 27.0% F_2 and 73.0% He by volume at 714 mm Hg and 27.5 °C?

10.70 An unknown gas is placed in a 1.500 L bulb at a pressure of 356 mm Hg and a temperature of 22.5 °C, and is found to weigh 0.9847 g. What is the molecular weight of the gas?

10.71 What are the molecular weights of the gases with the following densities:

(a) 1.342 g/L at STP

(b) 1.053 g/L at 25 °C and 752 mm Hg

10.72 Pure oxygen gas was first prepared by heating mercury(II) oxide, HgO:

$$2\,HgO(s) \longrightarrow 2\,Hg(l) + O_2(g)$$

What volume in liters of oxygen at STP is released by heating 10.57 g of HgO?

10.73 How many grams of HgO would you need to heat if you wanted to prepare 0.0155 mol of O_2 according to the equation in Problem 10.72?

10.74 Hydrogen gas can be prepared by reaction of zinc metal with aqueous HCl:

$$Zn(s) + 2\,HCl(aq) \longrightarrow ZnCl_2(aq) + H_2(g)$$

(a) How many liters of H_2 would be formed at 742 mm Hg and 15 °C if 25.5 g of zinc was allowed to react?

(b) How many grams of zinc would you start with if you wanted to prepare 5.00 L of H_2 at 350 mm Hg and 30.0 °C?

10.75 Ammonium nitrate can decompose explosively when heated according to the equation

$$2\,NH_4NO_3(s) \longrightarrow 2\,N_2(g) + 4\,H_2O(g) + O_2(g)$$

How many liters of gas would be formed at 450 °C and 1.00 atm pressure by explosion of 450 g of NH_4NO_3?

10.76 The reaction of sodium peroxide (Na_2O_2) with CO_2 is used in space vehicles to remove CO_2 from the air and generate O_2 for breathing:

$$2\,Na_2O_2(s) + 2\,CO_2(g) \longrightarrow 2\,Na_2CO_3(s) + O_2(g)$$

(a) Assuming that air is breathed at an average rate of 4.50 L/min (25 °C; 735 mm Hg) and that the concentration of CO_2 in expelled air is 3.4% by volume, how many grams of CO_2 are produced in 24 h?

(b) How many days would a 3.65 kg supply of Na_2O_2 last?

10.77 Titanium(III) chloride, a substance used in catalysts for preparing polyethylene, is made by high-temperature reaction of $TiCl_4$ vapor with H_2:

$$2\,TiCl_4(g) + H_2(g) \longrightarrow 2\,TiCl_3(s) + 2\,HCl(g)$$

(a) How many grams of $TiCl_4$ are needed for complete reaction with 155 L of H_2 at 435 °C and 795 mm Hg pressure?

(b) How many liters of HCl gas at STP will result from the reaction described in part (a)?

10.78 A typical high-pressure tire on a bicycle might have a volume of 365 mL and a pressure of 7.80 atm at 25 °C. Suppose the rider filled the tire with helium to minimize weight. What is the mass of the helium in the tire?

10.79 Propane gas (C_3H_8) is often used as a fuel in rural areas. How many liters of CO_2 are formed at STP by the complete combustion of the propane in a container with a volume of 15.0 L and a pressure of 4.5 atm at 25 °C? The unbalanced equation is

$$C_3H_8(g) + O_2(g) \longrightarrow CO_2(g) + H_2O(l)$$

Dalton's Law and Mole Fraction (Section 10.5)

10.80 Use the information in Table 10.1 to calculate the partial pressure in atmospheres of each gas in dry air at STP.

10.81 Natural gas is a mixture of many substances, primarily CH_4, C_2H_6, C_3H_8, and C_4H_{10}. Assuming that the total pressure of the gases is 1.48 atm and that their mole ratio is 94:4.0:1.5:0.50, calculate the partial pressure in atmospheres of each gas.

10.82 A special gas mixture used in bacterial growth chambers contains 1.00% by weight CO_2 and 99.0% O_2. What is the partial pressure in atmospheres of each gas at a total pressure of 0.977 atm?

10.83 A gas mixture for use in some lasers contains 5.00% by weight HCl, 1.00% H_2, and 94% Ne. The mixture is sold in cylinders that have a volume of 49.0 L and a pressure of 13,800 kPa at 210 °C. What is the partial pressure in kilopascals of each gas in the mixture?

10.84 What is the mole fraction of each gas in the mixture described in Problem 10.83?

10.85 A mixture of Ar and N_2 gases has a density of 1.413 g/L at STP. What is the mole fraction of each gas?

10.86 A mixture of 14.2 g of H_2 and 36.7 g of Ar is placed in a 100.0 L container at 290 K.

(a) What is the partial pressure of H_2 in atmospheres?

(b) What is the partial pressure of Ar in atmospheres?

10.87 A 20.0 L flask contains 0.776 g of He and 3.61 g of CO_2 at 300 K.

(a) What is the partial pressure of He in mm Hg?

(b) What is the partial pressure of CO_2 in mm Hg?

10.88 A sample of magnesium metal reacts with aqueous HCl to yield H_2 gas:

$$Mg(s) + 2 HCl(aq) \longrightarrow MgCl_2(aq) + H_2(g)$$

The gas that forms is found to have a volume of 3.557 L at 25 °C and a pressure of 747 mm Hg. Assuming that the gas is saturated with water vapor at a partial pressure of 23.8 mm Hg, what is the partial pressure in millimeters of mercury of the H_2? How many grams of magnesium metal were used in the reaction?

10.89 Chlorine gas was first prepared in 1774 by the oxidation of NaCl with MnO_2:

$$2 NaCl(s) + 2 H_2SO_4(l) + MnO_2(s) \longrightarrow$$
$$Na_2SO_4(s) + MnSO_4(s) + 2 H_2O(g) + Cl_2(g)$$

Assume that the gas produced is saturated with water vapor at a partial pressure of 28.7 mm Hg and that it has a volume of 0.597 L at 27 °C and 755 mm Hg pressure.

(a) What is the mole fraction of Cl_2 in the gas?

(b) How many grams of NaCl were used in the experiment, assuming complete reaction?

Kinetic–Molecular Theory and Graham's Law (Sections 10.6 and 10.7)

10.90 What are the basic assumptions of the kinetic–molecular theory?

10.91 What is the difference between effusion and diffusion?

10.92 What is the difference between heat and temperature?

10.93 Why does a helium-filled balloon lose pressure faster than an air-filled balloon?

10.94 The average temperature at an altitude of 20 km is 220 K. What is the average speed in m/s of an N_2 molecule at this altitude?

10.95 Calculate the average speed of a nitrogen molecule in m/s on a hot day in summer ($T = 37$ °C) and on a cold day in winter ($T = -25$ °C).

10.96 At what temperature (°C) will xenon atoms have the same average speed that Br_2 molecules have at 20 °C?

10.97 At what temperature does the average speed of an oxygen molecule equal that of an airplane moving at 580 mph?

10.98 Which has a higher average speed, H_2 at 150 K or He at 375 °C?

10.99 Which has a higher average speed, a Ferrari at 145 mph or a gaseous UF_6 molecule at 145 °C?

10.100 An unknown gas is found to diffuse through a porous membrane 2.92 times more slowly than H_2. What is the molecular weight of the gas?

10.101 What is the molecular weight of a gas that diffuses through a porous membrane 1.86 times faster than Xe? What might the gas be?

10.102 Rank the following gases in order of their speed of diffusion through a membrane, and calculate the ratio of their diffusion rates: HCl, F_2, Ar.

10.103 Which will diffuse through a membrane more rapidly, CO or N_2? Assume that the samples contain only the most abundant isotopes of each element, ^{12}C, ^{16}O, and ^{14}N.

10.104 A big-league fastball travels at about 45 m/s. At what temperature (°C) do helium atoms have this same average speed?

10.105 Traffic on the German autobahns reaches speeds of up to 23? km/h. At what temperature (°C) do oxygen molecules have thi? same average speed?

Real Gases (Section 10.8)

10.106 (a) The volume of the gas particles themselves most affect the overall volume of the gas sample at _____ (high or low) pressure.

(b) The volume of each particle causes the true volume of the ga? sample to be _____ (larger or smaller) than the volume calculated by the ideal gas law.

10.107 (a) The attractive forces between particles most affect the overall volume of the gas sample at _____ (high or low) pressure.

(b) The attractive forces between particles causes the true volume of the gas sample to be _____ (larger or smaller) than the volume calculated by the ideal gas law.

10.108 Assume that you have 0.500 mol of N_2 in a volume of 0.600 L at 300 K. Calculate the pressure in atmospheres using both the ideal gas law and the van der Waals equation. For N_2, $a = 1.35 (L^2 \cdot atm)/mol^2$ and $b = 0.0387$ L/mol.

10.109 Assume that you have 15.00 mol of N_2 in a volume of 0.600 L at 300 K. Calculate the pressure in atmospheres using both the ideal gas law and the van der Waals equation. For N_2, $a = 1.35 (L^2 \cdot atm)/mol^2$ and $b = 0.0387$ L/mol.

The Earth's Atmosphere and Pollution (Section 10.9)

10.110 Name the regions of the atmosphere. What property is used to distinguish between different regions of the atmosphere?

10.111 The Earth's atmosphere has a mass of approximately 5.15×10^{15} kg. If the average molar mass of air is 28.8 g/mol, how many moles of gas make up the atmosphere? What is the volume of the atmosphere in liters under conditions of STP? (Note: The average molar mass of air is the weighted average of the molar mass of nitrogen and oxygen 0.20 (32.0 g/mol) + 0.80(28.0 g/mol) = 28.8 g/mol.)

10.112 The troposphere contains about three quarters of the mass of the entire atmosphere. The troposphere is only 12 km thick while the whole atmosphere is about 120 km thick. Explain why the troposphere contains such a large fraction of the total mass.

10.113 The percent by volume of oxygen (20.95%) is constant throughout the troposphere.

(a) Express this percentage as a mole fraction.

(b) Give the partial pressure of oxygen at sea level where the total atmospheric pressure = 1.0 atm.

(c) Give the partial pressure oxygen at 11 km, the altitude where airplanes fly, if the total atmospheric pressure is 0.20 atm.

10.114 Based on the national ambient air quality standards provided in Table 10.7, which size particle pollution is more toxic for humans ($PM_{2.5}$ or PM_{10})? Explain.

10.115 Based on the national ambient air quality standards provided in Table 10.7, which pollutant is more toxic for humans (SO_2 or CO)? Explain.

10.116 Table 10.1 gives the percentage of CO_2 in the atmosphere as 0.040%. What is the concentration of CO_2 in units of ppm?

0.117 The concentration of nitrous oxide (N_2O) in dry air at sea level is 5×10^{-5}%. What is the concentration of N_2O in units of ppm and ppb?

0.118 The concentration of SO_2 in the air was measured to be 0.26 $\mu g/L$ at a pressure of 1 atm and temperature of 25 °C. Does this concentration exceed the 1-hour standard of 75 ppb?

0.119 A person inhales an average of 15,000 L per day. If the nitrogen dioxide concentration in air is 100 ppb, what mass of nitrogen dioxide is inhaled over the course of one day? (Assume the pressure is 1 atm and the temperature is 25 °C.)

0.120 One of the largest sources of SO_2 to the atmosphere is coal-fired power plants. Calculate the volume of SO_2 in L produced by burning 1 kg of coal that contains 2% sulfur. Assume all the sulfur is converted to SO_2.

0.121 Smelting of ores to produce pure metals is an atmospheric source of sulfur dioxide.

(a) Galena, the most common mineral of lead is primarily lead(II) sulfide (PbS). The first step in the production of pure iron is to oxidize lead sulfide into lead(II) sulfite ($PbSO_3$). Lead(II) sulfite is then thermally decomposed into lead(II) oxide and sulfur dioxide gas. Balance the following equation.

$$PbSO_3(s) \xrightarrow{\text{heat}} PbO(s) + SO_2(g)$$

(b) How many liters of SO_2 are produced at 1 atm and 300 °C, if 250 g of $PbSO_3$ is decomposed?

0.122 Which of the EPA criteria air pollutants are emitted

(a) directly from cars?

(b) directly from a coal-fired power plant?

0.123 What is a primary pollutant? Which of the criteria air pollutants are primary pollutants?

0.124 What is a secondary pollutant? Which of the criteria air pollutants are primary pollutants?

0.125 During ozone alerts, citizens are often asked to refrain from driving and from fueling their vehicles. Which primary pollutants that lead to ozone production are decreased by each eliminating each activity?

0.126 Photons with wavelengths less than 400 nm will break the nitrogen–oxygen bond in NO_2 in the series of reaction to form ozone. Calculate the bond energy in units of kJ/mol.

0.127 Key pollutants in a photochemical smog event are O_3, NO_2, and NO. Order the pollutants by the time of day that they reach their peak concentration. Explain their order of appearance.

0.128 The United States Environmental Protection Agency AirNow program reports an Air Quality Index (AQI) to protect human health. Colors are used to represent different levels of air quality (green = good, yellow = moderate, orange = unhealthy for sensitive groups, and red = unhealthy). The two maps shown below show air quality on June 17 and June 28, 2012. Propose an explanation for the drastically different levels of air quality in the Midwest and East.

Daily Peak AQI (Combined $PM_{2.5}$ and O_3)

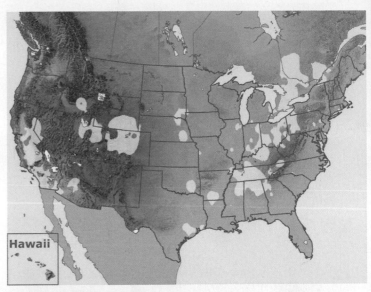

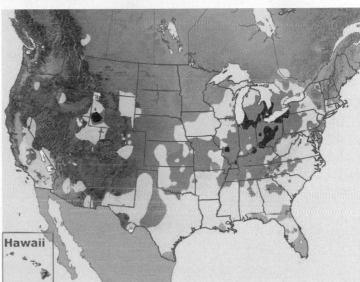

10.129 The United States Environmental Protection Agency AirNow program reports an Air Quality Index (AQI) to protect human health. Colors are used to represent different levels of air quality (green = good, yellow = moderate, orange = unhealthy for sensitive groups, and red = unhealthy). One map is from August 8, and the other is from December 5 in the year 2012. Which map corresponds to each day? Explain.

Daily Peak AQI (Combined PM$_{2.5}$ and O$_3$) August 8, 2012

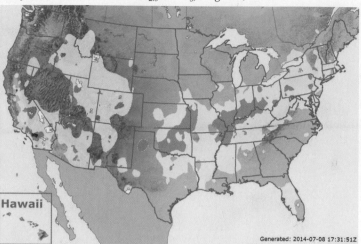

Generated: 2014-07-08 17:31:51Z

Daily Peak AQI (Combined PM$_{2.5}$ and O$_3$) December 5, 2012

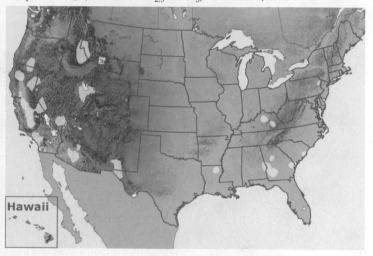

The Greenhouse Effect and Climate Change (Sections 10.10 and 10.11)

10.130 Fill in the blanks with the appropriate region of electromagnetic radiation. (UV, Visible, Infrared)

   (a) The sun most strongly emits in the _____ and _____ regions of electromagnetic radiation.

   (b) The atmosphere filters out biologically damaging _____ radiation from incoming solar radiation and prevents it from reaching Earth.

   (c) The Earth most strongly emits _____ radiation.

   (d) Greenhouse gases absorb _____ radiation.

10.131 **(a)** The wavelength of maximum emission of solar radiation is 483 nm. Calculate the energy of one mole of photons with a wavelength of 483 nm.

   (b) The wavelength of maximum emission of the Earth's radiation is 10,000 nm. Calculate the energy of one mole of photons with a wavelength of 10,000 nm.

   (c) Which emits higher energy radiation, the Earth or the Sun?

10.132 Why do the Earth and Sun have different emission spectra?

10.133 Ozone (O$_3$) is a harmful pollutant in the troposphere. Draw th electron-dot structure for ozone (O$_3$) and evaluate the symmet ric and asymmetric stretch to determine whether ozone is also greenhouse gas.

10.134 What major greenhouse gases are associated with human activities?

10.135 What is the trend in atmospheric CO$_2$ and CH$_4$ concentration over the past 150 years? Over several hundred thousand years?

10.136 Has the Earth's surface experienced warming as a result o increasing levels of greenhouse gases? If so, by how much?

10.137 Name several indicators that scientists use to evaluate the effects of climate change.

CHAPTER PROBLEMS

10.138 Match each of the gases to the correct atmospheric problem: ai quality, climate change, or both.

   (a) O$_3$ **(b)** SO$_2$

   (c) CO$_2$ **(d)** CH$_4$

   (e) NO$_2$

10.139 Chlorine occurs as a mixture of two isotopes, ^{35}Cl and ^{37}Cl What is the ratio of the diffusion rates of the three species (^{35}Cl)$_2$, ^{35}Cl^{37}Cl, and (^{37}Cl)$_2$?

10.140 What would the atmospheric pressure be in millimeters of mer cury if our atmosphere were composed of pure CO$_2$ gas?

10.141 The surface temperature of Venus is about 1050 K, and the pres sure is about 75 Earth atmospheres. Assuming that these con ditions represent a Venusian "STP," what is the standard mola volume in liters of a gas on Venus?

10.142 When you look directly up at the sky, you are actually looking through a very tall, transparent column of air that extends from the surface of the Earth thousands of kilometers into space. If the air in this column were liquefied, how tall would it be? The den sity of liquid air is 0.89 g/mL.

10.143 Uranium hexafluoride, a molecular solid used for purification of the uranium isotope needed to fuel nuclear power plants, sub limes at 56.5 °C. Assume that you have a 22.9 L vessel that con tains 512.9 g of UF$_6$ at 70.0 °C.

   (a) What is the pressure in the vessel calculated using the ideal gas law?

   (b) What is the pressure in the vessel calculated using the van der Waals equation? (For UF$_6$, $a = 15.80$ (L^2 · atm)/mol^2; $b = 0.1128$ L/mol.)

10.144 A driver with a nearly empty fuel tank may say she is "running on fumes." If a 15.0 gallon automobile gas tank had only gasoline va por remaining in it, what is the farthest the vehicle could travel if it gets 20.0 miles per gallon on liquid gasoline? Assume the aver age molar mass of molecules in gasoline is 105 g/mol, the density of liquid gasoline is 0.75 g/mL, the pressure is 743 mm Hg, and the temperature is 25 °C.

10.145 Two 112 L tanks are filled with gas at 330 K. One contains 5.00 mol of Kr, and the other contains 5.00 mol of O$_2$. Consider ing the assumptions of kinetic–molecular theory, rank the gases from low to high for each of the following properties:

   (a) collision frequency   **(b)** density (g/L)

   (c) average speed   **(d)** pressure

0.146 Two identical 732.0 L tanks each contain 212.0 g of gas at 293 K, with neon in one tank and nitrogen in the other. Based on the assumptions of kinetic–molecular theory, rank the gases from low to high for each of the following properties:

(a) average speed   (b) pressure

(c) collision frequency   (d) density (g/L)

0.147 Pakistan's K2 is the world's second tallest mountain, with an altitude of 28,251 ft. Its base camp, where climbers stop to acclimate, is located about 16,400 ft above sea level.

(a) Approximate atmospheric pressure P at different altitudes is given by the equation $P = e^{-h/7000}$, where P is in atmospheres and h is the altitude in meters. What is the approximate atmospheric pressure in mm Hg at K2 base camp?

(b) What is the atmospheric pressure in mm Hg at the summit of K2?

(c) Assuming the mole fraction of oxygen in air is 0.2095, what is the partial pressure of oxygen in mm Hg at the summit of K2?

0.148 When a 10.00 L vessel containing 42.189 g of I_2 is heated to 1173 K, some I_2 dissociates: $I_2(g) \longrightarrow 2\,I(g)$. If the final pressure in the vessel is 1.733 atm, what are the mole fractions of the two components $I_2(g)$ and $I(g)$ after the reaction?

0.149 Assume that you take a flask, evacuate it to remove all the air, and find its mass to be 478.1 g. You then fill the flask with argon to a pressure of 2.15 atm and reweigh it. What would the balance read in grams if the flask has a volume of 7.35 L and the temperature is 20.0 °C?

0.150 The apparatus shown consists of three bulbs connected by stopcocks. What is the pressure inside the system when the stopcocks are opened? Assume that the lines connecting the bulbs have zero volume and that the temperature remains constant.

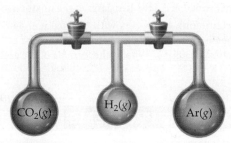

$CO_2(g)$	$H_2(g)$	$Ar(g)$
P = 2.13 atm	P = 0.861 atm	P = 1.15 atm
V = 1.50 L	V = 1.00 L	V = 2.00 L

0.151 The apparatus shown consists of three temperature-jacketed 1.000 L bulbs connected by stopcocks. Bulb **A** contains a mixture of $H_2O(g)$, $CO_2(g)$, and $N_2(g)$ at 25 °C and a total pressure of 564 mm Hg. Bulb **B** is empty and is held at a temperature of -70 °C. Bulb **C** is also empty and is held at a temperature of -190 °C. The stopcocks are closed, and the volume of the lines connecting the bulbs is zero. CO_2 sublimes at -78 °C, and N_2 boils at -196 °C.

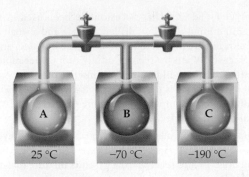

(a) The stopcock between **A** and **B** is opened, and the system is allowed to come to equilibrium. The pressure in **A** and **B** is now 219 mm Hg. What do bulbs **A** and **B** contain?

(b) How many moles of H_2O are in the system?

(c) Both stopcocks are opened, and the system is again allowed to come to equilibrium. The pressure throughout the system is 33.5 mm Hg. What do bulbs **A**, **B**, and **C** contain?

(d) How many moles of N_2 are in the system?

(e) How many moles of CO_2 are in the system?

10.152 Assume that you have 1.00 g of nitroglycerin in a 500.0 mL steel container at 20.0 °C and 1.00 atm pressure. An explosion occurs, raising the temperature of the container and its contents to 425 °C. The balanced equation is

$$4\,C_3H_5N_3O_9(l) \longrightarrow$$
$$12\,CO_2(g) + 10\,H_2O(g) + 6\,N_2(g) + O_2(g)$$

(a) How many moles of nitroglycerin and how many moles of gas (air) were in the container originally?

(b) How many moles of gas are in the container after the explosion?

(c) What is the pressure in atmospheres inside the container after the explosion according to the ideal gas law?

10.153 Use both the ideal gas law and the van der Waals equation to calculate the pressure in atmospheres of 45.0 g of NH_3 gas in a 1.000 L container at 0 °C, 50 °C, and 100 °C. For NH_3, $a = 4.17$ $(L^2 \cdot atm)/mol^2$ and $b = 0.0371$ L/mol.

10.154 When solid mercury(I) carbonate, Hg_2CO_3, is added to nitric acid, HNO_3, a reaction occurs to give mercury(II) nitrate, $Hg(NO_3)_2$, water, and two gases **A** and **B**:

$$Hg_2CO_3(s) + HNO_3(aq) \longrightarrow$$
$$Hg(NO_3)_2(aq) + H_2O(l) + A(g) + B(g)$$

(a) When the gases are placed in a 500.0 mL bulb at 20 °C, the pressure is 258 mm Hg. How many moles of gas are present?

(b) When the gas mixture is passed over $CaO(s)$, gas **A** reacts, forming $CaCO_3(s)$:

$$CaO(s) + A(g) + B(g) \longrightarrow CaCO_3(s) + B(g)$$

The remaining gas **B** is collected in a 250.0 mL container at 20 °C and found to have a pressure of 344 mm Hg. How many moles of **B** are present?

(c) The mass of gas **B** collected in part (b) was found to be 0.218 g. What is the density of **B** in g/L?

(d) What is the molecular weight of **B**, and what is its formula?

(e) Write a balanced equation for the reaction of mercury(I) carbonate with nitric acid.

10.155 Dry ice (solid CO_2) has occasionally been used as an "explosive" in mining. A hole is drilled, dry ice and a small amount of gunpowder are placed in the hole, a fuse is added, and the hole is plugged. When lit, the exploding gunpowder rapidly vaporizes the dry ice, building up an immense pressure. Assume that 500.0 g of dry ice is placed in a cavity with a volume of 0.800 L and the ignited gunpowder heats the CO_2 to 700 K. What is the final pressure inside the hole?

10.156 Consider the combustion reaction of 0.148 g of a hydrocarbon having formula C_nH_{2n+2} with an excess of O_2 in a 400.0 mL steel container. Before reaction, the gaseous mixture had a temperature of 25.0 °C and a pressure of 2.000 atm. After complete combustion and loss of considerable heat, the mixture of products and excess O_2 had a temperature of 125.0 °C and a pressure of 2.983 atm.

 (a) What is the formula and molar mass of the hydrocarbon?

 (b) What are the partial pressures in atmospheres of the reactants?

 (c) What are the partial pressures in atmospheres of the products and the excess O_2?

10.157 Natural gas is a mixture of hydrocarbons, primarily methane (CH_4) and ethane (C_2H_6). A typical mixture might have $X_{methane} = 0.915$ and $X_{ethane} = 0.085$. Let's assume that we have a 15.50 g sample of natural gas in a volume of 15.00 L at a temperature of 20.00 °C.

 (a) How many total moles of gas are in the sample?

 (b) What is the pressure of the sample in atmospheres?

 (c) What is the partial pressure of each component in the sample in atmospheres?

 (d) When the sample is burned in an excess of oxygen, how much heat in kilojoules is liberated?

10.158 A mixture of $CS_2(g)$ and excess $O_2(g)$ is placed in a 10.0 L reaction vessel at 100.0 °C and a pressure of 3.00 atm. A spark causes the CS_2 to ignite, burning it completely, according to the equation

$$CS_2(g) + 3 O_2(g) \longrightarrow CO_2(g) + 2 SO_2(g)$$

After reaction, the temperature returns to 100.0 °C, and the mixture of product gases (CO_2, SO_2, and unreacted O_2) is found to have a pressure of 2.40 atm. What is the partial pressure of each gas in the product mixture?

10.159 Gaseous compound **Q** contains only xenon and oxygen. When 0.100 g of **Q** is placed in a 50.0 mL steel vessel at 0 °C, the pressure is 0.229 atm.

 (a) What is the molar mass of **Q**, and what is a likely formula?

 (b) When the vessel and its contents are warmed to 100 °C, **Q** decomposes into its constituent elements. What is the total pressure, and what are the partial pressures of xenon and oxygen in the container?

10.160 When 10.0 g of a mixture of $Ca(ClO_3)_2$ and $Ca(ClO)_2$ is heated to 700 °C in a 10.0 L vessel, both compounds decompose, forming $O_2(g)$ and $CaCl_2(s)$. The final pressure inside the vessel is 1.00 atm.

 (a) Write balanced equations for the decomposition reactions.

 (b) What is the mass of each compound in the original mixture?

10.161 A 5.00 L vessel contains 25.0 g of PCl_3 and 3.00 g of O_2 at 15 °C. The vessel is heated to 200.0 °C, and the contents react to give $POCl_3$. What is the final pressure in the vessel, assuming that the reaction goes to completion and that all reactants and products are in the gas phase?

10.162 When 2.00 mol of NOCl(g) was heated to 225 °C in a 400.0 [L] steel reaction vessel, the NOCl partially decomposed according to the equation 2 NOCl(g) $\longrightarrow$ 2 NO(g) + $Cl_2(g)$. The pressure in the vessel after reaction is 0.246 atm.

 (a) What is the partial pressure of each gas in the vessel after reaction?

 (b) What percent of the NOCl decomposed?

10.163 Ozone (O_3) can be prepared in the laboratory by passing an electrical discharge through oxygen gas: 3 $O_2(g)$ $\longrightarrow$ 2 $O_3(g)$. Assume that an evacuated steel vessel with a volume of 10.00 L is filled with 32.00 atm of O_2 at 25 °C and an electric discharge is passed through the vessel, causing some of the oxygen to be converted into ozone. As a result, the pressure inside the vessel drops to 30.64 atm at 25.0 °C. What is the final mass percent of ozone in the vessel?

10.164 A steel container with a volume of 500.0 mL is evacuated, and 25.0 g of $CaCO_3$ is added. The container and contents are then heated to 1500 K, causing the $CaCO_3$ to decompose completely, according to the equation $CaCO_3(s) \longrightarrow CaO(s) + CO_2(g)$.

 (a) Using the ideal gas law and ignoring the volume of any solid remaining in the container, calculate the pressure inside the container at 1500 K.

 (b) Now make a more accurate calculation of the pressure inside the container. Take into account the volume of solid CaO (density = 3.34 g/mL) in the container, and use the van der Waals equation to calculate the pressure. The van der Waals constants for $CO_2(g)$ are $a = 3.59$ $(L^2 \cdot atm)/mol$ and $b = 0.0427$ L/mol.

10.165 Nitrogen dioxide dimerizes to give dinitrogen tetroxide: 2 $NO_2(g)$ $\longrightarrow$ $N_2O_4(g)$. At 298 K, 9.66 g of an NO_2/N_2O_4 mixture exerts a pressure of 0.487 atm in a volume of 6.51 L. What are the mole fractions of the two gases in the mixture?

10.166 A certain nonmetal reacts with hydrogen at 440 °C to form a poisonous, foul-smelling gas. The density of the gas at 25 °C and 1.00 atm is 3.309 g/L. What is the formula of the gas?

MULTICONCEPT PROBLEMS

10.167 An empty 4.00 L steel vessel is filled with 1.00 atm of $CH_4(g)$ and 4.00 atm of $O_2(g)$ at 300 °C. A spark causes the CH_4 to burn completely, according to the equation:

$$CH_4(g) + 2 O_2(g) \longrightarrow CO_2(g) + 2 H_2O(g) \quad \Delta H° = -802 [kJ]$$

 (a) What mass of $CO_2(g)$ is produced in the reaction?

 (b) What is the final temperature inside the vessel after combustion, assuming that the steel vessel has a mass of 14.500 kg, the mixture of gases has an average molar heat capacity of 21 J/(mol·°C), and the heat capacity of steel is 0.449 J/(g·°C)?

 (c) What is the partial pressure of $CO_2(g)$ in the vessel after combustion?

10.168 When a gaseous compound **X** containing only C, H, and O is burned in O_2, 1 volume of the unknown gas reacts with 3 volumes of O_2 to give 2 volumes of CO_2 and 3 volumes of gaseous H_2O. Assume all volumes are measured at the same temperature and pressure.

(a) Calculate a formula for the unknown gas, and write a balanced equation for the combustion reaction.

(b) Is the formula you calculated an empirical formula or a molecular formula? Explain.

(c) Draw two different possible electron-dot structures for the compound **X**.

(d) Combustion of 5.000 g of **X** releases 144.2 kJ heat. Look up $\Delta H°_f$ values for $CO_2(g)$ and $H_2O(g)$ in Appendix B, and calculate $\Delta H°_f$ for compound **X**.

0.169 Isooctane, C_8H_{18}, is the component of gasoline from which the term *octane rating* derives.

(a) Write a balanced equation for the combustion of isooctane to yield CO_2 and H_2O.

(b) Assuming that gasoline is 100% isooctane, that isooctane burns to produce only CO_2 and H_2O, and that the density of isooctane is 0.792 g/mL, what mass of CO_2 in kilograms is produced each year by the annual U.S. gasoline consumption of 4.6×10^{10} L?

(c) What is the volume in liters of this CO_2 at STP?

(d) How many moles of air are necessary for the combustion of 1 mol of isooctane, assuming that air is 21.0% O_2 by volume? What is the volume in liters of this air at STP?

0.170 The *Rankine* temperature scale used in engineering is to the Fahrenheit scale as the Kelvin scale is to the Celsius scale. That is, 1 *Rankine* degree is the same size as 1 Fahrenheit degree, and 0 °R = absolute zero.

(a) What temperature corresponds to the freezing point of water on the Rankine scale?

(b) What is the value of the gas constant R on the Rankine scale in (L·atm)/(°R·mol)?

(c) Use the van der Waals equation to determine the pressure inside a 400.0 mL vessel that contains 2.50 mol of CH_4 at a temperature of 525 °R. For CH_4, a = 2.253 $(L^2 \cdot atm)/mol^2$ and b = 0.04278 L/mol.

0.171 Chemical explosions are characterized by the instantaneous release of large quantities of hot gases, which set up a shock wave of enormous pressure (up to 700,000 atm) and velocity (up to 20,000 mi/h). For example, explosion of nitroglycerin ($C_3H_5N_3O_9$) releases four gases, **A**, **B**, **C**, and **D**:

$$n\, C_3H_5N_3O_9(l) \longrightarrow a\, A(g) + b\, B(g) + c\, C(g) + d\, D(g)$$

Assume that the explosion of 1 mol (227 g) of nitroglycerin releases gases with a temperature of 1950 °C and a volume of 1323 L at 1.00 atm pressure.

(a) How many moles of hot gas are released by the explosion of 0.004 00 mol of nitroglycerin?

(b) When the products released by explosion of 0.004 00 mol of nitroglycerin were placed in a 500.0 mL flask and the flask was cooled to −10 °C, product **A** solidified and the pressure inside the flask was 623 mm Hg. How many moles of **A** were present, and what is its likely identity?

(c) When gases **B**, **C**, and **D** were passed through a tube of powdered Li_2O, gas **B** reacted to form Li_2CO_3. The remaining gases, **C** and **D**, were collected in another 500.0 mL flask and found to have a pressure of 260 mm Hg at 25 °C. How many moles of **B** were present, and what is its likely identity?

(d) When gases **C** and **D** were passed through a hot tube of powdered copper, gas **C** reacted to form CuO. The remaining gas, **D**, was collected in a third 500.0 mL flask and found to have a mass of 0.168 g and a pressure of 223 mm Hg at 25 °C. How many moles each of **C** and **D** were present, and what are their likely identities?

(e) Write a balanced equation for the explosion of nitroglycerin.

10.172 Combustion analysis of 0.1500 g of methyl *tert*-butyl ether, an octane booster used in gasoline, gave 0.3744 g of CO_2 and 0.1838 g of H_2O. When a flask having a volume of 1.00 L was evacuated and then filled with methyl *tert*-butyl ether vapor at a pressure of 100.0 kPa and a temperature of 54.8 °C, the mass of the flask increased by 3.233 g.

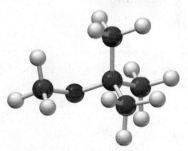

Methyl *tert*-butyl ether

(a) What is the empirical formula of methyl *tert*-butyl ether?

(b) What is the molecular weight and molecular formula of methyl *tert*-butyl ether?

(c) Write a balanced equation for the combustion reaction.

(d) The enthalpy of combustion for methyl *tert*-butyl ether is $\Delta H°_{combustion} = -3368.7$ kJ/mol. What is its standard enthalpy of formation, $\Delta H°_f$?

CHAPTER

11

Liquids, Solids, and Phase Changes

"Dry ice" is a solid form of carbon dioxide (CO_2) that sublimes at $-78.5\ °C$ at 1 atm. Different phases of CO_2 can be made by varying temperature and pressure. The liquid and supercritical phases of CO_2 are used as alternatives to harmful solvents in industrial applications.

 How is caffeine removed from coffee?

The answer to this question can be found in the **INQUIRY** ▸▸▸ on page 437.

CONTENTS

The kinetic–molecular theory developed in the previous chapter accounts for the properties of gases by assuming that gas particles act independently of one another. Because the intermolecular forces between them are so weak, the particles in gases are free to move about at random and occupy whatever volume is available. The same is not true in liquids and solids. Liquids and solids are distinguished from gases by the presence of substantial attractive forces between particles leading to small distances between particles. In this chapter, we'll examine how **intermolecular forces** influence the properties of liquids and solids. We'll pay particular attention to the ordering of particles in solids and to the different kinds of solids that result. In addition, we'll look at what happens during transitions between solid, liquid, and gaseous states and at the effects of temperature and pressure on these transitions.

11.1 ▶ PROPERTIES OF LIQUIDS

Many familiar and observable properties of liquids can be explained by the presence of intermolecular forces. We all know, for instance, that some liquids, such as water or gasoline, flow easily when poured, whereas others, such as motor oil or maple syrup, flow sluggishly.

The measure of a liquid's resistance to flow is called its **viscosity**. Viscosity is related to the ease with which individual molecules move around in the liquid and thus to the intermolecular forces present. Substances with weak intermolecular forces have relatively low viscosities, while substances with strong intermolecular forces have relatively high viscosities. Let's compare pentane (C_5H_{12}) with glycerol [$C_3H_5(OH)_3$]. Pentane, a nonpolar molecule, has London dispersion forces holding molecules together. The London dispersion forces are relatively weak and therefore pentane has a low viscosity. In contrast, glycerol has three hydroxyl (OH) groups that can form hydrogen bonds with other glycerol molecules. The intermolecular forces between glycerol molecules are strong resulting in a high viscosity. Glycerol is a clear liquid with a viscosity similar to honey.

Pentane

Glycerol

Another familiar property of liquids is **surface tension**, the resistance of a liquid to spread out and increase its surface area. Surface tension is caused by the difference in intermolecular forces experienced by molecules at the surface of a liquid and those experienced by molecules in the interior. Molecules at the surface feel attractive forces on only one side and are thus pulled in toward the liquid, while molecules in the interior are surrounded and are pulled equally in all directions (**FIGURE 11.1**). The ability of a water strider to walk on water and the beading up of water on a newly waxed car are both due to surface tension.

▲ Surface tension allows a water strider to walk on a pond without penetrating the surface.

◀ **FIGURE 11.1**

Surface tension. Surface tension is the resistance of a liquid to spread out and increase its surface area.

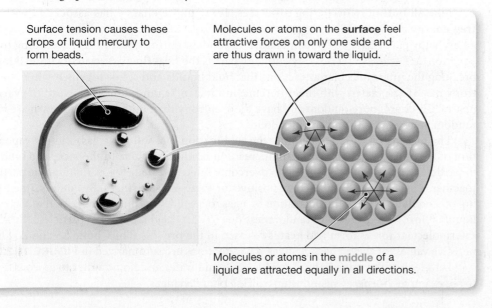

Surface tension causes these drops of liquid mercury to form beads.

Molecules or atoms on the **surface** feel attractive forces on only one side and are thus drawn in toward the liquid.

Molecules or atoms in the middle of a liquid are attracted equally in all directions.

TABLE 11.1 Viscosities and Surface Tensions of Some Common Substances at 20 °C

Name	Formula	Viscosity (N·s/m^2)	Surface Tension (J/m^2)
Pentane	C_5H_{12}	2.4×10^{-4}	1.61×10^{-2}
Benzene	C_6H_6	6.5×10^{-4}	2.89×10^{-2}
Water	H_2O	1.00×10^{-3}	7.29×10^{-2}
Ethanol	C_2H_5OH	1.20×10^{-3}	2.23×10^{-2}
Mercury	Hg	1.55×10^{-3}	4.6×10^{-1}
Glycerol	$C_3H_5(OH)_3$	1.49	6.34×10^{-2}

Surface tension, like viscosity, is generally higher in liquids that have stronger intermolecular forces. Both properties are also temperature-dependent because molecules at higher temperatures have more kinetic energy to counteract the attractive forces holding them together. Data for some common substances are given in **TABLE 11.1**. Note that mercury has a particularly large surface tension, causing droplets to form beads (Figure 11.1) and giving the top of the mercury column in a barometer a rounded shape called a *meniscus*.

11.2 ▶ PHASE CHANGES BETWEEN SOLIDS, LIQUIDS, AND GASES

We're all familiar with seeing solid ice melt to liquid water, liquid water freeze to solid ice or evaporate to gaseous steam, and gaseous steam condense to liquid water. Such processes, in which the physical form but not the chemical identity of a substance changes, are called **phase changes**, or *changes of state*. Matter in any one state, or **phase**, can change into either of the other two. Solids can even change directly into gases, as occurs when dry ice (solid CO_2) undergoes *sublimation*. The names of the various phase changes are:

Fusion (melting)	solid → liquid
Freezing	liquid → solid
Vaporization	liquid → gas
Condensation	gas → liquid
Sublimation	solid → gas
Deposition	gas → solid

Like all naturally occurring processes, every phase change has associated with it a free-energy change, ΔG. As we saw in Section 9.13, ΔG is made up of two contributions, an enthalpy part (ΔH) and a temperature-dependent entropy part ($T\Delta S$) according to the equation $\Delta G = \Delta H - T\Delta S$. The enthalpy part is the heat flow associated with making or breaking the intermolecular attractions that hold liquids and solids together, while the entropy part is associated with the difference in molecular randomness between the various phases. Gases are more random and have more entropy than liquids, which in turn are more random and have more entropy than solids.

The melting of a solid to a liquid, the sublimation of a solid to a gas, and the vaporization of a liquid to a gas all involve an increase in randomness due to the increased mobility of particles. Heat must be absorbed to overcome intermolecular forces holding the particles together. Thus, both ΔS and ΔH are positive for these phase changes. By contrast, the freezing of a liquid to a solid, the deposition of a gas to a solid, and the condensation of a gas to a liquid all involve a decrease in randomness due to decreased mobility of particles. Attractive intermolecular forces form and heat is released in the process. Thus, both ΔS and ΔH have negative values for these phase changes. The situations are summarized in **FIGURE 11.2**.

Let's look at the transitions of solid ice to liquid water and liquid water to gaseous steam to see examples of energy relationships during phase changes.

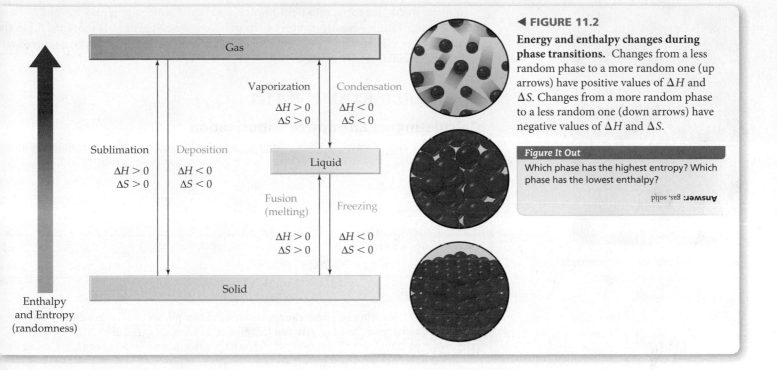

◀ **FIGURE 11.2**

Energy and enthalpy changes during phase transitions. Changes from a less random phase to a more random one (up arrows) have positive values of ΔH and ΔS. Changes from a more random phase to a less random one (down arrows) have negative values of ΔH and ΔS.

Figure It Out

Which phase has the highest entropy? Which phase has the lowest enthalpy?

Answer: gas, solid

Melting of ice to liquid water (fusion): $\Delta H = +6.01$ kJ/mol and $\Delta S = +22.0$ J/(K • mol)

Vaporization of liquid water to steam: $\Delta H = +40.67$ kJ/mol and $\Delta S = +109$ J/(K • mol)

Both ΔH and ΔS are larger for the liquid $\rightarrow$ vapor change than for the solid $\rightarrow$ liquid change because many more intermolecular attractions need to be overcome and much more randomness is gained in the change of liquid to vapor. This highly endothermic conversion of liquid water to gaseous water vapor is used by many organisms as a cooling mechanism. When our bodies perspire on a warm day, evaporation of the perspiration absorbs heat and leaves the skin feeling cooler.

For phase changes in the opposite direction, the numbers have the same absolute values but opposite signs.

Freezing of liquid water to ice: $\Delta H = -6.01$ kJ/mol and $\Delta S = -22.0$ J/(K • mol)

Condensation of water vapor to liquid water: $\Delta H = -40.67$ kJ/mol and $\Delta S = -109$ J/(K • mol)

Citrus growers take advantage of the exothermic freezing of water when they spray their trees with water on cold nights to prevent frost damage. As water freezes on the leaves, it releases heat that protects the tree.

Knowing the values of ΔH and ΔS for a phase transition makes it possible to calculate the temperature at which the change occurs. Recall from Section 9.13 that ΔG is negative for a spontaneous process, positive for a nonspontaneous process, and zero for a process at equilibrium. Thus, by setting $\Delta G = 0$ and solving for T in the free-energy equation, we can calculate the temperature at which two phases are in equilibrium. For the solid $\rightarrow$ liquid phase change in water, for instance, we have

$$\Delta G = \Delta H - T\Delta S = 0 \qquad \text{at equilibrium}$$

or $\qquad T = \Delta H / \Delta S$

where $\Delta H = +6.01$ kJ/mol and $\Delta S = +22.0$ J/(K • mol)

so $\qquad T = \dfrac{6.01 \dfrac{\text{kJ}}{\text{mol}}}{0.0220 \dfrac{\text{kJ}}{\text{K} \cdot \text{mol}}} = 273 \text{ K}$

▲ Evaporation of perspiration carries away heat and cools the body after exertion.

▲ Why do citrus growers spray their trees with water on cold nights?

In other words, ice turns into liquid water, and liquid water turns into ice, at 273 K, o 0 °C, at 1 atm pressure—hardly a surprise. In practice, the calculation is more useful in th opposite direction. That is, the temperature at which a phase change occurs is measured an then used to calculate $\Delta S(= \Delta H/T)$.

WORKED EXAMPLE 11.1

Calculating an Entropy of Vaporization

The boiling point of water is 100 °C, and the enthalpy change for the conversion of water to steam is ΔH_{vap} = 40.67 kJ/mol. What is the entropy change for vaporization, ΔS_{vap}, in J/(K • mol)?

IDENTIFY

Known	Unknown
Boiling point of water (T = 100 °C)	Entropy change (ΔS_{vap})
Enthalpy change (ΔH_{vap} = 40.67 kJ/mol)	

STRATEGY

At the temperature where a phase change occurs, the two phases coexist in equilibrium and ΔG, the free-energy difference between the phases, is zero: $\Delta G = \Delta H - T\Delta S = 0$. Rearranging gives $\Delta S = \Delta H/T$, where both ΔH and T are known. Remember that T must be expressed in kelvin.

SOLUTION

$$\Delta S_{vap} = \frac{\Delta H_{vap}}{T} = \frac{40.67 \frac{kJ}{mol}}{373.15 \text{ K}} = 0.1090 \text{ kJ}/(\text{K} \cdot \text{mol}) = 109.0 \text{ J}/(\text{K} \cdot \text{mol})$$

$/_{1,000}$

CHECK

Converting water from liquid to a gas should result in a rather large and positive entropy change. The calculation for ΔS_{vap} agrees with the expected sign and magnitude and is therefore a reasonable answer.

▶ **PRACTICE 11.1** The boiling point of ethanol is 78.4 °C, and the enthalpy change for the conversion of liquid to vapor is ΔH_{vap} = 38.56 kJ/mol. What is the entropy change for vaporization, ΔS_{vap}, in J/(K • mol)?

▶ **APPLY 11.2** Chloroform ($CHCl_3$) has ΔH_{vap} = 29.2 kJ/mol and ΔS_{vap} = 87.5 J/(K• mol). What is the boiling point of chloroform in kelvin?

The results of continuously adding heat to a substance can be displayed on a *heatin curve* like that shown in **FIGURE 11.3** for H_2O. The curve has five different regions associate with five different processes involved when heating solid H_2O from −25.0 °C to gaseous H_2 at 125.0 °C.

(1) **Heating solid H_2O:** Starting at an arbitrary temperature below the freezing point of wa ter, say −25.0 °C, addition of heat raises the ice's temperature until it reaches 0 °C. Be cause the **molar heat capacity** of ice is 36.57 J/(mol • °C), and because we need to rais the temperature 25.0 °C, 914 J/mol is required:

REMEMBER...

The **molar heat capacity (C_m)** of a substance is the amount of heat necessary to raise the temperature of 1 mol of the substance by 1 °C. (Section 9.7)

$$\text{Energy to heat ice from } -25 °C \text{ to } 0 °C = \left(36.57 \frac{J}{mol \cdot °C}\right)(25.0 °C)$$

$$= 914 \text{ J/mol} = 0.914 \text{ kJ/mol}$$

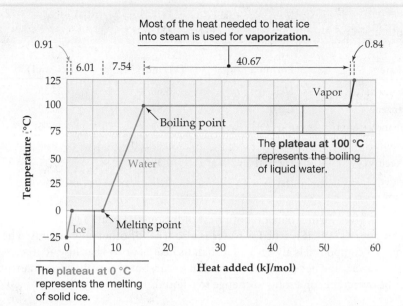

A heating curve for 1 mol of H_2O, showing the temperature changes and phase transitions that occur when heat is added.

Figure It Out

What processes are occurring in the flat regions of the heating curve? Heat is added in these regions, but the temperature does not increase. Explain.

Answer: The flat regions correspond to phase changes. The added energy as heat goes into breaking intermolecular forces and does not increase the average speed of particles which is an increase in temperature.

2) **Melting solid H_2O (fusion):** Once the temperature of the ice reaches 0 °C, addition of further heat goes into disrupting hydrogen bonds and other intermolecular forces rather than into increasing the temperature, as indicated by the plateau at 0 °C on the heating curve in Figure 11.3. At this temperature—the *melting point*—solid and liquid coexist in equilibrium as molecules break free from their positions in the ice crystals and enter the liquid phase. Not until the solid turns completely to liquid does the temperature again rise. The amount of energy required to overcome enough intermolecular forces to convert a solid into a liquid is the *enthalpy of fusion*, or **heat of fusion** (ΔH_{fusion}). For ice, $\Delta H_{fusion} = +6.01$ kJ/mol.

3) **Heating liquid H_2O:** Continued addition of heat to liquid water raises the temperature until it reaches 100 °C. Because the molar heat capacity of liquid water is 75.4 J/(mol · °C), 7.54 kJ/mol is required:

$$\text{Energy to heat water from 0 °C to 100 °C} = \left(75.4 \frac{J}{\text{mol} \cdot °C}\right)(100 °C)$$
$$= 7.54 \times 10^3 \text{ J/mol} = 7.54 \text{ kJ/mol}$$

4) **Vaporizing liquid H_2O:** Once the temperature of the water reaches 100 °C, addition of further heat again goes into overcoming intermolecular forces rather than into increasing the temperature, as indicated by the second plateau at 100 °C on the heating curve. At this temperature—the *boiling point*—liquid and vapor coexist in equilibrium as molecules break free from the surface of the liquid and enter the gas phase. The amount of energy necessary to convert a liquid into a gas is called the *enthalpy of vaporization*, or **heat of vaporization** (ΔH_{vap}). For water, $\Delta H_{vap} = +40.67$ kJ/mol.

5) **Heating H_2O vapor:** Only after the liquid has been completely vaporized does the temperature again rise. Because the molar heat capacity of water vapor is 33.6 J/(mol · °C), 0.840 kJ/mol is required:

$$\text{Energy to heat steam from 100 °C to 125 °C} = \left(33.6 \frac{J}{\text{mol} \cdot °C}\right)(25 °C)$$
$$= 840 \text{ J/mol} = 0.840 \text{ kJ/mol}$$

Notice that the largest part (40.67 kJ/mol) of the 56.05 kJ/mol required to convert solid ice at −25 °C to gaseous steam at 125 °C is used for vaporization. The heat of vaporization for water is larger than the heat of fusion because only a relatively small number of hydrogen bonds must be broken to convert the solid to the liquid, but *all* hydrogen bonds must be broken to convert the liquid to the vapor.

TABLE 11.2 Heats of Fusion and Heats of Vaporization for Some Common Compounds

Name	Formula	ΔH_{fusion} (kJ/mol)	ΔH_{vap} (kJ/mol)
Ammonia	NH_3	5.66	23.33
Benzene	C_6H_6	9.87	30.72
Ethanol	C_2H_5OH	4.93	38.56
Helium	He	0.02	0.08
Mercury	Hg	2.30	59.11
Water	H_2O	6.01	40.67

TABLE 11.2 gives further data on both heat of fusion and heat of vaporization for some common compounds. What is true for water is also true for other compounds: The heat of vaporization of a compound is always larger than its heat of fusion because all intermolecular forces must be overcome before vaporization can occur, but relatively fewer intermolecular forces must be overcome for a solid to change to a liquid.

WORKED EXAMPLE 11.2

Calculating the Amount of Heat of a Temperature Change

How much heat is required to convert 50.0 g of water at 25 °C to steam at 150 °C? The boiling point of water is 100 °C and $C_m[H_2O(l)] = 75.4$ J/(mol · °C), $\Delta H_{vap} = 40.67$ kJ/mol, $C_m[H_2O(g)] = 33.6$ J/(mol · °C).

IDENTIFY

Known	Unknown
Temperature change (25 °C to 150 °C)	
Amount of water (50.0 g)	Heat (q in kJ)
Boiling point (100 °C)	
$\Delta H_{vap} = 40.67$ kJ/mol	
$C_m[H_2O(l)] = 75.4$ J/(mol · °C)	
$C_m[H_2O(g)] = 33.6$ J/(mol · °C)	

STRATEGY

There are three separate steps involved in heating the sample. Step 1 is heating the liquid water to its boiling point, step 2 is vaporizing the liquid water, and step 3 is heating the water vapor. First convert from grams to moles since the molar heat capacities are given and then calculate q for each step.

SOLUTION

First find the number of moles of water:

$$50.0 \text{ g } H_2O \times \frac{1 \text{ mol } H_2O}{18.0 \text{ g } H_2O} = 2.78 \text{ mol } H_2O$$

Step 1. Heating liquid H_2O from 25 °C to 100 °C:

$$q = C_m \times \text{Moles of a substance} \times \Delta T$$

$$= \left(75.4 \frac{J}{\text{mol} \cdot °C}\right)(2.78 \text{ mol})(75 °C) = 1.57 \times 10^4 \text{ J} = 15.7 \text{ kJ}$$

Step 2. Vaporizing liquid H_2O:

$$q = \Delta H_{vap} \times \text{Moles of a substance} = (40.67 \text{ kJ/mol})(2.78 \text{ mol})$$

$$= 113.0 \text{ kJ}$$

Step 3. Heating gaseous H_2O from 100 °C to 150 °C:

$$q = C_m \times \text{Moles of a substance} \times \Delta T$$

$$= \left(33.6 \frac{J}{\text{mol} \cdot °C}\right)(2.78 \text{ mol})(50 °C) = 4.67 \times 10^3 \text{ J} = 4.67 \text{ kJ}$$

The total heat required is a sum of all three steps:
15.7 kJ + 113.0 kJ + 4.67 kJ = 133 kJ

CHECK

The amount of heat is large and positive, which is reasonable for a phase change. Also the magnitude of the heat for each step is reasonable. The phase change from liquid to vapor requires the most heat because strong hydrogen bonds between water molecules must be broken. Raising the temperature of liquid water requires more heat than raising the temperature of steam because the molar heat capacity and the temperature change are larger for the liquid.

▶ **PRACTICE 11.3** How much heat is required to convert 15.0 g of liquid benzene (C_6H_6) at 50 °C to gaseous benzene at 100 °C? The boiling point of benzene is 80.1 °C and $C_m[C_6H_6(l)] = 136.0$ J/(mol · °C), $\Delta H_{vap} = 30.72$ kJ/mol, $C_m[C_6H_6(g)] = 82.4$ J/(mol · °C)

▶ **APPLY 11.4** What is the sign and magnitude of q when 10.0 g of liquid water at 25 °C cools and freezes to form ice at −10 °C? The freezing point of water is 0 °C and $C_m[H_2O(l)] = 75.4$ J/(mol · °C), $\Delta H_{fus} = +6.01$ kJ/mol, $C_m[H_2O(s)] = 36.6$ J/(mol · °C).

11.3 ▶ EVAPORATION, VAPOR PRESSURE, AND BOILING POINT

The conversion of a liquid to a vapor is visible when the liquid boils, but it occurs under other conditions as well. Let's imagine the two experiments illustrated in **FIGURE 11.4**. In one experiment, we place a liquid in an open container; in the other experiment, we place the liquid in a closed container connected to a mercury **manometer**. After a certain amount of time has passed, the liquid in the first container has evaporated, while the liquid in the second container remains but the pressure has risen. At equilibrium and at a constant temperature, the pressure increase has a constant value called the **vapor pressure** of the liquid.

Evaporation and vapor pressure are both explained on a molecular level by the **kinetic–molecular theory**, developed in Section 10.6 to account for the behavior of gases. The molecules in a liquid are in constant motion but at a variety of speeds depending on the amount of kinetic energy they have. In considering a large sample, molecular kinetic energies follow a distribution curve like that shown in **FIGURE 11.5**, with the exact shape of the curve dependent on the temperature. The higher the temperature and the lower the boiling point of the substance, the greater the fraction of molecules in the sample that have sufficient kinetic energy to break free from the surface of the liquid and escape into the vapor.

Molecules that enter the vapor phase in an open container can escape from the liquid and drift away until the liquid evaporates entirely, but molecules in a closed container are trapped. As more and more molecules pass from the liquid to the vapor, the chances increase that random motion will cause some of them to return occasionally to the liquid. Ultimately, the number of molecules returning to the liquid becomes equal to the number escaping, at which point a dynamic equilibrium exists. Although *individual* molecules are constantly passing back and forth from one phase to the other, the *total numbers* of molecules in both liquid and vapor phases remain constant.

The numerical value of a liquid's vapor pressure depends on the magnitude of the intermolecular forces present and on the temperature. The smaller the intermolecular forces, the higher the vapor pressure, because loosely held molecules escape more easily. The higher the temperature, the higher the vapor pressure, because a larger fraction of molecules have sufficient kinetic energy to escape.

> **REMEMBER...**
>
> The gas pressure inside a container can be measured using an open-end **manometer**, which consists of a U-tube filled with mercury. The difference between the pressure of the gas and the pressure of the atmosphere is equal to the difference between the heights of the mercury levels in the two arms of the U-tube. (Section 10.1)

> **REMEMBER...**
>
> The **kinetic–molecular theory** is a group of five postulates that can be used to account for the behavior of gases and to derive the ideal gas law. Temperature and kinetic energy are related according to the equation $E_K = (3/2)RT$. (Section 10.6)

▲ Because bromine is colored, it's possible to see its reddish vapor above the liquid.

◀ **FIGURE 11.4**
The origin of vapor pressure.

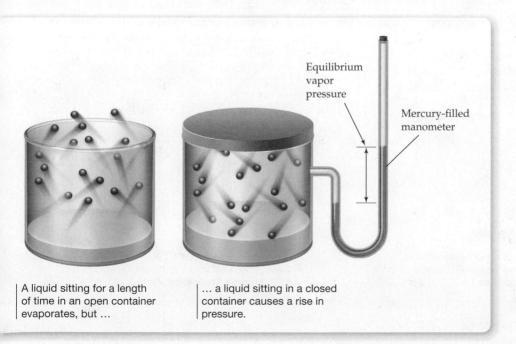

Equilibrium vapor pressure

Mercury-filled manometer

| A liquid sitting for a length of time in an open container evaporates, but ... | ... a liquid sitting in a closed container causes a rise in pressure. |

▶ **FIGURE 11.5**

The distribution of molecular kinetic energies in a liquid.

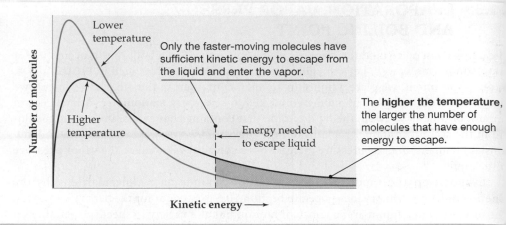

Lower temperature

Only the faster-moving molecules have sufficient kinetic energy to escape from the liquid and enter the vapor.

Higher temperature

Energy needed to escape liquid

The **higher the temperature,** the larger the number of molecules that have enough energy to escape.

Number of molecules

Kinetic energy ⟶

Figure It Out

How does the area under the curve to the right of the dashed line change as temperature increases? How does an increase in temperature affect vapor pressure?

Answer: The area under the curve to the right of the dashed line increases with temperature. Compare the area under the red line (higher *T*) with the area under the blue line (lower *T*). At higher temperature, more molecules have sufficient kinetic energy to move from the liquid to the gas phase and vapor pressure increases.

The Clausius–Clapeyron Equation

As indicated in **FIGURE 11.6**, the vapor pressure of a liquid rises with temperature in a non linear way. A linear relationship *is* found, however, when the natural logarithm of the vapor pressure, $\ln P_{vap}$, is plotted against the inverse of the Kelvin temperature, $1/T$. **TABLE 11.3** give the appropriate data for water, and Figure 11.6 shows the plot. A linear graph is characteristic of mathematical equations of the form $y = mx + b$. In this case, $y = \ln P_{vap}$, $x = 1/T$, m the slope of the line $(-\Delta H_{vap}/R)$, and b is the y-intercept (a constant, C). Thus, the data fit a expression known as the **Clausius–Clapeyron equation:**

Clausius–Clapeyron equation

natural logarithm

$$\ln P_{vap} = \left(-\frac{\Delta H_{vap}}{R}\right)\frac{1}{T} + C$$

$$y \quad = \quad m \quad x \; + \; b$$

where ΔH_{vap} is the heat of vaporization of the liquid, R is the gas constant (Section 10.3), and C is a constant characteristic of each specific substance.

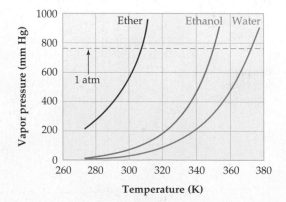

The vapor pressures of **ether**, **ethanol**, and **water** show a nonlinear rise when plotted as a function of temperature.

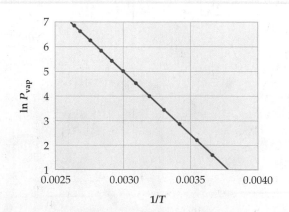

A plot of $\ln P_{vap}$ versus $1/T$ (kelvin) for water, prepared from the data in Table 11.3, shows a linear relationship.

▲ **FIGURE 11.6**

Vapor pressure of liquids at different temperatures.

TABLE 11.3	Vapor Pressure of Water at Various Temperatures						
Temp (K)	P_{vap} (mm Hg)	ln P_{vap}	$1/T$	Temp (K)	P_{vap} (mm Hg)	ln P_{vap}	$1/T$
273	4.58	1.522	0.003 66	333	149.4	5.007	0.003 00
283	9.21	2.220	0.003 53	343	233.7	5.454	0.002 92
293	17.5	2.862	0.003 41	353	355.1	5.872	0.002 83
303	31.8	3.159	0.003 30	363	525.9	6.265	0.002 75
313	55.3	4.013	0.003 19	373	760.0	6.633	0.002 68
323	92.5	4.527	0.003 10	378	906.0	6.809	0.002 65

The Clausius–Clapeyron equation makes it possible to calculate the heat of vaporization of a liquid by measuring its vapor pressure at several temperatures and then plotting the results to obtain the slope of the line. Alternatively, once the heat of vaporization and the vapor pressure at one temperature are known, the vapor pressure of the liquid at any other temperature can be calculated. A two-point form of the Clausius–Clapeyron equation that uses measurements of vapor pressure at two different temperatures to calculate ΔH_{vap} can be derived. The following expression can be written because C is a constant (its value is the same at any two pressures and temperatures). That is:

$$C = \ln P_1 + \frac{\Delta H_{vap}}{RT_1} = \ln P_2 + \frac{\Delta H_{vap}}{RT_2}$$

This equation can be rearranged so pressure is on one side and temperature is on the other.

$$\ln P_1 - \ln P_2 = \frac{\Delta H_{vap}}{RT_2} - \frac{\Delta H_{vap}}{RT_1}$$

Further mathematical rearrangement yields a two-point form of the Clausius–Clapeyron equation:

$$\ln\left(\frac{P_1}{P_2}\right) = \frac{\Delta H_{vap}}{R}\left(\frac{1}{T_2} - \frac{1}{T_1}\right)$$

If ΔH_{vap} and the vapor pressure at one temperature is known, then the vapor pressure of a liquid at a new temperature can be calculated as shown in Worked Example 11.3.

When the vapor pressure of a liquid rises to the point where it becomes equal to the external pressure, the liquid reaches its boiling point and changes into vapor. On a molecular level, you might picture boiling in the following way: Imagine that a few molecules in the interior of the liquid momentarily break free from their neighbors and form a microscopic bubble. If the external pressure from the atmosphere is greater than the vapor pressure inside the bubble, the bubble is immediately crushed. At the temperature where the external pressure and the vapor pressure in the bubble are the same, however, the bubble is not crushed. Instead, it rises through the denser liquid, grows larger as more molecules join it, and appears as part of the vigorous action we associate with boiling.

The temperature at which a liquid boils when the external pressure is exactly 1 atm (760 mm Hg) is called the **normal boiling point**. On the plots in Figure 11.6, the normal boiling points of the three liquids are reached when the curves cross the dashed line representing 760 mm Hg: for ether, 34.6 °C (307.8 K); for ethanol, 78.3 °C (351.5 K); and for water, 100.0 °C (373.15 K).

If the external pressure is less than 1 atm, then the vapor pressure necessary for boiling is reached earlier than 1 atm and the liquid boils at a lower than normal temperature. On top of Mt. Everest, for example, where the atmospheric pressure is only about 260 mm Hg, water boils at approximately 71 °C rather than 100 °C. Conversely, if the external pressure on a liquid is greater than 1 atm, the vapor pressure necessary for boiling is reached later and the liquid boils at a greater than normal temperature. Pressure cookers take advantage of this effect by raising the boiling point of water, thereby allowing food to cook more rapidly.

▲ What is the vapor pressure of the liquid at its boiling point?

WORKED EXAMPLE 11.3

Calculating a Vapor Pressure Using the Clausius–Clapeyron Equation

The vapor pressure of ethanol at 34.7 °C is 100.0 mm Hg, and the heat of vaporization of ethanol is 38.6 kJ/mol. What is the vapor pressure of ethanol in millimeters of mercury at 65.0 °C?

IDENTIFY

Known	Unknown
$\Delta H_{vap} = 38.6$ kJ/mol	Vapor pressure at 65.0 °C (P_2)

Vapor pressure at 34.7 °C ($P_1 = 100.0$ mmHg)

$T_1 = 34.7$ °C

$T_2 = 65.0$ °C

STRATEGY

Use the rearranged form of the Clausius–Clapeyron equation that enables the calculation of vapor pressure of the liquid at any other temperature if the heat of vaporization and the vapor pressure at one temperature are known. Convert temperature to Kelvin and ΔH_{vap} to units of joules.

$$\ln\left(\frac{P_1}{P_2}\right) = \frac{\Delta H_{vap}}{R}\left(\frac{1}{T_2} - \frac{1}{T_1}\right)$$

SOLUTION

Substitute all known values into the equation and solve for P_2.

$$\ln\left(\frac{100.0 \text{ mmHg}}{P_2}\right) = \frac{38,600 \text{ J/mol}}{8.314 \text{ J/mol} \cdot \text{K}}\left(\frac{1}{338.0} - \frac{1}{307.7}\right)$$

$$\ln\left(\frac{100.0 \text{ mmHg}}{P_2}\right) = -1.3527$$

$$\frac{100.0 \text{ mm Hg}}{P_2} = e^{-1.3527} = 0.2585$$

$$P_2 = 387 \text{ mmHg}$$

CHECK

We would expect the vapor pressure to be larger at a higher temperature so the answer is physically reasonable. It is difficult to estimate the magnitude of the new vapor pressure due to the complexity of the calculation.

▶ **PRACTICE 11.5** Bromine has $P_{vap} = 400$ mm Hg at 41.0 °C and a normal boiling point of 331.9 K. What is the heat of vaporization, ΔH_{vap}, of bromine in kJ/mol?

▶ **APPLY 11.6** The normal boiling point of water is 100.0 °C, and the heat of vaporization is $\Delta H_{vap} = 40.7$ kJ/mol. What is the boiling point of water in °C on top of Pikes Peak in Colorado, where $P = 407$ mm Hg?

11.4 ▶ KINDS OF SOLIDS

A brief look around tells you that most substances are solids rather than liquids or gases room temperature. That brief look also shows that there are many different kinds of solid Some solids, such as iron and aluminum, are hard and metallic. Others, such as sugar an table salt, are crystalline and easily broken. And still others, such as rubber and many plastic are soft and amorphous.

The most fundamental distinction between kinds of solids is that some are crystalli and others are amorphous. **Crystalline solids** are those whose constituent particles—atom ions, or molecules—have an ordered arrangement extending over a long range. This ord on the atomic level is also seen on the visible level because crystalline solids usually ha flat faces and distinct angles (**FIGURE 11.7a**). **Amorphous solids**, by contrast, are tho whose constituent particles are randomly arranged and have no ordered long-range stru ture (**FIGURE 11.7b**). Rubber is an example.

Crystalline solids can be further categorized as *ionic, molecular, covalent network,* metallic.

Ionic solids are those like sodium chloride, whose constituent particles are ions. A cry tal of sodium chloride is composed of alternating Na^+ and Cl^- ions ordered in a regular thre dimensional arrangement and held together by ionic bonds, as discussed in Section 2.11.

(a) A crystalline solid, such as this amethyst, has flat faces and distinct angles. These regular macroscopic features reflect a similarly ordered arrangement of particles at the atomic level.

(b) An amorphous solid like rubber has a disordered arrangement of its constituent particles.

▲ **FIGURE 11.7**
Crystalline and amorphous solids.

Molecular solids are those like sucrose or ice, whose constituent particles are molecules held together by the intermolecular forces discussed in Section 8.6. A crystal of ice, for instance, is composed of H_2O molecules held together in a regular way by hydrogen bonding (**FIGURE 11.8a**).

Covalent network solids are those like quartz (**FIGURE 11.8b**) or diamond, whose atoms are linked together by covalent bonds into a giant three-dimensional array. In effect, a covalent network solid is one *very* large molecule.

Metallic solids, such as silver or iron, also consist of large arrays of atoms, but their crystals have metallic properties such as electrical conductivity. We'll discuss metals in Chapter 21.

A summary of the different types of crystalline solids and their characteristics is given in **TABLE 11.4**.

(a) Ice consists of individual H_2O molecules held together in a regular manner by hydrogen bonds.

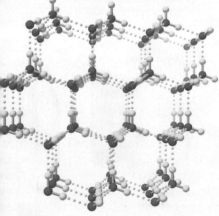

(b) Quartz (SiO_2) is essentially one very large molecule with Si–O covalent bonds. Each silicon atom has tetrahedral geometry and is bonded to four oxygens; each oxygen has approximately linear geometry and is bonded to two silicons.

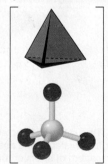

(c) This shorthand representation shows how SiO_4 tetrahedra join at their corners to share oxygen atoms.

▲ **FIGURE 11.8**
Crystal structures of ice, a molecular solid, and quartz, a covalent network solid.

Figure It Out
Quartz has a much higher melting point than water. Explain the difference based on the types of forces that hold individual units together.

Answer: In ice, water molecules are held together by hydrogen bonds. In quartz, individual SiO_2 units are held together by covalent bonds. Since covalent bonds are stronger than hydrogen bonds, quartz has a higher melting point.

TABLE 11.4	Types of Crystalline Solids and Their Characteristics		
Type of Solid	**Intermolecular Forces**	**Properties**	**Examples**
Ionic	Ion–ion forces	Brittle, hard, high-melting	NaCl, KBr, MgCl$_2$
Molecular	Dispersion forces, dipole–dipole forces, hydrogen bonds	Soft, low-melting, nonconducting	H$_2$O, Br$_2$, CO$_2$, CH$_4$
Covalent network	Covalent bonds	Hard, high-melting	C (diamond), SiO$_2$
Metallic	Metallic bonds	Variable hardness and melting point, conducting	Na, Zn, Cu, Fe

11.5 ▶ PROBING THE STRUCTURE OF SOLIDS: X-RAY CRYSTALLOGRAPHY

REMEMBER...

The **electromagnetic spectrum** is made up of all wavelengths of radiant energy, including visible light, infrared radiation, microwaves, radio waves, X rays, and so on. (Section 5.1)

How can the structure of a solid be found experimentally? According to a principle of optic the wavelength of light used to observe an object must be less than twice the length of th object itself. Since atoms have diameters of around 2×10^{-10} m and the visible light detecte by our eyes has wavelengths of $4 - 7 \times 10^{-7}$ m, it's impossible to see atoms using even th finest optical microscope. To "see" atoms, we must use "light" with a wavelength of approx mately 10^{-10} m, which is in the X-ray region of the **electromagnetic spectrum.**

The origins of X-ray crystallography go back to the work of Max von Laue in 1912. O passing X rays through a crystal of sodium chloride and letting them strike a photographi plate, Laue noticed that a pattern of spots was produced on the plate, indicating that th X rays were being *diffracted* by the atoms in the crystal. A typical diffraction pattern is show in **FIGURE 11.9**.

▶ **FIGURE 11.9**

An X-ray diffraction experiment.

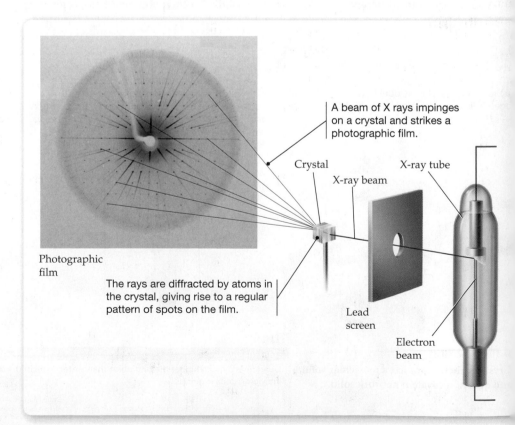

A beam of X rays impinges on a crystal and strikes a photographic film.

Crystal

X-ray tube

X-ray beam

Photographic film

The rays are diffracted by atoms in the crystal, giving rise to a regular pattern of spots on the film.

Lead screen

Electron beam

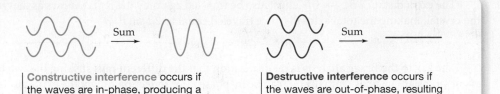

◀ **FIGURE 11.10**
Interference of electromagnetic waves.

Constructive interference occurs if the waves are in-phase, producing a wave with increased intensity.

Destructive interference occurs if the waves are out-of-phase, resulting in cancellation.

Diffraction of electromagnetic radiation occurs when a beam is scattered by an object containing regularly spaced lines (such as those in a diffraction grating) or points (such as the atoms in a crystal). This scattering happens only if the spacing between the lines or points is comparable to the wavelength of the radiation.

As shown schematically in **FIGURE 11.10**, diffraction is due to *interference* between two waves passing through the same region of space at the same time. If the waves are in-phase, peak to peak and trough to trough, the interference is constructive and the combined wave is increased in intensity. If the waves are out-of-phase, however, the interference is destructive and the wave is canceled. Constructive interference gives rise to the intense spots observed on Lauc's photographic plate, while destructive interference causes the surrounding light areas.

How does the diffraction of X rays by atoms in a crystal give rise to the observed pattern of spots on a photographic plate? According to an explanation advanced in 1913 by the English physicist William H. Bragg and his 22-year-old son, William L. Bragg, the X rays are diffracted by different layers of atoms in the crystal, leading to constructive interference in some instances but destructive interference in others.

To understand the Bragg analysis, imagine that incoming X rays with wavelength λ strike a crystal face at an angle θ and then bounce off at the same angle, just as light bounces off a mirror (**FIGURE 11.11**). Those rays that strike an atom in the top layer are all reflected at the same angle θ, and those rays that strike an atom in the second layer are also reflected at the angle θ. But because the second layer of atoms is farther from the X-ray source, the distance that the X rays have to travel to reach the second layer is farther than the distance they have to travel to reach the first layer by an amount indicated as BC in Figure 11.11. Using trigonometry, you can show that the extra distance BC is equal to the distance between atomic layers $d (= AC)$ times the sine of the angle θ:

$$\sin\theta = \frac{BC}{d} \qquad \text{so} \qquad BC = d\sin\theta$$

◀ **FIGURE 11.11**
Diffraction of X rays of wavelength λ from atoms in the top two layers of a crystal.

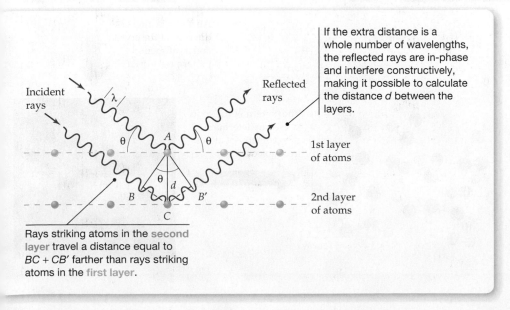

If the extra distance is a whole number of wavelengths, the reflected rays are in-phase and interfere constructively, making it possible to calculate the distance d between the layers.

Incident rays

Reflected rays

1st layer of atoms

2nd layer of atoms

Rays striking atoms in the **second layer** travel a distance equal to $BC + CB'$ farther than rays striking atoms in the **first layer**.

The extra distance $BC = CB'$ must also be traveled again by the *reflected* rays as they exit the crystal, making the total extra distance traveled equal to $2d \sin \theta$.

$$BC + CB' = 2d \sin \theta$$

The key to the Bragg analysis is the realization that the different rays striking the two layers of atoms are in-phase initially but can be in-phase after reflection only if the extra distance $BC + CB'$ is equal to a whole number of wavelengths $n\lambda$, where n is an integer (1, 2, 3, ...). If the extra distance is not a whole number of wavelengths, then the reflected rays will be out-of-phase and will cancel. Setting the extra distance $2d \sin \theta = n\lambda$ and rearranging to solve for d gives the **Bragg equation**:

$$BC + CB' = 2d \sin \theta = n\lambda$$

Bragg equation $\quad d = \dfrac{n\lambda}{2 \sin \theta}$

Of the variables in the Bragg equation, the value of the wavelength λ is known, the value of $\sin \theta$ can be measured, and the value of n is a small integer, usually 1. Thus, the distance between layers of atoms in a crystal can be calculated. For their work, the Braggs shared the 1915 Nobel Prize in Physics. The younger Bragg was 25 years old at the time.

Computer-controlled X-ray diffractometers are now available that automatically rotate crystal and measure the diffraction from all angles. Analysis of the X-ray diffraction pattern then makes it possible to measure the interatomic distance between any two nearby atoms in a crystal. For molecular substances, this knowledge of interatomic distances indicates which atoms are close enough to form a bond. X-ray analysis thus provides a means for determining the three dimensional structures of molecules (**FIGURE 11.12**).

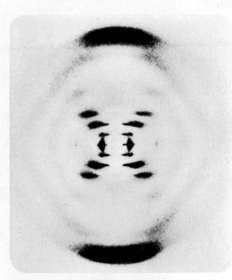

▲ Rosalind Franklin's X-ray diffraction image of DNA taken in 1952 provided crucial data that led to the double helical model of DNA proposed by James Watson and Francis Crick.

▶ **FIGURE 11.12**

A computer-generated structure of sucrose (table sugar), $C_{12}H_{22}O_{11}$, as determined by X-ray crystallography.

(Black = C, red = O, white = H)

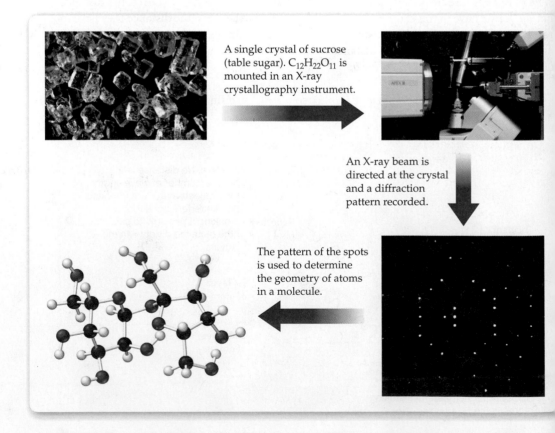

A single crystal of sucrose (table sugar). $C_{12}H_{22}O_{11}$ is mounted in an X-ray crystallography instrument.

An X-ray beam is directed at the crystal and a diffraction pattern recorded.

The pattern of the spots is used to determine the geometry of atoms in a molecule.

11.6 ▶ THE PACKING OF SPHERES IN CRYSTALLINE SOLIDS: UNIT CELLS

How do particles—whether atoms, ions, or molecules—pack together in crystals? Let's look at metals, which are the simplest examples of crystal packing because the individual metal atoms are spheres. Not surprisingly, metal atoms (and other kinds of particles as well) generally pack together in crystals so that they can be as close as possible and maximize intermolecular attractions.

If you were to take a large number of uniformly sized marbles and arrange them in a box in some orderly way, there are four possibilities you might come up with. One way to arrange the marbles is in orderly rows and stacks, with the spheres in one layer sitting directly on top of those in the previous layer so that all layers are identical (**FIGURE 11.13a**). In this arrangement, called **simple cubic packing**, each sphere is touched by six neighbors—four in its own layer, one above, and one below—and is thus said to have a **coordination number** of 6. Only 52% of the available volume is occupied by the spheres in simple cubic packing, making inefficient use of space and minimizing attractive forces. Of all the metals in the periodic table, only polonium crystallizes in this way.

Alternatively, space could be used more efficiently if, instead of stacking the spheres directly on top of one another, you slightly separate the spheres in a given layer and offset the alternating layers in an *a-b-a-b* arrangement so that the spheres in the *b* layers fit into the depressions between spheres in the *a* layers, and vice versa (**FIGURE 11.13b**). In this arrangement, called **body-centered cubic packing**, each sphere has a coordination number of 8—four neighbors above and four below—and space is used quite efficiently: 68% of the available volume is occupied. Iron, sodium, and 14 other metals crystallize in this way.

The remaining two packing arrangements of spheres are both said to be *closest-packed*. The **hexagonal closest-packing** arrangement (**FIGURE 11.14a**) has two alternating layers, *a-b-a-b*. Each layer has a hexagonal arrangement of touching spheres, which are offset so that spheres in a *b* layer fit into the small triangular depressions between spheres in an *a* layer. Zinc, magnesium, and 19 other metals crystallize in this way.

The **cubic closest-packing** arrangement (**FIGURE 11.14b**) has *three* alternating layers, *a-b-c-a-b-c*. The *a-b* layers are identical to those in the hexagonal closest-packed arrangement, but the third layer is offset from both *a* and *b* layers. Silver, copper, and 16 other metals crystallize with this arrangement.

In both kinds of closest-packed arrangements, each sphere has a coordination number of 12—six neighbors in the same layer, three above, and three below—and 74% of the available

▲ What kind of packing arrangement do these oranges have?

(a) Simple Cubic Packing:
All **layers are identical**, and all atoms are lined up in stacks and rows.

Simple cubic

Coordination Number 6:
Each sphere is touched by six neighbors, four in the same layer, one directly above, and one directly below.

(b) Body-Centered Cubic Packing:
The spheres in **layer a** are separated slightly and the spheres in **layer b** are offset so that they fit into the depressions between atoms in layer *a*. The third layer is a repeat of the first.

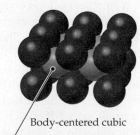

Body-centered cubic

Coordination Number 8:
Each sphere is touched by eight neighbors, four in the layer below, and four in the layer above.

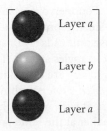

Layer *a*

Layer *b*

Layer *a*

◀ **FIGURE 11.13**

Simple cubic packing and body-centered cubic packing.

Figure It Out

Which type of packing would result in a more dense substance, simple cubic packing or body-centered cubic packing?

Answer: Density is defined as mass per volume. In body-centered cubic packing the atoms have less space between them, resulting in a greater amount of mass per volume.

▶ **FIGURE 11.14**

Hexagonal closest-packing and cubic closest-packing. In both kinds of packing, each sphere is touched by 12 neighbors, 6 in the same layer, 3 in the layer above, and 3 in the layer below.

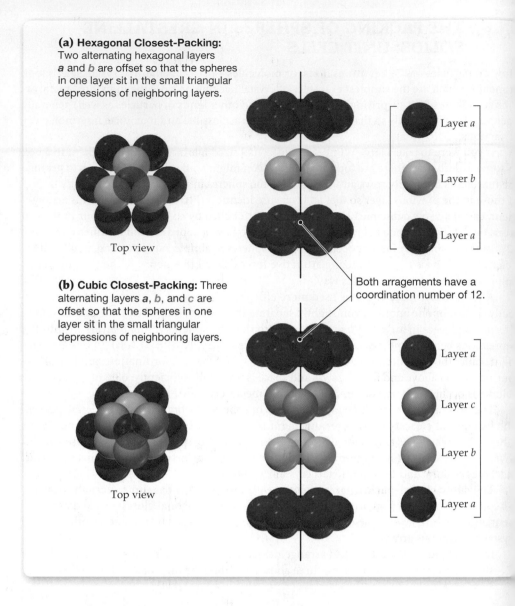

(a) Hexagonal Closest-Packing: Two alternating hexagonal layers *a* and *b* are offset so that the spheres in one layer sit in the small triangular depressions of neighboring layers.

Top view

(b) Cubic Closest-Packing: Three alternating layers *a*, *b*, and *c* are offset so that the spheres in one layer sit in the small triangular depressions of neighboring layers.

Top view

Layer *a*

Layer *b*

Layer *a*

Both arragements have a coordination number of 12.

Layer *a*

Layer *c*

Layer *b*

Layer *a*

▲ Just as these bricks are stacked together in a regular way on the pallet, a crystal is made of many small repeating units called unit cells that stack together in a regular way.

volume is filled. The next time you're in a grocery store, look to see how the oranges or appl are stacked in their display box. They'll almost certainly have a closest-packed arrangement

Having just taken a bulk view of how spheres can pack in a crystal, let's now take a clos up view. Just as a large brick wall is made up of many identical bricks stacked together in repeating pattern, a crystal is made up of many small repeat units, called **unit cells**, stacke together in three dimensions.

Fourteen different unit-cell geometries occur in crystalline solids. All are parallelepipeds- six-sided geometric solids whose faces are parallelograms. We'll be concerned here only wi those unit cells that have cubic symmetry; that is, cells whose edges are equal in length ar whose angles are 90°. The other 11 unit cells have noncubic geometry.

There are three kinds of cubic unit cells: *primitive-cubic*, *body-centered cubic*, and *fac centered cubic*. As shown in **FIGURE 11.15a**, a **primitive-cubic unit cell** for a metal has a atom at each of its eight corners, where it is shared with seven neighboring cubes that con together at the same point. As a result, only 1/8 of each corner atom "belongs to" a given c bic unit. This primitive-cubic unit cell, with all atoms arranged in orderly rows and stacks, the repeat unit found in simple cubic packing.

A **body-centered cubic unit cell** has eight corner atoms plus an additional atom in th center of the cube (**FIGURE 11.15b**). This body-centered cubic unit cell, with two repeatin offset layers and with the spheres in a given layer slightly separated, is the repeat unit found body-centered cubic packing.

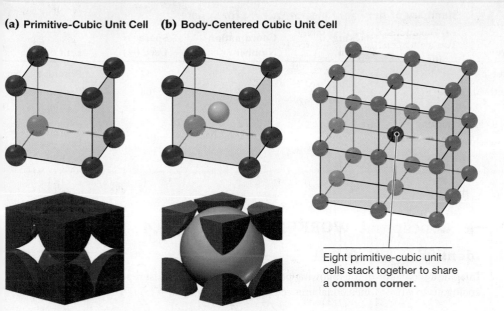

(a) Primitive-Cubic Unit Cell   **(b) Body-Centered Cubic Unit Cell**

Eight primitive-cubic unit cells stack together to share a **common corner**.

◄ **FIGURE 11.15**

Geometries of (a) primitive-cubic and (b) body-centered cubic unit cells. Both skeletal (top) and space-filling views (bottom) are shown.

Figure It Out

How many atoms are present in a primitive-cubic and body-centered cubic unit cell?

Answer: A primitive-cubic unit cell has an atom at each of its 8 corners, and each corner atom is shared by 8 cubes, so that only 1/8 of each atom "belongs" to a given unit cell. A primitive cubic unit cell has $1/8 \times 8$ (corner atoms) $= 1$ atom. A body-centered unit cell has one atom inside and 8 corner atoms; $[(1/8 \times 8 \text{ (corner atoms)}] + [1 \text{ (center atom)}] = 2$ atoms.

A **face-centered cubic unit cell** has eight corner atoms plus an additional atom in the center of each of its six faces that is shared with one other neighboring cube (**FIGURE 11.16a**). Thus, 1/2 of each face atom belongs to a given unit cell. This face-centered cubic unit cell is the repeat unit found in cubic closest-packing, as can be seen by looking down the body diagonal of a unit cell (**FIGURE 11.16b**). The faces of the unit-cell cube are at 54.7° angles to the layers of the atoms.

A summary of stacking patterns, coordination numbers, amount of space used, and unit cells for the four kinds of packing of spheres is given in **TABLE 11.5**. Hexagonal closest-packing is the only one of the four that has a noncubic unit cell.

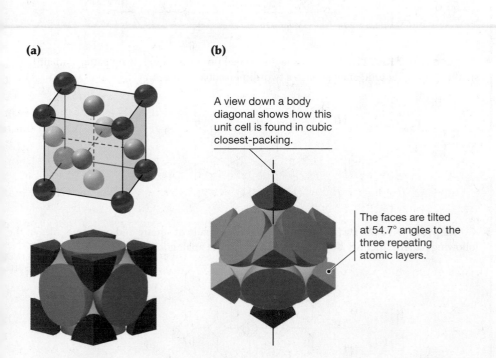

(a)   **(b)**

A view down a body diagonal shows how this unit cell is found in cubic closest-packing.

The faces are tilted at 54.7° angles to the three repeating atomic layers.

◄ **FIGURE 11.16**

Geometry of a face-centered cubic unit cell.

Figure It Out

How many atoms are present in a face-centered cubic unit cell?

Answer: A face-centered unit cell has 6 atoms on the faces that contribute 1/2 of an atom and 8 corner atoms that contribute 1/8 of an atom; $[(1/8 \times 8 \text{ (corner atoms)}] + [(1/2 \times 6 \text{ (face atoms)}] = 4$ atoms.

TABLE 11.5 **Summary of the Four Kinds of Packing for Spheres**

Structure	Stacking Pattern	Coordination Number	Space Used (%)	Unit Cell
Simple cubic	a-a-a-a-	6	52	Primitive-cubic
Body-centered cubic	a-b-a-b-	8	68	Body-centered cubic
Hexagonal closest-packing	a-b-a-b-	12	74	(Noncubic)
Cubic closest-packing	a-b-c-a-b-c-	12	74	Face-centered cubic

Conceptual WORKED EXAMPLE 11.4

Identifying a Unit Cell

Imagine a tiled floor in the following pattern. Identify the smallest repeating rectangular unit, analogous to a two-dimensional unit cell.

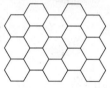

STRATEGY

Using trial and error, the idea is to draw two perpendicular sets of parallel lines that define a repeating rectangular unit. There may be more than one possibility.

SOLUTION

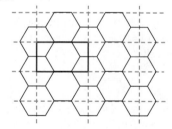 or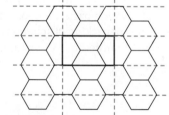

▶ **Conceptual PRACTICE 11.7** Imagine a tiled floor in the following pattern. Identify the smallest repeating unit, analogous to a two-dimensional unit cell.

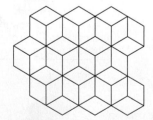

▶ **Conceptual APPLY 11.8** Imagine a tiled floor made of square and octagonal tiles in the following pattern. Identify the smallest repeating unit, analogous to a unit cell.

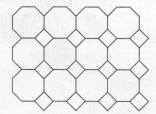

WORKED EXAMPLE 11.5

Using Unit-Cell Dimensions to Calculate the Radius of an Atom

Silver metal crystallizes in a cubic closest-packed arrangement with the edge of the unit cell having a length $d = 407$ pm. What is the radius in picometers of a silver atom?

STRATEGY AND SOLUTION

Cubic closest-packing uses a face-centered cubic unit cell. Looking at any one face of the cube head-on shows that the face atoms touch the corner atoms along the diagonal of the face but that corner atoms do not touch one another along the edges. Each diagonal is therefore equal to four atomic radii, $4r$:

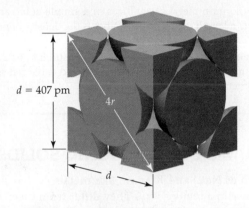

Because the diagonal and two edges of the cube form a right triangle, we can use the Pythagorean theorem to set the sum of the squares of the two edges equal to the square of the diagonal, $d^2 + d^2 = (4r)^2$ and then solve for r, the radius of one atom:

$$d^2 + d^2 = (4r)^2$$

$$2d^2 = 16r^2 \quad \text{and} \quad r^2 = \frac{d^2}{8}$$

$$\text{thus} \quad r = \sqrt{\frac{d^2}{8}} = \sqrt{\frac{(407 \text{ pm})^2}{8}} = 144 \text{ pm}$$

The radius of a silver atom is 144 pm.

▶ **PRACTICE 11.9** Calcium metal crystallizes in a cubic closest-packed arrangement. Calcium has an atomic radius of 197 pm. What is the edge length of a unit cell in pm?

▶ **APPLY 11.10** Polonium metal crystallizes in a simple cubic arrangement, with the edge of a unit cell having a length $d = 334$ pm. What is the radius in picometers of a polonium atom?

WORKED EXAMPLE 11.6

Using Unit-Cell Dimensions to Calculate the Density of a Metal

Nickel has a face-centered cubic unit cell with a length of 352.4 pm along an edge. What is the density of nickel in g/cm^3?

STRATEGY

Density is mass divided by volume. The mass of a single unit cell can be calculated by counting the number of atoms in the cell and multiplying by the mass of a single atom. The volume of a single cubic unit cell with edge d is $d^3 = (3.524 \times 10^{-8} \text{ cm})^3 = 4.376 \times 10^{-23} \text{ cm}^3$.

SOLUTION

Each of the eight corner atoms in a face-centered cubic unit cell is shared by eight unit cells, so that only $1/8 \times 8 = 1$ atom belongs to a single cell. In addition, each of the six face atoms is shared by two unit cells, so that $1/2 \times 6 = 3$ atoms belong to a single cell. Thus, a single cell

continued on next page

has 1 corner atom and 3 face atoms, for a total of 4, and each atom has a mass equal to the molar mass of nickel (58.69 g/mol) divided by Avogadro's number (6.022×10^{23} atoms/mol). We can now calculate the density:

$$\text{Density} = \frac{\text{Mass}}{\text{Volume}} = \frac{(4\text{ atoms})\left(\dfrac{58.69\ \dfrac{\text{g}}{\text{mol}}}{6.022 \times 10^{23}\ \dfrac{\text{atoms}}{\text{mol}}}\right)}{4.376 \times 10^{-23}\text{ cm}^3} = 8.909\text{ g/cm}^3$$

The calculated density of nickel is 8.909 g/cm³. (The measured value is 8.90 g/cm³.)

▶ **PRACTICE 11.11**   Polonium metal crystallizes in a simple cubic arrangement, with the edge of a unit cell having a length $d = 334$ pm. What is the density of polonium?

▶ **APPLY 11.12**   The density of a sample of metal was measured to be 22.67 g/cm³. An X-ray diffraction experiment measures the edge of a face-centered cubic cell as 383.3 pm. Calculate the atomic mass of the metal and identify it.

11.7 ▶ STRUCTURES OF SOME IONIC SOLIDS

Simple ionic solids such as NaCl and KBr are like metals in that the individual ions are spheres that pack together in a regular way. They differ from metals, however, in that the spheres are not all the same size—anions generally have larger **ionic radii** than cations. As a result, ionic solids adopt a variety of different unit cells, depending on the size and charge of the ions. NaCl, KCl, and a number of other salts have a face-centered cubic unit cell in which the larger Cl⁻ anions occupy corners and faces while the smaller Na⁺ cations fit into the holes between adjacent anions (**FIGURE 11.17**).

It's necessary, of course, that the unit cell of an ionic substance be electrically neutral, with equal numbers of positive and negative charges. In the NaCl unit cell, for instance, there are four Cl⁻ anions ($1/8 \times 8 = 1$ corner atom, plus $1/2 \times 6 = 3$ face atoms) and also four Na⁺ cations ($1/4 \times 12 = 3$ edge atoms, plus 1 center atom). (Remember that each corner atom in a cubic unit cell is shared by eight cells, each face atom is shared by two cells, and each edge atom is shared by four cells.)

Two other common ionic unit cells are shown in **FIGURE 11.18**. Copper(I) chloride has a face-centered cubic arrangement of the larger Cl⁻ anions, with the smaller Cu⁺ cations in holes so that each is surrounded by a tetrahedron of four anions. Barium chloride, by contrast, has a face-centered cubic arrangement of the smaller Ba²⁺ *cations*, with the larger Cl⁻ anions surrounded tetrahedrally. As required for charge neutrality, there are twice as many Cl⁻ anions as Ba²⁺ cations.

▶ **FIGURE 11.17**

The unit cell of NaCl. Both a skeletal view (a) and a space-filling view (b) in which the unit cell is viewed edge-on are shown.

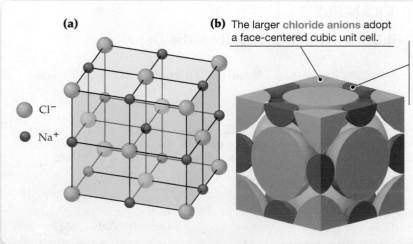

(a)

(b) The larger chloride anions adopt a face-centered cubic unit cell.

The smaller **sodium cations** fit into the holes between adjacent anions.

Cl⁻

Na⁺

(a)

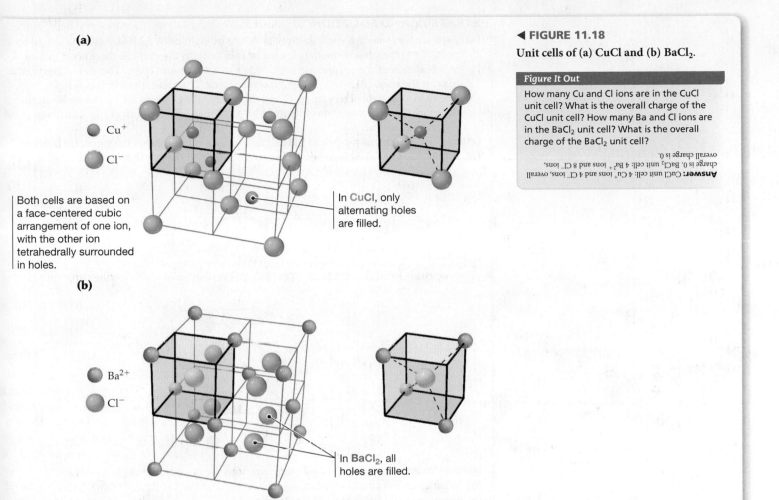

Cu$^+$

Cl$^-$

Both cells are based on a face-centered cubic arrangement of one ion, with the other ion tetrahedrally surrounded in holes.

In **CuCl**, only alternating holes are filled.

(b)

Ba^{2+}

Cl$^-$

In **BaCl$_2$**, all holes are filled.

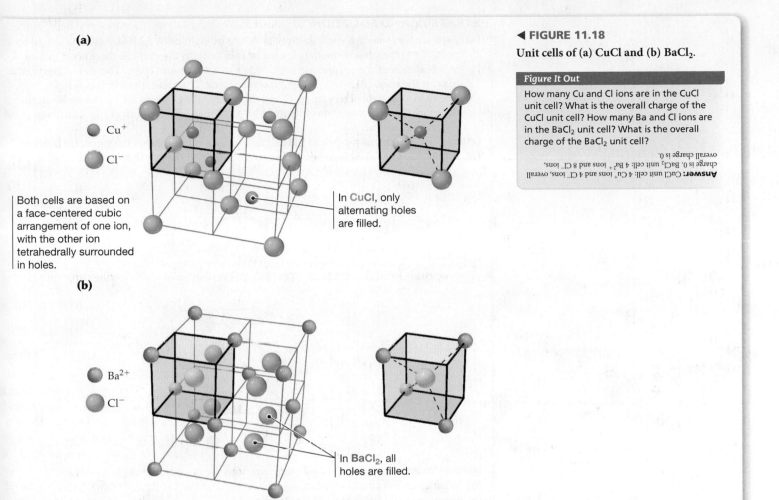

◀ **FIGURE 11.18**

Unit cells of (a) CuCl and (b) BaCl$_2$.

Figure It Out

How many Cu and Cl ions are in the CuCl unit cell? What is the overall charge of the CuCl unit cell? How many Ba and Cl ions are in the BaCl$_2$ unit cell? What is the overall charge of the BaCl$_2$ unit cell?

Answer: CuCl unit cell: 4 Cu$^+$ ions and 4 Cl$^-$ ions, overall charge is 0. BaCl$_2$ unit cell: 4 Ba^{2+} ions and 8 Cl$^-$ ions, overall charge is 0.

Conceptual WORKED EXAMPLE 11.7

Using the Unit-Cell to Determine Chemical Formula and Geometry

Rhenium oxide crystallizes in the following cubic unit cell:

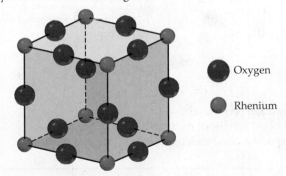

Oxygen

Rhenium

(a) How many rhenium ions and how many oxygen ions are in each unit cell?
(b) What is the formula of rhenium oxide?
(c) What is the oxidation state of rhenium?
(d) What is the geometry around each oxygen atom?
(e) What is the geometry around each rhenium atom?

continued on next page

STRATEGY AND SOLUTION

(a) Each corner atom in a cubic unit cell is shared by eight cells, each face atom is shared by two cells, and each edge atom is shared by four cells. In the unit cell there are rhenium atoms at each of the eight corners; $(8 \times 1/8) = 1$ rhenium atom. There are 12 oxygen atoms on the edges; $(12 \times 1/4) = 3$ oxygen atoms. Therefore, the formula is ReO_3.

(b) The oxidation state of oxygen is -2 and there are three O atoms for every one Re atom, therefore the oxidation state of Re must be $+6$ to make the unit cell of the ionic compound electrically neutral.

(c) Each oxygen forms two bonds the diagram of the unit cell shows the geometry is linear.

(d) If unit cells are linked together, Re forms 6 bonds with oxygen in an octahedral geometry.

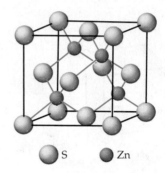

▶ **Conceptual** **PRACTICE 11.13** Zinc sulfide crystallizes in the following cubic unit cell:

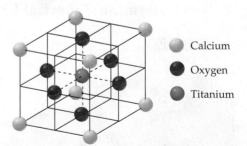

(a) How many sulfur ions and how many zinc ions are in each unit cell?

(b) What is the formula of zinc sulfide?

(c) What is the oxidation state of zinc?

(d) What is the geometry around each zinc atom?

▶ **Conceptual** **APPLY 11.14** Perovskite, a mineral containing calcium, oxygen, and titanium, crystallizes in the following cubic unit cell:

Calcium

Oxygen

Titanium

(a) What is the formula of perovskite?

(b) What is the oxidation number of the titanium ion in perovskite?

(c) What is the geometry around each titanium, oxygen, and calcium ion?

11.8 ▶ STRUCTURES OF SOME COVALENT NETWORK SOLIDS

The atoms in covalent network solids are held together by covalent bonds in a giant three-dimensional array. Covalent network solids are hard and have high melting points due to the strong covalent bonds between all atoms. Carbon and silicon form several different covalent network solids and familiar examples include graphite, diamond, and glass.

Carbon

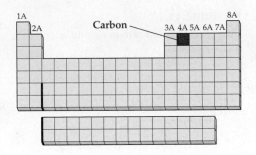

Carbon exists in more than 40 known structural forms, or **allotropes**, several of which are crystalline but most of which are amorphous. Graphite, the most common allotrope of carbon and the most stable under normal conditions, is a crystalline covalent network solid that consists of two-dimensional sheets of fused six-membered rings (**FIGURE 11.19a**). Each carbon atom is sp^2-hybridized and is bonded with trigonal planar geometry to three other carbons. The diamond form of elemental carbon is a covalent network solid in which each carbon atom is sp^3-hybridized and is bonded with tetrahedral geometry to four other carbons (**FIGURE 11.19b**).

In addition to graphite and diamond, a third crystalline allotrope of carbon called *fullerene* was discovered in 1985 as a constituent of soot. Fullerene consists of spherical C_{60} molecules with the extraordinary shape of a soccer ball. The C_{60} ball has 12 pentagonal and 20 hexagonal faces, with each atom sp^2-hybridized and bonded to three other atoms (**FIGURE 11.20a**). Closely related to both graphite and fullerene are a group of carbon allotropes called *nanotubes*—tubular structures made of repeating six-membered carbon rings, as if a sheet of graphite were rolled up (**FIGURE 11.20b**). Typically, the tubes have a diameter of about 2–30 nm and a length of up to 1 mm.

The different structures of the carbon allotropes lead to widely different properties. Diamond is the hardest known substance because of its three-dimensional network of strong single bonds that tie all atoms in a crystal together. In addition to its use in jewelry, diamond is widely used industrially for the tips of saw blades and drilling bits. It is an electrical insulator and has a melting point of about 8700 °C at a pressure of 6–10 million atm. Clear, colorless, and highly crystalline, diamonds are very rare and are found in only a few places in the world, particularly in central and southern Africa.

Graphite is the black, slippery substance used as the "lead" in pencils, as an electrode material in batteries, and as a lubricant in locks. All these properties result from its sheetlike

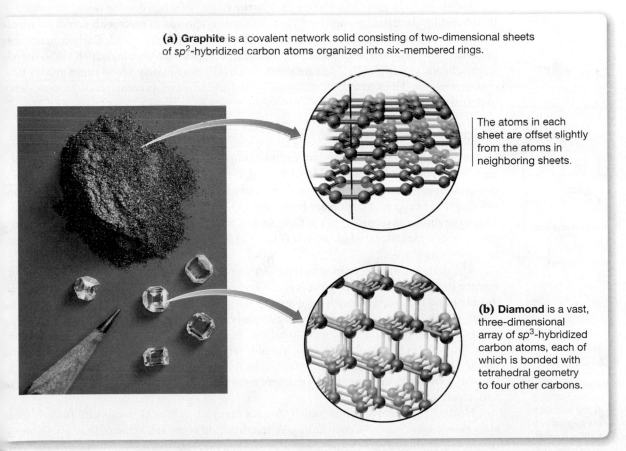

(a) Graphite is a covalent network solid consisting of two-dimensional sheets of sp^2-hybridized carbon atoms organized into six-membered rings.

The atoms in each sheet are offset slightly from the atoms in neighboring sheets.

(b) Diamond is a vast, three-dimensional array of sp^3-hybridized carbon atoms, each of which is bonded with tetrahedral geometry to four other carbons.

▲ **FIGURE 11.19**

Two crystalline allotropes of carbon, (a) graphite and (b) diamond.

▶ **FIGURE 11.20**

Fullerene, C₆₀, and carbon nanotubes.

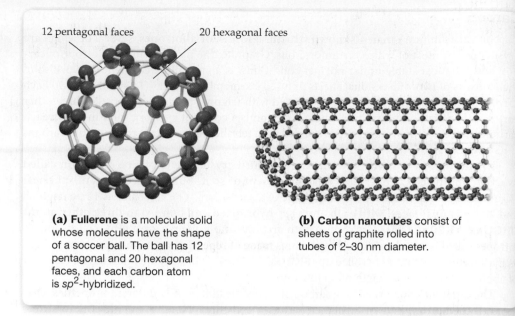

12 pentagonal faces   20 hexagonal faces

(a) Fullerene is a molecular solid whose molecules have the shape of a soccer ball. The ball has 12 pentagonal and 20 hexagonal faces, and each carbon atom is sp^2-hybridized.

(b) Carbon nanotubes consist of sheets of graphite rolled into tubes of 2–30 nm diameter.

structure. Air and water molecules can adsorb onto the flat faces of the sheets, allowing the sheets to slide over one another and giving graphite its greasy feeling and lubricating properties. Graphite is more stable than diamond at normal pressures but can be converted into diamond at very high pressure and temperature. In fact, approximately 120,000 kg of industrial diamonds are synthesized annually by applying 150,000 atm pressure to graphite at high temperature.

Fullerene, black and shiny like graphite, is the subject of much current research because of its interesting electronic properties. When fullerene is allowed to react with rubidium metal, a superconducting material called rubidium fulleride, Rb_3C_{60}, is formed. Carbon nanotubes are being studied for use as fibers in the structural composites used to make golf clubs, bicycle frames, boats, and airplanes. Their tensile strength is approximately 50–60 times greater than that of steel. We'll look at the chemistry of carbon and some of its compounds in more detail in Section 22.7.

Silica

Just as living organisms are based on carbon compounds, most rocks and minerals are based on silicon compounds. Quartz and much sand, for instance, are nearly pure *silica*, SiO_2. Silicon and oxygen together, in fact, make up nearly 75% of the mass of the Earth's crust. Considering that silicon and carbon are both in group 4A of the periodic table, you might expect SiO_2 to be similar in its properties to CO_2. In fact, though, CO_2 is a molecular substance and a gas at room temperature, whereas SiO_2 (Figure 11.8b) is a covalent network solid with a melting point over 1600 °C.

The dramatic difference in properties between CO_2 and SiO_2 is due primarily to the difference in electronic structure between carbon and silicon. The π part of a *carbon–oxygen* **double bond** is formed by sideways overlap of a carbon 2p orbital with an oxygen 2p orbital (Section 8.4). If a similar *silicon–oxygen* double bond were to form, it would require overlap of an oxygen 2p orbital and a silicon 3p orbital. But because the Si—O bond distance is longer than the C—O distance and a 3p orbital is larger than a 2p orbital, overlap between the silicon and oxygen p orbitals is not as favorable. As a result, silicon forms four single bonds to four oxygens in a covalent network structure rather than two double bonds to two oxygens in a molecular structure.

Heating silica above about 1600 °C breaks many of its Si—O bonds and turns it from a crystalline solid into a viscous liquid. When this fluid is cooled, some of the Si—O bonds re-form in a random arrangement and a noncrystalline, amorphous solid called *quartz glass* is formed. If additives are mixed in before cooling, a wide variety of glasses can be prepared.

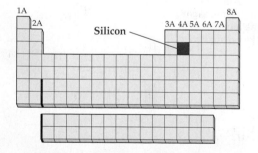

1A   8A
2A   3A 4A 5A 6A 7A
Silicon

REMEMBER...

Two atoms form a **double bond** when they approach each other with their hybrid orbitals aligned head-on for σ bonding and with their unhybridized p orbitals aligned in a parallel, sideways manner to form a π bond. (Section 8.4)

Common window glass, for instance, is prepared by adding $CaCO_3$ and Na_2CO_3. Addition of various transition metal ions results in the preparation of colored glasses, and addition of B_2O_3 produces a high-melting *borosilicate glass* that is sold under the trade name Pyrex. Borosilicate glass is particularly useful for cooking utensils and laboratory glassware because it expands very little when heated and is thus unlikely to crack. We'll look further at the chemistry of silicon and silicon-containing minerals in Section 22.8.

11.9 ▶ PHASE DIAGRAMS

Now that we've looked at the three phases of matter individually, let's draw everything together by taking an overall view. As noted previously, any one phase of matter can change spontaneously into either of the other two, depending on the temperature and pressure. A convenient way to picture this pressure–temperature dependency of a pure substance in a closed system without air present is to use what is called a **phase diagram**. As illustrated for water in **FIGURE 11.21**, a typical phase diagram shows which phase is stable at different combinations of pressure and temperature. When a boundary line between phases is crossed by changing either the temperature or the pressure, a phase change occurs.

The simplest way to interpret a phase diagram is to begin at the origin in the lower left corner of Figure 11.21 and travel up and right along the boundary line between solid on the left and gas on the right. Points on this line represent pressure/temperature combinations at which the two phases are in equilibrium in a closed system and a direct phase transition between solid ice and gaseous water vapor occurs. At some point along the solid/gas line, an intersection is reached where two lines diverge to form the bounds of the liquid region. The solid/liquid boundary for H_2O goes up and slightly left, while the liquid/gas boundary continues curving up and to the right. Called the **triple point**, this three-way intersection represents a unique combination of pressure and temperature at which all three phases coexist in equilibrium. For water, the triple-point temperature T_t is 0.0098 °C, and the triple-point pressure P_t is 6.0×10^{-3} atm.

Continuing up and slightly left from the triple point, the solid/liquid boundary line represents the melting point of solid ice (or the freezing point of liquid water) at various pressures. When the pressure is 1 atm, the melting point—called the **normal melting point**—is exactly 0 °C. There is a slight negative slope to the line, indicating that the melting point of ice decreases as pressure increases. Water is unusual in this respect, because most substances have a positive slope to their solid/liquid line, indicating that their melting points *increase* with pressure. For most substances, the solid phase is denser than the liquid because

▲ Colored glasses contain transition metal ions.

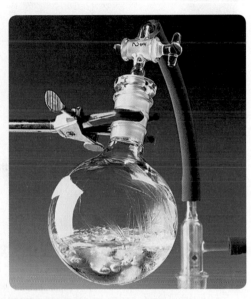
▲ At the triple point, solid exists in the boiling liquid. That is, solid, liquid, and gas coexist in equilibrium.

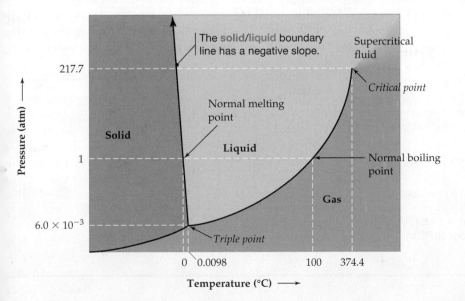

◀ **FIGURE 11.21**

A phase diagram for H_2O. Various features of the diagram are discussed in the text. Note that the pressure and temperature axes are not drawn to scale.

Figure It Out

(a) What phase of water exists at a pressure of 1 atm and temperature of −1 °C?
(b) What phase change occurs if the pressure is decreased from 1 atm to 1×10^{-4} atm at a constant temperature of −1 °C?

Answer: (a) $H_2O(s)$; **(b)** sublimation $H_2O(s) \rightarrow H_2O(g)$

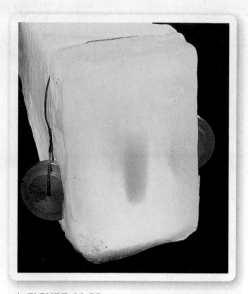

▲ **FIGURE 11.22**

Pressure and melting point. Why does the weighted wire cut through this block of ice?

particles are packed closer together in the solid. Increasing the pressure pushes the molecules even closer together, thereby favoring the solid phase even more and giving the solid/liquid boundary line a positive slope. Water, however, becomes less dense when it freezes to a solid because large empty spaces are left between molecules due to the ordered three-dimensional network of hydrogen bonds in ice (Figure 11.8a). As a result, increasing the pressure favors the liquid phase, giving the solid/liquid boundary a negative slope.

FIGURE 11.22 shows a simple demonstration of the effect of pressure on melting point. If a thin wire with heavy weights at each end is draped over a block of ice near 0 °C, the wire rapidly cuts through the block because the increased pressure lowers the melting point of the ice under the wire, causing the ice to liquefy.

Continuing up and right from the triple point, the liquid/gas boundary line represents the pressure/temperature combinations at which liquid and gas coexist and water vaporizes (or steam condenses). In fact, the part of the curve up to 1 atm pressure is simply the vapor pressure curve we saw previously in Figure 11.6. When the pressure is 1 atm, water is at its normal boiling point of 100 °C. Continuing along the liquid/gas boundary line, we suddenly reach the **critical point**, where the line abruptly ends. The critical temperature T_c is the temperature beyond which a gas cannot be liquefied, no matter how great the pressure, and the critical pressure P_c is the pressure beyond which a liquid cannot be vaporized, no matter how high the temperature. For water, $T_c = 374.4 °C$ and $P_c = 217.7$ atm.

We're all used to seeing solid/liquid and liquid/gas phase transitions, but behavior at the critical point lies so far outside our normal experiences that it's hard to imagine. Think of it this way: A *gas* at the critical point is under such high pressure, and its molecules are pressed so close together, that it becomes almost like a liquid. A *liquid* at the critical point is at such a high temperature, and its molecules are so relatively far apart, that it becomes almost like a gas. Thus, the two phases simply become the same and form a **supercritical fluid** that is neither true liquid nor true gas. No distinct physical phase change occurs on going beyond the critical point. Rather, a whitish, pearly sheen momentarily appears, and the visible boundary between liquid and gas suddenly vanishes. You really have to see it to believe it.

Conceptual WORKED EXAMPLE 11.8

Interpreting a Phase Diagram

Freeze-dried foods are prepared by freezing the food and removing water by subliming the ice at low pressure. Look at the phase diagram of water in Figure 11.21, and give the maximum pressure in mm Hg at which ice and water vapor are in equilibrium.

STRATEGY

Solid and vapor are in equilibrium only below the triple-point pressure, $P_t = 6.0 \times 10^{-3}$ atm, which needs to be converted to millimeters of mercury.

SOLUTION

$$6.0 \times 10^{-3} \text{ atm} \times \frac{760 \text{ mm Hg}}{1 \text{ atm}} = 4.6 \text{ mm Hg}$$

▶ Conceptual **PRACTICE 11.15** Look at the phase diagram of H_2O in Figure 11.21, and describe what happens to an H_2O sample when the following changes are made:

(a) The temperature is increased from $-25 °C$ to $50 °C$ at a constant pressure of 10 atm.
(b) The pressure is reduced from 50 atm to 0.5 atm at a constant temperature of $100 °C$.
(c) The pressure is first increased from 6×10^{-3} atm to 5 atm at $50°C$, and the temperature is then increased from $50 °C$ to $375 °C$.

▶ Conceptual **APPLY 11.16** Gallium metal has the following phase diagram (the pressure axis is not to scale). In the region shown, gallium has two different solid phases.

(a) Where on the diagram are the solid, liquid, and vapor regions?
(b) How many triple points does gallium have? Circle each on the diagram.
(c) At 1 atm pressure, which phase is more dense, solid or liquid? Explain.

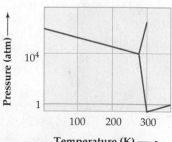

INQUIRY ▶▶▶ HOW IS CAFFEINE REMOVED FROM COFFEE?

Organic compounds with carbon-hydrogen bonds are nonpolar. Caffeine has high solubility in the nonpolar solvent benzene because a significant portion of the molecule is nonpolar.

Caffeine

Benzene

Caffeine ($C_8H_{10}N_4O_2$) is a pesticide found naturally in seeds and leaves of plants that kills or paralyzes certain insects that ingest it. In humans, caffeine acts a stimulant, and for this reason it is sometimes removed from coffee beans or tea leaves. *Extraction* is a process that refers to the separation of a substance from its surroundings, such as the removal of the caffeine molecule from a coffee bean. In 1905, Ludwig Roselius developed a method to extract caffeine from coffee using benzene (C_6H_6) as a solvent. Caffeine dissolves readily in the nonpolar solvent benzene because a significant portion of the molecule is nonpolar. If the polarity of solute and solvent are matched, then solubility will be high. In other words, nonpolar solvents dissolve nonpolar solutes and polar solvents dissolve polar solutes. However, in the food industry benzene is a poor choice for a solvent because it is highly toxic and carcinogenic (cancer causing). Residual benzene in the coffee can pose a severe health threat to those that consume it.

A much safer method uses supercritical CO_2 to extract caffeine from coffee beans. CO_2 is nontoxic, nonflammable, easily separated from a food sample, and recyclable. It is a nonpolar molecule and dissolves nonpolar solutes such as caffeine. However, at room temperature and pressure (25°C and 1 atm), CO_2 is a gas and cannot be used as a solvent. Raising the temperature and pressure produces the supercritical phase of CO_2, which has unique properties between those of gases and liquids. Supercritical CO_2 has solvent properties like the liquid phase, but the extraction can be performed faster than with a conventional organic solvent because it diffuses rapidly and flows easily like a gas. Supercritical CO_2 also has low surface tension allowing it to permeate into tiny pores in the coffee beans and dissolve caffeine on the inside.

The phase diagram of CO_2 shown in **FIGURE 11.23** shows that the supercritical phase of CO_2 can be reached at a relatively

◀ **FIGURE 11.23**

A phase diagram for CO_2. The pressure and temperature axes are not to scale.

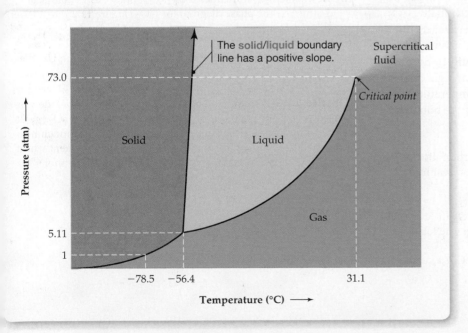

continued on next page

continued from previous page

At temperatures below the critical temperature, there is a clear boundary between liquid CO_2 and the gas phase.

Increasing temperature decreases the density of liquid CO_2, blurring the distinction between the liquid and gas phase.

At temperatures above the critical temperature (31.1°C), CO_2 is in the supercritical phase and the boundary disappears.

◀ **FIGURE 11.24**

Visualization of phase transition between liquid and supercritical CO_2 in a high pressure cell.

moderate temperature and pressure (31.1°C and 73.0 atm). The easily attainable critical point for CO_2 makes it the most widely used supercritical fluid. Industrial and research applications use supercritical CO_2 as a solvent in environmentally friendly dry cleaning, analytical separations, and polymerization reactions. The phase diagram for CO_2 has many of the same features as that of water (Figure 11.21) but differs in several interesting respects. First, the triple point is at $P_t = 5.11$ atm, meaning that CO_2 can't be a liquid below this pressure, no matter what the temperature. At 1 atm pressure, CO_2 is a solid below −78.5 °C but a gas above this temperature. This means that carbon dioxide never exists in the liquid form at standard pressure. Second, the slope of the solid/liquid boundary is positive, meaning that the solid phase is favored as the pressure rises and that the melting point of solid CO_2 therefore increases with pressure.

The transition between a liquid and a supercritical fluid can be observed using a high pressure cell (**FIGURE 11.24**). Initially, CO_2 is present in the cell in the liquid phase and there is clear distinction between the gas and liquid phase. In the high pressure cell at 75 atm, increasing the temperature causes the liquid to become less dense, so that the separation between the liquid and gas phases becomes less distinct. Upon reaching the critical temperature, the density of the gas and liquid phase are identical and the boundary between them no longer exists.

PROBLEM 11.17   A fire extinguisher containing carbon dioxide has a pressure of 70 atm at 75°F. What phase of CO_2 is present in the tank?

PROBLEM 11.18   Look at the phase diagram of CO_2 in Figure 11.23, and describe what happens to a CO_2 sample when the following changes are made:
  (a) The temperature is increased from −100 °C to 0 °C at a constant pressure of 2 atm.

  (b) The pressure is reduced from 72 atm to 5.0 atm at a constant temperature of 30 °C.
  (c) The pressure is first increased from 3.5 atm to 76 atm at −10 °C, and the temperature is then increased from −10 °C to 45 °C.

PROBLEM 11.19   Liquid carbon dioxide is also used as nontoxic solvent in dry cleaning. Refer to the phase diagram for CO_2 (Figure 11.23) to answer the following questions.
  (a) What is the minimum pressure at which liquid CO_2 can exist?
  (b) What is the minimum temperature at which liquid CO_2 can exist?
  (c) What is the maximum temperature at which liquid CO_2 can exist?

PROBLEM 11.20
  (a) For the phase transition $CO_2(s) \longrightarrow CO_2(g)$, predict the sign of ΔS.
  (b) At what temperature does CO_2 (s) spontaneously sublime at 1 atm? Use the phase diagram for CO_2 (Figure 11.23) to answer this question.
  (c) If ΔH for the sublimation of 1 mol of CO_2 (s) is 26.1 kJ, calculate ΔS in (J/K•mol) for this phase transition. (*Hint:* Use the temperature found in part b to calculate the answer.)

PROBLEM 11.21   A sample of supercritical carbon dioxide was prepared by heating 100.0 g of $CO_2(s)$ at −78.5°C to CO_2 (g) at 33°C. Then the pressure was increased to 75.0 atm. How much heat was required to sublime the sample of $CO_2(s)$ and subsequently heat CO_2 (g)? ($\Delta H_{sub} = 26.1$ kJ/mol; C_m for $CO_2(g) = 35.0$ J/mol • °C)

STUDY GUIDE

Section	Concept Summary	Learning Objectives	Test Your Understanding
11.1 ▸ Properties of Liquids	Two important properties of liquids are **viscosity**, a measure of a liquid's resistance to flow and **surface tension**, the resistance of a liquid to spread out and increase its surface area. These properties are influenced by the strength of the intermolecular forces in a liquid sample.	**11.1** Predict which substance has higher viscosity or surface tension based on its molecular structure.	Problems 11.30–11.33
11.2 ▸ Phase Changes between Solids, Liquids, and Gases	Matter in any one **phase**—solid, liquid, or gas—can undergo a **phase change** to either of the other two phases. Like all naturally occurring processes, a phase change has an associated free-energy change, $\Delta G = \Delta H - T\Delta S$. The enthalpy component, ΔH, is a measure of the change in intermolecular forces; the entropy component, ΔS, is a measure of the change in molecular randomness accompanying the phase transition.	**11.2** Use the equation for Gibbs free energy to calculate the temperature or ΔS for a phase change.	Worked Example 11.1; Problems 11.2, 11.46, 11.47
		11.3 Calculate the amount of heat associated with phase changes and draw heating curves.	Worked Example 11.2; Problems 11.3, 11.4, 11.40–11.45
11.3 ▸ Evaporation, Vapor Pressure, and Boiling Point	In a closed container, liquid molecules evaporate and the resulting increase in pressure is called the **vapor pressure**. Vapor pressure increases as temperature increases because more molecules have sufficient energy to overcome intermolecular forces and enter the gas phase.	**11.4** Use the Clausius–Clapeyron equation to calculate vapor pressure at varying temperatures or ΔH_{vap}.	Worked Example 11.3; Problems 11.48–11.57
11.4 ▸ Types of Solids	Solids can be characterized as **amorphous solids** if their particles are randomly arranged or **crystalline solids** if their particles are ordered. Crystalline solids can be further characterized as **ionic solids** if their particles are ions, **molecular solids** if their particles are molecules, **covalent network solids** if they consist of a covalently bonded array of atoms without discrete molecules, or **metallic solids** if their particles are metal atoms.	**11.5** Classify types of solids based on their chemical composition and properties.	Problems 11.58–11.63
11.5–11.8 ▸ Structures of Solids	The structures of atoms in a solid can be measured by directing X rays at a crystal and analyzing the resulting diffraction pattern. The regular three-dimensional network of particles in a crystal is made up of small repeating units called **unit cells**. **Simple cubic packing** uses a **primitive-cubic unit cell**, with an atom at each corner of the cube. **Body-centered cubic packing** uses a **body-centered cubic unit cell**, with an atom at the center and at each corner of the cube. **Cubic closest-packing** uses a **face-centered cubic unit cell**, with an atom at the center of each face and at each corner of the cube. A fourth kind of packing, called **hexagonal closest-packing**, uses a noncubic unit cell.	**11.6** Use the Bragg equation to calculate spacing between atomic layers in a crystal.	Problems 11.64, 11.65
		11.7 Identify a unit cell from a pattern	Worked Example 11.4; Problems 11.7, 11.8
		11.8 Identify the four kinds of spherical packing arrangements in crystalline solids and the three kinds of cubic unit cells.	Problems 11.23, 11.66, 11.67
		11.9 Calculate the density of a substance, atomic radii of its atoms, or molecular mass given its unit cell dimensions.	Worked Examples 11.5, 11.6; Problems 11.68–11.75, 11.80, 11.81
		11.10 Use the unit cell to determine the formula and geometry of ionic compounds.	Worked Example 11.7; Problems 11.13, 11.14, 11.24, 11.25

Section	Concept Summary	Learning Objectives	Test Your Understanding
11.9 ▸ Phase Diagrams	The effects of temperature and pressure on phase changes can be displayed graphically on a **phase diagram**. A typical phase diagram has three regions—solid, liquid, and gas—separated by three boundary lines that represent pressure/temperature combinations at which two phases are in equilibrium and phase changes occur. At exactly 1 atm pressure, the temperature at the solid/liquid boundary corresponds to the **normal melting point** of the substance, and the temperature at the liquid/gas boundary corresponds to the **normal boiling point**. The three lines meet at the **triple point**, a unique combination of temperature and pressure at which all three phases coexist in equilibrium. The liquid/gas line runs from the triple point to the **critical point**, a pressure/temperature combination beyond which liquid and gas phases become a **supercritical fluid** that is neither a true liquid nor a true gas.	**11.11** Use a phase diagram to identify the phase of a substance at a given temperature and pressure. **11.12** Sketch a phase diagram given the appropriate data.	Worked Example 11.8; Problems 11.26, 11.27, 11.82, 11.83, 11.87–11.93 Problems 11.84, 11.85

KEY TERMS

allotrope *433*
amorphous solid *420*
body-centered cubic
  packing *425*
body-centered cubic unit
  cell *426*
Bragg equation *424*
Clausius–Clapeyron
  equation *418*

coordination number *425*
covalent network solid *421*
critical point *436*
crystalline solid *420*
cubic closest-packing *425*
diffraction *423*
face-centered cubic unit cell *427*
heat of fusion (ΔH_{fusion}) *415*
heat of vaporization(ΔH_{vap}) *415*

hexagonal closest-packing *425*
ionic solid *420*
metallic solid *421*
molecular solid *421*
normal boiling point *419*
normal melting point *435*
phase *412*
phase change *412*
phase diagram *435*

primitive-cubic unit cell *426*
simple cubic packing *425*
supercritical fluid *436*
surface tension *411*
triple point *435*
unit cell *426*
vapor pressure *417*
viscosity *411*

KEY EQUATIONS

• **Clausius–Clapeyron equation (Section 11.3)**

$$\ln P_{vap} = \left(-\frac{\Delta H_{vap}}{R} \right)\frac{1}{T} + C$$ where ΔH_{vap} is the heat of vaporization, R is the gas constant, and C is a constant characteristic of the specific substance.

• **Two-point form of the Clausius–Clapeyron equation (Section 11.3)**

$$\ln\left(\frac{P_1}{P_2}\right) = \frac{\Delta H_{vap}}{R}\left(\frac{1}{T_2} - \frac{1}{T_1}\right)$$

• **Bragg equation (Section 11.5)**

$$d = \frac{n\lambda}{2\sin\theta}$$ where d is the distance between layers of atoms, λ is the wavelength of the X rays, and θ is the angle at which X rays strike the atomic layer.

CONCEPTUAL PROBLEMS

Problems 11.1–11.21 appear within the chapter.

11.22 Assume that you have a liquid in a cylinder equipped with a movable piston. There is no air in the cylinder, the volume of space above the liquid is 200 mL, and the equilibrium vapor pressure above the liquid is 28.0 mm Hg. What is the equilibrium pressure above the liquid when the volume of space is decreased from 200 mL to 100 mL at constant temperature?

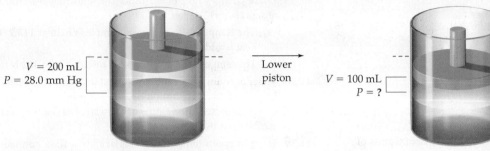

$V = 200$ mL
$P = 28.0$ mm Hg

Lower piston →

$V = 100$ mL
$P = ?$

11.23 Identify each of the following kinds of packing:

(a)   **(b)**   **(c)**   **(d)**

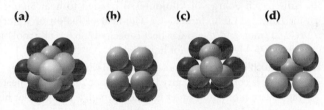

11.24 Zinc sulfide, or sphalerite, crystallizes in the following cubic unit cell:

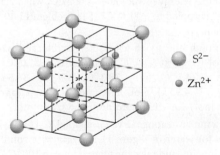

○ S^{2-}

● Zn^{2+}

(a) What kind of packing do the sulfide ions adopt?

(b) How many S^{2-} ions and how many Zn^{2+} ions are in the unit cell?

11.25 Titanium oxide crystallizes in the following cubic unit cell:

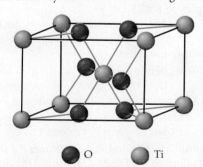

● O   ○ Ti

(a) How many titanium ions and how many oxygen ions are in each unit cell?

(b) What is the formula of titanium oxide?

(c) What is the oxidation state of titanium?

11.26 The phase diagram of a substance is shown below.

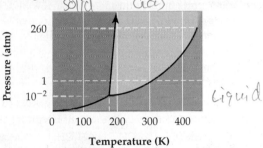

(a) Approximately what is the normal boiling point and what is the normal melting point of the substance?

(b) What is the physical state of the substance under the following conditions?

(i) $T = 150$ K, $P = 0.5$ atm

(ii) $T = 325$ K, $P = 0.9$ atm

(iii) $T = 450$ K, $P = 265$ atm

11.27 The following phase diagram of elemental carbon has three different solid phases in the region shown.

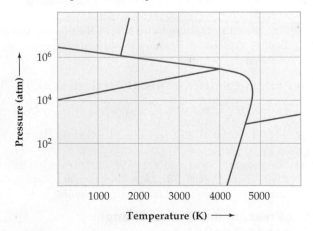

(a) Show where the solid, liquid, and vapor regions are on the diagram.

(b) How many triple points does carbon have? Circle each on the diagram.

(c) Graphite is the most stable solid phase under normal conditions. Identify the graphite phase on the diagram.

(d) On heating graphite to 2500 K at a pressure of 100,000 atm, it can be converted into diamond. Identify the diamond phase on the graph.

(e) Which phase is more dense, graphite or diamond? Explain.

SECTION PROBLEMS

Properties of Liquids (Section 11.1)

11.28 A magnetized needle gently placed on the surface of a glass of water acts like a makeshift compass. Is it water's viscosity or its surface tension that keeps the needle on top?

11.29 Water flows quickly through the narrow neck of a bottle, but maple syrup flows sluggishly. Is this different behavior due to a difference in viscosity or in surface tension for the liquids?

11.30 Predict which substance in each pair has the highest surface tension.
(a) CCl_4 or CH_2Br_2
(b) ethanol (CH_3CH_2OH) or ethylene glycol ($HOCH_2CH_2OH$)

11.31 Predict which substance in each pair has the highest viscosity.
(a) hexane ($CH_3CH_2CH_2CH_2CH_2CH_3$) or 1-hexanol ($CH_3CH_2CH_2CH_2CH_2CH_2OH$)
(b) pentane or neopentane

Pentane

Neopentane

11.32 The chemical structure for oleic acid, the primary component of olive oil is shown. Explain why olive oil has a higher viscosity than water.

11.33 The viscosity of water at 20 °C $1.00 \times 10^{-3}(N \cdot s/m^2)$ is higher than dimethyl sulfide $[(CH_3)_2S]$ $2.8 \times 10^{-4}(N \cdot s/m^2)$. Explain the difference in viscosity based on chemical structure.

Vapor Pressure and Phase Changes (Sections 11.2 and 11.3)

11.34 Why is ΔH_{vap} usually larger than ΔH_{fusion}?

11.35 Why is the heat of sublimation, ΔH_{subl}, equal to the sum of ΔH_{vap} and ΔH_{fusion} at the same temperature?

11.36 Mercury has mp $= -38.8$ °C and bp $= 356.6$ °C. What, if any, phase changes take place under the following conditions at 1.0 atm pressure?
(a) The temperature of a sample is raised from -30 °C to 365 °C.
(b) The temperature of a sample is lowered from 291 K to 238 K.
(c) The temperature of a sample is lowered from 638 K to 231 K.

11.37 Iodine has mp $= 113.7$ °C and bp $= 184.4$ °C. What, if any, phase changes take place under the following conditions at 1.0 atm pressure?
(a) The temperature of a solid sample is held at 113.7 °C while heat is added.
(b) The temperature of a sample is lowered from 452 K to 389 K.

11.38 Water at room temperature is placed in a flask connected by rubber tubing to a vacuum pump, and the pump is turned on. After several minutes, the volume of the water has decreased and what remains has turned to ice. Explain.

11.39 Ether at room temperature is placed in a flask connected by a rubber tube to a vacuum pump, the pump is turned on, and the ether begins boiling. Explain.

11.40 How much energy in kilojoules is needed to heat 5.00 g of ice from -11.0 °C to 30.0 °C? The heat of fusion of water is 6.01 kJ/mol, and the molar heat capacity is 36.6 J/(K·mol) for ice and 75.4 J/(K·mol) for liquid water.

11.41 How much energy in kilojoules is released when 15.3 g of steam at 115.0 °C is condensed to give liquid water at 75.0 °C? The heat of vaporization of liquid water is 40.67 kJ/mol, and the molar heat capacity is 75.4 J/(K·mol) for the liquid and 33.6 J/(K·mol) for the vapor.

11.42 How much energy in kilojoules is released when 7.55 g of water at 33.5 °C is cooled to -11.0 °C? (See Problem 11.40 for the necessary data.)

11.43 How much energy in kilojoules is released when 25.0 g of ethanol vapor at 93.0 °C is cooled to -11.0 °C? Ethanol has mp $= -114.1$ °C, bp $= 78.3$ °C, $\Delta H_{vap} = 38.56$ kJ/mol, and $\Delta H_{fusion} = 4.93$ kJ/mol. The molar heat capacity is 112.3 J/(K·mol) for the liquid and 65.6 J/(K·mol) for the vapor.

11.44 Draw a molar heating curve for ethanol, C_2H_5OH, similar to that shown for water in Figure 11.3. Begin with solid ethanol at its melting point, and raise the temperature to 100 °C. The necessary data are given in Problem 11.43.

11.45 Draw a molar heating curve for sodium similar to that shown for water in Figure 11.3. Begin with solid sodium at its melting point, and raise the temperature to 1000 °C. The necessary data are mp $= 97.8$ °C, bp $= 883$ °C, $\Delta H_{vap} = 89.6$ kJ/mol, and $\Delta H_{fusion} = 2.64$ kJ/mol. Assume that the molar heat capacity is 20.8 J/(K·mol) for both liquid and vapor phases and does not change with temperature.

11.46 Naphthalene, better known as "mothballs," has bp $= 218$ °C and $\Delta H_{vap} = 43.3$ kJ/mol. What is the entropy of vaporization ΔS_{vap} in J/(K·mol) for naphthalene?

11.47 What is the entropy of fusion, ΔS_{fusion} in J/(K·mol) for sodium? The necessary data are given in Problem 11.45.

11.48 Carbon disulfide, CS_2, has $P_{vap} = 100$ mm Hg at -5.1 °C and a normal boiling point of 46.5 °C. What is ΔH_{vap} for carbon disulfide in kJ/mol?

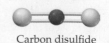

Carbon disulfide

11.49 The vapor pressure of $SiCl_4$ is 100 mm Hg at 5.4 °C, and the normal boiling point is 57.7 °C. What is ΔH_{vap} for $SiCl_4$ in kJ/mol?

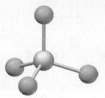

Silicon tetrachloride

11.50 What is the vapor pressure of CS_2 in mm Hg at 20.0 °C? (See Problem 11.48.)

11.51 What is the vapor pressure of $SiCl_4$ in mm Hg at 30.0 °C? (See Problem 11.49.)

11.52 Dichloromethane, CH_2Cl_2, is an organic solvent used for removing caffeine from coffee beans. The following table gives the vapor pressure of dichloromethane at various temperatures. Fill in the rest of the table, and use the data to plot curves of P_{vap} versus T and ln P_{vap} versus $1/T$.

Temp (K)	P_{vap}(mm Hg)	ln P_{vap}	$1/T$
263	80.1	?	?
273	133.6	?	?
283	213.3	?	?
293	329.6	?	?
303	495.4	?	?
313	724.4	?	?

11.53 The following table gives the vapor pressure of mercury at various temperatures. Fill in the rest of the table, and use the data to plot curves of P_{vap} versus T and ln P_{vap} versus $1/T$.

Temp (K)	P_{vap}(mm Hg)	ln P_{vap}	$1/T$
500	39.3	?	?
520	68.5	?	?
540	114.4	?	?
560	191.6	?	?
580	286.4	?	?
600	432.3	?	?

11.54 Use the plot you made in Problem 11.52 to find a value in kJ/mol for ΔH_{vap} for dichloromethane.

11.55 Use the plot you made in Problem 11.53 to find a value in kJ/mol for ΔH_{vap} for mercury. The normal boiling point of mercury is 630 K.

11.56 Choose any two temperatures and corresponding vapor pressures in the table given in Problem 11.52, and use those values to calculate ΔH_{vap} for dichloromethane in kJ/mol. How does the value you calculated compare to the value you read from your plot in Problem 11.54?

11.57 Choose any two temperatures and corresponding vapor pressures in the table given in Problem 11.53, and use those values to calculate ΔH_{vap} for mercury in kJ/mol. How does the value you calculated compare to the value you read from your plot in Problem 11.55?

Structures of Solids (Sections 11.4–11.8)

11.58 List the four main classes of crystalline solids, and give a specific example of each.

11.59 What kinds of particles are present in each of the four main classes of crystalline solids?

11.60 Which of the substances Na_3PO_4, CBr_4, rubber, Au, and quartz best fits each of the following descriptions?
(a) Amorphous solid
(b) Ionic solid
(c) Molecular solid
(d) Covalent network solid
(e) Metallic solid

11.61 Which of the substances diamond, Hg, Cl_2, glass, and KCl best fits each of the following descriptions?
(a) Amorphous solid
(b) Ionic solid
(c) Molecular solid
(d) Covalent network solid
(e) Metallic solid

11.62 Silicon carbide is very hard, has no known melting point, and diffracts X rays. What type of solid is it: amorphous, ionic, molecular, covalent network, or metallic?

11.63 Arsenic tribromide melts at 31.1 °C, diffracts X rays, and does not conduct electricity in either the solid or liquid phase. What type of solid is it: amorphous, ionic, molecular, covalent network, or metallic?

11.64 Diffraction of X rays with $\lambda = 154.2$ pm occurred at an angle $\theta = 22.5°$ from a metal surface. What is the spacing (in pm) between the layers of atoms that diffracted the X rays?

11.65 Diffraction of X rays with $\lambda = 154.2$ pm occurred at an angle $\theta = 76.84°$ from a metal surface. What is the spacing (in pm) between layers of atoms that diffracted the X rays?

11.66 Which of the four kinds of packing used by metals makes the most efficient use of space, and which makes the least efficient use?

11.67 (a) What is a unit cell?
(b) How many atoms are in one primitive cubic unit cell, one body-centered unit cell, and one face-centered unit cell?

11.68 Copper crystallizes in a face-centered cubic unit cell with an edge length of 362 pm. What is the radius of a copper atom in picometers? What is the density of copper in g/cm³?

11.69 Lead crystallizes in a face-centered cubic unit cell with an edge length of 495 pm. What is the radius of a lead atom in picometers? What is the density of lead in g/cm³?

11.70 Aluminum has a density of 2.699 g/cm³ and crystallizes with a face-centered cubic unit cell. What is the edge length of a unit cell in picometers?

11.71 Tungsten crystallizes in a body-centered cubic unit cell with an edge length of 317 pm. What is the length in picometers of a unit-cell diagonal that passes through the center atom?

11.72 In light of your answer to Problem 11.71, what is the radius in picometers of a tungsten atom?

11.73 Sodium has a density of 0.971 g/cm³ and crystallizes with a body-centered cubic unit cell. What is the radius of a sodium atom, and what is the edge length of the cell in picometers?

11.74 Titanium metal has a density of 4.506 g/cm³ and an atomic radius of 144.8 pm. In what cubic unit cell does titanium crystallize?

11.75 Calcium metal has a density of 1.55 g/cm^3 and crystallizes in a cubic unit cell with an edge length of 558.2 pm.

(a) How many Ca atoms are in one unit cell?

(b) In which of the three cubic unit cells does calcium crystallize?

11.76 Sodium hydride, NaH, crystallizes in a face-centered cubic unit cell similar to that of NaCl (Figure 11.17). How many Na^+ ions touch each H^- ion, and how many H^- ions touch each Na^+ ion?

11.77 Cesium chloride crystallizes in a cubic unit cell with Cl^- ions at the corners and a Cs^+ ion in the center. Count the numbers of $+$ and $-$ charges, and show that the unit cell is electrically neutral.

11.78 If the edge length of an NaH unit cell is 488 pm, what is the length in picometers of an Na—H bond? (See Problem 11.76.)

11.79 The edge length of a CsCl unit cell (Problem 10.93) is 412.3 pm. What is the length in picometers of the Cs—Cl bond? If the ionic radius of a Cl^- ion is 181 pm, what is the ionic radius in picometers of a Cs^+ ion?

11.80 The atomic radius of Pb is 175 pm and the density is 11.34 g/cm^3. Does lead have a primitive cubic structure or a face-centered cubic structure?

11.81 The density of a sample of metal was measured to be 6.84 g/cm^3. An X-ray diffraction experiment measures the edge of a face-centered cubic cell as 350.7 pm. What is the atomic weight, atomic radius, and identity of the metal?

Phase Diagrams (Section 11.9)

11.82 Look at the phase diagram of CO_2 in Figure 11.23, and tell what phases are present under the following conditions:

(a) $T = -60\,°C, P = 0.75$ atm

(b) $T = -35\,°C, P = 18.6$ atm

(c) $T = -80\,°C, P = 5.42$ atm

11.83 Look at the phase diagram of H_2O in Figure 11.21, and tell what happens to an H_2O sample when the following changes are made:

(a) The temperature is reduced from $48\,°C$ to $-4.4\,°C$ at a constant pressure of 6.5 atm.

(b) The pressure is increased from 85 atm to 226 atm at a constant temperature of $380\,°C$.

11.84 Bromine has $T_t = -7.3\,°C, P_t = 44 \text{ mm Hg}, T_c = 315\,°C$, and $P_c = 102$ atm. The density of the liquid is 3.1 g/cm^3, and the density of the solid is 3.4 g/cm^3. Sketch a phase diagram for bromine, and label all points of interest.

11.85 Oxygen has $T_t = 54.3 \text{ K}, P_t = 1.14 \text{ mm Hg}, T_c = 154.6 \text{ K}$, and $P_c = 49.77$ atm. The density of the liquid is 1.14 g/cm^3, and the density of the solid is 1.33 g/cm^3. Sketch a phase diagram for oxygen, and label all points of interest.

11.86 Refer to the bromine phase diagram you sketched in Problem 11.84, and tell what phases are present under the following conditions:

(a) $T = -10\,°C, P = 0.0075$ atm

(b) $T = 25\,°C, P = 16$ atm

11.87 Refer to the oxygen phase diagram you sketched in Problem 11.85, and tell what phases are present under the following conditions:

(a) $T = -210\,°C, P = 1.5$ atm

(b) $T = -100\,°C, P = 66$ atm

11.88 Does solid oxygen (Problem 11.85) melt when pressure is applied as water does? Explain.

11.89 Assume that you have samples of the following three gases at $25\,°C$. Which of the three can be liquefied by applying pressure, and which cannot? Explain.

Ammonia: $T_c = 132.5\,°C$ and $P_c = 112.5$ atm

Methane: $T_c = -82.1\,°C$ and $P_c = 45.8$ atm

Sulfur dioxide: $T_c = 157.8\,°C$ and $P_c = 77.7$ atm

11.90 Benzene has a melting point of $5.53\,°C$ and a boiling point of $80.09\,°C$ at atmospheric pressure. Its density is 0.8787 g/cm^3 when liquid and 0.899 g/cm^3 when solid; it has $T_c = 289.01\,°C$, $P_c = 48.34$ atm, $T_t = 5.52\,°C$, and $P_t = 0.0473$ atm. Starting from a point at 200 K and 66.5 atm, trace the following path on a phase diagram:

(1) First, increase T to 585 K while keeping P constant.

(2) Next, decrease P to 38.5 atm while keeping T constant.

(3) Then, decrease T to 278.66 K while keeping P constant.

(4) Finally, decrease P to 0.0025 atm while keeping T constant.

What is your starting phase, and what is your final phase?

11.91 Refer to the oxygen phase diagram you drew in Problem 11.85 and trace the following path starting from a point at 0.0011 atm and $-225\,°C$:

(1) First, increase P to 35 atm while keeping T constant.

(2) Next, increase T to $-150\,°C$ while keeping P constant.

(3) Then, decrease P to 1.0 atm while keeping T constant.

(4) Finally, decrease T to $-215\,°C$ while keeping P constant.

What is your starting phase, and what is your final phase?

11.92 How many phase transitions did you pass through in Problem 11.90, and what are they?

11.93 What phase transitions did you pass through in Problem 11.91?

CHAPTER PROBLEMS

11.94 What is the atomic radius in picometers of an argon atom if solid argon has a density of 1.623 g/cm^3 and crystallizes at low temperature in a face-centered cubic unit cell?

11.95 Mercury has $mp = -38.8\,°C$, a molar heat capacity of $27.9 \text{ J/(K} \cdot \text{mol)}$ for the liquid and $28.2 \text{ J/(K} \cdot \text{mol)}$ for the solid and $\Delta H_{fusion} = 2.33 \text{ kJ/mol}$. Assuming that the heat capacities don't change with temperature, how much energy in kilojoules is needed to heat 7.50 g of Hg from a temperature of $-50.0\,°C$ to $+50.0\,°C$?

11.96 Silicon carbide, SiC, is a covalent network solid with a structure similar to that of diamond. Sketch a small portion of the SiC structure.

11.97 In Denver, the Mile-High City, water boils at $95\,°C$. What is atmospheric pressure in atmospheres in Denver? ΔH_{vap} for H_2O is 40.67 kJ/mol.

11.98 If a protein can be induced to crystallize, its molecular structure can be determined by X-ray crystallography. Protein crystals, though solid, contain a large amount of water molecules along with the protein. The protein chicken egg-white lysozyme, for instance, crystallizes with a unit cell having angles of $90°$ and with edge lengths of $7.9 \times 10^3 \text{ pm}, 7.9 \times 10^3 \text{ pm}$, and $3.8 \times 10^3 \text{ pm}$. There are eight molecules in the unit cell. If the lysozyme molecule has a molecular weight of 1.44×10^4 and a density of 1.35 g/cm^3, what percent of the unit cell is occupied by the protein?

11.99 The molecular structure of a scorpion toxin, a small protein, was determined by X-ray crystallography. The unit cell has angles of 90°, contains 16 molecules, and has a volume of 1.019×10^2 nm³. If the molecular weight of the toxin is 3336 and the density is about 1.35 g/cm³, what percent of the unit cell is occupied by protein?

1.100 Magnesium metal has $\Delta H_{fusion} = 9.037$ kJ/mol and $\Delta S_{fusion} = 9.79$ J/(K·mol). What is the melting point in °C of magnesium?

1.101 Titanium tetrachloride, $TiCl_4$, has a melting point of -23.2 °C and has $\Delta H_{fusion} = 9.37$ kJ/mol. What is the entropy of fusion, ΔS_{fusion} in J/(K·mol), for $TiCl_4$?

1.102 Dichlorodifluoromethane, CCl_2F_2, one of the chlorofluorocarbon refrigerants responsible for destroying part of the Earth's ozone layer, has $P_{vap} = 40.0$ mm Hg at -81.6 °C and $P_{vap} = 400$ mm Hg at -43.9 °C. What is the normal boiling point of CCl_2F_2 in °C?

Dichlorodifluoromethane

1.103 The chlorofluorocarbon refrigerant trichlorofluoromethane, CCl_3F, has $P_{vap} = 100.0$ mm Hg at -23 °C and $\Delta H_{vap} = 24.77$ kJ/mol.
(a) What is the normal boiling point of trichlorofluoromethane in °C?
(b) What is ΔS_{vap} for trichlorofluoromethane?

1.104 Nitrous oxide, N_2O, occasionally used as an anesthetic by dentists under the name "laughing gas," has $P_{vap} = 100$ mm Hg at -111.3 °C and a normal boiling point of -88.5 °C. What is the heat of vaporization of nitrous oxide in kJ/mol?

1.105 Acetone, a common laboratory solvent, has $\Delta H_{vap} = 29.1$ kJ/mol and a normal boiling point of 56.1 °C. At what temperature in °C does acetone have $P_{vap} = 105$ mm Hg?

Acetone

1.106 Use the following data to sketch a phase diagram for krypton: $T_t = -169$ °C, $P_t = 133$ mm Hg, $T_c = -63$ °C, $P_c = 54$ atm, mp $= -156.6$ °C, bp $= -152.3$ °C. The density of solid krypton is 2.8 g/cm³, and the density of the liquid is 2.4 g/cm³. Can a sample of gaseous krypton at room temperature be liquefied by raising the pressure?

1.107 What is the physical phase of krypton (Problem 11.106) under the following conditions:
(a) $P = 5.3$ atm, $T = -153$ °C
(b) $P = 65$ atm, $T = 250$ K

1.108 Calculate the percent volume occupied by the spheres in a body-centered cubic unit cell.

1.109 Iron crystallizes in a body-centered cubic unit cell with an edge length of 287 pm. What is the radius of an iron atom in picometers?

11.110 Iron metal has a density of 7.86 g/cm³ and a molar mass of 55.85 g. Use this information together with the data in Problem 11.109 to calculate a value for Avogadro's number.

11.111 Silver metal crystallizes in a face-centered cubic unit cell with an edge length of 408 pm. The molar mass of silver is 107.9 g/mol, and its density is 11.50 g/cm³. Use these data to calculate a value for Avogadro's number.

11.112 A drawing of the NaCl unit cell is shown in Figure 11.17.
(a) What is the edge length in picometers of the NaCl unit cell? The ionic radius of Na^+ is 97 pm, and the ionic radius of Cl^- is 181 pm.
(b) What is the density of NaCl in g/cm³?

11.113 Niobium oxide crystallizes in the following cubic unit cell:

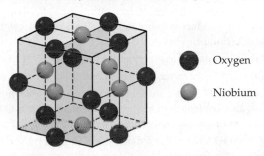

Oxygen

Niobium

(a) How many niobium atoms and how many oxygen atoms are in each unit cell?
(b) What is the formula of niobium oxide?
(c) What is the oxidation state of niobium?

11.114 One form of silver telluride (Ag_2Te) crystallizes with a cubic unit cell and a density of 7.70 g/cm³. X-ray crystallography shows that the edge of the cubic unit cell has a length of 529 pm. How many Ag atoms are in the unit cell?

11.115 Substance **X** has a vapor pressure of 100 mm Hg at its triple point (48 °C). When 1 mol of **X** is heated at 1 atm pressure with a constant rate of heat input, the following heating curve is obtained:

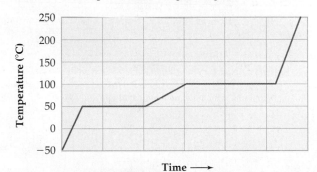

(a) Sketch the phase diagram for **X**, including labels for different phases, triple point, melting point, and boiling point.
(b) For each of the following, choose which phase of **X** (solid, liquid, or gas) fits the description:
(i) Is the most dense at 50 °C
(ii) Is the least dense at 50 °C
(iii) Has the greatest specific heat
(iv) Predominates at 80 °C and 1 atm
(v) Can have a vapor pressure of 20 mm Hg

MULTICONCEPT PROBLEMS

11.116 For each of the following substances, identify the intermolecular force or forces that predominate. Using your knowledge of the relative strengths of the various forces, rank the substances in order of their normal boiling points.

$$Al_2O_3, F_2, H_2O, Br_2, ICl, NaCl$$

11.117 Look up thermodynamic data for ethanol (C_2H_5OH) in Appendix B, estimate the normal boiling point of ethanol, and calculate the vapor pressure of ethanol at 25 °C.

11.118 The mineral *magnetite* is an iron oxide ore that has a density of 5.20 g/cm^3. At high temperature, magnetite reacts with carbon monoxide to yield iron metal and carbon dioxide. When 2.660 g of magnetite is allowed to react with sufficient carbon monoxide, the CO_2 product is found to have a volume of 1.136 L at 298 K and 751 mm Hg pressure.

(a) What mass of iron in grams is formed in the reaction?

(b) What is the formula of magnetite?

(c) Magnetite has a somewhat complicated cubic unit cell with an edge length of 839 pm. How many Fe and O atoms are present in each unit cell?

11.119 A group 3A metal has a density of 2.70 g/cm^3 and a cubic unit cell with an edge length of 404 pm. Reaction of a 1.07 cm³ chunk of the metal with an excess of hydrochloric acid gives a colorless gas that occupies 4.00 L at 23.0 °C and a pressure of 740 mm Hg.

(a) Identify the metal.

(b) Is the unit cell primitive, body-centered, or face-centered?

(c) What is the atomic radius of the metal atom in picometers?

11.120 A cube-shaped crystal of an alkali metal, 1.62 mm on an edge, was vaporized in a 500.0 mL evacuated flask. The resulting vapor pressure was 12.5 mm Hg at 802 °C. The structure of the solid metal is known to be body-centered cubic.

(a) What is the atomic radius of the metal atom in picometers?

(b) Use the data in Figure 5.19 to identify the alkali metal.

(c) What are the densities of the solid and the vapor in g/cm^3?

11.121 Assume that 1.588 g of an alkali metal undergoes complete reaction with the amount of gaseous halogen contained in a 0.500 flask at 298 K and 755 mm Hg pressure. In the reaction, 22.83 is released ($\Delta H = -22.83 \text{ kJ}$). The product, a binary ionic compound, crystallizes in a unit cell with anions in a face-centered cubic arrangement and with cations centered along each edge between anions. In addition, there is a cation in the center of the cube.

(a) What is the identity of the alkali metal?

(b) The edge length of the unit cell is 535 pm. Find the radius of the alkali metal cation from the data in Figure 6.1, and then calculate the radius of the halide anion. Identify the anion from the data in Figure 6.2.

(c) Sketch a space-filling, head-on view of the unit cell, labeling the ions. Are the anions in contact with one another?

(d) What is the density of the compound in g/cm^3?

(e) What is the standard heat of formation for the compound?

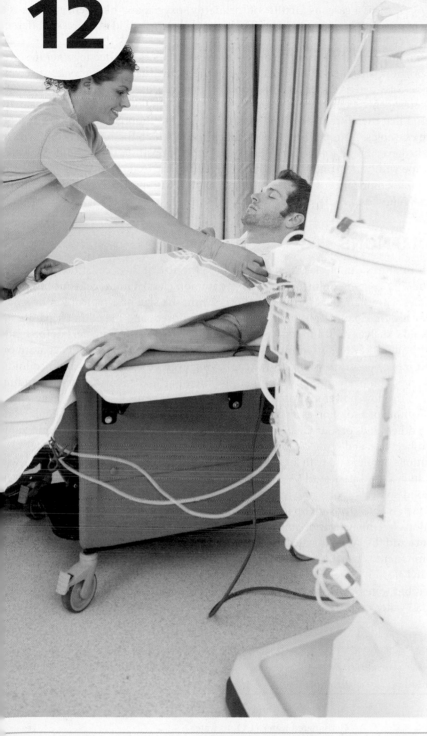

Solutions and Their Properties

Hemodialysis on this artificial kidney machine cleanses the blood of individuals whose kidneys no longer function. The principles of osmosis discussed in this chapter are the basis for hemodialysis.

 How does hemodialysis cleanse the blood of patients with kidney failure?

The answer to this question can be found in the **INQUIRY** ▸▸▸ on page 479.

CONTENTS

Thus far, we've been concerned only with pure substances, both elements and compounds. If you look around, though, most of the substances you see in day-to-day life are *mixtures*. Air is a gaseous mixture of (primarily) oxygen and nitrogen, gasoline is a liquid mixture of many different organic compounds, and rocks are solid mixtures of different minerals.

A **mixture** is any combination of two or more pure substances blended together in some arbitrary proportion without chemically changing the individual substances themselves. Mixtures can be classified as either *heterogeneous* or *homogeneous*, depending on their appearance. A heterogeneous mixture is one in which the mixing of components is visually nonuniform and that therefore has regions of different composition. Sugar with salt and oil with water are examples. A homogeneous mixture is one in which the mixing *is* uniform, at least to the naked eye, and that therefore has a constant composition throughout. Seawater (sodium chloride with water) and brass (copper with zinc) are examples. We'll explore the properties of some homogeneous mixtures in this chapter, with particular emphasis on the mixtures we call *solutions*.

12.1 ▶ SOLUTIONS

Homogeneous mixtures can be classified according to the size of their constituent particles as either *solutions* or *colloids*. **Solutions**, the most common class of homogeneous mixtures, contain particles with diameters in the range 0.1–2 nm, the size of a typical ion or small molecule. They are transparent, although they may be colored, and they don't separate on standing. **Colloids**, such as milk and fog, contain larger particles, with diameters in the range 2–500 nm. Although they are often murky or opaque to light, colloids don't separate on standing. Mixtures called *suspensions* also exist, having even larger particles than colloids. These are not truly homogeneous, however, because their particles separate out on standing and are visible with a low-power microscope. Blood, paint, and aerosol sprays are examples.

We usually think of a solution as a solid dissolved in a liquid or as a mixture of liquids, but there are many other kinds of solutions as well. In fact, any one state of matter can form a solution with any other state, and seven different kinds of solutions are possible (TABLE 12.1).

Even solutions of one solid with another and solutions of a gas in a solid are well known. Metal alloys, such as stainless steel (4–30% chromium in iron) and brass (10–40% zinc in copper), are solid/solid solutions, and hydrogen in palladium is a gas/solid solution. Metallic palladium, in fact, is able to absorb up to 935 times its own volume of H_2 gas!

For solutions in which a gas or solid is dissolved in a liquid, the dissolved substance is called the **solute** and the liquid is called the **solvent**. When one liquid is dissolved in another, the minor component is usually considered the solute and the major component is the solvent. Thus, ethyl alcohol is the solute and water the solvent in a mixture of 10% ethyl alcohol and 90% water, but water is the solute and ethyl alcohol the solvent in a mixture of 90% ethyl alcohol and 10% water.

▲ The pewter in this 200-year-old jug is a solid–solid solution of approximately 96% tin and 4% copper.

TABLE 12.1 Some Different Kinds of Solutions	
Kind of Solution	**Example**
Gas in gas	Air (O_2, N_2, Ar, and other gases)
Gas in liquid	Carbonated water (CO_2 in water)
Gas in solid	H_2 in palladium metal
Liquid in liquid	Gasoline (mixture of hydrocarbons)
Liquid in solid	Dental amalgam (mercury in silver)
Solid in liquid	Seawater (NaCl and other salts in water)
Solid in solid	Metal alloys, such as sterling silver (92.5% Ag, 7.5% Cu)

12.2 ▶ ENERGY CHANGES AND THE SOLUTION PROCESS

With the exception of gas–gas mixtures, such as air, the different kinds of solutions listed in Table 12.1 involve *condensed phases*, either liquid or solid. Thus, all the **intermolecular forces** described in Section 8.6 to explain the properties of pure liquids and solids are also important for explaining the properties of solutions. The situation is more complex for solutions than for pure substances, though, because three types of interactions among particles have to be taken into account: solvent–solvent interactions, solvent–solute interactions, and solute–solute interactions.

A good rule of thumb, often summarized in the phrase "like dissolves like," is that solutions will form when the three types of interactions are similar in kind and in magnitude. Thus, ionic solids like NaCl dissolve in polar solvents like water because the strong ion–dipole attractions between Na$^+$ and Cl$^-$ ions and polar H$_2$O molecules are similar in magnitude to the strong hydrogen bonding attractions between water molecules and to the strong ion–ion attractions between Na$^+$ and Cl$^-$ ions. In the same way, nonpolar organic substances like cholesterol, C$_{27}$H$_{46}$O, dissolve in nonpolar organic solvents like benzene, C$_6$H$_6$, because of the similar London dispersion forces present among both kinds of molecules. Oil, however, does not dissolve appreciably in water because the two liquids have different kinds of intermolecular forces.

REMEMBER...

Intermolecular forces include ion–dipole, dipole–dipole, and London dispersion forces, as well as hydrogen bonds. (Section 8.6)

Like dissolves like. Solutions form when the intermolecular forces involved in these three kinds of interactions are similar.

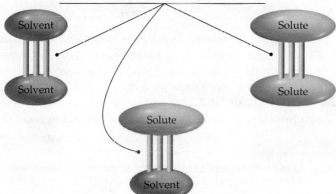

The dissolution of a solid in a liquid can be visualized as shown in **FIGURE 12.1** for NaCl. When solid NaCl is placed in water, those ions that are less tightly held because of their position at a corner or an edge of the crystal are exposed to water molecules, which collide with them until an ion happens to break free. More water molecules then cluster around the ion, stabilizing it by means of ion–dipole attractions. A new edge or corner is thereby exposed on the crystal, and the process continues until the entire crystal has dissolved. The ions in solution are said to be *solvated*—more specifically, *hydrated*, when water is the solvent—meaning that they are surrounded and stabilized by an ordered shell of solvent molecules.

Like all chemical and physical processes, the dissolution of a solute in a solvent has associated with it a free-energy change, $\Delta G = \Delta H - T\Delta S$, whose value describes its spontaneity (Section 9.13). If ΔG is negative, the process is spontaneous and the substance dissolves; if ΔG is positive, the process is nonspontaneous and the substance does not dissolve. The enthalpy term ΔH measures the heat flow into or out of the system during dissolution, and the temperature-dependent entropy term $T\Delta S$ measures the change in the amount of molecular randomness in the system. The enthalpy change is called the *heat of solution*, or **enthalpy of solution** (ΔH_{soln}), and the entropy change is called the **entropy of solution** (ΔS_{soln}).

What values might we expect for ΔH_{soln} and ΔS_{soln}? Let's take the entropy change first. Entropies of solution are usually positive because molecular randomness usually increases during dissolution: $+43.4$ J/(K • mol) for NaCl in water, for example. When a solid dissolves in a liquid, randomness increases upon going from a well-ordered crystal to a less-ordered state in which solvated ions or molecules are able to move freely in solution. When

▲ Why don't oil and water mix?

▶ **FIGURE 12.1**

Dissolution of NaCl crystals in water.

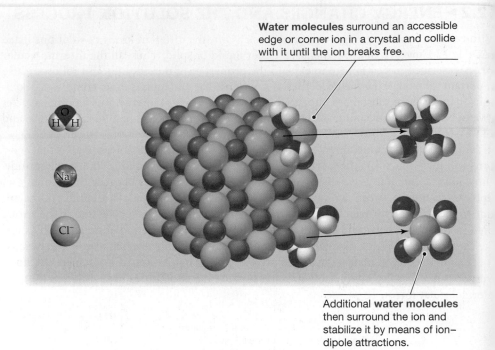

Water molecules surround an accessible edge or corner ion in a crystal and collide with it until the ion breaks free.

Additional water molecules then surround the ion and stabilize it by means of ion–dipole attractions.

one liquid dissolves in another, randomness increases as the different molecules intermingl▶ (**FIGURE 12.2**). **TABLE 12.2** lists values of ΔS_{soln} for some common ionic substances.

Values for enthalpies of solution, ΔH_{soln}, are more difficult to predict (Table 12.2). Som▶ solids dissolve exothermically and have a negative ΔH_{soln} (-37.0 kJ/mol for LiCl in water▶ but others dissolve endothermically and have a positive ΔH_{soln} ($+17.2$ kJ/mol for KCl i▶ water). Athletes benefit from both situations when they use instant hot packs or cold packs t▶

▶ **FIGURE 12.2**

Entropy of solution. Entropies of solution are usually positive because molecular randomness usually increases when a solid dissolves in a liquid or one liquid dissolves in another.

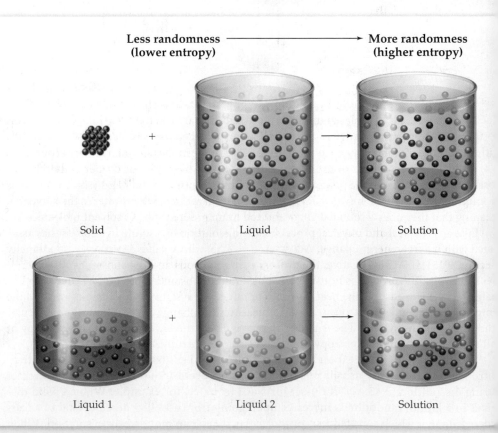

Less randomness (lower entropy) ⟶ More randomness (higher entropy)

Solid   Liquid   Solution

Liquid 1   Liquid 2   Solution

TABLE 12.2 Some Enthalpies and Entropies of Solution in Water at 25 °C

Substance	ΔH_{soln}(kJ/mol)	ΔS_{soln}[J/(K · mol)]
LiCl	−37.0	10.5
NaCl	3.9	43.4
KCl	17.2	75.0
LiBr	−48.8	21.5
NaBr	−0.6	54.6
KBr	19.9	89.0
KOH	−57.6	12.9

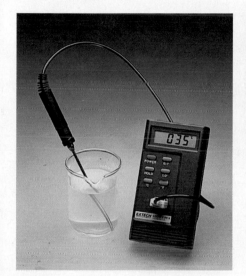
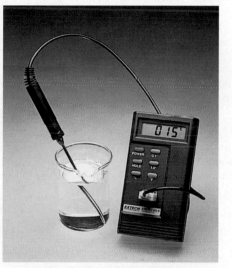

Dissolution of $CaCl_2$ in water is **exothermic**, causing the temperature of the water to rise from its initial 25 °C value.

Dissolution of NH_4NO_3 is endothermic, causing the temperature of the water to fall from its initial 25 °C value.

◀ **FIGURE 12.3**

Enthalpy of solution. Enthalpies of solution can be either negative (exothermic) or positive (endothermic).

eat injuries. Both kinds of instant packs consist of a pouch of water and a dry chemical, either $CaCl_2$ or $MgSO_4$ for hot packs, and NH_4NO_3 for cold packs. When the pack is squeezed, he pouch breaks and the solid dissolves, either raising or lowering the temperature FIGURE 12.3).

Hot packs:	$CaCl_2(s)$	$\Delta H_{soln} = -81.3\,kJ/mol$
	$MgSO_4(s)$	$\Delta H_{soln} = -91.2\,kJ/mol$
Cold pack:	$NH_4NO_3(s)$	$\Delta H_{soln} = +25.7\,kJ/mol$

The exact value of ΔH_{soln} for a given substance results from an interplay of the three nds of interactions mentioned at the beginning of this section:

- **Solvent–solvent interactions**: Energy is absorbed (positive ΔH) to overcome intermolecular forces between solvent molecules because the molecules must be separated and pushed apart to make room for solute particles.
- **Solute–solute interactions**: Energy is absorbed (positive ΔH) to overcome intermolecular forces holding solute particles together in a crystal. For an ionic solid, this is related to the **lattice energy**. As a result, substances with higher lattice energies tend to be less soluble than substances with lower lattice energies. Compounds with singly charged ions

REMEMBER...

Lattice energy (U) is the sum of the electrostatic interaction energies between ions in a crystal. It is the amount of energy that must be supplied to break up an ionic solid into its individual gaseous ions. (Section 6.8)

are thus more soluble than compounds with doubly or triply charged ions, as we saw when discussing solubility guidelines in Section 4.6.

- **Solvent–solute interactions**: Energy is released (negative ΔH) when solvent molecules cluster around solute particles and solvate them. For ionic substances in water, the amount of hydration energy released is generally greater for smaller cations than for larger ones because water molecules can approach the positive nuclei of smaller ions more closely and thus bind more tightly. In addition, hydration energy generally increases as the charge on the ion increases.

The first two kinds of interactions are endothermic, requiring an input of energy to spread apart solvent molecules and to break apart crystals. Only the third interaction is exothermic, as attractive intermolecular forces develop between solvent and solute particles. The sum of the three interactions determines whether ΔH_{soln} is endothermic or exothermic. For some substances, the one exothermic interaction is sufficiently large to outweigh the two endothermic interactions, but for other substances, the opposite is true (**FIGURE 12.4**).

▶ **FIGURE 12.4**

Components of ΔH_{soln}. The value of ΔH_{soln} is the sum of three terms: solvent–solvent, solute–solute, and solvent–solute.

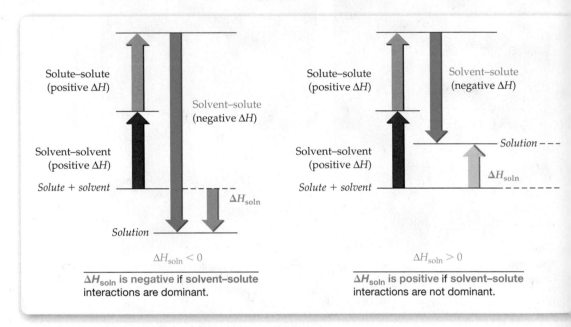

ΔH_{soln} is negative if solvent–solute interactions are dominant.

ΔH_{soln} is positive if solvent–solute interactions are not dominant.

Conceptual WORKED EXAMPLE 12.1

Predicting Solubility from Chemical Structure

Pentane (C_5H_{12}) and 1-butanol (C_4H_9OH) are organic liquids with similar molecular weights but substantially different solubility behavior. Which of the two would you expect to be more soluble in water? Explain.

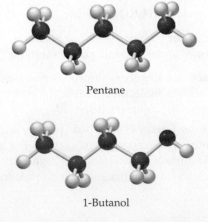

Pentane

1-Butanol

STRATEGY

Look at the two structures and decide on the kinds of intermolecular forces present between molecules in each case. If the intermolecular forces between the three types of interactions; solvent–solvent interactions, solvent–solute interactions, and solute–solute interactions are similar in magnitude, then a solution will likely form. The substance with intermolecular forces more like those in water will probably be more soluble in water.

SOLUTION

London dispersion forces are the intermolecular forces between two pentane molecules because pentane is nonpolar. In contrast, water is polar and forms hydrogen bonds. Pentane is unlikely to have strong intermolecular interactions with water. 1-Butanol, however, has an —OH group and can hydrogen bond to itself and to water. 1-Butanol forms 3 hydrogen bonds with water, resulting in strong solvent–solute interactions as shown. As a result, 1-butanol is more soluble in water.

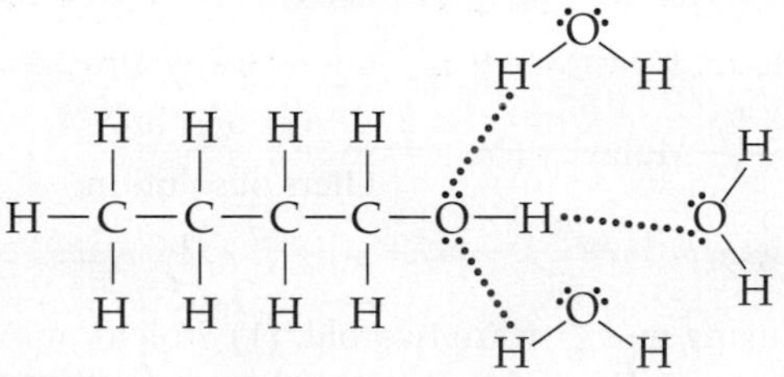

Conceptual PRACTICE 12.1 Arrange the following three compounds in order of their expected increasing solubility in pentane (C_5H_{12}): 1,5 pentanediol ($C_5H_{12}O_2$), KBr, toluene (C_7H_8, a constituent of gasoline).

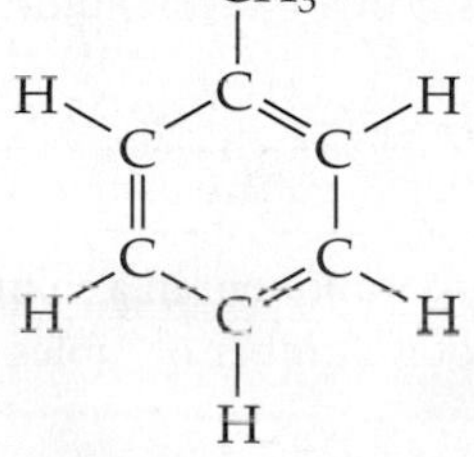

Toluene

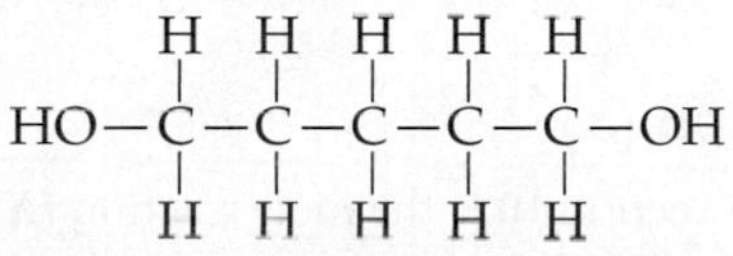

1,5 Pentanediol

Conceptual APPLY 12.2 Many people take vitamin supplements to promote health and the potential for overdose depends on whether a particular vitamin is fat soluble or water soluble. If a vitamin is soluble in nonpolar fatty tissues it will accumulate in the body, whereas water-soluble vitamins are readily excreted. Use the ball-and-stick models of vitamin C and vitamin E to determine which molecule is more fat soluble.

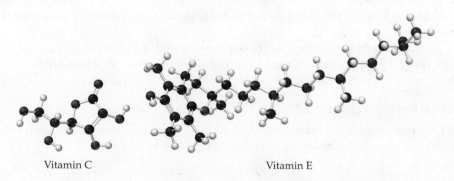

Vitamin C   Vitamin E

12.3 ▶ CONCENTRATION UNITS FOR SOLUTIONS

In daily life, it's often sufficient to describe a solution as either *dilute* or *concentrated*. In scien tific work, though, it's usually necessary to know the exact concentration of a solution—that is, t know the exact amount of solute dissolved in a certain amount of solvent. There are many way of expressing concentration, each of which has its own advantages and disadvantages. The four o the most common units for concentration are **molarity**, *mole fraction*, *mass percent*, and *molality*

Molarity (M)

The most common way of expressing concentration in a chemistry laboratory is to use molarity (M As discussed in Section 4.1, a solution's molarity is given by the number of moles of solute per lite of solution (mol/L, abbreviated M). If, for example, you dissolve 0.500 mol (20.0 g) of NaOH i enough water to give 1.000 L of solution, then the solution has a concentration of 0.500 M.

$$\text{Molarity (M)} = \frac{\text{Moles of solute}}{\text{Liters of solution}}$$

The advantages of using molarity are twofold: (1) Stoichiometry calculations are sim plified because numbers of moles are used rather than mass, and (2) amounts of solutio (and therefore of solute) are conveniently measured by volume (pouring) rather than by mas (weighing). As a result, **titrations** are particularly easy.

The disadvantages of using molarity are also twofold: (1) The exact solute concentratio depends on the temperature because the volume of a solution expands or contracts as th temperature changes, and (2) the exact amount of solvent in a given volume can't be deter mined unless the density of the solution is known.

Mole Fraction (X)

As discussed in Section 10.5, the mole fraction (X) of any component in a solution is given b the number of moles of the component divided by the total number of moles making up th solution (including solvent):

$$\text{Mole fraction } (X) = \frac{\text{Moles of component}}{\text{Total moles making up the solution}}$$

For example, a solution prepared by dissolving 1.00 mol (32.0 g) of methyl alcoho (CH_3OH) in 5.00 mol (90.0 g) of water has a methyl alcohol concentration $X = 1.00$ mol (1.00 mol + 5.00 mol) = 0.167. Note that mole fractions are dimensionless because the uni cancel.

Mole fractions are independent of temperature and are particularly useful for calcula tions involving gas mixtures. Except in special situations, mole fractions are not often use for liquid solutions because other units are generally more convenient.

Mass Percent (Mass %)

As the name suggests, the **mass percent (mass %)** of any component in a solution is the ma of that component divided by the total mass of the solution times 100%:

$$\text{Mass percent} = \frac{\text{Mass of component}}{\text{Total mass of solution}} \times 100\%$$

For example, a solution prepared by dissolving 10.0 g of glucose in 100.0 g of water has a glu cose concentration of 9.09 mass %:

$$\text{Mass \% glucose} = \frac{10.0 \text{ g}}{10.0 \text{ g} + 100.0 \text{ g}} \times 100\% = 9.09 \text{ mass \%}$$

Closely related to mass percent, and particularly useful for very dilute solutions, are the concentration units **parts per million (ppm)** and **parts per billion (ppb)**:

$$\text{Parts per million (ppm)} = \frac{\text{Mass of component}}{\text{Total mass of solution}} \times 10^6$$

$$\text{Parts per billion (ppb)} = \frac{\text{Mass of component}}{\text{Total mass of solution}} \times 10^9$$

A concentration of 1 ppm for a substance means that each kilogram (1 million mg) of solution contains 1 mg of solute. For dilute aqueous solutions near room temperature, where 1 kg has a volume of 1 L, 1 ppm also means that each liter of solution contains 1 mg of solute. In the same way, a concentration of 1 ppb means that each liter of an aqueous solution contains 0.001 mg of solute.

Values in ppm and ppb are frequently used for expressing the concentrations of trace amounts of impurities in air or water. Thus, you might express the maximum allowable concentration of lead in drinking water as 15 ppb, or about 1 g per 67,000 L.

The advantage of using mass percent (or ppm) for expressing concentration is that the values are independent of temperature because masses don't change when substances are heated or cooled. The disadvantage of using mass percent is that it is generally less convenient when working with liquid solutions to measure amounts by mass rather than by volume. Furthermore, the density of a solution must be known before a concentration in mass percent can be converted into molarity. Worked Example 12.5 shows how to make the conversion.

▲ Dissolving a quarter teaspoon of table sugar in this backyard swimming pool would give a sugar concentration of about 1 ppb.

─● WORKED EXAMPLE 12.2

Mass Percent Concentration

Assume that you have a 5.75 mass % solution of LiCl in water. What mass of solution in grams contains 1.60 g of LiCl?

IDENTIFY

Known	Unknown
Mass percent (5.75%)	Mass of solution (g)
Mass of solute (1.60 g LiCl)	

STRATEGY

Describing this concentration as 5.75 mass % means that 100.0 g of aqueous solution contains 5.75 g of LiCl (and 94.25 g of H_2O), a relationship that can be used as a conversion factor.

SOLUTION

$$\text{Mass of soln} = 1.60 \text{ g LiCl} \times \frac{100 \text{ g soln}}{5.75 \text{ g LiCl}} = 27.8 \text{ g soln}$$

▸**PRACTICE 12.3** What is the mass percent concentration of a saline solution prepared by dissolving 1.00 mol of NaCl in 1.00 L of water?

▸**APPLY 12.4** Calculate the amount of water (in grams) that must be added to 15.0 g of sucrose ($C_{12}H_{22}O_{11}$) to make a solution with a concentration of 7.50 mass %.

─● WORKED EXAMPLE 12.3

Using Parts Per Million (ppm) as a Concentration Unit

Nitrate (NO_3^-) enters the water supply from fertilizer runoff or leaking septic tanks. The legal limit in drinking water is 10.0 ppm because infants that drink water exceeding this level can become seriously ill. What mass of nitrate (mg) is present in 1.5 L of water with a concentration of 10.0 ppm? (Assume the density of water is 1.0 g/mL.)

continued on next page

IDENTIFY

Known	Unknown
Concentration of nitrate (10.0 ppm)	Mass of nitrate (mg)
Volume of water (1.5 L)	

STRATEGY

Use density to convert from volume of water to mass of water (total mass of solution). Substitute known values into the equation for parts per million and solve for the mass of NO_3^-.

SOLUTION

Find the mass of water: $1.5 \text{ L} \times \dfrac{1 \text{ mL}}{1 \times 10^{-3} \text{ L}} \times \dfrac{1.0 \text{ g}}{1 \text{ mL}} = 1500 \text{ g}$

Use the equation for ppm: $10.0 \text{ ppm} = \dfrac{\text{Mass of } NO_3^-}{1500 \text{ g}} \times 10^6$

Mass of $NO_3^- = 0.015 \text{ g} = 15 \text{ mg}$

CHECK

The mass of nitrate is very small, which is reasonable because ppm is used to express trace amounts of solutes. Since the multiplication factor for ppm is 10^6, we would expect the amount of solute to be about 10^6 times less than the total mass of the solution.

▶ **PRACTICE 12.5** A 50.0 mL sample of drinking water was found to contain 1.25 μg of arsenic. Calculate the concentration in units of ppb and determine whether the sample exceeds the legal limit of 10.0 ppb.

▶ **APPLY 12.6** The legal limit for human exposure to carbon monoxide in the workplace is 35 ppm. Assuming that the density of air is 1.3 g/L, how many grams of carbon monoxide are in 1.0 L of air at the maximum allowable concentration?

Molality (*m*)

The **molality** (*m*) of a solution is defined as the number of moles of solute per kilogram of solvent (mol/kg):

$$\text{Molality } (m) = \dfrac{\text{Moles of solute}}{\text{Mass of solvent (kg)}}$$

To prepare a 1.000 *m* solution of KBr in water, for example, you might dissolve 1.000 mol of KBr (119.0 g) in 1.000 kg (1000 mL) of water. You can't say for sure what the final volume of the solution will be, although it will probably be a bit larger than 1000 mL. Although the names sound similar, note the differences between molarity and molality. Molarity is the number of moles of solute per *volume* (liter) of *solution*, whereas molality is the number of moles of solute per *mass* (kilogram) of *solvent*.

The main advantage of using molality is that it is temperature-independent because masses don't change when substances are heated or cooled. Thus, it is well suited for calculating certain properties of solutions that we'll discuss later in this chapter. The disadvantages of using molality are that amounts of solution must be measured by mass rather than by volume and that the density of the solution must be known to convert molality into molarity (see Worked Example 12.5).

A summary of the four most common units for expressing concentration, together with a comparison of their relative advantages and disadvantages, is given in **TABLE 12.3**.

TABLE 12.3 A Comparison of Various Concentration Units

Name	Units	Advantages	Disadvantages
Molarity (M)	$\dfrac{\text{mol solute}}{\text{L solution}}$	Useful in stoichiometry; by volume	Temperature-dependent; must know density to find solvent mass
Mole fraction (X)	none	Temperature-independent; useful in special applications	Measure by mass; must know density to convert to molarity
Mass %	%	Temperature-independent; useful for small amounts	Measure by mass; must know density to convert to molarity
Molality (m)	$\dfrac{\text{mol solute}}{\text{kg solvent}}$	Temperature-independent; useful in special applications	Measure by mass; must know density to convert to molarity

─• WORKED EXAMPLE 12.4

Calculating the Molality of a Solution

What is the molality of a solution made by dissolving 1.45 g of table sugar (sucrose, $C_{12}H_{22}O_{11}$) in 30.0 mL of water? The molar mass of sucrose is 342.3 g/mol.

IDENTIFY

Known	Unknown
Mass of sucrose (1.45 g)	Molality (m)
Volume of water (30.0 mL)	
Molar mass of sucrose (342.3 g/mol)	

STRATEGY

Molality is the number of moles of solute per kilogram of solvent. Thus, we need to find how many moles are in 1.45 g of sucrose and how many kilograms are in 30.0 mL of water.

SOLUTION

The number of moles of sucrose is

$$1.45 \text{ g sucrose} \times \frac{1 \text{ mol sucrose}}{342.3 \text{ g sucrose}} = 4.24 \times 10^{-3} \text{ mol sucrose}$$

Since the density of water is 1.00 g/mL, 30.0 mL of water has a mass of 30.0 g, or 0.0300 kg. Thus, the molality of the solution is

$$\text{Molality} = \frac{4.24 \times 10^{-3} \text{ mol}}{0.0300 \text{ kg}} = 0.141 \, m$$

▸ **PRACTICE 12.7** What mass in grams of a 0.500 m solution of sodium acetate, CH_3CO_2Na, in water would you use to obtain 0.150 mol of sodium acetate?

▸ **APPLY 12.8** What is the molality of a solution prepared by dissolving 0.385 g of cholesterol, $C_{27}H_{46}O$, in 40.0 g of chloroform, $CHCl_3$? What is the mole fraction of cholesterol in the solution?

─• WORKED EXAMPLE 12.5

Using Density to Convert Molarity to Other Measures of Concentration

A 0.750 M solution of H_2SO_4 in water has a density of 1.046 g/mL at 20 °C. What is the concentration of this solution in (a) mole fraction, (b) mass percent, and (c) molality? The molar mass of H_2SO_4 is 98.1 g/mol.

IDENTIFY

Known	Unknown
Concentration of H_2SO_4 solution (0.750M)	Mole fraction, mass percent, molality
Density of H_2SO_4 solution (1.046 g/mL)	
Molar mass of H_2SO_4 (98.1 g/mol)	

continued on next page

STRATEGY

Write the expressions for both the beginning and ending concentrations labeled with units. This will help you see how to use the known quantities of density and molar mass to change from the starting unit into the ending unit.

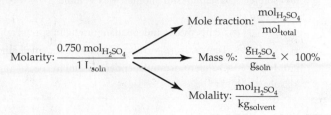

Molarity: $\dfrac{0.750 \text{ mol}_{H_2SO_4}}{1 \text{ L}_{soln}}$

Mole fraction: $\dfrac{\text{mol}_{H_2SO_4}}{\text{mol}_{total}}$

Mass %: $\dfrac{g_{H_2SO_4}}{g_{soln}} \times 100\%$

Molality: $\dfrac{\text{mol}_{H_2SO_4}}{kg_{solvent}}$

SOLUTION

(a) Mole fraction: To calculate mole fraction, the 1.00 L of solution needs to be converted to total number of moles making up the solution (mol H_2SO_4 + mol H_2O).
The number of moles of H_2SO_4 in 1.00 L is 0.750, because the molarity is 0.750 M.
To find the number of moles of H_2O, first find the total mass of the solution using the density.

$$\text{Mass of 1.00 L soln} = 1.00 \text{ L} \times \frac{1 \text{ mL}}{1 \times 10^{-3} \text{ L}} \times \frac{1.046 \text{ g}}{1 \text{ mL}} = 1046 \text{ g}$$

The total mass of solution is the mass of H_2SO_4 + the mass of H_2O. The mass of H_2SO_4 is found by converting from moles to grams:

$$\text{Mass of } H_2SO_4 \text{ in 1.00 L soln} = 0.750 \text{ mol} \times \frac{98.1 \text{ g}}{\text{mol}} = 73.6 \text{ g}$$

The mass of water is: 1046 g solution − 73.6 g H_2SO_4 = 972.4 g H_2O
The number of moles of water is found by converting mass into moles.

$$972.4 \text{ g } H_2O \times \frac{1 \text{ mol } H_2O}{18.0 \text{ g } H_2O} = 54.0 \text{ mol } H_2O$$

Thus, the mole fraction of H_2SO_4 is

$$X_{H_2SO_4} = \frac{0.750 \text{ mol } H_2SO_4}{0.750 \text{ mol } H_2SO_4 + 54.0 \text{ mol } H_2O} = 0.0137$$

(b) Mass percent: To calculate mass percent, convert 0.750 mol H_2SO_4 into grams using molar mass and convert 1.00 L of solution into grams using density. Both of these quantities were calculated in part (a). The mass % of H_2SO_4 is

$$\text{Mass \% of } H_2SO_4 = \frac{73.6 \text{ g } H_2SO_4}{1046 \text{ g solution}} \times 100\% = 7.04\%$$

(c) Molality: To calculate molality, convert 1.00 L of solution to kg of solvent. The mass of water was found in part **(a)** to be 972.4 g or 0.9724 kg.

$$\text{Molality of } H_2SO_4 = \frac{0.750 \text{ mol } H_2SO_4}{0.972 \text{ kg } H_2O} = 0.772 \text{ } m$$

▶**PRACTICE 12.9**   The density at 20 °C of a 0.500 M solution of acetic acid in water is 1.0042 g/mL. What is the mol fraction, mass %, and molality of the solution? The molar mass of acetic acid, CH_3CO_2H, is 60.05 g/mol.

▶**APPLY 12.10**   The density at 20 °C of a 0.258 m solution of glucose in water is 1.0173 g/mL, and the molar mass of glucose is 180.2 g/mol. What is the molarity of the solution?

12.4 ▶ SOME FACTORS THAT AFFECT SOLUBILITY

If you take solid NaCl and add it to water, dissolution occurs rapidly at first but then slow down as more and more NaCl is added. Eventually the dissolution stops because a dynam equilibrium is reached where the number of Na^+ and Cl^- ions leaving a crystal to go int solution is equal to the number of ions returning from the solution to the crystal. At th

| A supersaturated solution of sodium acetate in water. | When a tiny seed crystal is added, larger crystals begin to grow and precipitate from the solution until equilibrium is reached. |

oint, the maximum possible amount of NaCl has dissolved and the solution is said to be **aturated** in that solute.

$$\text{Solute} + \text{Solvent} \underset{\text{crystallize}}{\overset{\text{dissolve}}{\rightleftharpoons}} \text{Solution}$$

Note that this definition requires a saturated solution to be at *equilibrium* with undis-
olved solid at a given temperature. Some substances, however, can form what are called
upersaturated solutions, which contain a greater-than-equilibrium amount of solute. For
xample, when a saturated solution of sodium acetate is prepared at high temperature and
en cooled slowly, a supersaturated solution results, as shown in **FIGURE 12.5**. Such a solu-
on is unstable, however, and precipitation occurs when a tiny seed crystal of sodium acetate
added to initiate crystallization.

ffect of Temperature on Solubility

he solubility of a substance—that is, the amount of the substance per unit volume of solvent
eeded to form a saturated solution at a given temperature—is a physical property character-
tic of that substance. Different substances can have greatly different solubilities, as shown
 FIGURE 12.6. Sodium chloride, for instance, has a solubility of 35.9 g/100 mL of water
 20 °C, and sodium nitrate has a solubility of 87.3 g/100 mL of water at 20 °C. Sometimes,
articularly when two liquids are involved, the solvent and solute are **miscible**, meaning
at they are mutually soluble in all proportions. A solution of ethyl alcohol and water is an
xample.

Solubilities are temperature-dependent, so the temperature at which a specific mea-
urement is made must be reported. As Figure 12.6 shows, there is no obvious correlation
etween structure and solubility or between solubility and temperature. The solubilities of most
olecular and ionic solids increase with increasing temperature, but the solubilities of some
NaCl) are almost unchanged and the solubilities of others [$Ce_2(SO_4)_3$] actually decrease.

▶ **FIGURE 12.6**

Solubilities of some common solids in water as a function of temperature.

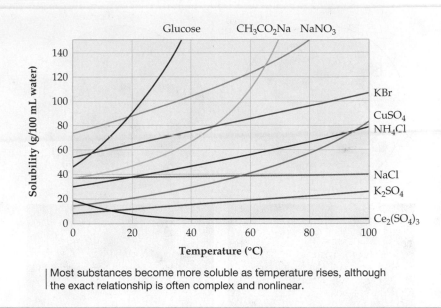

Most substances become more soluble as temperature rises, although the exact relationship is often complex and nonlinear.

The effect of temperature on the solubility of gases in water is more predictable than its effect on the solubility of solids: Gases become less soluble in water as the temperature increase (**FIGURE 12.7**). One consequence of this decreased solubility is that carbonated drinks bubbl continuously as they warm up to room temperature after being refrigerated. Soon, they lose s much dissolved CO_2 that they go flat. A much more important consequence is the damage t aquatic life that can result from the decrease in concentration of dissolved oxygen in lakes an rivers when hot water is discharged from industrial plants, an effect known as thermal pollutior

Effect of Pressure on Solubility

Pressure has practically no effect on the solubility of liquids and solids but has a profound e fect on the solubility of gases. According to **Henry's law**, the solubility of a gas in a liquid at given temperature is directly proportional to the partial pressure of the gas over the solution

Henry's law   Solubility $= k \cdot P$

The constant k in this expression is characteristic of a specific gas, and P is the parti pressure of the gas over the solution. Doubling the partial pressure doubles the solubility, tr pling the partial pressure triples the solubility, and so forth. Henry's law constants are usuall

▶ **FIGURE 12.7**

Solubilities of some gases in water as a function of temperature.

Figure It Out

Is the amount of dissolved oxygen higher in water at 10 °C or at 30 °C?

Answer: 10 °C

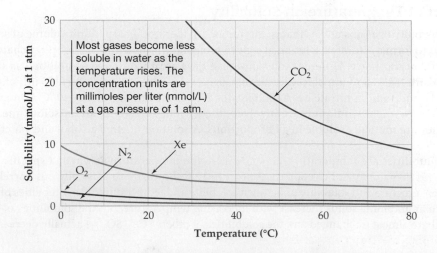

Most gases become less soluble in water as the temperature rises. The concentration units are millimoles per liter (mmol/L) at a gas pressure of 1 atm.

Equilibrium	Pressure increase	Equilibrium restored
At a given pressure, an equilibrium exists in which equal numbers of gas particles enter and leave the solution.	When **pressure is increased** by pushing on the piston, more gas particles are temporarily forced into solution than are able to leave.	Solubility therefore increases until a new equilibrium is reached.

◀ **FIGURE 12.8**

A molecular view of Henry's law.

Figure It Out

After pressure is increased and equilibrium is restored, how does the concentration of gas in solution compare to its original level?

Answer: The concentration is higher since the pressure is higher.

iven in units of mol/(L·atm), and measurements are reported at 25 °C. Note that when the as partial pressure P is 1 atm, the Henry's law constant k is numerically equal to the solubility of the gas in moles per liter.

Perhaps the most common example of Henry's law behavior occurs when you open a can f soda or other carbonated drink. Bubbles of gas immediately come fizzing out of solution ecause the pressure of CO_2 in the can drops and CO_2 suddenly becomes less soluble. A more erious example of Henry's law behavior occurs when a deep-sea diver surfaces too quickly nd develops a painful and life-threatening condition called *decompression sickness (DCS)* or he *bends*. The bends occur because large amounts of nitrogen dissolve in the blood at high nderwater pressures. When the diver ascends and pressure decreases too rapidly, bubbles f nitrogen form in the blood, blocking capillaries and inhibiting blood flow. The condition an be prevented by using an oxygen/helium mixture for breathing rather than air (oxygen/ itrogen), because helium has a much lower solubility in blood than nitrogen.

On a molecular level, the increase in gas solubility with increasing pressure occurs because f a change in the position of the equilibrium between dissolved and undissolved gas. At a given ressure, an equilibrium is established in which equal numbers of gas particles enter and leave he solution. When the pressure is increased, however, more particles are forced into solution han leave it, so gas solubility increases until a new equilibrium is established (**FIGURE 12.8**).

▲ Divers who ascend too quickly can develop the bends, a condition caused by the formation of nitrogen bubbles in the blood. Treatment involves placing the diver into a decompression chamber like this one and slowly changing the pressure.

─● WORKED EXAMPLE 12.6

Using Henry's Law to Calculate Gas Solubility

The Henry's law constant of methyl bromide (CH_3Br), a gas used as a soil fumigating agent, is $k = 0.159$ mol/(L·atm) at 25 °C. What is the solubility in mol/L of methyl bromide in water at 25 °C and a partial pressure of 125 mm Hg?

IDENTIFY

Known	Unknown
$k = 0.159$ mol/(L·atm)	Solubility (mol/L)
T = 25 °C	
Partial pressure of CH_3Br = 125 mm Hg	

STRATEGY

According to Henry's law, solubility $= k \cdot P$.

continued on next page

SOLUTION

$$k = 0.159 \; \text{mol}/(\text{L} \cdot \text{atm})$$

$$P = 125 \; \text{mm Hg} \times \frac{1 \; \text{atm}}{760 \; \text{mm Hg}} = 0.164 \; \text{atm}$$

$$\text{Solubility} = k \cdot P = 0.159 \; \frac{\text{mol}}{\text{L} \cdot \text{atm}} \times 0.164 \; \text{atm} = 0.0261 \; \text{M}$$

The solubility of methyl bromide in water at a partial pressure of 125 mm Hg is 0.0261 M.

▶ **PRACTICE 12.11** The solubility of CO_2 in water is 3.2×10^{-2} M at 25 °C and 1 atm pressure. What is the Henry's law constant for CO_2 in $\text{mol}/(\text{L} \cdot \text{atm})$?

▶ **APPLY 12.12** Use the Henry's law constant you calculated in Problem 12.11 to find the concentration of CO_2 in:

(a) A can of soda under a CO_2 pressure of 2.5 atm at 25 °C

(b) A can of soda open to the atmosphere at 25 °C (CO_2 is approximately 0.04% by volume in the atmosphere.)

12.5 ▶ PHYSICAL BEHAVIOR OF SOLUTIONS: COLLIGATIVE PROPERTIES

The behavior of solutions is qualitatively similar to that of pure solvents but is quantitatively different. Pure water boils at 100.0 °C and freezes at 0.0 °C, for instance, but a 1.00 *m* (molal) solution of NaCl in water boils at 101.0 °C and freezes at −3.7 °C.

The higher boiling point and lower freezing point observed for a solution compared to a pure solvent are examples of **colligative properties**, which depend only on the amount of dissolved solute but not on the solute's chemical identity. The word *colligative* means "bound together in a collection" and is used because a collection of solute particles is responsible for the observed effects. Other colligative properties are a lower vapor pressure for a solution compared with the pure solvent and *osmosis*, the migration of solvent and other small molecules through a semipermeable membrane.

▶ **Colligative properties** In comparing the properties of a pure solvent with those of a solution:

- The vapor pressure of the solution is lower.
- The boiling point of the solution is higher.
- The freezing (or melting) point of the solution is lower.
- The solution gives rise to *osmosis*, the migration of solvent molecules through a semipermeable membrane.

We'll look at each of the four colligative properties in more detail in Sections 12.6–12.8

12.6 ▶ VAPOR-PRESSURE LOWERING OF SOLUTIONS: RAOULT'S LAW

Recall from Section 11.3 that a liquid in a closed container is in equilibrium with its vapor and that the amount of pressure exerted by the vapor is called the **vapor pressure**. When you compare the vapor pressure of a pure solvent with that of a solution at the same temperature however, you find that the two values are different. If the solute is nonvolatile and has no appreciable vapor pressure of its own, as occurs when a solid is dissolved, then the vapor pressure of the solution is always lower than that of the pure solvent. If the solute *is* volatile and has a significant vapor pressure of its own, as occurs in a mixture of two liquids, then the vapor pressure of the mixture is intermediate between the vapor pressures of the two pure liquids.

REMEMBER...

At equilibrium, the rate of evaporation from the liquid to the vapor is equal to the rate of condensation from the vapor back to the liquid. The resulting **vapor pressure** is the partial pressure of the gas in the equilibrium. (Section 11.3)

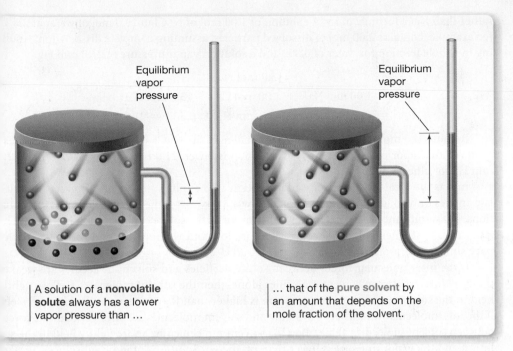

Equilibrium
vapor
pressure

Equilibrium
vapor
pressure

| A solution of a **nonvolatile solute** always has a lower vapor pressure than … | … that of the **pure solvent** by an amount that depends on the mole fraction of the solvent. |

Solutions with a Nonvolatile Solute

It's easy to demonstrate with **manometers** that a solution of a nonvolatile solute has a lower vapor pressure than the pure solvent (**FIGURE 12.9**). Alternatively, you can show the same effect by comparing the evaporation rate of a pure solvent with the evaporation rate of a solution. A solution always evaporates more slowly than a pure solvent does because its vapor pressure is lower and its molecules therefore escape less readily.

According to **Raoult's law**, the vapor pressure of a solution containing a nonvolatile solute is equal to the vapor pressure of the pure solvent times the mole fraction of the solvent. That is,

> **REMEMBER…**
>
> A **manometer** is a mercury-filled U-tube used for reading pressure. One end of the tube is connected to the sample container and the other end is open to the atmosphere. (Section 10.1)

Raoult's law
$$P_{soln} = P_{solv} \times X_{solv}$$
where P_{soln} is the vapor pressure of the solution, P_{solv} is the vapor pressure of pure solvent at the same temperature, and X_{solv} is the mole fraction of the solvent in the solution.

What is the vapor pressure of a solution of 1.00 mol of glucose in 15.0 mol of water at 25 °C? The vapor pressure of pure water at 25 °C is 23.76 mm Hg so we expect the vapor pressure of the solution to be lower. The mole fraction of water in the solution is:

$$X_{solv} = \frac{15.0 \text{ mol H}_2\text{O}}{15.0 \text{ mol H}_2\text{O} + 1.0 \text{ mol glucose}} = 0.938$$

Thus, Raoult's law predicts a vapor pressure for the solution of 22.3 mm Hg, which corresponds to a vapor-pressure lowering, ΔP_{soln}, of 1.5 mm Hg:

$$P_{soln} = P_{solv} \times X_{solv} = 23.76 \text{ mm Hg} \times 0.938 = 22.3 \text{ mm Hg}$$
$$\Delta P_{soln} = P_{solv} - P_{soln} = 23.76 \text{ mm Hg} - 22.3 \text{ mm Hg} = 1.5 \text{ mm Hg}$$

Alternatively, the amount of vapor-pressure lowering can be calculated directly by multiplying the mole fraction of the *solute* times the vapor pressure of the pure solvent. That is,

$$\Delta P_{soln} = P_{solv} \times X_{solute} = 23.76 \text{ mm Hg} \times \frac{1.00 \text{ mol glucose}}{1.00 \text{ mol glucose} + 15.0 \text{ mol H}_2\text{O}}$$
$$= 1.49 \text{ mm Hg}$$

If an ionic substance such as NaCl is the solute rather than a molecular substance, we have to calculate mole fractions based on the total concentration of solute *particles* (ions) rather than NaCl formula units. A solution of 1.00 mol of NaCl in 15.0 mol of water at 25 °C, for example, contains 2.00 mol of dissolved particles, assuming complete dissociation, resulting in a mole fraction for water of 0.882 and a solution vapor pressure of 21.0 mm Hg.

$$X_{water} = \frac{15.0 \text{ mol H}_2\text{O}}{1.00 \text{ mol Na}^+ + 1.00 \text{ mol Cl}^- + 15.0 \text{ mol H}_2\text{O}} = 0.882$$

$$P_{soln} = P_{solv} \times X_{solv} = 23.76 \text{ mm Hg} \times 0.882 = 21.0 \text{ mm Hg}$$

Because the mole fraction of water is smaller in the NaCl solution than in the glucose solution, the vapor pressure of the NaCl solution is lower: 21.0 mm Hg for NaCl versus 22. mm Hg for glucose at 25 °C.

Just as the ideal gas law discussed in Section 10.3 applies only to "ideal" gases, Raoult's law applies only to ideal solutions. Raoult's law approximates the behavior of most real solutions, but significant deviations from ideality occur as the solute concentration increases. The law works best when solute concentrations are low and when solute and solvent particles have similar intermolecular forces.

If the intermolecular forces between solute particles and solvent molecules are weaker than the forces between solvent molecules alone, then the solvent molecules are less tightly held in the solution and the vapor pressure is higher than Raoult's law predicts. Conversely, if the intermolecular forces between solute and solvent molecules are stronger than the forces between solvent molecules alone, then the solvent molecules are more tightly held in the solution and the vapor pressure is lower than predicted. Solutions of ionic substances, in particular, often have a vapor pressure significantly lower than predicted, because the ion–dipole forces between dissolved ions and polar water molecules are so strong.

A further complication is that ionic substances rarely dissociate completely, so a solution of an ionic compound usually contains fewer particles than the formula of the compound would suggest. The actual extent of dissociation can be expressed as a **van't Hoff factor** (i):

van't Hoff factor $i = \dfrac{\text{Moles of particles in solution}}{\text{Moles of solute dissolved}}$

Rearranging this equation shows that the number of moles of particles dissolved in a solution is equal to number of moles of solute dissolved times the van't Hoff factor.

$$\text{Moles of particles in solution} = i \times \text{Moles of solute dissolved}$$

For any nonelectrolyte such as sucrose ($C_{12}H_{22}O_{11}$), we can always assume $i = 1$. To take solution of an electrolyte such as NaCl as an example, the experimentally determined van't Hoff factor for 0.05 m NaCl is 1.9, meaning that each mole of NaCl gives only 1.9 mol of particles rather than the 2.0 mol expected for complete dissociation. Of the 1.9 mol of particles, 0.9 mol is Cl$^-$, 0.9 mol is Na$^+$, and 0.1 mol is undissociated NaCl. Thus, NaCl is only $(0.9/1.0) \times 100\% = 90\%$ dissociated in a 0.05 m solution and the amount of vapor pressure lowering is less than expected.

What accounts for the lowering of the vapor pressure when a nonvolatile solute is dissolved in a solvent? As we've noted on many prior occasions, a physical process such as the vaporization of a liquid to a gas is accompanied by a free-energy change $\Delta G_{vap} = \Delta H_{vap} - T\Delta S_{vap}$. The more negative the value of ΔG_{vap}, the more favored the vaporization process. Thus, if we want to compare the ease of vaporization of a pure solvent with that of the solvent in a solution, we have to compare the signs and relative magnitudes of the ΔH_{vap} and ΔS_{vap} terms in the two cases.

The vaporization of a liquid to a gas is disfavored by enthalpy (positive ΔH_{vap}) because energy is required to overcome intermolecular attractions in the liquid. At the same time, however, vaporization is favored by entropy (positive ΔS_{vap}) because randomness increases when molecules go from a liquid state to a gaseous state.

Comparing a pure solvent with a solution, the *enthalpies* of vaporization for the pure solvent and solvent in a solution are similar because similar intermolecular forces must be

Because the entropy of the solvent in a solution is higher than that of pure solvent to begin with, ΔS_{vap} **is smaller for the solution** than for the pure solvent.

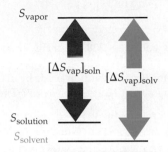

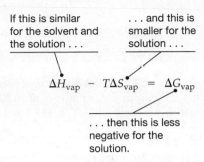

If this is similar for the solvent and the solution . . .

. . . and this is smaller for the solution . . .

$$\Delta H_{vap} - T\Delta S_{vap} = \Delta G_{vap}$$

. . . then this is less negative for the solution.

As a result, vaporization of the solvent from the solution is less favored (less negative ΔG_{vap}), and the vapor pressure of the solution is lower.

◀ **FIGURE 12.10**

Vapor-pressure lowering. The lower vapor pressure of a solution relative to that of a pure solvent is due to the difference in their entropies of vaporization, ΔS_{vap}.

> **Figure It Out**
>
> Which has a larger ΔS_{vap}, a pure solvent or a solution? How does a larger value of ΔS_{vap} affect the value of ΔG_{vap}?
>
> **Answer:** ΔS_{vap} is larger for the pure solvent. A larger value of ΔS_{vap} makes ΔG_{vap} more negative, meaning the process is more product favored.

vercome in both cases for solvent molecules to escape from the liquid. The *entropies* of va-orization for a pure solvent and a solvent in a solution are *not* similar, however. Because a olvent in a solution has more molecular randomness and higher entropy than a pure solvent oes, the entropy *change* on going from liquid to vapor is smaller for the solvent in a solution han for the pure solvent. Subtracting a smaller $T\Delta S_{vap}$ from ΔH_{vap} thus results in a larger less negative) ΔG_{vap} for the solution. As a result, vaporization is less favored for the solution nd the vapor pressure of the solution at equilibrium is lower (**FIGURE 12.10**).

―● WORKED EXAMPLE 12.7

Calculating the Vapor Pressure of an Electrolyte Solution

What is the vapor pressure in mm Hg of a solution made by dissolving 18.30 g of NaCl in 500.0 g of H_2O at 70 °C, assuming a van't Hoff factor of 1.9? The vapor pressure of pure water at 70 °C is 233.7 mm Hg.

IDENTIFY

Known	Unknown
Mass of solute (18.3 g NaCl)	P_{soln}
Mass of solvent (500.0 g H_2O)	
$i = 1.9$	
$P_{H_2O} = 233.7$ mm Hg	

STRATEGY

According to Raoult's law, the vapor pressure of the solution (P_{soln}) equals the vapor pressure of pure solvent (P_{H_2O}) times the mole fraction of the solvent (X_{solv}) in the solution. Thus, we have to find the numbers of moles of solvent and solute particles and then calculate the mole fraction of solvent. Electrolytes dissociate in solution and the van't Hoff factor is used to find the number of solute particles.

SOLUTION

First, use molar mass to calculate the number of moles of NaCl and H_2O.

$$\text{Moles of NaCl} = 18.3 \text{ g NaCl} \times \frac{1 \text{ mol NaCl}}{58.44 \text{ g NaCl}} = 0.313 \text{ mol NaCl}$$

$$\text{Moles of } H_2O = 500.0 \text{ g } H_2O \times \frac{1 \text{ mol } H_2O}{18.02 \text{ g } H_2O} = 27.75 \text{ mol } H_2O$$

continued on next page

Next, calculate the mole fraction of water in the solution. A van't Hoff factor of 1.9 means that the NaCl dissociates incompletely and gives only 1.9 particles per formula unit. Thus, the solution contains $1.9 \times 0.313 \text{ mol} = 0.59 \text{ mol}$ of dissolved particles and the mole fraction of water is

$$\text{Mole fraction of } H_2O = \frac{27.75 \text{ mol}}{0.59 \text{ mol} + 27.75 \text{ mol}} = 0.9792$$

From Raoult's law, the vapor pressure of the solution is

$$P_{\text{soln}} = P_{\text{solv}} \times X_{\text{solv}} = 233.7 \text{ mm Hg} \times 0.9792 = 228.8 \text{ mm Hg}$$

▶ **PRACTICE 12.13**   What is the vapor pressure in mm Hg of a solution made by dissolving 10.00 g of $CaCl_2$ in 100.0 g of H_2O at 70 °C, assuming a van't Hoff factor of 2.7? The vapor pressure of pure water at 70 °C is 233.7 mm Hg.

▶ **APPLY 12.14**   A solution made by dissolving 8.110 g of $MgCl_2$ in 100.0 g of water has a vapor pressure 224.7 mm Hg at 70 °C. The vapor pressure of pure water at 70 °C is 233.7 mm Hg. What is the van't Hoff factor for $MgCl_2$?

● WORKED EXAMPLE 12.8

Calculating Vapor-Pressure Lowering of a Solution

How many grams of sucrose must be added to 320 g of water to lower the vapor pressure by 1.5 mm Hg at 25 °C? The vapor pressure of water at 25 °C is 23.8 mm Hg, and the molar mass of sucrose is 342.3 g/mol.

IDENTIFY

Known	Unknown
Mass of water (320 g)	Mass of sucrose (g)
$P_{H_2O} = 23.8$ mm Hg	
Molar mass of sucrose (342.3 g/mol)	
Vapor pressure lowering (1.5 mm Hg)	

STRATEGY

According to Raoult's law, $P_{\text{soln}} = P_{\text{solv}} \times X_{\text{solv}}$, which can be rearranged to the solve for mole fraction of solvent, $X_{\text{solv}} = P_{\text{soln}}/P_{\text{solv}}$. The equation $X_{\text{solv}} = \dfrac{\text{mol } H_2O}{\text{mol sucrose} + \text{mol } H_2O}$ can be used to find the moles of sucrose.

SOLUTION

First, calculate the vapor pressure of the solution, P_{soln}, by subtracting the amount of vapor-pressure lowering from the vapor pressure of the pure solvent, P_{solv}:

$$P_{\text{soln}} = 23.8 \text{ mm Hg} - 1.5 \text{ mm Hg} = 22.3 \text{ mm Hg}$$

Now calculate the mole fraction of water, X_{solv}.

Since $\qquad P_{\text{soln}} = P_{\text{solv}} \times X_{\text{solv}}$

then $\qquad X_{\text{solv}} = \dfrac{P_{\text{soln}}}{P_{\text{solv}}} = \dfrac{22.3 \text{ mm Hg}}{23.8 \text{ mm Hg}} = 0.937$

This mole fraction of water is the number of moles of water divided by the total number of moles of sucrose plus water:

$$X_{\text{solv}} = \frac{\text{Moles of water}}{\text{Total moles}}$$

Since the number of moles of water is

$$\text{Moles of water} = 320 \text{ g} \times \frac{1 \text{ mol}}{18.0 \text{ g}} = 17.8 \text{ mol}$$

then the total number of moles of sucrose plus water is

$$\text{Total moles} = \frac{\text{moles of water}}{X_{\text{solv}}} = \frac{17.8 \text{ mol}}{0.937} = 19.0 \text{ mol}$$

Subtracting the number of moles of water from the total number of moles gives the number of moles of sucrose needed:

$$\text{Moles of sucrose} = 19.0 \text{ mol} - 17.8 \text{ mol} = 1.2 \text{ mol}$$

Converting moles into grams then gives the mass of sucrose needed.

$$\text{Grams of sucrose} = 1.2 \text{ mol} \times 342.3 \frac{\text{g}}{\text{mol}} = 4.1 \times 10^2 \text{ g}$$

▶ **PRACTICE 12.15** How many grams of NaBr must be added to 250 g of water to lower the vapor pressure by 1.30 mm Hg at 40 °C assuming complete dissociation? The vapor pressure of water at 40 °C is 55.3 mm Hg.

▶ **Conceptual APPLY 12.16** The following diagram shows a close-up view of part of the vapor-pressure curve for a pure solvent and a solution of a nonvolatile solute. Which curve represents the pure solvent, and which the solution?

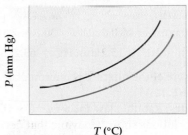

Solutions with a Volatile Solute

According to **Dalton's law of partial pressures**, the overall vapor pressure P_{total} of a mixture of two volatile liquids, A and B, is the sum of the vapor-pressure contributions of the individual components, P_A and P_B:

$$P_{\text{total}} = P_A + P_B$$

The individual vapor pressures P_A and P_B are calculated by Raoult's law. That is, the vapor pressure of A is equal to the mole fraction of A (X_A) times the vapor pressure of pure A $(P°_A)$, and the vapor pressure of B is equal to the mole fraction of B (X_B) times the vapor pressure of pure B $(P°_B)$. Thus, the total vapor pressure of the solution is

$$P_{\text{total}} = P_A + P_B = (P°_A \cdot X_A) + (P°_B \cdot X_B)$$

Take a mixture of the two similar organic liquids benzene (C_6H_6, bp $= 80.1$ °C) and toluene (C_7H_8, bp $= 110.6$ °C), as an example. Pure benzene has a vapor pressure $P° = 96.0$ mm Hg at 25 °C, and pure toluene has $P° = 30.3$ mm Hg at the same temperature. In a 1:1 molar mixture of the two, where the mole fraction of each is $X = 0.500$, the vapor pressure of the solution is 63.2 mm Hg:

REMEMBER...

Dalton's law of partial pressures states that the total pressure exerted by a mixture of gases in a container at constant volume and temperature is equal to the sum of the pressures of each individual gas in the container. (Section 10.5)

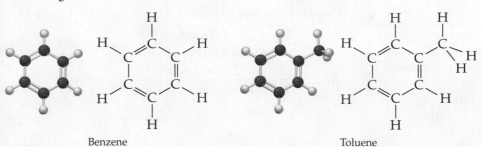

Benzene   Toluene

▶ **FIGURE 12.11**

Raoult's law for a mixture of volatile liquids. The vapor pressure of a solution of the two volatile liquids benzene and **toluene** at 25 °C is the sum of the two individual contributions, each calculated by Raoult's law.

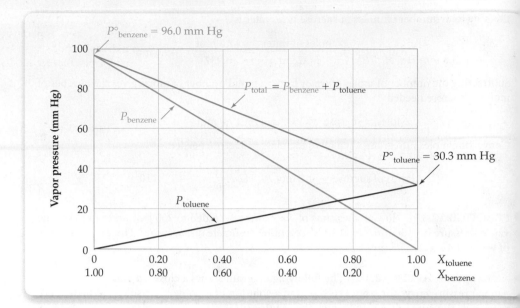

$$P_{total} = (P°_{benzene})(X_{benzene}) + (P°_{toluene})(X_{toluene})$$
$$= (96.0 \text{ mm Hg} \times 0.500) + (30.3 \text{ mm Hg} \times 0.500)$$
$$= 48.0 \text{ mm Hg} + 15.2 \text{ mm Hg} = 63.2 \text{ mm Hg}$$

Note that the vapor pressure of the mixture is intermediate between the vapor pressure of the two pure liquids (**FIGURE 12.11**).

As with nonvolatile solutes, Raoult's law for a mixture of volatile liquids applies onl to ideal solutions. Most real solutions show behaviors that deviate slightly from the ideal i either a positive or negative way, depending on the kinds and strengths of intermolecula forces present in the solution.

⟶● WORKED CONCEPTUAL EXAMPLE 12.9

Calculating the Vapor Pressure in a Solution with a Volatile Solute

What is the vapor pressure in mm Hg of a solution prepared by dissolving 25.0 g of ethyl alcohol (C_2H_5OH) in 100.0 g of water at 25 °C? The vapor pressure of pure water is 23.8 mm Hg, and the vapor pressure of ethyl alcohol is 61.2 mm Hg at 25 °C.

IDENTIFY

Known	Unknown
Mass of C_2H_5OH (25.0 g)	P_{soln}
Mass of H_2O (100.0 g)	
$P_{H_2O} = 23.8$ mm Hg	
$P_{C_2H_5OH} = 61.2$ mm Hg	

STRATEGY

The total vapor pressure of a solution with a volatile solute is calculated using the general equation $P_{total} = P_A + P_B = (P°_A \cdot X_A) + (P°_B \cdot X_B)$. The mole fraction of each liquid is calculated by converting from grams to moles and then dividing by the total number of moles.

SOLUTION

Convert from grams to moles of each liquid:

$$\text{Mol } C_2H_5OH = 25.0 \text{ g } C_2H_5OH \times \frac{1 \text{ mol } C_2H_5OH}{46.0 \text{ g } C_2H_5OH} = 0.543 \text{ mol C}$$

$$\text{Mol } H_2O = 100.0 \text{ g } H_2O \times \frac{1 \text{ mol } H_2O}{18.0 \text{ g } H_2O} = 5.56 \text{ mol } H_2O$$

Calculate the mole fraction of each liquid:

$$X_{C_2H_5OH} = \frac{0.543 \text{ mol } C_2H_5OH}{0.543 \text{ mol } C_2H_5OH + 5.56 \text{ mol } H_2O} = 0.089$$

$$X_{H_2O} = \frac{5.56 \text{ mol } H_2O}{0.543 \text{ mol } C_2H_5OH + 5.56 \text{ mol } H_2O} = 0.911$$

The vapor pressure of the solution is:

$$P_{soln} = P_{H_2O} + P_{C_2H_5OH}$$
$$= (23.8 \text{ mm Hg} \times 0.911) + (61.2 \text{ mm Hg} \times 0.089)$$
$$= 27.1 \text{ mm Hg}$$

CHECK

The answer is reasonable because the vapor pressure of the solution, 27.1 mm Hg, is in between the vapor pressure of each pure liquid. The vapor pressure of the solution is closer to the vapor pressure of water than ethyl alcohol, which makes sense because water has a much higher mole fraction than ethyl alcohol in the mixture.

▶ **PRACTICE 12.17** What is the vapor pressure of the solution if 25.0 g of water is dissolved in 100.0 g of ethyl alcohol at 25 °C?

▶ **Conceptual APPLY 12.18** The following diagram shows a close-up view of part of the vapor-pressure curves for two pure liquids and a mixture of the two. Which curves represent pure liquids, and which represents the mixture?

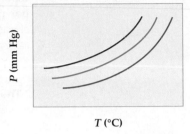

12.7 ▶ BOILING-POINT ELEVATION AND FREEZING-POINT DEPRESSION OF SOLUTIONS

We saw in Section 11.3 that the vapor pressure of a liquid rises with increasing temperature and that the liquid boils when its vapor pressure equals atmospheric pressure. Because a solution of a nonvolatile solute has a lower vapor pressure than a pure solvent has at a given temperature, the solution must be heated to a higher temperature to cause it to boil. Furthermore, the lower vapor pressure of the solution means that the liquid–vapor phase transition line on a **phase diagram** is always lower for the solution than for the pure solvent. As a result, the triple-point temperature T_t is lower for the solution, the solid–liquid phase transition line is shifted to a lower temperature for the solution, and the solution must be cooled to a lower temperature to freeze. **FIGURE 12.12** shows the situation.

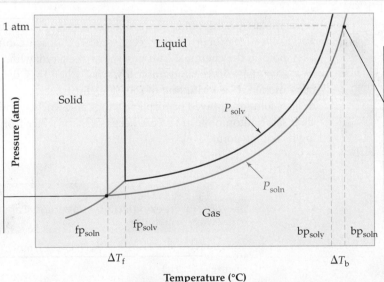

Because the liquid/vapor phase transition line is lower for the **solution** than for the **pure solvent**, the triple-point temperature T_t is lower and the solid/liquid phase transition line is shifted to a lower temperature. As a result, the freezing point of the solution is lower than that of the pure solvent by an amount ΔT_f.

Because the vapor pressure of the **solution** is lower than that of the **pure solvent** at a given temperature, the temperature at which the vapor pressure reaches atmospheric pressure is higher for the solution than for the solvent. Thus, the boiling point of the solution is higher by an amount ΔT_b.

▲ **FIGURE 12.12**

Phase diagrams for a pure solvent (red) and a solution of a nonvolatile solute (green).

Figure It Out

How does the vapor pressure of a solution compare to the vapor pressure of a pure solvent? How is the freezing point and boiling point of the solution affected by the change in vapor pressure?

Answer: The vapor pressure of a solution is lower than the pure solvent. This causes the freezing point of the solution to be lower than that of the pure solvent and the boiling point of the solution to be higher than that of the pure solvent.

The boiling point of a solution depends on the concentration of dissolved particles, just as vapor pressure does. Thus, a 1.00 m solution of glucose in water boils at approximately 100.51 °C at 1 atm pressure (0.51 °C above normal), but a 1.00 m solution of NaCl in water boils at approximately 101.02 °C (1.02 °C above normal) because there are twice as many particles (ions) dissolved in the NaCl solution as there are in the glucose solution. As with vapor-pressure lowering described in the previous section, the actual amount of boiling-point elevation observed for a solution of an ionic substance depends on the extent of dissociation, as given by the van't Hoff factor:

> **The change in boiling point for a solution**
>
> $$\Delta T_b = K_b \cdot m \cdot i$$
>
> where K_b is the **molal boiling-point-elevation constant (K_b)** characteristic of a given solvent, m is the molal concentration of the solute, and i is the van't Hoff factor.

The concentration must be expressed in molality—the number of moles of solute particles per kilogram of solvent—rather than molarity so that the solute concentration is independent of temperature. Molal boiling-point-elevation constants are given in **TABLE 12.4** for some common substances.

The freezing point of a solution depends on the concentration of solute particles, just as the boiling point does. For example, a 1.00 m solution of glucose in water freezes at −1.86 °C and a 1.00 m solution of NaCl in water freezes at approximately −3.72 °C.

> **The change in freezing point ΔT_f for a solution**
>
> $$\Delta T_f = -K_f \cdot m \cdot i$$
>
> where K_f is the **molal freezing-point-depression constant (K_f)** characteristic of a given solvent, m is the molal concentration of the solute, and i is the van't Hoff factor.

Some molal freezing-point-depression constants are also given in Table 12.4.

Colligative properties may seem somewhat obscure, but in fact they have many practical uses, both in the chemical laboratory and in everyday life. Motorists in winter, for instance, take advantage of freezing-point lowering when they drive on streets where the snow has been melted by a sprinkling of salt. The antifreeze added to automobile radiators and the de-icer solution sprayed on airplane wings also work by lowering the freezing point of water. That same automobile antifreeze keeps radiator water from boiling over in summer by raising its boiling point.

TABLE 12.4   Molal Boiling-Point-Elevation Constants (K_b) and Molal Freezing-Point-Depression Constants (K_f) for Some Common Substances

Substance	$K_b[(°C \cdot kg)/mol]$	$K_f[(°C \cdot kg)/mol]$
Benzene (C_6H_6)	2.64	5.07
Camphor ($C_{10}H_{16}O$)	5.95	37.8
Chloroform ($CHCl_3$)	3.63	4.70
Diethyl ether ($C_4H_{10}O$)	2.02	1.79
Ethyl alcohol (C_2H_6O)	1.22	1.99
Water (H_2O)	0.51	1.86

The fundamental cause of boiling-point elevation and freezing-point depression in solutions is the same as the cause of vapor-pressure lowering: the entropy difference between the pure solvent and the solvent in solution. Let's take boiling-point elevations first. We know that liquid and vapor phases are in equilibrium at the boiling point (T_b) and that the free-energy difference between the two phases (ΔG_{vap}) is therefore zero.

Since
$$\Delta G_{vap} = \Delta H_{vap} - T_b\Delta S_{vap} = 0$$

then
$$\Delta H_{vap} = T_b\Delta S_{vap} \quad \text{and} \quad T_b = \frac{\Delta H_{vap}}{\Delta S_{vap}}$$

In comparing the enthalpies of vaporization (ΔH_{vap}) for the pure solvent and for the solvent in a solution, the values are similar because similar intermolecular forces holding the solvent molecules must be overcome in both cases. In comparing the entropies of vaporization, however, the values are not similar. Because the solvent in solution has more molecular randomness than the pure solvent has, the entropy change between solution and vapor is smaller than the entropy change between pure solvent and vapor. But if ΔS_{vap} is smaller for the solution, then T_b must be correspondingly larger. In other words, the boiling point of the solution (T_b) is higher than that of the pure solvent (**FIGURE 12.13**).

A similar explanation accounts for freezing-point depression. Because liquid and solid phases are in equilibrium at the freezing point, the free-energy difference between the phases (ΔG_{fusion}) is zero:

Since
$$\Delta G_{fusion} = \Delta H_{fusion} - T_f\Delta S_{fusion} = 0$$

then
$$\Delta H_{fusion} = T_f\Delta S_{fusion} \quad \text{and} \quad T_f = \frac{\Delta H_{fusion}}{\Delta S_{fusion}}$$

In comparing the solvent in solution with pure solvent, the enthalpies of fusion (ΔH_{fusion}) are similar because similar intermolecular forces between solvent molecules are involved. The entropies of fusion (ΔS_{fusion}), however, are not similar. Because the solvent in solution has more molecular randomness than the pure solvent has, the entropy change between the solvent in the solution and the solid is larger than the entropy change between pure solvent and the solid. With ΔS_{fusion} larger for the solution, T_f must be correspondingly smaller, meaning that the freezing point of the solution (T_f) is lower than that of the pure solvent (**FIGURE 12.14**).

◀ **FIGURE 12.13**

Boiling-point elevation. The higher boiling point of a solution relative to that of a pure solvent is due to a difference in their entropies of vaporization, ΔS_{vap}.

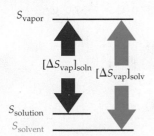

Because the solvent in a solution has a higher entropy to begin with, **ΔS_{vap} is smaller for the solution** than for the pure solvent.

S_{vapor}

$[\Delta S_{vap}]_{soln}$ $[\Delta S_{vap}]_{solv}$

$S_{solution}$
$S_{solvent}$

As a result, the boiling point of the solution T_b is higher than that of the pure solvent.

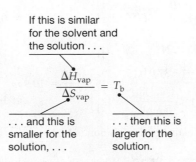

If this is similar for the solvent and the solution . . .

$$\frac{\Delta H_{vap}}{\Delta S_{vap}} = T_b$$

. . . and this is smaller for the solution, . . .

. . . then this is larger for the solution.

▶ **FIGURE 12.14**

Freezing-point lowering. The lower freezing point of a solution relative to that of a pure solvent is due to a difference in their entropies of fusion, ΔS_{fusion}.

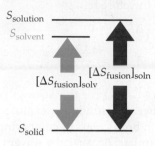

Because the solvent in a solution has a higher entropy level to begin with, ΔS_{fusion} **is larger for the solution** than for the pure solvent.

$S_{solution}$ _____
$S_{solvent}$ _____

$[\Delta S_{fusion}]_{solv}$ $[\Delta S_{fusion}]_{soln}$

S_{solid} _____

As a result, the freezing point of the solution T_f is lower than that of the pure solvent.

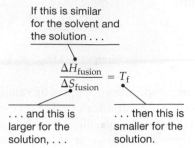

If this is similar for the solvent and the solution . . .

$$\frac{\Delta H_{fusion}}{\Delta S_{fusion}} = T_f$$

. . . and this is larger for the solution, ... then this is smaller for the solution.

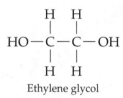

Ethylene glycol

WORKED EXAMPLE 12.10

Using Freezing-Point Depression to Calculate the Molality of a Solution

Ethylene glycol $(C_2H_6O_2)$ is commonly dissolved in water to make antifreeze for your car's engine. What is the molality of an aqueous solution of ethylene glycol if the freezing point of the solution at 1 atm pressure is $-37.0\,°C$? The molal freezing-point-depression constant for water is given in Table 12.4.

IDENTIFY

Known	Unknown
Freezing point $(-37.0\,°C)$	Molality (m)
$K_f = 1.86\ (°C \cdot kg)/mol$	

STRATEGY

Ethylene glycol is a nonelectrolyte, therefore $i = 1$. Rearrange the equation for molal freezing-point depression to solve for m:

$$\Delta T_f = -K_f \cdot m \cdot i \quad so \quad m = \frac{-\Delta T_f}{K_f \cdot i}$$

where $K_f = 1.86(°C \cdot kg)/mol$ and $\Delta T_f = 0.0\,°C - 37.0\,°C = -37.0\,°C$.

SOLUTION

$$m = \frac{-(-37.0\,°C)}{1.86\dfrac{(°C \cdot kg)}{mol}} = 19.9\frac{mol}{kg} = 19.9\ m$$

The molality of the solution is $19.9\ m$.

▶ **PRACTICE 12.19**   What is the normal boiling point in °C of an antifreeze solution prepared by dissolving 616.9 g of ethylene glycol $(C_2H_6O_2)$ in 500.0 g of water? The molal boiling-point-elevation constant for water is given in Table 12.4.

▶ **APPLY 12.20**   The following phase diagram shows a close-up view of the liquid–vapor phase transition boundaries for pure chloroform and a solution of a nonvolatile solute in chloroform.

(a) What is the approximate boiling point of pure chloroform?
(b) What is the approximate molal concentration of the nonvolatile solute? See Table 12.4 to find K_b for chloroform.

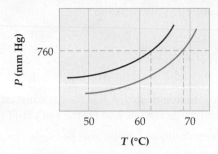

12.8 ▶ OSMOSIS AND OSMOTIC PRESSURE

Certain materials, including those that make up the membranes around living cells, are *semipermeable*. That is, they allow water or other smaller molecules to pass through, but they block the passage of larger solute molecules or solvated ions. When a solution and a pure solvent, or two solutions of different concentration, are separated by the right kind of semipermeable membrane, solvent molecules pass through the membrane in a process called **osmosis**. Although the passage of solvent through the membrane takes place in both directions, passage from the pure solvent side to the solution side is more favored and occurs at a greater rate. As a result, the amount of liquid on the pure solvent side of the membrane decreases, the amount of liquid on the solution side increases, and the concentration of the solution decreases.

Osmosis can be demonstrated with the experimental setup shown in **FIGURE 12.15**, in which a solution inside the bulb is separated from pure solvent in the beaker by a semipermeable membrane. Solvent passes through the membrane from the beaker to the bulb, causing the liquid level in the attached tube to rise. The increased weight of liquid in the tube creates an increased pressure that pushes solvent back through the membrane until the rates of forward and reverse passage become equal and the liquid level stops rising.

The amount of pressure necessary to achieve this equilibrium passage of solvent molecules through the membrane is called the solution's **osmotic pressure** (Π) (Greek capital pi). Osmotic pressures can be extremely high, even for relatively dilute solutions. The osmotic pressure of a 0.15 M NaCl solution at 25 °C, for example, is 7.3 atm, a value that will support a difference in water level of approximately 250 ft! The rise in water from the roots to leaves

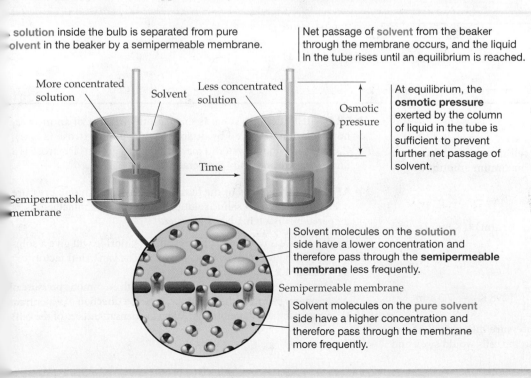

solution inside the bulb is separated from pure solvent in the beaker by a semipermeable membrane.

Net passage of solvent from the beaker through the membrane occurs, and the liquid In the tube rises until an equilibrium is reached.

More concentrated solution

Solvent

Less concentrated solution

Osmotic pressure

Time

Semipermeable membrane

At equilibrium, the **osmotic pressure** exerted by the column of liquid in the tube is sufficient to prevent further net passage of solvent.

Solvent molecules on the **solution** side have a lower concentration and therefore pass through the **semipermeable membrane** less frequently.

Semipermeable membrane

Solvent molecules on the **pure solvent** side have a higher concentration and therefore pass through the membrane more frequently.

◀ FIGURE 12.15
The phenomenon of osmosis.

Figure It Out

When does the rate of solvent crossing the semipermeable membrane from both the solution and the pure solvent sides become equal?

Answer: When the pressure exerted by the column of liquid in the tube is equal to the osmotic pressure of the solution.

of trees, such as the giant coastal California redwoods, is due to osmotic pressure. The fluid inside the tree has a higher solute concentration than groundwater.

The amount of osmotic pressure at equilibrium depends on the concentration of solute particles in the solution according to the equation

Osmotic pressure

$$\Pi = MRTi$$

where M is the molar concentration of solute, R is the gas constant $[0.082\,06\,(L \cdot atm)/(K \cdot mol)]$, T is the temperature in kelvins, and i is the van't Hoff factor.

For example, a 1.00 M solution of glucose, a nonelectrolyte, in water at 300 K has an osmotic pressure of 24.6 atm:

$$\Pi = MRTi = \left(1.00\,\frac{mol}{L}\right)\left(0.082\,06\,\frac{L \cdot atm}{K \cdot mol}\right)(300\,K)(1) = 24.6\,atm$$

Note that the solute concentration is given in *molarity* when calculating osmotic pressure rather than in molality as for other colligative properties. Because osmotic-pressure measurements are made at the specific temperature given in the equation $\Pi = MRTi$, it's not necessary to express concentration in a temperature-independent unit like molality.

Osmosis, like all colligative properties, results from a favorable increase in entropy of the pure solvent as it passes through the membrane and mixes with the solution. We can also explain osmosis on a molecular level by noting that molecules on the solvent side of the membrane, because of their greater concentration, approach the membrane a bit more frequently than molecules on the solution side, thereby passing through more often (Figure 12.15).

WORKED EXAMPLE 12.11

Calculating the Osmotic Pressure of a Solution

The total concentration of dissolved particles inside red blood cells and in the surrounding plasma is approximately 0.30 M, and the membrane surrounding the cells is semipermeable. **FIGURE 12.16** shows the transport of solvent across the cell membrane in solutions of varying concentration. What would the maximum osmotic pressure in atmospheres inside the cells be if the cells were removed from blood plasma and placed in pure water at 298 K?

IDENTIFY

Known	Unknown
Molarity of particles in solution (0.30 M)	Osmotic pressure (Π)
Temperature (298 K)	

STRATEGY

If red blood cells were removed from plasma and placed in pure water, water would pass through the cell membrane, causing a pressure increase inside the cells. The maximum amount of this pressure would be

$$\Pi = MRTi$$

where $Mi = 0.30\,mol/L$, $R = 0.082\,06\,(L \cdot atm)/(K \cdot mol)$, and $T = 298\,K$.

SOLUTION

$$\Pi = \left(0.30\,\frac{mol}{L}\right)\left(0.082\,06\,\frac{L \cdot atm}{K \cdot mol}\right)(298\,K) = 7.3\,atm$$

In fact, red blood cells rupture if the pressure differential across their membrane is greater than 5 atm, so the cells would swell and burst if placed in pure water.

▶ **PRACTICE 12.21** What is the osmotic pressure of an intravenous solution prepared by dissolving 50.0 g of dextrose (sugar), $C_6H_{12}O_6$, in enough water to make 1.00 L of solution? Dextrose is a nonelectrolyte and body temperature is 37.0 °C.

▶ **APPLY 12.22** Cells in the human eye have an osmotic pressure of 8.0 atm at 25 °C. A saline solution, used to store contact lenses, is prepared by dissolving NaCl in water.

(a) What molar concentration of sodium chloride will give a solution with an equal osmotic pressure? The van't Hoff factor for NaCl is 1.9.

(b) A contact lens stored in saline solution with an osmotic pressure of 5.5 atm is placed in the eye. What is the net direction of water transfer across the cell membrane? What happens to the size of the cell?

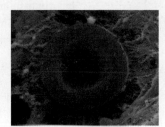

◀ **FIGURE 12.16**
The effect of solute concentration
on red blood cells in plasma.

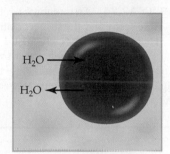

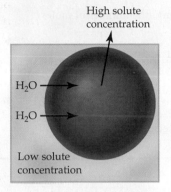

In a *hypertonic solution*, the solute concentration in the solution is greater than that inside the cell. There is a net transfer of water from the cell to the solution causing the cell to shrivel and shrink.

In an *isotonic solution*, the solute concentration in the solution is the same as that inside the cell. There is no net transfer of water across the cell membrane.

In a *hypotonic solution*, the solute concentration in the solution is less than that inside the cell. There is a net transfer of water across the cell membrane into the cell causing the cell to swell and burst if the internal pressure becomes too high.

In the laboratory, colligative properties are sometimes used for determining the molecular weight of an unknown substance. Any of the four colligative properties we've discussed can be used, but the most accurate values are obtained from osmotic-pressure measurements because the magnitude of the osmosis effect is so great. For example, a solution of 0.0200 M glucose in water at 300 K will give an osmotic-pressure reading of 374.2 mm Hg, a value that can easily be read to four significant figures. The same solution, however, will lower the freezing point by only 0.04 °C, a value that can be read to only one significant figure. Worked Example 12.12 shows how osmotic pressure can be used to find molecular weight.

─● WORKED EXAMPLE 12.12

Using Osmotic Pressure to Calculate the Molecular Weight of a Solute

A solution prepared by dissolving 20.0 mg of insulin, a nonelectrolyte, in water and diluting to a volume of 5.00 mL gives an osmotic pressure of 12.5 mm Hg at 300 K. What is the molecular weight of insulin?

IDENTIFY

Known	Unknown
Osmotic pressure (Π = 12.5 mm Hg)	Mol. wt. of insulin
Mass of insulin (20.0 mg)	
Volume of solution (5.00 mL)	

STRATEGY

To determine molecular weight, we need to know the number of moles of insulin represented by the 20.0 mg sample. We can do this by first rearranging the equation for osmotic pressure to find the molar concentration of the insulin solution and then multiplying by the volume of the solution to obtain the number of moles of insulin.

continued on next page

SOLUTION

Since $\Pi = MRTi$, then $M = \dfrac{\Pi}{RTi}$

$$M = \dfrac{12.5 \text{ mm Hg} \times \dfrac{1 \text{ atm}}{760 \text{ mm Hg}}}{0.082\,06 \dfrac{\text{L} \cdot \text{atm}}{\text{K} \cdot \text{mol}} \times 300 \text{ K} \times 1} = 6.68 \times 10^{-4} \text{ M}$$

Since the volume of the solution is 5.00 mL, the number of moles of insulin is

$$\text{Moles insulin} = 6.68 \times 10^{-4} \dfrac{\text{mol}}{\text{L}} \times \dfrac{1 \times 10^{-3} \text{ L}}{1 \text{ mL}} \times 5.00 \text{ mL}$$

$$= 3.34 \times 10^{-6} \text{ mol}$$

Knowing both the mass and the number of moles of insulin, we can calculate the molar mass and thus the molecular weight:

$$\text{Molar mass} = \dfrac{\text{mass of insulin}}{\text{moles of insulin}} = \dfrac{0.0200 \text{ g insulin}}{3.34 \times 10^{-6} \text{ mol insulin}}$$

$$= 5990 \text{ g/mol}$$

The molecular weight of insulin is 5990.

▶ **PRACTICE 12.23** A solution prepared by dissolving 0.8220 g of glucose, a nonelectrolyte, in enough water to produce 200.0 mL of solution has an osmotic pressure of 423.1 mm Hg at 298 K. What is the molecular weight of glucose?

▶ **APPLY 12.24** An unknown white powder is found on the table at a crime scene and the suspect claims that it is table sugar (sucrose, $C_{12}H_{22}O_{11}$). A forensic chemist dissolves 0.512 g of the unknown white powder in enough water to produce 100.0 mL of solution. The osmotic pressure is measured at 25 °C and found to be 278 mm Hg. Does the osmotic pressure measurement support the claim that the powder is sucrose?

▲ The Hadera seawater reverse-osmosis (SWRO) desalination plant in Israel is the world's largest, producing approximately 4.4 *billion* gallons per year of desalinated water.

One of the more interesting uses of colligative osmotic pressure is the desalination seawater by *reverse osmosis*. When pure water and seawater are separated by a suitable men brane, the passage of water molecules from the pure side to the solution side is faster tha passage in the reverse direction. As osmotic pressure builds up, though, the rates of forwa and reverse water passage eventually become equal at an osmotic pressure of about 30 at at 25 °C. If, however, a pressure even *greater* than 30 atm is applied to the solution side, the the reverse passage of water becomes favored. As a result, pure water can be obtained fro seawater (**FIGURE 12.17**).

Desalination of seawater has many applications, from use on cruise ships, to use troops on battlefields, to use for providing drinking water to entire countries. Israel, f instance, now obtains more than 35% of its domestic consumer water demand by rever osmosis.

▶ **FIGURE 12.17**

Desalination of seawater by reverse osmosis at high pressure.

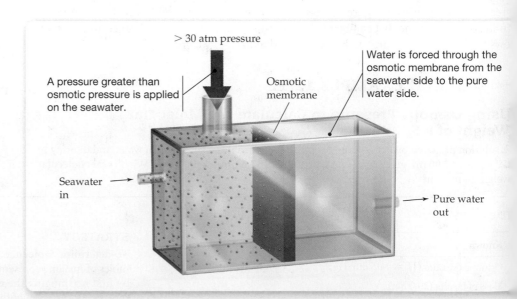

12.9 ▶ FRACTIONAL DISTILLATION OF LIQUID MIXTURES

Undoubtedly the most important of all commercial applications of colligative properties is in the refining of petroleum to make gasoline. Although petroleum refineries appear as a vast maze of pipes, tanks, and towers, the pipes are just for transferring the petroleum or its products and the tanks are just for storage. It's in the towers that the separation of crude petroleum into usable fractions takes place. As we saw in Section 9.11, petroleum is a complex mixture of hydrocarbon molecules that are refined by distillation into different fractions: straight-run gasoline (bp 30 − 200 °C), kerosene (bp 175   300 °C), and gas oil (bp 275 − 400 °C).

Called **fractional distillation**, the separation of a mixture of different liquids into fractions with different boiling points occurs when the mixture is boiled and the vapors are condensed. Because the vapor is enriched in the component with the higher vapor pressure according to Raoult's law (Section 12.6), the condensed vapors are also enriched in that component and a partial purification can be effected. If the boil/condense cycle is repeated a large number of times, complete purification of the more volatile liquid component can be achieved.

As an example, let's look at the separation by fractional distillation of a 1:1 molar mixture of benzene and toluene ($X_{benzene}$ and $X_{toluene}$ are both 0.500). If we begin by heating the mixture, boiling occurs when the sum of the vapor pressures equals atmospheric pressure—that is, when $X \cdot P°_{benzene} + X \cdot P°_{toluene} = 760$ mm Hg. Reading from the vapor-pressure curves in **FIGURE 12.18** (or calculating values with the **Clausius–Clapeyron equation** as discussed in Section 11.3), we find that boiling occurs at 365.3 K (92.2 °C), where $P°_{benzene} = 1084$ mm Hg and $P°_{toluene} = 436$ mm Hg:

$$P_{mixt} = X \cdot P°_{benzene} + X \cdot P°_{toluene}$$
$$= (0.500)(1084 \text{ mm Hg}) + (0.500)(436 \text{ mm Hg})$$
$$= 542 \text{ mm Hg} + 218 \text{ mm Hg}$$
$$= 760 \text{ mm Hg}$$

Although the starting *liquid* mixture of benzene and toluene has a 1:1 molar composition, the composition of the *vapor* is not 1:1. Of the 760 mm Hg total vapor pressure for the boiling mixture, 542/760 = 71.3% is due to benzene and 218/760 = 28.7% is due to toluene. If we now condense the vapor, the liquid we get has this same 71.3:28.7 composition. On boiling this new liquid mixture, the composition of the vapor now becomes 86.4% benzene and 13.6% toluene. A third condense/boil cycle brings the composition of the vapor to 94.4% benzene/5.6% toluene, and so on through further cycles until the desired level of purity is reached.

▲ Distillation of petroleum into fractions according to boiling point is carried out in the large towers of this refinery.

REMEMBER…

The **Clausius–Clapeyron equation** describes the relationship between vapor pressure and heat of vaporization, according to the expression $\ln P_{vap} = (-\Delta H_{vap}/R)(1/T) + C$. (Section 11.3)

◀ **FIGURE 12.18**

Vapor-pressure curves for pure benzene (blue), pure toluene (red), and a 1:1 mixture of the two (green).

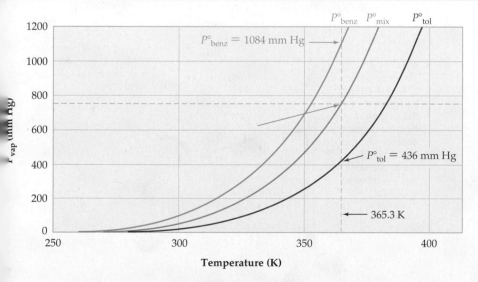

The mixture of benzene and toluene boils at 92.2 °C (365.3 K) at atmospheric pressure, intermediate between the boiling points of the two pure liquids.

▶ **FIGURE 12.19**

A phase diagram of temperature versus composition (mole fraction) for a mixture of benzene and toluene.

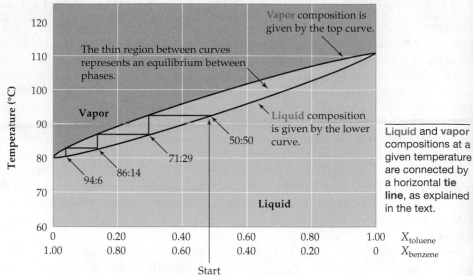

Figure It Out

A mixture that is 80% toluene and 20% benzene is distilled. **(a)** At what approximate temperature will the mixture begin to boil? **(b)** If the vapor is collected and condensed into a liquid, what is the approximate percent composition of toluene and benzene?

Answer: (a) 103 °C (b) 63% toluene, 37% benzene

Fractional distillation can be represented on a liquid–vapor phase diagram by plotting temperature versus composition, as shown in **FIGURE 12.19**. The lower region of the diagram represents the liquid phase, and the upper region represents the vapor phase. Between the two is a thin equilibrium region where liquid and vapor coexist.

To understand how the diagram works, let's imagine starting in the middle of Figure 12.19 with the 50:50 benzene/toluene mixture and heating it to its boiling point (92.2 °C on the diagram). The lower curve represents the liquid composition (50:50), but the upper curve represents the vapor composition (approximately 71:29 at 92.2 °C). The two points are connected by a short horizontal line called a *tie line* to indicate that the temperature is the same at both points. Condensing the 71:29 vapor mixture by lowering the temperature gives a 71:29 liquid mixture that, when heated to *its* boiling point (86.6 °C), has an 86:14 vapor composition as represented by another tie line. In essence, fractional distillation is simply a stairstep walk across successive tie lines in whatever number of steps is necessary to reach the desired purity.

In practice, the successive boil/condense cycles occur naturally in the distillation column, and there is no need to isolate liquid mixtures at intermediate stages of purification. Fractional distillation is therefore relatively simple to carry out and is used on a daily basis in refineries, chemical plants, and laboratories throughout the world (**FIGURE 12.20**).

▶ **FIGURE 12.20**

A simple fractional distillation column used in a chemistry laboratory.

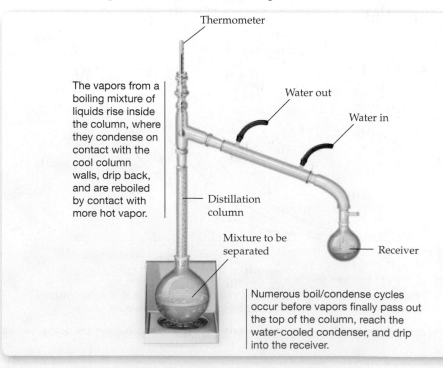

INQUIRY ▶▶▶ HOW DOES HEMODIALYSIS CLEANSE THE BLOOD OF PATIENTS WITH KIDNEY FAILURE?

The primary function of the kidneys is to filter the blood and remove wastes by producing urine. Kidneys also have a number of other important functions on which life depends, including regulation of blood pressure, body pH, and electrolyte balance. In addition, they produce several important hormones and are responsible for the reabsorption from filtered blood of ions and small molecules such as glucose, amino acids, and water. Even though these functions are vital, more than 330,000 people in the United States currently suffer from end-stage renal disease, or kidney failure.

There is no cure for kidney failure. For those afflicted, the only choices are regular dialysis treatments to replace some of the kidney's functions or organ transplantation. Unfortunately, more than 1,000 patients are currently awaiting a kidney transplant according to the United Network for Organ Sharing (UNOS), and the number of available organs is nowhere close to sufficient. Thus, dialysis is the only realistic choice for many people.

The process of *dialysis* is similar to osmosis, except that *both* solvent molecules and small solute particles pass through the semipermeable dialysis membrane. Recall that in osmosis, only solvent molecules pass through a semipermeable membrane. In dialysis, only large colloidal particles such as cells and large molecules such as proteins can't pass (**FIGURE 12.21**). (The exact dividing line between a small molecule and a large one is imprecise, and dialysis membranes with a variety of pore sizes are available.) Because they can't dialyze, proteins can be separated from small ions and molecules, making dialysis a valuable procedure for purification of the proteins needed in laboratory studies.

The most important medical use of dialysis is in artificial kidney machines, where *hemodialysis* is used to cleanse the blood of patients, removing waste products like urea and controlling the potassium/sodium ion balance (**FIGURE 12.22**). Blood is diverted from the body and pumped through a dialysis tube suspended in a solution formulated to contain many of the same components as blood plasma. These substances—NaCl, $NaHCO_3$, KCl, and glucose—can have the same concentrations in the dialysis solution as they do in blood, so that they have no net passage through the membrane. However, it is sometimes desirable to adjust dialysis solutions to cause an increase in the concentration of a critical substance in the blood. For example, excess bicarbonate (HCO_3^-) in dialysis solution can be used to correct acid imbalances (that is, to adjust blood pH). **TABLE 12.5** shows the concentration and various components of a hemodialysis solution.

Small waste products pass through the dialysis membrane from the blood to the solution side, where they are washed away, but cells, proteins, and other important blood components are prevented by their size from passing through the membrane. A typical hemodialysis treatment lasts for approximately 3–4 hours, and treatments are usually repeated 3–4 times per week.

The water used to make dialysis solutions must be extremely pure because if a solute has a higher concentration in dialysis solution than in blood, it will pass through the membrane and enter the patient's body. Fluoride ion is of particular concern because over half of the drinking water in the United States is enriched with

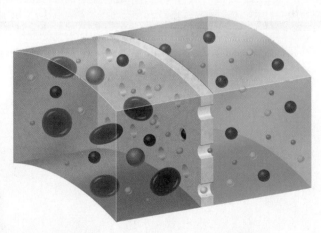

▲ **FIGURE 12.21**

Dialysis membrane. Solvent and small solute molecules (represented by pink, green, and yellow spheres) pass through tiny pores in the membrane. Cells and proteins (represented by red discs and blue spheres respectively) are retained because they are too large to pass through the membrane.

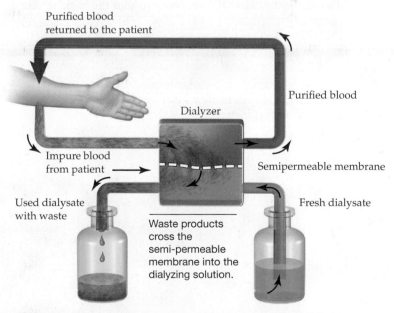

Purified blood returned to the patient

Purified blood

Dialyzer

Impure blood from patient

Semipermeable membrane

Used dialysate with waste

Fresh dialysate

Waste products cross the semi-permeable membrane into the dialyzing solution.

▲ **FIGURE 12.22**

Representation of hemodialysis, an artificial kidney

continued on next page

continued from previous page

TABLE 12.5 Composition of a Typical Hemodialysis Solution

Substance	Concentration
Na^+	137 mmol/L
Cl^-	105 mmol/L
Ca^{2+}	3.0 mmol/L
CH_3COO^-	4.0 mmol/L
K^+	2.0 mmol/L
HCO_3^-	33 mmol/L
Mg^{2+}	0.75 mmol/L
Dextrose $(C_6H_{12}O_6)$	1.11 mmol/dL

fluoride to promote dental health. The optimal level of fluoride in drinking water is 1.0 ppm, and the maximum legal limit is 4.0 ppm. Water from a municipal supply must be purified and the fluoride ion concentration reduced to less than 0.2 ppm before it can be used in dialysis. Fluoride poisoning in dialysis treatment has occurred when water purification systems malfunctioned. In a Chicago hospital in 1993, 12 patients became ill with symptoms of pain, vomiting, and itching, and 3 patients died from cardiac arrest. New standards are now in place in dialysis units to prevent cases of fluoride intoxication.

And by the way: Please consider signing the necessary authorization card indicating your willingness to become an organ donor. In most states, all that's needed is to sign the back of your driver's license.

PROBLEM 12.25   What is the difference between a dialysis membrane and the typical semipermeable membrane used for osmosis?

PROBLEM 12.26   Urea has a high solubility in blood serum and one waste product filtered from blood in a dialysis treatment. Classify the strongest type of intermolecular force in the following interactions; solvent–solvent, solvent–solute, and solute–solute when urea is dissolved in aqueous blood serum (H_2O).

$$H_2N-\overset{\overset{\textstyle O}{\|}}{C}-NH_2$$

Urea

PROBLEM 12.27
(a) Use Table 12.5 to calculate the osmotic pressure of the hemodialysis solution at 25 °C.

(b) If the osmotic pressure of blood at 25 °C is 7.70 atm, what the direction of solvent movement across the semipermeable membrane in dialysis? (Blood to dialysis solution or dialysis solution blood)

PROBLEM 12.28   The presence of fluoride ion in dialysis solution is dangerous because patients are exposed to much greater volume of fluid than the average person. A typical 4-hour dialysis treatment exposes a patient to as much fluid as the average person drinks one week (~12 L). What mass of fluoride ion is present in 12.0 L water with a fluoride concentration of 2.0 ppm?

STUDY GUIDE

Section	Concept Summary	Learning Objectives	Test Your Understanding
12.1–12.2 ▶ Solutions and Energy Changes	**Solutions** are homogeneous mixtures that contain particles the size of a typical ion or small molecule. Any one state of matter can mix with any other state, leading to seven possible kinds of solutions. For solutions in which a gas or solid is dissolved in a liquid, the dissolved substance is called the **solute** and the liquid is called the **solvent**.	**12.1** Describe the types of intermolecular forces involved when a solute dissolves in a solvent.	Problems 12.44, 12.45, 12.52, 12.53
		12.2 Predict solubility based on the chemical structure of the solute and solvent.	Worked Example 12.1; Problems 12.29, 12.46–12.51, 12.54
	The dissolution of a solute in a solvent has an associated free-energy change, $\Delta G = \Delta H - T\Delta S$. The enthalpy change is the **enthalpy of solution** (ΔH_{soln}), and the entropy change is the **entropy of solution** (ΔS_{soln}). Enthalpies of solution can be either positive or negative, depending on the relative strengths of solvent–solvent, solute–solute, and solvent–solute intermolecular forces. Entropies of solution are usually positive because molecular randomness increases when a pure solute dissolves in a pure solvent.	**12.3** Describe the effect of ion size and lattice energy on the solubility of ionic compounds.	Problems 12.41–12.43
12.3 ▶ Concentration Units for Solutions	The concentration of a solution can be expressed in many ways, including molarity (moles of solute per liter of solution), mole fraction (moles of solute per mole of solution), mass percent (mass of solute per mass of solution times 100%), and molality (moles of solute per kilogram of solvent).	**12.4** Calculate the concentration of a solution in units of mass percent.	Worked Example 12.2; Problems, 12.68, 12.140
		12.5 Calculate the concentration of a solution in units of parts per million (ppm) or parts per billion (ppb).	Worked Example 12.3; Problems, 12.5, 12.6, 12.63, 12.70, 12.71,
		12.6 Calculate the concentration of a solution in units of molality.	Worked Example 12.4; Problems, 12.7, 12.8, 12.65, 12.69, 12.78
		12.7 Convert from one unit of concentration to another.	Worked Example 12.5; Problems, 12.58, 12.60, 12.66, 12.74, 12.80
12.4 ▶ Some Factors That Affect Solubility	When equilibrium is reached and no further solute dissolves in a given amount of solvent, a solution is said to be **saturated**. Solubilities are usually temperature-dependent, although often not in a simple way. Gas solubilities usually decrease in water with increasing temperature, but the solubilities of solids can either increase or decrease. The solubilities of gases also depend on pressure. According to **Henry's law**, the solubility of a gas in a liquid at a given temperature is proportional to the partial pressure of the gas over the solution.	**12.8** Analyze a solubility graph to determine the solubility of a solute at a given temperature.	Problems 12.82, 12.83
		12.9 Use Henry's Law to calculate the solubility of a gas.	Worked Example 12.6; Problems 12.84–12.89

Section	Concept Summary	Learning Objectives	Test Your Understanding
12.5–12.8 ▶ Physical Properties of Solutions: Colligative Properties (Vapor Pressure Lowering, Boiling-Point Elevation, Freezing-Point Depression, Osmotic Pressure)	In comparison with a pure solvent, a solution has a lower vapor pressure at a given temperature, a lower freezing point, and a higher boiling point. In addition, a solution that is separated from solvent by a semipermeable membrane gives rise to the phenomenon of **osmosis**. All four of these properties of solutions depend only on the concentration of dissolved solute particles rather than on the chemical identity of those particles and are therefore called **colligative properties**. The fundamental cause of all colligative properties is the same: the higher entropy of the solvent in a solution relative to that of the pure solvent.	**12.10** Qualitatively predict changes to the vapor pressure of a solution when a solute is added.	Problems 12.16, 12.18, 12.31, 12.32, 12.33, 12.100
		12.11 Calculate the vapor pressure of a solution containing a nonvolatile solute.	Worked Examples 12.7, 12.8; Problems 12.13, 12.14, 12.15, 12.16, 12.92, 12.93, 12.102
		12.12 Calculate the vapor pressure of a solution containing a volatile solute.	Worked Example 12.9; Problems 12.109–12.113
		12.13 Calculate the amount of boiling-point elevation and freezing-point depression for a solution.	Worked Example 12.10; Problems, 12.90, 12.94, 12.98, 12.104, 12.106
		12.14 Calculate osmotic pressure for a solution.	Worked Example 12.11; Problems, 12.118–12.121
		12.15 Use osmotic pressure or other colligative property to calculate the molecular weight of a solute.	Worked Example 12.12; Problems, 12.124, 12.128, 12.129, 12.164
12.9 ▶ Fractional Distillation of Liquid Mixtures	**Fractional distillation** separates liquids in a mixture based on differences in boiling point. The vapor above a boiling liquid contains a higher mole percent of the more volatile liquid than the original liquid mixture. Successive boil and condense cycles will purify the more volatile liquid.	**12.16** Interpret the liquid–liquid phase diagram of a solution and apply the diagram to fractional distillation.	Problems 12.38, 12.39

KEY TERMS

colligative property 462
colloid 448
enthalpy of solution (ΔH_{soln}) 449
entropy of solution (ΔS_{soln}) 449
fractional distillation 477
Henry's law 460

mass percent (mass %) 454
miscible 459
mixture 448
molal boiling-point-elevation constant (K_b) 470
molal freezing-point-depression constant (K_f) 470

molality (m) 456
osmosis 473
osmotic pressure (Π) 473
parts per billion (ppb) 455
parts per million (ppm) 455
Raoult's law 463
saturated 459

solute 448
solution 448
solvent 448
supersaturated 459
van't Hoff factor (i) 464

KEY EQUATIONS

• **Units of Solution Concentration (Section 12.3)**

$$\text{Molarity (M)} = \frac{\text{Moles of solute}}{\text{Liters of solution}}$$

$$\text{Mole fraction } (X) = \frac{\text{Moles of component}}{\text{Total moles making up the solution}}$$

$$\text{Mass percent} = \frac{\text{Mass of component}}{\text{Total mass of solution}} \times 100\%$$

$$\text{Parts per million (ppm)} = \frac{\text{Mass of component}}{\text{Total mass of solution}} \times 10^6$$

$$\text{Parts per billion (ppb)} = \frac{\text{Mass of component}}{\text{Total mass of solution}} \times 10^9$$

Henry's law (Section 12.4)

$$\text{Solubility} = k \cdot P \qquad \text{where the constant } k \text{ is characteristic of a specific gas and } P$$
is the partial pressure of the gas over the solution.

Raoult's law and vapor pressure lowering (Section 12.6)

$$P_{soln} = P_{solv} \times X_{solv}$$
$$\Delta P_{soln} = P_{solv} \times X_{solute}$$

van't Hoff factor

$$i = \frac{\text{Moles of particles in solution}}{\text{Moles of solute dissolved}} \quad (i = 1 \text{ for nonelectrolytes})$$

Molal boiling-point elevation (Section 12.7)

$$\Delta T_b = K_b \cdot m \cdot i \qquad \text{where } K_b \text{ is the molal boiling-point-elevation constant and } i \text{ is the van't Hoff factor}$$

Molal freezing point depression (Section 12.7)

$$\Delta T_f = -K_f \cdot m \cdot i \qquad \text{where } K_b \text{ is the molal freezing-point-depression constant and } i \text{ is the van't Hoff factor}$$

Osmotic pressure (Section 12.8)

$$\Pi = MRTi$$

CONCEPTUAL PROBLEMS

Problems 12.1–12.28 appear within the chapter.

12.29 Many people take vitamin supplements to promote health and the potential for overdose depends on whether a particular vitamin is fat soluble or water soluble. If a vitamin is soluble in nonpolar fatty tissues it will accumulate in the body, whereas water-soluble vitamins are readily excreted. Use the ball-and-stick models of niacin and vitamin D to determine which is more fat soluble.

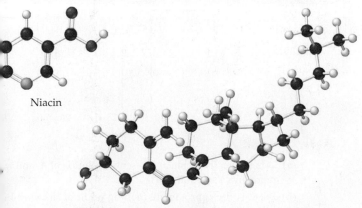

Niacin

Vitamin D

12.30 Rank the situations represented by the following drawings according to increasing entropy.

(a)   (b)

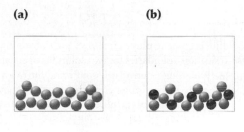

(c)

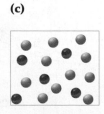

12.31 The following phase diagram shows part of the vapor-pressure curves for a pure liquid (green curve) and a solution of the first liquid with a second volatile liquid (red curve).

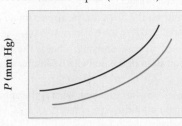

P (mm Hg)

T (°C)

(a) Is the boiling point of the second liquid higher or lower than that of the first liquid?

(b) Draw on the diagram the approximate position of the vapor-pressure curve for the second liquid

12.32 The following phase diagram shows part of the liquid–vapor phase-transition boundaries for pure ether and a solution of a nonvolatile solute in ether.

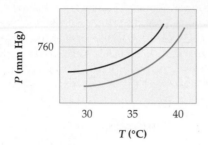

(a) What is the approximate normal boiling point of pure ether?

(b) What is the approximate molal concentration of the solute? [K_b for ether is 2.02 ($°C \cdot kg$)/mol.]

12.33 The following diagram shows a close-up view of part of the vapor-pressure curves for a solvent (red curve) and a solution of the solvent with a second liquid (green curve). Is the second liquid more volatile or less volatile than the solvent?

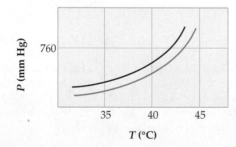

12.34 The following phase diagram shows part of the liquid–vapor phase-transition boundaries for two solutions of equal concentration, one containing a nonvolatile solute and the other containing a volatile solute whose vapor pressure at a given temperature is approximately half that of the pure solvent.

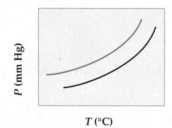

(a) Which curve, red or green, represents the solution of the nonvolatile solute and which represents the solution of the volatile solute?

(b) Draw on the diagram the approximate position of the vapor-pressure curve for the pure solvent.

(c) Based on your drawing, what is the approximate molal concentration of the nonvolatile solute, assuming the solvent has $K_b = 2.0 °C/m$?

(d) Based on your drawing, what is the approximate normal boiling point of the pure solvent?

12.35 Two beakers, one with pure water (blue) and the other with a solution of NaCl in water (green), are placed in a closed container represented by drawing **(a)**. Which of the drawings **(b)**–**(d)** represents what the beakers will look like after a substantial amount time has passed?

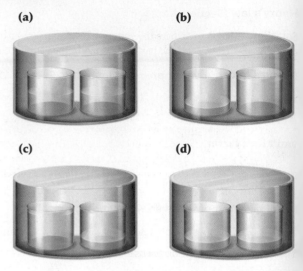

12.36 Assume that two liquids are separated by a semipermeable membrane. Make a drawing that shows the situation after equilibrium is reached.

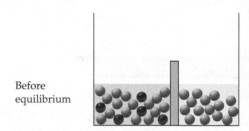

12.37 The following phase diagram shows a very small part of solid–liquid phase-transition boundaries for two solution equal concentration. Substance **A** has $i = 1$, and substance **B** $i = 3$.

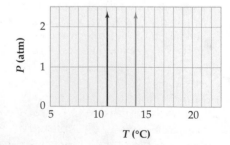

(a) Which line, red or blue, represents a solution of **A**, and w represents a solution of **B**?

(b) What is the approximate melting point of the pure li solvent?

(c) What is the approximate molal concentration of each s tion, assuming the solvent has $K_f = 3.0 °C/m$?

12.38 A phase diagram of temperature versus composition for a ture of the two volatile liquids hexane (bp = 69 °C) and de (bp = 126 °C) is shown. Assume that you begin with a mi containing 0.60 mol of decane and 0.40 mol of hexane.

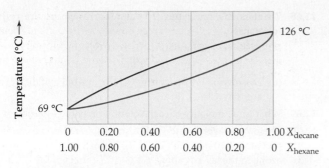

(a) What region on the diagram corresponds to vapor, and what region corresponds to liquid?

(b) At what approximate temperature will the mixture begin to boil? Mark as point **A** on the diagram the liquid composition at the boiling point, and mark as point **B** the vapor composition at the boiling point.

(c) Assume that the vapor at point **B** condenses and is reboiled. Mark as point **C** on the diagram the liquid composition of the condensed vapor and as point **D** on the diagram the vapor composition of the reboiled material.

2.39 The following graph is a phase diagram of temperature versus composition for mixtures of the two liquids chloroform and dichloromethane.

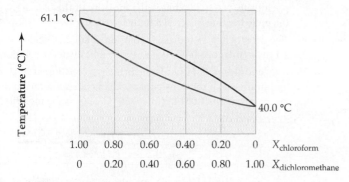

(a) Label the regions on the diagram corresponding to liquid and vapor.

(b) Assume that you begin with a mixture of 60% chloroform and 40% dichloromethane. At what approximate temperature will the mixture begin to boil? Mark as point **A** on the diagram the liquid composition at the boiling point, and mark as point **B** the vapor composition at the boiling point.

(c) Assume that the vapor at point **B** condenses and is reboiled. Mark as point **C** on the diagram the liquid composition of the condensed vapor and as point **D** on the diagram the vapor composition of the reboiled material.

(d) What will the approximate composition of the liquid be after carrying out two cycles of boiling and condensing?

ECTION PROBLEMS

utions and Energy Changes (Sections 12.1 and 12.2)

40 If a single 5-g block of NaCl is placed in water, it dissolves slowly, but if 5 g of powdered NaCl is placed in water, it dissolves rapidly. Explain.

41 Why do ionic substances with higher lattice energies tend to be less soluble in water than substances with lower lattice energies?

12.42 Which would you expect to have the larger (more negative) hydration energy?
(a) Na^+ or Cs^+
(b) K^+ or Ba^{2+}

12.43 Which would you expect to have the larger hydration energy, SO_4^{2-} or ClO_4^-? Explain.

12.44 Classify the strongest type of intermolecular force in the following interactions: solvent–solvent, solvent–solute, and solute–solute when solid iodine (I_2) is placed in the water. Based on these interactions predict whether I_2 is soluble in water.

12.45 Classify the strongest type of intermolecular force in the following interactions: solvent–solvent, solvent–solute, and solute–solute when solid glucose ($C_6H_{12}O_6$) is placed in the water. Based on these interactions predict whether glucose is soluble in water.

$$HO-\overset{\overset{\displaystyle H}{|}}{C}-\overset{\overset{\displaystyle OH}{|}}{C}-\overset{\overset{\displaystyle H}{|}}{C}-\overset{\overset{\displaystyle OH}{|}}{C}-\overset{\overset{\displaystyle H}{|}}{C}-\overset{\overset{\displaystyle O}{||}}{C}-H$$

Glucose

12.46 Br_2 is much more soluble in tetrachloromethane, CCl_4, than in water. Explain.

12.47 Predict whether the solubility of formaldehyde, CH_2O, is greater in water or tetrachloromethane, CCl_4.

12.48 Predict whether the solubility of acetic acid, CH_3COOH, is greater in water or benzene, C_6H_6.

12.49 Predict whether the solubility of butane, C_4H_{10}, is greater in water or benzene, C_6H_6.

12.50 Arrange the following compounds in order of their expected increasing solubility in water: Br_2, KBr, toluene (C_7H_8, a constituent of gasoline).

12.51 Arrange the following compounds in order of their expected increasing solubility in hexane (C_6H_{14}): NaBr, CCl_4, CH_3COOH.

12.52 When ethyl alcohol, CH_3CH_2OH, dissolves in water, how many hydrogen bonds are formed between one ethyl alcohol molecule and surrounding water molecules? Sketch the hydrogen bonding interactions. (Hint: Add lone pairs of electrons to the structure before drawing hydrogen bonds.)

$$H-\overset{\overset{\displaystyle H}{|}}{\underset{\underset{\displaystyle H}{|}}{C}}-\overset{\overset{\displaystyle H}{|}}{\underset{\underset{\displaystyle H}{|}}{C}}-O-H$$

12.53 When the vitamin niacin ($C_6H_5O_2N$) dissolves in water, how many hydrogen bonds are formed between one niacin molecule and surrounding water molecules? Sketch the hydrogen bonding interactions. (Hint: Add lone pairs of electrons to the structure before drawing hydrogen bonds.)

Niacin

12.54 Ethyl alcohol, CH_3CH_2OH, is miscible with water at 20 °C but pentyl alcohol, $CH_3CH_2CH_2CH_2CH_2OH$, is soluble in water only to the extent of 2.7 g/100 mL. Explain.

Pentyl alcohol (Problem 12.54) is miscible with octane, C_8H_{18}, but methyl alcohol, CH_3OH, is insoluble in octane. Explain.

Units of Concentration (Section 12.3)

12.56 The dissolution of $CaCl_2(s)$ in water is exothermic, with $\Delta H_{soln} = -81.3$ kJ/mol. If you were to prepare a 1.00 m solution of $CaCl_2$ beginning with water at 25.0 °C, what would the final temperature of the solution be in °C? Assume that the specific heats of both pure H_2O and the solution are the same, 4.18 J/(K·g).

12.57 The dissolution of $NH_4ClO_4(s)$ in water is endothermic, with $\Delta H_{soln} = +33.5$ kJ/mol. If you prepare a 1.00 m solution of NH_4ClO_4 beginning with water at 25.0 °C, what is the final temperature of the solution in °C? Assume that the specific heats of both pure H_2O and the solution are the same, 4.18 J/(K·g).

12.58 Assuming that seawater is an aqueous solution of NaCl, what is its molarity? The density of seawater is 1.025 g/mL at 20 °C, and the NaCl concentration is 3.50 mass %.

12.59 Assuming that seawater is a 3.50 mass % aqueous solution of NaCl, what is the molality of seawater?

12.60 Propranolol ($C_{16}H_{21}NO_2$), a so-called beta-blocker that is used for treatment of high blood pressure, is effective at a blood plasma concentration of 50 ng/L. Express this concentration of propranolol in the following units:
(a) Parts per billion (assume a plasma density of 1.025 g/mL)
(b) Molarity

12.61 Residues of the herbicide atrazine ($C_8H_{14}ClN_5$) in water can be detected at concentrations as low as 0.050 µg/L. Express this concentration of atrazine in the following units:
(a) Parts per billion (assume a solution density of 1.00 g/mL)
(b) Molarity

12.62 How would you prepare each of the following solutions?
(a) A 0.150 M solution of glucose in water
(b) A 1.135 m solution of KBr in water
(c) A solution of methyl alcohol (methanol) and water in which $X_{methanol} = 0.15$ and $X_{water} = 0.85$

12.63 How would you prepare each of the following solutions?
(a) 100 mL of a 155 ppm solution of urea, CH_4N_2O, in water
(b) 100 mL of an aqueous solution whose K^+ concentration is 0.075 M

12.64 How would you prepare 165 mL of a 0.0268 M solution of benzoic acid ($C_7H_6O_2$) in chloroform ($CHCl_3$)?

12.65 How would you prepare 250 mL of a 0.325 M solution of benzoic acid ($C_7H_6O_2$) in chloroform ($CHCl_3$)?

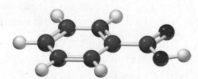

Benzoic acid

12.66 Which of the following solutions is more concentrated?
(a) 0.500 M KCl or 0.500 mass % KCl in water
(b) 1.75 M glucose or 1.75 m glucose in water

12.67 Which of the following solutions has the higher molarity?
(a) 10 ppm KI in water or 10,000 ppb KBr in water
(b) 0.25 mass % KCl in water or 0.25 mass % citric acid ($C_6H_8O_7$) in water

12.68 What is the mass percent concentration of the followin solutions?
(a) Dissolve 0.655 mol of citric acid, $C_6H_8O_7$, in 1.00 kg of wate
(b) Dissolve 0.135 mg of KBr in 5.00 mL of water.
(c) Dissolve 5.50 g of aspirin, $C_9H_8O_4$, in 145 g of dichlorometh ane, CH_2Cl_2.

12.69 What is the molality of each solution prepared in Problem 12.6

12.70 The ozone layer in the Earth's stratosphere has an average to pressure of 10 mm Hg (1.3×10^{-2} atm). The partial pressure ozone in the layer is about 1.2×10^{-6} mm Hg (1.6×10^{-9} atm What is the concentration of ozone in parts per million, assumi that the average molar mass of air is 29 g/mol?

12.71 Persons are medically considered to have lead poisoning if th have a concentration of greater than 10 µg of lead per decili of blood. What is this concentration in parts per billion? Assur that the density of blood is the same as that of water.

12.72 What is the concentration of each of the following solutions?
(a) The molality of a solution prepared by dissolving 25.0 g H_2SO_4 in 1.30 L of water
(b) The mole fraction of each component of a solution pr pared by dissolving 2.25 g of nicotine, $C_{10}H_{14}N_2$, in 80.0 g CH_2Cl_2

12.73 Household bleach is a 5.0 mass % aqueous solution of sodium pochlorite, NaOCl. What is the molality of the bleach? Wha the mole fraction of NaOCl in the bleach?

12.74 The density of a 16.0 mass % solution of sulfuric acid in wate 1.1094 g/mL at 25.0 °C. What is the molarity of the solution?

12.75 Ethylene glycol, $C_2H_6O_2$, is the principal constituent of auton bile antifreeze. If the density of a 40.0 mass % solution of ethyl glycol in water is 1.0514 g/mL at 20 °C, what is the molarity?

Ethylene glycol

12.76 What is the molality of the 40.0 mass % ethylene glycol solut used for automobile antifreeze (Problem 12.75)?

12.77 Ethylene glycol, $C_2H_6O_2$, is a colorless liquid used as automo antifreeze. If the density at 20 °C of a 4.028 m solution of ethyl glycol in water is 1.0241 g/mL, what is the molarity of the so tion? The molar mass of ethylene glycol is 62.07 g/mol.

12.78 Nalorphine ($C_{19}H_{21}NO_3$), a relative of morphine, is used to co bat withdrawal symptoms in narcotics users. How many gr of a 1.3×10^{-3} m aqueous solution of nalorphine are neede obtain a dose of 1.5 mg?

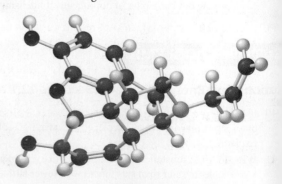

Nalorphine

2.79 How many grams of water should you add to 32.5 g of sucrose, $C_{12}H_{22}O_{11}$, to get a 0.850 m solution?

2.80 A 0.944 M solution of glucose, $C_6H_{12}O_6$, in water has a density of 1.0624 g/mL at 20 °C. What is the concentration of this solution in the following units?

(a) Mole fraction

(b) Mass percent

(c) Molality

2.81 Lactose, $C_{12}H_{22}O_{11}$, is a naturally occurring sugar found in mammalian milk. A 0.335 M solution of lactose in water has a density of 1.0432 g/L at 20 °C. What is the concentration of this solution in the following units?

(a) Mole fraction

(b) Mass percent

(c) Molality

olubility and Henry's Law (Section 12.4)

.82 Look at the solubility graph in Figure 12.6, and estimate which member of each of the following pairs has the higher molar solubility at the given temperature:

(a) $CuSO_4$ or NH_4Cl at 60 °C

(b) CH_3CO_2Na or glucose at 20 °C

.83 Look at the solubility graph in Figure 12.6, and estimate which member of each of the following pairs has the higher molar solubility at the given temperature:

(a) NaCl or NH_4Cl at 40 °C

(b) K_2SO_4 or $CuSO_4$ at 20 °C

.84 Vinyl chloride ($H_2C = CHCl$), the starting material from which PVC polymer is made, has a Henry's law constant of 0.091 mol/(L·atm) at 25 °C. What is the solubility of vinyl chloride in water in mol/L at 25 °C and a partial pressure of 0.75 atm?

.85 Hydrogen sulfide, H_2S, is a toxic gas responsible for the odor of rotten eggs. The solubility of $H_2S(g)$ in water at STP is 0.195 M. What is the Henry's law constant of H_2S at 0 °C? What is the solubility of H_2S in water at 0 °C and a partial pressure of 25.5 mm Hg?

.86 Fish generally need an O_2 concentration in water of at least 4 mg/L for survival. What partial pressure of oxygen above the water in atmospheres at 0 °C is needed to obtain this concentration? The solubility of O_2 in water at 0 °C and 1 atm partial pressure is 2.21×10^{-3} mol/L.

.87 At an altitude of 10,000 ft, the partial pressure of oxygen in the lungs is about 68 mm Hg. What is the concentration in mg/L of dissolved O_2 in blood (or water) at this partial pressure and a normal body temperature of 37 °C? The solubility of O_2 in water at 37 °C and 1 atm partial pressure is 1.93×10^{-3} mol/L.

.88 Sulfur hexafluoride, which is used as a nonflammable insulator in high-voltage transformers, has a Henry's law constant of 2.4×10^{-4} mol/(L·atm) at 25 °C. What is the solubility in mol/L of sulfur hexafluoride in water at 25 °C and a partial pressure of 2.00 atm?

.89 The nonstick polymer Teflon is made from tetrafluoroethylene, C_2F_4. If C_2F_4 is a gas that dissolves in water at 298 K to the extent of 1.01×10^{-3} M with a partial pressure of 0.63 atm, what is its Henry's law constant at 298 K?

Colligative Properties (Sections 12.5–12.8)

12.90 Rank the following aqueous solutions from lowest to highest freezing point: 0.10 m $FeCl_3$, 0.30 m glucose ($C_6H_{12}O_6$), 0.15 m $CaCl_2$. Assume complete dissociation.

12.91 Which of the following aqueous solutions has the (a) higher freezing point, (b) higher boiling point, (c) lower vapor pressure: 0.50 m sucrose ($C_{12}H_{22}O_{11}$) or 0.35 m HNO_3?

12.92 What is the vapor pressure in mm Hg of a solution prepared by dissolving 5.00 g of benzoic acid ($C_7H_6O_2$) in 100.00 g of ethyl alcohol (C_2H_6O) at 35 °C? The vapor pressure of pure ethyl alcohol at 35 °C is 100.5 mm Hg.

12.93 What is the normal boiling point in °C of a solution prepared by dissolving 1.50 g of aspirin (acetylsalicylic acid, $C_9H_8O_4$) in 75.00 g of chloroform ($CHCl_3$)? The normal boiling point of chloroform is 61.7 °C, and K_b for chloroform is given in Table 12.4.

12.94 What is the freezing point in °C of a solution prepared by dissolving 7.40 g of $MgCl_2$ in 110 g of water? The value of K_f for water is given in Table 12.4, and the van't Hoff factor for $MgCl_2$ is $i = 2.7$.

12.95 Assuming complete dissociation, what is the molality of an aqueous solution of KBr whose freezing point is −2.95 °C? The molal freezing-point-depression constant of water is given in Table 12.4.

12.96 When 9.12 g of HCl was dissolved in 190 g of water, the freezing point of the solution was −4.65 °C. What is the value of the van't Hoff factor for HCl?

12.97 The observed osmotic pressure for a 0.125 M solution of $MgCl_2$ at 310 K is 8.57 atm. Calculate the value of the van't Hoff factor for $MgCl_2$ under these conditions.

12.98 When 1 mol of NaCl is added to 1 L of water, the boiling point increases. When 1 mol of methyl alcohol is added to 1 L of water, the boiling point decreases. Explain.

12.99 When 100 mL of 9 M H_2SO_4 at 0 °C is added to 100 mL of liquid water at 0 °C, the temperature rises to 12 °C. When 100 mL of 9 M H_2SO_4 at 0 °C is added to 100 g of solid ice at 0 °C, the temperature falls to −12 °C. Explain the difference in behavior.

12.100 Draw a phase diagram showing how the phase boundaries differ for a pure solvent compared with a solution.

12.101 A solution concentration must be expressed in molality when considering boiling-point elevation or freezing-point depression but can be expressed in molarity when considering osmotic pressure. Why?

12.102 What is the vapor pressure in mm Hg of the following solutions, each of which contains a nonvolatile solute? The vapor pressure of water at 45.0 °C is 71.93 mm Hg.

(a) A solution of 10.0 g of urea, CH_4N_2O, in 150.0 g of water at 45.0 °C

(b) A solution of 10.0 g of LiCl in 150.0 g of water at 45.0 °C, assuming complete dissociation

12.103 What is the vapor pressure in mm Hg of a solution of 16.0 g of glucose ($C_6H_{12}O_6$) in 80.0 g of methanol (CH_3OH) at 27 °C? The vapor pressure of pure methanol at 27 °C is 140 mm Hg.

12.104 What is the boiling point in °C of each of the solutions in Problem 12.102? For water, $K_b = 0.51$ (°C·kg)/mol.

12.105 What is the freezing point in °C of each of the solutions in Problem 12.102? For water, $K_f = 1.86$ (°C·kg)/mol.

12.106 A 1.0 m solution of K_2SO_4 in water has a freezing point of −4.3 °C. What is the value of the van't Hoff factor i for K_2SO_4?

The van't Hoff factor for KCl is $i = 1.85$. What is the boiling point of a 0.75 m solution of KCl in water? For water, $K_b = 0.51(°C \cdot kg)/mol$.

12.108 Heptane (C_7H_{16}) and octane (C_8H_{18}) are constituents of gasoline. At 80.0 °C, the vapor pressure of heptane is 428 mm Hg and the vapor pressure of octane is 175 mm Hg. What is $X_{heptane}$ in a mixture of heptane and octane that has a vapor pressure of 305 mm Hg at 80.0 °C?

12.109 Cyclopentane (C_5H_{10}) and cyclohexane (C_6H_{12}) are volatile, nonpolar hydrocarbons. At 30.0 °C, the vapor pressure of cyclopentane is 385 mm Hg and the vapor pressure of cyclohexane is 122 mm Hg. What is $X_{pentane}$ in a mixture of C_5H_{10} and C_6H_{12} that has a vapor pressure of 212 mm Hg at 30.0 °C?

12.110 Acetone, C_3H_6O, and ethyl acetate, $C_4H_8O_2$, are organic liquids often used as solvents. At 30 °C, the vapor pressure of acetone is 285 mm Hg and the vapor pressure of ethyl acetate is 118 mm Hg. What is the vapor pressure in mm Hg at 30 °C of a solution prepared by dissolving 25.0 g of acetone in 25.0 g of ethyl acetate?

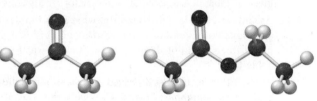

Acetone · · · · · · · · · · · · · Ethyl acetate

12.111 The industrial solvents chloroform, $CHCl_3$, and dichloromethane, CH_2Cl_2, are prepared commercially by reaction of methane with chlorine, followed by fractional distillation of the product mixture. At 25 °C, the vapor pressure of $CHCl_3$ is 205 mm Hg and the vapor pressure of CH_2Cl_2 is 415 mm Hg. What is the vapor pressure in mm Hg at 25 °C of a mixture of 15.0 g of $CHCl_3$ and 37.5 g of CH_2Cl_2?

12.112 What is the mole fraction of each component in the liquid mixture in Problem 12.110, and what is the mole fraction of each component in the vapor at 30 °C?

12.113 What is the mole fraction of each component in the liquid mixture in Problem 12.111, and what is the mole fraction of each component in the vapor at 25 °C?

12.114 A solution prepared by dissolving 5.00 g of aspirin, $C_9H_8O_4$, in 215 g of chloroform has a normal boiling point that is elevated by $\Delta T = 0.47 °C$ over that of pure chloroform. What is the value of the molal boiling-point-elevation constant for chloroform?

12.115 A solution prepared by dissolving 3.00 g of ascorbic acid (vitamin C, $C_6H_8O_6$) in 50.0 g of acetic acid has a freezing point that is depressed by $\Delta T = 1.33 °C$ below that of pure acetic acid. What is the value of the molal freezing-point-depression constant for acetic acid?

12.116 A solution of citric acid, $C_6H_8O_7$, in 50.0 g of acetic acid has a boiling point elevation of $\Delta T = 1.76 °C$. What is the molality of the solution if the molal boiling-point-elevation constant for acetic acid is $K_b = 3.07(°C \cdot kg)/mol$?

12.117 What is the normal boiling point in °C of ethyl alcohol if a solution prepared by dissolving 26.0 g of glucose ($C_6H_{12}O_6$) in 285 g of ethyl alcohol has a boiling point of 79.1 °C? See Table 12.4 to find K_b for ethyl alcohol.

12.118 What osmotic pressure in atmospheres would you expect for each of the following solutions?

(a) 5.00 g of NaCl in 350.0 mL of aqueous solution at 50 °C

(b) 6.33 g of sodium acetate, CH_3CO_2Na, in 55.0 mL of aqueous solution at 10 °C

12.119 What osmotic pressure in mm Hg would you expect for an aqueous solution of 11.5 mg of insulin (mol. weight = 5990) in 6.60 mL solution at 298 K? What would the height of the water column be meters? The density of mercury is 13.534 g/mL at 298 K.

12.120 A solution of an unknown molecular substance in water at 300 gives rise to an osmotic pressure of 4.85 atm. What is the molar of the solution?

12.121 Human blood gives rise to an osmotic pressure of approximat 7.7 atm at body temperature, 37.0 °C. What must the molarity an intravenous glucose solution be to give rise to the same motic pressure as blood?

12.122 When salt is spread on snow-covered roads at −2 °C, the sn melts. When salt is spread on snow-covered roads at −30 nothing happens. Explain.

12.123 If cost per gram were not a concern, which of the following su stances would be the most efficient per unit mass for melti snow from sidewalks and roads: glucose ($C_6H_{12}O_6$), LiCl, Na or $CaCl_2$? Explain.

12.124 Cellobiose is a sugar obtained by degradation of cellulose. If 20 mL of an aqueous solution containing 1.500 g of cellobiose 25.0 °C gives rise to an osmotic pressure of 407.2 mm Hg, wha the molecular weight of cellobiose?

12.125 Met-enkephalin is one of the so-called endorphins, a class of n urally occurring morphine-like chemicals in the brain. Wha the molecular weight of met-enkephalin if 20.0 mL of an aque solution containing 15.0 mg of met-enkephalin at 298 K supp a column of water 32.9 cm high? The density of mercury at 29 is 13.534 g/mL.

12.126 The freezing point of a solution prepared by dissolving 1 mol of hydrogen fluoride, HF, in 500 g of water is −3.8 °C, the freezing point of a solution prepared by dissolving 1 mol of hydrogen chloride, HCl, in 500 g of water is −7.4 Explain.

12.127 The boiling point of a solution prepared by dissolving 71 Na_2SO_4 in 1.00 kg of water is 100.8 °C. Explain.

12.128 Elemental analysis of β-carotene, a dietary source of vitami shows that it contains 10.51% H and 89.49% C. Dissolving 0.0 g of β-carotene in 1.50 g of camphor gives a freezing-point pression of 1.17 °C. What are the molecular weight and form of β-carotene? [K_f for camphor is 37.7 (°C · kg)/mol.]

12.129 Lysine, one of the amino acid building blocks found in prote contains 49.29% C, 9.65% H, 19.16% N, and 21.89% O by emental analysis. A solution prepared by dissolving 30.0 m lysine in 1.200 g of the organic solvent biphenyl gives a freez point depression of 1.37 °C. What are the molecular weight formula of lysine? [K_f for biphenyl is 8.00 (°C · kg)/mol.]

CHAPTER PROBLEMS

12.130 How many grams of ethylene glycol (automobile antifre $C_2H_6O_2$) dissolved in 3.55 kg of water are needed to lower freezing point of water in an automobile radiator to −22.0 °C

12.131 When 1 mL of toluene is added to 100 mL of benzene (bp 80.1 the boiling point of the benzene solution rises, but when 1 m benzene is added to 100 mL of toluene (bp 110.6 °C), the bo point of the toluene solution falls. Explain.

When solid $CaCl_2$ is added to liquid water, the temperature rises. When solid $CaCl_2$ is added to ice at 0 °C, the temperature falls. Explain.

Silver chloride has a solubility of 0.007 mg/mL in water at 5 °C. What is the osmotic pressure in atmospheres of a saturated solution of AgCl at 5 °C?

When a 2.850 g mixture of the sugars sucrose ($C_{12}H_{22}O_{11}$) and fructose ($C_6H_{12}O_6$) was dissolved in water to a volume of 1.50 L, the resultant solution gave an osmotic pressure of 0.1843 atm at 298.0 K. What is $X_{sucrose}$ of the mixture?

Glycerol ($C_3H_8O_3$) and diethylformamide ($C_5H_{11}NO$) are non-volatile, miscible liquids. If the volume of a solution made by dissolving 10.208 g of a glycerol–diethylformamide mixture in water is 1.75 L and the solution has an osmotic pressure of 1.466 atm at 298.0 K, what is $X_{glycerol}$ of the mixture?

How many grams of naphthalene, $C_{10}H_8$ (commonly used as household mothballs), should be added to 150.0 g of benzene to depress its freezing point by 0.35 °C? See Table 12.4 to find K_f for benzene.

Bromine is sometimes used as a solution in tetrachloromethane, CCl_4. What is the vapor pressure in mm Hg of a solution of 1.50 g of Br_2 in 145.0 g of CCl_4 at 300 K? The vapor pressure of pure bromine at 300 K is 30.5 kPa, and the vapor pressure of CCl_4 is 16.5 kPa.

Assuming that seawater is a 3.5 mass % solution of NaCl and that its density is 1.00 g/mL, calculate both its boiling point and its freezing point in °C.

There's actually much more in seawater than just dissolved NaCl. Major ions present include 19,000 ppm Cl^-, 10,500 ppm Na^+, 2650 ppm SO_4^{2-}, 1350 ppm Mg^{2+}, 400 ppm Ca^{2+}, 380 ppm K^+, 140 ppm HCO_3^-, and 65 ppm Br^-.

(a) What is the total molality of all ions present in seawater?

(b) Assuming molality and molarity to be equal, what amount of osmotic pressure in atmospheres would seawater give rise to at 300 K?

Rubbing alcohol is a 90 mass % solution of isopropyl alcohol, C_3H_8O, in water.

(a) How many grams of rubbing alcohol contains 10.5 g of isopropyl alcohol?

(b) How many moles of isopropyl alcohol are in 50.0 g of rubbing alcohol?

Although not particularly convenient, it's possible to use osmotic pressure to measure temperature. What is the Kelvin temperature if a solution prepared by dissolving 17.5 mg of glucose ($C_6H_{12}O_6$) in 50.0 mL of aqueous solution gives rise to an osmotic pressure of 37.8 mm Hg?

The van't Hoff factor for $CaCl_2$ is 2.71. What is its mass % in an aqueous solution that has $T_f = -1.14$ °C?

What is the van't Hoff factor for K_2SO_4 in an aqueous solution that is 5.00% K_2SO_4 by mass and freezes at −1.21 °C?

If the van't Hoff factor for LiCl in a 0.62 m solution is 1.96, what is the vapor pressure depression in mm Hg of the solution at 298 K? (The vapor pressure of water at 298 K is 23.76 mm Hg.)

What is the value of the van't Hoff factor for KCl if a 1.00 m aqueous solution shows a vapor pressure depression of 0.734 mm Hg at 298 °C? (The vapor pressure of water at 298 K is 23.76 mm Hg.)

The steroid hormone *estradiol* contains only C, H, and O; combustion analysis of a 3.47 mg sample yields 10.10 mg CO_2 and 2.76 mg H_2O. On dissolving 7.55 mg of estradiol in 0.500 g of

camphor, the melting point of camphor is depressed by 2.10 °C. What is the molecular weight of estradiol, and what is a probable formula? [For camphor, $K_f = 37.7$ (°C · kg)/mol.]

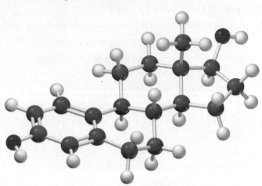

Estradiol

12.147 Many acids are partially dissociated into ions in aqueous solution. Trichloroacetic acid (CCl_3CO_2H), for instance, is partially dissociated in water according to the equation

$$CCl_3CO_2H(aq) \rightleftharpoons H^+(aq) + CCl_3CO_2^-(aq)$$

What is the percentage of molecules dissociated if the freezing point of a 1.00 m solution of trichloroacetic acid in water is −2.53 °C?

12.148 Addition of 50.00 mL of 2.238 m H_2SO_4 (solution density = 1.1243 g/mL) to 50.00 mL of 2.238 M $BaCl_2$ gives a white precipitate.

(a) What is the mass of the precipitate in grams?

(b) If you filter the mixture and add more H_2SO_4 solution to the filtrate, would you obtain more precipitate? Explain.

12.149 A solid mixture of KCl, KNO_3, and $Ba(NO_3)_2$ is 20.92 mass % chlorine, and a 1.000 g sample of the mixture in 500.0 mL of aqueous solution at 25 °C has an osmotic pressure of 744.7 mm Hg. What are the mass percents of KCl, KNO_3, and $Ba(NO_3)_2$ in the mixture?

12.150 A solution of LiCl in a mixture of water and methanol (CH_3OH) has a vapor pressure of 39.4 mm Hg at 17 °C and 68.2 mm Hg at 27 °C. The vapor pressure of pure water is 14.5 mm Hg at 17 °C and 26.8 mm Hg at 27 °C, and the vapor pressure of pure methanol is 82.5 mm Hg at 17 °C and 140.3 mm Hg at 27 °C. What is the composition of the solution in mass percent?

12.151 An aqueous solution of KI has a freezing point of −1.95 °C and an osmotic pressure of 25.0 atm at 25.0 °C. Assuming that the KI completely dissociates in water, what is the density of the solution?

12.152 An aqueous solution of a certain organic compound has a density of 1.063 g/mL, an osmotic pressure of 12.16 atm at 25.0 °C, and a freezing point of −1.03 °C. The compound is known not to dissociate in water. What is the molar mass of the compound?

12.153 At 60 °C, compound X has a vapor pressure of 96 mm Hg, benzene (C_6H_6) has a vapor pressure of 395 mm Hg, and a 50:50 mixture by mass of benzene and X has a vapor pressure of 299 mm Hg. What is the molar mass of X?

12.154 An aqueous solution containing 100.0 g of NaCl and 100.0 g of $CaCl_2$ has a volume of 1.00 L and a density of 1.15 g/mL. The vapor pressure of pure water at 25 °C is 23.8 mm Hg, and you can assume complete dissociation for both solutes.

(a) What is the boiling point of the solution?

(b) What is the vapor pressure of the solution at 25 °C?

.55 Iodic acid, HIO_3, is a weak acid that undergoes only partial dissociation in water. If a 1.00 M solution of HIO_3 has a density of 1.07 g/mL and a freezing point of $-2.78\,°C$, what percent of the HIO_3 is dissociated?

12.156 A 1.24 M solution of KI has a density of 1.15 g/cm^3.

(a) What is the molality of the solution?

(b) What is the freezing point of the solution, assuming complete dissociation of KI?

(c) The actual freezing point of the solution is $-4.46\,°C$. What percent of the KI is dissociated?

12.157 Desert countries like Saudi Arabia have built reverse osmosis plants to produce freshwater from seawater. Assume that seawater has the composition 0.470 M NaCl and 0.068 M $MgCl_2$ and that both compounds are completely dissociated.

(a) What is the osmotic pressure of seawater at 25 °C?

(b) If the reverse osmosis equipment can exert a maximum pressure of 100.0 atm at 25.0 °C, what is the maximum volume of freshwater that can be obtained from 1.00 L of seawater?

12.158 A solution prepared by dissolving 100.0 g of a mixture of sugar ($C_{12}H_{22}O_{11}$) and table salt (NaCl) in 500.0 g of water has a freezing point of $-2.25\,°C$. What is the mass of each individual solute? Assume that NaCl is completely dissociated.

12.159 A solution of 0.250 g of naphthalene (mothballs) in 35.00 g of camphor lowers the freezing point by 2.10 °C. What is the molar mass of naphthalene? The freezing-point-depression constant for camphor is 37.7 $(°C \cdot kg)/mol$.

MULTICONCEPT PROBLEMS

12.160 Treatment of 1.385 g of an unknown metal M with an excess of aqueous HCl evolved a gas that was found to have a volume of 382.6 mL at 20.0 °C and 755 mm Hg pressure. Heating the reaction mixture to evaporate the water and remaining HCl then gave a white crystalline compound, MCl_x. After dissolving the compound in 25.0 g of water, the melting point of the resulting solution was $-3.53\,°C$.

(a) How many moles of H_2 gas are evolved?

(b) What mass of MCl_x is formed?

(c) What is the molality of particles (ions) in the sol MCl_x?

(d) How many moles of ions are in solution?

(e) What are the formula and molecular weight of MCl_x?

(f) What is the identity of the metal M?

12.161 A compound that contains only C and H was burned i O_2 to give CO_2 and H_2O. When 0.270 g of the compo burned, the amount of CO_2 formed reacted completely v mL of 2.00 M NaOH solution according to the equation

$$2\,OH^-(aq) + CO_2(g) \rightarrow CO_3{}^{2-}(aq) + H_2O(l$$

When 0.270 g of the compound was dissolved in 50.0 g phor, the resulting solution had a freezing point of 177.9 ° camphor freezes at 179.8 °C and has $K_f = 37.7$ $(°C \cdot kg)$,

(a) What is the empirical formula of the compound?

(b) What is the molecular weight of the compound?

(c) What is the molecular formula of the compound?

12.162 Combustion analysis of a 36.72 mg sample of the male h testosterone gave 106.43 mg CO_2 and 32.100 mg H_2C only combustion products. When 5.00 mg of testoster dissolved in 15.0 mL of a suitable solvent at 25 °C, an pressure of 21.5 mm Hg was measured. What is the m formula of testosterone?

12.163 When 8.900 g of a mixture of an alkali metal chloride (X an alkaline earth chloride (YCl_2) was dissolved in 15 water, the freezing point of the resultant solution was $-$ Addition of an excess of aqueous $AgNO_3$ to the solution a white precipitate with a mass of 27.575 g. How much metal chloride was present in the original mixture, and the identities of the two metals **X** and **Y**?

12.164 Combustion analysis of a 3.0078 g sample of digitoxin, pound used for the treatment of congestive heart failu 7.0950 g of CO_2 and 2.2668 g of H_2O. When 0.6617 g of d was dissolved in water to a total volume of 0.800 L, the pressure of the solution at 298 K was 0.026 44 atm. Wh molecular formula of digitoxin, which contains only C, H